Hydrology *and* Hydraulic Systems

Third Edition

Hydrology *and* Hydraulic Systems

Third Edition

Ram S. Gupta, Ph.D., P.E.

Roger Williams University, Bristol, RI
Delta Engineers, Inc., Bristol, RI

WAVELAND

PRESS, INC.

Long Grove, Illinois

For information about this book, contact:
    Waveland Press, Inc.
    4180 IL Route 83, Suite 101
    Long Grove, IL 60047-9580
    (847) 634-0081
    info@waveland.com
    www.waveland.com

Copyright © 2008 by Ram S. Gupta

10-digit ISBN 1-57766-455-8
13-digit ISBN 978-1-57766-455-0

All rights reserved. No part of this book may be reproduced, stored in a retrieval system, or transmitted in any form or by any means without permission in writing from the publisher.

Printed in the United States of America

7 6 5 4 3 2 1

Table of Contents

Preface

◆◆◆

Hydrology and Hydraulic Systems, Third Edition, provides a comprehensive, up-to-date treatment of the field of hydrology. It offers a broad selection of subject material and can thus accommodate the high degree of variability in course content among various universities. This text is ideal for at least two undergraduate or graduate level courses—a basic course in hydrology and an applied course on hydrology and hydrologic systems. It also works well with minimal augmentation by the instructor for courses in groundwater hydrology and for courses encompassing the disciplines of surface water hydrology and water resources management.

In the *Second Edition,* the major thrust of the revision was the hydrologic content. A strong emphasis was placed on practical applications and designs. The addition of a new chapter on "Development and Monitoring of Groundwater" solidified the treatment of groundwater, while inclusion of relatively new advancements in kinematic hydrology and contaminant hydrology broadened the scope of the book.

The *Third Edition* features a major revision of hydraulics. All hydraulic-related material has been revised and consolidated. The principles of hydraulics and the basic concepts underlying the application of those principles have been included in a new chapter, while hydraulic devices for measuring flows are now consolidated into a single chapter on hydraulic structures. The chapter on conveyance systems contains a completely rewritten treatment of channel section design, and the new concept of combining the regime theory and the power function laws is presented. The chapter on pressure flow systems is enhanced with a revised discussion of pipeline analysis.

Additionally, computer software applications have been updated throughout the text. More examples are presented in SI units, and the end-of-chapter problems have been expanded. The *Third Edition* retains the important pedagogical features of the previous edition, in particular its development of subject matter in the same sequence in which water resource development activities progress on an actual project.

Key features of the *Third Edition* include the following:

- More than 300 illustrations and 175 tables
- More than 225 fully solved example problems in both FPS and SI units, well integrated within the text
- 520 end-of-chapter problems for student assignment
- Application of EPA-recommended statistical procedures for groundwater monitoring
- Discussion of theory of contaminant transport through porous medium
- Detailed treatment of hydrologic field investigation and analytical procedures for data assessment
- Systematic presentation of various surface-water-flow assessment techniques
- Presentation of theory and design of loose-boundary channels
- Complete treatment of storm, sanitary, agricultural, and urban drainage systems

In a revision of this magnitude, assistance came from many sources. My former professor, Dr. Alvin S. Goodman at the Polytechnic University of New York, and Dr. Erhard F. Joeres of the University of Wisconsin-Madison, reviewed the previous edition manuscript and offered many valuable suggestions.

My students at Roger Williams University provided very useful input. Laurence Mace helped me develop the regime-power functions. Kristy Oak, Laurence Curan, and Lisa Gheringhelli reviewed individual chapters and end-of-chapter problems.

My wife, Saroj Gupta, helped in the preparation of the manuscript. Laurie Prossnitz of Waveland Press, who edited the previous edition, once again very ably handled the manuscript through all its various stages. She received invaluable assistance from Bonnie Highsmith, who redrew the illustrations in digital format, and from Deborah Underwood, who formatted the manuscript, rebuilt hundreds of equations, and prepped the entire document for typesetting. Thanks also go to Gayle Zawilla for creating the index. The Waveland staff provided the editorial support that led to the completion of the project. I extend my sincere thanks to these individuals and to my colleagues at Roger Williams University who provided a helping hand from time to time.

1

Demand for Water

◆◆

1.1 DEVELOPMENT OF WATER RESOURCES

Water resources are developed either to use or to control groundwater and surface water flows. A developmental activity for the use of water comprises studying the availability of and demand for water and contemplating a project that can meet the expected needs from available supplies by means of engineering works and nonstructural measures. Water control activities are concerned with regulating the rate of water supply in order to improve conditions within an area. The purpose of the facilities provided by a project is to vary the quantity and quality, time of supply, and place of use of the water resources in accordance with need. Developing a project involves the design and arrangement of structural components, along with nonstructural measures, if any, to serve the intended purpose.

This book aims to provide a basic understanding of all the quantitative elements that are involved in the formulation of a water development project. The process of development can be explained by the following sequence of questions. The book is devoted to answering these questions.

1. How much water is needed? In order to develop a water-use project, planners must know the current requirements and be able to predict future requirements for the duration of the plan period for all of the anticipated uses. This is dealt with later in this chapter.

2. How much water is available? This question reflects the prime objective of the study of hydrology—a science related to the occurrence and distribution of natural water on and below the earth's surface and in the atmosphere. Obtaining reliable estimates of fresh surface water and groundwater entering a basin and estimating their spatial and temporal distributions are the most difficult problems for the hydrologist. In this book we study comprehensively all aspects of hydrology. This includes the fundamentals of natural water circulation through the hydrologic cycle and discussion of the elements of the cycle (Chapter 2); the theory, assessment, and monitoring of groundwater supplies (Chapters 3, 4, and 5); the field measurements of streamflows (Chapter 6); and the assessment of surface water supplies with their distribution (Chapter 7). Estimation of peak flows (Chapter 8) and handling of peak flows (Chapters 9, 13, and 14) are important components of water control projects.

3. How are the requirements satisfied by the supplies? The formulation of a project is the answer to this question. A development project may encompass a single unit, a regional development, or a basinwide development. It may be for a single purpose or may involve multipurpose uses. The magnitude and distribution of supplies (Chapter 7) are compared with the variation in demands (Chapter 1). When the supplies always exceed

the demands, a run-of-river project (direct withdrawal from the source) is indicated. However, storage of adequate size (Chapter 10) is needed in the case of seasonal or overall water deficiencies. Dams and control structures (Chapter 10) are designed to store as well as regulate water flows. Other facilities depend on the purpose to be served by the project. For example, a water supply project needs pumps and conveyance (channel) and distribution (pipe) systems; an irrigation project requires conveyance (canal) and drainage systems; a hydropower project includes channels and penstocks (conduits); a flood control project may require embankments, channel improvements, drainage works, and nonstructural measures; and a navigation project involves improvements to existing channels as well as construction of artificial channels.

Hydraulic principles (Chapter 9) and the basic hydraulic structures (Chapter 10), conveyance systems or channels (Chapter 11), distribution systems or pipes (Chapter 12), pumps (Chapter 12), and drainage systems (Chapters 13 and 14) are included in the book. In addition to these structures, there are certain specialized facilities such as treatment plants for water supply, headworks and outlets for irrigation, powerhouses for hydropower, detention basins for flood control, and shiplocks for navigation. These structures are not covered in this book.

4. How is the used-up water disposed of? The used-up waters or excess stormflows from urban areas, irrigated lands, highways, and airports are disposed of through the drainage system. Chapters 13 and 14 deal with this.

In addition to the preceding questions there are other pertinent questions that relate to the quality of water, such as: What type of water is needed? What kind of water is available? What treatment is required? What are the quality and effects of the wastewater discharged or drained into a river system? These are equally important questions in water resources development; however, these qualitative aspects lie outside the scope of this book.

It should be recognized that planning and design provide only the conceptual framework of a water development project. The complete process of development includes the construction, operation, and management of the project as well.

1.2 ASSESSMENT OF DEMANDS

The need for a water development activity arises from the demand for water for some purpose. After demand is established, the availability of water resources in the vicinity of the demand center is assessed through application of the principles of hydrology. The reconciliation of the demand with the available resources in an optimal manner is the objective of water resources planning. Where the resources are restricted compared to the demand, as for irrigation in some regions, the problem is approached by considering how much demand can be satisfied with the available water resources. There may be conflicting demands when more than one purpose is involved. These have to be resolved by establishing a priority ranking among water uses. The location, type, and components of a project, as well as its functional characteristics, are dependent on the purpose and magnitude of demands. Thus the project cost is a function of the demand. But the demand for water is affected to some extent by the cost of providing water via the project. Therefore, demand is not a static problem that can be conclusively determined at one point in time. Rather, the initial estimates are reviewed at a later stage in the planning process. The project estimates also are revised accordingly. The procedures for making tentative estimates of demand for each major purpose of development are discussed in subsequent sections.

Demand for Water   Chapter 1

1.3 DEMAND FOR WATER SUPPLY

Withdrawal uses involve diversion of surface water or groundwater from its source of supply, such as for irrigation and water supply. *Nonwithdrawal uses* are on-site uses such as navigation and recreation. *Consumptive uses* involve that portion of withdrawn quantity that is no longer available for further use because it is used up by crops, human beings, industrial plant processes, evapotranspiration, and so on. For example, in the context of agriculture, the consumptive use requirement of a crop is defined as the amount of water needed for crop growth.

Water supply requirements usually have the highest priority among the developmental uses, and water of good quality is needed. Although the total quantity of withdrawal in big cities may be relatively large, the consumptive water use is small since 80 to 90% of the total intake is returned to the river system (of course, its quality is degraded). The water requirements of a city can be divided into three broad categories:

- Municipal requirements
- Large industrial requirements
- Waste dilution requirements

The order of magnitude of these requirements is indicated in Figure 1.1 for a typical city with a population of about 150,000. If the sewage and industrial waste are discharged after treatment, the waste dilution requirements can be reduced substantially as discussed subsequently. The total water supply requirement in a river system for a number of cities is not equal to the sum of the requirements of the individual cities because the dilution requirement is a nonwithdrawal use that is available to all cities on the same river, and the consumptive requirement is only a small fraction of the total. Thus, if all cities are situated on the same river with sufficient distance in between for purification of the discharged wastes, the total requirements will be only slightly more than the largest city requirements.

1.4 MUNICIPAL REQUIREMENTS

This includes (1) such domestic uses as drinking, cooking, washing, watering lawns, and air conditioning, (2) public uses such as in public buildings and for firefighting, (3) commercial use in shopping centers, hotels, and laundries, (4) small industrial use by industries not having a separate system, and (5) losses in the distribution system. The municipal requirements are highly variable, depending on such factors as size of city, characteristics of the population, nature and size of commercial and industrial establishments, climatic conditions, and cost of supply.

Municipal requirements are estimated by the following simple equation:

$$\text{required quantity} = \begin{pmatrix} \text{population at} \\ \text{the end of} \\ \text{design period} \end{pmatrix} \times \begin{pmatrix} \text{per} \\ \text{capita} \\ \text{usage} \end{pmatrix} \left[L^3 \right] \tag{1.1}$$

The two parameters for assessing the municipal requirements in eq. (1.1)—population estimates and per capita water usage—are discussed in the following sections.

Figure 1.1 Water requirements of a city of 150,000 population.

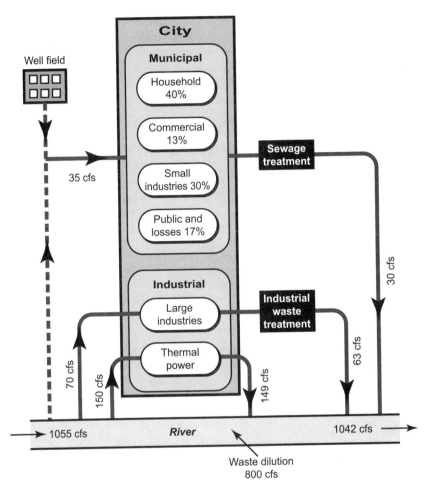

1.5 POPULATION FORECASTING

Like any other natural phenomenon, the prediction of future population is quite complex. Any sophisticated model has several implicit and explicit assumptions. The success of forecasting lies in the judgment of the forecaster and the reliability of these assumptions.

Most methods pertain to trend analysis wherein future changes in population are expected to follow the pattern of the past. However, dynamic human growth involves continuous deviations from past trends, which are difficult to assess. Long-term projection methods consider probable shifts in trends. There are four broad categories of population forecasting techniques: (1) graphical, (2) mathematical, (3) ratio and correlation, and (4) component methods.

The two types of population predictions used are (1) short-term estimates for 1 to 10 years, and (2) long-term forecasting for 10 to 50 or more years. The methods used for these two types of estimates are different, as described in the following sections.

1.6 SHORT-TERM ESTIMATES

Certain techniques from the categories of graphical and mathematical methods are used for short-term estimates. These methods are essentially trend analyses in graphic or mathematical form. The mathematical approach assumes three forms of population growth: geometric growth, arithmetic growth, and declining rate of growth. These are shown by three segments in Figure 1.2. Each segment has a separate relation. The historic population data of the study area may be plotted on a regular graph. Depending on whether the shape is similar to *ab*, *bc*, or *cd*, the relationship of that segment should be used for population projection. The short-term methods are used for either intercensal estimates for any year between two censuses or postcensal estimates from the last census until the next census.

1.6.1 Graphical Extension Method

This method consists of plotting the population of past census years against time, sketching a curve that fits the data, and extending this curve into the future to obtain the projected population. Since it is convenient and more accurate to project a straight line, it is aimed to get a straight-line fit to the past data by making a semilog or logarithmic plot, as necessary. The forecast may vary widely depending on whether the last two known points are joined and extended or other points are joined and extended.

1.6.2 Arithmetic Growth Method

This method considers that the same population increase takes place in a given period. Mathematically,

$$\frac{dP}{dt} = K_a$$

where

P = population

t = time, years

K_a = uniform growth-rate constant

By integrating the equation above, we obtain

$$P_t = P_0 + K_a t \tag{1.2}$$

where

P_t = projected population t years after P_0

P_0 = present population

t = period of projection

and

$$K_a = \frac{P_2 - P_1}{\Delta t} \quad [\text{T}^{-1}] \tag{1.3}$$

where P_1 and P_2 are recorded populations at some t interval apart.

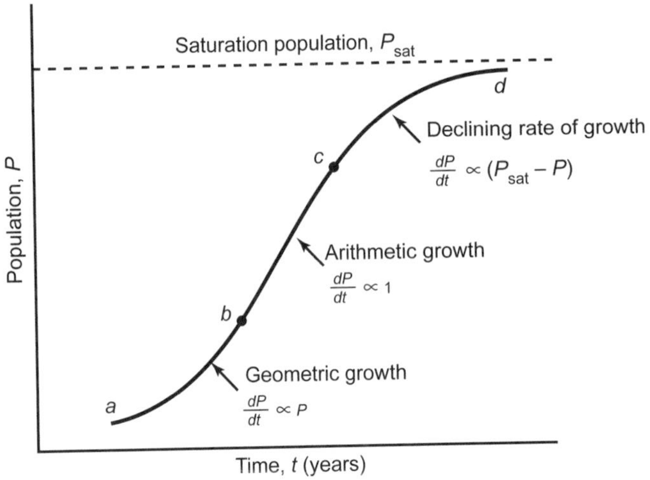

Figure 1.2 Population growth curve.

EXAMPLE 1.1

The population of a city has been recorded in 1990 and 2005 as 100,000 and 110,000, respectively. Estimate the 2015 population, assuming arithmetic growth.

SOLUTION   From eq. (1.3),

$$K_a = \frac{110,000 - 100,000}{15} = 667$$

From eq. (1.2),

$$P_{2015} = 110,000 + 667(10) = 116,667 \text{ persons}$$

1.6.3 Geometric Growth Method

This method considers that the increase in population takes place at a constant percent of the current population. Mathematically,

$$\frac{dP}{dt} = K_p P$$

By integrating we obtain

$$\ln P_t = \ln P_0 + K_p t \tag{1.4}$$

where

$$K_p = \frac{\ln P_2 - \ln P_1}{\Delta t} \quad [\text{T}^{-1}] \tag{1.5}$$

EXAMPLE 1.2

Using the information from Example 1.1, estimate the population assuming geometric growth.

Demand for Water   Chapter 1

SOLUTION From eq. (1.5),

$$K_p = \frac{\ln 110{,}000 - \ln 100{,}000}{15} = 0.0064$$

From eq. (1.4),

$$\ln P_{2015} = \ln 110{,}000 + 0.0064(10) = 11.672$$
$$P_{2015} = e^{11.672} = 117{,}240 \text{ persons}$$

1.6.4 Declining Growth Rate Method

This method assumes that the city has a saturation population and the rate of growth becomes less as the population approaches the saturation level. In other words, the rate of increase is a function of the population deficit $(P_{sat} - P)$, that is,

$$\frac{dP}{dt} = K_D \left(P_{sat} - P \right)$$

Upon integration, we have

$$P_t = P_{sat} - \left(P_{sat} - P_0 \right) e^{-K_D t} \tag{1.6}$$

Rearranging eq. (1.6) gives

$$K_D = -\frac{1}{\Delta t} \ln \left(\frac{P_{sat} - P_2}{P_{sat} - P_1} \right) \quad [\text{T}^{-1}] \tag{1.7}$$

EXAMPLE 1.3

In Example 1.1, if the saturation population of the city is 200,000, estimate the 2015 population. Assume a declining rate of growth.

SOLUTION From eq. (1.7),

$$K_D = -\frac{1}{15} \ln \left(\frac{200{,}000 - 110{,}000}{200{,}000 - 100{,}000} \right) = 0.007$$

From eq. (1.6),

$$P_{2015} = 200{,}000 - (200{,}000 - 110{,}000)e^{-(.007)(10)}$$
$$= 116{,}085 \text{ persons}$$

EXAMPLE 1.4

A city has a present population of 200,000, which is estimated to increase geometrically to 220,000 in the next 15 years. The existing treatment plant capacity is 51 mgd. The rate of input to the treatment plant is 165 gallons per person per day. For how many years from now will the treatment plant be adequate?

SOLUTION

1. From eq. (1.5),

$$K_p = \frac{\ln 220{,}000 - 200{,}000}{15} = 0.00635$$

2. Population that can be served by the plant:

$$\frac{51.0\left(10^6\right)}{165} = 309{,}090 \text{ persons}$$

3. Time to reach the design population:

From eq. (1.4),

$$\ln 309{,}090 = \ln 200{,}000 + 0.00635(t)$$

$$t = 68.55 \text{ years}$$

1.7 Long-Term Forecasting

Long-term predictions are made using techniques from all four categories. The entire past record of historic population data is used in long-term predictions. The mathematical curve-fitting approach is popular because it is relatively easy to apply. McJunkin (1964) indicates, however, that the component and ratio methods offer greater reliability than the traditional graphical-mathematical methods.

1.7.1 Graphic Comparison Method

Several larger cities in the vicinity are selected whose earlier growth exhibited characteristics similar to those of the study area. The population-time curves for these cities and for the study area are plotted in Figure 1.3. From point O' corresponding to the last known population for study area A, a horizontal line is drawn intersecting the other curves at O_1, O_2, O_3, etc. At O', lines parallel to O_1B, O_2C, O_3D, etc. are drawn as $O'B'$, $O'C'$, $O'D'$, etc., respectively. These lines establish a range of future growth within which $O'A'$ is extended. This method has a shortcoming since it is not certain that the future growth of the study area will be similar to the past growth of the other areas.

1.7.2 Mathematical Logistic Curve Method

This method is suitable for the study of large population centers such as large cities, states, or nations. On the basis of the study of the growth curve of Figure 1.2, certain mathematical equations of an empirical curve conforming to this shape (S-shape) were proposed. One of the best known functions is the logistic curve in the form

$$P_t = \frac{P_{sat}}{1 + ae^{bt}} \tag{1.8}$$

where

P_t = population at any time t from an assumed origin

P_{sat} = saturation population

a, b = constants

Figure 1.3 Graphical projection by comparison.

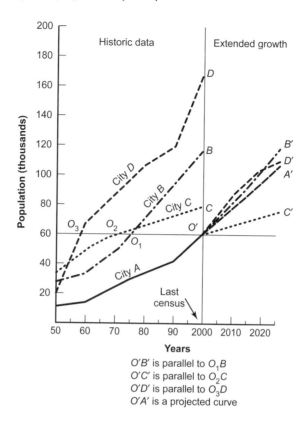

O'B' is parallel to O_1B
O'C' is parallel to O_2C
O'D' is parallel to O_3D
O'A' is a projected curve

The constants are determined by selecting three populations from the record: one in the beginning, P_0, one in the middle, P_1, and one near the end of the record, P_2, associated with the years T_0, T_1, and T_2 such that the number of years (interval) between T_0 and T_1 is the same as that between T_1 and T_2. This interval between T_0 and T_1 is designated as n. The constants are given by

$$P_{sat} = \frac{2P_0 P_1 P_2 - P_1^2 (P_0 + P_2)}{P_0 P_2 - P_1^2} \qquad (1.9)$$

$$a = \frac{P_{sat} - P_0}{P_0} \qquad (1.10)$$

$$b = \frac{1}{n} \ln \left[\frac{P_0 (P_{sat} - P_1)}{P_1 (P_{sat} - P_0)} \right] \qquad (1.11)$$

In eq. (1.8), time t is counted from the year T_0. From eq. (1.9), P_{sat} must be positive and must exceed the latest known population. A test of the validity of logistic growth is that the population data plot as a straight line on specially scaled (logistic) graph paper. (The graph of $\log [(P_{sat} - P) / P]$ versus t is a straight line.)

Section 1.7 Long-Term Forecasting

EXAMPLE 1.5

In two 20-year periods, a city grew from 45,000 to 258,000 to 438,000. Estimate **(a)** the saturation population, **(b)** the equation of the logistic curve, and **(c)** the population 40 years after the last period.

SOLUTION

$$P_0 = 45,000$$
$$P_1 = 258,000$$
$$P_2 = 438,000$$
$$n = 20$$

(a) From eq. (1.9) (expressing numbers in thousands),

$$P_{sat} = \frac{2(45)(258)(438) - (258)^2(45+438)}{(45)(438) - (258)^2} = 469,000$$

(b) From eq. (1.10),

$$a = \frac{469 - 45}{45} = 9.42$$

From eq. (1.11),

$$b = \frac{1}{20}\ln\left[\frac{45(469-258)}{258(469-45)}\right] = -0.122$$

The equation of the logistic curve:

$$P_t = \frac{469,000}{1 + 9.42e^{-0.122t}}$$

(c) The time from the beginning, $t = 40 + 40 = 80$ years; thus

$$P = \frac{469,000}{1 + 9.42e^{-0.122(80)}} = 468,745 \text{ persons}$$

1.7.3 Ratio and Correlation Methods

A city or smaller area is a part of a region, state, nation, or other larger area. Many factors and influences affecting population growth occur throughout a region. Thus the growth of the smaller area has some relation to the growth of the larger area. Because a careful projection of the future population of the nation and/or state (larger area) is made by an authoritative organization, these may be used to forecast the growth of the smaller area.

In the simplest technique, a constant ratio obtained from the most recent data is used as follows:

$$K_r = \frac{P_i}{P_i'}$$

$$P_t = K_r P_t' \tag{1.12}$$

where

P_i = population of study area at last census

P_i' = population of larger area at last census representing the same year

P_t = future population for study area

P_t' = estimated future population of larger area

K_r = constant

In a refined technique using variation in ratios, the ratios of the population of the smaller area to the larger area are calculated for a series of census years. By using any of the graphical or mathematical methods of short-term estimates (Section 1.6), the trend line of the ratios is projected. The projected ratio in the year of interest is applied to an estimate of the study area population.

In another statistical method, the correlation technique described in Chapter 7 is applied. Between the two series of census data relating to the study area and larger area, a relation is established through regression analysis. For example, a simple regression equation may be of the following form:

$$P_f = aP_f' + b \qquad (1.13)$$

where

P_f = population of the study area

P_f' = population of the region (larger area)

a,b = regression constants

The future population is projected from eq. (1.13).

1.7.4 Component Methods

A population change can occur in only three ways: (1) by birth, (2) by death, and (3) through migration. These components of population change can be linked by the balance equation:

$$P_t = P_0 + B - D \pm M \qquad (1.14)$$

where

P_t = forecast population at the end of time t

P_0 = existing population

B = number of births during time t

D = number of deaths during time t

M = net number of migrants during time t

(positive value indicates migration into the study area)

Because migration affects the births and deaths in an area, the estimates of net migration are made before estimating the natural change due to births and deaths. The migratory trends may be estimated by applying eq. (1.14) backward to the past census data on population, births, and deaths during a selected period. The school attendance method, comparing the actual children enrolled to the children from birth records, is also used for estimating migration.

For determining the natural change due to births and deaths, the simplest procedure is to multiply the existing population by the expected birth and death rates, that is,

$$B = K_1 P_0 \Delta t \tag{1.15}$$

$$D = K_2 P_0 \Delta t \tag{1.16}$$

where

K_1, K_2 = birth and death rate, respectively

Δt = forecast period

Better estimates of natural change (births and deaths) are made by the cohort-survival technique, which makes projections for each subcomponent related to the natural change.

1.8 PER CAPITA WATER USAGE

Per capita use is normally expressed as the average daily rate, which is the mean annual usage of water averaged for a day in terms of gallons (or liters) per capita per day (gpcd). The seasonal, monthly, daily, and hourly variations in the rate are given in percentages of the average. Deciding which of these should be used for the design capacity depends on the components of the water supply system. The layouts of two water supply systems—one for direct pumping from a river or from a well field and one for an impounding reservoir—are shown in Figure 1.4. The period of design for which the population projection is to be made and the design capacity criteria of different component structures of the systems are indicated in Table 1.1.

Figure 1.4 Layout of typical water supply systems (from Fair, Geyer, and Okun, 1966).

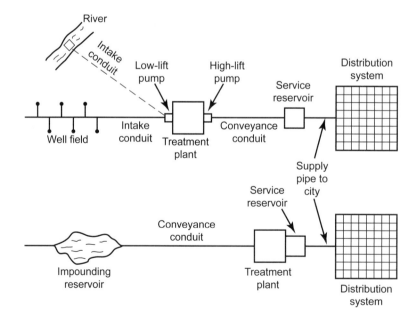

Demand for Water   Chapter 1

Table 1.1   Design Periods and Capacity Criteria for Constituent Structures

Structure	Design Period[a] (years)	Required Capacity
1. Source of supply		
a. River	Indefinite	Maximum daily (requirements)
b. Well field	10–25	Maximum daily
c. Reservoir	25–50	Average annual demand
2. Conveyance		
a. Intake conduit	25–50	Maximum daily
b. Conduit to treatment plant	25–50	Maximum daily
3. Pumps		
a. Low-lift	10	Maximum daily plus one reserve unit
b. High-lift	10	Maximum hourly plus one reserve unit
4. Treatment plant	10–15	Maximum daily
5. Service reservoir	20–25	Working storage (storage capacity computation, Sect. 10.10) plus fire demand for a specified duration (Table 1.3) plus emergency storage
6. Distribution		
a. Supply pipe or conduit	25–50	Greater of (1) maximum daily plus fire demand of a day or (2) maximum hourly requirement
b. Distribution grid	Full development	

[a] "Design period" does not necessarily indicate the life of the structure. A design period takes into account other factors, such as subsequent ease of extension, rate of population growth and shifts in community, and industrial/commercial developments.

1.8.1 Average Daily per Capita per Day Usage

For household use, the per capita water consumption ranges between 60 and 180 gallons per day, with a reasonable average of 101 gallons per day. Municipal water use should, however, also include commercial use, small industrial use, public use, and losses in the system. According to the U.S. Geological Survey (USGS) the estimated usage of public water supplies in the U.S. in 2000 was 179 gpcd. A typical distribution for an average city is given in Table 1.2. When industrial requirements become relatively important in a city, they should be considered separately.

1.8.2 Variations in Usage

The values in Table 1.2 refer to the daily average of the long-term (many years) usage. In actuality, consumption changes with the seasons, varies from day to day in the week, and fluctuates from hour to hour in a day. Knowledge of these variations is important for the design of project components, as indicated in Table 1.1. There are two common trends: (1) the smaller the city the more variable the demand; and (2) the shorter the period of flow, the wider is the variation from the average (i.e., the hourly peak flow is much higher than the daily peak).

Typical variation in daily usage from a city water supply is shown in Figure 1.5. The variations are commonly indicated in terms of the percentage of the long-term average value. There are no fixed ratios; each city has its own trend. However, in the absence of

Table 1.2   Typical Average Usage

Use	Average Use (gpcd)[a]	Percent of Total
Household	101	56
Commercial	30	17
Industrial	22	12
Public	17	9
Loss	10	6
Total	180	100

[a] Liter = gallons × 3.8.

Figure 1.5   Typical variation in usage in a day.

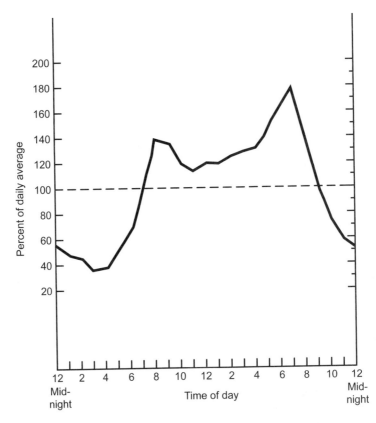

data, the following formula devised by R.O. Goodrich is very convenient for estimating the maximum usage from 2 hours (2/24 day) to a year (365 days) for small cities.

$$p = \frac{180}{t^{0.1}} \quad \text{[unbalanced]} \tag{1.17}$$

Demand for Water   Chapter 1

where

p = percentage of annual average daily usage

t = time, days

From eq. (1.17), the maximum daily use is 180% of the (long-term) average daily usage, and the maximum monthly use is 128%. Larger cities may have smaller peaks.

The maximum hourly consumption in any day is likely to be 150% of the usage for that day (McGhee, 1991). For a distribution system, the fire demands also have to be added. It is unlikely that water will be drawn at the maximum hourly rate while a serious fire is raging. Hence, the capacity is based on the maximum daily usage plus fire demand or maximum hourly usage, whichever is greater.

For pump design, the information on minimum flow rate, which is considered to be 25 to 50% of the average daily flow, also is important.

EXAMPLE 1.6

For a city having an average daily water usage of 180 gpcd from the municipal supply, determine the maximum hourly requirement (excluding the fire demands).

SOLUTION

1. Maximum daily usage = 180% of average daily

$$= \frac{180}{100}(180) = 324 \text{ gpcd}$$

2. For a maximum day, maximum hourly usage = 150% of daily use

$$= \frac{150}{100}(324) = 486 \text{ gpcd}$$

1.9 FIRE DEMANDS

The quantity of water included under the category of "public use" for firefighting purposes is quite small but the rate of withdrawal is high. The distribution system and service reservoir thus should incorporate provisions for the fire demands in their capacities.

Many methods are available for determining the requirements of fire flow and duration. These include Appendix H of the Uniform Fire Code (2006), the Insurance Services Office (ISO) Guide, the National Fire Protection Association (NFPA) Standards, the Iowa State Method, the Illinois Institute of Technology Research Method, and the American Water Works Association (AWWA) Manual. The Uniform Code adopts a table to simplify the procedure.

The Code utilizes the type of building construction and the floor area to determine the required flows. Buildings are classified based on the type of building material and the fire resistance rating. The *fire resistance rating* is defined as the time in hours or fraction thereof that a material or assembly of materials will withstand the passage of flame and the transmission of heat when exposed to fire under specified conditions of test and performance criteria.

1.9.1 Types of Construction

The two basic categories are:

1. *Noncombustible* (concrete, masonry, steel): Type I and II. Type II is further subdivided into II-FR (fire resistive), II-1 hr, and II-N (no resistance rating).

2. *Combustible:* Type III, IV, and V. Types III and IV have noncombustible exterior and combustible interior. The interior of Type IV has heavy timber (HT) construction of solid or laminated wood without concealed spaces with a minimum 8 in. nominal size. Type III, entirely or partially wood construction of smaller dimensions than Type IV, is further subdivided into III-1 hr and III-N (no rating). Type V has combustible exterior and interior. It is subdivided into V-1 hr and V-N. The majority of private homes fall into the last category.

1.9.2 Fire Area

The fire area is the total floor area in square feet of all floor levels excluding the basement. For Type I and Type II constructions, the fire area is the area of three largest successive floors due to inherent fire resistance.

1.9.3 Fire Flow Requirements and Duration

Buildings are classified into (1) one- and two-family dwellings, and (2) other buildings. For one- and two-family dwellings having a fire area of 3600 ft^2 or less, the minimum fire flow is 1000 gallons per minute. The minimum flow duration is 2 hours. For one- and two-family dwellings of area in excess of 3600 ft^2, use Table 1.3 to determine the required flow and duration. Under the column for type of construction, find the row corresponding to the fire area and read across the last two columns for fire flow in gpm and duration in hours, respectively.

For buildings other than one- and two-family dwellings, use Table 1.3 similarly. In this case the minimum of 3600 ft^2 does not apply.

In buildings with sprinkler systems, fire demands could be reduced up to 50% for one- and two-family dwellings and up to 75% but not less than 1500 gpm for other buildings.

For most development projects, the building with the largest fire flow determines the flow requirements for the entire project.

EXAMPLE 1.7

A city with a population of 20,000 has an average daily usage of 180 gpcd. The fire demand is dictated by a three-story building, each floor of which has an area of 2400 ft^2 of ordinary (V-N) construction. Determine the fire demand and the design capacity for different components of a water supply project. The working service storage is 1.5 mgd.

SOLUTION

1. From Example 1.6

    maximum daily usage = 324 gpcd

    maximum hourly usage = 486 gpcd

Table 1.3   Minimum Required Fire Flow and Flow Duration for Buildings[a]

Fire Area ft^2 ($\times$ 0.0929 for m^2)					Fire Flow gpm[b] ($\times$ 3.785 for L/min)	Flow duration (hours)
I, II-FR	II-1 hr., III-1 hr.	IV-HT, V-1 hr.	II-N, III-N	V-N		
0–22,700	0–12,700	0–8,200	0–5,900	0–3,600	1,500	
22,701–30,200	12,701–17,000	8,201–10,900	5,901–7,900	3,601–4,800	1,750	
30,201–38,700	17,001–21,800	10,901–12,900	7,901–9,800	4,801–6,200	2,000	
38,701–48,300	21,801–24,200	12,901–17,400	9,801–12,600	6,201–7,700	2,250	2
48,301–59,000	24,201–33,200	17,401–21,300	12,601–15,400	7,701–9,400	2,500	
59,001–70,900	33,201–39,700	21,301–25,500	15,401–18,400	9,401–11,300	2,750	
70,901–83,700	39,701–47,100	25,501–30,100	18,401–21,800	11,301–13,400	3,000	
83,701–97,700	47,101–54,900	30,101–35,200	21,801–25,900	13,401–15,600	3,250	
97,701–112,700	54,901–63,400	35,201–40,600	25,901–29,300	15,601–18,000	3,500	3
112,701–128,700	63,401–72,400	40,601–46,400	29,301–33,500	18,001–20,600	3,750	
128,701–145,900	72,401–82,100	46,401–52,500	33,501–37,900	20,601–23,300	4,000	
145,901–164,200	82,101–92,400	52,501–59,100	37,901–12,700	23,301–26,300	4,250	
164,201–183,400	92,401–103,100	59,101–66,000	42,701–47,700	26,301–29,300	4,500	
183,401–203,700	103,101–114,600	66,001–73,300	47,701–53,000	29,301–32,600	4,750	
203,701–225,200	114,601–126,700	73,301–81,100	53,001–58,600	32,601–36,000	5,000	
225,201–247,700	126,701–139,400	81,101–89,200	58,601–65,400	36,001–39,600	5,250	
247,701–271,200	139,401–152,600	89,201–97,700	65,401–70,600	39,601–43,400	5,500	
271,201–295,900	152,601–166,500	97,701–106,500	70,601–77,000	43,401–47,400	5,750	
295,901–Greater	166,501–Greater	106,501–115,800	77,001–83,700	47,401–51,500	6,000	4
295,901–Greater	166,501–Greater	115,801–125,500	83,701–90,600	51,501–55,700	6,250	
295,901–Greater	166,501–Greater	125,501–135,500	90,601–97,900	55,701–60,200	6,500	
295,901–Greater	166,501–Greater	135,501–145,800	97,901–106,800	60,201–64,800	6,750	
295,901–Greater	166,501–Greater	145,801–156,700	106,801–113,200	64,801–69,600	7,000	
295,901–Greater	166,501–Greater	156,701–167,900	113,201–121,300	69,601–74,600	7,250	
295,901–Greater	166,501–Greater	167,901–179,400	121,301–129,600	74,601–79,800	7,500	
295,901–Greater	166,501–Greater	179,401–191,400	129,601–138,300	79,801–85,100	7,750	
295,901–Greater	166,501–Greater	191,401–Greater	138,301–Greater	85,101–Greater	8,000	

[a] Types of construction are based on NFPA 220.
[b] Measured at 20 psi (139.9 kPa).
Note: The fire flow requirements in Table 1.3 are based on buildings without fire sprinkler protection. Buildings with fire sprinkler protection should receive a fire sprinkler credit to the required fire flow.
Source: Reprinted with permission from the *Uniform Fire Code Handbook* (2006). Copyright © 2006 National Fire Protection Association. All rights reserved.

2. Average daily draft $= \dfrac{180(20,000)}{1,000,000} = 3.6$ mgd

3. Maximum daily draft $= \dfrac{324(20,000)}{1,000,000} = 6.48$ mgd

Section 1.9   Fire Demands

4. Maximum hourly draft $= \dfrac{486(20{,}000)}{1{,}000{,}000} = 9.72$ mgd

5. Fire flow

   Total floor area $= 3(2400) = 7200$ ft^2

   From Table 1.3,

   fire flow $= 2250$ gpm or 3.24 mgd

   duration $= 2$ hours

6. Maximum daily + fire flow $= 6.48 + 3.24 = 9.72$ mgd

7. Pumps: Assume that the required flow is handled by three units and that one reserve unit is installed.

   a. Low-lift pumps $= \dfrac{4}{3}$ (maximum daily) $= \dfrac{4}{3}(6.48) = 8.64$ mgd

   b. High-lift pumps $= \dfrac{4}{3}$ (maximum hourly) $= \dfrac{4}{3}(9.72) = 12.96$ mgd

8. Service reservoir:

   a. Fire flow duration $= 2$ hr

   b. Total quantity of fire flow $= \dfrac{2}{24}$ (3.24) $= 0.27$ mgd

   c. Working storage (given) $= 1.5$ mgd

   d. Emergency storage $=$ (days) (average daily draft) $= 3(3.6) = 10.8$ mgd

   e. Service storage $= 0.27 + 1.5 + 10.8 = 12.57$ mgd

9. Design capacities:

Structure	Basis	Capacity (mgd)
River flow	Maximum daily	6.48
Intake conduit, conduit to treatment	Maximum daily	6.48
Low-lift pumps	Maximum daily plus reserve	8.64
High-lift pumps	Maximum hourly plus reserve	12.96
Treatment plant	Maximum daily	6.48
Service storage	Working storage plus fire plus emergency	12.57
Distribution system	Maximum daily plus fire or maximum hourly, whichever is greater	9.72

1.10 INDUSTRIAL REQUIREMENTS

Table 1.2 included a typical industrial water component from a public water supply system. This is not adequate for a community with large water-using industries. Large industries are usually served by separate supplies. About 5% of the industries use 80% of the industrial water demand; 70% of industrial plants use as little as 2% of the demand. Thermal power stations are the heaviest water users, with a requirement of 700 gpcd or 20 gal/kWh on the average. Other major water-using industries are steel, paper, chemicals, textiles, and petroleum refining. The average industrial requirement for the entire country (excluding thermal power stations) is 100 gpcd. Table 1.4 lists the water intake of major industries, which is the amount of water pumped into the establishment per unit of production. The actual consumption by unit of production for an individual industry is anywhere from 2 to 20%; the balance is discharged. Thus, by recycling wastewater, industries can substantially reduce the quantities of fresh water required. In the foreseeable future, improved plant efficiencies due to new technologies in the industrial sector, increased water recycling, higher energy prices, and changes in environmental regulations will contribute to a decline in industrial water use.

Table 1.4 Requirements of Major Industries

Industry	Average Water Use
Thermoelectric power	20 gal/kWh
Steel	38,000 gal/ton
Paper	39,000 gal/ton
Organic chemicals	55,000 gal/ton
Woolens	140,000 gal/ton
Coke (coal byproduct)	3,600 gal/ton
Petroleum refining	770 gal/barrel

1.11 WASTE DILUTION REQUIREMENTS

In earlier times it was a common practice to dump raw municipal and industrial wastes into the same river that served as the source of supply, thus relying on the self-purification properties of the stream. As long as the streamflow is at least 40 times that of the wasteflow and there is a sufficiently long reach of river to the next city, both nuisance and unsafe conditions can be avoided. But with the rapid growth of cities and industrial activities and with increased demand placed on water supplies, the dumping of raw wastes into rivers is no longer permitted. The problem now is to what extent the municipal and industrial wastes should be treated before discharge into the river (industrial wastes have to be considered even if the industrial supplies are developed from a different source than the municipal supplies). The amount of streamflow required for sufficient natural treatment of municipal and industrial wasteflow is a function of the pollutant characteristics of the waste and streamflow properties with regard to oxygen content, dissolved minerals, water temperature, and length of the available downstream reach. This relationship is depicted in many models, the most common being the oxygen sag curve. For average conditions it has been found that the raw (fully untreated) waste from municipal and industrial sources, excluding thermal power plants, requires a ratio of streamflow to wasteflow of 40, and thoroughly treated waste requires a ratio of 2, with a linear variation in between as shown in Figure 1.6 (Kuiper, 1965).

If water supply is being planned from a reservoir project, there are three annual cost components to be considered: (1) cost of storage to provide the municipal (and industrial)

Figure 1.6 Requirements for waste dilution.

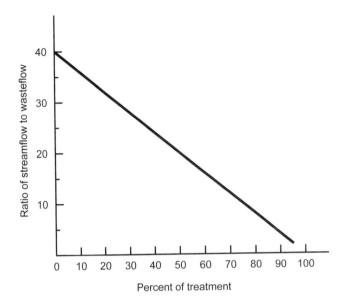

requirements, (2) cost of storage to produce the required quantity for waste dilution, and (3) cost of waste treatment. Items (2) and (3) act opposite to each other; that is, when the degree of treatment is increased, the cost of treatment rises but the cost of waste dilution storage decreases, and vice versa. The various degrees of treatment and the annual costs associated with them are considered until the lowest cost is found that indicates the most economic treatment of the city's waste.

EXAMPLE 1.8

A city had a total withdrawal (excluding in-stream dilution requirements) of 140 mgd in 1997 distributed as follows: municipal usage, 30 mgd; manufacturing industries, 35 mgd; and thermal power, 75 mgd. The city had a population of 200,000, which is expected to rise to 220,000 in the year 2010. An industrial expansion of 20% and a thermal power increase of 80 MW is expected in the city by 2010. Estimate the total withdrawal in the year 2010. If the waste is to be discharged after 80% of the treatment, determine the total water requirements.

SOLUTION

1. Existing per capita municipal usage $= \dfrac{30\left(10^{6}\right)}{200{,}000} = 150$ gpcd

    In the year 2010:

2. Municipal requirements $= \dfrac{(220{,}000)(150)}{1{,}000{,}000} = 33$ mgd

3. Manufacturing industry requirements $= \dfrac{120}{100}(35) = 42$ mgd (20% expansion)

4. Thermal power requirements:

    a. Assuming a plant capacity factor of 0.6, the additional energy produced per day is

$$(80 \text{ MW})(1 \text{ day})(0.6)\left[\frac{(24 \text{ hr})}{(1 \text{ day})}\right]\left[\frac{(1000 \text{ kW})}{(1 \text{ MW})}\right] = 1.15 \times 10^6 \text{ kWh}$$

    b. From Table 1.4, water usage = 20 gal/kWh

    c. Additional water required $= (20)\left(1.15 \times 10^6\right)\left(\dfrac{1}{10^6}\right) = 23$ mgd

    d. Total thermal power requirements = 75 + 23 = 98 mgd (existing plus new)

5. Total for municipal and manufacturing usage = 33 + 42 = 75 mgd

6. Manufacturing and thermal (total industrial) = 42 + 98 = 140 mgd

7. Waste dilution requirements:

    a. From Fig. 1.6, for 80% treatment the streamflow/wasteflow ratio (excluding thermal plant effluent) = 9

    b. Streamflow requirement = 9(75) = 675 mgd

    c. Waste dilution = 675 − (33 + 42 + 98) = 502 mgd

8. City water supply requirements are summarized as follows:

Sector	mgd
Municipal requirements	33
Industrial (including thermal) requirements	140
Waste dilution requirements	502

1.12 DEMAND FOR IRRIGATION WATER

The demand for irrigation water depends on several factors, including the method of irrigation, type of crop to be grown, soil conditions, and prevailing climate. The gross irrigation water requirement is indicated in terms of the depth of water over the irrigable area. The basic quantity of interest is the consumptive use of the crop, from which the water available through rainfall is subtracted and various losses are added to establish the *gross irrigation demand*.

A considerable quantity of water applied to farmland returns to the river system. This includes surface runoff during irrigation, wasted water, canal seepage, and deep percolation. This may be 30 to 60% of the gross requirement. About one-half of this reaches the river through groundwater flow. The remainder reaches the river as surface runoff during the irrigation season and, thus, becomes available for use to downstream projects.

Irrigation projects must include subsurface drainage facilities as a measure against waterlogging and salinity. Drainage aspects are described in Chapter 14.

1.13 CONSUMPTIVE USE OF CROPS

The consumptive use and evapotranspiration from a cropped area are considered synonymous. Numerous procedures have been devised to estimate the consumptive use of water. Two methods commonly applied to cropped areas are described here. The data from actual farm experience or experimental basins, wherever available, are given preference over the computational procedures because of the empirical constants involved in the latter.

1.13.1 Penman-Monteith Method

Currently this is a very powerful tool used in hydrologic practice to assess evapotranspiration. The method is described in detail in Section 2.12.2 and should be studied along with the Blaney-Criddle method.

1.13.2 Blaney-Criddle Method

For the conditions in the arid western regions of the United States, Blaney and Criddle (1945) proposed a relation that determined consumptive use as the multiplication of the mean monthly temperatures, monthly percent of annual daytime hours, and a coefficient for individual crops that varied monthly and seasonally. The coefficients presented originally were the seasonal values for the entire growing season of crops. Subsequently, monthly coefficient values were suggested (Blaney and Criddle, 1962). However, these coefficients did not include the effects of humidity, wind movement, and other climatological factors. The modified Blaney-Criddle method [U.S. SCS, 1970; now the Natural Resources Conservation Service (NRCS)] split the coefficient in two parts to consider these factors indirectly. The modified formula is

$$U = \Sigma K_t \, K_c \, t_m \, \frac{p}{100} \quad \text{[unbalanced]} \tag{1.18}$$

where

U = consumptive use in./month

K_t = climatic coefficient related to mean monthly temperatures

K_c = growth stage coefficient

t_m = mean monthly temperature, °F

p = monthly percentage of annual daytime hours (Table 1.5)

The values of K_t are based on the formula

$$K_t = 0.0173 t_m - 0.314 \quad \text{[unbalanced]} \tag{1.19}$$

For $t_m < 36$ °F, use $K_t = 0.30$.

The monthly values of K_c are obtained from Table 1.6(a) for a perennial crop with a year-round growing season. For other seasonal crops, Table 1.6(b) is used, based on the percentage of the growing season covered by the month in question. The monthly consumptive amounts are summed over the growing season to obtain the seasonal consumptive use.

Table 1.5   Monthly Percentage of Daytime Hours of the Year

Month	Latitude (degrees north of equator)													
	24	26	28	30	32	34	36	38	40	42	44	46	48	50
Jan.	7.58	7.49	7.40	7.30	7.20	7.10	6.99	6.87	6.76	6.62	6.49	6.33	6.17	5.98
Feb.	7.17	7.12	7.07	7.03	6.97	6.91	6.86	6.79	6.73	6.65	6.58	6.50	6.42	6.32
Mar.	8.40	8.40	8.39	8.38	8.37	8.36	8.35	8.34	8.33	8.31	8.30	8.29	8.27	8.25
Apr.	8.60	8.64	8.68	8.72	8.75	8.80	8.85	8.90	8.95	9.00	9.05	9.12	9.18	9.25
May	9.30	9.38	9.46	9.53	9.63	9.72	9.81	9.92	10.02	10.14	10.26	10.39	10.53	10.69
June	9.20	9.30	9.38	9.49	9.60	9.70	9.83	9.95	10.08	10.21	10.38	10.54	10.71	10.93
July	9.41	9.49	9.58	9.67	9.77	9.88	9.99	10.10	10.22	10.35	10.49	10.64	10.80	10.99
Aug.	9.05	9.10	9.16	9.22	9.28	9.33	9.40	9.47	9.54	9.62	9.70	9.79	9.89	10.00
Sept.	8.31	8.31	8.32	8.34	8.34	8.36	8.36	8.38	8.38	8.40	8.41	8.42	8.44	8.44
Oct.	8.09	8.06	8.02	7.99	7.93	7.90	7.85	7.80	7.75	7.70	7.63	7.58	7.51	7.43
Nov.	7.43	7.36	7.27	7.19	7.11	7.02	6.92	6.82	6.72	6.62	6.49	6.36	6.22	6.07
Dec.	7.46	7.35	7.27	7.14	7.05	6.92	6.79	6.66	6.52	6.38	6.22	6.04	5.86	5.65
Annual	100.00	100.00	100.00	100.00	100.00	100.00	100.00	100.00	100.00	100.00	100.00	100.00	100.00	100.00

Table 1.6(a) Crop-Growth-Stage Coefficient K_c (Modified Blaney-Criddle Method) Perennial Crops (Northern Hemisphere)

Crop	Average K_c values by months											
	Jan.	Feb.	Mar.	Apr.	May	June	July	Aug.	Sept.	Oct.	Nov.	Dec.
Alfalfa	0.63	0.74	0.86	0.99	1.09	1.13	1.11	1.06	0.99	0.90	0.78	0.65
Grass pasture	0.48	0.58	0.74	0.85	0.90	0.92	0.92	0.91	0.87	0.79	0.67	0.55
Grapes	0.20	0.23	0.32	0.49	0.70	0.80	0.81	0.76	0.66	0.50	0.35	0.25
Citrus orchards	0.64	0.66	0.68	0.70	0.71	0.72	0.72	0.71	0.70	0.68	0.66	0.64
Deciduous, with cover	0.63	0.74	0.86	0.98	1.09	1.13	1.12	1.06	0.99	0.90	0.78	0.65
Deciduous, no cover	0.17	0.25	0.39	0.63	0.87	0.96	0.95	0.82	0.53	0.30	0.20	0.16
Avocados	0.27	0.42	0.58	0.71	0.78	0.81	0.78	0.71	0.63	0.54	0.43	0.36
Walnuts	0.10	0.14	0.23	0.43	0.68	0.92	0.98	0.88	0.69	0.49	0.31	0.15

Table 1.6(b) Annual Crops

Crop	K_c values at listed % of growing season										
	0	10	20	30	40	50	60	70	80	90	100
Field corn (grain)	0.44	0.49	0.58	0.71	0.93	1.05	1.08	1.06	1.01	0.93	0.85
Field corn (silage)	0.44	0.48	0.55	0.65	0.80	0.97	1.06	1.08	1.06	1.02	0.96
Grain sorghum	0.30	0.38	0.60	0.83	1.01	1.07	0.99	0.88	0.76	0.65	0.56
Winter wheat[a]	1.46	1.44	1.42	1.39	1.35	1.30	1.23	1.15	1.03	0.86	0.78
Spring grains	0.29	0.45	0.67	0.89	1.09	1.28	1.31	1.17	0.90	0.55	0.20
Cotton	0.20	0.25	0.33	0.50	0.79	0.97	1.02	0.95	0.81	0.65	0.29
Dry beans	0.50	0.59	0.71	0.87	1.02	1.10	1.12	1.06	0.94	0.81	0.67
Sugar beets	0.45	0.50	0.61	0.79	0.95	1.10	1.20	1.25	1.21	1.13	1.04
Potatoes	0.33	0.40	0.51	0.72	0.98	1.17	1.31	1.37	1.36	1.31	1.23
Tomatoes	0.45	0.45	0.47	0.56	0.75	0.95	1.03	0.99	0.90	0.80	0.70
Melons and cantaloupe	0.44	0.48	0.56	0.65	0.76	0.81	0.81	0.78	0.75	0.71	0.67
Small vegetables	0.29	0.40	0.57	0.69	0.77	0.81	0.82	0.79	0.72	0.58	0.38

[a] Data given only for springtime season of 70 days prior to harvest (after last frost). K_c increases from 0.50 at seeding to 1.46 during period with average temperature below 32°F.
Source: Davis and Sorensen (1969).

EXAMPLE 1.9

For the growing season of sugar beets at Limberly, Idaho, located at latitude 42.4° N, the long-term mean monthly air temperatures are given in column 2 of Table 1.7. The crop is planted April 11 and harvested October 15. Estimate the seasonal consumptive use of water.

SOLUTION The monthly consumptive values are arranged in Table 1.7.
Using values from col. 8 of Table 1.7, the seasonal consumptive use is:

$$U = \frac{20}{30}(0.89) + 2.15 + 4.01 + 7.06 + 7.13 + 3.96 + \frac{15}{31}(2.03)$$
$$= 25.92 \text{ in.}$$

Table 1.7 Consumptive Use Computation

(1)	(2)	(3)	(4)	(5)	(6)	(7)	(8)
Period	Mean Monthly Temp. (°F)	Number of Days in Period	Mid-period Percent of Total Season[a]	K_c from Table 1.6(b)	K_t[b]	Percent p from Table 1.5	U^c (in./ month)
Apr. 11–30	44.5	20	5	0.48	0.46	9.01	0.89
May	55.2	31	19	0.60	0.64	10.16	2.15
June	60.8	30	35[d]	0.87	0.74	10.24	4.01
July	69.5	31	51	1.10	0.89	10.38	7.06
Aug.	68.6	31	68	1.24	0.87	9.64	7.13
Sept.	57.9	30	84	1.18	0.69	8.40	3.96
Oct. 1–15	47.9	15	96	1.08	0.51	7.69	2.03
Total		188					

a $\dfrac{\text{Number of days up to middle of the period}}{\text{Total days in growing season}}$

b $K_t = 0.0173(\text{col. 2}) - 0.314.$

c $U = \dfrac{(\text{col. 2})(\text{col. 5})(\text{col. 6})(\text{col. 7})}{100}$

d $\dfrac{20 + 31 + 15}{188}$

1.14 EFFECTIVE RAINFALL

The portion of the rainfall during the growing season that is utilized in meeting the requirements of crops is termed the *effective rainfall*. The remainder is lost through surface runoff and deep percolation. In humid areas, this may provide a major portion of the requirements, whereas in arid areas it may constitute only a small part. The necessity of irrigation in humid regions may arise due to the unbalanced distribution of rainfall.

Section 1.14 Effective Rainfall   **25**

Effective rainfall is influenced by many factors relating to (1) soil moisture, (2) cropping pattern, (3) application of irrigation, and (4) rainfall characteristics. Based on the study of extensive data, the Soil Conservation Service (1964) suggested the relationship shown in Table 1.8. The limitation on use is given at the bottom of the table.

Whereas the crop consumptive-use requirements vary from year to year by a small margin, the variations in rainfall are large. As such, the frequency analysis of effective rainfall is made as follows:

1. For the region under consideration, available data on monthly rainfall are collected.

2. Using Table 1.8, the effective rainfall figures for each month of record are determined.

3. For each year on record, the total effective precipitation for all the months of the growing season is determined.

4. From the resultant values—one for each year on record—a frequency curve is prepared by the method of Section 8.8.

5. If an irrigation supply is desired that will be adequate 90% of the time (9 of 10 years), the effective rainfall corresponding to the 90% value of the frequency curve is observed.

6. The total effective rainfall is distributed over the months of the growing season in the ratio indicated by the 10 driest years on record.

Table 1.8   Average Monthly Effective Rainfall Related to Mean Monthly Rainfall and Average Monthly Consumptive Use

Monthly mean rainfall (in.)	Average monthly consumptive use, U (in.)									
	1.0	2.0	3.0	4.0	5.0	6.0	7.0	8.0	9.0	10.0
	Average monthly effective rainfall (in.)[a]									
0.5	0.20	0.25	0.30	0.30	0.30	0.35	0.40	0.45	0.50	0.50
1.0	0.55	0.60	0.65	0.70	0.70	0.75	0.80	0.85	0.95	1.00
2.0	1.00	1.25	1.35	1.55	1.55	1.55	1.60	1.70	1.85	2.00
3.0	1.00	1.85	1.95	2.10	2.20	2.30	2.40	2.55	2.70	2.90
4.0	1.00	2.00	2.55	2.70	2.90	2.95	3.15	3.30	3.50	3.80
5.0	1.00	2.00	3.00	3.25	3.50	3.60	3.85	4.05	4.30	4.60
6.0	1.00	2.00	3.00	3.80	4.10	4.25	4.50	4.80	5.10	5.40
7.0	1.00	2.00	3.00	4.00	4.60	4.80	5.05	5.40	5.70	6.05
8.0	1.00	2.00	3.00	4.00	5.00	5.30	5.60	5.90	6.20	
9.0	1.00	2.00	3.00	4.00	5.00	5.75	6.05	6.35		

[a] Based on 3-in. net depth of application for irrigation. For other net depths of application, multiply by the factors shown below.

Net depth of application	0.75	1.0	1.5	2.0	2.5	3.0	4.0	5.0	6.0	7.0
Factor	0.72	0.77	0.86	0.93	0.97	1.00	1.02	1.04	1.06	1.07

Note: Average monthly effective rainfall cannot exceed average monthly rainfall or average monthly consumptive use. When the application of the factors above results in a value of effective rainfall exceeding either, this value must be reduced to a value equal to the lesser of the two.

Source: U.S. SCS (1964), now the Natural Resources Conservation Service (NRCS).

1.15 Farm Losses

The losses that take place from the water delivered to the farm are measured by the *on-farm efficiency*. Thus

$$\text{on-farm efficiency} = \frac{\text{water utilized in crop growth}}{\text{water delivered to farm}} \qquad (1.20)$$

The principal factors that affect efficiency include (1) the method of applying the water, (2) the texture and condition of the soil, (3) the slope of the land, (4) the preparation of the land (i.e., ditched or bordered), (4) the rate of irrigation flow in relation to the farm size, and (5) the management practices of the irrigator. Farm losses take place due to (1) deep percolation beyond the root zone of crops, (2) surface wastes from the fields, and (3) on-farm distribution losses and nonproductive consumption.

For deep percolation, a minimum allowance of about 20% of the applied water is made. This ensures adequate leaching if the applied water does not contain more than 1300 ppm of dissolved salts and the drained water can be accepted with 6000 ppm of dissolved salts. It may be necessary to pass additional water for leaching if the applied irrigation water is more saline, the drained water has to have a lower salinity, or the soil requires reclamation. Refer to Hansen et al. (1980) for the leaching requirements.

Surface waste or runoff is inherent in most irrigation systems. This quantity is, however, recovered and becomes available for use within a project or elsewhere in the basin downstream. Surface wastes usually amount to 6 to 10% of the quantity delivered to the fields. Distribution losses vary from a very small quantity in pipelines to about 15% in unlined ditches.

On-farm efficiency, which is a product of the efficiency of each of the items above, can be achieved in the range of 40 to 70% for a properly designed efficient irrigation system. The higher value in the range above is associated with the sprinkler system.

1.16 Conveyance Losses and Waste

Conveyance losses and waste relate to the water lost between the point of diversion from a stream or reservoir to the points of delivery to farms and is measured by the *off-farm efficiency*. The losses comprise the evaporation through the canal system, the water seeped through the canal system, and the operational wastes discharged into drains or streams. Evaporation losses from the canal water surface are not too large unless the canal is very shallow and wide. Usually, these are less than 1% of the canal flow.

Seepage losses from canals depend on (1) the permeability of the soil, (2) the wetted surface of the canal, and (3) the difference in level of water in the canal and the adjacent groundwater table. The average unit seepage rates for the western United States are similar to those cited by Hart (Worstell, 1975); see Table 1.9.

The loss rates from lined canals are between about 0.1 and 1.0 ft/day (Worstell, 1975). To determine the total quantity of seepage from a canal system, the data required are (1) the predominant soil texture to ascertain the average seepage rates, (2) the widths, and (3) the lengths of the canals. The chart developed by Worstell (1975), illustrated in Figure 1.7, can be used to estimate the seepage loss in cfs per mile for different canal widths. Broadly, the seepage losses range from 15 to 45% of diverted flow on unlined canals and from 5 to 15% on lined canals.

Table 1.9 Average Seepage Loss from Canals in Southern Idaho

Type of Soil	Seepage Loss (ft/day)
1. Medium clay and loam	0.5–1.5
2. Impervious clay	0.5
3. Medium soils	1.0
4. Somewhat pervious soils	1.5–2.0
5. Gravel	2.5–5.0
Source: Hart (1963).	

Operational wastes are unavoidable. These result from the inability to release into the canal system the quantity to match exactly all the requirements, operation of the canals at high levels to reduce siltation, unexpected rainfalls, and breaches in the system. These losses range from 5 to 30% of diversions on projects with ample supplies and from 1 to 10% with limited supplies.

Off-farm efficiency, comprising the foregoing items of conveyance losses, ranges between 50 and 90%. In cases where water originates on the farm itself, such as from a well, the off-farm efficiency is 100%. The average irrigation efficiency for the entire United States is indicated in Table 1.10.

Figure 1.7 Chart to estimate the seepage losses from a canal (from Worstell, 1975).

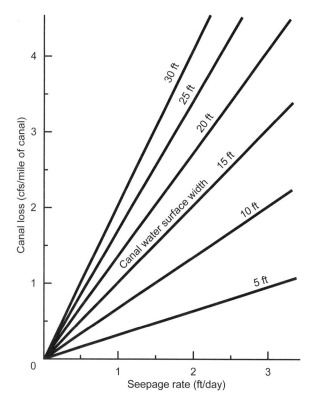

Table 1.10   Average Irrigation Efficiency in the United States

	Trend Efficiency %	High Efficiency %
On-farm	59	66
Off-farm	83	88
System	49	58

System efficiency is the overall efficiency obtained by multiplying the on-farm efficiency by the off-farm efficiency. Trend efficiency reflects irrigation/water improvements following the present trend in upgrading of systems. High efficiency considers an accelerated program of improving irrigation systems and water management using the best practical technology available.

1.17 COMPUTATION OF IRRIGATION DEMANDS

The computation for gross water requirements proceeds as follows in accordance with the procedure for each item described in previous sections:

1. Compute the monthwise *consumptive use or crop requirements or evapotranspiration* for the specified crop.

2. Estimate the *irrigation requirements* by subtracting from the consumptive use the *effective rainfall* available for plant growth. To this quantity the following items are included (if applicable): (a) irrigation applied prior to crop growth should be added; (b) water required for leaching should be added; (c) miscellaneous requirements of germination, frost protection, plant cooling, and so on, should be added; and (d) decrease in soil moisture should be subtracted.

3. Assess the *farm delivery requirements* by dividing the irrigation requirements of item 2 by the on-farm efficiency.

4. Compute the *gross water requirements* by dividing the farm delivery requirements of item 3 by the off-farm efficiency.

EXAMPLE 1.10

An irrigation project serves an area of 50,000 acres. The cropping pattern* is: alfalfa, 30%; wheat, 50%; rice, 30%; cotton, 20%. The monthly consumptive use values for these crops, which are calculated by the procedure of Section 1.13.2, are given in Table 1.11. The monthly effective rainfall values, which are calculated by the procedure of Section 1.14, are also given in the table. The irrigation water applied prior to crop growth and the soil moisture withdrawal in certain cases are also indicated in the table. On-farm efficiency is 60% and off-farm efficiency is 90%. Determine the monthly and total irrigation demands.

SOLUTION   Refer to Table 1.11.

---

* The cropping pattern is defined as the percent of the total irrigable area devoted to each crop during each of the two principal growing seasons of a year. Each area used for crops in both seasons will be counted twice. The perennial crops using water in all 12 months will also be counted twice. Complete utilization of land in both seasons will sum up to 200%.

Table 1.11 Computation of Monthly Irrigation Demands

Item	Jan.	Feb.	Mar.	Apr.	May	June	July	Aug.	Sept.	Oct.	Nov.	Dec.	Total
R (in.)	0.8	1.0	1.9	2.0	1.4	1.9	3.9	2.3	1.8	1.0	0.8	0.7	19.5
1. Alfalfa (30% of irrigable area)													
U (in.)	1.40	1.71	2.0	2.33	4.14	6.06	7.94	6.65	3.77	3.00	2.10	1.50	42.6
PP (in.)(+)													0
SM (in.)(−)													0
IR, gross (in.)	1.40	1.71	2.0	2.33	4.14	6.06	7.94	6.65	3.77	3.00	2.10	1.50	42.6
IR, net (in.)	0.6	0.71	0.10	0.33	2.74	4.16	4.04	4.35	1.97	2.00	1.30	0.80	23.1
IR, eff. (in.)	0.18	0.21	0.03	0.10	0.82	1.25	1.21	1.31	0.59	0.60	0.39	0.24	6.93
2. Wheat (50%)													
U (in.)	2.11	3.43	6.29	5.90						1.30	1.75	1.50	22.28
PP (in.)(+)									2.10				2.10
SM (in.)(−)				2.20									2.20
IR, gross (in.)	2.11	3.43	6.29	3.70					2.10	1.30	1.75	1.50	22.18
IR, net (in.)	1.31	2.43	4.39	1.70					0.30	0.30	0.95	0.80	12.18
IR, eff. (in.)	0.66	1.21	2.20	0.85					0.15	0.15	0.48	0.40	6.10
3. Rice (30%)													
U (in.)					6.80	10.42	12.53	9.50					39.25
PP (in.)(+)				9.00									9.00
SM (in.)(−)								2.50					2.50
IR, gross (in.)				9.00	6.80	10.42	12.53	7.00					45.75
IR, net (in.)				7.00	5.40	8.52	8.63	4.70					34.25
IR, eff. (in.)				2.10	1.62	2.56	2.59	1.41					10.28

Item	Jan.	Feb.	Mar.	Apr.	May	June	July	Aug.	Sept.	Oct.	Nov.	Dec.	Total
4. Cotton (20%)													
U (in.)			1.34	2.82	2.96	5.36	6.86	4.14	1.14				24.62
PP (in.)(+)		1.00											1.00
SM (in.)(−)								1.5	0.5				2.00
IR, gross (in.)		1.00	1.34	2.82	2.96	5.36	6.86	2.64	0.64				23.62
IR, net (in.)		0	0	0.82	1.56	3.46	2.96	0.34	0				9.14
IR, eff. (in.)		0	0	0.16	0.31	0.69	0.59	0.07	0				1.82
Total, IR (eff.)(in.)	0.84	1.42	2.23	3.21	2.75	4.50	4.39	2.79	0.74	0.75	0.87	0.64	25.13
IR for area (acre-ft × 10^3)	3.50	5.92	9.29	13.38	11.46	18.75	18.29	11.62	3.08	3.13	3.62	2.67	104.71
Farm delivery @ 60% efficiency (acre-ft × 10^3)	5.83	9.87	15.48	22.30	19.10	31.25	30.48	19.37	5.13	5.21	6.03	4.45	174.52
Gross requirements @ 90% (acre-ft × 10^3)	6.48	10.97	17.20	24.78	21.22	34.72	33.87	21.52	5.70	5.79	6.70	4.94	193.91

Abbreviations:

R = effective rainfall
U = consumptive use
PP = irrigation applied prior to crop growth
SM = soil moisture withdrawal
IR, gross = gross irrigation required = $U + PP - SM$
IR, net = net irrigation required = IR(gross) − R
IR, effective = IR(net) × percent irrigable area
IR for area = Farm area × Total IR (eff.)

$$\text{Farm delivery} = \frac{\text{IR for area}}{\text{farm efficiency}}$$

$$\text{Gross requirement} = \frac{\text{farm delivery}}{\text{off-farm conveyance efficiency}}$$

1.18 DEMAND FOR HYDROPOWER

Water power involves the nonconsumptive use of water. This feature makes the water utilization distinct in two respects: (1) hydropower generation can readily be integrated with other development objectives, and (2) all resources (streamflows) available at a site are evaluated from the consideration of power-producing potential. With proper planning, a very high percent of the total available streamflow in a river basin may be used for hydropower through a series of power plants. The challenge lies in locating potential hydropower sites that are within a reasonable transmission distance of the power market under consideration. Since hydroenergy is the product of the available head and the available flow (multiplied by a certain constant), sites that offer a good combination of head and flow are investigated.

In terms of producing water head, rapids, falls, and dams offer good hydropower potential. Whereas the head at a site is practically constant, the available flows are highly variable. The study of maximum flows is important from the viewpoint of the design or installed capacity of the power plant; the average flows are important from the consideration of the energy output, and minimum flows are required to predict the dependable plant capacity. Since the entire quantity available at a site (except the flood flows) is utilized in power production, the study of water demands for hydropower amounts to the collection of streamflow data and their analysis. Usually, the analysis relates to the preparation of the flow-duration curve discussed in Section 7.27.2, which indicates the magnitude of discharge against the percentage of time that discharge is exceeded at a site.

There are two types of hydropower plants: (1) a run-of-river plant uses direct streamflows, and its energy output is subjected to the instantaneous flow of the river; and (2) a storage plant with a reservoir is able to produce increased dependable energy on the basis of a controlled water release. If the reservoir serves only to smooth out the weekly fluctuations in streamflows, the plant is said to have a pondage capacity. On the other hand, a reservoir that serves to store water from the wet season to the dry season is said to have a storage capacity.

1.18.1 Power and Energy Production from Available Streamflows

When the number of pounds or Newtons of fluid flowing per second (γQ) is lowered by level h, it releases energy at a rate of (γQh). Including the efficiency term, an equation for power (rate of energy) can be given by

$$P = \gamma Qhe \quad [\text{FLT}^{-1}] \tag{1.21}$$

where

P = plant capacity

Q = discharge through the turbines

h = net head on the turbines

e = combined efficiency for turbines and generators

Flow-duration curves developed from long-term monthly streamflow records offer a convenient tool in plant capacity design. The procedure for preparation of a flow-duration curve is described in Section 7.27.2. A typical curve is shown in Figure 1.8.

In eq. (1.21), with an average value of head, the efficiency and head are practically constant for a plant. Thus the power is directly proportional to the flow. In other words, the curve in Figure 1.8 indicates the power production with a suitable modification of the vertical scale. The design or installed capacity of a plant is based on the maximum flow, which is usually taken to be Q_{15} (i.e., flow exceeded 15% of the time). Floodflows above this magnitude are allowed to overflow without producing power. In metric units, using $\gamma = 9.81$ kN/m³ and taking $e = 0.84$, the installed capacity in kW is given by

$$P_{instal} = 8.24 Q_{15} h \text{ (metric units)} \quad [FLT^{-1}] \tag{1.22a}$$

In FPS units, using $\gamma = 62.4$ lbs/ft³ and taking $e = 0.84$ and converting units, the installed capacity is

$$P_{instal} = \frac{Q_{15} h}{14} \text{ (English units)} \quad [FLT^{-1}] \tag{1.22b}$$

where

$$P_{instal} = \text{installed capacity, kW}$$
$$Q_{15} = \text{discharge with 15% exceedance, cfs}$$
$$h = \text{net head, ft}$$

If the time scale (abscissa) in Figure 1.8 is expressed in terms of hours in a year, the area under the curve will provide the annual energy production. Mathematically,

$$E = 8.24 Q_{av} h \, (8760) \text{ (metric units)} \quad [FL] \tag{1.23a}$$

$$E = \frac{Q_{av} h}{14} (8760) \text{ (English units)} \quad [FL] \tag{1.23b}$$

where

$$E = \text{annual energy, kWh}$$
$$Q_{av} = \text{average discharge, cfs}$$
$$8760 = \text{number of hours in a year}$$

Q_{av} is the average discharge under the curve in Figure 1.8 taking Q_{15} as the highest magnitude of discharge, similar to eq. (7.62).

A plant capacity factor is the ratio of the average power production to the installed capacity. This is practically equal to the ratio Q_{av}/Q_{15}, assuming that the head and the efficiency are essentially constant. By reservoir storage, both Q_{av} and Q_{min} are improved, and thus the annual energy production and the dependable (firm) power are enhanced. The plant capacity factor also increases, resulting in more efficient use of a plant. A plant capacity factor of 0.6 is common for storage-type power plants.

Energy computations assume that an adequate number and adequate sizes of turbine units have been installed to utilize the minimum available flow. If only one turbine unit is provided, its operative range is generally from 30 to 110% of the turbine design flow, which means that the turbine will be inoperative during the times the flow is less than 30% of the design value. Thus the energy production will be for a shorter period in a year and the total annual generation will, accordingly, be less.

Similarly, depending on the turbine type, there is an operating limitation on the head. Usually, a turbine can operate in the range of 60 to 120% of the design head. It is considered that the available head is fairly constant or that an average value of head is used in energy computations by eq. (1.23) when there are small fluctuations, which is the case with

Section 1.18   Demand for Hydropower

Figure 1.8 Flow-duration curve.

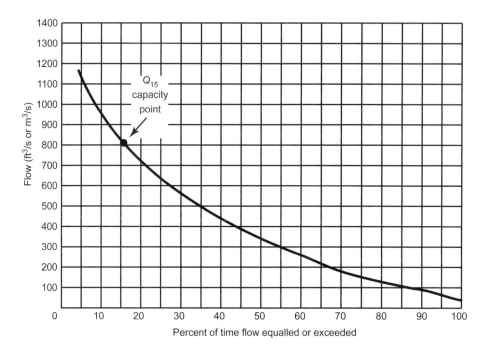

run-of-river projects and projects with remote location of power plants. If variations in head are substantial, a sequential analysis is made where the energy calculations are made in steps at different intervals.

EXAMPLE 1.11

At the Rimmon Pond site on the Naugatuck River near Seymourtown, Connecticut, in the Housatonic basin (drainage area 300 mi^2), the flow-duration data from the monthly flow records are as given in Figure 1.8. The average head is 30 ft. Assess the site for its hydropower potential.

SOLUTION

1. From Figure 1.8, $Q_{15} = 810$ cfs

2. From eq. (1.22b),

$$P_{instal} = \frac{810(30)}{14} = 1736 \text{ kW}$$

3. $Q_{av} = 0.175Q_{15} + 0.075Q_{20} + 0.10(Q_{30} + Q_{40} + Q_{50}$
$\qquad + \ Q_{60} + Q_{70} + Q_{80} + Q_{90}) + 0.05\tilde{Q}_{100}{}^*$
$\qquad = 0.175(810) + 0.075(705) + 0.10(550 + 430 + 340 + 260 + 180$
$\qquad + \ 130 + 90) + 0.05(40)$
$\qquad = 395$ cfs

* Approximate 100% flow.

4. From eq. (1.23b),

$$E = \frac{395(30)(8760)}{14} = 7.41 \times 10^6 \text{ kWh}$$

5. Plant capacity factor $= \dfrac{Q_{av}}{Q_{15}} = \dfrac{395}{810} = 0.49$

1.19 DEMAND FOR NAVIGATION

There are three different methods to provide navigable waterways: (1) river regulation, (2) lock-and-dam, and (3) artificial canalization. In the first method, a river channel is improved by means of river training works and dredging. In some sections of the river channel, the natural depth of water often is not sufficient to maintain navigability, which requires release of water from upstream reservoirs. This demand from the reservoirs is likely to be on the order of several thousand cubic feet per second successively for several months. Thus huge reservoir capacities of several million acre-feet are needed for navigation purposes. One of the shortcomings of this method is that water deficiencies are usually in the lower reaches of a river, while the reservoir sites are in its upper part. This results in many technical, operational, and legal difficulties in maintaining the navigable flow in downstream reaches.

In the second method, the depth of water for low streamflow is increased behind a series of dams through a succession of backwater curves. At each dam, a shiplock is provided to negotiate the difference in water levels upstream and downstream of the dam. The water demand relates to the (1) evaporation losses from the reservoir pools, (2) water requirements for locking operations, and (3) leakages at shiplocks.

Each locking operation requires the release of water in the downstream direction equivalent to the volume of the lock between the upstream and downstream levels. This might involve a flow of over 1000 acre-ft/day (500 cfs). The water for lockages is not accumulative since the water displaced by one lock subsequently can be used by the next lock downstream. Compared to the locking requirements, the evaporation and leakages are insignificant.

The third method provides for an artificially constructed new channel with a number of shiplocks. This method is adopted either to connect two different river systems or in situations where the other two methods are not suitable. As regards the water demand, a flow of several hundred cubic feet per second has to be maintained through the channel. This is supplied from a stream with a natural dependable flow, or by a reservoir. In addition, provisions must be made for the requirements of evaporation, lockage operation, and leakage as discussed for the second method. If an unlined channel is constructed, the seepage losses also have to be included.

PROBLEMS

1.1 A community had a population of 12,000 in 1980, which is increased to 20,000 in 2005. The saturation population is 80,000. Estimate the 2015 population by (a) arithmetic growth, (b) constant percent increase, and (c) decreasing rate of increase.

1.2 Using the following census figures, estimate the population for 2010 by (a) the graphical method, and (b) the most appropriate mathematical method.

Year	Population (thousands)
1970	35.8
1980	38.3
1990	40.8
2000	43.3

1.3 From the following census data, estimate the 1995 and 2010 population by (a) the graphical method, and (b) the most appropriate mathematical method.

Year	Population
1970	25,000
1980	30,500
1990	37,250
2000	45,500

1.4 A water supply reservoir has an annual capacity of 25×10^3 acre-ft. It is serving a city with a present population of 60,000, which is expected to increase to 100,000 in 20 years. How many years from now will the city's needs exceed the reservoir's capacity? Assume arithmetic growth of the population and an average daily draft from the reservoir of 160 gallons per person. Disregard the losses.

1.5 It is estimated that a community will have a population of 15,000 fifteen years from today. The water treatment plant of the community has a capacity of 7.0 mgd, which is adequate for 35 years from the present, with an input rate to the plant of 175 gallons per person per day. If the community is growing at a geometric rate, what is the current population?

1.6 A community has a current population of 28,000. It is estimated that in 20 years its population will be 38,000. The saturation population is expected to be 80,000. The total water consumption at present has been estimated to be 4.0 million gallons per day. The treatment plant has a capacity of 9.2 million gallons per day. Determine how many years from now the consumption will reach its design capacity with current usage rates if the community has a declining growth rate.

1.7 The continental United States registered the following populations. Determine (a) the saturation population, (b) the equation of the logistic curve, and (c) the projected population in the year 2010.

Year	Population (millions)
1840	9.6
1920	76.0
2000	225.1

1.8 A city has the following census data. Fit a logistic curve to the data and determine **(a)** the saturation population, and **(b)** the population in the year 2020.

Year	Population (thousands)	Year	Population (thousands)
1880	15.0	1950	67.8
1890	20.0	1960	78.0
1900	25.0	1970	83.4
1910	32.0	1980	91.8
1920	40.0	1990	96.6
1930	47.5	2000	103.2
1940	58.8		

1.9 A community located in the city of Problem 1.8 has the following census data. Estimate the population for 2000 by **(a)** the constant ratio method, and **(b)** the changing ratio method using the graphical extension.

Year	Population (thousands)	Year	Population (thousands)
1920	6.5	1960	16.0
1930	8.4	1970	17.4
1940	11.2	1980	19.5
1950	13.5		

1.10 Estimate the 2000 and 2020 population of the community in Problem 1.9 by simple regression analysis from the data in Problems 1.8 and 1.9.

1.11 Average daily usage of water in a city is 175 gallons per capita per day (gpcd), which excludes the fire demands. Determine **(a)** the maximum monthly usage in gpcd, **(b)** the maximum weekly usage in gpcd, **(c)** the maximum daily usage in gpcd, and **(d)** the maximum hourly requirement for water.

1.12 A residential area has typical single-family timber dwellings of 3000 ft^2 floor area. Determine the fire demand and its duration for the area.

1.13 The fire demand of a community is dictated by a two-story building of ordinary construction with each floor area equal to 2,500 ft^2. Determine the daily requirement (quantity) for firefighting purposes.

1.14 An industrial four-story building of noncombustible fire resistive construction measures 5000 ft^2 for each floor area. Determine the total fire demand.

1.15 Determine the fire flow for a four-story wood frame building with each floor area of 100 m^2, which is connected with a six-story building of noncombustible nonresistance construction rating that has each floor area of 90 m^2.

1.16 The population of a community is 50,000 and the average daily usage of municipal supply is 175 gpcd. The largest building in the development is a three-story timber framework dwelling of 1-hour rating having a total floor area of 5000 ft^2. Determine the design flow for the following:

a. Groundwater source development

b. Conduit to the treatment plant

Problems

    c. Water treatment plant

    d. Pumping plant

    e. Distribution system

    f. Service reservoir given working storage of 2.5 mgd and three days of emergency storage

1.17 In 1996 the water requirements of a city of 500,000 population were as follows:

Municipal:	87 mgd
Industries:	
Manufacturing	115 mgd
Thermal power	210 mgd
Waste dilution:	6.59 bgd (billion gallons/day)

In the year 2010 it is expected that the population will increase by 10%, industries by 15%, and the thermal power by 100 MW. Determine the requirements by each sector assuming the same level of waste treatment as at present. Assume a plant capacity factor of 0.6.

1.18 At Boise, Idaho, latitude 43°54'N, the long-term mean monthly temperatures are as follows:

Month	Temp. (°F)	Month	Temp. (°F)
Jan.	27.9	July	72.5
Feb.	33.6	Aug.	71.0
Mar.	41.4	Sept.	61.2
Apr.	49.1	Oct.	50.1
May	56.1	Nov.	39.7
June	64.5	Dec.	30.4

Compute the seasonal consumptive use of water for an alfalfa crop having a growing season of April 1 to September 15.

1.19 For the Boise, Idaho, climate in Problem 1.18, compute the seasonal consumptive water use for potatoes. The growing season is May 10 to September 15.

1.20 For the Boise, Idaho, climate in Problem 1.18, compute the seasonal consumptive water use for grain sorghum. The growing season is June 5 to November 2.

1.21 An irrigation project serves an area of 100,000 acres. The cropping pattern is: wheat, 40%; potatoes, 30%; grain sorghum, 35%; and citrus, 25%. The monthly consumptive use and the effective irrigation for these crops are given in the following table. The irrigation water applied prior to crop growth and the soil moisture withdrawals for certain months are also indicated. The on-farm irrigation efficiency is 65% and the off-farm conveyance efficiency is 85%. Determine the monthly diversions and total demand for irrigation.

Item	Jan.	Feb.	Mar.	Apr.	May	June	July	Aug.	Sept.	Oct.	Nov.	Dec.
R (in.)	0.8	0.9	1.5	1.9	1.7	2.5	4.5	3.2	2.1	1.4	0.6	0.5
Wheat												
U (in.)	2.1	3.2	5.95	5.70						1.5	1.75	1.40
PP (in.)									2.2			
SM (in.)				1.5								
Potatoes												
U (in.)					1.52	3.65	8.58	8.53	4.95			
PP (in.)			3.00									
SM (in.)								0.50	2.0			
Grain sorghum												
U (in.)						2.04	5.36	6.59	3.43	1.39	0.49	
PP (in.)					1.0							
SM (in.)												
Citrus												
U (in.)	1.35	1.75	2.75	4.1	4.4	5.0	6.8	7.5	5.9	4.9	2.5	1.6
PP (in.)												
SM (in.)												

Abbreviations:

R = effective rainfall

U = consumptive use

PP = irrigation applied prior to crop growth

SM = soil moisture withdrawal

1.22 Flow-duration data for the Housatonic River near New Milford Town, Connecticut, are indicated below. The average head at the site is 25 ft. Assess the site with respect to (a) potential capacity, (b) annual energy generation, and (c) plant capacity factor.

Flow (cfs)	450	930	1180	1370	1680	1950	2180	2360
Percent of time flow exceeded	100	90	80	70	50	30	20	15

1.23 At the Harrisville, New York, site on the West Oswegatchie River, the flow-duration data are as given below. The average head is 15 m. Assess the site for (a) potential capacity, (b) annual energy generation, and (c) plant capacity factor.

Flow (m^3/s)	40	30	21	15.5	10	7	2.2	1
Percent of time flow exceeded	9	12	15	19	28	47	85	100

1.24 In Problem 1.23, if the plant capacity factor is increased to 0.65 by the storage capacity, determine the percent increase in the annual energy generation.

2

Availability of Water

◆◆◆

2.1 ASSESSMENT OF WATER QUANTITY

Designing a water resources project is essentially an exercise in matching the demand for water with its supply. Two obvious sources of supply are surface water and groundwater. Measurement techniques and issues related to quantitative assessment are the basic elements of hydrology that are covered in Chapters 3 through 8. In this chapter we provide a summary of the fundamental processes that contribute to the formation of surface and groundwater flows and a discussion of the key parameters of the process. An understanding of these parameters facilitates hydrologic analyses and planning.

2.2 HYDROLOGIC CYCLE

Both surface and groundwater flows originate from precipitation, which includes all forms of moisture falling on the ground from clouds, including rain, snow, dew, hail, and sleet. Precipitation at any place is distributed as follows:

1. A portion known as the *interception* is retained on buildings, trees, shrubs, and plants. This is eventually evaporated.

2. Some of the remaining precipitation is evaporated back into the atmosphere directly.

3. Another portion infiltrates into the ground. A part of the infiltration in the root zone is consumed by plants and trees and ultimately transpired into the atmosphere.

4. The water that percolates deeper into the ground constitutes the groundwater flow. It may ultimately appear as the *baseflow* in streams.

5. If the precipitation exceeds the combined evaporation and infiltration, puddles known as *depression storage* are formed. Evaporation takes place from these puddles.

6. After the puddles are filled, the water begins flowing over the surface to join a stream channel. With reference to precipitation this is called the *precipitation excess*. From a consideration of the surface water flow, this is known as *direct runoff*. Some evaporation takes place from the stream surface.

7. A layer of water is formed as runoff occurs. The water in this layer is known as *detention storage*. Evaporation takes place from this storage as well. When precipitation ceases, the water in detention storage eventually joins the stream channel.

8. The destination of all streams is open bodies of water, such as oceans, seas, and lakes, which are subject to extensive evaporation.

9. The evaporation from all the sources above, together with the transpiration, carries moisture into the atmosphere. This results in the formation of clouds that contribute to precipitation, through which steps 1 through 9 repeat.

This chain process, driven principally by energy from the sun, is known as the *hydrologic cycle*. The complete cycle is global in nature. Subcycles with smaller boundary limits also exist.

2.3 WATER BALANCE EQUATION

In quantitative terms the hydrologic cycle can be represented by a closed equation which represents the principle of conservation of mass, often referred to in hydraulics as the *continuity equation*. Many forms of this expression, called the *water balance equation*, are possible by subdividing, consolidating, or eliminating some of the terms, depending on the purpose of computation. The water balance can be expressed (1) for a short interval or for a long duration; (2) for a natural drainage basin or an artificially separated boundary or with respect to water bodies such as lakes, reservoirs, and groundwater basins; and (3) for the phase above the ground surface, that below the surface, or the entire phase. Three applications of the water balance equation are common: (1) a water balance equation for large basin areas, (2) a water balance equation for water bodies, and (3) a water balance equation for direct runoff. In the first two cases the entire phase above and below the ground surface is considered in the equation in terms of the streamflows. The infiltration term, I, that drops out in the entire phase appears in the direct runoff case above the ground surface. In its general form, the equation may be represented by

$$P + Q_{SI} + Q_{GI} - E - Q_{SO} - Q_{GO} - \Delta s - n = 0 \quad [\text{L}^3 \text{ or } \text{L}^3\text{T}^{-1}] \qquad (2.1)$$

where

$$P = \text{precipitation}$$
$$Q_{SI}, Q_{GI} = \text{surface and groundwater inflow into the boundary from outside}$$
$$E = \text{evaporation (including transpiration)}$$
$$Q_{SO}, Q_{GO} = \text{surface and groundwater outflow from the boundary}$$
$$\Delta s = \text{change of storage volume within the boundary}$$
$$n = \text{discrepancy term}$$

Since all water balance components are subject to errors of measurement and estimation, a discrepancy term has been included. The components of eq. (2.1) are expressed as a volume of water or in the form of flow rates or as a mean depth over the basin. The last form is convenient for the balance equation of direct runoff.

2.3.1 Balance Equation for Water Bodies for Short Duration

The water balance equation for reservoirs, lakes, streams, and groundwater reservoirs is used to predict the consequences of the prevailing inflow and outflow conditions on the body of water. The equation is relevant for day-to-day operation of the water body. A short time period is involved in these studies and the term Δs must be considered. If the inflow terms of eq. (2.1), Q_{SI} and Q_{GI} are combined into one term, Q_i, and the outflow terms Q_{SO} and Q_{GO} into one term, Q_o, eq. (2.1) can be represented, ignoring the discrepancy term, by

$$P + Q_i - E - Q_o - \Delta s = 0 \quad [\text{L}^3] \qquad (2.2)$$

where

Q_i = inflow (volume) into water body

Q_o = outflow (volume) from water body

Δs = change of storage volume during Δt

When only a segment of a river is involved, the terms P and E within the river reach can be dropped.

EXAMPLE 2.1

At a particular time the storage in a river reach is 55.3 acre-ft. At that instant, the inflow into the reach is 375 cfs and the outflow is 563 cfs. After 2 hours, the inflow and outflow are 600 cfs and 675 cfs, respectively. Determine (a) the change of storage during 2 hours and (b) the storage volume after 2 hours.

SOLUTION

(a) 1. Average inflow rate = $\dfrac{375 + 600}{2}$ = 487.5 cfs

2. Inflow during 2 hrs = 487.5×2 = 975 cfs-hr

3. Average outflow rate = $\dfrac{563 + 675}{2}$ = 619 cfs

4. Outflow during 2 hrs = 619×2 = 1238 cfs-hr

5. From eq. (2.2) without P and E,

$$975 - 1238 - \Delta s = 0$$

or

$$\Delta s = -263 \text{ cfs-hr}$$

$$= \left(-263 \frac{\text{ft}^3}{\text{sec}} \text{ hr} \right) \left[\frac{60 \times 60 \text{ sec}}{1 \text{ hr}} \right] \left[\frac{1 \text{ acre-ft}}{43{,}560 \text{ ft}^3} \right]$$

$$= -21.73 \text{ acre-ft}$$

(b) $S_2 = S_1 + \Delta s = 55.30 - 21.73 = 33.57$ acre-ft

2.3.2 Balance Equation for Large River Basins for Long Duration

In large river basins, the water balance equation is used for the quantitative evaluation of basin resources and for substantiation of projects for their intended use and proposed modifications. The study of mean water balances is usually performed on a long-duration basis (for an annual cycle). Over a long period, positive and negative water storage variations tend to balance, and the change in storage, Δs, may be disregarded. The groundwater exchange in large basins with neighboring basins is ignored (i.e., $Q_{GI} - Q_{GO} = 0$). There is no surface water inflow into a basin with a distinct watershed divide (i.e., $Q_{SI} = 0$). Ignoring the discrepancy term, eq. (2.1) reduces to

$$P - E - Q = 0 \quad [\text{L}^3 \text{ or L}] \tag{2.3}$$

where Q is the discharge volume from the basin into the river.

2.3.3 Balance Equation for Direct Runoff within a Basin during a Storm

Surface contribution to streamflow and direct runoff are synonymous terms. In terms of runoff, the water balance from a storm over the ground surface is

$$P - E - I - S_D - R = 0 \quad [\text{L}] \tag{2.4}$$

where

P = precipitation

E = evapotranspiration

R = direct surface runoff or precipitation excess

I = infiltration

S_D = interception and depression storage

The storm evaporation during the short period is small and can be disregarded. If the interception and depression storage can also be ignored in comparison with the infiltration (in a more exact determination, these terms are estimated separately), eq. (2.4) reduces to

$$R = P - I \quad [\text{L}] \tag{2.5}$$

The application of eq. (2.5) is discussed in Section 2.13.

2.3.4 Water Balance Equation for Direct Runoff within a Basin for Longer than Storm Duration

The long duration in this balance equation means a period longer than the storm duration, for which the evapotranspiration component cannot be neglected. The values of the water balance components are averaged for this period. This can be a daily, weekly, monthly, or yearly duration. Models of this type have been developed by Thornthwaite and Mather (1955), Palmer (1965), and Haan (1972), in which the input of water from precipitation has been equated to the outflow of water by evapotranspiration, infiltration, and runoff. Conceptually, these models consider that moisture is either added to or subtracted from the soil, depending on whether precipitation for a period is greater than or less than the potential evapotranspiration.

When precipitation is less than the potential evapotranspiration, actual evapotranspiration in these models is treated as a function of the soil moisture content. This results in the loss of soil moisture and an increased *moisture deficit*; that is, the difference between the soil moisture capacity and the soil moisture storage at a given time.

When precipitation for a period exceeds the potential evapotranspiration, moisture is added to the soil until it attains its capacity. Any excess water contributes to runoff.

A model presented by Thomas (1981), known as the *abcd* model, places an upper limit on the sum of evapotranspiration and soil moisture storage rather than only on the soil moisture storage to its capacity. This provides a value of actual evapotranspiration less than the potential evapotranspiration and can simulate a decrease in soil moisture storage even when precipitation is in excess of potential evapotranspiration.

The Thomas model defines two state variables. One, known as the *available water*, is the sum of the precipitation to the end of a period i and the soil moisture storage to the end of the previous period $(i - 1)$; that is,

$$W_i = P_i + S_{i-1} \quad [\text{L}^3 \text{ or L}] \tag{2.6}$$

The other state variable, Y_i, is the sum of actual evapotranspiration and soil moisture storage at the end of period i; that is,

$$Y_i = E_i + S_i \quad [\text{L}^3 \text{ or L}] \tag{2.7}$$

Thomas has suggested the following nonlinear relation between the two state variables.

$$Y_i = \frac{W_i + b}{2a} - \left[\left(\frac{W_i + b}{2a} \right)^2 - \frac{W_i b}{a} \right]^{0.5} \quad [\text{L}^3 \text{ or L}] \tag{2.8}$$

where a and b are the model parameters. Parameter a, according to Thomas, reflects *the propensity of runoff to occur before the soil is fully saturated*. Its value of less than 1 results in runoff for $W_i < b$. Parameter b is an upper limit on the sum of evapotranspiration and soil moisture storage. Equation (2.8) assures that $Y_i < W_i$.

In order to allocate Y_i of eq. (2.7) between evapotranspiration and soil moisture storage at the end of the period, it is assumed that the rate of loss of soil moisture due to evapotranspiration is proportional to the soil moisture storage and potential evapotranspiration (PE), which leads to the relation

$$S_i = Y_i e^{-PE_i / b} \quad [\text{L}^3 \text{ or L}] \tag{2.9}$$

The difference $W_i - Y_i$ represents the sum of direct runoff $(DR)_i$ and infiltration contributing to groundwater recharge $(GR)_i$, since a part of the infiltrated water results in a change of soil moisture storage $(S_i - S_{i-1})$. The allocation between the direct runoff and groundwater recharge is suggested as follows:

$$(GR)_i = c(W_i - Y_i) \quad [\text{L}^3 \text{ or L}] \tag{2.10}$$

$$(DR)_i = (1 - c)(W_i - Y_i) \quad [\text{L}^3 \text{ or L}] \tag{2.11}$$

where c is a model parameter that is related to the fraction of mean runoff that comes from groundwater.

If G_i denotes the groundwater storage at the end of period i, then

$$G_i = \frac{(GR)_i + G_{i-1}}{d + 1} \quad [\text{L}^3 \text{ or L}] \tag{2.12}$$

The groundwater discharge is given by

$$(QG)_i = dG_i \quad [\text{L}^3 \text{ or L}] \tag{2.13}$$

where d is a model parameter for the fraction of groundwater storage discharged.

The streamflow at the end of period i is equal to $(DR)_i + (QG)_i$. Thus the model is applied to determine the averaged streamflow.

The values of parameters a, b, c, and d are obtained by calibrating the model from the known data for the water balance components. Alley (1984) estimated the following mean monthly values from the study of 10 sites in New Jersey, each having a record of 50 years: $a = 0.992$, $b = 30$, $c = 0.16$, and $d = 0.26$. Runoff estimates are very sensitive to parameter a.

The application of the model requires initial estimates of soil moisture storage, S_0, and groundwater storage, G_0. Thomas suggests the use of optimized values of S_0 and G_0 from the basin study. Alley (1984) suggests assuming some initial values of S_0 and G_0 and simulating data for some period (for a year in monthly data) prior to the beginning of the period of interest. The potential evapotranspiration for use in eq. (2.9) is computed by the Penman-Monteith method, described in Section 2.12.2.

EXAMPLE 2.2

The average monthly precipitation data recorded at the Whippany River Basin (drainage area 29.4 mi^2) at Morristown, New Jersey, during 2005 are given below. The monthly computed potential evapotranspiration values for the basin are also indicated. The model parameters are: $a = 0.98$, $b = 25$, $c = 0.10$, and $d = 0.35$. Initial soil moisture storage and groundwater storage are ascertained to be 7.8 and 1.5 in., respectively. For each month, determine **(a)** moisture storage, **(b)** direct runoff, **(c)** groundwater recharge, **(d)** groundwater storage, **(e)** groundwater discharge, and **(f)** streamflow.

Month:	J	F	M	A	M	J	J	A	S	O	N	D
Precipitation (in.)	1.2	2.2	0.2	0.75	0.1	0.6	7.0	7.8	5.7	0.88	0.48	5.5
Potential evapotranspiration (in.)	1.92	1.96	2.3	2.4	3.3	3.4	3.9	4.0	3.0	2.1	1.93	1.95

SOLUTION

1. The computations are made for successive periods; monthly for this problem.

2. For January 2005:

    From eq. (2.6),

    $$W_1 = P_1 + S_0 = 1.2 + 7.8 = 9.0 \text{ in.}$$

    From eq. (2.8),

    $$Y_1 = \frac{9.0 + 25.0}{2(0.98)} - \left\{ \left[\frac{9.0 + 25.0}{2(0.98)} \right]^2 - \frac{(9.0)(25.0)}{(0.98)} \right\}^{0.5} = 8.90 \text{ in.}$$

    From eq. (2.9), soil moisture storage,

    $$S_1 = 8.90 e^{-1.92/25.0} = 8.24 \text{ in.}$$

    From eq. (2.10), groundwater recharge,

    $$(GR)_1 = 0.10(9.0 - 8.9) = 0.01 \text{ in.}$$

    From eq. (2.11), direct runoff,

    $$(DR)_1 = (1 - 0.10)(9.0 - 8.9) = 0.09 \text{ in.}$$

    or

    $$(0.09 \text{ in.})(29.4 \text{ mi}^2)\left(\frac{1 \text{ ft}}{12 \text{ in.}}\right)\left(\frac{5280^2 \text{ ft}^2}{1 \text{ mi}^2}\right) = 6.15 \times 10^6 \text{ ft}^3$$

    or

    $$\left(6.15 \times 10^6 \frac{\text{ft}^3}{\text{month}}\right)\left(\frac{1 \text{ month}}{31 \times 24 \times 60 \times 60 \text{ sec}}\right) = 2.30 \text{ cfs}$$

From eq. (2.12), groundwater storage,

$$G_1 = \frac{0.01 + 1.5}{1 + 0.35} = 1.12 \text{ in.}$$

or

$$G_1 = (\text{depth})(\text{drainage area})$$

$$= (1.12 \text{ in.})(29.4 \text{ mi}^2)\left(\frac{1 \text{ ft}}{12 \text{ in.}}\right)\left(\frac{5280^2 \text{ ft}^2}{1 \text{ mi}^2}\right)$$

$$= 76.5 \times 10^6 \text{ ft}^3$$

From eq. (2.13), groundwater discharge,

$$(QG)_1 = G_1 d = \left(76.5 \times 10^6 \frac{\text{ft}^3}{\text{month}}\right)(0.35)\left(\frac{1 \text{ month}}{31 \times 24 \times 60 \times 60 \text{ sec}}\right)$$

$$= 10 \text{ cfs}$$

$$\text{streamflow} = (DR)_1 + (QG)_1 = 2.30 + 10.0 = 12.3 \text{ cfs}$$

3. Similar computations are performed for the month of February with starting values of S and G as 8.24 and 1.12, respectively, and so on.

2.4 ERRORS OF COMPUTATION OF WATER BALANCE COMPONENTS

Evaluation of water balance components always involves errors due to measurement and interpretation (Winter, 1981). Precipitation and streamflow are the only components of the balance equation that are observed extensively from the network of stations. The data for evaporation are observed on a limited scale and for infiltration from the experimental basins. The water storage variations are obtained from water-level and soil-moisture observations and by snow surveys. Also, empirical formulas are used for the computation of evaporation, infiltration, and water storage. Winter (1981) discussed various types of errors involved in the measurement and computation of components of the water balance equation. The time frame is very important since the long-term averages have smaller errors than the short-term values. The errors associated with annual and monthly estimates of water balance components are indicated in Table 2.1 based on commonly used methodologies.

The water balance equation usually does not balance out; accordingly, a residual term for the discrepancy has been included in eq. (2.1). When a component is estimated through an empirical formula, the error of imperfection of the formula is added into the residual term. If a component is evaluated indirectly by substituting other known terms in the water balance equation, this component includes the errors of computation of other variables. Winter (1981) computed the groundwater inflow of lakes in three different geometric and climatic settings as the balance term of the water budget equation. The errors involved in the other three components of the equation (i.e., precipitation, evaporation, and streamflow) propagate as the sum of variances and covariances. Thus the variance of the error of the residual term is equal to the summation of variances of error of each of the other components of the equation, plus twice the covariances of the error of components with each other. Covariance terms relate to interrelationships of the measurement error of the components, not to the interrelationships of the components themselves. As indicated by Winter, overall error is the standard deviation of the error of the residual. For a worst possible estimate of the total error, the error due to each component is considered additive (of the same sign). Due to these errors, the groundwater inflow has been overestimated by 60% in one case and about 400% in two cases on an annual basis. Thus the errors of measurement and estimation have a significant impact on water balance calculations. This is particularly serious when a component is computed as the residual quantity. To minimize the error it is desirable to measure or compute all components using the best independent methods.

Table 2.1 Errors in Hydrologic Components by Commonly Used Methodologies

	Percent Error	
	Annual Estimate	Monthly Estimate
1. Precipitation		
Gage observation	2	2
Gage placement (height)	5	5
No windshield		20
Areal averaging	10	15
Gage density	13	20
2. Streamflow		
Current-meter measurement	5	5
Stage-discharge relationship	20	30
Channel bias	5	5
Regionalization of discharge	70	
3. Evaporation		
Energy budget	10	
Class A plan	10	10
Pan to lake coefficient	15	50
Areal averaging	15	15

Source: Based on Winter (1981).

2.5 PRECIPITATION

Precipitation, largely in the form of rain and snow, is the source of moisture coming to the earth. It is a key source parameter in the water balance equation. The ability to accurately measure and compute precipitation determines to a considerable extent the reliability of all water balance computations (Sokolov and Chapman, 1974). The rainfall and snowfall at any location are measured by self-recording or manual observation gages. These gages record the depth of rainfall or snowfall in inches or millimeters at any place within a given time frame. Snow measurements are made by standard rain gages equipped with shields to reduce the effect of wind. Snow boards and stakes also are used. The direct method of measuring snowfall is, however, not entirely satisfactory. It is supplemented by snow surveying. For this purpose, snow courses are established. Each course comprises a series of sampling points from which the samples of snowpack are taken by core-cutting equipment on a regular basis. In addition to surveying on ground, aerial snow surveying is performed in remote places. Snow surveying provides information on snow depth variation, water equivalent, density, and snow quality. Each gage catches the precipitation falling within a circle 8 in. in diameter and hence indicates only a point measurement. To obtain the precipitation for an area on a representative basis, a number of gages are needed. A network of over 15,000 gages (stations) exists throughout the country, about 80% of which are operated by the National Weather Service and the remainder by other government and private agencies.

The measured precipitation data are subject to errors due to (1) water displacement by the dipstick, (2) amount for wetting of the gage collector, (3) evaporation during precipitation and manual readings, (4) height of the gage above the ground, and (5) the effect of wind.

The density and arrangement of the network and the method of analysis influence the estimate of areal distribution of rainfall from point data. Numerous papers have been published on precipitation measurement errors. Studies indicate that wind is the major cause of error in precipitation gage measurements (Larson and Peck, 1974). The errors increase with wind speed and are much greater for snowfall than for rainfall. A properly selected and well-protected site can reduce wind errors considerably. Gage shields can further reduce catch deficiency* for snow, although they have very little effect on rain measurement. However, shields are not effective at wind speeds above 20 mph. Larson and Peck have summarized the catch deficiency versus wind speed based on studies at two National Weather Service gages, as shown in Figure 2.1. At 20 mph, a catch deficiency of 70% can be experienced in snow measurement by an unshielded gage. A shield can reduce the error to about 50%. For rainfall, a deficiency of about 20% is expected with or without the shield at 20 mph. The curves of Figure 2.1 can be used to apply corrections to observed precipitation data.

2.6 ANALYSIS OF POINT PRECIPITATION DATA

The point observations from a precipitation gage are subject to two regular problems. A gage site (station) may have a short break in the record because of instrument failure or absence of the observer. It is often necessary to estimate the missing record. Another problem is that the recording conditions at a gage site may have changed significantly some time during the period of record, due to relocation or upgrading of a station in the same vicinity, difference in observational procedure, or any other reason. The problem is resolved in both cases by comparison with neighboring gage sites.

2.6.1 Estimating Missing Data

The precipitation value missing at a site can be estimated from concurrent observations at three or more neighboring stations, known as *index stations*, located as close to and evenly spaced from the missing data station as possible. The normal-ratio method is used, according to which

$$\frac{P_x}{N_x} = \frac{1}{n}\left(\frac{P_1}{N_1} + \frac{P_2}{N_2} + \frac{P_3}{N_3} + \ldots + \frac{P_n}{N_n}\right) \quad \text{[dimensionless]} \qquad (2.14)$$

---

* Catch deficiency $= 1 - \dfrac{\text{gage catch}}{\text{true catch}}$

Figure 2.1 Gage catch deficiency versus wind speed: 1, rain gage; 2, snow gage with a shield; 3, snow gage without a shield (from Larson and Peck, 1974).

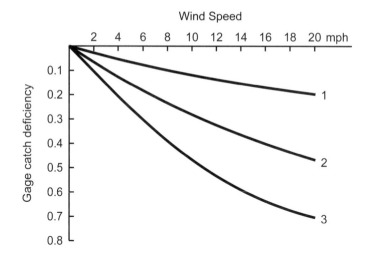

where

$$P_x = \text{missing precipitation value for station X}$$

$$P_1, P_2, P_3, \ldots, P_n = \text{precipitation values at the neighboring stations for the concurrent period}$$

$$N_x = \text{normal long-term, usually annual, precipitation at station X}$$

$$N_1, N_2, \ldots, N_n = \text{normal long-term precipitation for neighboring stations}$$

$$n = \text{number of index (neighboring) stations}$$

Equation (2.14) also can be applied for estimating the missing storm depth at a site by treating P_1, P_2, ... as related to a particular storm. Also, N_1, N_2, ... can be taken as the average values for a particular month for all years on record for index stations.

2.6.2 Checking Consistency of Data: Double-Mass Analysis

Double-mass analysis is a consistency check used to detect whether the data at a site have been subjected to significant change in magnitude due to external factors such as tampering with the instrument, a change in recording conditions, or a shift in observation practices. The change due to meteorological factors will equally affect all stations involved in the test and thus will not cause a lack of consistency. The analysis also provides a means of adjusting the inconsistent data. In the analysis, a plot is made of accumulated annual or seasonal precipitation values at the site in question (being checked for consistency) against the concurrent accumulated values of several surrounding stations. More conveniently, the mean of the surrounding stations is used in the accumulation and the plot, as shown in Figure 2.2. If the data are consistent, the plot will be a straight line. On the other hand, inconsistent data will exhibit a change in slope or break at the point where the inconsistency occurred. This is shown by point V in Figure 2.2 for the year 1981. If the slope of the line UV is a and of the line VW is b, the adjustment of the inconsistent data is made by the ratio of the slopes of the two line segments. Two ways of adjustment are possible.

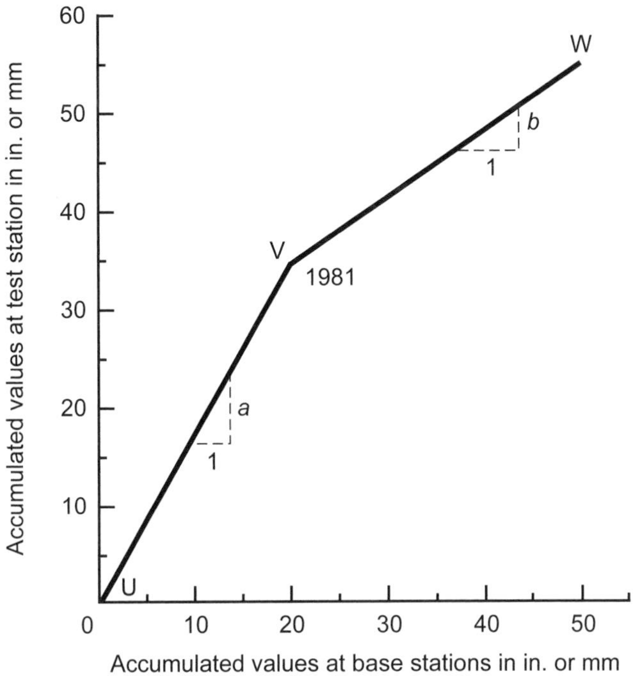

Figure 2.2 Double-mass curve analysis.

1. The data are adjusted to reflect the conditions that existed prior to the indicated break. This is done by multiplying each recent precipitation value after breakpoint V of station X (being tested) by the ratio a/b.

2. The data are adjusted to reflect recent conditions following the break. This is achieved by multiplying each value of the precipitation before the breakpoint by the ratio of b/a.

An adjustment of the second type usually is made. In certain cases, more than one break (change in the slope) in the data is observed. Sometimes an apparent change in slope is noticed because of a natural variation in the data and is unassociated with changes in gage location, gage environment, or observation procedure. If doubt exists, a test of hypothesis should be performed by the Fisher distribution (Section 8.5.2) on the two sets of data (before and after the apparent break) to check whether the data are homogeneous and the break is purely by chance. Searcy and Hardison of the U.S. Geological Survey (1960) recommended that if fewer than 10 stations are grouped together to check the consistency of a station, the record of each station should be tested by double-mass analysis for consistency by plotting it against the group of all other stations, and those records that are inconsistent should be eliminated from the group.

The double-mass curve technique should seldom be used in mountainous areas. It also is not suitable for adjusting daily or storm precipitation. When the records of the stations have different starting dates, the mass curve can be formed by accumulating the data starting with the most recent values, in the reverse order from that indicated earlier.

EXAMPLE 2.3

The annual records of five precipitation stations are given in Table 2.2. Check the consistency of station A. Adjust the record if it is inconsistent.

Table 2.2 Annual Precipitation for Double-Mass Analysis

	(1)	(2)	(3)	(4)	(5)	(6)	(7)	(8)	(9)
		Annual Precipitation for Station (in.)					Mean of Stations B,C,D,E	Cumulated Precipitation for Station A	Cumulated Precipitation for Mean of B,C,D,E
	Year	A	B	C	D	E			
	1990	26.28	29.89	24.55	36.56	31.80	30.70	26.28	30.70
	1991	22.46	24.70	32.79	30.82	31.66	29.99	48.74	60.69
	1992	26.81	33.60	32.35	38.61	33.61	34.54	75.55	95.23
	1993	23.66	31.94	25.99	27.71	33.11	29.69	99.21	124.92
	1994	19.00	29.06	29.38	36.10	25.24	29.95	118.21	154.87
	1995	46.71	29.29	49.88	42.62	44.43	41.56	164.92	196.43
	1996	36.99	30.89	38.28	32.06	38.49	34.93	201.91	231.36
	1997	24.27	21.51	26.19	23.66	31.88	25.81	226.18	257.17
	1998	37.42	25.95	28.90	33.34	36.32	31.13	263.60	288.30
	1999	30.45	33.25	24.58	38.50	35.91	33.06	294.05	321.36
	2000	34.26	25.06	28.32	31.58	26.11	27.77	328.31	349.13
	2001	30.34	35.31	31.33	35.29	36.70	34.66	358.65	383.79
	2002	40.53	40.50	34.62	31.15	36.84	35.78	399.18	419.57
	2003	37.48	32.87	39.88	33.26	39.81	36.46	436.66	456.03
	2004	40.42	31.21	38.29	39.73	37.81	36.76	477.08	492.79
	2005	27.50	27.56	25.72	25.54	29.66	27.12	504.58	519.91

SOLUTION

1. The mean of a group of stations (B, C, D, and E) is computed in column 7.
2. The accumulated values for station A and the group of stations are given in columns 8 and 9.
3. Column 8 is plotted against column 9 in Figure 2.3. The breakpoint is observed at 1994.
4. The ratio of recent to past slope = 1.06/0.78 = 1.36.
5. The data prior to the breakpoint (1990–1993) are corrected by the factor 1.36, as indicated in Table 2.3.

Table 2.3 Adjusted Precipitation of Station A

Year	Recorded Precipitation (in.)	Adjusted Precipitation (in.)
1990	26.28	35.74
1991	22.46	30.55
1992	26.81	36.46
1993	23.66	32.18
1994–2005		Same as recorded

Figure 2.3 Double-mass curve for Example 2.3.

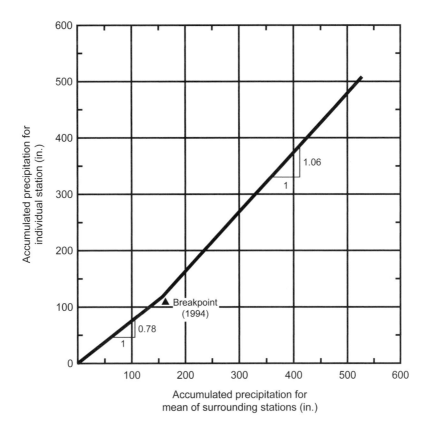

2.7 CONVERSION OF POINT PRECIPITATION TO AREAL PRECIPITATION

The representative precipitation over a defined area is required in engineering applications, whereas the gaged observation pertains to point precipitation. The areal precipitation is computed from the record of a group of rain gages within the area by the following methods.

1. Arithmetic or station average method
2. Weighted average method
   a. Thiessen polygon method
   b. Isohyetal method

2.7.1 Arithmetic Average Method

This simple method consists of computing the arithmetic average of the values of the precipitation for all stations within and in proximity to the area. This method assigns equal weight to all stations irrespective of their relative spacings and other factors.

2.7.2 Thiessen Polygon Method

In this method, weight is assigned to each station in proportion to its representative area defined by a polygon. These polygons are formed as follows:

1. The stations are plotted on a map of the area drawn to a scale (Figure 2.4).

2. The adjoining stations are connected by dashed lines.

3. Perpendicular bisectors are constructed on each of these dashed lines, as shown by the solid lines in Figure 2.4.

4. These bisectors form polygons around each station. Each polygon is representative of the effective area for the station within the polygon. For stations close to the boundary, the boundary forms the closing limit of the polygons.

5. The area of each polygon is determined* and then multiplied by the rainfall value for the station within the polygon.

6. The sum of item 5 divided by the total drainage area provides the weighted average precipitation.

EXAMPLE 2.4

The rain gages located in and around a drainage area are shown in Figure 2.4 along with the rainfall recorded at these stations due to a storm. Determine the average precipitation for the drainage area by **(a)** the arithmetic average method, and **(b)** the Thiessen polygon method.

SOLUTION

(a) Arithmetic average $= \dfrac{2.1+3.1+5.2+3.8+5.4+3.5+4.5}{7}$

$\bar{p} = 3.91$ in.

(b) Thiessen polygon method: Refer to Table 2.4.

2.7.3 Isohyetal Method

This is the most accurate of the three methods and provides a means of considering the orographic (mountains) effect. The procedure is as follows:

1. The stations and rainfall values are plotted on a map to a suitable scale.

2. The contours of equal precipitation (isohyets) are drawn as shown in Figure 2.5. The accuracy depends on the construction of the isohyets and their intervals.

3. The area between successive isohyets is computed and multiplied by the numerical average of the two contour (isohyets) values.

4. The sum of item 3 divided by the drainage area provides the weighted average precipitation.

---

* This is done by a graphic tool like AutoCAD or a planimeter or, alternatively, by drawing the figure to a scale on graph paper, counting the total number of squares covered by the polygon, and multiplying by the square of the map scale.

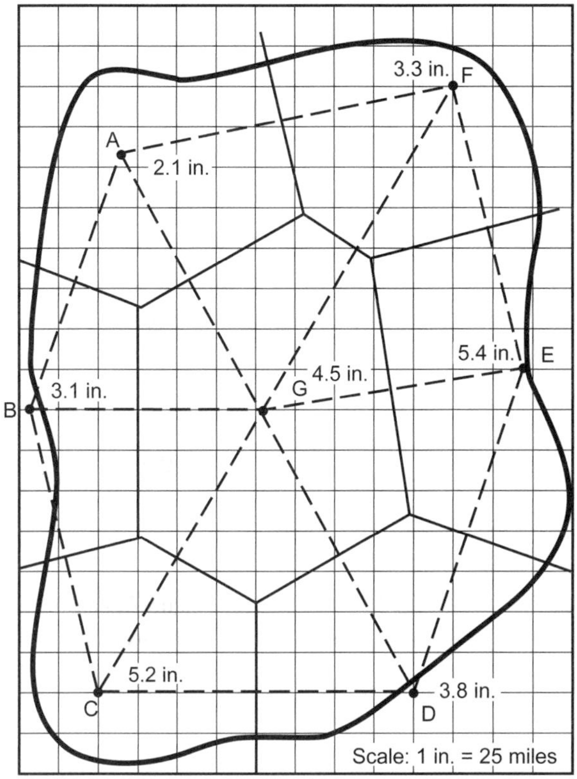
3.3 in.
F
A
2.1 in.
5.4 in.
E
4.5 in.
G
3.1 in.
B
5.2 in.
C
3.8 in.
D
Scale: 1 in. = 25 miles

Figure 2.4 Thiessen polygon method of computing average areal precipitation.

Table 2.4 Average Precipitation Computation by Thiessen Polygon Method

(1) Observed Precipitation (in.)	(2) Area of Polygon (mi^2)	(3) Precipitation × area (col. 1 × col. 2)
2.1	735	1,543.5
3.1	475	1,472.5
5.2	640	3,328.0
3.8	620	2,356.0
5.4	740	3,996.0
3.3	685	2,260.5
4.5	1,210	5,445.0
Total	5,105	20,401.5

$$\text{Average } \overline{p} = \frac{20{,}401.5}{5{,}105} = 4.0 \text{ in.}$$

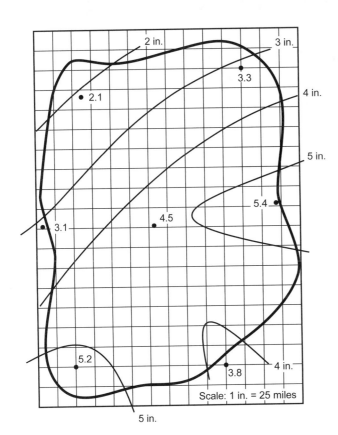

Figure 2.5 Isohyetal method of computing average areal precipitation.

Inside figure labels: 2 in., 3 in., 3.3, 2.1, 4 in., 5 in., 5.4, 4.5, 3.1, 5.2, 3.8, 4 in., Scale: 1 in. = 25 miles, 5 in.

EXAMPLE 2.5

Solve Example 2.4 by the isohyetal method.

SOLUTION

Refer to Table 2.5.

2.8 INTENSITY-DURATION-FREQUENCY (IDF) ANALYSIS OF POINT PRECIPITATION

The point precipitation data of various storms are analyzed in an IDF study. Since the precipitation data serve the purpose of estimating the streamflows in many instances, not only the total quantity of precipitation but its rate, known as the *intensity* (expressed in in./hr or mm/hr), and duration (in min or hr) are important in a peak-flow study. A point or gaged observation can be considered to be representative of a 10-mi² drainage area. Hence the studies of point-rainfall extremes are extensively used in the design of small-area drainage systems comprising storm sewers, drains, culverts, and so on. The application of intensity, duration, and frequency data in the rational method is discussed in Chapter 13. The intensity-duration-frequency analysis can be carried out only where the data from a recording rain gage are available. The procedure of analysis is as follows:

1. A specific duration of rainfall, such as 5 min, is selected.

Table 2.5 Average Precipitation Computation by Isohyetal Method

(1)	(2)	(3)	(4)	(5)
Isohyet (in.)	Area Covered by the Isohyet (mi²)	Area between Two Isohyets (mi²)	Average of Two Isohyets (in.)	Precipitation × Area (col. 3 × col. 4)
<2	0			
		125	1.8 (est)	225
2	125			
		820	2.5	2,050
3	945			
		1,100	3.5	3,850
4	2,045			
		2,555	4.5	11,498
5	4,600			
		500	5.2	260
>5	5,100			
Total		$\overline{5,100}$		$\overline{20,233}$

$$\text{Average } \overline{p} = \frac{20,233}{5,100} = 3.96 \text{ in.}$$

2. From the record of the rain gage, which indicates the accumulated amount of precipitation with respect to time, the maximum rainfall of this duration in each year is noted. This is the maximum incremental precipitation (difference between accumulated precipitation values) for the selected (5 min) duration obtained from the gage record. For a partial duration series, all values in the record that exceed a level given by the *excessive precipitation* for the selected duration are noted. The excessive precipitation as defined by the National Weather Service is precipitation that falls at a rate equaling or exceeding that indicated by the following formula:

$$p = \frac{t + 20}{100} \quad \text{[unbalanced]}$$

where

p = precipitation, in.

t = precipitation duration, min

3. The precipitation values are arranged in descending order and the return period for each value is obtained using the formula $T = n + 1/m$, where m is the rank of the data and n is the total number of years of data in the record (Section 8.8). For partial duration series (Section 8.5), the adjustment in the precipitation values is made by applying the following empirical multiplication factors:

Return Period (years)	Conversion Factor
2	0.88
5	0.96
10	0.99
>10	1.0

Section 2.8 Intensity-Duration-Frequency (IDF) Analysis of Point Precipitation   57

4. Similar analyses are carried out for other selected durations (10, 15, 20, . . . min), as shown in Example 2.6.

5. For each frequency level computed, the precipitation amounts (depths) are plotted for different durations. These are the depth-duration-frequency curves. The precipitation depths can be converted to intensities by $i = 60\ p/t$. For instance, a precipitation of 0.5 in. of 30 min duration has an intensity of 1 in. /hr. These values are plotted as the intensity-duration-frequency curve on arithmetic (ordinary) graph paper as in Figure 2.6, or on log-log paper. Interpolation between the return periods can be done from the curves of the lower and higher return periods.

EXAMPLE 2.6

For the precipitation data arranged for different durations in Table 2.6, prepare intensity-duration-frequency curves for 20-year and 10-year frequencies.

SOLUTION

1. For each duration, the precipitation depths are arranged in descending order. The highest value has been assigned a rank of 1 and the lowest a rank of 22. The return periods are obtained in column 8 of Table 2.6.

2. The depths of different duration corresponding to 20-year frequency are interpolated from the values for 23- and 11.5-year frequencies and converted to intensities in Table 2.7. Similar calculations are done for 10-year frequency. These are plotted in Figure 2.6.

Table 2.6 Frequency Analysis of Different Duration of Precipitation Depths

(1)	(2)	(3)	(4)	(5)	(6)	(7)	(8)
			Precipitation (in.)				
Rank			of Duration				Return Period
m	5 min.	10 min.	15 min.	20 min.	30 min.	60 min.	$T = (n + 1)/m$
1	0.40	0.66	0.89	1.07	1.48	2.15	23
2	0.38	0.63	0.83	0.97	1.29	1.92	11.5
3	0.37	0.62	0.79	0.91	1.26	1.48	7.7
4	0.36	0.60	0.76	0.86	0.91	1.06	5.8
5	0.35	0.60	0.73	0.80	0.83	0.96	4.6
6	0.33	0.58	0.72	0.77	0.82	0.94	3.8
7	0.33	0.50	0.72	0.77	0.78	0.90	3.3
8	0.31	0.50	0.63	0.70	0.75	0.87	2.9
9	0.30	0.49	0.57	0.65	0.67	0.77	2.6
10	0.28	0.44	0.56	0.62	0.66	0.75	2.3
.							
.							
.							
22	0.13	0.23	0.32	0.40	0.40	0.43	1.05

← 20 yr (pointing to Return Period 23)
← 10 yr (pointing to Return Period 11.5)

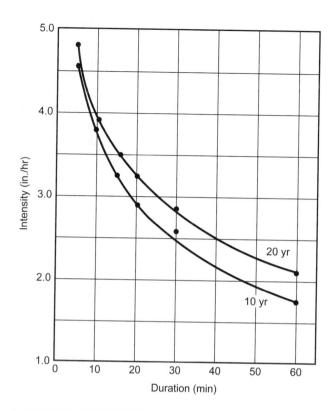

Figure 2.6 Intensity-duration-frequency curve.

Table 2.7 Values for 20-Year and 10-Year Precipitation Intensities of Different Duration

Return Period (years)	Intensity (in./hr) of Duration					
	5 min.	10 min.	15 min.	20 min.	30 min.	60 min.
20	4.74	3.90	3.50	3.13	2.86	2.09
10	4.51	3.78	3.24	2.85	2.58	1.75

The National Weather Service has prepared a series of intensity-duration-frequency maps for the United States for several combinations of return period and duration of precipitation. These generalized maps are used in the absence of recording gage data.

The intensity-duration-frequency curves are also prepared using empirical relations of the type:

$$i = \frac{A}{t+B} \quad \text{[unbalanced]} \tag{2.15}$$

where

$\qquad i$ = intensity, in./hr

$\qquad t$ = duration, min

$\qquad A, B$ = constants that depend on the return period and climatic factors

The constants for different parts of the country in Figure 2.7 are given in Table 2.8.

Section 2.8 Intensity-Duration-Frequency (IDF) Analysis of Point Precipitation 59

Figure 2.7   Map of similar rainfall characteristics (from Steel and McGhee, 1979).

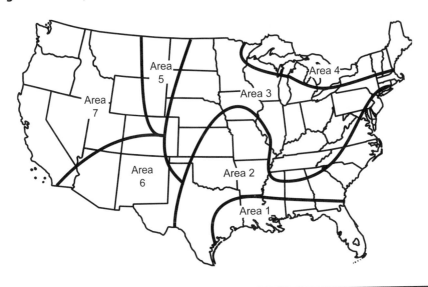

Table 2.8   **Intensity–Duration Constants for Various Regions[a]**

Frequency (years)		Area 1	Area 2	Area 3	Area 4	Area 5	Area 6	Area 7
2	A =	206	139.75	102	70	70	68	31.9
	B =	30	21	17	13	16	14	11
5	A =	246.9	190.2	131.1	96.9	81.1	74.8	48
	B =	29	25	19	16	13	12	12
10	A =	300	229.9	170	111	111	122	59.8
	B =	36	29	23	16	17	23	13
25	A =	326.8	259.8	229.9	170	129.9	155.1	66.9
	B =	33	32	30	27	17	26	10
50	A =	315	350	250	187	187	159.8	65
	B =	28	38	27	24	25	21	8
100	A =	366.9	374.8	290	220	240.2	209.8	77.2
	B =	33	36	31	28	29	26	10

[a] Conversion of units by the author.
Source: Steel and McGhee (1979).

2.9   DEPTH-AREA-DURATION (DAD) ANALYSIS OF A STORM

DAD is an areal precipitation analysis of a single storm. The analysis is performed to determine the maximum amounts of precipitation of various durations over areas of various sizes. The procedure is, as such, applied to a storm that produces an excessive depth of precipitation. In a study of the *probable maximum precipitation* (PMP), which is defined as a rational upper limit of precipitation of a given duration over a particular basin, several severe storms are analyzed and the maximum values for various durations are selected for each size of area. The generalized depth-area-duration values of the probable maximum precipitation for the eastern United States, as prepared by the National Weather Service, are illustrated in Table 8.14.

The information from a recording gage is needed for this study. The procedure consists of first determining the depth-area relation for the total depth of a storm and then breaking down the overall depth relating to each area among different durations. The points of equal duration on the depth and area graph produce curves of the depth-area-duration. The steps are explained below.

1. Prepare the accumulated precipitation or mass curves resulting from a storm for each station in the basin, which, in fact, are the records from the rain gages.

2. From the total amounts of precipitation from the storm at various stations, prepare an isohyetal map. The simpler storms present a single isohyetal pattern. The complex storms that are produced by two or more closely spaced bursts of rainfall have closed isohyetal patterns divided into zones.

3. The isohyets are assumed to be the boundaries of individual areas. Determine the average depth of precipitation for the areas enclosed by successive isohyets. This provides the total storm depth and area relation.

4. Start with the smallest isohyet. Within an area enclosed by this isohyet, there will be a certain number of gage stations. Determine the weight of each station by drawing the Thiessen polygons for these stations.

5. For each of the above stations, using the mass curve from step 1 determine the incremental (difference of) precipitation values for various durations. Multiply these values by the respective weight of each station.

6. For all stations within the area enclosed by the smallest isohyet, sum up the values of step 5 for different durations separately, with the last duration equal to the storm period.

7. The ratio of the average depth of precipitation from step 3 to the value from step 6 for the total storm period, corresponding to the area within the smallest isohyet, is the factor by which all station values of different durations in step 6 are multiplied to derive the adjusted values for this isohyet.

8. Steps 4 through 7 are repeated for successive isohyetal areas of the map created in step 2. The values are plotted on semilog paper and the lines through the similar durations are drawn as shown in Figure 2.8.

2.10 EVAPORATION AND TRANSPIRATION

Precipitation that does not ultimately become available as surface or subsurface runoff is called *water loss*. It consists of *evaporation*, which is the amount of water vaporized into the atmosphere from free water surface and land areas, and *transpiration*, which is the water absorbed by plants and crops and eventually discharged into the atmosphere. As discussed in Section 2.2, evaporation and transpiration take place in each stage of the hydrologic cycle. They form a major segment of the hydrologic cycle since about 70% of the precipitation in the United States is returned back to the atmosphere as evaporation and transpiration. From open water bodies such as lakes, reservoirs, seas, and oceans, the loss is by direct evaporation. From a drainage basin the loss is due to (1) evaporation from the soil, (2) evaporation of the intercepted water, (3) evaporation from the depression storage, and (4) transpiration of water by plants and trees. This combined total loss from the drainage

Figure 2.8 Depth-area-duration curve.

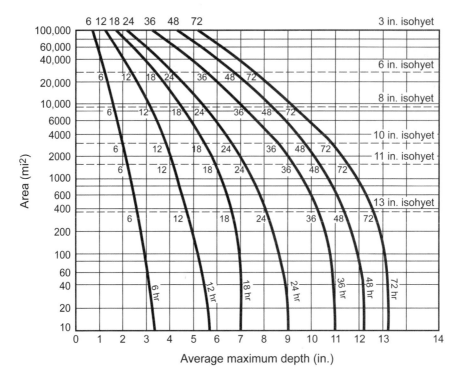

basin is also called *evapotranspiration*. The methodologies for estimation are grouped according to the type of surface from which evaporation/transpiration occurs.

2.11 EVAPORATION FROM FREE-WATER BODIES

Evaporation is a two-phase process. The first phase occurs when water molecules acquire sufficient energy to break through the water surface and escape into the atmosphere. This energy is provided principally by solar radiation. The second phase consists of transporting the vapor molecules from the vicinity of the water surface into the atmosphere. This is controlled by the difference between the vapor pressure of the body of water and that of the air for molecular diffusion, and by the wind speed for evaporation due to convection. There are at least eight factors on which the rate of evaporation depends: (1) solar radiation, (2) air temperature, (3) atmospheric pressure, (4) relative humidity, (5) water temperature, (6) wind speed, (7) quality of water, and (8) geometry of the evaporating surface.

The methods of estimating evaporation include (1) comparative methods, such as pan evaporation and atmometers; (2) aerodynamic methods, such as eddy correlation, gradient, and mass transfer; (3) balance methods, such as water budget and energy budget; and (4) combination methods such as the Penman method (1948). These are discussed in detail in the *National Handbook of Recommended Methods for Water Data Acquisition* (U.S. Geological Survey, 1977*). Four widely used methods, one from each category, are described next.

---

* The chapter on evaporation was prepared in 1982.

2.11.1 Evaporation Using Pans

The most common method of estimating evaporation from a free-water body is by means of an evaporation pan. The Standard National Weather Service Class A pan is widely used. This pan, built of unpainted galvanized iron, is 4 ft in diameter by 10 in. in depth, and is mounted on a wooden frame 12 in. above the ground, to circulate the air beneath the pan. It is filled to a depth of 8 in. The water surface level is measured daily by a hook gage in a stilling well. The evaporation is computed as the difference in the observed levels adjusted for any precipitation during observation intervals. It has been observed that evaporation occurs more rapidly from a pan than from larger water bodies. A coefficient is accordingly applied to pan observations to derive the equivalent lake or reservoir evaporation.

$$E_L = KE_p \quad [L] \tag{2.16}$$

where

E_L = evaporation from a water body

E_p = evaporation from the pan

K = pan coefficient

Pan coefficients are normally computed on an annual basis, but monthly coefficients also have been used. Gray (1973) and Linsley et al. (1982) have summarized the coefficient values, which range from 0.6 to 0.8 with an average of 0.7.

Kohler et al. (1955) proposed a formula to be used before applying the pan coefficient, to adjust the pan observation for heat exchange through the pan and the effect of wind (advected energy). The *National Handbook* suggests another correction for splashout and blowout whenever precipitation is greater than 8 mm. Kohler and Parmele (1967) indicate that the corrected pan data represent the "free water evaporation" that will occur in a very shallow water body. For a natural water body, it is necessary to consider significant heat storage and energy advected by water coming in or going out. An equation for this adjustment has been suggested by Kohler and Parmele (1967).

A refinement to eq. (2.16), which is suitable for estimating monthly or even daily evaporation losses, considers the saturation vapor pressure of the lake (water body) and the pan.

$$E_L = K' \frac{e_{sL} - e_Z}{e_{sp} - e_Z} E_p \quad [L] \tag{2.17}$$

where

e_{sL} = saturation vapor pressure for maximum temperature just below lake surface

e_{sp} = saturation vapor pressure for maximum temperature in evaporation pan

e_Z = mean vapor pressure of air at a height Z above the lake surface

K' = a coefficient, mainly a function of the type of pan, equal to 1.5 for U.S. class A pan at $Z = 4$ m (Webb, 1966)

Daily computed values of E_L are summed up to get the monthly evaporation. For direct monthly computation, mean monthly values of e_{sL}, e_{sp}, and e_Z are used to compute E_L. Saturation vapor pressure is given in Appendix C.

The pan-to-lake coefficient is a major cause of error in the pan method. Errors in the range of 10 to 15% of annual estimates and up to 50% for monthly estimates have been reported. The method thus is preferred for long-duration estimates.

EXAMPLE 2.7

Find the daily evaporation from a lake for a day on which the following data were observed:

1. Air temperature at 4 m (mean) = 30 °C
2. Lake temperature
   Maximum = 20 °C
   Minimum = 18 °C
3. Pan temperature
   Maximum = 28 °C
   Minimum = 25 °C
4. Relative humidity = 25%
5. Wind speed at 4 m = 6 m/s
6. Pan evaporation = 9 mm

Compute the results by **(a)** the simple relation, and **(b)** the refined formula.

SOLUTION

(a) From eq. (2.16),

$$E_L = K E_p = 0.7(9.0) = 6.3 \text{ mm}$$

(b) Saturation vapor pressure at maximum temperature for lake (from Appendix C), $e_{sL} = 2.337$ kPa

Saturation vapor pressure at maximum temperature for pan, $e_{sp} = 3.781$ kPa.

Saturation vapor pressure at mean air temperature = 4.243 kPa.

Moisture in air = 25%.

Vapor pressure of the air,

$$e_Z = \left(\frac{25}{100}\right)(4.243) = 1.061 \text{ kPa}$$

From eq. (2.17),

$$E_L = (1.5)\left(\frac{2.337 - 1.061}{3.781 - 1.061}\right)(9.0) = 6.33 \text{ mm}$$

2.11.2 Evaporation by the Aerodynamic Method

This is a very widely used method to determine evaporation from lakes and reservoirs. The method is based on a diffusion equation suggested by John Dalton in 1802, according to which the evaporation is proportional to the difference between saturated vapor pressure at the surface water temperature and the vapor pressure due to the moisture in the air:

$$E_a = M(e_s - e_Z)u_Z \quad [\text{LT}^{-1}] \tag{2.18}$$

where

E_a = evaporation by the aerodynamic method

M = mass-transfer coefficient, the dimension is the inverse of pressure

e_s = saturation vapor pressure at water temperature

e_Z = vapor pressure of the air at level Z, $e_Z = \text{RH}\left(e_Z^0\right)$

e_Z^0 = saturation vapor pressure at air temperature at level Z

RH = relative humidity (fraction)

u_Z = wind velocity at level Z

The mass-transfer coefficient, M, is commonly determined by calibration with reference to the energy budget evaporation method. If a plot is made of the product $(e_s - e_Z)u_Z$ against the independent estimate of evaporation, the slope of the line is M.

Thornthwaite and Holzman in 1939 developed a relation of the form of equation (2.18) from mass and momentum transfer of water vapor that provided an expression to represent M. Since then research has been directed to many forms of aerodynamic equations and mass-transfer coefficients. An expression for the mass-transfer coefficient, M, in terms of the bulk evaporation coefficient is:

$$M = 0.622\frac{\rho_a C_E}{\rho_w P} \quad [L^2/F] \tag{2.19}$$

where

ρ_w = density of water (Appendix C)

ρ_a = density of air (Appendix D)

P = atmospheric pressure at level Z

C_E = bulk evaporation coefficient, dimensionless

Extensive research has been conducted for estimation of C_E. If the pressure-related terms, e_s, e_Z and P are measured in kPa, the velocity and evaporation in m/s, and the air and water density in kg/m^3, then the values of C_E range from 1.15×10^{-3} to 1.4×10^{-3} at the 8-m-high level, according to various studies. It is about 10% higher at the 4-m height.

When the wind speed is observed at a different elevation than the level for computing evaporation, it can be adjusted from one level to another by the logarithmic law that accounts for roughness of surface. A simplified form of the adjustment is

$$u_2 = u_1 \frac{\ln\dfrac{Z_2}{Z_0}}{\ln\dfrac{Z_1}{Z_0}} \quad [L/T] \tag{2.20}$$

where Z_1 and Z_2 are measurement heights for levels 1 and 2, respectively. Z_0 is the reference height where velocity is zero, as follows:

Roughness class	Roughness length, Z_0, m	Landscape
0	0.0002	Water surface
0.5	0.0024	Open terrain
1	0.03	Open agricultural area
1.5–2.5	0.055–0.2	Agricultural land with houses
3.0	0.4	Village, small town, forests
3.5	0.8	Larger cities
4.0	1.6	Very large cities

EXAMPLE 2.8

Solve Example 2.7 by the aerodynamic method. $C_E = 1.4 \times 10^{-3}$ at 4 m.

SOLUTION

1. No height adjustment is needed as all measurements are at 4 m.

2. Mean lake temperature = 19 °C
   Saturation vapor pressure at 19 °C, e_s = 2.198 kPa (Appendix C)

3. Mean air temperature = 30 °C
   Saturation vapor pressure at 30 °C = 4.243 kPa
   Vapor pressure of air at 25% RH, $e_Z = (0.25)(4.243) = 1.061$ kPa

4. At 30 °C, atmospheric pressure = 101.3 kPa, air density = 1.16 kg/m³ (Appendix D)

5. From equation (2.19),

$$M = \frac{(0.622)(1.16)(1.4 \times 10^{-3})}{(1000)(101.3)} = 1.0 \times 10^{-8} \, \text{kPa}^{-1}$$

6. From equation (2.18),

$$E_a = (1 \times 10^{-8})(2.198 - 1.061)(6.0) = 6.82 \times 10^{-8} \, \text{m/s}$$

or

$$(6.82 \times 10^{-8} \, \text{m/s}) \left[\frac{1000 \, \text{mm}}{1 \, \text{m}} \right] \left[\frac{86,400 \, \text{s}}{1 \, \text{d}} \right] = 5.90 \, \text{mm/day}$$

2.11.3 Evaporation by the Energy Balance Method

The method is based on accounting for all heat energy received and dissipated by a water body. The procedure is highly data intensive. It is accurate in application for daily or longer periods when energy terms tend to stabilize. The major incoming and outgoing heat energy components in a controlled volume of a water body or a cropped area to just below the land surface are shown in Figure 2.9.

Energy required to evaporate a unit mass of water is called the *latent heat of vaporization* of water, λ, as given in Table 2.9. Thus, the energy absorbed per unit area to evaporate E is $\rho_w \lambda E$. The radiant energy of the sun captured at the earth's surface is a dominating factor on evaporation rates. A portion of the radiant energy input onto the earth is not

Figure 2.9 Energy inflow and outflow from a water body and a cropped area.

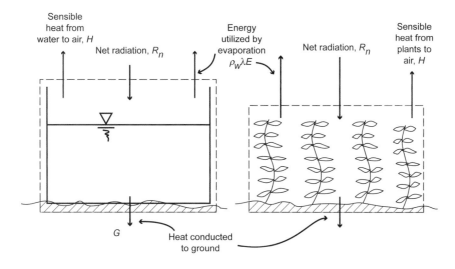

used up in direct evaporation. The air in contact with the ground or water surface is warmed and then moves upward. The associated flow of energy through the air is termed the *sensible heat flux* since this involves a change of air temperature, which is a property that can be sensed or measured. Also, there is a sensible heat exchange (heat conduction), G, from the soil or water surface to the layer of soil or water below.

In addition to these, there are other small energy terms such as the energy temporarily stored within the controlled volume, energy associated with horizontal air movement into and out of the controlled volume, energy possessed by the evaporated water, and energy involved in photosynthesis and respiration. However, these are generally insignificant. Energy balance for the controlled volume is

$$R_n - H - G - \rho_w \lambda E_r = 0 \quad [\text{FL}^{-1}\text{T}^{-1}] \tag{2.21}$$

where

$$R_n = \text{net radiation flux at the surface, J m}^{-2}\text{ s}^{-1}$$
$$\rho_w \lambda E_r = \text{latent heat flux (+ during evaporation)}$$
$$H = \text{sensible heat flux to air (+ when air is warming)}$$
$$G = \text{sensible heat flux to soil or water (+ if soil or water is warming)}$$
$$E_r = \text{Evaporation by the energy balance method}$$

There are two approaches to represent the sensible heat flux. It can be expressed as an aerodynamic relation in terms of the difference of the temperatures at water surface and air level above. Alternatively, it can be expressed as a ratio to the latent heat, called the Bowen ratio, $\beta = H/\rho_w \lambda E$. The Bowen ratio can be derived from temperatures and vapor pressures at two heights above the water surface:

$$\beta = \gamma \frac{T_2 - T_1}{e_2 - e_1} \quad [\text{dimensionless}] \tag{2.22}$$

Section 2.11 Evaporation from Free-Water Bodies 67

Table 2.9 Specific Heat, Saturated Vapor Pressure, Gradient, Psychrometric Constant, and Latent Heat of Vaporization at Standard Atmospheric Pressure

Temperature °C	Specific Heat c_p, kJ/kg °C	Saturated vapor pressure e_s, kPa	Gradient of saturated vapor pressure Δ, kPa °C^{-1}	Psychrometric constant γ, kPa °C^{-1}	Latent heat of vaporization λ, MJ/kg
0	4.218	0.611	0.044	0.0654	2.501
1	4.215	0.657	0.047	0.0655	2.499
2	4.211	0.706	0.051	0.0656	2.496
3	4.208	0.758	0.054	0.0656	2.494
4	4.205	0.814	0.057	0.0657	2.492
5	4.202	0.873	0.061	0.0658	2.489
6	4.200	0.935	0.065	0.0659	2.487
7	4.198	1.002	0.069	0.0659	2.484
8	4.196	1.073	0.073	0.0660	2.482
9	4.194	1.148	0.078	0.0660	2.480
10	4.192	1.228	0.082	0.0661	2.478
11	4.191	1.313	0.087	0.0661	2.475
12	4.190	1.403	0.093	0.0662	2.473
13	4.188	1.498	0.098	0.0663	2.470
14	4.187	1.599	0.104	0.0663	2.468
15	4.186	1.706	0.110	0.0664	2.466
16	4.185	1.819	0.116	0.0665	2.463
17	4.184	1.938	0.123	0.0665	2.461
18	4.184	2.065	0.130	0.0666	2.459
19	4.183	2.198	0.137	0.0666	2.456
20	4.182	2.337	0.145	0.0667	2.454
21	4.182	2.488	0.153	0.0668	2.451
22	4.181	2.645	0.161	0.0668	2.449
23	4.181	2.810	0.170	0.0669	2.447
24	4.180	2.985	0.179	0.0670	2.444
25	4.180	3.169	0.189	0.0670	2.442
26	4.180	3.363	0.199	0.0671	2.440
27	4.179	3.567	0.209	0.0672	2.437
28	4.179	3.781	0.220	0.0672	2.435
29	4.178	4.007	0.232	0.0673	2.433
30	4.178	4.243	0.243	0.0674	2.430
31	4.178	4.494	0.256	0.0674	2.428
32	4.178	4.756	0.269	0.0675	2.425
33	4.178	5.032	0.282	0.0676	2.423
34	4.178	5.321	0.296	0.0676	2.421
35	4.178	5.625	0.311	0.0677	2.418
36	4.178	5.943	0.326	0.0678	2.416
37	4.178	6.277	0.342	0.0678	2.414
38	4.178	6.627	0.358	0.0679	2.411
39	4.178	6.994	0.375	0.0680	2.409

The psychrometric constant, γ, represents a balance between the specific heat of moist air, c_p, and the latent heat of vaporization, λ. The values of c_p, λ, and γ are listed in Table 2.9. Incorporating the Bowen ratio in equation (2.21) and rearranging:

$$E_r = \frac{R_n - G}{\rho_w \lambda (1 + \beta)} \quad [\text{LT}^{-1}] \tag{2.23}$$

Instruments are available commercially that measure air temperature and vapor pressure at two elevations above the surface in addition to R_n and G to determine E_r by eq. (2.23). The sign of β often changes in morning and evening since H is positive (upward) during the day and negative at night.

The magnitude of soil (water) sensible heat flux, G, is usually small on a daily or longer basis and can be neglected since the heat gained early in the day is lost at night. However, for hourly periods, G can be significant. One method used to ascertain the value of G is to install heat flux measuring plates below the soil surface or inside the lake.

Net radiant energy, R_n, is the balance of the solar energy available at an evaporating surface. Of the total extraterrestrial radiation, R_A, received at the top of the earth's atmosphere, a portion, R_S, reaches the earth in short 0.3 to 3 μm wavelengths. A portion of the short-wave radiation S_n is reflected back to the atmosphere as *albedo*, α, leaving a net short-wave radiation of $(1 - \alpha)R_S$ captured at the ground surface.

Doorenbos and Pruitt (1977) recommended the following empirical relation for net short-wave radiation:

$$S_n = (1 - \alpha)\left(0.25 + 0.5\frac{n}{N}\right)R_A \quad [\text{FL}^{-1}\text{T}^{-1}] \tag{2.24}$$

where

$\quad\quad S_n$ = net short-wave radiation, MJ/m^2 day

$\quad\quad \alpha$ = albedo, 0.23 for crops and grass, and 0.08 for open water

$\quad\quad R_A$ = extraterrestrial radiation, see Table 2.10

$\quad\quad n/N$ = ratio of actual measured bright sunshine hours and maximum possible sunshine hours

Hargreaves and Samani (1982) proposed a relation based on air temperature for net short-wave radiation:

$$S_n = (1 - \alpha)K_{RS}\left(T_{\max} - T_{\min}\right)^{0.5} R_A \quad [\text{FL}^{-1}\text{T}^{-1}] \tag{2.25}$$

where

$\quad\quad T_{\max}$ = maximum daily air temperature, °C

$\quad\quad T_{\min}$ = minimum daily air temperature, °C

$\quad\quad K_{RS}$ = coefficient, 0.16 for interior regions and 0.19 for coastal regions

Both the ground and the atmosphere emit long 3 to 100 μm wavelength radiation characteristic of their temperatures, in accordance with Stefan-Boltzmann's law of back body radiation. The net long-wave radiation, which is the difference between the thermal radiation from the ground and emission from atmosphere and clouds, is given by an empirical relation of the form:

$$R_b = -\left(0.1 + 0.9\frac{n}{N}\right)\left(0.34 - 0.14\sqrt{e_d}\right)\sigma T^4 \quad [\text{FL}^{-1}\text{T}^{-1}] \tag{2.26}$$

where

R_b = net long-wave radiation, MJ/m^2 day

e_d = vapor pressure at air temperature, kPa

or (saturation vapor pressure at mean air temperature × relative humidity)

σ = Stefan-Boltzmann constant $\left(4.903 \times 10^{-9}\ \dfrac{\text{MJ}}{\text{m}^2\text{K}^4\text{day}}\right)$

T = mean air temperature, °K = 273.2 + °C

The negative sign in eq. (2.26) appears since there is a net loss of energy from the ground.

Net radiation, R_n, in eq. (2.23) can be measured using instrumentation or satellite data. Often, however, it has to be determined from the climatic data. Then R_n is equal to the sum of S_n and R_b, as follows:

$$R_n = \text{eq. (2.24) or eq. (2.25)} + \text{eq. (2.26)} \quad [\text{FL}^{-1}\text{T}^{-1}] \tag{2.27}$$

EXAMPLE 2.9

Solve Example 2.7 by the energy balance method. Daily Bowen ratio = 0.1, $n/N = 0.7$ and the extraterrestrial radiation at the location = 40 MJ/m^2 day. Disregard the soil heat flux.

SOLUTION

1. Net short-wave radiation: From eq. (2.24),

$$S_n = (1 - 0.08)[0.25 + 0.5(0.7)](40) = 22.08 \text{ MJ/m}^2 \text{ day}$$

2. Net long-wave radiation:

Saturation vapor pressure at air temperature, $e_a = 4.243$ kPa

Vapor pressure at 25% relative humidity, $e_d = 0.25(4.243) = 1.061$ kPa

$T = 30 + 273.2 = 303.2$ °K

From eq. (2.26),

$$R_b = -[0.1 + 0.9(0.7)]\left(0.34 - 0.14\sqrt{1.061}\right)(4.903 \times 10^{-9})(303.2)^4$$

$$= -5.92 \text{ MJ/m}^2 \text{ day}$$

3. Net radiation: From eq. (2.27),

$$R_n = 22.08 - 5.92 = 16.16 \text{ MJ/m}^2 \text{ day}$$

4. Evaporation:

At 30 °C, $\lambda = 2.43$ MJ/kg (from Table 2.9)

From eq. (2.23),

$$E_r = \frac{16.16}{(1000)(2.43)(1+0.1)} = 0.006 \text{ m/day or 6 mm/day}$$

Table 2.10 Total Daily Solar Radiation Received on a Horizontal Surface at the Top of the Atmosphere

								Approximate Date								
Latitude	Jan 13	Feb 4	Feb 26	Mar 21	Apr 13	May 6	May 29	June 22	July 15	Aug 8	Aug 31	Sep 23	Oct 16	Nov 8	Nov 30	Dec 22
								($MJ\,m^{-2}d^{-1}$)								
70		1.01	5.54	12.98	22.20	31.67	39.28	42.79	39.49	31.37	21.94	12.81	5.45	1.01		
60	3.09	6.17	11.67	18.90	26.85	34.20	39.49	41.39	39.28	33.86	26.59	18.69	11.54	6.17	3.04	2.07
50	8.67	12.22	17.71	24.31	30.95	36.66	40.50	41.81	40.33	36.32	30.65	24.01	17.50	12.09	8.62	7.44
40	14.80	18.35	23.38	29.00	34.12	38.47	41.09	41.90	40.88	38.09	33.74	28.62	23.04	18.14	14.71	13.40
30	20.89	24.01	28.33	32.77	36.57	39.28	40.88	41.22	40.59	38.94	36.19	32.34	28.03	23.84	20.80	19.70
20	26.64	29.21	32.51	35.56	37.80	39.02	39.53	39.53	39.32	38.73	37.37	35.13	32.13	28.96	26.51	25.58
10	31.79	33.61	35.73	37.29	37.92	37.75	37.25	36.91	37.08	37.46	37.50	36.82	35.30	33.36	31.62	30.99
0	36.15	37.12	37.88	37.84	36.91	35.39	33.99	33.40	33.82	35.09	36.49	37.42	37.46	36.78	35.98	35.64
-10	39.57	39.57	38.94	37.29	34.84	32.13	29.89	29.05	29.76	31.84	34.41	36.82	38.47	39.19	39.36	39.45
-20	41.98	40.93	38.81	35.56	31.71	27.90	25.07	23.97	24.94	27.65	31.33	35.13	38.35	40.54	41.77	42.24
-30	43.34	41.14	37.54	32.77	27.65	22.96	19.66	18.43	19.57	22.75	27.31	32.34	37.08	40.76	43.12	44.01
-40	43.63	40.29	35.01	29.00	22.75	17.46	13.91	12.56	13.87	17.29	22.53	28.62	34.63	39.91	43.42	44.77
-50	43.04	38.43	31.79	24.31	17.25	11.67	8.16	6.98	8.12	11.58	17.08	24.01	31.41	38.09	42.87	44.65
-60	41.94	35.81	27.57	18.90	11.37	5.92	2.87	1.99	2.87	5.88	11.25	18.69	27.23	35.51	41.73	44.22
-70	42.19	33.19	22.79	12.98	5.37	0.97				0.97	5.33	12.81	22.49	32.89	41.98	45.70

Northern Hemisphere (latitudes 70 to 0)

Southern Hemisphere (latitudes -10 to -70)

Source: Jensen, Burman, and Allen (1990). Used with permission of ASCE.

2.11.4 Combination Method of Penman

Penman (1948) combined evaporation computed by the aerodynamic method and evaporation computed by the energy balance method to estimate a weighted evaporation. The weighing factors used were $\gamma/(\Delta + \gamma)$, and $\Delta/(\Delta + \gamma)$, where γ is the psychrometric constant and Δ is the gradient of the saturated vapor pressure to the air temperature, de_s / dT. The values of both of these are listed in Table 2.9. The total heat content of an air mass is the sum of the sensible heat and latent heat represented by the terms $\gamma/(\Delta + \gamma)$, and $\Delta/(\Delta + \gamma)$, respectively. Thus, the basic equation for weighted evaporation is

$$E = \frac{\Delta}{(\Delta + \gamma)} E_r + \frac{\gamma}{(\Delta + \gamma)} E_a \quad [LT^{-1}] \qquad (2.28)$$

where

E_r = evaporation by the energy balance method, eq. (2.23)

E_a = evaporation by the aerodynamic method, eq. (2.18)

γ = psychrometric constant, Table 2.9

Δ = gradient of saturated vapor pressure, Table 2.9

EXAMPLE 2.10

Use the combination method to solve Example 2.7.

SOLUTION

1. At 30 °C air temperature, $\gamma = 0.0674$ kPa/°C, $\Delta = 0.243$ kPa/°C from Table 2.9

2. $E_r = 6$ mm/day from Example 2.9

3. $E_a = 5.9$ mm/day from Example 2.8

4. From eq. (2.28),

$$E = \frac{0.243(6)}{(0.243 + 0.0674)} + \frac{0.0674(5.9)}{(0.243 + 0.0674)}$$
$$= 0.78(6) + 0.22(5.9) = 5.98 \text{ mm/day}$$

2.12 EVAPOTRANSPIRATION FROM A DRAINAGE BASIN

Evapotranspiration considers evaporation from natural surfaces whether the water source is in the soil, plants, or a combination of both. As defined in Chapter 1, *consumptive use* is the amount of water required to support the optimum growth of a particular crop under field conditions. With respect to the cropped area, the consumptive use denotes the total evaporation from an area including the water used by plant tissues. Thus, evapotranspiration is the same as consumptive use except the latter includes water in the plant tissue, which is a minor quantity. The term "evapotranspiration" is more common than "consumptive use." The determination of evaporation and transpiration as separate elements for a drainage basin is unreliable. Moreover, their separate evaluation is not required for most studies.

Evapotranspiration is one of the most popular subjects of research in the field of hydrology and irrigation. Numerous procedures have been developed to estimate evapotranspiration. These fall in the categories of (1) water balance methods, such as evapotranspirometers, hydraulic budget on field plots, and soil moisture depletion; (2) energy balance methods; (3) mass-transfer methods, such as wind speed function, eddy flux, and use of enclosures; (4) a combination of energy and mass-transfer methods, such as the Penman-Monteith method; (5) prediction methods, such as empirical equations and the indices applied to pan-evaporation data; and (6) methods for specific crops. Currently, the most preferred approach to evapotranspiration study is based on the Penman-Monteith method.

In the context of evapotranspiration, Thornthwaite in 1948 introduced the term *potential evapotranspiration* to define the evapotranspiration that will occur when the soil contains an adequate moisture supply at all times (i.e., when moisture is not a limiting factor in evapotranspiration). For application to a cropped area, the *reference crop evapotranspiration* concept was introduced whereby the potential evapotranspiration was considered on the basis of an idealized crop of a uniform height, completely covering the ground, growing actively, and not experiencing any shortage of water. For the reference crop, clipped grass of 0.12 m height is used as a standard. Alfalfa of 0.5 m height is used as a standard crop specifically for arid regions. Evapotranspiration from any crop is determined from the reference crop evapotranspiration by applying a crop coefficient that simulates the condition of the specific crop.

2.12.1 Evapotranspirometers

A properly constructed and installed evapotranspirometer provides the most accurate estimates of evapotranspiration and is a reliable means of calibrating other methods. It is an instrument consisting of a block of soil with some planted vegetation enclosed in a container. If there is a provision for drainage of the soil water, it is referred to as a *lysimeter.* Evapotranspiration is ascertained by maintaining a water budget for the container; that is, accounting for the water applied, the water drained from the bottom, and the change in moisture content of the soil in the lysimeter. However, these instruments are rare and expensive and are applicable to a particular place, soil type, and vegetation.

2.12.2 Penman-Monteith Method

The aerodynamic process for determining evaporation considers the transport of water vapor by the turbulence of the wind blowing over a natural surface. In the aerodynamic eq. (2.18), with M defined by eq. (2.19), the factor $C_E u_Z$ is referred to as the *transport function.* Its inverse value is recognized as the *aerodynamic resistance* to water vapor transfer, r_a. By substituting $r_a = 1/C_E u_Z$ in eqs. (2.18) and (2.19), the aerodynamic equation takes the form

$$E_a = 0.622 \frac{\rho_a}{\rho_w P} \frac{1}{r_a} (e_s - e_Z) \quad [\text{LT}^{-1}] \tag{2.29}$$

Since $\rho_a = 3.486 P/(273 + T)$, with T in °C, eq. (2.29) becomes

$$E_a = \frac{2.17}{\rho_w (T + 273)} \frac{1}{r_a} (e_s - e_Z) \quad [\text{LT}^{-1}] \tag{2.30}$$

Another resistance function relevant to plants and vegetation is associated with the movement of water vapor from inside plant leaves to the air outside through small apertures in the leaves called *stomata*. This stomatal resistance for the entire plant canopy is called the *surface resistance*, r_s. Monteith (1965) combined the aerodynamic resistance and surface resistance into a function of the form $(1 + r_s / r_a)$ and applied it to the Penman combination equation (2.28). The result is what is recognized as the Penman-Monteith equation, which currently is the most widely used relationship in evapotranspiration study. The full expanded form of the equation without the Bowen ratio is

$$E_{to} = \frac{\Delta}{\Delta + \gamma \left(1 + \dfrac{r_s}{r_a}\right)} \frac{R_n - G}{\rho_w \lambda}$$

$$+ \frac{\gamma}{\Delta + \gamma \left(1 + \dfrac{r_s}{r_a}\right)} \frac{2.17}{\rho_w (T + 273)} \frac{1}{r_a} \left(e_Z^o - e_Z\right)(86{,}400) \quad [LT^{-1}] \qquad (2.31)$$

where

E_{to} = potential or reference crop evapotranspiration

e_Z^o = mean saturation vapor pressure at air temperature to which vegetation is exposed

e_Z = vapor pressure at air temperature for given relative humidity

All other terms were defined earlier. The factor (86,400) converts the second term to per day in line with the first term.

There are two approaches for applying the Penman-Monteith equation (2.31) to determine the actual evapotranspiration from any specific crop under a limited soil moisture condition. Because of its simplicity, the first method is more commonly applied.

1. The reference crop approach determines the potential evapotranspiration for a reference crop that is either clipped grass or alfalfa of a standard height. A crop coefficient, K_c, and a soil moisture coefficient, K_a, are applied to estimate the actual evapotranspiration.

2. In the direct application, the resistance factors are estimated from known field observations to directly represent the characteristics of the surface and vegetation type in question.

2.12.3 Reference Crop Evapotranspiration by the Penman-Monteith Method

For a water body or a fully wetted surface, $r_s = 0$. Then eq. (2.31) becomes identical to the Penman combination equation (2.28). For vegetated surfaces, r_s is related to the *leaf area index* (LAI), which is the area of (one side of) leaves of a crop growing on each unit area of ground surface. Allen et al. (1989) approximated the LAI to the height of grass and alfalfa reference crops, thus leading to the following relations:

$$\text{For grass reference crop: } r_s = 69 \quad \text{s/m}$$
$$\text{For alfalfa reference crop: } r_s = 45 \quad \text{s/m} \tag{2.32a}$$

Though the aerodynamic resistance, r_a, has an involved form of relation, it is simplified for reference crops for wind measurements at a standardized height of 2 m as follows:

$$\text{For grass reference crop: } r_a = 208/u_2 \quad \text{s/m}$$
$$\text{For alfalfa reference crop: } r_a = 110/u_2 \quad \text{s/m} \tag{2.32b}$$

If the wind measurements are made at heights other than 2 m above the ground surface, they should be adjusted to 2 m by eq. (2.20) for use in eq. (2.32b).

Using the values of r_s and r_a as discussed above, the resistance function for reference crops reduces as follows:

$$\text{For grass reference crop: } \left(1 + \frac{r_s}{r_a}\right) = \left(1 + 0.33\, u_2\right)$$
$$\text{For alfalfa reference crop: } \left(1 + \frac{r_s}{r_a}\right) = \left(1 + 0.40\, u_2\right) \tag{2.33}$$

EXAMPLE 2.11

Estimate the daily reference crop (alfalfa) evapotranspiration for July at Kimberly, Idaho; latitude 42.4°N; elevation 3922 ft. The mean values of the meteorological data for July from 1966 to 1985 are as follows:

1. Mean air temperature = 20.7 °C
2. Relative humidity = 49%
3. Wind velocity at 3.66 m height = 2.48 m/s
4. Mean percentage sunshine = 82.5%

SOLUTION   Net radiation estimate:

1. Extraterrestrial radiation for July at 42.4 °N, R_A = 40.8 MJ/m² day (from Table 2.10)
2. From eq. (2.24),

$$S_n = \left(1 - 0.23\right)\left[0.25 + 0.5\left(\frac{82.5}{100}\right)\right]40.8 = 20.8 \text{ MJ/m}^2 \text{ day}$$

3. Saturation vapor pressure at 20.7 °C, $e_Z^o = 2.44$ kPa

$$e_Z = e_d = \text{RH} \times 2.44 = (0.49)(2.44) = 1.2 \text{ kPa}$$

4. From eq. (2.26),

$$T = 273.2 + 20.7 = 293.90 \text{ °K}$$
$$R_b = -\left[0.1 + 0.9\left(\frac{82.5}{100}\right)\right]\left(0.34 - 0.14\sqrt{1.2}\right)\left(4.903 \times 10^{-9}\right)(293.9)^4$$
$$= -5.75 \text{ MJ/m}^2 \text{ day}$$

5. From eq. (2.27),

$$R_n = 20.8 - 5.75 = 15.05 \text{ MJ/m}^2 \text{ day}$$

6. Vapor pressure deficit

$$(e_z^o - e_z) = 2.44 - 1.20 + 1.24 \text{ kPa}$$

7. Common parameters

At 20.7 °C from Table 2.9, $\lambda = 2.45$ MJ/kg

$$\gamma = 0.0668 \text{ kPa/°C}$$
$$\Delta = 0.151 \text{ kPa/°C}$$

8. From eq. (2.20), for $Z_0 = .03$ m

$$\text{Wind velocity at 2 m, } u_2 = u_{3.66} \left(\frac{\ln \dfrac{2}{0.03}}{\ln \dfrac{3.66}{0.03}} \right) = 2.2 \text{ m/s}$$

9. From eq. (2.32b):

$$r_a = 110/2.2 = 50 \text{ s/m}$$

From eq. (2.33):

$$\left(1 + \frac{r_s}{r_a} \right) = \left[1 + 0.4(2.2) \right] = 1.88$$

10. Ratio

$$\frac{\Delta}{\Delta + \gamma \left(1 + \dfrac{r_s}{r_a} \right)} = \frac{0.151}{0.151 + 0.0668(1.88)} = 0.55$$

$$\frac{\gamma}{\Delta + \gamma \left(1 + \dfrac{r_s}{r_a} \right)} = \frac{0.0668}{0.151 + 0.0668(1.88)} = 0.24$$

11. From eq. (2.31):

$$E_{to} = 0.55 \frac{15.05}{(1000)(2.45)} + 0.24 \frac{(2.17)(1.24)(86,400)}{(1000)(20.7 + 273)(50)}$$
$$= 3.38 \times 10^{-3} + 3.80 \times 10^{-3} = 7.18 \times 10^{-3} \text{ m/day}$$

2.12.4 Actual Evapotranspiration from Any Surface

The reference crop evapotranspiration determined in the previous section is multiplied by the coefficients to determine actual evapotranspiration; thus

$$E_t = K_c K_a E_{to} \quad [LT^{-1}] \tag{2.34}$$

where

E_t = actual evapotranspiration from a surface

E_{to} = reference crop evapotranspiration

K_c = cover or crop coefficient at a specified growth stage

K_a = coefficient dependent on available soil water

The term "crop coefficient" applies to nonagricultural vegetation and bare soil as well. It is a complex coefficient to describe since it has to encompass different stages of crop growth, many climatic factors, and a variety of crops. Extensive research has been conducted on crop coefficients (Jensen, 1974; Doorenbos and Pruitt, 1977; Wright, 1982; Synder et al., 1989; Allen et al., 1991; Slack et al., 1996; Snyder, 2000). Because of the two reference crop definitions with grass and alfalfa, two families of K_c curves exist for various crops. The coefficients should be clearly associated with their reference crop. As an approximate conversion, the alfalfa-based K_c values can be changed to grass reference values by multiplying by a factor ranging from 1.1 (for humid conditions) to 1.25 (for arid conditions).

Doorenbos and Pruitt (1977) presented a relatively simple procedure for constructing the grass reference crop coefficients, K_c. A K_c curve, as shown in Figure 2.10, is constructed as follows:

1. From local information determine the total growing season and divide this into four growing stages—initial stage, crop development stage, mid-season stage, and late-season stage. If this information is not known, use Table 2.11. Mark the end point of each stage as D_1, D_2, D_3, and D_4 on Figure 2.10.

2. The coefficient for stage 1, K_{c1}, is a function of the reference evapotranspiration at the time of planting and the period between occurrence of irrigation or significant rain. Determine this from Figure 2.11. Draw a horizontal line at this K_{c1} value from the beginning to the end of stage 1 (D_1).

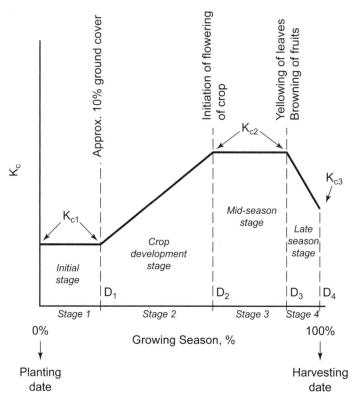

Figure 2.10
Schematic diagram of the crop coefficient curve.

Table 2.11 Growing Season, Distribution among Stages, and Mid- and Late-Season Crop Coefficients for Selected Crops

Crop	Typical growing season, days	Fraction of stage time at growth stage				K_{c2}^a	K_{c3}^a
		1	2	3	4		
Artichokes (perennial)	310–360	0.09	0.12	0.70	0.09	1.00	0.95
Barley	120–150	0.12	0.20	0.44	0.24	1.15	0.25
Beans (dry)/Pulses	95–110	0.16	0.25	0.40	0.19	1.15	0.35
Carrots	100–150	0.18	0.27	0.39	0.16	1.05	0.80
Celery	125–180	0.16	0.27	0.46	0.11	1.05	1.00
Corn (sweet)	80–110	0.23	0.29	0.37	0.11	1.15	1.05
Corn (grain)	125–180	0.17	0.28	0.33	0.22	1.15	0.55[b]
Cotton	180–195	0.16	0.27	0.31	0.26	1.20	0.70
Cucumber	105–130	0.19	0.28	0.38	0.15	1.00	0.75
Eggplant	130–140	0.21	0.32	0.30	0.17	1.05	0.90
Grain (small)	150–165	0.14	0.20	0.40	0.26		
Lentil	150–170	0.14	0.20	0.41	0.25	1.15	0.30
Lettuce	75–140	0.26	0.37	0.27	0.10	1.05	0.95
Melons	120–160	0.20	0.28	0.37	0.15	1.05	0.75
Millet	105–140	0.14	0.23	0.39	0.24	1.10	0.30
Oats	120–150	0.12	0.20	0.44	0.24	1.15	0.25
Onion (dry)	150–210	0.10	0.17	0.49	0.24	1.05	0.85
Groundnuts (peanuts)	130–140	0.22	0.30	0.30	0.18	1.05	0.60
Peas	90–100	0.21	0.26	0.37	0.16	1.15	1.10[c]
Peppers (fresh)	120–125	0.22	0.29	0.33	0.16	1.05	0.90
Potato	105–145	0.21	0.25	0.33	0.21	1.15	0.75
Radishes	35–40	0.20	0.27	0.40	0.13	0.90	0.85
Rice	150–180	0.20	0.20	0.40	0.20	1.20	0.90
Safflower	125–190	0.17	0.27	0.35	0.21	1.15	0.25
Sorghum	120–130	0.16	0.27	0.33	0.24	1.00	0.55
Soybeans	135–150	0.14	0.21	0.46	0.19	1.15	0.50
Spinach	60–100	0.27	0.31	0.34	0.08	1.00	0.95
Squash	90–100	0.24	0.34	0.26	0.16	0.95	0.75
Sugar beet	160–230	0.18	0.27	0.33	0.22	1.20	0.95
Sunflower	125–130	0.17	0.28	0.36	0.19	1.15	0.35
Tomato	135–180	0.20	0.28	0.33	0.19	1.20	0.65
Wheat	120–150	0.12	0.20	0.44	0.24	1.15	0.25

[a] K_{c2} and K_{c3} values are for a subhumid climate (minimum daytime humidity 45%) and moderate wind speed (averaging 2 m/s). For adjustments to other conditions refer to ASCE (1996).

[b] (high moist), 0.35 (dry)

[c] (fresh), 0.3 (dry)

Source: Derived from Doorenbos and Pruitt (1977).

Figure 2.11 Crop coefficient during initial stage (derived from Doorenbos and Pruitt, 1977).

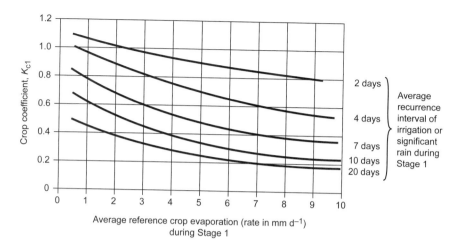

3. Determine K_{c2} and K_{c3} from Table 2.11. These values are for a minimum daytime humidity of about 45% and a wind speed of 2 m/s at 2 m above the crop level. Refer to ASCE (1996) to make corrections for a different climate. Draw a horizontal line at K_{c2} from the beginning to the end of stage 3 (from D_2 to D_3). Mark a point at K_{c3} at the end of stage 4 (D_4).

4. Place straight line segments through end points to complete the diagram similar to Figure 2.10. The crop coefficients are read from this figure and applied to the reference crop evapotranspiration.

For well-watered crops, the value of the coefficient K_a is 1. It is less than 1 when available soil moisture limits transpiration; it then depends on the percentage of available soil water compared to the field capacity (Jensen et al., 1990).

EXAMPLE 2.12

Prepare the crop coefficient curve (grass reference) for sugar beets. The crop is planted April 11. The grass reference evapotranspiration for April, E_{to}, is 3.5 mm/day. The soil is wetted in 10-day intervals during the initial stage. The crop is well watered.

SOLUTION

1. From Table 2.11, for sugar beets, $D_1 = 18\%$, $D_2 = 27\%$, $D_3 = 33\%$, $D_4 = 22\%$

2. From Figure 2.11, for sugar beets, $E_{to} = 3.5$ and 10-day wetting, $K_{c1} = 0.4$

3. From Table 2.11, for sugar beets, $K_{c2} = 1.2$ and $K_{c3} = 0.95$

4. The curve is drawn in Figure 2.12. Draw a horizontal line AB at $K_{c1} = 0.4$ for the first 18% (D_1), another horizontal line CD at $K_{c2} = 1.2$ starting at $D_1 + D_2$ or 45% and ending at $D_1 + D_2 + D_3$ or 78%. Mark a point, E, at $K_{c3} = 0.95$ at 100% growth at the end of stage D_4. Join BC and DE as shown in the figure.

EXAMPLE 2.13

The sugar beet crop in Example 2.12 is planted on April 11 and harvested on Oct. 15. Estimate monthly evapotranspiration. Compare with the Blaney-Criddle Method (Example 1.9). The estimated grass reference evapotranspiration is

Month	April	May	June	July	Aug	Sept	Oct
E_{to}, mm/day	3.5	4.0	4.5	5.0	5.0	3.0	2.0

SOLUTION   The computations are arranged in the table below.

(1)	(2)	(3)	(4)	(5)	(6)	(7)	
		Midperiod[a] percent of total season	K_c value from Fig. 2.12	Ref. Crop E_{to} mm/day	$K_c \times E_{to}$[c] mm/day	\multicolumn{2}{c}{U from Example 1.9}
Period	Number of days					in./month	mm/day
Apr. 11–30	20	5	0.4	3.5	1.4	0.89	0.75
May	31	19[b]	0.42	4.0	1.68	2.15	1.76
June	30	35	0.90	4.5	4.05	4.01	3.40
July	31	51	1.2	5.0	6.0	7.06	5.78
Aug.	31	68	1.2	5.0	6.0	7.13	5.84
Sept.	30	84	1.13	3.0	3.39	3.96	3.35
Oct. 1–15	15	96	1.0	2.0	2.0	2.03	1.66
Total	188						

a $\dfrac{\text{Number of days to middle of period}}{\text{Total days in growing season}} \times 100$

b $\dfrac{20 + 31/2}{188} = 19\%$

c column 4 × column 5

2.12.5 Blaney-Criddle Method

There is another category of methods suitable for the cropped area. The Blaney-Criddle method, developed for conditions in the arid western regions of the United States, is described in section 1.13.2. The method directly provides the monthly evapotranspiration for a specific crop. The crop coefficients used in the Blaney-Criddle method are different from the reference crop coefficients. The climatic effects and crop parameters are not separated in the Blaney-Criddle relations. Application of the method is provided in Example 1.9 and a comparison with the Penman-Monteith method is shown in Example 2.13.

2.13 DIRECT RUNOFF FROM RAINFALL OR RAIN EXCESS

Information on rainfall excess is necessary in hydrograph analysis, discussed in Chapter 7. As indicated by the water balance equation (2.5), the direct runoff or rainfall excess contributing to immediate streamflow is assessed by subtracting the infiltration from the total rainfall. A simple model, a homogeneous soil column with a uniform initial water content, is considered. There are three distinct cases of infiltration.

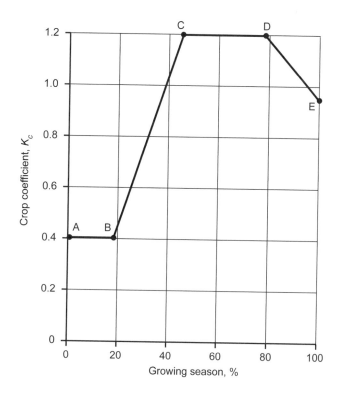

Figure 2.12 Crop coefficient curve for Example 2.12.

1. When a rainfall intensity, i, is less than the saturated hydraulic conductivity, K_s,* all the rainfall infiltrates, as shown by line I in Figure 2.13.

2. The effect of the rainfall rate, which is greater than the saturated conductivity $(i > K_s)$ is shown by curve II. Initially, water infiltrates at the application rate. After a time t_p, the capacity of soil to infiltrate water falls below the rainfall rate. Surface ponding begins, which results in depression storage and runoff.

3. For a rainfall intensity that exceeds the capacity of soil to infiltrate water from the beginning, water is always ponded on the surface. In this case, the rate of infiltration is controlled only by soil-related factors. This rate, shown by curve III in Figure 2.13, is called the infiltration capacity of a given soil, f_p.

The infiltration capacity, f_p, decreases with time, due primarily to reduction in the hydraulic gradient between the surface and the *wetting front*, which is the limit of water penetration into the soil. The front separates the wet soil from the dry soil. The infiltration capacity eventually approaches a constant rate, f_c, which is considered to be equal to the apparent saturated hydraulic conductivity, K_s.

After the surface ponding (beyond time t_p for case 2 and from the beginning for case 3), for a continuous uniform rain of intensity i, the surface runoff hydrograph has a shape indicated by q in Figure 2.14. The difference between rainfall and runoff appears as the curve marked $(i - q)$. The curve f_p relates to the infiltration rate. The difference between

---

* Natural soils are usually not completely saturated, even below the water table, due to air entrapment during the wetting process. The hydraulic conductivity, K_s, is taken to be the residual air saturation conductivity and is sometimes referred to as the apparent saturated hydraulic conductivity. For a definition of hydraulic conductivity, refer to Chapter 3.

Section 2.13 Direct Runoff from Rainfall or Rain Excess

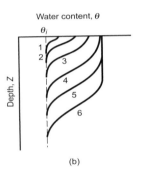

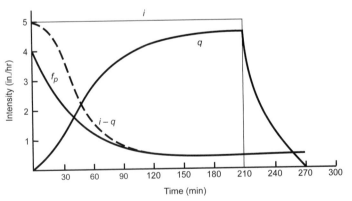

Figure 2.14
Water balance components of overland flow.

the dotted $(i - q)$ curve and the f_p curve signifies interception and other minor losses (storages) at the beginning. After the surface storage is filled in, the two curves coincide (i.e., direct runoff results from subtracting the infiltration from the rainfall). If there is knowledge of the minor losses,[*] these are deducted from the first part of the precipitation after ponding. Ordinarily, these are ignored because they are relatively minor and cannot be assessed reliably. The basic problem thus relates to determination of the infiltration loss rate under different conditions. This is known as the infiltration approach to surface runoff assessment, as compared to the direct rainfall-runoff correlation (Section 7.15) and multivariate runoff relation (Linsley et al., 1982, pp. 244–249).

There are four approaches to determine the rainfall excess using the infiltration concept. Two of these, the infiltration capacity curve and the nonlinear loss rate function, are detailed methods that consider the time-varying infiltration rates. In the simplified index approach, the average rate of infiltration for the storm period is used. The NRCS (formerly SCS) method uses the time-averaged parameters and indirectly considers the infiltration rate through the soil characteristics.

2.13.1 Infiltration Capacity Curve Approach

Green and Ampt proposed in 1911 a relation for infiltration capacity based on Darcy's law of soil water movement. Extensive research on the theory of infiltration was carried out during the 1930s and mid-1940s. Kostiakov and Horton suggested empirical relations for the infiltration capacity that became quite popular because of their simplicity. Subsequent empirical equations were formulated by Philip in 1957 and Holton in 1961.

For unsaturated soil, the equation for flux (volume of water moving per unit area per unit time) is given by Darcy's law, in which the hydraulic conductivity is a function of water content. When combined with the equation of conservation of mass, this relation yields the following:

$$\frac{\partial \theta}{\partial t} = \frac{\partial}{\partial z}\left(K \frac{dh}{d\theta}\frac{\partial \theta}{\partial z} \right) - \frac{\partial K}{\partial z} \quad [\mathrm{T}^{-1}] \tag{2.35}$$

where

θ = water content of soil

K = hydraulic conductivity

h = pressure head on soil medium

z = distance measured positively downward from the surface

Equation (2.35), known as the Richards equation, is the governing equation of infiltration through saturated and unsaturated soil. Exact analytical solutions to the Richards equation are limited to a few cases. Numerical solutions have been developed for various initial and boundary conditions of interest.

The elaborate procedures of the numerical method have been of limited value in practice because of computational cost, time, and soil properties data requirements. On the other hand, the simple equation of Green and Ampt has been a focus of renewed interest by many researchers.

---

[*] The minor losses are considered in several ways, depending on the available information: (1) only interception is excluded from the precipitation; (2) surface retention is excluded, comprising interception, depression storage, and evaporation during the storm; or (3) initial storm loss is subtracted, which is the interception and only a small fraction of the depression storage. Other depressions are considered as a part of the drainage.

In addition, there are empirical models for the infiltration capacity curve. These empirical models are of two types. In the Kostiakov, Horton, and Philip models, the infiltration capacity is expressed as a function of time; it decreases rapidly with time during the early part of the infiltration process if rainfall intensity is greater than the infiltration capacity. However, if the rainfall intensity is less than the capacity, the decay in the curve is less in the ratio of actual infiltration to potential (capacity) infiltration. This necessitates an adjustment of the capacity curve. The Holton model removes the problem of capacity curve adjustment for the rainfall application rate by relating the infiltration capacity to the soil moisture deficiency. The moisture deficiency (available storage) is reduced with time due to infiltrated water, and so is the infiltration capacity. The Green-Ampt model, though having a theoretical basis, is also based on the storage concept.

2.13.2 Horton Model

Horton (1939) presented a three-parameter equation expressed as

$$f_p = (f_0 - f_c) \, e^{-kt} + f_c \quad [LT^{-1}] \tag{2.36}$$

where

f_0 = initial infiltration capacity, in./hr

f_c = final constant infiltration capacity (equal to apparent saturated conductivity), in./hr

k = factor representing the rate of decrease in the capacity, 1/time

The parameters f_0 and k have no physical basis; that is, they cannot be determined from soil water properties and must be ascertained from experimental data.

The plot of eq. (2.36) is an asymptotic curve that starts at f_0 and attains a constant value of f_c as shown by ABD in Figure 2.15. The portion of the precipitation intensity above this $f_p - t$ curve during different time intervals designates the runoff and the depression storage, if any.

At the beginning of a storm, if the precipitation for a certain duration occurs at a rate less than the infiltration capacity, a soil moisture deficiency exists and the capacity for infiltration remains higher to a point C in Figure 2.15 rather than falling to point B according to eq. (2.36). This is illustrated in Example 2.14.

For calculating runoff, if there is any depression storage capacity, S, depression storage is accounted for first until S is completely full. For each time interval, the depression storage, ΔS, is the *minimum* of the following two values:

$$\Delta S = i\Delta t - \Delta F \quad [L] \tag{2.37a}$$

or

$$\Delta S = \underset{\begin{pmatrix} \text{depression} \\ \text{storage} \\ \text{capacity} \end{pmatrix}}{S} - \underset{\begin{pmatrix} \text{cumulated} \\ \text{storage from} \\ \text{previous step} \end{pmatrix}}{\Sigma \Delta S} \quad [L] \tag{2.37b}$$

After ΔS is computed, the runoff is calculated by

$$RO = i\Delta t - \Delta F - \Delta S \quad [L] \tag{2.38}$$

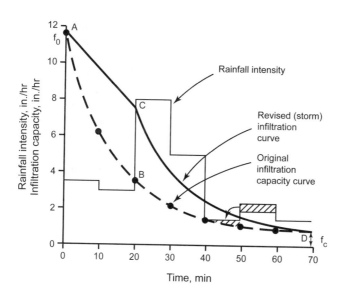

Figure 2.15 Rainfall intensity and infiltration capacity curves.

where

$\quad$ RO = rainfall excess or runoff during time Δt

$\quad$ ΔF = difference in cumulated infiltration F during Δt

$\quad$ ΔS = change in depression storage during Δt

When there is no depression storage capacity, ΔS is zero.

$\quad$ After runoff ensues, if the precipitation intensity in a certain period falls below the infiltration capacity curve, the moisture deficiency for this period has to be met from the subsequent excessive precipitation as explained in the example.

EXAMPLE 2.14

The infiltration capacity curve for a watershed is given by $f_p = (11.66 - 0.83)\, e^{-0.07t} + 0.83$ where t is in min and f_p in in./hr. The storm pattern is:

t, min	Intensity, in./hr
0–10	3.5
10–20	3.0
20–30	8.0
30–40	5.0
40–50	1.5
50–60	2.4
60–70	1.5

Determine the rainfall excess for the successive 10-min period.

SOLUTION

1. The infiltration capacity f_p is computed in col. 2 of Table 2.12 at various times by the given formula.

Section 2.13 Direct Runoff from Rainfall or Rain Excess

2. During the first 20 min, the rainfall intensity is less than the infiltration capacity; hence all rainfall is infiltrated.

$$\text{Total rainfall during 20 min} = 3.5\left(\frac{10}{60}\right) + 3\left(\frac{10}{60}\right) = 1.08 \text{ in.}$$

3. To prepare the revised infiltration curve, the infiltration is cumulated in col. 6 of Table 2.12 for various time intervals. The infiltration capacity at time zero (col. 2) is plotted against the zero cumulated infiltration, and the infiltration capacity of each subsequent period is plotted against the cumulated infiltration (col. 6) of the corresponding period in Figure 2.16, i.e., f_p of 11.66 plotted against F of 0.0, f_p of 6.21 plotted against F of 1.49, and so on.

4. From Figure 2.16, corresponding to $F = 1.08$ in. from step 2, f_p is 7.5 in./hr.

5. Set 7.5 in./hr as the initial value of f_0 at 20 min. From this point onward, the storm infiltration curve is given by

$$f_p' = [7.5 - 0.83]e^{-0.07t'} + 0.83$$

where t' is the time counted 20 min after the start of the storm.

Table 2.12 Infiltration Capacity and Cumulated Infiltration

(1) Time min	(2) f_p in./hr	(3) Δt[a] min	(4) Average f_p[b] in./hr	(5) ΔF[c] in.	(6) F[d] in.
0	11.66				
		10	8.94	1.49	1.49
10	6.21				
		10	4.86	0.81	2.30
20	3.50				
		10	2.83	0.47	2.77
30	2.16				
		10	1.83	0.31	3.08
40	1.49				
		10	1.33	0.22	3.30
50	1.16				
		10	1.08	0.18	3.48
60	0.99				
		10	0.95	0.16	3.64
70	0.91				

[a] Successive difference col. 1

[b] Average of two successive values of col. 2

[c] col. 3 × col. 4 × $\left[\dfrac{1 \text{ hr}}{60 \text{ min}}\right]$

[d] Cumulation of col. 5

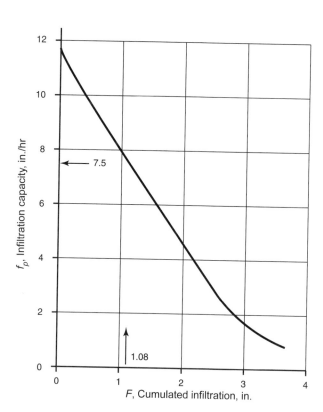

Figure 2.16 Cumulated infiltration curve.

Table 2.13 Revised Infiltration Capacity for the Storm of Example 2.14

(1) t' min	(2) Time from beginning of storm t min	(3) f_p using t' in./hr
0	20	7.5
10	30	4.14
20	40	2.47
30	50	1.65
40	60	1.24
50	70	1.03

6. Using the equation of step 5, the revised storm infiltration capacity at different times is calculated in Table 2.13 and plotted on Figure 2.15, shown as the revised curve.

7. The computations of rainfall excess during different time intervals are arranged in Table 2.14. From eq. (2.38), runoff is $i\Delta t - \Delta F$ since there is no depression storage.

8. During the time interval 40–50 min, the rainfall amount is less than the cumulated infiltration during this period by 0.09 in. The runoff for this period is zero (it can not be negative). This deficiency of 0.09 in. is met from the rainfall of the succeeding period (50–60 min.), as shown in Figure 2.15 and Table 2.14.

Section 2.13 Direct Runoff from Rainfall or Rain Excess

Table 2.14 Computations of Rainfall Excess by the Horton Method

(1)	(2)	(3)	(4)	(5)	(6)	(7)	(8)
Time min	Revised infiltration capacity f'_p in./hr	Δt min	Average f'_p in./hr	Cumulated ΔF in.	Rainfall intensity i in./hr	$\Delta P = i\Delta t^a$ in.	$RO^b = \Delta P - \Delta F$ in.
0							
10							
20	7.5						
		10	5.82	0.97	8.0	1.33	0.36
30	4.14						
		10	3.31	0.55	5.0	0.83	0.28
40	2.47						0
		10	2.06	0.34	1.5	0.25	− 0.09
50	1.65						
		10	1.45	0.24	2.4	0.4	0.16
							(−).09^c
							= 0.07
60	1.24						
		10	1.13	0.19	1.5	0.25	0.06
70	1.03						

Note: Computations of col. (3), (4) and (5) are similar to Table 2.13.

[a] Col. 6 × col. 3 × $\left[\dfrac{1 \text{ hr}}{60 \text{ min}} \right]$

[b] Col. 7 − col. 5

[c] Negative value of the previous step.

2.13.3 Holton Model

For agriculture watersheds, Holton and others in the Agriculture Research Service of the U.S. Department of Agriculture developed infiltration models during the mid-1960s and 1970s. The modified equation used in the USDAHL-70 Watershed Model has the form:

$$f_p = GI \cdot aS^{1.4} + f_c \qquad (2.39)$$

where

f_p = infiltration capacity, in./hr

GI = growth index of crop, percent of maturity

a = index of surface connected porosity

S = available storage in the surface layer, in.

f_c = constant rate of infiltration after long wetting, in./hr

The Agriculture Research Service has developed experimental GI curves for several crops (see for example, Holton et al., 1975).

Table 2.15   Estimates of Index *a* in the Holton Equation

| Land Use or Cover | Basal Area Rating[a] | |
	Poor Condition	Good Condition
Fallow[b]	0.10	0.30
Row crops	0.10	0.20
Small grains	0.20	0.30
Hay (legumes)	0.20	0.40
Hay (sod)	0.40	0.60
Pasture (bunch grass)	0.20	0.40
Temporary pasture (sod)	0.20	0.60
Permanent pasture (sod)	0.80	1.00
Woods and forests	0.80	1.00

[a] Adjustments needed for "weeds" and "grazing."
[b] For fallow land only, poor condition means "after row crop" and good condition means "after sod."
Source: Skaggs and Khaleel (1982).

Table 2.16   Estimates of Final Infiltration Rate

Hydrologic Soil Grade	f_c (in./hr)
A	0.45–0.30
B	0.30–0.15
C	0.15–0.05
D	0.05–0

Source: Skaggs and Khaleel (1982).

Index *a* is a function of surface conditions and the density of plant roots. Estimates of *a* are given in Table 2.15.

The values for f_c are based on the hydrologic soil groups, as categorized in the *SCS National Engineering Handbook* and explained in Section 2.15. Estimates of f_c are given in Table 2.16.

Available storage, *S*, is computed by $S = (\theta_s - \theta)d$, where θ_s is the water content at saturation that equals porosity (in fact, it is the water content at residual air saturation), θ is the water content at any instant, and *d* is the surface-layer depth. For control depth, *d*, Holton and Creitz have suggested using the depth of the plow layer or the depth to the first impeding layer. However, Huggins and Monke (1966) consider that determining the depth is uncertain since it is highly dependent on surface condition and practices of preparing the seedbed.

The procedure for applying the Holton model is as follows:

1. First measure or estimate the initial moisture content, θ_i.
2. Compute the initial available storage by $S_0 = (\theta_s - \theta_i)d$.
3. Determine the initial infiltration capacity f_p from eq. (2.39).

Section 2.13   Direct Runoff from Rainfall or Rain Excess

4. Determine S after a period of time Δt by $S = S_0 - F + f_c \Delta t + ET\Delta t$, where F is the minimum of $f_p \Delta t$ and $i\Delta t$ (the available storage is reduced by the infiltration water but partly recovered due to drainage from the surface layer at the rate of f_c and by evapotranspiration, ET, through plants).

5. Determine f_p after period Δt by eq. (2.39).

6. Repeat the process.

2.13.4 Approximate Infiltration Model of Green-Ampt

The Green-Ampt model (1911) has received considerable renewed research attention and has found favor in field applications because (1) it is a simple model, (2) it has a theoretical base on Darcy's law (it is not strictly empirical), (3) its parameters have physical significance that can be computed from soil properties, and (4) it has been used with good results for profiles that become dense with depth, for profiles where hydraulic conductivity increases with depth, for soils with partially sealed surfaces, and for soils having nonuniform initial water contents. The model is developed as follows.

Consider a column of homogeneous soil of unlimited depth with an initial uniform water content θ_i. It is assumed that a ponding depth H is maintained over the surface from time 0 and a sharply defined wetting front is formed as shown in Figure 2.17. The length of the wet zone increases as infiltration progresses.

The application of Darcy's law results in the following form of the Green-Ampt equation:

$$f_p = K_s(H + S_f + L)/L \quad [LT^{-1}] \tag{2.40}$$

where

K_s = effective hydraulic conductivity

H = ponding depth

S_f = suction (capillary) head at the wetting front

L = depth to the wetting front

If the total (cumulative) infiltration is expressed as $F = (\theta_s - \theta_i)L$ or ML, θ_s being porosity, and the ponding depth is very shallow, $H \approx 0$, then eq. (2.40) can be written as

$$f_p = K_s + \frac{K_s MS_f}{F} \quad [LT^{-1}] \tag{2.41}$$

where $M = (\theta_s - \theta_i)$ is the initial soil water deficit.

Since $f_p = dF/dt$, the integration of eq. (2.41) with the condition $F = 0$ at $t = 0$ provides a cumulative infiltration as

$$K_s t = F - MS_f \ln\left(1 + \frac{F}{MS_f}\right) \quad [L] \tag{2.42}$$

Morel-Seytoux and Khanji (1974) indicated that the form of eqs. (2.40) through (2.42) remains the same when simultaneous movement of both water and air take place. The terms on the right side of the equation would, however, have to be divided by a viscous resistance correction factor, ranging from 1.1 to 1.7.

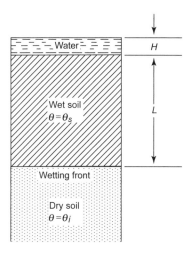

Figure 2.17 Simplified wetting front in the Green-Ampt model.

The effective suction at the wetting front, S_f, has been a subject of further research (refer to Skaggs and Khaleel, 1982). Many suction-related terms have been used to represent it. Mein and Larson (1973) suggested the average suction at the wetting front, S_{av}, to represent S_f and used the ratio of unsaturated hydraulic conductivity to effective conductivity as a weighting factor to define it, as given by eq. (2.46) subsequently. Many investigators found the application of S_{av} satisfactory.

Equations (2.40) through (2.42) apply when a ponding exists $(i \geq f_p)$ from the beginning. If $i < f_p$,* the surface ponding effect will not take place until time t_p. Under this condition, for a steady rainfall, the actual infiltration rate, f, can be summarized as follows:

1. For $t < t_p$,

$$f = i \quad [\text{LT}^{-1}] \tag{2.43}$$

2. For $t = t_p$,

$$f = f_p = i \quad [\text{LT}^{-1}] \tag{2.44a}$$

The cumulative infiltration at the time of surface ponding, F_p, can be obtained from eq. (2.41) after substituting S_{av} for S_f.

$$F_p = \frac{S_{av} M}{i / K_s - 1} \quad [\text{L}] \tag{2.44b}$$

$$t_p = \frac{F_p}{i} \quad [\text{T}] \tag{2.44c}$$

3. For $t > t_p$, as given in eq. (2.41),

$$f = f_p = K_s + \frac{K_s S_{av} M}{F} \quad [\text{LT}^{-1}] \tag{2.45a}$$

For cumulative infiltration, Mein and Larson suggested an equation analogous to eq. (2.42):

$$K_s \left(t - t_p + t_p' \right) = F - M S_{av} \ln \left(1 + \frac{F}{M S_{av}} \right) \quad [\text{L}] \tag{2.45b}$$

* It is assumed that $i > K_s$. If not, surface ponding will not occur at all as discussed in Section 2.13.

Section 2.13   Direct Runoff from Rainfall or Rain Excess   **91**

where t_p' is the equivalent time to infiltrate F_p under the condition of surface ponding from the beginning as obtained from eq. (2.42) after substituting S_{av} for S_f.

For unsteady rainfall, the Green-Ampt model provides good results if the rainfall variations are not excessive and the rainfall contributes to an extension of the wetted profile. However, if there are relatively long periods of low or zero rainfall, the model predictions are less accurate, due to redistribution of the soil water. For rainfall after a long dry period, a new soil water distribution should be considered.

The depression storage and runoff computations should proceed according to equations (2.37) and (2.38).

2.13.5 Determination of Parameters in the Green-Ampt Model

As stated earlier, an advantage of the Green-Ampt model as compared to the empirical models is that its parameters can be ascertained from the physical properties of soil. The saturated volumetric water content, θ_s, is measured by the porosity of soil, although it is somewhat less due to entrapped air even below the water table. Similarly, the value of K_s is less than the saturated hydraulic conductivity, K_0. Bouwer (1966) described an air-entry parameter to measure K_s. In the absence of a field-measured value, he suggested that K_s be estimated as $K_s \approx 0.5 K_0$.

Mein and Larson (1973) provided the relations of capillary suction (S_f) versus relative conductivity (K/K_s) for selected soils. The parameter of average capillary suction is defined as

$$S_{av} = \int_0^1 S_f \, dK_r \quad [\text{L}] \tag{2.46}$$

where K_r = relative hydraulic conductivity = K/K_s. Thus S_{av} is the area under the S_f vs K_r curve of a particular soil. The values of porosity, saturated hydraulic conductivity, and average wetting front suction are given in Table 2.17. Usually, it proves advantageous to determine the parameters of the model from field measurements by fitting measured infiltration data into the equation (Skaggs and Khaleel, 1982).

EXAMPLE 2.15

Rainfall at a constant intensity of 6 mm/hr falls on a homogeneous soil which has an initial uniform moisture content of 0.23. The soil property data obtained are $K_s = 1.24$ mm/hr and θ_s (porosity) = 0.48. The estimated value of S_{av} is 150 mm. Determine the rainfall excess. Assume no interception and depression storage.

SOLUTION

(a) Time to surface ponding, t_p:

    1. $M = \theta_s - \theta_i = 0.48 - 0.23 = 0.25$

    2. From eq. (2.44b),

$$F_p = \frac{150(0.25)}{6/1.24 - 1} = 9.8 \text{ mm}$$

    3. From eq. (2.44c), $t_p = 9.8/6 = 1.63$ hr.

(b) Infiltration after the ponding:

    4. First determine t_p' from eq. (2.42) as if ponding is from the beginning

Table 2.17 Green-Ampt Infiltration Parameters

Soil texture class	Porosity, θ_s	Wetting front soil suction head, S_f, cm	Effective hydraulic conductivity, K_s, cm/h
Sand	0.437	4.95	11.78
	(0.374–0.500)	(0.97–25.36)	
Loamy sand	0.437	6.13	2.99
	(0.363–0.506)	(1.35–27.94)	
Sandy loam	0.453	11.01	1.09
	(0.351–0.555)	(2.67–45.47)	
Loam	0.463	8.89	0.66
	(0.375–0.551)	(1.33–59.38)	
Silt loam	0.501	16.68	0.34
	(0.420–0.582)	(2.92–95.39)	
Sandy clay loam	0.398	21.85	0.15
	(0.332–0.464)	(4.42–108.0)	
Clay loam	0.464	20.88	0.10
	(0.409–0.519)	(4.79–91.10)	
Silty clay loam	0.471	27.30	0.10
	(0.418–0.524)	(5.67–131.50)	
Sandy clay	0.430	23.90	0.06
	(0.370–0.490)	(4.08–140.2)	
Silty clay	0.479	29.22	0.05
	(0.425–0.533)	(6.13–139.4)	
Clay	0.475	31.63	0.03
	(0.427–0.523)	(6.39–156.5)	

Source: Rawls and Brakensiek (1983).

$$(1.24)t_p' = (9.8)-(0.25)150 \ \ln\left[1+\frac{9.8}{0.25(150)}\right]$$

$$t_p' = 0.88 \ \text{hr}$$

5. From eq. (2.45a),

$$f = 1.24+\frac{1.24(150)(0.25)}{F}$$

or

$$f = 1.24+\frac{46.5}{F} \tag{a}$$

6. From eq. (2.45b),

$$1.24(t-1.63+0.88) = F-0.25(150)\ln\left[1+\frac{F}{0.25(150)}\right]$$

Section 2.13   Direct Runoff from Rainfall or Rain Excess

or

$$t = 0.75 + 0.81F - 30.24 \ln(1 + 0.0267F) \qquad \text{(b)}$$

A graph of F versus t for eq. (b) is plotted in Figure 2.18.

7. At successive time levels, the value of F is noted in col. 2 of Table 2.18 (derived from Figure 2.18). Using eq. (2.38), since there is no storage, the rainfall excess for successive intervals is computed in the table. Actual infiltration, f, for various values of F can be computed from eq. (a), although it is not required to calculate the net rainfall.

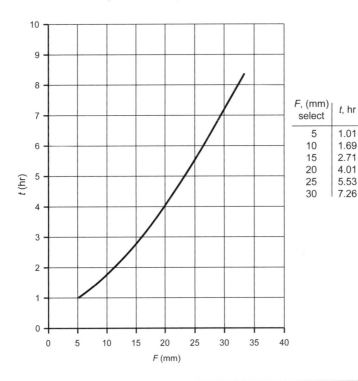

F, (mm) select	t, hr
5	1.01
10	1.69
15	2.71
20	4.01
25	5.53
30	7.26

Figure 2.18
Plot for graphical solution of infiltration equation in Example 2.15.

Table 2.18   Infiltration and Rainfall Excess Computations with the Green-Ampt Model[a]

(1)	(2)	(3)	(4)	(5)	(6)
Time t	F	Δt	ΔF	$i\Delta t$[b]	RO (i.e., $i\Delta t - \Delta F$)
(hr)	(mm)	(hr)	(mm)	(mm)	(mm)
1.63	9.8 (F_p)				
		0.37	2.0	2.2	0.2
2.0	11.8				
		1.00	4.5	6.0	1.5
3.0	16.3				
		1.00	3.7	6.0	2.3
4.0	20.0				

[a] Columns 3 and 4 indicate the difference in successive values (increment) of columns 1 and 2, respectively.
[b] $i = 6$ mm/hr

Availability of Water   Chapter 2

EXAMPLE 2.16

A storm pattern for a watershed is as follows:

t (min)	Intensity (in./hr)
0–10	0.5
10–20	2.0
20–30	6.5
30–40	5.0
40–50	0.9
50–60	2.0
60–70	3.0

The soil texture is sandy with a saturated moisture content (porosity) of 0.50, an effective hydraulic conductivity of 1.0 in./hr, and an average capillary suction of 6 in. The initial moisture content is 0.3. Determine the rainfall excess for successive 10-min periods. Assume a depression storage of 0.5 in.

SOLUTION

(a) First rainfall period (0–10 min)

1. Since $i < K_s$, there is no ponding and the entire rain infiltrates.

2. Cumulated infiltration, $F_1 = 0.5(10/60) = 0.08$ in.

3. The values to determine the rainfall excess are listed in Table 2.19.

(b) Second rainfall period (10–20 min)

4. Cumulative infiltration,

$$F_p = \frac{6(0.2)}{2/1 - 1} = 1.2 \text{ in.}$$

5. Incremental cumulative infiltration,

$$\Delta F_p = 1.2 - 0.08 = 1.12 \text{ in.}$$

6. Ponding time,

$$t_p = \frac{1.12}{2} = 0.56 \text{ hr or } 33.6 \text{ min.}$$

Hence there is no ponding in the second period.

7. Infiltration during second period, $\Delta F_2 = 2(10/60) = 0.33$ in.

8. Cumulative infiltration to end of the period, $F_2 = 0.33 + 0.08 = 0.41$ in.

9. Infiltration capacity, f_p [from eq. (2.41)]; $M = 0.5 - 0.3 = 0.2$

$$f_p = 1 + \frac{(1)(0.2)(6)}{0.41} = 3.93 \text{ in./hr}$$

(c) Third rainfall period (20–30 min)

Section 2.13   Direct Runoff from Rainfall or Rain Excess

10. Rainfall rate increases to 6.5 in./hr, but f_p is 3.93 in. /hr, so the surface ponding occurs at the outset of this period (i.e., $t_p = 20$ min or 0.33 hr).

11. From eq. (2.42), computing t'_p;

$$1 \cdot t'_p = 0.41 - 6(0.2)\ln\left[1 + \frac{0.41}{0.2(6)}\right]$$

$$t'_p = 0.06 \text{ hr}$$

12. From eq. (2.45b),

$$1(t - 0.33 + 0.06) = F - 0.2(6)\ln\left[1 + \frac{F}{0.2(6)}\right]$$

or

$$t = 0.27 + F - 1.2 \ln(1 + 0.83F)$$

A plot of F versus t for this equation is given in Figure 2.19.

13. At the end of the period, when $t = 30$ min or 0.5 hr, $F = 0.90$ in. from Figure 2.19.

14. Ponding will accumulate up to depression storage capacity of 0.5 in., according to eq. (2.37b), then runoff will commence to be computed by eq. (2.38) as explained in Table 2.19.

(d) Fourth rainfall period (30–40 min)

15. Surface ponding continues from the third period; hence the equation of step 12 still holds.

16. At the end of the period, when $t = 40$ min or 0.67 hr, $F = 1.25$ in., from Figure 2.19. Depression storage is full, so the runoff is computed by eq. (2.38) in Table 2.19.

(e) Fifth rainfall period (40–50 min)

17. Infiltration capacity at the beginning of the period from eq. (2.45a):

$$f_p = (1) + \frac{(1)(6)(0.2)}{1.25} = 1.96 \text{ in./hr}$$

18. Since the rainfall rate of 0.9 in./hr is less than f_p but the depression storage is full at 0.5 in. in the previous period (see Table 2.19), the infiltration at full capacity will continue meeting the deficiency of rainfall from the depression storage.

19. If the infiltration continues at full capacity (i.e., the ponding effect continues throughout the period), the cumulative infiltration at the end of the period can be computed from the same equation of step 12 plotted in Figure 2.19. For $t = 50$ min or 0.83 hr, $F = 1.60$ in.

20. Depression storage will reduce by 0.2 in. equal to ($i\Delta t - \Delta F$).

(f) Sixth and seventh rainfall periods

21. Computation continues similar to the fourth period by applying the equation of step 12 as shown in Table 2.19.

Availability of Water   Chapter 2

Figure 2.19 Graphical solution of cumulated infiltration equation in Example 2.16.

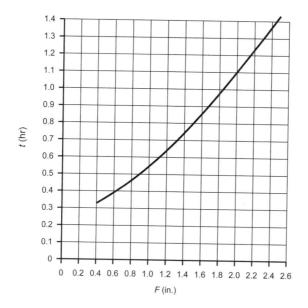

F, (in.) select	t, hr
0.4	0.33
0.6	0.39
0.8	0.46
1.0	0.54
1.5	0.80
2.0	1.10
2.5	1.42

Table 2.19 Computations for Unsteady Rainfall with the Green-Ampt Model

(1) Time		(2) F	(3) Δt	(4) ΔF^a	(5) i	(6) $i\Delta t$	(7) $i\Delta t - \Delta F^b$	(8) ΔS^c	(9) $\Sigma\Delta S$	(10) RO^d
(min)	(hr)	(in.)	(hr)	(in.)	(in./hr)	(in.)	(in.)	(in.)	(in.)	(in.)
0	0	0								
			0.167	0.08	0.5	0.08	0	0	0	0
10	0.167	0.08								
			0.166	0.33	2.0	0.33	0	0	0	0
20	0.333	0.41								
			0.167	0.49	6.5	1.09	0.6	0.5	0.5	0.10
30	0.50	0.90								
			0.167	0.35	5.0	0.84	0.49	0	0.5	0.49
40	0.667	1.25								
			0.166	0.35	0.9	0.15	−0.20	−0.20	0.3	0
50	0.833	1.60								
			0.167	0.25	2.0	0.33	0.08	0.08	0.38	0
60	10	1.85								
			0.167	0.25	3.0	0.50	0.25	0.12	0.50	0.13
70	1.167	2.10								

[a] Successive difference of column 2.
[b] Col. 6 – Col. 4.
[c] Eq. (2.37a) or eq. (2.37b), whichever is the minimum.
[d] Eq. (2.38), after computing ΔS.

Section 2.13   Direct Runoff from Rainfall or Rain Excess

2.14 HEC'S NONLINEAR LOSS-RATE FUNCTION APPROACH FOR DIRECT RUNOFF

The Hydrologic Engineering Center (HEC) of the U.S. Army Corps of Engineers has used the term *loss* for the precipitation not available to direct runoff and has indicated that the rate of loss is related nonlinearly to the rainfall intensity and a *loss-rate function* that decreases with increased ground wetness. The HEC studies indicated that a fairly definite quantity of water loss by interception and infiltration is required to satisfy initial moisture deficiencies before runoff can occur. An allowance for this initial loss or initial abstraction is made according to various antecedent soil moisture conditions. After the initial abstraction, the loss takes place at the following rate, which does not exceed the amount of precipitation for any time interval:

$$f = Kp^E \quad [LT^{-1}] \tag{2.47a}$$

with

$$K = K_0 C^{-0.1L} \quad [\text{dimensionless}] \tag{2.47b}$$

where

f = loss rate, in. or mm per hour

K = loss rate function

p = rainfall intensity, in. or mm per hour

E = exponent ranging between 0.3 and 0.9, with a frequent value of 0.7

K_0 = loss coefficient at the start of a storm

C = coefficient controlling the rate of decrease of the loss-rate function

L = accumulated loss during the storm, in. or mm

A typical plot for the loss-rate function is shown in Figure 2.20. The loss-rate coefficients are determined from the rainfall and runoff data. The HEC has developed a loss-rate optimization program for this purpose (HEC, 1973).

EXAMPLE 2.17

Determine the rainfall excess for successive periods for the storm of Example 2.16 by the loss-rate function approach. Assume that Figure 2.20 applies for the loss-rate function. Use $E = 0.7$.

SOLUTION

1. Initial accumulated loss from Figure 2.20 is 1.0 in.

2. Assuming a uniform distribution of the rain within 10-min observation periods, the rainfall of the first 25 min will be abstracted in the initial loss of about 1.0 in., as follows:

$$0-10 \text{ min} \quad 0.5\left(\frac{10}{60}\right) = 0.08 \text{ in.}$$

$$10-20 \text{ min} \quad 2.0\left(\frac{10}{60}\right) = 0.33 \text{ in.}$$

$$20-25 \text{ min} \quad 6.5\left(\frac{5}{60}\right) = 0.54 \text{ in.}$$

$$\overline{ 0.95 \text{ in.}}$$

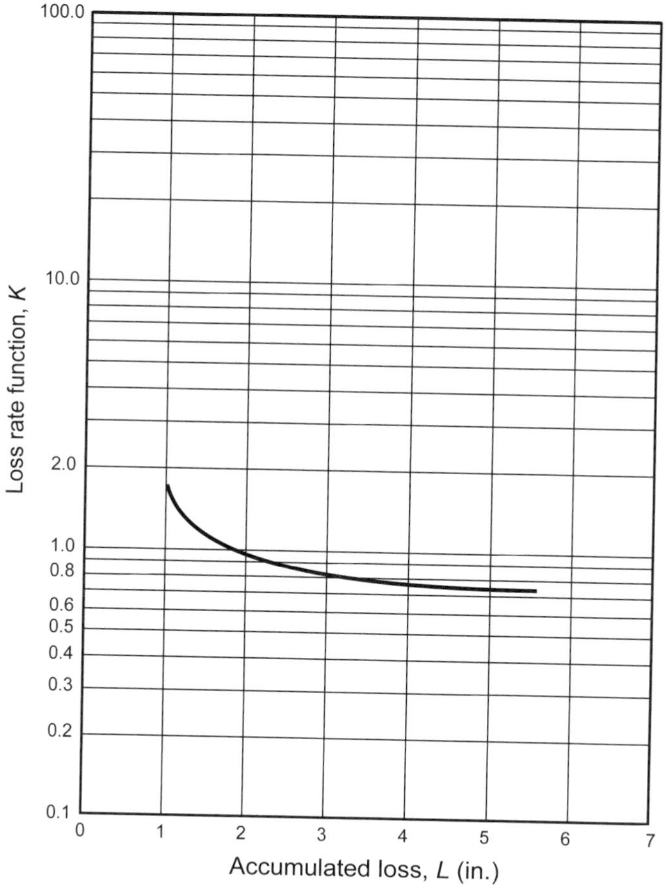

Figure 2.20
Loss-rate function.

3. The direct runoff will appear after 25 min. The computation is shown in Table 2.20.

2.15 THE NRCS APPROACH FOR DIRECT RUNOFF

By studying the infiltration behavior of different types of soils, the Natural Resources Conservation Service (formerly SCS) developed a method to compute the direct runoff resulting from a rainfall storm (U.S. SCS, 1972). The factors affecting infiltration are: hydrologic soil group, type of land cover, hydrologic condition and antecedent (pre-storm) moisture condition, and cropping practice in the case of cultivated agricultural land. Each of these factors is subdivided into many classes. Hydrologically, soils are assigned four groups on the basis of intake of water on bare soil when thoroughly wetted, as shown in Table 2.21. With urbanization the soil profile is disturbed considerably. The group classification can be based on the texture of disturbed soil.

The type of land cover, such as bare soil, vegetation, impervious surface, and so on, establishes runoff production potential. Important cover types for urban areas, cultivated agricultural lands, other agricultural lands, and arid rangelands are given in Table 2.22. Cultivated agricultural lands are further subdivided by treatment or cropping practice, such as straight row, contoured, and contoured and terraced. The hydrologic conditions

Table 2.20 Computation of Rainfall Excess by the Loss-Rate Function Method

(1) Time (min)	(2) Rainfall Intensity (in./hr)	(3) Rainfall during Period (in.)	(4) K Value (Fig. 2.20)	(5) Loss Rate, $f = Kp^{0.7}$ (in./hr)	(6) Loss during Period (in.)	(7) Accumulated Loss (in.)	(8) Rainfall Excess (in.) (col. 3–col. 6)
0–10	0.50	0.08					
10–20	2.00	0.33			0.95 initial loss (≈1)		
20–25	6.50	0.54					
25–30	6.50	0.54	1.70[a]	6.30[b]	0.53[c]	1.48	0.01
30–40	5.0	0.83	1.17	3.61	0.60	2.08	0.23
40–50	0.9	0.15	0.93	0.86	0.14	2.22	0.01
50–60	2.0	0.33	0.90	1.46	0.24	2.46	0.09
60–70	3.0	0.50	0.87	1.88	0.31	2.77	0.19

[a] The value from Fig. 2.20 corresponding to accumulated loss in the preceding step.
[b] For example, $f = (1.7)(6.5)^{0.7} = 6.30$.
[c] (Column 5)(duration in hours): i.e., $(6.3)(5/60) = 0.53$.

Table 2.21 Hydrologic Soil Groups

Group	Minimum Infiltration Rate (in./hr)	Texture[a]
A	0.3–0.45	Sand, loamy sand, or sandy loam
B	0.15–0.30	Silt loam or loam
C	0.05–0.15	Sandy clay loam
D	0–0.05	Clay loam, silty clay loam, sandy clay, silty clay, or clay

[a] Reproduced from U.S. Soil Conservation Service (1986).

reflect the level of land management. Hydrologically poor conditions indicate a state of land use that will provide higher runoff compared to good conditions. The antecedent moisture condition (AMC) is the index of the soil condition with respect to runoff potential before a storm event. It has three categories as shown in Table 2.23.

The NRCS (SCS) has evolved a system of curve numbers. A distinct curve number (CN) is assigned on the basis of the combination of the factors above. Table 2.22 gives curve numbers for antecedent moisture condition II. Table 2.24 provides conversion of CNs to other conditions. For an area with many different subareas, the composite CN is determined by adding the product of CN and respective area and dividing by the total area.

Table 2.22 Curve Numbers for Antecedent Moisture Condition II

Use	Cover Type	Treatment	Hydrologic Condition	Hydrologic soil group			
				A	B	C	D
Urban	Fully developed						
	Open space (lawns, parks)		Poor (cover < 50%)	68	79	86	89
			Fair	49	69	79	84
			Good (grass cover > 75%)	39	61	74	80
	Impervious areas (paved parking, roofs, driveways, paved roads)			98	98	98	98
	Dirt roads			72	82	87	89
	Urban districts						
	Commercial and business			89	92	94	95
	Industrial			81	88	91	93
	Developing areas			77	86	91	94
Cultivated agriculture lands	Fallow	Bare soil	...	77	86	91	94
	Row crops	Straight row	Poor	72	81	88	91
		Straight row	Good	67	78	85	89
		Contoured	Poor	70	79	84	88
		Contoured	Good	65	75	82	86
		Contoured and terraced	Poor	66	74	80	82
		Contoured and terraced	Good	62	71	78	81
	Small grain	Straight row	Poor	65	76	84	88
		Straight row	Good	63	75	83	87
		Contoured	Poor	63	74	82	85
		Contoured	Good	61	73	81	84
		Contoured and terraced	Poor	61	72	79	82
		Contoured and terraced	Good	59	70	78	81
	Close-seeded legumes or rotation meadow	Straight row	Poor	66	77	85	89
		Straight row	Good	58	72	81	85
		Contoured	Poor	64	75	83	85
		Contoured	Good	55	69	78	83
		Contoured and terraced	Poor	63	73	80	83
		Contoured and terraced	Good	51	67	76	80

(continued)

Table 2.22 Curve Numbers for Antecedent Moisture Condition II (Continued)

Use	Cover Type	Treatment	Hydrologic Condition	Hydrologic soil group			
				A	B	C	D
Agriculture lands	Pasture or range		Poor	68	79	86	89
			Fair	49	69	79	84
			Good	39	61	74	80
	Meadow			30	58	71	78
	Woods		Poor	45	66	77	83
			Fair	36	60	73	79
			Good	30	55	70	77
	Farmsteads (building, lanes, driveways)			59	74	82	86
Arid and semiarid rangelands	Herbaceous (mixture of grass, weeds, and low-growing brush)		Poor (< 30% ground cover)		80	87	93
			Fair		71	81	89
			Good (> 70% cover)		62	74	85
	Oak-aspen (mountain brush mixture)		Poor		66	74	79
			Fair		48	57	63
			Good		30	41	48
	Pinyon-juniper		Poor		75	85	89
			Good		41	61	71
	Sagebrush with grass understory		Poor		67	80	85
			Good		35	47	55
	Desert shrub		Poor	63	77	85	88
			Fair	55	72	81	86
			Good	49	68	79	84

Source: Condensed from U.S. Soil Conservation Service (1986).

Table 2.23   Antecedent Moisture Condition

Category	Condition
I	Dry soil but not to the wilting point
II	Average conditions
III	Saturated soils; heavy rainfall or light rainfall with low temperatures has occurred in the last 5 days

The NRCS runoff equation is

$$Q = \frac{(P - 0.2S)^2}{(P + 0.8S)} \quad [L] \tag{2.48}$$

where

Q = accumulated runoff, in. or mm depth over the drainage area

P = accumulated rainfall depth, in. or mm

S = potential maximum retention* of water by the soil, in. or mm

The potential maximum retention, S, is related to the curve number, CN, by the following relation:

$$CN = \frac{1000}{10 + S} \quad [\text{unbalanced}] \tag{2.49}$$

Thus, once a curve number is ascertained from Table 2.22 and Table 2.24 for the known conditions, the direct runoff can be computed from eqs. (2.48) and (2.49). The TR-55 (SCS, 1986) contains a graph and a table that solve eq. (2.48) directly. The tabular solution is reproduced in Table 2.25 for a certain range of CNs and rainfall values.

To use the method for sequential rainfall, the intensity is converted to the rainfall depth for each period of sequence and accumulated to the end of each period. From the accumulated rainfall to the end of successive rain periods, the accumulated direct runoff or rainfall excess is derived for each time period using the NRCS equation (2.48). The accumulated direct runoff is then converted to the increments of the runoff.

EXAMPLE 2.18

Determine the direct runoff (rainfall excess) for successive 10-min periods of the storm of Example 2.16. The soil in the basin belongs to hydrologic group B. The basin is mostly wooded in good hydrologic condition. The saturated soil condition (condition III) exists in the basin.

SOLUTION

1. For given hydrologic characteristics and for moisture condition II, CN = 55, from Table 2.22.

2. Corresponding CN = 75 for condition III, from Table 2.24.

3. Computations for accumulated rain and runoff are given in Table 2.26.

---

* This is mostly the infiltration. The term is distinct from the surface retention, which does not include the infiltration.

Table 2.24   Cross-Linking of Curve Numbers for Various Antecedent Moisture Conditions

Curve Number for Condition	Corresponding Curve Number for Condition	
II	I	III
100	100	100
95	87	99
90	78	98
85	70	97
80	63	94
75	57	91
65	45	83
60	40	79
55	35	75
50	31	70
45	27	65
40	23	60
35	19	55
30	15	50
25	12	45
20	9	39
15	7	33
10	4	26
5	2	17
0	0	0

Source: U.S. Soil Conservation Service (1972).

2.16 INFILTRATION-INDEX APPROACH FOR DIRECT RUNOFF

The index approach is the simplest procedure to estimate the total volume of storm runoff. The object of this method is to obtain a coefficient that may be applied to an entire rain period, or to an entire storm if it is made up of several rain periods, to arrive at an estimate of the direct runoff (Cook, 1946). Three types of indices are common: (1) the ϕ index, which represents a level (horizontal line) of intensity that divides the rainfall intensity diagram in such a manner that the depth of rain above the index line is equivalent to surface runoff depth over the basin, as illustrated in Figure 2.21; (2) the f_{av} index, which indicates the average rate of infiltration during a period in which the rainfall intensity is equal to or more than the infiltration capacity, f_p; and (3) the W index, which is a mean of f_{av} when it varies across a watershed. The value of W for a rain occurring after the watershed is wetted and the infiltration capacity is reduced to the minimum is known as the W_{min} index.

The ϕ index is the simplest of these indices. For its determination, the storm rainfall is measured and the amount of runoff is obtained from the corresponding direct hydrograph. The difference is the ϕ index. The value of ϕ increases with the increase of

Table 2.25   Runoff Depth for Selected CNs and Rainfall Amounts[a]

Rainfall	Runoff Depth (in.) for Curve Number of:												
	40	45	50	55	60	65	70	75	80	85	90	95	98
0.5	0.00	0.00	0.00	0.00	0.00	0.00	0.00	0.00	0.00	0.01	0.06	0.17	0.32
1.0	0.00	0.00	0.00	0.00	0.00	0.00	0.00	0.03	0.08	0.17	0.32	0.56	0.79
1.2	0.00	0.00	0.00	0.00	0.00	0.00	0.03	0.07	0.15	0.27	0.46	0.74	0.99
1.4	0.00	0.00	0.00	0.00	0.00	0.02	0.06	0.13	0.24	0.39	0.61	0.92	1.18
1.6	0.00	0.00	0.00	0.00	0.01	0.05	0.11	0.20	0.34	0.52	0.76	1.11	1.38
1.8	0.00	0.00	0.00	0.00	0.03	0.09	0.17	0.29	0.44	0.65	0.93	1.29	1.58
2.0	0.00	0.00	0.00	0.02	0.06	0.14	0.24	0.38	0.56	0.80	1.09	1.48	1.77
2.5	0.00	0.00	0.02	0.08	0.17	0.30	0.46	0.65	0.89	1.18	1.53	1.96	2.27
3.0	0.00	0.02	0.09	0.19	0.33	0.51	0.71	0.96	1.25	1.59	1.98	2.45	2.77
3.5	0.02	0.08	0.20	0.35	0.53	0.75	1.01	1.30	1.64	2.02	2.45	2.94	3.27
4.0	0.06	0.18	0.33	0.53	0.76	1.03	1.33	1.67	2.04	2.46	2.92	3.43	3.77
4.5	0.14	0.30	0.50	0.74	1.02	1.33	1.67	2.05	2.46	2.91	3.40	3.92	4.26
5.0	0.24	0.44	0.69	0.98	1.30	1.65	2.04	2.45	2.89	3.37	3.88	4.42	4.76
6.0	0.50	0.80	1.14	1.52	1.92	2.35	2.81	3.28	3.78	4.30	4.85	5.41	5.76
7.0	0.84	1.24	1.68	2.12	2.60	3.10	3.62	4.15	4.69	5.25	5.82	6.41	6.76
8.0	1.25	1.74	2.25	2.78	3.33	3.89	4.46	5.04	5.63	6.21	6.81	7.40	7.76
9.0	1.71	2.29	2.88	3.49	4.10	4.72	5.33	5.95	6.57	7.18	7.79	8.40	8.76
10.0	2.23	2.89	3.56	4.23	4.90	5.56	6.22	6.88	7.52	8.16	8.78	9.40	9.76
11.0	2.78	3.52	4.26	5.00	5.72	6.43	7.13	7.81	8.48	9.13	9.77	10.39	10.76
12.0	3.38	4.19	5.00	5.79	6.56	7.32	8.05	8.76	9.45	10.11	10.76	11.39	11.76
13.0	4.00	4.89	5.76	6.61	7.42	8.21	8.98	9.71	10.42	11.10	11.76	12.39	12.76
14.0	4.65	5.62	6.55	7.44	8.30	9.12	9.91	10.67	11.39	12.08	12.75	13.39	13.76
15.0	5.33	6.36	7.35	8.29	9.19	10.04	10.85	11.63	12.37	13.07	13.74	14.39	14.76

[a] Interpolate the values shown to obtain runoff depths for CNs or rainfall amounts not shown.
Source: U.S. Soil Conservation Service (1986).

rain intensity up to a certain level and then approaches a constant number. It also is affected by the rainfall pattern (Cook, 1946). Whenever possible, the ϕ index should be applied to a similar storm from which it is derived. However, it must be appreciated that the use of indices does not constitute a rational application of the infiltration theory. Rainfall intensities less than the ϕ index are not considered in determining ϕ. Trial and error is involved.

EXAMPLE 2.19

The rainfall intensities during each 30 min of a 150-min storm over a 500-acre basin are 4.5, 3, 1, 3.5, and 2 in./hr, respectively. The direct runoff from the basin is 105 acre-ft. Determine the ϕ index for the basin.

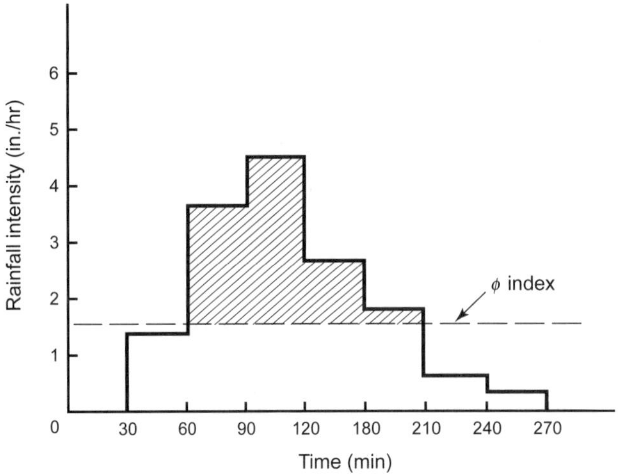

Figure 2.21
Representation
of the ϕ index.

Table 2.26   Computation of Runoff by the NRCS (SCS) Method

(1)	(2)	(3)	(4)	(5)	(6)
				Accumulated	
	Rainfall	Amount of	Accumulated	Direct Runoff	Runoff
Time	Intensity	Rain[a]	Rainfall	(Table 2.25)	Increments[b]
(min)	(in./hr)	(in.)	(in.)	(in.)	(in.)
0–10	0.5	0.08	0.08	0	0
10–20	2.0	0.33	0.41	0	0
20–30	6.5	1.08	1.49	0.16	0.16
30–40	5.0	0.83	2.32	0.55	0.39
40–50	0.9	0.15	2.47	0.65	0.10
50–60	2.0	0.33	2.80	0.84	0.19
60–70	3.0	0.50	3.30	1.17	0.33

[a] For example, (0.5 in./hr × 10 min)(1 hr/60 min).
[b] Difference between successive values, 0.55 − 0.16 = 0.39.

SOLUTION

1. Total rainfall $= \left(4.5\dfrac{\text{in.}}{\text{hr}}\right)(30 \text{ min})\left(\dfrac{1 \text{ hr}}{60 \text{ min}}\right)$

    $+\ 3\left(\dfrac{30}{60}\right)+1\left(\dfrac{30}{60}\right)+3.5\left(\dfrac{30}{60}\right)+2\left(\dfrac{30}{60}\right)$

    $= 7.0$ in. or 0.583 ft.

2. Rainfall volume $= (500)\,(0.583) = 291.5$ acre-ft.

3. Runoff volume $= 105$ acre-ft (given).

4. Volume under ϕ index $= 291.5 - 105 = 186.5$ acre-ft.

5. Infiltration depth $= \dfrac{186.5}{500} = 0.373$ ft or 4.48 in.

6. ϕ index $= (4.48 \text{ in.}) \left(\dfrac{1}{150 \text{ min}} \right) \left(\dfrac{60 \text{ min}}{1 \text{ hr}} \right)$

$\qquad = 1.79$ in./hr.

7. The revised ϕ index should be computed excluding the rainfall of 1 in./hr intensity since it is less than the computed ϕ index.

8. Total rainfall (excluding 1 in./hr)

$$= 4.5 \left(\frac{30}{60} \right) + 3 \left(\frac{30}{60} \right) + 3.5 \left(\frac{30}{60} \right) + 2 \left(\frac{30}{60} \right)$$

$$= 6.5 \text{ in. or } 0.542 \text{ ft}$$

9. Rainfall volume $= (500)(0.542) = 271$ acre-ft

10. Volume under ϕ index $= 271 - 105 = 166$ acre-ft

11. Infiltration depth $= \dfrac{166}{500} = 0.332$ ft or 4 in.

12. ϕ index $= (3.98) \left(\dfrac{1}{120 \text{ min}} \right) \left(\dfrac{60 \text{ min}}{1 \text{ hr}} \right)$

$\qquad = 2$ in./hr

EXAMPLE 2.20

Using the ϕ index of 2 in. /hr, determine the rainfall excess from the storm of Example 2.16.

SOLUTION

1. Runoff intensities = rainfall intensities $- \phi$ index, but not less than zero.

2. Hence the runoff intensities for successive 10-min periods are 0, 0, 4.5, 3.0, 0, 0, and 1.0, respectively.

3. Total runoff $= 4.5 \left(\dfrac{10}{60} \right) = 3.0 \left(\dfrac{10}{60} \right) + 1.0 \left(\dfrac{10}{60} \right)$

$\qquad = 1.42$ in.

2.17 DIRECT RUNOFF FROM SNOWMELT

Snowfall is the second major form of precipitation after rainfall. In mountainous regions it is the primary form of precipitation. It has a particular relevance in mountainous and high plains basins, where it accumulates throughout the season and produces substantial runoff during the spring melt. Whether a precipitation will fall as rain or snow depends on surface temperature, with snowfall occurring wherever the temperatures are below 34 to 36 °F.

The direct measurement of snowfall is obtained through the use of rain gages equipped with shields for better catch or by using snow boards and snow stakes. However, measurement at points (gage sites) over large areas does not always serve very useful purposes, due to drift and blowing effects. Snowpack surveys are accordingly performed to ascertain the areal extent of the snow-covered area, the water equivalent, and the depth of the snowpack at selected points in the basin before and during melting of the snow. Where the snowpack varies considerably, the water equivalent should be measured at various elevations and exposures.

2.17.1 Snowmelt Process

To a hydrologist, it is the runoff produced from the snowpack that is of direct concern in floods and water supply studies. The snowmelt is essentially a process of converting ice into water by heat energy. The sources of heat energy for snowmelt are all natural, comprising (1) absorbed solar radiation; (2) net long-wave terrestrial and atmospheric radiation; (3) condensation from air, releasing latent heat of vaporization to the snowpack; (4) convection heat transfer by wind; (5) heat content of rain water; and (6) conduction of heat from the ground.

A snowpack accumulated for a prolonged cold period has a temperature well below the freezing point. Mild weather conditions cause melting of the snowpack surface. This initial meltwater moves slightly into the snowpack and again refreezes, through contact with the colder underlying snow. In the process, the temperature of the snowpack rises slightly, due to the release of heat of fusion by the meltwater. When the warm conditions persist, the temperature of the entire snowpack rises to 32 °F. Melt from the surface flows down through the pack. Although it does not refreeze, it is still retained by the snowpack because of the water-holding capacity or liquid-water deficiency. Once the holding capacity is completely satisfied, the snowpack is considered to be *ripe*. The total water stored within the snowpack has been elaborately discussed by the U.S. Army Corps of Engineers (1956). Any additional heat energy input to the snowpack causes the melt to reach the snow-soil interface. This is similar to rainfall reaching the ground. Infiltration and other losses take place from the snowmelt, as in the case of rainfall, and the melt excess appears as direct runoff.

Snowmelt quantities cannot be measured directly. A number of equations have been developed to estimate snowmelt. These are presented with examples of their application by the U.S. Army Corps of Engineers (1960, 1956). Two common procedures are described here. From the estimated snowmelt, deductions should be made for infiltration and other losses to compute melt excess.

2.17.2 Temperature Index or Degree-Day Method

The simplest approach for computing snowmelt is based on degree-days, or simply the air temperature, which is the single most important factor in the melting of snow. The relation is given by

$$M = C(T - T_B) \quad [LT^{-1}] \tag{2.50}$$

where

M = snowmelt, in./day

C = melt rate coefficient or degree-day factor, ranging from 0.015 to 0.2; a mean value of 0.08 is used in the western United States

T = mean daily air temperature or maximum daily air temperature, °F

T_B = base temperature above which snowmelt is assumed to occur, typically 32 °F for mean daily or 40 °F for maximum daily value

2.17.3 Generalized Equation of the Corps of Engineers

The U.S. Army Corps of Engineers conducted an extensive study of the snowmelt for regions in the western United States. The theory developed has been based on the theoretical process of heat transfer, as well as the empirical approach of experimental observations. The equations are summarized below.

Snowmelt during a Rainstorm

1. For open or partly forested (0 – 60% cover) basin

$$M = 0.09 + (0.029 + 0.0084KW + 0.007R)(T - 32) \quad \text{[unbalanced]} \quad (2.51)$$

2. For heavily forested (60 – 100% cover) basin

$$M = 0.05 + (0.074 + 0.007R)(T - 32) \tag{2.52}$$

Snowmelt during No-Rain Period

1. For partly forested (10 – 60% cover) basin

$$M = K'(1 - F)(0.004S)(1 - A) + K(0.0084W)\left(0.22T' + 0.78T'_D\right) + F(0.029T') \tag{2.53}$$

2. For heavily forested (≥ 80% cover) basin

$$M = 0.074\left(0.53T' + 0.47T'_D\right) \tag{2.54}$$

where

M = snowmelt, in./day

T = air temperature at 10-ft height, °F

T' = difference between air temperature at 10-ft height and at snow surface, °F

T'_D = difference between dew point temperature at 10-ft height and at the snow surface, °F

R = rainfall, in./day

F = estimated basin forest cover, fraction

W = wind speed at 50 ft above the snow, mph

S = solar radiation at the snow surface per day, langleys/day

A = albedo or reflectivity of snow determined by $0.7/D^{0.2}$ to $.75/D^{0.2}$, should be above 0.4

D = days since last snowfall

K = basin condensation – convection melt factor; ranges from 1.0 for an open basin to 0.3 for densely forested basin

K' = basin short-wave radiation melt factor; usually 0.9 to 1.1

In mountainous regions, the temperature drops on the average about 3 °F per 1000 ft. Therefore, for application of these equations, a number of zones that are 1000 ft or its multiple in height are selected. The melt in each zone by the appropriate equation is determined for each day and is multiplied by the area of that zone. The sum of all zones is then divided by the total basin area to obtain the basin-mean snowmelt.

EXAMPLE 2.21

The basin characteristics and initial conditions (by snow survey at the beginning of the melt) for a mountainous region are given in Table 2.27 and the meteorologic data are given in Table 2.28. Determine the snowmelt for each day of the record. Assume that $K = 0.6$, $K' = 1.0$, and forest cover $F = 0.5$. Consider that the last snowfall occurred on January 31.

SOLUTION

(a) Considering the lapse rate of 3 °F per 1000 ft, the air temperature and dew point temperature at various elevations are calculated in Table 2.29. The zone temperatures are averages of two respective elevation temperatures.

(b) Rainfall occurs on the first day (Feb. 3, 2005). The total melt is determined using eq. (2.51).

1. Zone 1

$$M = 0.09 + [.029 + 0.0084(0.6)(10.5) + 0.007(0.05)](38.5 - 32)$$
$$= 0.625 \text{ in./day}$$

2. Zone 2

$$M = 0.09 + [0.029 + 0.0084(0.6)(10.5) + 0.007(.05)](35.5 - 32)$$
$$= 0.378 \text{ in./day}$$

Table 2.27 Initial Conditions

Characteristics	Zone 1	Zone 2
Elevation (ft)	3000 – 4000	4000 – 5000
Area (mi^2)	315	730
Snowpack (in.)	18	13

Table 2.28 Meteorological Data

Date	Rainfall (in.)	Air Temperature at 3000 ft (°F)	Dew Point at 3000 ft (°F)	Wind Speed (mph)	Solar Radiation (langleys/day)
Feb. 3, 2005	0.05	40	31	10.5	400
Feb. 4, 2005	0	48	29.2	11.0	567
Feb. 5, 2005	0	42	28.5	8.2	397

Table 2.29 Air and Dew Point Temperatures at Various Elevations

Date	Air Temperature (°F)			Dew Point Temperature (°F)		
	At 3000 ft	At 4000 ft	At 5000 ft	At 3000 ft	At 4000 ft	At 5000 ft
Feb. 3, 2005	40	37	34	31	29	26
Feb. 4, 2005	48	45	42	29.2	26.2	23.2
Feb. 5, 2005	42	39	36	28.5	25.5	22.5

3. The weighted basin average snowmelt for the first day:

$$M = \frac{0.625(315) + 0.378(730)}{315 + 730} = 0.452 \text{ in./day}$$

4. The status of the snowpack at the end of first day is as follows:

	Zone 1	Zone 2
a. Snowpack at beginning (in.)	18	13
b. Snowmelt on first day (in.)	0.625	0.378
c. Snowfall during the day (in.)	0	0
d. Snowpack at the end of first day (in.)	17.375	12.622

(c) For the second day (Feb. 4, 2005), there is no rain. Equation (2.53) is used to compute the snowmelt.

1. Zone 1

$$D = 4 \qquad A = \frac{0.7}{4^{0.2}} = 0.53$$

$$T' = 46.5 - 32 = 14.5 \text{ °F}$$

$$T_D' = 27.7 - 32 = -4.3 \text{ °F}$$

$$M = 1(1 - 0.5)(0.004 \times 567)(1 - 0.53) + 0.6(0.0084 \times 11.0)$$
$$\times \left[0.22(14.5) + 0.78(-4.3) \right] + 0.5(0.029 \times 14.5)$$
$$= 0.73 \text{ in./day}$$

2. Zone 2

$$T' = 43.5 - 32 = 11.5$$

$$T_D' = 24.7 - 32 = -7.3$$

$$M = 1(1 - 0.5)(0.004 \times 567)(1 - 0.53) + 0.6(0.0084 \times 11)$$
$$\times \left[0.22(11.5) + 0.78(-7.3) \right] + 0.5(.029 \times 11.5)$$
$$= 0.52 \text{ in./day}$$

3. Weighted average snowmelt for the basin

$$M = \frac{0.73(315) + 0.52(730)}{315 + 730} = 0.58 \text{ in./day}$$

The melt can be similarly computed for the third day.

2.18 REMOTE SENSING OF HYDRAULIC COMPONENTS

Remote-sensing techniques, though relatively new to hydrology, have become important tools in hydrologic information gathering. They are mostly complementary to, rather than a replacement of, conventional methods. The electromagnetic spectrum comprised of gamma rays; x-rays; visible light; near, medium, and far infrared rays; microwaves; and radio waves is the basis of remote sensing. Remote sensing measures reflected energy or reflectance from the earth's surface. Some parts of the electromagnetic spectrum are blocked by the atmosphere. The regions of the spectrum that are efficiently transmitted, known as *atmospheric windows*, are used for remote sensing. Different earth features, i.e., green vegetation, dry soil, moist soil, clean lake water, and snow cover, produce different types of reflectance that are measured by sensors. The interpretation of the measured variations in the reflectance characteristics forms the basis of the hydrologic data collection. There are a variety of sensors designed to measure specific parts of the spectrum. Gamma radiation sensors measure the difference between the natural terrestrial gamma radiation and the radiation attenuated by soil water or snow cover. Other sensors are thermal sensors, microwave sensors, aerial photography cameras and video recorders, multispectral scanners, and lasers. The type of data to be collected dictates which parts of the spectrum have to be observed. A careful matching of the sensor to data needs is necessary. The sensors are mounted either on ground-based platforms, on aircraft, or on satellites. Generally, ground-based systems are used for developing sensors by verifying how a sensor responds to the targeted feature. Aircraft are effective for small areas—for aerial photography and radar surveys. Modern remote sensing essentially centers around satellite systems.

There are two basic satellite systems: (1) Polar-orbiting satellites orbit the entire earth over a fixed period of days, providing coverage everywhere on the planet; (2) geostationary satellites synchronize with the earth's rotation, providing continuous coverage of the same area. The major commercial satellites of hydrologic interest are the NASA Landsat, the TM series, the U.S. NOAA satellite, and the French SPOT. Each of these satellites provides measurements in four to seven spectral ranges. The choice of the satellite system depends on the specific spectral bands to be covered as dictated by the data needs.

Remote-sensing systems obtain data in three forms: images as photographs, analog signals, and digital signals. The first two also can be converted to digital data to be analyzed by computer systems. A number of software programs have been developed that perform intricate analyses of the data. The application of remote-sensing techniques to measurements of various water balance components is summarized here.

2.18.1 Rainfall

The direct measurement of rainfall from satellite has not been satisfactory because clouds prevent observation of raindrops by the sensors. However, visible and infrared imagery of clouds has led to many techniques for inference of rainfall information. Microwave techniques, which are in a research phase, appear very promising for measuring precipitation since clouds are transparent at some microwave frequencies. Remote sensing of rainfall is very useful as a complement to conventional rain-gaging procedures because the conventional gages have spatial limitations when measuring rainfall.

2.18.2 Snow Precipitation

Remote sensing is very useful in snow measurement. Data are derived from all regions of the electromagnetic spectrum. Data relevant to snow measurements are the extent of snowpack, the water equivalent of snow, and the presence of liquid water in the pack which signals the onset of the melting process.

The extent of snowpack can be easily ascertained from the visible band. The visible band provides contrast between snow and non-snow areas. Infrared satellite imagery determines the water content of snowpacks by measuring gamma emissions before and during snow covers. To gather complete information on the extent of snowpack, the water equivalent, and the presence of water in the pack, microwave remote sensing appears very promising.

2.18.3 Evapotranspiration

Remote sensing cannot measure evaporation or evapotranspiration directly. However, several variables related to evaporation and evapotranspiration can be measured or estimated by remote sensing, e.g., the surface temperature, incoming solar radiation, surface albedo, soil cover, and moisture conditions. Special empirical formulas have been derived that are appropriate to remote-sensing inputs.

2.18.4 Runoff

Remote sensing cannot measure runoff directly. However, the information derived from remote sensing is applied to estimate runoff in the following ways:

1. The Natural Resources Conservation Service utilizes a curve number to estimate runoff. The curve number is based on the land cover, land use, soil type, and moisture condition. These data are conveniently obtained by remote sensing.

2. Watershed topographic features and drainage patterns are easily ascertained from good imagery. This information can be used in empirical formulas to estimate runoff.

3. Formulas relate peak discharge to drainage area as discussed in Chapter 8. Remote sensing is quite convenient to ascertain basin area and drainage characteristics. The mapping of the areal extent of flooding and delineation of flood plains has effectively been done using remote-sensed data.

PROBLEMS

2.1 The evaporation losses from a reservoir of constant surface area of 500 acres are 150 acre-ft per day. If the outflow from the reservoir is 50 cfs, determine the change in the water level of the reservoir in a day without inflow.

2.2 In Problem 2.1, a precipitation of 3 in. falls on the reservoir surface in a day. What is the change in the reservoir depth?

2.3 In a river reach, the rate of inflow at any time is 350 cfs and the rate of outflow is 285 cfs. After 90 min, the inflow and outflow are 250 cfs and 200 cfs, respectively, and storage of 10.8 acre-ft has been observed. Determine the change in storage in 90 min and the initial storage volume.

2.4 At an upstream location on a river the flow is 100 m³/s and at a downstream location it is 120 m³/s. In one hour the storage in the river reduces by 70,000 m³. If the upstream flow after one hour is 120 m³/s, what is the downstream flow after one hour?

2.5 It is estimated that 60% of the annual precipitation in a basin with a drainage area of 20,000 acres is evaporated. If the average annual river flow at the outlet of the basin has been observed to be 2.5 cfs, determine the annual (long-term) precipitation in the basin.

2.6 During a two-hour storm the precipitation over a 100-acre area is 1.2 in. There is a constant infiltration at a rate of 0.4 in./hour. The depression storage is 20 acre-ft. Is there any runoff? If yes, how much? Disregard evaporation during the storm.

2.7 Water from a 100 ft × 200 ft lawn converges into a gully. During a 1-hour storm, discharge into the gully is 1.05 ft³/s. The interception by grass is 0.35 acre-ft. Overall, 70% of the rain is infiltrated. What is the amount of rainfall?

2.8 The average monthly rainfall recorded at the Metapoisset Basin (drainage area 30.5 mi^2) near New Bedford, Massachusetts, during 2005 and the computed potential evapotranspiration are given below. The initial moisture storage and groundwater storage are assessed to be 7.5 in. and 2.5 in., respectively. The fitted parameters of the Thomas *abcd* model are $a = 0.96$, $b = 20$, $c = 0.15$, and $d = 0.11$. Determine the monthly values of the (**a**) moisture storage, (**b**) direct runoff, (**c**) groundwater recharge, (**d**) groundwater storage, (**e**) groundwater discharge, and (**f**) streamflow.

Month:	J	F	M	A	M	J	J	A	S	O	N	D
Rainfall (in.)	0	1.7	2.66	1.24	0.36	2.12	4.44	7.48	4.66	2.22	0.6	0.24
Evapotranspiration (in.)	1.05	1.48	1.95	2.10	2.80	2.80	3.10	3.20	2.80	1.90	1.50	1.30

2.9 During a storm, gage station A was inoperative but the surrounding stations, B, C, and D, recorded rainfall of 5.2, 4.5, and 6.1 in., respectively. The average annual precipitation for stations A, B, C, and D are 48, 53, 59, and 67 in., respectively. Estimate the storm precipitation for station A.

2.10 The following rainfall record exists for five stations in a basin for the month of August. Estimate the missing rainfall for August 1997 for station A.

Station	All Years' Average for August (in.)	Measured in August 1997 (in.)
A	4.5	?
B	3.0	3.5
C	2.0	1.9
D	2.5	2.7
E	4.8	5.0

2.11 The precipitation data from October to April for six stations are given as follows. Check the consistency of data of station A and make an adjustment to reflect the recent conditions.

		Precipitation for October to April (in.)				
Year	Station A	Station B	Station C	Station D	Station E	Station F
1990	13.75	16.31	14.04	14.15	15.16	15.13
1991	20.22	20.79	21.85	21.06	17.16	22.23
1992	18.74	21.89	18.30	21.93	18.08	20.61
1993	20.27	20.26	17.13	19.04	22.27	22.29
1994	15.17	20.18	19.41	17.23	19.42	16.69
1995	17.13	14.36	17.37	15.58	13.78	18.84
1996	15.22	19.75	21.18	13.52	18.29	16.76
1997	18.71	19.98	18.34	15.90	14.27	20.58
1998	12.14	13.57	13.01	14.40	15.83	13.35
1999	18.50	21.17	17.64	21.05	16.99	20.35
2000	23.36	24.43	23.44	27.43	16.11	25.69
2001	9.50	13.88	19.86	16.16	15.98	10.45
2002	11.83	18.21	15.24	14.30	17.56	13.02
2003	13.40	18.48	21.23	17.65	18.48	14.74
2004	11.23	17.41	16.95	18.03	13.58	12.36
2005	13.14	17.49	20.11	13.05	16.44	14.45

2.12 The annual precipitation at station X and the mean annual precipitation at 12 surrounding stations are given below.

   a. Determine the consistency of the data of station X.

   b. In what year did the breakpoint (change in regime) occur?

   c. What is the unadjusted mean annual flow at station X?

   d. What is the adjusted mean annual flow according to recent regime conditions?

   e. Plot the double-mass curve of the adjusted data.

	Annual Precipitation (mm)			Annual Precipitation (mm)	
Year	Station X	12-Station Mean	Year	Station X	12-Station Mean
			1983	447	325
1969	517		1984	396	442
1970	264		1985	385	364
1971	462		1986	308	299
1972	397	338	1987	319	429
1973	275	299	1988	352	430
1974	385	390	1989	429	455
1975	378	338	1990	275	338
1976	337	340	1991	330	377
1977	407	339	1992	253	364
1978	385	364	1993	407	442
1979	638	520	1994	374	429
1980	358	338	1995	330	455
1981	350	312	1996	308	438
1982	290	286	1997	297	325

2.13 The rainfall recorded at the network of stations due to a storm over a trapezoidal watershed is shown in Figure P2.13. Determine the average areal rainfall by the **(a)** average method, and **(b)** Thiessen polygon method.

2.14 Solve Problem 2.13 by the isohyetal method.

2.15 The annual precipitation (in mm) observed on a network of stations is shown in Fig. P2.15. Trace the drainage basin along with the location and values of the rainfall on paper and compute the average areal rainfall by **(a)** the arithmetic average, and **(b)** the Thiessen polygon method. The scale of the figure in x- and y-directions is 1 cm = 16 km.

2.16 Solve Problem 2.15 by the isohyetal method.

2.17 From the precipitation data of various durations given in Example 2.6, prepare the intensity-duration-frequency curves for 5-year and 3-year return periods.

2.18 Construct the depth-area-duration curves of maximum probable precipitation for a basin located in New York City. Obtain the values of precipitation for different durations and areas from Table 8.14 and Figure 8.9. [*Hint*: multiply the factors for region VIIa in Table 8.14 by the PMP for New York from Figure 8.9 for different areas and durations.]

Figure P2.13

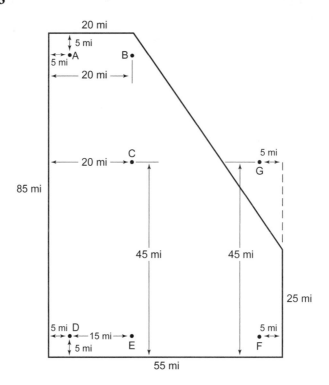

Station	Precipitation (in.)
A	2.8
B	3.5
C	4.2
D	5.1
E	6.0
F	4.5
G	3.9

2.19 In pan measurements close to a reservoir, the following data were observed on a certain day. Determine the daily evaporation by **(a)** the simple formula assuming that $K = 0.7$, and **(b)** the revised formula, assuming that $K' = 1.5$ at 4 m.

1. Pan evaporation = 15 mm
2. Mean air temperature at 4 m = 23°C
3. Maximum reservoir temperature = 18°C
4. Minimum reservoir temperature = 15°C
5. Maximum pan temperature = 24°C
6. Minimum pan temperature = 18°C
7. Relative humidity = 47%

2.20 The temperature and relative humidity averaged for the month of October have the following values. During the month, the evaporation measured from the pan was 19 cm. Determine the monthly evaporation from the lake by **(a)** the simple formula, assuming that $K = 0.75$, and **(b)** the revised formula, assuming that $K' = 1.5$.

Figure P2.15

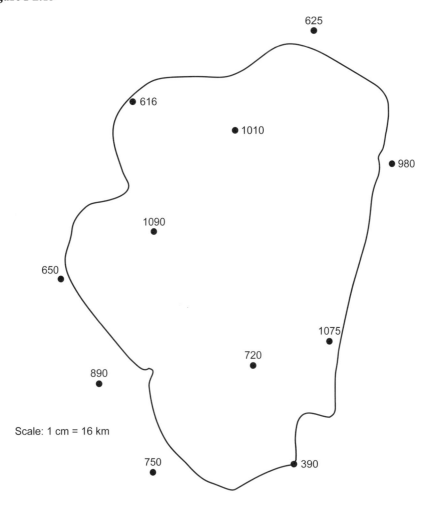

1. Monthly average mean air temperature at 4 m = 17 °C
2. Monthly average maximum lake temperature = 13 °C
3. Monthly average maximum pan temperature = 16 °C
4. Monthly average relative humidity = 60%

2.21 Solve Problem 2.19 by the aerodynamic method of evaporation at 4 m if the wind speed is 23 m/s and $C_E = 1.3 \times 10^{-3}$ at 4 m.

2.22 Solve Problem 2.20 by the aerodynamic method for evaporation at 4 m. The wind speed is 47 m/s at 8 m level. Assume the mean lake temperature to be 85% of the maximum. C_E is 1.1×10^{-3}. Use the reference height Z_0 corresponding to water surface.

2.23 The following data are recorded at a meteorological station. Determine the average pan coefficient for the simple pan equation. (*Hint:* Compute E_{day} by the aerodynamic method for each day and substitute in the pan equation to determine the daily coefficient.) $C_E = 1.3 \times 10^{-3}$.

Date	Mean Water Temperature (°C)	Mean Air Temperature (°C)	Relative Humidity (%)	Wind Speed (m/s)	Pan Evaporation (mm)
June 8, 2007	17	25	45	18	10.9
June 15, 2007	18	31	20	10	13.4
June 21, 2007	15	23	47	40	16.8
June 27, 2007	23	32	35	30	36.6

2.24 Solve Problem 2.19 for evaporation by the energy balance method. The Bowen ratio is 0.1, $n/N = 0.85$, and the place is located at 70 °N latitude, having an extraterrestrial radiation of 42.5 MJ/m^2 day. Disregard the soil heat index.

2.25 Solve Problem 2.20 for evaporation by the energy balance method. The lake is located at 30 °N latitude. The Bowen ratio = 0.05 and $n/N = 0.8$. Disregard the soil heat flux.

2.26 Solve Problem 2.19 for evaporation by the combination method using information from Problems 2.21 and 2.24.

2.27 Solve Problem 2.20 for evaporation by the combination method using information from Problems 2.22 and 2.25.

2.28 Estimate the daily reference crop (alfalfa) evapotranspiration at Boston, MA, for June; latitude ~42 °N. The mean values of the meteorological data are as follows:

1. Mean air temperature (June) = 24.2 °C
2. Relative humidity = 65%
3. Wind velocity at 2 m = 1.8 m/s
4. Mean % sunshine = 70%

Disregard the soil heat flux.

2.29 For Boise, Idaho, latitude ~44 °N, determine the daily reference crop (grass) evapotranspiration for July. The meteorological data are as follows:

1. Mean July air temperature = 22.5 °C
2. Relative humidity = 45%
3. Wind velocity at 4 m = 2.54 m/s
4. Mean % sunshine = 82%

Disregard the soil heat flux. Use a reference height of 0.03 m.

2.30 Prepare the crop coefficient curve (grass reference) for grain sorghum planted on June 5. For June the grass reference E_{to} is 3.5 mm/day. The soil is wetted in 10-day intervals during the initial period. The crop is well watered.

2.31 Prepare the crop coefficient curve (grass reference) for potatoes planted on May 10. The grass reference E_{to} for May is 5 mm/day. The soil is watered every 10 days in the initial stage. The crop is well watered.

2.32 The grain sorghum of Problem 2.30 is planted on June 5 and harvested on Nov. 2. Estimate the monthly evapotranspirations. Compare the results with Problem 1.20.

Month	June	July	August	Sept.	Oct.	Nov.
E_{to} mm/day	3.5	6.5	5.5	3.0	1.5	1.0

2.33 The potato crop of Problem 2.31 is planted on May 10 and harvested on Sept. 15. Estimate the monthly evapotranspirations. Compare the results with Problem 1.19. The grass reference evapotranspirations are:

Month	May	June	July	Aug.	Sept.
E_{to} mm/day	5.0	5.5	6.5	6.0	6.0

2.34 The standard f_p curve can be given by the equation

$$f_p = 1.2 + (9 - 1.2)e^{-0.076t}$$

where f_p is in./hr and t is in min. Plot the **(a)** standard infiltration capacity curve, **(b)** mass (cumulated) infiltration curve, **(c)** revised infiltration capacity curve if the rainfall intensity is 1.5 in./hr during the first 40 min and then intensities are 6 in./hr, 1.5 in./hr and 2.5 in./hr for three successive intervals of 20 min each.

2.35 For the rainfall intensities of Problem 2.34, determine the rainfall excess by the Horton method.

2.36 Rainfall occurs at a constant rate of 10.5 mm/day on a sandy loam having an initial moisture content of 0.20. At saturation, the moisture content is 0.52 and the hydraulic conductivity is 2×10^{-4} mm/sec. The average capillary suction is 250 mm. Determine the rainfall excess for every hour up to 4 hours with the Green-Ampt model. Assume no depression storage.

2.37 Rainfall rates for successive 20-min periods of a 140-min storm are 1.5, 1.5, 6.0, 4.0, 1.0, 0.8, and 3.2 in./hr, totaling 6.0 in. Determine the rainfall excesses for successive periods by the Green-Ampt model. Assume a saturated hydraulic conductivity of 0.8 in./hr, a porosity of 0.62, an initial moisture content of 0.28, and an average capillary suction of 7.2 in. The depression storage is 0.5 in.

2.38 Following are the rates of rainfall in in./hr for successive 30-min periods from a storm of 180 min: 0.5, 0, 0, 4.5, 5.0, and 3.0. Determine the rainfall excess in each successive period with the Green-Ampt model. The soil is sandy with a porosity of 0.59 and a saturated conductivity of 3.3×10^{-4} in./sec. The initial moisture content is 0.15 and the average capillary suction is 6.5 in. Disregard the depression storage.

2.39 From the rainfall sequence of Problem 2.37, determine the rainfall excess in successive 20-min periods by the loss-rate function approach. Assume an initial loss of 1.5 in. The following values may be used for the loss-rate function. Use $E = 0.7$.

Accumulated Loss (in.)	K
1.5	1.50
2.0	0.95
3.0	0.72
4.0	0.65
5.0	0.55

2.40 For the storm of Problem 2.38, determine the successive rainfall excess by the loss rate function approach. Use Figure 2.20 for the loss-rate function. Initial loss = 1 in., $E = 0.7$.

2.41 From the rainfall sequence of Problem 2.37, determine the rainfall excesses during successive 20-min periods by the NRCS method. The soil in the basin belongs to group A. It is an agriculture row crop land with contoured pattern in good hydrologic condition. The soil is in average condition before the storm (moisture condition II).

2.42 From the storm of Problem 2.38, determine the rainfall excesses during successive 30-min periods by the NRCS method. The watershed is part of an urban impervious area. The hydrologic soil group is C and the soil is in a dry condition (condition I).

2.43 The rainfall intensity for each 20-min period is given below for a storm. The direct runoff from the basin is 2.5 in. Determine **(a)** the total storm rain, and **(b)** the ϕ index for the basin.

20-minute period	1	2	3	4	5
Rain intensity (in./hr)	2.5	5.0	6.5	3.5	1.0

2.44 Determine the volume of direct surface runoff in acre-ft that will result from the following storm. The basin area is 1000 acres. The ϕ index is 0.5 in./hr.

30-minute period	1	2	3	4
Rain intensity (in./hr)	0.9	0.5	0.60	0.5

2.45 The direct surface runoff volume computed from the hydrograph of a 5-mi^2 basin area is 13,040 cfs-hr. The hydrograph was produced by a 5-hour storm of uniform rainfall intensity of 2.05 in./hr. Determine **(a)** the ϕ index, and **(b)** the direct runoff.

2.46 For a plain area, the following climatic data are observed on a certain rain-free day in a region covered with snow. Determine the snowmelt rate by **(a)** the degree-day method, $C = 0.08$, and **(b)** the energy-budget method, $K' = 1.0$, $K = 0.5$, forest cover = 0.45. The snowfall did not occur for the last 4 days from the date of measurement. Snow surface is at 32 °F.

1. Air temperature (°F) = 53
2. Dew point (°F) = 28
3. Average wind (mph) = 11.0
4. Solar radiation (langleys/day) = 560

2.47 For a mountainous region, the following snow survey and climatic data are observed on three consecutive days. Determine the mean snowmelt rates for each day. The last snowfall occurred on February 27, 2007. Assume that $K = 0.5$, $K' = 1.0$, and $F = 0.55$. Snow surface is at 32 °F.

Initial Conditions

	Zone 1	Zone 2	Zone 3
Elevation (ft)	2000–3000	3000–4000	4000–5000
Area (mi^2)	250	730	400
Snowpack (in.)	9	11	15

Meteorological Conditions

Date	Rainfall (in)	Air Temp. (°F) at 2000 ft	Dew Point (°F) at 2000 ft	Wind Speed (mph)	Solar Radiation (langleys/day)
Mar. 1	0	45	29	8.8	450
Mar. 2	0.1	40	28.5	6.5	560
Mar. 3	0	50	31	10.2	470

3

Theory of Groundwater Flow

◆◆◆

3.1 SCOPE

The terms *subsurface water* and *groundwater* have been given different meanings by different researchers all over the world. These terms have been used in a broader sense to include all waters below the surface of the earth in liquid, solid, or vapor forms, appearing as physically or chemically bound waters, as free water in the zones of aeration and saturation, and in a supercritical state in the zone of dense fluids extending below the zone of saturation having a pressure greater than 218 atm and temperature higher than 374 °C. These terms also have been used to refer only to the free water that can move through rock and soil, comprising water in the capillary fringe, gravitational water infiltered through the zone of aeration, and moving groundwater in the zone of saturation. Further, these terms have been used when referring to water in the zone of saturation only. The use of these terms in the United States, however, almost stabilized when in 1923 Meinzer defined subsurface water to designate all waters that occur below the earth's surface, and groundwater to designate the water in the zone of saturation. The *International Glossary of Hydrology*, prepared by WMO and UNESCO (1974), adopted the same meanings for these terms. Meinzer's concept of subsurface water as all varieties of water in the interior of the earth is very broad, something very similar to the present definition of subsurface hydrosphere. In the common sense, subsurface flow is meant to indicate water moving in the zones of aeration and saturation and the deep percolation.

Hydrogeology covers the study of subsurface water in all its phases: origin, manner of deposition, laws of motion, distribution, physical and chemical properties, interrelationship with atmospheric and surface waters, effects of human activities, economic values, and so on. On the other hand, civil engineers are more concerned about the movement and distribution of groundwater and its application, which is the subject matter of this chapter.

3.2 CLASSIFICATION OF SUBSURFACE WATER

Hydrogeologists have classified subsurface water based on the fundamental ideas of geological structures containing such water. However, this classification scheme based on the manner in which water is deposited is widely accepted by both hydrogeologists and engineers. This scheme takes into consideration the fact that physical, geographical, geological, and thermodynamic conditions are responsible factors in the deposition of water in the interior of the earth. In this classification, zonal divisions of subsurface water are made. Nineteenth-century scientists noted that there existed a law of zonation of natural phenomena such as climatic zonation, soil zonation, and vertical zonation of the material of the globe. All natural water supply is distributed in three zones: atmospheric, surface, and subsurface waters.

The principle of zonation was traced in subsurface water as well. From 1910 onward, a number of classification schemes based on the manner in which water is deposited were proposed by Soviet and American scientists as well as by scientists in France, Germany, and other western European nations. A very original classification system suggested by Meinzer (1923) is still widely accepted today. This book adopts the same classification. This classification, shown in Figure 3.1, established two broader divisions: interstitial (rock voids) water and internal (deep-lying) water. Interstitial water is subdivided into suspended (vadose) water in the zone of aeration and groundwater in the zone of saturation. Suspended water has three further subdivisions: soil water zone, intermediate zone, and capillary zone. The water in the zone of saturation was divided by Meinzer into free water and pressure water.

Figure 3.1   Meinzer's classification and modification suggested by Davis and DeWiest (1966).

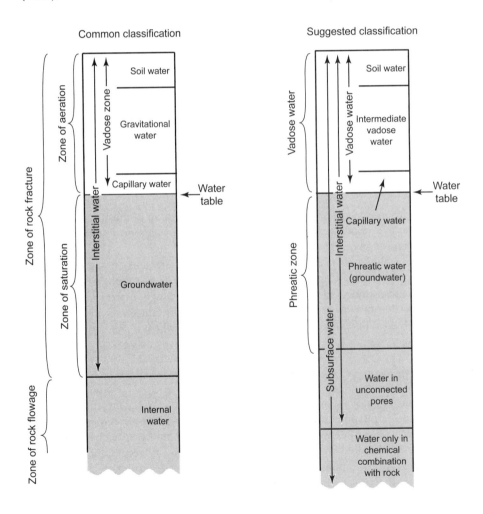

Theory of Groundwater Flow   Chapter 3

The French scientist Schoeller, in 1962, distinguished the following zones beneath the surface: (1) the evaporation zone, (2) the infiltration zone, (3) the capillary fringe, and (4) the zone of groundwater accumulation. In the last zone, free surface and pressure surface are recognized.

In Lange's classification of 1969, often used by hydrogeologists in the former USSR, three basic groups of water are recognized: soil water, subsurface water (in the former USSR this term is commonly used in the sense of groundwater) and interstratal water.

Davis and DeWiest (1966) of the United States suggested certain minor changes in Meinzer's classification. The original classification and suggested changes are shown schematically in Figure 3.1. Davis and DeWiest combined the collecting-rock properties, thus describing groundwater as (1) water of igneous and metamorphic rocks, (2) water of hard sedimentary rocks, (3) water of unconsolidated sediments, and (4) water of regions of extreme climatic conditions.

Pinneker (1983) considered that present classifications are concerned only with the distribution of water pertaining to land masses. Groundwater below the oceans and seas is not covered. Also, the deep-lying water in the zone of saturation that is acted upon by geostatic pressure or other internal forces is not identified in these classifications, although artesian water pressured by hydrostatic pressure has been recognized. Pinneker thus suggested a classification that included the process by which groundwater deposits are formed, while retaining Meinzer's concept (see Table 3.1). This classification has the following scheme:

- *Groups:* depending on the position of groundwater in the earth's crust
- *Sections:* according to the degree of saturation of rock formation with water
- *Types:* on the basis of hydraulic features
- *Classes:* basic varieties of groundwater according to the way in which they are formed
- *Subclasses:* on the basis of water-collecting properties of rocks
- *Special conditions:* specific nature of surroundings

3.3 WATER-BEARING FORMATIONS

Formations that can yield significant quantities of water are known as *aquifers*. This characteristic is imparted to the formations by interconnected openings or pores through which water can move. Alluvial deposits are thus the best form of aquifers: probably 90% of all developed aquifers are in such formations. Such aquifers often have the advantage of direct replenishment by seepage from streams and land. Rock formations of a volcanic nature, limestone, and sandstone possess cracks, fissures, cavities, faults, caverns, and joints through which they yield water. The quality of such rocks as an aquifer depends on the extent of such openings; sometimes they form highly permeable aquifers. Generally, metamorphic and igneous rocks are in solid forms and serve as poor aquifers. Similarly, clays, although having a high level of porosity, prove to be poor aquifers because their pores are too small and not well connected.

Types of aquifers are shown schematically in Figure 3.2. Mainly there are two types: unconfined aquifers and confined aquifers. In unconfined aquifers, the upper surface of the groundwater body is exposed to atmospheric pressure or a *water table* exists. Confined aquifers, also known as pressure or artesian aquifers, occur where groundwater is under

Table 3.1   Groundwater Classification According to the Manner in Which it Has Been Formed

Group	Section	Type	Class	Subclass		Special Conditions	
				Water in strata of porous rocks (pore and stratal water)	Water in fissured cavernous rocks (fissure and vein-fissure water)	Water in permafrost regions	Water in volcanically active regions
Continental groundwater	Groundwater of the zone of aeration	Suspended water	Perched water (in the broad sense)	Salt water and infiltrating water, perched water	Salt water and infiltrating water, perched water	Active layer	Upper part of lava cover
	Groundwater of the zone of saturation on continents	Mainly nonpressure water	Groundwater	Aquifer nearest to the surface on stable impermeable layer	Upper parts of the zone of intensive fissuring and karst massif	Suprapermafrost / Interpermafrost and intrapermafrost	Lower part of lava cover / Water of hydrothermal systems under hydrostatic pressure
		Pressure water	Artesian water	Industrial water under hydrostatic pressure	Buried fissured zone under hydrostatic-pressure	Subpermafrost	
			Deep-lying	Sedimentary layers, which are subjected to the action of geostatic pressure and endogenic forces	Water of deep-lying faults within the sphere of activity of endogenic forces	Absent	Water of volcanic structures and hot spring systems, connected with a rising stream from the magma chamber
Groundwater below seas and oceans	Groundwater of the submarine zone of saturation	Mainly pressure water	Water connected with the land mass	Shelf and marine deposits	Karsted rock of the shelf and fault zones	Subpermafrost shelf of the northern seas	Submarine volcanic structures and marine hot spring systems
			Water not connected with the land mass	Water of deep basins	Trenches and mid-oceanic rifts	Absent	Submarine volcanic structures and marine hot spring systems

Source: Pinneker (1983). Used with permission of Cambridge University Press.

Figure 3.2 Types of aquifers.

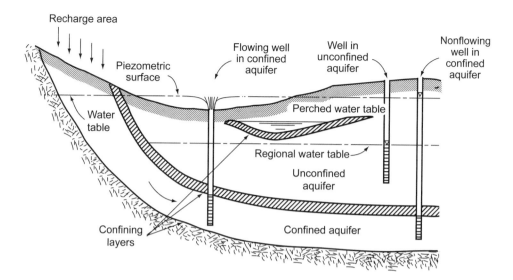

greater-than-atmospheric pressure due to an overlaying confined layer of a relatively impermeable medium.

A special case of unconfined aquifers involves perched aquifers, where a stratum of relatively impermeable material exists above the main body of groundwater. The downward-percolating water is intercepted by this stratum and a groundwater body of limited areal extent is thus formed. The zone of aeration is present between the bottom of the perching bed and the main water table.

A special case of confined aquifers is the leaky aquifer, also known as a semi-confined aquifer. Such an aquifer is either overlain or underlain by a semipervious layer through which water leaks in or out of the confined aquifer.

3.4 FLUID POTENTIAL AND HYDRAULIC HEAD

Just as the energy head is the energy quantity per unit weight of fluid, the fluid potential is the energy quantity per unit mass. Thus,

$$\Phi = gh \quad [\text{L}^2\text{T}^{-2}] \tag{3.1}$$

where

Φ = fluid potential at any point

h = hydraulic head at that point

Flow of any kind occurs from a region in which the potential has a higher value towards the region of a lower value. Thus, the flow of heat, electricity, or water requires existence of a potential gradient. For a porous medium, the flow velocity and, hence, the kinetic energy are extremely small and can be ignored. Thus, the fluid potential and the hydraulic head are comprised of the elevation head and the pressure head. With reference to Figure 3.3:

$$h = Z + \psi \quad [\text{L}] \tag{3.2}$$

where

Z = elevation of a point with reference to a datum

$\psi = p/\gamma$; p being the pressure above atmosphere at the point

3.5 Basic Equation of Groundwater Flow: Darcy's Law

The fundamental law of groundwater movement was discovered by Henri Darcy in 1856. He ran an experiment on a pipe filled with sand under conditions simulated by Figure 3.3. He concluded that the flow rate Q was proportional to the cross-sectional area A, inversely proportional to the length L of the sand-filter flow path, and proportional to head drop $(h_1 - h_2)$. This provided the famous Darcy equation

$$Q = \frac{KA(h_1 - h_2)}{L} \quad [\mathrm{L^3T^{-1}}] \tag{3.3}$$

where K is the hydraulic conductivity, which represents the constant of proportionality, as discussed in Section 3.6.

The ratio $h_1 - h_2/L$ is known as the *hydraulic gradient*. Defining specific discharge, q, or discharge velocity, v, as discharge per unit cross-sectional area, eq. (3.3) becomes

$$q = v = \frac{K\Delta h}{L} \quad [\mathrm{LT^{-1}}] \; ^* \tag{3.4}$$

where

q = specific discharge

v = Darcy velocity or discharge velocity

Δh = drop of head $(h_1 - h_2)$ in length L

3.5.1 Darcy Velocity and Seepage Velocity

In eq. (3.4), v, known as the Darcy velocity, is a fictitious velocity since it assumes that flow occurs through the entire cross section of the material, whereas the flow is actually limited to the pore space only. If v_v is the seepage velocity and A_v is the area of voids, then, from the continuity equation,

$$Q = Av = A_v v_v$$

or

$$v_v = v\frac{A}{A_v}$$

Multiplying both sides by the length of the medium,

$$v_v = v\frac{AL}{A_v L} = v\frac{V}{V_v}$$

---

* A negative sign is inserted on the right side of eq. (3.4) if Δh is taken as $(h_2 - h_1)$, thus indicating flow in the direction of decreasing head.

Figure 3.3 Simulation of Darcy's experiment.

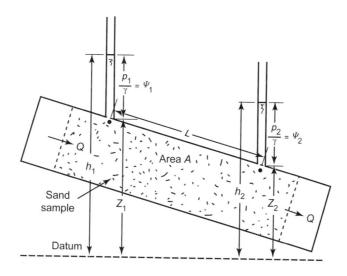

The ratio of volume of voids to total volume is porosity, η by definition, as discussed in 3.7.1, i.e., $\eta = \dfrac{v_v}{v}$ and $v_v = v/\eta$. Thus,

$$v_v = \frac{K}{\eta}\frac{\Delta h}{L} \quad [\mathrm{LT^{-1}}] \tag{3.5}$$

where

v = Darcy velocity

v_v = seepage or interstitial velocity

η = porosity

EXAMPLE 3.1

A channel runs almost parallel to a river as shown in Figure 3.4. The water level in the river is at an elevation of 120 ft and in the channel at an elevation of 110 ft. The river and channel are 2000 ft apart, and a pervious formation with an average thickness of 30 ft and hydraulic conductivity of 0.25 ft/hr joins them. Determine the rate of seepage flow each day from the river to the channel.

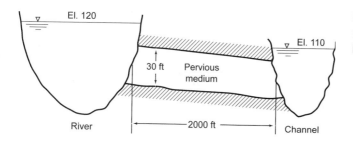

Figure 3.4 Model of river and channel in Example 3.1.

Section 3.5 Basic Equation of Groundwater Flow: Darcy's Law

SOLUTION

1. Consider a 1-ft width of river and channel perpendicular to the paper. The area of cross section of the aquifer normal to flow

$$A = (30\ \text{ft} \times 1\ \text{ft}) = 30\ \text{ft}^2$$

2. $K = \left(0.25\dfrac{\text{ft}}{\text{hr}}\right)\left[\dfrac{24\ \text{hr}}{1\ \text{day}}\right] = 6\ \text{ft/day}$

3. From eq. (3.3)

$$Q = \dfrac{K\,A\left(h_1 - h_2\right)}{L}$$

or

$$Q = \left(6\dfrac{\text{ft}}{\text{day}}\right)\left(30\ \text{ft}^2\right)\dfrac{(120 - 110\ \text{ft})}{2000\ \text{ft}}$$

$$= 0.9\ \text{ft}^3/\text{day/ft width}$$

EXAMPLE 3.2

A semipervious layer (aquitard) separates an overlying water-table aquifer from an underlying confined aquifer as shown in Figure 3.5. Determine the rate of flow taking place between the aquifers.

SOLUTION

1. Since the water table is above the piezometric surface and a semipervious (leaky) layer exists, flow will take place from the water-table aquifer to the confined aquifer.

2. Head loss will take place when water moves through the water-table (top) aquifer, which is not known.

3. Assume that the head at point b shown in the figure is h_b and consider the unit horizontal area through which flow takes place.

4. Between points a and b, from eq. (3.4),

$$q = \dfrac{40\left(90 - h_b\right)}{90} \tag{1}$$

5. Between points b and c, with reference to datum at c: The head at c equals the piezometric level above c, i.e., $(80 + 15)$. The head at b equals h_b plus datum head of 15 ft above c, i.e., $(h_b + 15)$. From eq. (3.4),

$$q = (1)\dfrac{\left(h_b + 15\right) - (80 + 15)}{15} \tag{2}$$

6. Equating (1) and (2) provides $h_b = 88.70$ ft

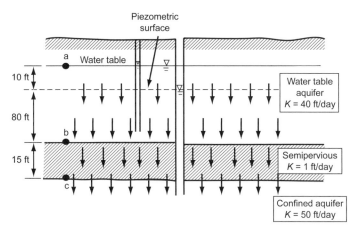

From eq. (1), $q = 0.58$ ft^3/day per square foot

EXAMPLE 3.3

A confined aquifer has a source of recharge as shown in Figure 3.6. The hydraulic conductivity of the aquifer is 50 m/day and its porosity is 0.2. The piezometric head in two wells 1000 m apart is 55 m and 50 m, respectively, from a common datum. The average thickness of the aquifer is 30 m and the average width is 5 km. **(a)** Determine the rate of flow through the aquifer. **(b)** Determine the time of travel from the head of the aquifer to a point 4 km downstream (assume no dispersion or diffusion).

SOLUTION

1. Area of cross section of flow $= (30 \text{ m})(5 \text{ km})\left[\dfrac{1000 \text{ m}}{1 \text{ km}}\right] = 15 \times 10^4 \text{ m}^2$

2. Hydraulic gradient $= \dfrac{55 - 50}{1000} = 5 \times 10^{-3}$

3. Rate of flow, from eq. (3.3):

$$Q = (50 \text{ m/day})(15 \times 10^4 \text{ m}^2)(5 \times 10^{-3}) = 37,500 \text{ m}^3\text{/day}$$

4. Darcy velocity:

$$v = \frac{Q}{A} = \left(37,500 \frac{\text{m}^3}{\text{day}}\right)\left(\frac{1}{15 \times 10^4 \text{ m}^2}\right) = 0.25 \text{ m/day}$$

5. Seepage velocity:

$$v_v = \frac{v}{\eta} = \frac{0.25}{0.2} = 1.25 \text{ m/day}$$

Figure 3.6 Travel time in a uniform-sized aquifer.

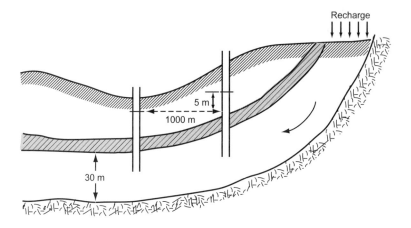

6. Time to travel 4 km or 4000 m downstream:

$$t = \frac{4000 \text{ m}}{1.25 \text{ m/day}} = 3200 \text{ days} \text{ or } 8.77 \text{ years}$$

This demonstrates that water moves very slowly underground.

3.6 PARAMETERS OF GROUNDWATER MOVEMENT

Besides serving as an underground storage reservoir, an aquifer acts as a conduit through which water is transmitted from a higher level to a lower level of energy. The difference in energies at various locations is caused by a continuous process of infiltration and extraction of water underground. A basic parameter connected with water movement through a porous medium is the coefficient of permeability or hydraulic conductivity, which is a constant of proportionality between the rate of flow and the energy gradient causing that flow in accordance with Darcy's law.

3.6.1 Hydraulic Conductivity

Hydraulic conductivity combines the properties of a porous medium and the fluid flowing through it. The relevant fluid properties are dynamic viscosity, μ, and specific weight, γ. The medium properties comprise porosity, grain-size distribution, and shape of the grains. A term used to communicate the effectiveness of the porous medium alone as a transmitting medium is the *intrinsic* (specific) *permeability* k, which has a dimension of L^2. When fluid and medium properties are combined, the term is called the coefficient of permeability or hydraulic conductivity, expressed as:

$$K = k\frac{\gamma}{\mu} \quad [LT^{-1}] \tag{3.6}$$

where

K = hydraulic conductivity

k = intrinsic permeability

γ = specific weight of fluid

μ = dynamic viscosity of fluid

Darcy's law is used to define hydraulic conductivity. From relation (3.3), a medium has a hydraulic conductivity of 1 (having a dimension of length per unit time) if it transmits a unit discharge through a cross-section of unit area under a hydraulic gradient of unit change in head through a unit length of flow.

Similarly, to define the intrinsic permeability, eq. (3.6) is substituted in eq. (3.4), and k thus becomes

$$k = \frac{qv}{gi} \quad [L^2] \tag{3.7}$$

where v = kinematic viscosity = μ/ρ and $i = \Delta h/L$. Accordingly, a medium is said to have an intrinsic permeability of 1 (unit of length squared) if it transmits a unit discharge of fluid of unit kinematic viscosity through a cross-section of unit area under a unit potential gradient. The units of cm^2, ft^2, and darcy are used for intrinsic permeability. Their equivalence is shown in Table 3.2.

Table 3.2   Equivalence of Intrinsic Permeability, Hydraulic Conductivity, and Transmissivity

	Intrinsic Permeability	
darcy	cm^2	ft^2
1	0.987×10^{-8}	1.062×10^{-11}
	Hydraulic Conductivity	
meinzer or gpd/ft^2	ft/day	m/day
1	0.134	0.041
	Transmissivity	
gpd/ft	ft^2/day	m^2/day
1	0.134	0.0124

The units used for hydraulic conductivity are ft/day, m/day, and gallons per day/ft^2. The last unit, also known as *meinzers*, has been adopted by the U.S. Geological Survey. For laboratory measurement, a water temperature of 60 °F is considered standard, whereas the actual temperature in the field is used to measure the field coefficient of permeability. The equivalence of these terms is also indicated in Table 3.2.

Values of hydraulic conductivity can be obtained from empirical formulas, from laboratory measurements, and from field tests. The representative values for various aquifer mediums are given in Table 3.3.

EXAMPLE 3.4

Determine the hydraulic conductivity of a medium with intrinsic permeability of 1 darcy and through which water flows at 60 °F.

Table 3.3 Representative Values of Hydraulic Conductivity for Soils and Rocks

Formation	Hydraulic Conductivity, m/day
Unconsolidated Formations	
Gravel, coarse	1000–8600
Gravel, medium	20–1000
Gravel, fine	20–50
Sand, coarse	0.1–500
Sand, medium	0.1–50
Sand, fine	0.01–20
Silt, sandy	1–4
Silt, clayey	0.2–1
Till, gavel	30
Till, sandy	0.2
Till, clayey	$\leq 10^{-5}$
Clay	≤ 0.0005
Sedimentary Rocks	
Limestone	10^{-4}–800
Sandstone	10^{-5}–3
Siltstone	10^{-6}–0.001

SOLUTION

At 60 °F or 15.6 °C, $\rho = 0.999$ g/cm^3 and $\mu = 1.12$ cP or 1.12×10^{-2} P or g/cm-sec. From eq. (3.6),

$$K = k\frac{\gamma}{\mu} = k\frac{\rho g}{\mu}$$

$$K = (1 \text{ darcy})\left(0.999\frac{\text{g}}{\text{cm}^3}\right)\left(980\frac{\text{cm}}{\text{sec}^2}\right)\left(\frac{1}{1.12\times10^{-2}}\frac{\text{cm-sec}}{\text{g}}\right)$$

$$\times \left[\frac{0.987\times10^{-8}}{1}\frac{\text{cm}^2}{\text{darcy}}\right] = 862.8\times10^{-6} \text{ cm/sec}$$

$$\uparrow$$

from Table 3.2

Conversion to meinzers:

$$K = \left(862.8\times10^{-6}\frac{\text{cm}}{\text{sec}}\right)\left[\frac{1 \text{ m}}{100 \text{ cm}}\right]\left[\frac{24\times60\times60 \text{ sec}}{1 \text{ day}}\right]\left[\frac{1 \text{ meinzer}}{0.041 \text{ m/day}}\right]$$

$$= 18.2 \text{ meinzers}$$

Thus 1 darcy = 18.2 meinzers for water at 60 °F.

EXAMPLE 3.5

At station A the water-table elevation is 650 ft above sea level, and at B, which is 1000 ft apart from A, the elevation is 632 ft. The average velocity of flow is observed to be 0.1 ft/day. Determine the coefficient of permeability in meinzers.

SOLUTION

From eq. (3.4),

$$K = \frac{v}{\Delta h / L}$$

where

$$v = \text{specific discharge} = \text{velocity} = 0.1 \text{ ft/day}$$
$$\Delta h / L = \text{hydraulic gradient} = (650 - 632)/1000 = 0.018$$
$$\text{thus, } K = 0.1/0.018 = 5.56 \text{ ft/day}$$

Conversion to meinzers:

$$K = \left(5.56 \frac{\text{ft}}{\text{day}} \right) \left[\frac{1 \text{ meinzer}}{0.134 \text{ ft/day}} \right]$$
$$= 41.9 \text{ meinzers}$$

3.6.2 Variation of Hydraulic Conductivity

Hydraulic conductivity varies from aquifer to aquifer, from liquid to liquid, from location to location, from direction to direction, and from temperature to temperature. When K is the same in all places (space), the medium is *homogeneous*. When it varies in space, the medium is said to be *heterogeneous*. Even in a homogeneous medium, K can vary with the direction of flow. The medium is then called *anisotropic*.

When the hydraulic conductivity is a continuous function of depth,

$$\overline{K} = \frac{1}{b} \int_0^b K_z \, dz \quad [\text{LT}^{-1}] \tag{3.8}$$

where b is the thickness of the medium.

When a medium is stratified, two conditions can exist: the direction of flow is either parallel to the stratifications or normal to it. When flow direction is parallel to the stratifications, as shown in Figure 3.7, the average value of hydraulic conductivity can be given by

$$\overline{K} = \frac{1}{b} \left(K_1 b_1 + K_2 b_2 + K_3 b_3 + \cdots + K_n b_n \right) \quad [\text{LT}^{-1}] \tag{3.9}$$

For flow perpendicular to stratifications, as shown in Figure 3.8,

$$\overline{K} = \frac{b}{b_1 / K_1 + b_2 / K_2 + b_3 / K_3 \cdots + b_n / K_n} \quad [\text{LT}^{-1}] \tag{3.10}$$

Figure 3.7 Flow parallel to stratifications.

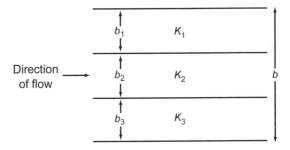

Figure 3.8 Flow normal to stratifications.

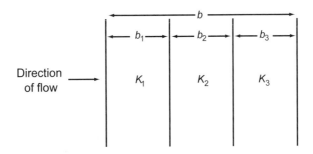

EXAMPLE 3.6

The soil under a dam consists of four layers as follows:

Layer	Hydraulic Conductivity (cm/hr)	Depth (m)
1	5	4.8
2	2	8.0
3	0.6	18.0
4	1.0	3.0

What is the average vertical (perpendicular to flow) conductivity of the soil?

SOLUTION

From eq. (3.10),

$$\overline{K} = \frac{4.8 + 8.0 + 18.0 + 3.0}{4.8/5 + 8.0/2 + 18.0/0.6 + 3.0/1} = 0.89 \text{ cm/hr}$$

$$= \left(0.89 \frac{\text{cm}}{\text{hr}}\right)\left[\frac{1 \text{ m}}{100 \text{ cm}}\right]\left[\frac{24 \text{ hr}}{1 \text{ day}}\right]$$

$$= 0.214 \text{ m/day}$$

EXAMPLE 3.7

In a soil stratum, the hydraulic conductivity at the surface is 2×10^{-3} cm/sec. It uniformly reduces to 4×10^{-4} cm/sec at a depth of 22 m, as shown in Figure 3.9. If the water table is 3 m below the surface, determine the average hydraulic conductivity of the stratum.

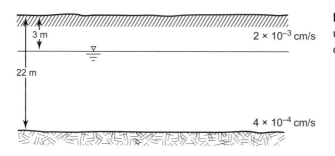

Figure 3.9 Stratum with uniformly varying hydraulic conductivity, Example 3.7.

SOLUTION For linear variation, the hydraulic conductivity at a height x (bottom as datum) can be expressed as

$$K = 4 \times 10^{-4} + \left(\frac{2 \times 10^{-3} - 4 \times 10^{-4}}{22} \right) x$$

or

$$K = 4 \times 10^{-4} + 0.727 \times 10^{-4} x$$

From eq. (3.8),

$$\overline{K} = \frac{1}{19} \int_0^{19} \left(4 \times 10^{-4} + 0.727 \times 10^{-4} x \right) dx$$

$$= \frac{1}{19} \left\{ 4 \times 10^{-4} [x]_0^{19} + 0.727 \times 10^{-4} \left[\frac{x^2}{2} \right]_0^{19} \right\} = 10.9 \times 10^{-4} \text{ cm/sec}$$

or

$$\overline{K} = \left(10.9 \times 10^{-4} \frac{\text{cm}}{\text{sec}} \right) \left[\frac{1 \text{ m}}{100 \text{ cm}} \right] \left[\frac{24 \times 60 \times 60 \text{ sec}}{1 \text{ day}} \right] = 0.942 \text{ m/day}$$

3.6.3 Transmissivity

Transmissivity determines the ability of an aquifer to transmit water through its entire thickness. In an aquifer of uniform thickness,

$$T = \overline{K} b \quad [\text{L}^2 \text{T}^{-1}] \tag{3.11}$$

where

T = transmissivity

$\overline{K}$ = average hydraulic conductivity

b = thickness of aquifer

Field tests to determine the transmissivity of a medium are described in Chapter 4.

EXAMPLE 3.8

What is the transmissivity of the soil in Example 3.6 when the water table is at the ground surface?

SOLUTION

From Example 3.6, $\overline{K}$ = 0.214 m/day

From eq. (3.11), T = 0.214 (4.8 + 8.0 + 18.0 + 3.0) = 7.23 m^2/day

3.6.4 Leakance, Retardation Coefficient, and Leakage Factor (for Leaky Aquifer)

Hantush (1964) introduced leakance or coefficient of leakage as a term characteristic of the semipervious confining layer through which water leaks out from an aquifer. Defined, as follows, it is a measure of the ability of the confining layer to transmit vertical leakage:

$$L_e = \frac{K'}{b'} \quad [\mathrm{T}^{-1}]$$ (3.12)

where

L_e = leakance or coefficient of leakage

K' = coefficient of permeability of semipervious layer of thickness b'

Other factors, introduced by Hantush to indicate areal distribution of leakage and used for the solution of the equation of flow through a leaky aquifer, were the retardation coefficient, a, and leakage factor, B, defined as

$$a = \frac{K}{K'/b'} \quad [\mathrm{L}]$$ (3.13)

and

$$B = \sqrt{\frac{Kb}{K'/b'}} \quad [\mathrm{L}]$$ (3.14)

where K is the coefficient of permeability of aquifer of thickness b.

EXAMPLE 3.9

The banks and bottom of a stream consist of silty clay of hydraulic conductivity 0.008 m/day having an average depth of 150 cm. The underlying aquifer of fine sand has an average thickness of 20 m. Determine the (a) coefficient of leakage, (b) retardation coefficient, and (c) leakage factor of the stream bed. Hydraulic conductivity of fine sand = 2.5 m/day.

SOLUTION From eq. (3.12),

$$b' = (150 \text{ cm}) \left[\frac{1}{100} \frac{\text{cm}}{\text{m}} \right] = 1.5 \text{ m}$$

$$L_e = \frac{0.008}{1.5} = 0.0053 \text{ per day}$$

From eq. (3.13),

$$a = \frac{2.5}{0.0053} = 471.7 \text{ m}$$

From eq. (3.14),

$$B = \sqrt{\frac{2.5 \times 20}{0.0053}} = 97.1 \text{ m}$$

3.7 PARAMETERS OF GROUNDWATER STORAGE

Two important aspects of the study of groundwater are the movement of water underground to streams and wells and underground storage in which an aquifer serves as a storage reservoir. The volume of water taken or released from storage with changes in water levels is reflected in the parameters of *specific yield* or specific retention for water-table aquifers and by *specific storage* or storage coefficient for confined aquifers. The fundamental parameter of groundwater phenomena, however, is porosity, which allows soil to be considered a porous medium.

3.7.1 Porosity

An element of soil, separated in three phases, is shown schematically in Figure 3.10. Porosity is defined as the ratio of the volume of voids to the total volume, or

$$\eta = \frac{V_v}{V_t} \quad \text{[dimensionless]} \tag{3.15}$$

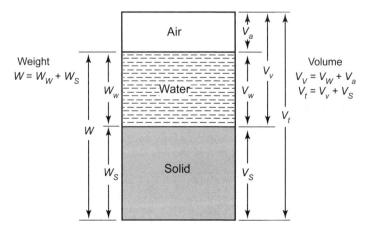

Figure 3.10 Three phases in a soil element.

The term *void ratio* is commonly used in soil mechanics to provide an indication of voids or pores in the soil. It is defined as the ratio of the volume of voids to the volume of solids in a soil sample, or

$$e = \frac{V_v}{V_s} \quad \text{[dimensionless]} \tag{3.16}$$

This term, however, is rarely used in groundwater flow.

Porosity and void ratio are interrelated by the expression

$$e = \frac{\eta}{1-\eta} \quad \text{[dimensionless]} \tag{3.17}$$

Bulk (dry) density, ρ_b, of soil is the mass of soil solids (dry soil) per unit gross volume of soil, and the density of soil particles (grains), ρ_s is equal to the mass of soil solids per unit volume of soil solids. For the same mass of soil solids,

$$\rho_b \propto \frac{1}{V_t} \text{ and } \rho_s \propto \frac{1}{V_s}$$

where

ρ_b = dry (bulk) density

V_t = total soil volume

ρ_s = grain density

V_s = dry soil volume

From these relations and eq. (3.15), the following relation emerges:

$$\frac{\rho_b}{\rho_s} = 1-\eta$$

or

$$\frac{G_b}{G_s} = 1-\eta \quad \text{[dimensionless]} \tag{3.18}$$

where G_b and G_s are bulk specific gravity and specific gravity of soil solids, respectively.

Porosity is a measure of the water-bearing capacity of a formation. However, it is not just the total magnitude of porosity that is important from the consideration of water extraction and transmission, but the size of voids and the extent to which they are interconnected since pores may be open (interconnected) or closed (isolated). For instance, a clay formation may have a very high porosity but it is a poor medium as an aquifer. Specific yield, or effective porosity, and specific retention, as discussed below, are important from this consideration.

EXAMPLE 3.10

A sample of sandy soil is collected from an aquifer. The sampler with a volume of 50 cm^3 is filled with the soil. When the soil is poured into a graduated cylinder, it displaces 30.5 cm^3 of water. What are the porosity and the void ratio of the sand?

SOLUTION

The volume of water displaced is equal to the volume of soil particles (solids); thus $V_s = 30.5$ cm³ while $V_t = 50$ cm³

Hence

$$V_v = V_t - V_s = 50 - 30.5 = 19.5 \text{ cm}^3$$

From eq. (3.15),

$$\eta = \frac{V_v}{V_t} = \frac{19.5}{50.0} = 0.39 \text{ or } 39\%$$

From eq. (3.17),

$$e = \frac{\eta}{1-\eta} = \frac{0.39}{1-0.39} = 0.64$$

EXAMPLE 3.11

A soil sample occupies 0.132 ft³. When dried, it weighs 15.8 lb. If the specific gravity of soil solids is 2.65, calculate (a) the bulk density of the soil, and (b) the porosity of the soil.

SOLUTION

1. Dry unit weight $= \dfrac{15.8}{0.132} = 119.7 \text{ lb/ft}^3$

2. Dry (bulk) density, $\rho_b = \dfrac{119.7}{g} = \dfrac{119.7}{32.2} = 3.72 \text{ slugs/ft}^3$

3. Unit weight of soil grains $= G_s \gamma_w = (2.65)(62.4)$

$$= 165.36 \text{ lb/ft}^3$$

4. Density of soil grains, $\rho_s = \dfrac{165.36}{g} = \dfrac{165.36}{32.2} = 5.14 \text{ slugs/ft}^3$

5. From eq. (3.18),

$$\eta = 1 - \frac{3.72}{5.14} = 0.276 \text{ or } 27.6\%$$

6. [Instead of dry weight, the natural (wet) weight of 18 lb/ft³ with a moisture content, ω, of 14% could have been given in the problem. In such a case,

$$W_s = \frac{W_t}{1+\omega/100} = \frac{18}{1+14/100} = 15.8 \text{ lb}$$

Other steps are the same as above.]

3.7.2 Specific Retention (of Water-Table Aquifer)

When the water table is lowered, water drains from the pore spaces of an aquifer and is replaced with air. This process occurs because the pressure of water inside the pores becomes less than the surrounding air pressure. However, a part of the water is retained within the pores, due to forces of *adhesion* (attraction between pore walls and adjacent water molecules) and *cohesion* (attraction between molecules of water), which are stronger than the pressure difference between the air pressure and the water pressure. The difference of air pressure and water pressure is known as *capillary pressure*, P_c. The volume of water thus retained against the force of gravity, compared to the total volume of rock (soil), is called the *specific retention*. It is also known as the *field capacity* or water-holding capacity. This is a measure of the water-retaining capacity of the porous medium. Specific retention is thus dependent on both pore characteristics and factors affecting the surface tension, such as temperature, viscosity, mineral composition of water, and so on.

As stated above, the amount of water drained from the saturated soil is a function of capillary pressure. A characteristic curve of this function is shown in Figure 3.11. As P_c increases, the volumetric-moisture content* decreases. At a large value of P_c, the volumetric-moisture content tends toward a constant value because of adhesion and cohesion (explained earlier) and the gradient $\Delta\omega/\Delta P_c$ approaches zero. The volumetric-moisture content at this state is equal to the specific retention, as shown in Figure 3.11.

A simple device consisting of a porous plate, capillary tube, and leveling bottles is used to measure volumetric-moisture content and capillary pressure head on a saturated sample. The data are plotted as in Figure 3.11 to obtain the specific retention of the representative sample.

The porewater pressure at any depth h below the water table is equal to γh like hydrostatic pressure, or simply h in terms of water head. Thus, pressure above the water table, with reference to the water table as a datum, will be negative and equal to the height of the point from the water table. This negative pressure is simply the capillary pressure, P_c. If we follow the relationship of Figure 3.11 between capillary pressure and moisture content, the same curve indicates moisture content (in volumetric terms) of the soil at various heights above the water table. Consider the water table in Figure 3.12a at level 1; the moisture distribution curve will be as shown by the outer solid curve in Figure 3.12b. Suppose that the water table drops down to level 2. When equilibrium is achieved, the moisture distribution curve will be similar to level 1 but will be displaced to level 2, as shown by the dotted curve in Figure 3.12b. The area under the curve represents the moisture in the soil. The shaded area between the two curves or at the base between the two water-table lines represents the amount of water drained from the soil with the reduction of water table from level 1 to level 2.

EXAMPLE 3.12

A 200-g dry soil sample is tested by a porous plate test. The negative (capillary) pressure head and the incremental amount of water released from the sample are indicated below. The bulk density of the soil is 1.5 g/cc and at saturation the weight moisture content of the soil is 29.33%. **(a)** Calculate the volumetric-moisture content of the soil for each capillary pressure

---

* Moisture content, ω, is a weight parameter. Here this term is used to indicate the quantity of water inside the pores in terms of volume. The following relation holds:

$$\text{volumetric moisture content} = \text{weight moisture content} \times \frac{\text{bulk density of soil}}{\text{density of water}}$$

or volumetric moisture content = weight moisture content × bulk specific gravity

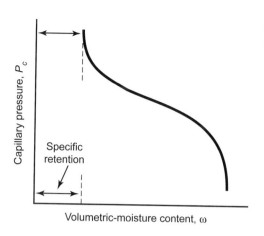

Figure 3.11 Soil-water retention curve.

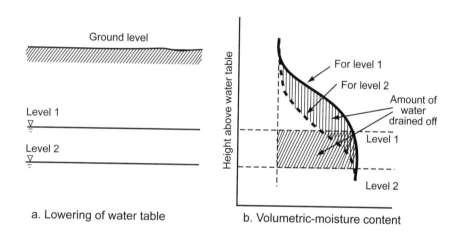

Figure 3.12 Water drained with lowering of the water table.

a. Lowering of water table

b. Volumetric-moisture content

head and plot the moisture distribution curve. **(b)** If the water table was initially located 300 cm below the surface and subsequently receded to 350 cm below the surface, calculate the volume of water removed from the soil per unit area and the specific retention for the soil.

Capillary head, cm	0	20	50	80	100	120	150	170	200	230	250	280	300
Incremental water release, cm³	0	0	1.34	3.33	3.34	4.00	6.66	4.00	6.67	5.33	3.33	2.00	0

SOLUTION

1. Volume of soil sample = $\dfrac{\text{dry mass}}{\text{bulk density}} = \dfrac{(200 \text{ g})}{(1.5 \text{g/cc})} = 133.33 \text{ cm}^3$

2. Volumetric moisture content (at saturation)

$$= (\text{weight moisture content})\frac{(\text{soil bulk density})}{(\text{water density})}$$

$$= (29.33)\left(\frac{1.5}{1}\right) = 44\% \quad \text{or} \quad 0.44$$

3. Volume of water in sample (at saturation) $= 0.44(133.33) = 58.67 \text{ cm}^3$

4. Volume of water retained in the soil, as indicated in col. 4 of Table 3.4, equals the volume of water at saturation minus the total water released of col. 3.

5. Volumetric moisture content at various capillary heads in col. 5 is the water retained in col. 4 divided by the volume of soil.

6. The values of capillary head and moisture content are plotted on a graph designated initial water table, as shown by the solid curve in Figure 3.13.

Figure 3.13   Amount of water drained with lowering of water level in Example 3.12.

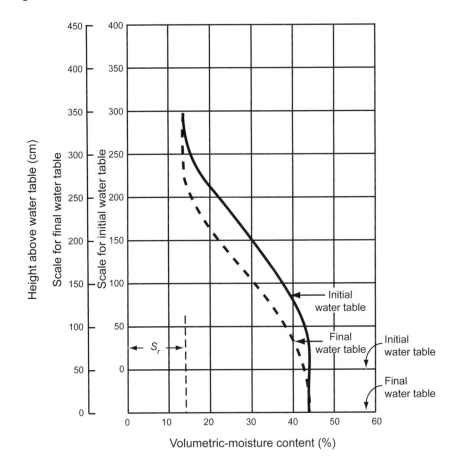

Theory of Groundwater Flow   Chapter 3

7. The scale on the y-axis is scaled down by 50 cm to represent the lowering of the water table. The values of capillary head and moisture content are now plotted on this revised scale as shown by the dotted curve on the figure. The volume of water removed per unit soil area is represented by the area between the two curves.

8. The area between the two curves = 1500 cm-%.

9. The volume of water removed = $\dfrac{1500}{100^*}$ = 15 cm^3 per cm^2 area

10. The total volume of soil per unit surface area between the two water tables = 1×50 = 50 cm^3.

11. The water removed per unit (1 cm^3) volume of soil = 15/50 = 0.30 or 30%

12. The specific retention, S_r, = 14% (from the figure).

Table 3.4 Moisture Contents for Various Capillary Heads

(1)	(2)	(3)	(4)	(5)
Capillary head, cm	Incremental water released, cm^3	Total water released, cm^3	Water retained, cm^3	Volumetric-moisture content, %
0	0	0	58.67	44
20	0	0	58.67	44
50	1.34	1.34	57.33	43
80	3.33	4.67	54.00	40.5
100	3.34	8.01	50.66	38
120	4.00	12.01	46.66	35
150	6.66	18.67	40.00	30
170	4.00	22.67	36.00	27
200	6.67	29.34	29.33	22
230	5.33	34.67	24.00	18
250	3.33	38.00	20.67	15.5
280	2.00	40.00	18.67	14
300	0	40.00	18.67	14

3.7.3 Specific Yield (of Water-Table Aquifer)

Specific yield, also known as *effective porosity*, is defined as the volume of water yielded by an unconfined aquifer by gravity as compared to the unit volume of the aquifer. As the water level falls, water is drained from the pores. Specific yield can not be determined for a confined aquifer since the aquifer is not drained. Specific yield is given by

$$S_y = \frac{\text{Volume of water yielded by gravity}}{\text{Volume of unconfined aquifer}} \times 100$$

---

* Since the moisture content is in %.

or

$$S_y = \frac{1}{A}\frac{dV}{dh} \quad \text{[dimensionless]} \tag{3.19}$$

where

S_y = specific yield

A = area of soil formation

dV = volume of water drained

dh = change in water table

Since some water remains in the soil, the sum of specific yield and specific retention is equal to the porosity, or

$$S_y = \eta - S_r \quad \text{[dimensionless]} \tag{3.20}$$

where

S_y = specific yield

S_r = specific retention

In addition to the capillary head moisture content technique discussed in the preceding section, there are other procedures for determining specific yield, including the well-pumping tests discussed in Chapter 4.

Table 3.5 indicates the representative values of specific yield for various types of soils and rocks. The specific yield of most aquifer formations ranges from about 0.10 to about 0.30 and averages 0.20.

Table 3.5   Representative Values of Specific Yield for Soils and Rocks

Formation	Range of values	Typical
Gravel, coarse	0.10–0.25	0.21
Gravel, medium	0.15–0.45	0.24
Gravel, fine	0.15–0.40	0.28
Sand, coarse	0.15–0.45	0.30
Sand, medium	0.15–0.45	0.32
Sand, fine	0.01–0.45	0.23
Silt	0.01–0.40	0.20
Till, gravel	0.05–0.20	0.16
Till, sand		0.16
Till, silt		0.06
Clay	0.01–0.20	0.06
Sandstone, medium grained	0.01–0.40	0.27
Sandstone, fine grained		0.21
Limestone	0.01–0.35	0.14
Siltstone	0.01–0.35	0.12

EXAMPLE 3.13

For Example 3.12, determine the porosity and specific yield of the soil.

SOLUTION

Porosity = volumetric-moisture content at saturation or zero capillary head

$$\eta = 44\% \text{ (from Figure 3.13)}$$

From eq. (3.20), $S_y = 44 - 14 = 30\%$
(Note that the water drained out in Example 3.12 is equal to the specific yield.)

EXAMPLE 3.14

A water table drops 5 ft over an area of 3.5 acres. If the soil has a specific yield of 4%, how much water has drained from the area?

SOLUTION

1. The area of 3.5 acres = $3.5 \times 43{,}560 = 152{,}460 \text{ ft}^2$
2. The total volume of soil drained off = $5 \times 152{,}460 = 762{,}300 \text{ ft}^3$
3. The volume of water = $S_y \times$ total volume of soil drained off (by definition)

$$= 0.04 \times 762{,}300$$
$$= 30{,}492 \text{ ft}^3$$

In water-table aquifers, some quantity is derived from compression of the aquifer and change of density of the water. Therefore the storage coefficient for water-table aquifers is the total specific yield plus the fraction attributable to compressibility. The latter is, however, negligible compared to gravity drainage; specific yield provides an indication of aquifer release. The term storage coefficient is generally used in relation to confined aquifers.

3.7.4 Specific Storage for Confined Aquifers

The term *specific storage* is defined as the volume of water released from storage per unit decline in pressure head within the unit volume of an aquifer. It is a constant property of an aquifer and, as such, is a more fundamental parameter. A confined aquifer remains saturated at all times and, as such, water release is not derived from drainage of the voids by gravity as in the case of unconfined aquifers. In confined aquifers, the release or addition of water is attained due to the change in pore pressure.

In an equilibrium condition, the forces due to the weight of the formations overlying the aquifer and all other loads from the top are balanced by the skeleton and water within the pores of the aquifer. Due to the pumping of a well, the water pressure inside the pores is reduced. This results in a slight compaction of the skeleton of the aquifer and expansion of the water permitted by its elasticity. A certain amount of water is thus released from storage. The reverse process takes place in response to recharge.

Jacob made the first attempt in 1940 to introduce an analytical expression for the *specific storage*. For an elastic confined aquifer, he defined

$$S_s = \eta \gamma_w \left(\frac{1}{\eta E_s} + \frac{1}{E_w} \right) \quad [\text{L}^{-1}] \tag{3.21a}$$

or

$$S_s = \rho g (\alpha + \eta \beta) \quad [\text{L}^{-1}] \tag{3.21b}$$

where

S_s = specific storage

E_w = bulk modulus of elasticity of water (3×10^5 psi at ordinary temperatures)

E_s = bulk modulus of elasticity of soil solids

α = aquifer compressibility $(1/E_s)$

β = water compressibility $(1/E_w)$

η = porosity

The first term of the expression in parentheses relates to the compressibility of the aquifer and the second term to the expansibility of water.

DeWiest (1966) criticized this derivation, which had considered deformation of the aquifer (one side of the volume element was considered deformable). Cooper (1966) made a further refinement considering flow rate relative to moving grains of the aquifer medium and the flow rate across the fixed boundaries of the control volume. For a very small grain velocity, his form reduces to Jacob's formulation. Table 3.6 indicates the range of values of specific storage of soils and rocks.

Table 3.6 Specific Storage Values

Formation	Specific storage, m^{-1}
Gravel, dense sandy	$1.0 \times 10^{-4} - 4.9 \times 10^{-5}$
Sand, dense	$2.0 \times 10^{-4} - 1.3 \times 10^{-4}$
Sand, loose	$1.0 \times 10^{-3} - 4.9 \times 10^{-4}$
Clay, medium hard	$1.3 \times 10^{-3} - 9.2 \times 10^{-4}$
Clay, stiff	$2.6 \times 10^{-3} - 1.3 \times 10^{-3}$
Clay, plastic	$2.0 \times 10^{-2} - 2.6 \times 10^{-3}$
Rock, fissured	$6.9 \times 10^{-5} - 3.3 \times 10^{-6}$
Rock, unfissured	$< 3.3 \times 10^{-6}$

3.7.5 Storage Coefficient or Storativity

The *storage coefficient* is the volume of water that is released or taken into storage by an aquifer per unit area of the aquifer per unit decline or rise in pressure head. Storage coefficient is expressed as

$$S = \frac{S_y}{100} + S_s b \quad [\text{dimensionless}] \tag{3.22}$$

In a confined aquifer, S_y is zero. Thus for a confined aquifer the relation is

$$S = S_s b \quad \text{[dimensionless]} \tag{3.23}$$

In contrast to the specific yield of an unconfined aquifer, the storage coefficient of the confined aquifer is much smaller, ranging from about 10^{-5} to 10^{-3}.

In an unconfined aquifer, S_s has no relevance. Hence, storage coefficient is similar to the specific yield in decimal points.

EXAMPLE 3.15

A confined aquifer of 40 m thickness has a porosity of 0.3. Determine the specific storage and storage coefficient. $\alpha = 1.5 \times 10^{-9}$ cm^2/dyn, $\beta = 5 \times 10^{-10}$ cm^2/dyn.

SOLUTION

1. $\rho g = (1 \text{ g/cm}^3)(980 \text{ cm/s}^2) = 980 \text{ dyn/cm}^3$
   $b = (40 \text{ m})[100 \text{ cm/m}] = 4000 \text{ cm}$

2. Specific storage [from eq. (3.21b)]:

$$S_s = \gamma_w(\alpha + \eta\beta)$$

$$= \left(980\frac{\text{dyn}}{\text{cm}^3}\right)\left(1.5\times10^{-9} + 0.3\times5\times10^{-10}\frac{\text{cm}^2}{\text{dyn}}\right)$$

$$= 1.62\times10^{-6} \text{ per cm}$$

3. Storage coefficient [from eq. (3.22)]:

$$S = 1.62\times10^{-6} \times 4000$$

$$= 6.47\times10^{-3}$$

EXAMPLE 3.16

The storage coefficient determined from a pumping test in an aquifer is 4×10^{-4} at a location where the aquifer depth is 100 ft. If the average volume of the aquifer per square foot is 80 ft^3, how much water will be released by the aquifer with a drop in head of 70 ft?

SOLUTION

1. $S_s = \dfrac{4\times10^{-4}}{100} = 4\times10^{-6}$ per foot per unit volume

2. The amount of water released is

$$(4 \times 10^{-6})(80)(70) = 0.022 \text{ ft}^3 \text{ per ft}^2 \text{ of area}$$

3. Note that if the storage coefficient is used directly, the volume of water released is

$$(4 \times 10^{-4})(1 \text{ ft}^2)(70) = 0.028 \text{ ft}^3 \text{ per ft}^2$$

This is incorrect because the average thickness of the aquifer is 70 ft, whereas the storage coefficient is based on a 100-ft depth.

4. To use the storage coefficient, the following procedure has to be followed:

$$S_s = 4 \times 10^{-6} \text{ per foot}$$

$$S \text{ for aquifer} = (4 \times 10^{-6})(70) = 2.8 \times 10^{-4}$$

The amount of water released is

$$(2.8 \times 10^{-4})(1)(80) = 0.022 \text{ ft}^3 \text{ per ft}^2$$

3.8 GENERALIZATION OF DARCY'S LAW

Darcy's law has been presented for one-dimensional flow in Section 3.5, which is the form in which it was empirically proposed by Darcy. However, at any point of a fluid in three-dimensional flow, there are three velocity components, a pressure component, and a density component. In groundwater flow, density is commonly considered constant (water is assumed to be incompressible unless specifically stated to the contrary, as in storage coefficient computation for an artesian aquifer). When water flows through an inclined medium as shown in Figure 3.3, its pressure (piezometric) head, which is a scalar quantity, is written as

$$h = Z + \frac{p}{\gamma} \quad [L] \tag{3.24}$$

There also is a kinetic energy term, which can be disregarded in considering the loss of head due to flow.

The velocity (specific discharge) component, however, is a vector quantity and can be expressed in three directions by Darcy's equation (3.4), with a negative sign for differential form of downward gradient:

$$v_x = -K \frac{\partial h}{\partial x} \quad [LT^{-1}]$$

$$v_y = -K \frac{\partial h}{\partial y} \tag{3.25}$$

$$v_z = -K \frac{\partial h}{\partial z}$$

If **i**, **j**, **k** represent standard unit vectors in the x, y, and z directions, respectively, the velocity (specific discharge) components in the three coordinate directions will be $\mathbf{i}v_x$, $\mathbf{j}v_y$, and $\mathbf{k}v_z$. The resultant velocity (specific discharge) vector will be given by

$$v = \mathbf{i}v_x + \mathbf{j}v_y + \mathbf{k}v_z$$

Treating K as constant and substituting eq. (3.25), we have

$$v = -K \left\{ \mathbf{i} \frac{\partial h}{\partial x} + \mathbf{j} \frac{\partial h}{\partial y} + \mathbf{k} \frac{\partial h}{\partial z} \right\}$$

or

$$v = -K \nabla h \quad [LT^{-1}] \tag{3.26}$$

where ∇h denotes the head-gradient vector.

Equation (3.26) is a generalized form of Darcy's law expressed in vector notation. This equation is for isotropic soil in which hydraulic conductivity K is constant in all directions. For anisotropic aquifers, in which a different hydraulic conductivity is assigned to each of the coordinate directions, eqs. (3.25) and (3.26) will have the following forms:

$$v_x = -K_x \frac{\partial h}{\partial x} \quad v_y = -K_y \frac{\partial h}{\partial y} \quad v_z = -K_z \frac{\partial h}{\partial z}$$

and

$$v = \left\{ -\mathbf{i}K_x \frac{\partial h}{\partial x} - \mathbf{j}K_y \frac{\partial h}{\partial y} - \mathbf{k}K_z \frac{\partial h}{\partial z} \right\} \quad [\mathrm{LT}^{-1}] \tag{3.27}$$

3.8.1 Velocity Potential

For the case when hydraulic conductivity K is constant, the velocity potential is defined as a scalar quantity having the following relation:

$$\phi = Kh \quad [\mathrm{L}^2\mathrm{T}^{-1}] \tag{3.28}$$

In terms of the velocity potential

$$v_x = -\frac{\partial \phi}{\partial x}$$
$$v_y = -\frac{\partial \phi}{\partial y} \quad [\mathrm{LT}^{-1}] \tag{3.29}$$
$$v_z = -\frac{\partial \phi}{\partial z}$$

and Darcy's law, eq. (3.26), takes the form:

$$v = -\nabla \phi \quad [\mathrm{LT}^{-1}] \tag{3.30}$$

EXAMPLE 3.17

A homogeneous but anisotropic aquifer has the following values of hydraulic conductivity and head gradient:

	K (cm/sec)	Gradient h
x direction	0.03	0.22
y direction	0.035	0
z direction	0.002	−0.98

(a) Calculate the Darcy velocity vector.

(b) Plot the vectors $K\nabla h$.

(c) Compute the magnitude of the Darcy velocity.

SOLUTION

(a) **1.** Positive coordinate directions are set up as shown in Figure 3.14(a).

2. $\mathbf{i}K_x\dfrac{\partial h}{\partial x} = \mathbf{i}(0.03)(0.22) = 6.6\times10^{-3}\,\mathbf{i}$ cm/s

$\mathbf{k}K_z\dfrac{\partial h}{\partial x} = \mathbf{k}(0.002)(-0.98) = -2.0\times10^{-3}\,\mathbf{k}$ cm/s

3. $v = -\{6.6\times10^{-3}\mathbf{i} - 2.0\times10^{-3}\mathbf{k}\}$ cm/s

(b) **4.** Suppose that standard unit vectors $\mathbf{i}, \mathbf{j}, \mathbf{k}$ are represented by 10^3 in./cm per second in the respective directions.

5. $6.6\times10^{-3}\,\mathbf{i} = (6.6\times10^{-3}\,\text{cm/s})(10^3\,\text{in./cm/s}) = 6.6$ in.

$-2.0\times10^{-3}\,\mathbf{k} = (-2.0\times10^{-3}\,\text{cm/s})(10^3\,\text{in./cm/s}) = -2.0$ in.

These are plotted in Figure 3.14b.

(c) **6.** Magnitude of the Darcy velocity,

$$\text{Resultant magnitude} = \sqrt{\left(6.6\times10^{-3}\right)^2 + \left(-2.0\times10^{-3}\right)^2}$$

$$= 6.9\times10^{-3}\ \text{cm/s}$$

$$\cos\alpha = \frac{6.6\times10^{-3}}{6.9\times10^{-3}} = 0.956 \quad \alpha = 17°$$

$$\cos\gamma = \frac{-2\times10^{-3}}{6.9\times10^{-3}} = -0.29 \quad \gamma = 107°$$

3.9 VALIDITY OF DARCY'S LAW

1. In laminar flow, velocity bears a linear relationship to the hydraulic gradient. Since Darcy's law states that the discharge velocity is proportional to the first power of the hydraulic gradient, it is valid only within a laminar flow condition. As for pipes, Reynolds number is used to distinguish laminar flow from turbulent flow. For a porous medium, Reynolds number is expressed as

$$\text{Re} = \frac{\rho v d_{10}}{\mu} \quad \text{[dimensionless]} \tag{3.31}$$

where

v = Darcy velocity

d_{10} = effective grain size (i.e., 10% of materials are finer than size indicated)

For Re < 1, laminar flow occurs, as in most cases; for Re ≥ 1 but < 10, there is no serious departure from laminar flow; and for Re >10, there is turbulent flow, as in the immediate vicinity of pumped wells.

2. Darcy's law is not valid where water flows through extremely fine-grained materials (e.g., colloidal clays).

3. Darcy's law is modified where the medium is not fully saturated.

Figure 3.14 Plot of velocity-vector components.

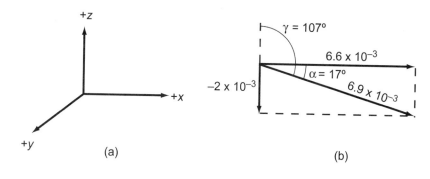

(a)   (b)

EXAMPLE 3.18

A 0.3-m well has a 25-m-long screen that covers the entire depth of an aquifer. The aquifer medium has an effective grain size of 1.5 mm. The well is pumped at a rate of 0.2 m³/s. Assess the validity of Darcy's law near the well.

SOLUTION

1. Area through which flow into well takes place $= (\pi \times d)(\text{thickness of aquifer})$

$$= (\pi \times 0.3)(25) = 23.55 \text{ m}^2$$

2. Darcy velocity $= \dfrac{Q}{A} = \dfrac{0.2}{23.55} = 8.5 \times 10^{-3}$ m/s or 0.85 cm/s

3. Assume that $\rho = 1$ g/cm³ and $\mu = 0.01$ P or g/cm·sec

4. From eq. (3.31),

$$\text{Re} = \frac{(1)(0.85)(0.15)}{0.01} = 12.75 > 10$$

5. Since Re > 10, flow is turbulent and Darcy's law is not applicable.

3.10 STEADY-STATE FLOW AND UNSTEADY-STATE FLOW

There are two types of groundwater flow. In a steady or equilibrium state of flow, the water table or piezometric surface is stabilized in a position and does not change with time as flow takes place, which means that the inflow of water matches the outflow of water. In an unsteady or nonequilibrium state, the water table or piezometric head varies with time; thus water is either added or withdrawn from the groundwater storage during the flow. The fundamental equation of groundwater flow is Darcy's law. For steady state, it can be used to determine the specific discharge (and discharge) if the head gradient is known, or to compute the gradient for a given specific discharge. Darcy's law by itself, however, does not provide all the necessary conditions to solve groundwater flow problems in general. It gives three relations among four variables: three velocity (specific discharge) vectors and the head. For a general equation of flow, a fourth relation is provided by the equation of continuity or conservation of mass.

3.11 GENERAL EQUATION OF GROUNDWATER FLOW

The groundwater flow equation contains three components: Darcy's law, the continuity equation, and the storage component. Conceptually, the flow equation is developed as follows. The continuity equation is written in terms of mass rate of inflow and outflow and accumulation of matter within an elemental volume situated in the field of flow. The terms of the mass rate are substituted by Darcy's law in the form of the piezometric (water table) head, and the term relating to matter accumulation is expressed through the storage coefficient. This equation, usually in terms of piezometric head, is recognized as the flow equation.

The form of the equation differs for confined and unconfined aquifers because of different expressions for Darcy's law in both cases. In the case of confined flow, the area of flow is constant. For unconfined flow, however, area is a function of head (saturated depth). Also, the storage coefficient has different meanings in both cases, although this does not affect the form of the equation. The final equation for a confined aquifer is a linear diffusion type of equation and for an unconfined aquifer it is the nonlinear Boussinesq equation. Derivations are presented in Sections 3.11.1 and 3.11.2 separately for both cases.

3.11.1 Equation for Confined Aquifers

I. Darcy's law:
    From eq. (3.25), three velocity (specific discharge) vectors:

$$v_x = -K_x \frac{\partial h}{\partial x}$$

$$v_y = -K_y \frac{\partial h}{\partial y}$$

$$v_z = -K_z \frac{\partial h}{\partial z}$$

II. Continuity equation:
   1. Mass discharge is equal to the water density times the volume discharge. The mass balance equation is written since the water density is considered variable in the storage coefficient. Mass discharge (flux) through face 1 into the element of Figure 3.15 in the x direction is

$$(\rho Q_x)_1 \quad \text{or} \quad (\rho v_x)_1 \Delta y \Delta z$$

   2. Mass discharge through face 2 out of the element is

$$(\rho Q_x)_2 \quad \text{or} \quad (\rho v_x)_2 \Delta y \Delta z$$

   3. If ρv_x is considered as a continuous function, the Taylor series may be used to expand $(\rho v_x)_2$ in terms of $(\rho v_x)_1$ as follows:

$$\left(\rho v_x\right)_2 = \left(\rho v_x\right)_1 + \frac{\partial \left(\rho v_x\right)}{\partial x}\Delta x + \frac{1}{2}\frac{\partial^2 \left(\rho v_x\right)}{\partial x^2}\left(\Delta x\right)^2 + \cdots$$

Taking the first two terms (neglecting the others) yields

$$\left(\rho v_x\right)_2 = \left(\rho v_x\right)_1 + \frac{\partial}{\partial x}\left(\rho v_x\right)\Delta x$$

Figure 3.15 Elemental volume in the field of flow.

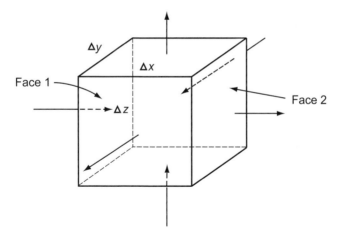

4. Net influx in the x direction
   Outflow rate – inflow rate

$$= \left[(\rho v_x)_1 + \frac{\partial (\rho v_x)}{\partial x} \Delta x \right] \Delta y \Delta z - (\rho v_x)_1 \Delta y \Delta z$$

$$= \frac{\partial (\rho v_x)}{\partial x} \Delta x \Delta y \Delta z$$

Similar terms could be written for the y and z directions.

5. The total net flux (outflow – inflow) will be the summation of fluxes in the x, y, z directions and will be equal to the rate of change of mass within the element:

$$\frac{\partial (\rho v_x)}{\partial x} \Delta x \Delta y \Delta z + \frac{\partial (\rho v_y)}{\partial y} \Delta x \Delta y \Delta z + \frac{\partial (\rho v_z)}{\partial z} \Delta x \Delta y \Delta z = -\frac{\partial M}{\partial t}$$

The negative sign is to make the net flux positive when the mass is depleted. The equation above can be rewritten as

$$\frac{\partial (\rho v_x)}{\partial x} + \frac{\partial (\rho v_y)}{\partial y} + \frac{\partial (\rho v_z)}{\partial z} + \frac{1}{\Delta x \Delta y \Delta z} \frac{\partial M}{\partial t} = 0 \quad [\mathrm{ML}^{-3}\mathrm{T}^{-1}] \quad (3.32)$$

This is the *continuity equation*.

III. Rate of change of mass:

6. Mass accumulated within the element

$$M = \rho \eta \, \Delta x \Delta y \Delta z$$

7. Considering that compression and expansion occur in the z direction only and Δx and Δy are constant:

$$\frac{\partial M}{\partial t} = \left[\eta \Delta z \frac{\partial \rho}{\partial t} + \rho \Delta z \frac{\partial \eta}{\partial t} + \rho \eta \frac{\partial (\Delta z)}{\partial t} \right] \Delta x \Delta y$$

8. The first term relates to the compressibility of water and the other two with compressibility of material. The expression above reduces to (Marino and Luthin, 1982, pp. 146–147)

$$\frac{\partial M}{\partial t} = (\alpha + \beta\eta)\rho\Delta x \Delta y \Delta z \frac{\partial p}{\partial t}$$

where

$$a = 1/E_s$$
$$\beta = 1/E_w$$
$$E_s = \text{bulk modulus of elasticity of aquifer solids}$$
$$E_w = \text{bulk modulus of elasticity of water}$$
$$\eta = \text{porosity}$$
$$p = \text{pressure}, \ p = \gamma h$$

In terms of h (since $p = \gamma h$),

$$\frac{\partial M}{\partial t} = (\alpha + \eta\beta)\rho\Delta x \Delta y \Delta z \gamma \frac{\partial h}{\partial t}$$

or

$$\frac{\partial M}{\partial t} = \rho S_s \Delta x \Delta y \Delta z \frac{\partial h}{\partial t} \quad \text{since } S_s = \gamma(\alpha + \eta\beta)$$

IV. Manipulation of continuity equation
9. Since

$$\frac{\partial (\rho v_x)}{\partial t} = \rho \frac{\partial v_x}{\partial x} + v_x \frac{\partial \rho}{\partial x}$$

10. If the second term on the right side relating to change of water density is dropped in comparison to the first term,

$$\frac{\partial (\rho v_x)}{\partial x} = \rho \frac{\partial v_x}{\partial x}$$

11. Substituting v_x by Darcy's law from eq. (3.25) as shown in item I:

$$\frac{\partial (\rho v_x)}{\partial x} = \rho \frac{\partial v_x}{\partial x} = -\rho \frac{\partial}{\partial x}\left(K_x \frac{\partial h}{\partial x} \right)$$

Similar terms could be written for the y and z directions.
12. Substituting these and the rate of mass term of step 8 into the continuity equation (3.32) of step 5:

$$\frac{\partial}{\partial x}\left(K_x \frac{\partial h}{\partial x} \right) + \frac{\partial}{\partial y}\left(K_y \frac{\partial h}{\partial y} \right) + \frac{\partial}{\partial z}\left(K_z \frac{\partial h}{\partial z} \right) = S_s \frac{\partial h}{\partial t} \quad [\text{T}^{-1}] \qquad (3.33)$$

13. Equation (3.33) is for *nonhomogeneous, anisotropic* confined aquifers. For *homogeneous, anisotropic* cases, when the hydraulic conductivity will be the same in space, eq. (3.33) will become

$$K_x \frac{\partial^2 h}{\partial x^2} + K_y \frac{\partial^2 h}{\partial y^2} + K_z \frac{\partial^2 h}{\partial z^2} = S_s \frac{\partial h}{\partial t} \quad [\text{T}^{-1}] \tag{3.34}$$

This is a linear second-order partial differential equation for unsteady-state flow in a confined aquifer. Equations of similar form appear in the flow of heat and electricity. The derivation above is based on the concept of Jacob (1950). DeWiest (1965) followed a different approach and in the process redefined the specific storage term, S_s, as discussed in Section 3.7.4.

3.11.2 Equation for Unconfined Aquifers

Development here is on the same lines as in the case of a confined aquifer. Consider an elemental volume cutting through the entire saturated thickness, as shown in Figure 3.16.

I. Darcy's laws: Assuming the Dupuit assumption of horizontal flow, we have

$$Q_x = -K_x \frac{\partial h}{\partial x} h \Delta y$$

$$Q_y = -K_y \frac{\partial h}{\partial y} h \Delta x$$

There is no vertical flow according to the assumption above.

II. Continuity Equation:
1. It can be written in volume-discharge form here since compressibility of water is not involved.
2. In the x direction, net flux:

$$\text{outflow} - \text{inflow} = Q_{x1} + \frac{\partial}{\partial x}(Q_x)\Delta x - Q_{x1}$$

$$= \frac{\partial}{\partial x}(Q_x)\Delta x$$

Figure 3.16 Elemental volume through the entire depth of an unconfined aquifer.

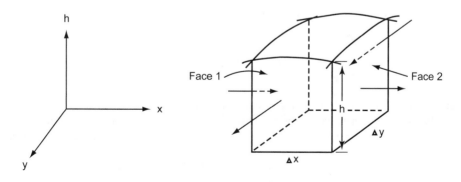

3. In the y direction $= \dfrac{\partial (Q_y)}{\partial y} \Delta y$

4. The total net flux is equal to the change in stored water volume.

$$\frac{\partial (Q_x)}{\partial x} \Delta x + \frac{\partial (Q_y)}{\partial y} \Delta y = -\frac{\partial V_W}{\partial t}$$

III. Rate of change of stored water volume:

$$V_W = S_y \, dh \Delta x \Delta y \quad \text{(by definition)}$$

$$\frac{\partial V_W}{\partial t} = S_y \, \frac{\partial h}{\partial t} \Delta x \Delta y$$

IV. Manipulation of the continuity equation: in the eq. of step II (4), substitute the values of Q from Darcy's law from Step I and V_W from step III.

$$-\frac{\partial}{\partial x}\left(K_x \, h \frac{\partial h}{\partial x} \right) \Delta x \Delta y = -\frac{\partial}{\partial y}\left(K_y \, h \frac{\partial h}{\partial y} \right) \Delta x \Delta y = -S_y \, \frac{\partial h}{\partial t} \Delta x \Delta y$$

or

$$\frac{\partial}{\partial x}\left(K_x \, h \frac{\partial h}{\partial x} \right) + \frac{\partial}{\partial y}\left(K_y \, h \frac{\partial h}{\partial y} \right) = S_y \, \frac{\partial h}{\partial t} \quad [\mathrm{LT}^{-1}] \tag{3.35}$$

This is the nonlinear Boussinesq equation. Linearization of this could be achieved if the change in water table is small compared to the water-table depth h. In that case, the average aquifer thickness, b, could be substituted for h. For a homogeneous case, then, eq. (3.35) will become

$$K_x \frac{\partial^2 h}{\partial x^2} + K_y \frac{\partial^2 h}{\partial y^2} = \frac{S_y}{b} \frac{\partial h}{\partial t} \quad [\mathrm{T}^{-1}] \tag{3.36}$$

Thus eq. (3.36) for an unconfined aquifer becomes exactly like eq. (3.34) in two dimensions for a confined aquifer. Only storage terms will have different meanings.

3.12 An Overview of the Groundwater Flow Equation

Incorporating a source or sink term in eq. (3.33), the most general form of the equation of saturated flow through a porous medium is

$$\frac{\partial}{\partial x}\left(K_x \frac{\partial h}{\partial x} \right) + \frac{\partial}{\partial y}\left(K_y \frac{\partial h}{\partial y} \right) + \frac{\partial}{\partial z}\left(K_z \frac{\partial h}{\partial z} \right) \pm W(x,y,z,t) = S_s \frac{\partial h}{\partial t} \quad [\mathrm{T}^{-1}] \tag{3.37}$$

where $W(x, y, z, t)$ is the source or sink term of discharge per unit volume representing recharge or discharge (well) point. Dropping out the source/sink term, for a homogeneous, anisotropic medium, the above equation becomes eq. (3.34), as reproduced again

$$K_x \frac{\partial^2 h}{\partial x^2} + K_y \frac{\partial^2 h}{\partial y^2} + K_z \frac{\partial^2 h}{\partial z^2} = S_s \frac{\partial h}{\partial t} \quad [\mathrm{T}^{-1}] \tag{3.34}$$

For a homogeneous, isotropic medium, this reduces to

$$\frac{\partial^2 h}{\partial x^2} + \frac{\partial^2 h}{\partial y^2} + \frac{\partial^2 h}{\partial z^2} = \frac{S_s}{K}\frac{\partial h}{\partial t} \quad [\text{L}^{-1}] \tag{3.38}$$

Equations (3.37), (3.34), and (3.38) are also expressed in terms of transmissivities instead of hydraulic conductivities. In such a case, the specific storage term is replaced by the storage coefficient or specific yield and the source/sink term is substituted by $bW(x, y, z, t)$.

Equation (3.38) is recognized as the *equation of heat conduction, equation of flow of electricity, diffusion equation,* and *nonequilibrium equation of groundwater flow.*

In steady flow, the pressure distribution does not change with time. Accordingly, eq. (3.38) becomes

$$\frac{\partial^2 h}{\partial x^2} + \frac{\partial^2 h}{\partial y^2} + \frac{\partial^2 h}{\partial z^2} = 0 \quad [\text{L}^{-1}] \tag{3.39a}$$

or

$$\nabla^2 h = 0 \tag{3.39b}$$

This is the *Laplace equation,* which appears in mathematics and many branches of physics: for steady-state conduction of heat and electricity, for steady diffusion, and in the elastic membrane theory. An equation identical in form to the Laplace equation for saturated flow was developed independently by Jules Dupuit (in 1863), P. Forchheimer (in 1886), and Charles Slichter (in 1899).

The equation of flow in an unconfined aquifer is the nonlinear Boussinesq equation, which is extremely difficult to solve. With certain permissible assumptions, it is linearized to the form of eq. (3.36). All discussion above, as such, holds true for an unconfined aquifer as well, with specific yield being substituted for the storage coefficient.

3.13 UNSATURATED FLOW AND TWO-PHASE FLOW

Darcy's law and the concepts of hydraulic head and hydraulic conductivity in this chapter have been developed with respect to flow through a saturated medium when all the voids are filled with water. However, above the water table through which flow takes place during infiltration, recharge, evaporation and transpiration, and capillary suction, the soil pores are only partly filled with water, the remaining pore spaces being occupied by air. The flow under such conditions is referred to as unsaturated flow. Darcy's law is extended to such flows by incorporating a relative hydraulic conductivity parameter in the relation to represent unsaturated hydraulic conductivity as a function of moisture content.

Similarly when another liquid is present that does not mix with water, the flow of that liquid takes place as a separate phase distinct from water flow. The rate of flow of each phase is controlled by the relative hydraulic conductivity of each liquid and the hydraulic gradient due to each liquid. Darcy's law is accordingly modified with respect to each phase. This topic is discussed in Section 5.6.1 in the context of movement of pollutants through soil and groundwater. Extensive literature both on unsaturated flow and on multiphase flow exists, most of which has been contributed by soil scientists, agricultural engineers, and chemical engineers. Groundwater hydrologists also have joined in these studies with their special interests in contaminant hydrogeology.

PROBLEMS

3.1 Piezometers have been installed at three sites, A, B, and C. The ground elevation at these sites, the length of piezometer tubes, and the height of water within the tubes are given below:

	At A	At B	At C
Ground elevation above mean sea level (m)	150	120	175
Length from ground level to bottom of piezometer (m)	100	82	125
Height of water from bottom of piezometer (m)	32	45	27

Determine the following:

a. Elevation head at A, B, and C above the mean sea level

b. Pressure head at A, B, and C

c. Hydraulic head at A, B, and C

d. Fluid potential at A, B, and C

e. Pressure at B in kPa

f. Hydraulic gradient between A and B, if the two sites are 500 m apart.

3.2 Three piezometers A, B, and C are 1500 m apart from each other. Piezometer B is located due north of piezometer A, and C to the west of line AB as shown in Figure P3.2. The ground surface elevation, the bottom elevation, and depth of water in each piezometer are given below:

	At A	At B	At C
Ground elevation (above mean sea level) (m)	95	125	110
Piezometer bottom elevation (m)	25	75	0
Depth to water level in piezometer from ground surface (m)	5	35	22

a. Determine the pressure heads at A, B, and C.

b. Determine the hydraulic heads above mean sea level at A, B, and C.

c. Draw the groundwater contours.

d. Determine the direction of the groundwater flow that is perpendicular to the contours.

e. Calculate the hydraulic gradient of groundwater.

Figure P3.2

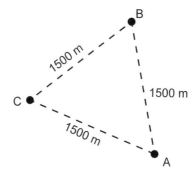

3.3 A confined aquifer slopes gradually from 12 m to 8 m thickness. The slope of piezometric surface is 0.25 m per kilometer. If the hydraulic conductivity of the aquifer is 25 m/day, how much water flows through the aquifer from its width of 2.5 km?

3.4 Flow in a valley takes place as shown in Figure P3.4. The formation in the valley has a hydraulic conductivity of 400 ft/day and a porosity of 0.25. The difference in the water levels in the two wells shown is 1 ft. Between the observation wells, the average depth of the aquifer is 100 ft.

Figure P3.4

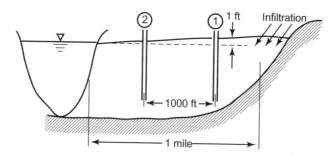

a. Determine the rate of flow per mile width of the aquifer.

b. How long will it take the groundwater to travel from the head of the valley to the stream bank?

3.5 A porous medium is oriented at an angle of 40° with the horizontal plane. If the hydraulic grade line is parallel to the orientation of the medium and the specific discharge is 5 m³/day per square meter, determine the hydraulic conductivity of the medium.

3.6 The hydraulic conductivity of a soil for water flow at 50 °F is 0.015 ft/sec. What is its intrinsic permeability?

3.7 A tracer element was introduced into an aquifer at an upstream location and from its appearance at a downstream location, the average flow velocity was found to be 0.5 in./day. The slope of the piezometric surface was 1 ft/mi. Determine the hydraulic conductivity of the aquifer. What is the intrinsic permeability for water flow in darcy at 40 °F?

3.8 A layered soil is shown in Figure P3.8. Estimate the transmissivity for the formation for vertically downward flow.

Figure P3.8

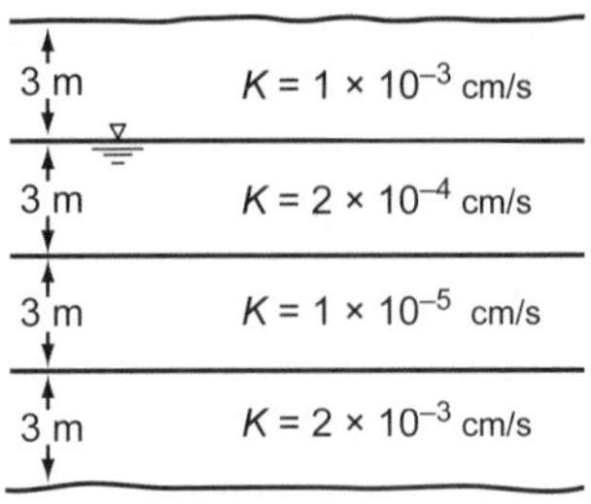

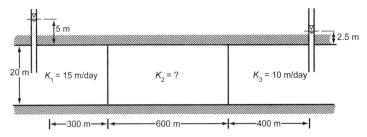

3.9 Two observation wells have been constructed in the formation shown in Figure P3.9. If the horizontal flow rate is 0.01 m³/hr per m width (perpendicular to the paper), determine K_2.

Figure P3.9

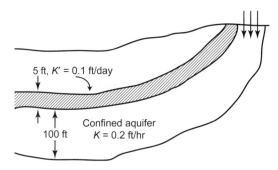

3.10 A semi-impervious layer with hydraulic conductivity of 1 ft/day separates an overlying water-table aquifer with hydraulic conductivity of 40 ft/day from an underlying confined aquifer of hydraulic conductivity 50 ft/day with heads in two aquifers similar to Example 3.2. If the rate of flow between aquifers is 0.02 ft³/hr per unit area of aquifer, determine the thickness of the semi-impervious layer.

3.11 For Figure P3.11, determine the coefficient of leakage and leakage factor.

Figure P3.11

5 ft, K' = 0.1 ft/day

Confined aquifer
100 ft K = 0.2 ft/hr

3.12 What volume of solid material is present in 1 ft³ of sandstone if the porosity of the sandstone is 0.35?

3.13 A soil has a bulk density of 1.7 g/cm³. The specific gravity of soil solids is 2.6. What is the porosity and the void ratio of the soil?

3.14 A wet soil weighs 300 g and occupies a volume of 200 cm³. If it contains 15% moisture by weight:

a. What is its bulk density?

b. What is its porosity for specific gravity of soil grains of 2.65?

c. What is its void ratio?

3.15 For the capillary pressure and volumetric moisture content distribution given below, if the water table was initially 170 cm below the surface and finally it went down to 220 cm, how much water drained from a unit area of the soil? [Use the procedure of Example 3.12 from step 6 onward]

Capillary pressure, cm	0	20	40	60	80	100	120	137	150	170
Volumetric moisture content, %	52	52	50	46.6	41.2	32	28	26	25	25

3.16 The capillary pressure heads and volumetric moisture contents for a soil sample are given below. The water table was initially at 280 cm. If it is lowered by 20 cm, determine the quantity of water released per unit area.

Capillary pressure, cm	0	20	50	80	100	120	150	170	200	230	250	280
Volumetric moisture content, %	44	44	43	40.5	38	35	30	27	22	18	15.5	14

3.17 A 100-g dry soil sample is tested for capillary head moisture distribution. The negative pressure head changes and the incremental amount of water released from the sample are indicated below. The bulk density of the soil is 1.4 g/cc, and at saturation it contained 38% moisture by weight.

Negative Pressure Head (cm)	Water Released (cm³)
0	0.0
10	0.2
20	0.3
30	0.6
40	0.9
50	1.3
60	1.7
70	1.4
80	0.8
90	0.4
100	0.2

a. Calculate the volumetric moisture content at each pressure head (ratio of volume of water retained to volume of soil).

b. Plot a moisture distribution curve.

c. If the water table drops from 100 cm to 150 cm below the surface, determine the amount of water drained from the soil.

Problems 161

3.18 In Problem 3.17,

   a. What is the porosity of the soil?

   b. What is the specific retention of the soil?

   c. What is the specific yield of the soil?

3.19 In a clayey formation, the water table drops by 0.6 m over an area of 8 hectares. What volume of water is drained off if the specific yield is 23%? (1 km^2 = 100 hectares)

3.20 An unconfined aquifer system covers an area of 20 million square meters. The water table is 22 m below the land surface. When 50 million m^3 of water is added to the aquifer, the water table rises by 10 m. What is the specific yield of the aquifer?

3.21 The specific storage of an artesian aquifer is 1.2×10^{-6} per ft and its porosity is 0.35. Determine the modulus of elasticity of soil solids. What percentage of the storage coefficient is due to the expansion of water within the aquifer and what percent by the compression of the aquifer? ($E_W = 3 \times 10^5$ psi)

3.22 From a pumping test on a confined formation, the storage coefficient is found to be 3×10^{-3} for a location having a water depth of 50 m. For an aquifer of an area of 3.2 km^2 with an average depth of 35 m, estimate the volume of water contributed by the area when the pressure head is dropped by 10 m.

3.23 A confined aquifer has an average thickness of 150 ft and a porosity of 0.35. If the compressibility of water is 3×10^{-6} in.^{2}/lb and the compressibility of material is 2.5×10^{-6} in.^{2}/lb, what is the storage coefficient of the aquifer?

3.24 For the confined aquifer in Problem 3.23 the compressibility of the material is not known but the storage coefficient is 0.0003. Determine the compressibility of the material, and also the storage coefficient components contributed by water and material, respectively.

3.25 A well of 6 in. radius is drilled through a confined aquifer of 100 ft thickness. The aquifer consists of uniform sand with average grain size of 0.3 in. The well is pumped at a rate of 100 ft^3/hr. Determine whether Darcy's flow conditions exist near the well at 60 °F.

3.26 A homogeneous, isotropic aquifer has a hydraulic conductivity of 0.02 ft/s. Its head gradients in the x, y, and z directions are 0, 0.19, and 0.87, respectively.

   a. Determine the velocity vector components.

   b. Plot the velocity vectors for the aquifer.

   c. Find the magnitude and direction of the velocity for the aquifer.

3.27 A homogeneous anisotropic aquifer has the following gradients and hydraulic conductivities:

	i	K(ft/min)
x direction	0	0.04
y direction	0.15	0.042
z direction	0.82	0.003

   a. Calculate the Darcy's velocity vector.

   b. Plot the velocity potential vectors.

   c. Compute the specific discharge.

4

Applications and Development of Groundwater Flow

◆◆◆

4.1 GRAPHICAL SOLUTION OF STEADY-STATE FLOW EQUATIONS: THE FLOW NET TECHNIQUE

As discussed in Section 3.12, the Laplace equation represents the steady-state groundwater flow through a homogeneous, isotropic medium. The geometric solution of the Laplace equation in two dimensions consists of two families of curves intersecting each other orthogonally. The lines of equal hydraulic head, called *equipotentials*, and the lines describing flow paths of water particles through the aquifer media, called *streamlines*, represent such a family of curves. Since water moves in the direction of steepest hydraulic gradient, streamlines are perpendicular to equipotential lines. Together they form a *flow net*. A flow net thus represents a solution of the steady-state flow equation for homogeneous, isotropic soil.

4.1.1 Flow Net in Isotropic Soil

The equation for flow nets originates from Darcy's law. However, where Darcy's law cannot be applied directly due to the irregularity of flow zones or directions of flow and difficulty in defining boundaries, the flow net method can be used.

Consider the portion of a flow net drawn in Figure 4.1(a). The flow taking place through a *flow channel* between equipotential lines 1 and 2 as shown in Figure 4.1(b) per unit width (perpendicular to the paper) is (from Darcy's law)

$$\Delta q = K\left(dm_1 \times 1\right)\left(\frac{\Delta h_1}{dl_1}\right) \tag{a}$$

The flow between equipotential lines 2 and 3 is

$$\Delta q = K\left(dm_2 \times 1\right)\left(\frac{\Delta h_2}{dl_2}\right) \tag{b}$$

Since the same flow continues, eq. (a) = eq. (b), and for $dl_1 = dm_1$, $dl_2 = dm_2$, etc.

$$\Delta h_1 = \Delta h_2 \tag{c}$$

that is, the head drop is the same in each potential drop (two successive equipotential lines). If there are n_d such drops, then

Figure 4.1 (a) Portion of a flow net; (b) flow through a single flow channel.

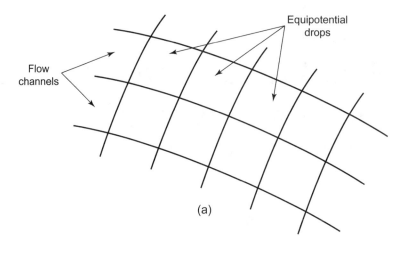

Equipotential drops

Flow channels

(a)

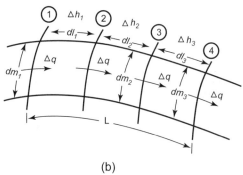

(b)

$$\Delta h = \frac{h}{n_d} \tag{d}$$

where h is the total head loss between the first and last equipotential lines. Substituting (d) in (a) yields

$$\Delta q = K \frac{dm_1}{dl_1} \frac{h}{n_d} \tag{e}$$

Equation (e) is for one flow channel. If there are n_f such channels, the total flow per unit width is

$$q = \frac{n_f}{n_d} K \frac{dm_1}{dl_1} h \tag{f}$$

If the flow net is drawn as squares, so that $dm_1 \approx dl_1$, eq. (f) becomes

$$q = \frac{n_f}{n_d} Kh \quad [\mathrm{L}^2\mathrm{T}^{-1}] \tag{4.1}$$

where

q = rate of flow or seepage per unit width

n_f = number of flow channels

n_d = number of equipotential drops

h = total head loss in flow system

K = hydraulic conductivity

Each potential line indicates the available head, which is equal to total head minus loss of head to that point. If all equipotential lines are assigned a serial number counting the last line on the downstream as zero, then the uplift pressure on the base at the position of the nth equipotential line will be

$$u = \left(\frac{n}{n_d} h + Z \right) \gamma_w \quad [\text{FL}^{-2}] \tag{4.2}$$

where

u = uplift water pressure

n = serial number of the equipotential line counting the
     last line on the downstream as zero

n_d = number of potential drops

h = total head loss in flow system

Z = depth of the base below the datum (if the
     base is above the datum, Z is negative)

4.1.2 Procedure to Draw a Flow Net

Drawing a flow net is a trial procedure. Certain rules to be observed are:

1. Equipotential lines cross flow lines at a right angle.
2. Shapes formed by equipotential lines and flow lines should be as close to square as possible.
3. Impermeable boundaries are flow lines.
4. The soil-water interface at upstream and downstream of a structure is an equipotential line.
5. The seepage surface is a flow line.

The method of sketching the flow net is as follows:

1. Draw to a convenient scale the cross section of the structure, water elevations, and soil profiles.
2. Establish boundary conditions and draw flow lines and equipotential lines for the boundaries.
3. Sketch in intermediate flow lines and equipotential lines by smooth curves adhering to right-angle intersections and square figures. Where the direction of flow is in a straight line, flow lines are an equal distance apart and parallel.

4. Continue sketching until an inconsistency develops. Each inconsistency will indicate changes to be made in flow lines and equipotential lines. Successive trials will result in a reasonably consistent flow net.

5. It is for the student to decide the number of flow lines to be drawn. Three to five flow lines are usually sufficient. Depending on the number of flow lines selected, the number of equipotential lines will be automatically fixed because of the requirement for the square figures of the net.

EXAMPLE 4.1

A dam is constructed on a permeable stratum underlain by an impermeable rock as shown in Figure 4.2. A row of sheet pile is installed at the upstream face. If the permeable soil has a hydraulic conductivity of 150 ft/day, determine **(a)** the rate of flow from upstream to downstream, and **(b)** the uplift pressure acting in the bottom of the dam.

SOLUTION

The flow net is drawn as shown in Figure 4.2. $n_f = 5$, $n_d = 17$.

(a) From eq. (4.1),

$$q = \frac{n_f}{n_d} Kh$$

$$= \frac{5}{17}(150)(35) = 1554 \text{ ft}^3/\text{day per foot} \quad \text{or} \quad 0.018 \text{ cfs per foot}$$

(b) From eq. (4.2),

$$u = \left(\frac{n}{n_d}h + Z\right)\gamma_w$$

$$= \left[\frac{n}{17}(35) + 2\right]62.4$$

$$= (2.06n + 2)(62.4) \text{ psf} \quad \text{or} \quad 0.0624(2.06n + 2)\text{ksf}$$

Position:	A	B	C	D	E	F	G	H	I	J
Distance from front toe (ft)	0	3	22	37.5	50	62.5	75	86	94	100
n		16.5	9	8	7	6	5	4	3	2	1.2
u (ksf)		2.25	1.28	1.15	1.02	0.90	0.77	0.64	0.51	0.38	0.28

4.1.3 Flow Net in Anisotropic Soil

In anisotropic soil when K_x is not equal to K_z, the groundwater flow equation is not of the Laplace form and the flow net is not the solution to the equation. However, by transposing $x' = \sqrt{K_z/K_x}\,x$, the flow equation can be converted to Laplace form in terms of x' and z. Thus, the flow net solution for an anisotropic soil can be obtained by distorting the horizontal scale for the system.

The procedure is as follows:

Figure 4.2 (a) Flow net for a dam with sheet pile; (b) distribution of uplift pressure.

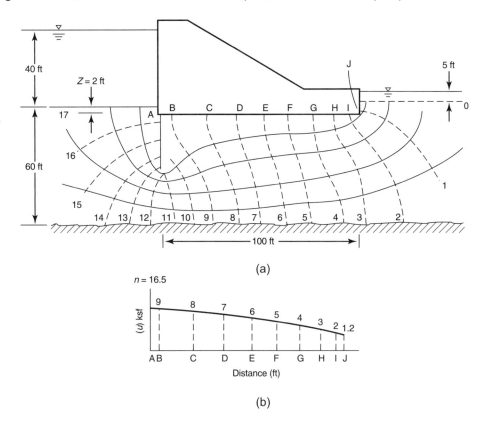

(a)

(b)

1. Transform all horizontal dimensions into notional dimensions using

$$x' = \sqrt{K_z / K_x}\, x \quad [\text{L}] \tag{4.3}$$

   where

   x = natural horizontal dimensions

   x' = notional horizontal dimensions

2. Draw the cross section of the structure in natural vertical dimensions and notional (distorted) horizontal dimensions, to a convenient scale.

3. Sketch the flow net as usual.

4. Determine the rate of flow using

$$q = \frac{n_f}{n_d}\sqrt{K_x K_z}\, h \quad [\text{L}^2\text{T}^{-1}] \tag{4.4}$$

EXAMPLE 4.2

A concrete dam is found on a permeable stratum having $K_x = 30$ m/day and $K_z = 6$ m/day, as shown in Figure 4.3(a). Determine **(a)** the rate of flow below the dam, and **(b)** the distribution of the uplift pressure.

SOLUTION

The factor for converting horizontal dimensions $= \sqrt{\dfrac{K_z}{K_x}} = \sqrt{\dfrac{6}{30}} = 0.45$

Length of dam $= 30 \times 0.45 = 13.5$ m

The profile of the structure with distorted x scale is shown in Figure 4.3(b).
The flow net has also been sketched in Figure 4.3(b).

(a) $n_f = 4$, $n_d = 8$. From eq. (4.4),

$$q = \frac{n_f}{n_d}\sqrt{K_x K_z}\, h$$

$$= \frac{4}{8}\sqrt{(30)(6)}\,(10) = 67 \text{ m}^3/\text{day per meter width}$$

(b) From eq. (4.2),

$$u = \left(\frac{n}{n_d}h + Z\right)\gamma_w$$

$$= \left[\frac{n}{8}(10) + 1\right](9.81)$$

$$= (1.25n + 1)(9.81) \text{ kN/m}^2$$

Position:	A	B	C	D	E	F	G
Distance on distorted model (m)	0	0.65	3.5	6.75	9.65	12.2	13.5
Natural distance (m)	0	1.45	7.8	15	21.4	27	30
n	7	6	5	4	3	2	1
u (kN/m²)	95.65	83.4	71.1	58.9	46.6	34.3	22.1

4.2 ANALYTICAL SOLUTION OF STEADY-STATE FLOW EQUATIONS

Groundwater flow equations are summarized in Section 3.12. When an appropriate equation is solved to satisfy the initial and boundary conditions of a given flow system, the distribution of head throughout the system is obtained. These equations are, however, difficult to solve for many flow problems since the theory of partial differential equations is very limited. The steady-state flow is represented by the Laplace equation, for which a general solution can be obtained from the theory of partial differential equations. For actual flow problems, however, the solutions should meet certain boundary conditions or specifications regarding the distribution and variation of head at one or more boundaries or the rate of flow across them. Solutions to some problems have been obtained either by adopting some

Figure 4.3 (a) Profile of a dam on an isotropic stratum; (b) transformed section and flow net for a dam.

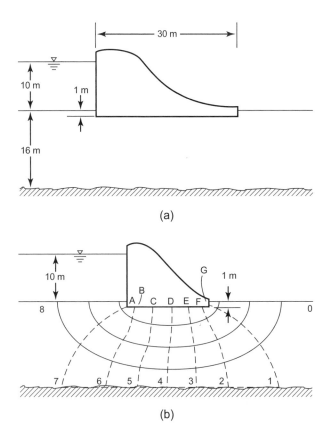

(a)

(b)

approximate differential equations that are easier to solve *or* by idealizing the conditions for the system. A few cases are analyzed below. The medium has been considered homogeneous and isotropic in all cases.

4.2.1 Groundwater Flow between Two Water Bodies

Figure 4.4 shows a one-dimensional confined flow model. Equation (3.39a) in one-dimensional steady-state form is:

$$\frac{\partial^2 h}{\partial x^2} = 0 \tag{a}$$

The general solution of this is

$$h = Bx + C \tag{b}$$

Substituting the boundary condition in (b): At $x = 0$, $h = h_0$:

$$h_0 = B(0) + C \quad \text{or} \quad C = h_0 \tag{c}$$

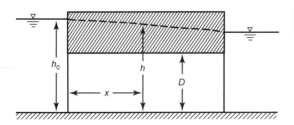

Figure 4.4 Confined flow between two water bodies.

Differentiating (b) yields

$$\frac{dh}{dx} = B \qquad \text{(d)}$$

Darcy's law for unit width (perpendicular to paper); negative sign since h is decreasing in the direction of x

$$Q = -KD\frac{dh}{dx} \qquad \text{(e)}$$

Substituting (d) in (e), we obtain

$$B = -Q/KD \qquad \text{(f)}$$

Substituting (c) and (f) in (b) gives us

$$Q = \frac{h_0 - h}{x}KD \quad [\text{L}^2\text{T}^{-1}] \qquad \text{(4.5)}$$

Equation (4.5) also can be applied to unconfined flow between two water bodies if the difference of water levels between two water bodies is small compared to the water table depth. In that case D will be the average wetted thickness of the aquifer. The equation is for the unit width of the aquifer.

EXAMPLE 4.3

A channel runs parallel to a river. The water level in the river is at an elevation of 120 ft and in the channel at an elevation of 110 ft. The river and channel are 2000 ft apart. A confined aquifer of 30 ft thickness joins them. The hydraulic conductivity is 0.25 ft/hr. Determine the rate of seepage from the river to the channel.

SOLUTION

From eq. (4.5),

$$Q = \frac{h_0 - h}{x}KD$$

$$= \frac{120 - 110}{2000}\frac{\text{ft}}{\text{ft}}\left(0.25\frac{\text{ft}}{\text{hr}}\right)(30 \text{ ft})$$

$$= 0.0375 \text{ ft}^3/\text{hr} \quad \text{or} \quad 0.9 \text{ ft}^3/\text{day per ft width}$$

EXAMPLE 4.4

Consider the cross section shown in Figure 4.5. Determine the rate of flow.

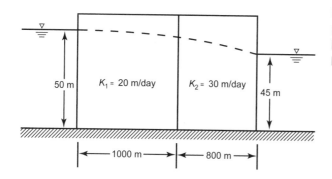

Figure 4.5 Unconfined nonhomogeneous aquifer between two water bodies, Example 4.4.

SOLUTION

$$\text{Average } \overline{K} = \frac{b}{b_1/K_1 + b_2/K_2} = \frac{1800}{1000/20 + 800/30} = 23.48 \text{ m/day}$$

From eq. (4.5),

$$Q = \left(23.48\frac{\text{m}}{\text{day}}\right)\left(\frac{50-45}{1800}\frac{\text{m}}{\text{m}}\right)\left(\frac{50+45}{2}\text{m}\right)$$

$$= 3.1 \text{ m}^3/\text{day per meter}$$

ALTERNATIVE SOLUTION

1. Suppose that the head is h at the interface of the two soil media.
2. Per meter width of the first medium, from eq. (4.5),

$$Q = (20)\left(\frac{50-h}{1000}\right)\left(\frac{50+h}{2}\right)$$

$$= \left(\frac{1}{100}\right)\left(50^2 - h^2\right) \qquad (a)$$

3. Per meter width of the second medium, from eq. (4.5),

$$Q = (30)\left(\frac{h-45}{800}\right)\left(\frac{h+45}{2}\right)$$

$$= \left(\frac{3}{160}\right)\left(h^2 - 45^2\right) \qquad (b)$$

4. Equating eqs. (a) and (b) gives us

$$\frac{1}{100}\left(50^2 - h^2\right) = \frac{3}{160}\left(h^2 - 45^2\right)$$

$$h = 46.8 \text{ m}$$

5. Substituting in eq. (a) yields

$$Q = 3.1 \text{ m}^3/\text{day per meter}$$

4.2.2 Confined Flow to a Well

Flow toward a well in a homogeneous and isotropic aquifer is radially symmetric. When the well screen, perforated pipe, or open well bore extends through the entire thickness of an aquifer, it is known as a *fully penetrating well*, as shown in Figure 4.6. Usually, a gravel pack is provided around the well screen. In such cases the well radius is considered from the center of the well to the outside of the gravel pack. Loss of head in the well and gravel pack is known as *well loss;* this quantity is generally very small.

Equation (3.39a) refers to the steady-state flow in a homogeneous, isotropic aquifer, which is reproduced below in two dimensions:

$$\frac{\partial^2 h}{\partial x^2} + \frac{\partial^2 h}{\partial y^2} = 0 \tag{a}$$

Polar coordinates are convenient for the problems concerning radial flow, as indicated in Figure 4.7.

$$x = r \cos\theta$$
$$y = r \sin\theta$$
$$r = \left(x^2 + y^2\right)^{1/2} \quad \theta = \tan^{-1}\frac{y}{x} \tag{b}$$

Using the chain rule of differentiation gives

Figure 4.6 Fully penetrating well in a confined aquifer.

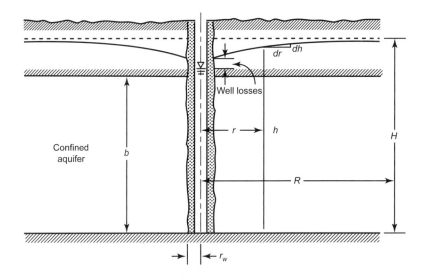

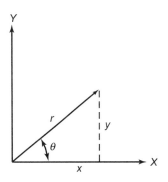

Figure 4.7   Polar coordinates.

$$\frac{\partial^2 h}{\partial x^2} = \frac{x^2}{r^2}\frac{\partial^2 h}{\partial r^2} + \frac{y^2}{r^3}\frac{\partial h}{\partial r} + \frac{y^2}{r^4}\frac{\partial^2 h}{\partial\theta^2} - \frac{2xy}{r^4}\frac{\partial h}{\partial\theta}$$

$$\frac{\partial^2 h}{\partial y^2} = \frac{y^2}{r^2}\frac{\partial^2 h}{\partial r^2} + \frac{x^2}{r^3}\frac{\partial h}{\partial r} + \frac{x^2}{r^4}\frac{\partial^2 h}{\partial\theta^2} + \frac{2xy}{r^4}\frac{\partial h}{\partial\theta} \qquad \text{(c)}$$

Adding the two equations above and substituting $x^2 + y^2 = r^2$ yield

$$\frac{\partial^2 h}{\partial r^2} + \frac{1}{r}\frac{\partial h}{\partial r} + \frac{1}{r^2}\frac{\partial^2 h}{\partial\theta^2} = 0 \qquad \text{(d)}$$

When flow is directed toward or originates from the origin of the coordinate system, it is independent of θ, and eq. (d) above reduces to the following ordinary differential equation, since the head is a function of the radial coordinate only:[*]

$$\frac{d^2 h}{dr^2} + \frac{1}{r}\frac{dh}{dr} = 0 \quad [\text{L}^{-1}] \qquad \text{(4.6)}$$

This is the groundwater flow equation in polar or cylindrical coordinates.

Equation (4.6) can be written in the following form:

$$\frac{1}{r}\frac{d}{dr}\left(r\frac{dh}{dr}\right) = 0 \qquad \text{(e)}$$

Integrating (e) yields

$$r\frac{dh}{dr} = C_1 = \text{constant} \qquad \text{(f)}$$

The constant C_1 has to be evaluated from the boundary condition. The periphery of a circle of radius r is $2\pi r$. For thickness b, the peripheral area will be $2\pi rb$. According to Darcy's law, the flow at the periphery will thus be

$$Q = K\left(\frac{dh}{dr}\right)(2\pi rb)$$

or

$$r\left(\frac{dh}{dr}\right) = \frac{Q}{2\pi bK} \qquad \text{(g)}$$

[*] For direct derivation of this equation, refer to Lohman (1972) or any other groundwater text.

Section 4.2   Analytical Solution of Steady-State Flow Equations

Comparing (f) and (g),

$$C_1 = \frac{Q}{2\pi bK}$$

Rearranging (f) after substituting C_1

$$dh = \frac{Q}{2\pi bK}\frac{dr}{r} \qquad \text{(h)}$$

Integrating (h) again, we have

$$h = \frac{Q}{2\pi bK}\ln r + C_2 \qquad \text{(i)}$$

For the initial boundary condition at a radial distance R from the well, the head is H (as the boundary on an island or at the end of the cone of depression). Substituting in (i) gives*

$$\left(H-h\right) = \frac{Q}{2\pi bK}\ln\frac{R}{r} \quad [\text{L}] \qquad \text{(4.7a)}$$

rearranging:

$$Q = \frac{2\pi bK\left(H-h\right)}{\ln\dfrac{R}{r}} \quad [\text{L}^3\text{T}^{-1}] \qquad \text{(4.7b)}$$

where

H = piezometric head at radial distance R

h = piezometric head at any distance r

Q = discharge from the well

b = thickness of confined aquifer

K = hydraulic conductivity of aquifer

bK = transmissivity of aquifer

This form of the equation is known as the Theim equation. Field tests for determining the coefficient of permeability are made under steady-state conditions wherein water levels in test and observation wells are stabilized after a long period of pumping. Equation (4.7b) is used to determine the coefficient of permeability. For a known value of K, this formula is

---

* Equation (4.7a) can be derived easily by direct application of Darcy's law. Referring to Fig. 4.6 and applying Darcy's law at a distance r, we have

$$Q = 2\pi rbK\frac{dh}{dr}$$

$$\int_h^H dh = \frac{Q}{2\pi bK}\int_r^R \frac{dr}{r}$$

$$H-h = \frac{Q}{2\pi bK}\ln\frac{R}{r}$$

the same as eq. (4.7a).

used to compute the discharge. Instead of pumping, the test is sometimes performed by adding water into the well. Equation (4.7b) holds good for that case also.

EXAMPLE 4.5

A well is pumped from a confined aquifer at a rate of 0.08 m³/s for a long time. In two observation wells located 50 m and 10 m away from the well, the difference in water elevation has been observed as 1.5 m. What is the transmissivity of the aquifer?

SOLUTION

1. $H - h =$ difference in water elevation

   $= 1.5$ m

2. Rearranging eq. (4.7b)

$$T = bK = \frac{Q}{2\pi(H-h)}\ln\frac{R}{r}$$

$$= \frac{(0.08 \text{ m}^3/\text{s})}{2\pi(1.5 \text{ m})}\ln\frac{50 \text{ m}}{10 \text{ m}}$$

$$= 0.0137 \text{ m}^2/\text{s}$$

EXAMPLE 4.6

An aquifer of 20 m average thickness is overlain by an impermeable layer of 30 m thickness. A test well of 0.5 m diameter and two observation wells at a distance of 10 m and 60 m from the test well are drilled through the aquifer. After pumping at a rate of 0.1 m³/s for a long time, the following *drawdowns* (amount that head is lowered from the initial position) are stabilized in these wells: first observation well, 4 m; second observation well, 3 m. Determine the hydraulic conductivity of the aquifer and the drawdown in the test well.

SOLUTION

1. Arrangement is shown in Figure 4.8.

   $H = (Z - 3)$ and $h = (Z - 4)$ as shown in Fig. 4.8

   hence, $(H - h) = (Z - 3) - (Z - 4) = 1$ m

2. From eq. (4.7a),

$$1 = \frac{0.1}{2\pi(20)K}\ln\frac{60}{10}$$

$$K = \frac{0.1 \text{ m}^3/\text{s}}{2\pi(20 \text{ m})(1 \text{ m})}\ln\frac{60 \text{ m}}{10 \text{ m}}$$

$$= 1.43\times10^{-3} \text{ m/s}$$

3. Apply eq. (4.7a) between the test well and the first observation well.

   $H = (Z - 4)$, $h = (Z - s)$

   hence, $(H - h) = (Z - 4) - (Z - s) = (s - 4)$

Figure 4.8 Testing of a confined aquifer, Example 4.6.

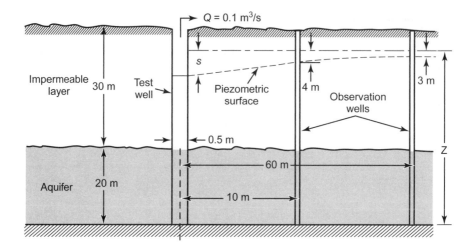

4. From eq. (4.7a),

$$(s-4) = \frac{0.1}{2\pi(20)(1.43 \times 10^{-3})} \ln\frac{10}{0.25}$$

$$= 2.05$$

$$\text{or } s = 6.05 \text{ m}$$

4.2.3 Unconfined Flow to a Well

Analysis of unconfined aquifers is made on the assumptions of Dupuit that consider (1) the flow to be horizontal, and (2) the velocity of flow to be proportional to the tangent of the hydraulic gradient instead of its sine. An essential difference between confined flow and unconfined flow is that in unconfined flow, as the water table slopes, the saturated thickness changes and the area of cross section of flow varies as shown in Figure 4.9.

Equation (4.6) represents the steady-state groundwater flow equation for an unconfined aquifer as well. As previously noted, integration of eq. (4.6) provides

$$r\frac{dh}{dr} = C_1 \tag{a}$$

Substituting a similar boundary condition as for the confined case, that is, radial flow at the periphery of a circle of radius r having saturated thickness, h,

$$Q = 2\pi rhK\frac{dh}{dr} \tag{b}$$

(Note that saturated thickness here is h, not the aquifer thickness b as for the confined aquifer.) Substituting in eq. (a) yields

Applications and Development of Groundwater Flow   Chapter 4

Figure 4.9 Flow in an unconfined aquifer.

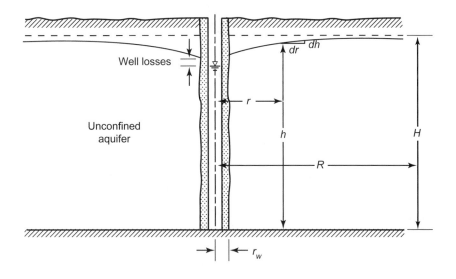

$$C_1 = \frac{Q}{2\pi hK}$$

Hence

$$r \frac{dh}{dr} = \frac{Q}{2\pi hK}$$

or

$$h\,dh = \frac{Q}{2\pi K}\frac{dr}{r} \tag{c}$$

Integrating eq. (c) gives

$$h^2 = \frac{Q}{\pi K}\ln r + C_2 \tag{d}$$

For boundary conditions in eq. (d), if at a radial distance R, the head is H, then*

$$H^2 - h^2 = \frac{Q}{\pi K}\ln \frac{R}{r} \quad [L^2] \tag{4.8a}$$

---

* For a direct derivation, apply Darcy's law at a distance r (see Fig. 4.9):

$$Q = 2\pi rhK\frac{dh}{dr}$$

$$\int_h^H h\,dh = \frac{Q}{2\pi K}\int_r^R \frac{dr}{r}$$

$$H^2 - h^2 = \frac{Q}{\pi K}\ln \frac{R}{r}$$

Rearranging

$$Q = \frac{\pi K\left(H^2 - h^2\right)}{\ln R/r} \quad \left[L^3 T^{-1}\right] \tag{4.8b}$$

Equation (4.8) is used in a similar manner as the confined aquifer eq. (4.7); that is, to assess the hydraulic coefficient by performing field tests or to compute the steady-state discharge.

EXAMPLE 4.7

A fully penetrating 12-in. diameter well has its bottom 80 ft below the static water table. After 24 hours of pumping at 1100 gpm, the water level in the test well stabilizes at 10 ft below the static water table. A drawdown of 3.65 ft is noticed in an observation well 320 ft away from the test well. Determine the hydraulic conductivity of the aquifer.

SOLUTION

1. Refer to Figure 4.10.

$$Q = 1100 \text{ gpm} = 2.45 \text{ cfs}$$

$$H = 80 - 3.65 = 76.35 \text{ ft}$$

$$h = 80 - 10 = 70 \text{ ft}$$

2. Rearranging eq. (4.8b),

$$K = \frac{Q}{\pi\left(H^2 - h^2\right)} \ln \frac{R}{r}$$

$$= \frac{2.45 \text{ ft}^3/\text{s}}{\pi\left(76.35^2 - 70.0^2 \text{ ft}^2\right)} \ln \frac{320 \text{ ft}}{0.5 \text{ ft}}$$

$$= 0.0054 \text{ ft/s} \quad \text{or} \quad 469 \text{ ft/day}$$

4.2.4 Groundwater Travel Time

Refer to Figure 4.6 or 4.9. Assume that a water particle takes time dt to move a distance dr:

$$\text{velocity of water movement}, \quad v_r = \frac{dr}{dt} \tag{a}$$

According to Darcy's law,

$$\text{Darcy velocity}, \quad v = -K\frac{dh}{dr}$$

$$\text{seepage velocity}, \quad v_s = \frac{v}{\eta} = -\frac{K}{\eta}\frac{dh}{dr} \tag{b}$$

Equating (a) and (b) yields

Figure 4.10 Test well in an unconfined aquifer.

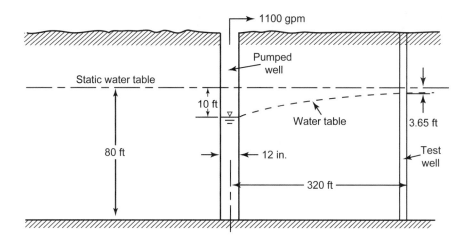

$$dt = -\frac{\eta}{K}\frac{dr}{dh}dr \qquad\qquad (c)$$

For flow toward a well, the radial discharge from the periphery of a circle of radius r and depth D is found by Darcy's law:

$$Q = K\frac{dh}{dr}2\pi r\, D$$

or

$$\frac{1}{K}\frac{dr}{dh} = \frac{2\pi Dr}{Q} \qquad\qquad (d)$$

Substituting eq. (d) in eq. (c) and integrating gives us

$$\int_0^t dt = \int_R^r -\frac{2\pi Dr}{Q}\eta dr$$

$$t = \frac{\pi D\eta}{Q}\left(R^2 - r^2\right)\ \ [T] \qquad\qquad (4.9)$$

where

t = time of travel from R to r

r = any radial distance

R = radial distance at the boundary from where the time
of travel is to be computed

D = thickness of the confined aquifer, b, or average saturated
thickness between radial distances R and r

η = porosity

EXAMPLE 4.8

In Example 4.6, determine the time of travel of groundwater from the observation well at a distance 60 m to the pumped well if the porosity of the aquifer is 0.3.

SOLUTION

From eq. (4.9),

$$t = \frac{\pi D \eta}{Q}\left(R^2 - r^2\right)$$

$$= \frac{\pi(20 \text{ m})(0.3)}{0.1 \text{ m}^3/\text{s}}\left(60^2 - 0.25^2 \text{ m}^2\right)$$

$$= 678 \times 10^3 \text{ sec} \quad \text{or} \quad 7.8 \text{ days}$$

4.3 ANALYTICAL SOLUTION OF UNSTEADY-STATE FLOW EQUATIONS

Equations (3.34), (3.36), and (3.38) represent transient or unsteady-state flow conditions. Certain boundary value problems for simplified cases have been solved using these equations. An extensive application of eq. (3.38), in terms of polar coordinates, has, however, been made on problems relating to radial flow toward wells; this has practical significance. The unconfined flow, in the form of the linear Boussinesq equation, becomes identical to the confined flow equation. Thus, for the mathematical solution of two kinds of flow, no distinction is required. The unconfined flow, however, imposes certain limitations that have to be recognized and specifically handled, as discussed subsequently.

4.3.1 Unsteady Flow to a Well: Theis Equation

Following the approach of Section 4.2.2 for conversion into polar coordinates, eq. (3.38) can be written in two dimensions as

$$\frac{\partial^2 h}{\partial r^2} + \frac{1}{r}\frac{\partial h}{\partial r} = \frac{S}{T}\frac{\partial h}{\partial t} \quad [\text{L}^{-1}] \tag{4.10}$$

or, in terms of drawdown,

$$\frac{\partial^2 s}{\partial r^2} + \frac{1}{r}\frac{\partial s}{\partial r} = \frac{S}{T}\frac{\partial s}{\partial t} \quad [\text{L}^{-1}] \tag{4.11}$$

Equation (4.11) has been solved by first converting it to base u defined by the Boltzmann variable (i.e., $u = r^2 S/4Tt$). This will reduce it to an ordinary differential equation of the type of eq. (4.6). Its double integration with constants evaluated by appropriate boundary conditions will yield the solution.

C. V. Theis, in 1935, was the first to obtain the solution for eq. (4.11) based on an analogy between groundwater flow and heat conduction. He based his solution on the following assumptions and for the initial and boundary conditions indicated below.

Assumptions

1. The aquifer is homogeneous, isotropic, and of infinite extent (this is a built-in assumption of the groundwater flow equation).

2. The transmissivity of the aquifer is practically constant.

3. The water derived is entirely from storage and is released instantaneously with decline of head.

4. The well penetrates the entire thickness of the aquifer, and its diameter is very small compared to pumping rates, so that storage in the well is negligible.

Initial and boundary conditions

1. At time = 0, drawdown = 0, at any distance.

2. At time > 0, drawdown = 0, at infinite distance.

3. At the well face, r_w, $\partial s/\partial r = -Q/2\pi r_w T$ (according to Darcy's law, i.e., flow into the well is equal to its discharge).

The solution to eq. (4.11) is

$$s = \frac{Q}{4\pi T} \int_u^\infty \frac{e^{-u}}{u} du \quad [\text{L}] \tag{4.12}$$

This is known as the Theis equation, and the exponential integral is referred to as the *well function*, $W(u)$; thus the set of equations is

$$s = \frac{Q}{4\pi T} W(u) \quad [\text{L}] \tag{4.13}$$

when

$$u = \frac{r^2 S}{4Tt} \tag{4.14}$$

$$W(u) = -0.5772 - \ln u + u - \frac{u^2}{2.2!} + \frac{u^3}{3.3!} - \cdots \tag{4.15}$$

where

r = any distance from the center of pumping well

t = any time when drawdown is s

S = storage coefficient

T = transmissivity

The set of equations above is generally used to determine the hydraulic properties of transmissivity and storage coefficient of an aquifer. The required values of other variables to solve the equations above are obtained from aquifer testing in the field. The aquifer test comprises pumping a well at a constant rate for a period ranging from several hours to several days and measuring the change in water levels at fixed time intervals in observation wells located at different distances from the pumped well.

The observation wells are used for measurements because it is easy to get stable readings in them. On the other hand, it is difficult to measure the level in a production well while it is being pumped. Also, flow conditions exist in the vicinity of a pumped or production well that cause additional drawdown. However, where observation wells are not available, production wells are used to obtain useful data as discussed in Section 4.10. The Theis equation is a significant contribution since it permits making tests in much less time in an unsteady state itself without waiting to achieve steady-state conditions.

Many procedures have been suggested to solve the preceding equations, including the one suggested by Theis himself. Earlier methods were related to procedures for solving the foregoing equations only. Subsequently, refinements were introduced by making modifications in the equations, particularly the well function $W(u)$, to reflect actual conditions of flow systems as they differ from the assumptions listed earlier.

4.3.2 Aquifer-Test Analysis

Aquifer-test analysis involves applying field data from an aquifer or pumping test to compute hydraulic properties of the aquifer. Equation (4.13) has a general form

$$s = \frac{Q}{4\pi T} W(u, \alpha, \beta, \ldots) \quad [\text{L}] \tag{4.16}$$

where

$$u = r^2 S / 4Tt$$

$\alpha, \beta =$ dimensionless factors to define particular aquifer-system conditions

In general procedure, a graph from the field data is prepared between s versus t/r^2 (or r^2/t). Instead of drawdown, the recovery data after pumping ceases could be used to prepare this curve. Standard curves are drawn between W and u for various controlled values of α, β, . . . , which are known as the *type curves*. Utilizing a curve-matching process between a type curve and a field data curve, as explained in the next section, eq. (4.16) is solved for the values of T and S.

There can be many site conditions in a well-aquifer system, as listed below.

I. Areal extent of aquifer
    1. Aquifer of infinite extent
    2. Aquifer bound by an impermeable boundary
    3. Aquifer bound by a recharge boundary
II. Depth of well
    1. Fully penetrating well
    2. Partially penetrating well
III. Confined aquifer
    1. Nonleaky aquifer
    2. Leaky confining bed releasing water from storage
    3. Leaky confining bed not yielding water from storage but transmitting water from overlying aquifer
    4. Leaky aquifer in which head in the overlying aquifer changes
IV. Unconfined Aquifer
    1. Aquifer in which significant dewatering (reduction in saturated thickness) occurs
    2. Aquifer in which vertical flow occurs near the well
    3. Aquifer with delayed yield (i.e., water from storage does not release quickly)

The combination of a condition from one category with any other condition of another category can lead to numerous site conditions for which a special type curve or set of curves has to be developed. The contributions in this regard have been significant; more than 100 papers have been written dealing with different situations. Theis began by introducing the concept of the type curve. Jacob and Hantush advanced the theory to cover leaky aquifer problems and produced many type curves for such cases. Neuman and Witherspoon also made contributions to leaky aquifers. Boulton did extensive research on unconfined aquifers and developed many tables and type curves for vertical flow and delayed yield. Neuman and Streltsova also made significant contributions in unconfined flow hydraulics.

Reed (1980) has compiled the type-curve solutions for 11 conditions of flow in confined aquifers. Marino and Luthin (1982) provided detailed coverage regarding unconfined aquifers.

The selection of a proper type curve or set of curves is imperative for the data analysis. A type curve that fits the site conditions should be used. Stallman (1971) found that an error of many orders of magnitude could be committed by improper use of type curves.

4.4 UNSTEADY-STATE ANALYSIS OF CONFINED AQUIFERS

This confined aquifer analysis is applicable to an impermeable (nonleaky) aquifer of infinite extent. Besides the type-curve method originally developed by Theis, another method of analysis is also described below.

4.4.1 Type-Curve Method

Taking the logarithm of eq. (4.13),

$$\log s = \left[\frac{Q}{4\pi T}\right] + \log W(u) \tag{a}$$

Rearranging eq. (4.14),

$$\frac{t}{r^2} = \frac{S}{4T}\frac{1}{u}$$

Taking the logarithm,

$$\log \frac{t}{r^2} = \left[\log \frac{S}{4T}\right] + \log \frac{1}{u} \tag{b}$$

For a constant Q, the bracketed parts of equations (a) and (b) are constant. Thus if a constant equal to $\log(Q/4\pi T)$ is added to $\log W(u)$, $\log s$ is obtained. Similarly, when $\log(S/4T)$ is added to $\log(1/u)$, the result is $\log(t/r^2)$. In other words, a graph between $\log W(u)$ and $\log 1/u$ is similar to a graph between $\log s$ and $\log t/r^2$. It is offset by constant amounts, as shown in Figure 4.11.

The procedure is summarized below.

1. Prepare a plot on log-log paper of $W(u)$ (on vertical coordinates) and $1/u$ (on horizontal coordinates). This is known as the type curve. For various values of u, $W(u)$ can be calculated from eq. (4.15) or more conveniently from Table 4.1. Three type curves are shown in Figure 4.12; curve A covers the range of $1/u$ from 10^{-1} to 10^2, curve B from 10^2 to 10^5, and curve C from 10^5 to 10^8.

2. From given pumping test data, prepare a plot, on transparent log-log paper, of drawdown, s versus t/r^2.* The length of each cycle of this log-log paper should be the same as that used for the type curve of step 1. This is known as the *data curve*.

---

* There are many other ways in which the type curve and the data curve are plotted. Two of these are: (1) plot $W(u)$ and u for the type curve and s and r^2/t for the data curve; and (2) plot $W(u)$ and $1/u$ for the type curve and s and t for the data curve. The plotting of $1/u$ and t/r^2 is, however, preferred because it eliminates the necessity for computing $1/t$ for various values of s.

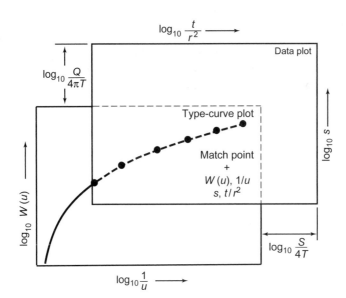

Figure 4.11 Relation of $W(u)$ versus $1/u$ and s versus t/r^2.

Data for this curve are obtained from a pumping test in which discharge is kept constant. Drawdowns can be observed in an observation well at any distance r for different time intervals; that is, r is constant and time varies. Thus the drawdown-time analysis is made or alternatively, drawdowns can be observed at the same time in wells located at different distances, thus involving drawdown-distance analysis. In both cases t/r^2 is computed and plotted against s in the form of a data curve.

3. The data plot is superimposed (placed) over the type-curve plot. The data curve plot is moved up or down, right or left, keeping its x and y axes parallel to the type-curve axes, until the data curve overlaps a certain portion of the type curve.

4. Any arbitrary point is selected on the overlapping part of two sheets (plots). This point need not be on the curves. It is often convenient to select a point on the type curve whose coordinates are a multiple of 10. Record $W(u)$ and $1/u$ coordinates, and the corresponding s and t/r^2 coordinates, of this matching point.

5. Transmissivity is computed from eq. (4.13), rearranged as

$$T = \frac{Q}{4\pi s}W(u) \quad [\text{L}^2\text{T}^{-1}] \tag{4.17}$$

and the storage coefficient from eq. (4.14), rearranged as

$$S = 4T\frac{t}{r^2}u \quad [\text{dimensionless}] \tag{4.18}$$

In the type curve, the maximum variation in $W(u)$ takes place in the range of $1/u$ of 10^{-1} to 10^2 (type curve A, Figure 4.12) when from almost vertical the curve becomes almost horizontal. Gradually, the curve becomes more flat (horizontal). In many instances, the data curve might match curve A. The data curve should be visually compared with Figure 4.12 to decide which of the type curves might be appropriate for a comparison.

Table 4.1 Values of Well Function $W(u)$ for Values of $1/u$

$1/u$	$1/u \times 10^{-1}$	1	10	10^2	10^3	10^4	10^5	10^6
1.0		0.21938	1.82292	4.03793	6.33154	8.63322	10.93572	13.23830
1.2	0.00003	0.29255	1.98932	4.21859	6.51369	8.81553	11.11804	13.42062
1.5	0.00017	0.39841	2.19641	4.44007	6.73667	9.03866	11.34118	13.64376
2.0	0.00115	0.55977	2.46790	4.72610	7.02419	9.32632	11.62886	13.93144
2.5	0.00378	0.70238	2.68126	4.94824	7.24723	9.54945	11.85201	14.15459
3.0	0.00857	0.82889	2.85704	5.12990	7.42949	9.73177	12.03433	14.33691
3.5	0.01566	0.94208	3.00650	5.28357	7.58359	9.88592	12.18847	14.49106
4.0	0.02491	1.04428	3.13651	5.41675	7.71708	10.01944	12.32201	14.62459
5.0	0.04890	1.22265	3.35471	5.63939	7.94018	10.24258	12.54515	14.84773
6.0	0.07833	1.37451	3.53372	5.82138	8.12247	10.42490	12.72747	15.03006
7.0	0.11131	1.50661	3.68551	5.97529	8.27659	10.57905	12.88162	15.18421
8.0	0.14641	1.62342	3.81727	6.10865	8.41011	10.71258	13.01515	15.31774
9.0	0.18266	1.72811	3.93367	6.22629	8.52787	10.83036	13.13294	15.43551

$1/u$	$1/u \times 10^7$	10^8	10^9	10^{10}	10^{11}	10^{12}	10^{13}	10^{14}
1.0	15.54087	17.84344	20.14604	22.44862	24.75121	27.05379	29.35638	31.65897
1.2	15.72320	18.02577	20.32835	22.63094	24.93353	27.23611	29.53870	31.84128
1.5	15.94634	18.24892	20.55150	22.85408	25.15668	27.45926	29.76184	32.06442
2.0	16.23401	18.53659	20.83919	23.14177	25.44435	27.74693	30.04953	32.35211
2.5	16.45715	18.75974	21.06233	23.36491	25.66750	27.97008	30.27267	32.57526
3.0	16.63948	18.94206	21.24464	23.54723	25.84982	28.15240	30.45499	32.75757
3.5	16.79362	19.09621	21.39880	23.70139	26.00397	28.30655	30.60915	32.91173
4.0	16.92715	19.22975	21.53233	23.83492	26.13750	28.44008	30.74268	33.04526
5.0	17.15030	19.45288	21.75548	24.05806	26.36064	28.66322	30.96582	33.26840
6.0	17.33263	19.63521	21.93779	24.24039	26.54297	28.84555	31.14813	33.45071
7.0	17.48677	19.78937	22.09195	24.39453	26.69711	28.99969	31.30229	33.60487
8.0	17.62030	19.92290	22.22548	24.52806	26.83064	29.13324	31.43582	33.73840
9.0	17.73808	20.04068	22.34326	24.64584	26.94843	29.25102	31.55360	33.85619

Example: when $\dfrac{1}{u} = 1.2 \times 10^{-1}$, $W(u) = .00003$

Source: Reed (1980).

Figure 4.12 Type curves.

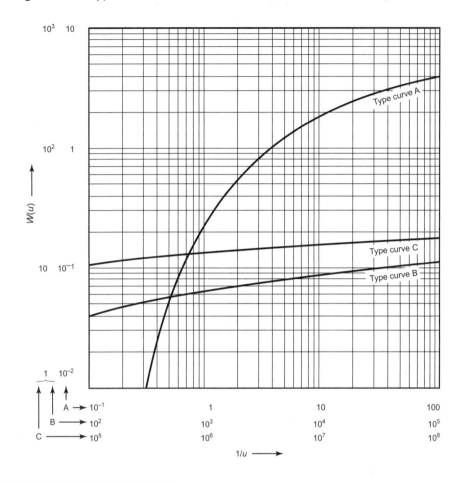

EXAMPLE 4.9

A confined aquifer is pumped at a rate of 1.11 ft³/sec. In an observation well a distance of 200 ft from the well, the following drawdown data were observed. Determine the transmissivity and storage coefficient of the aquifer.

Time since pumping started (min)	1	1.5	2.0	2.5	3.0	4.0	5.0	8.0
Observed drawdown (ft)	0.66	0.87	0.99	1.11	1.21	1.36	1.49	1.75
Time (min)	10.0	14.0	18.0	24.0	30.0	40.0	50.0	60.0
Drawdown (ft)	1.86	2.08	2.20	2.36	2.49	2.65	2.78	2.88
Time (min)	80.0	100.0	120.0	150.0	180.0	210.0	240.0	
Drawdown (ft)	3.04	3.16	3.28	3.42	3.51	3.61	3.67	

SOLUTION

1. From the pumping test data:

t/r^2	2.5×10^{-5}	3.75×10^{-5}	5×10^{-5}	6.25×10^{-5}	7.5×10^{-5}	1×10^{-4}	1.25×10^{-4}	2×10^{-4}
s	0.66	0.87	0.99	1.11	1.21	1.36	1.49	1.75
t/r^2	2.5×10^{-4}	3.5×10^{-4}	4.5×10^{-4}	6.0×10^{-4}	7.5×10^{-4}	1×10^{-3}	1.25×10^{-3}	1.5×10^{-3}
s	1.86	2.08	2.20	2.36	2.49	2.65	2.78	2.88
t/r^2	2×10^{-3}	2.5×10^{-3}	3×10^{-3}	3.75×10^{-3}	4.5×10^{-3}	5.25×10^{-3}	6×10^{-3}	
s	3.04	3.16	3.28	3.42	3.51	3.61	3.67	

2. These data (s versus t/r^2) are plotted in Figure 4.13. The curve matches with the type curve A. The match point corresponding to $W(u) = 1$ and $1/u = 10$ on the data curve is $s = 0.55$ ft and $t/r^2 = 5 \times 10^{-5}$ min/ft^2.

Figure 4.13   Data plot for Example 4.9 on drawdown-time analysis by Theis method.

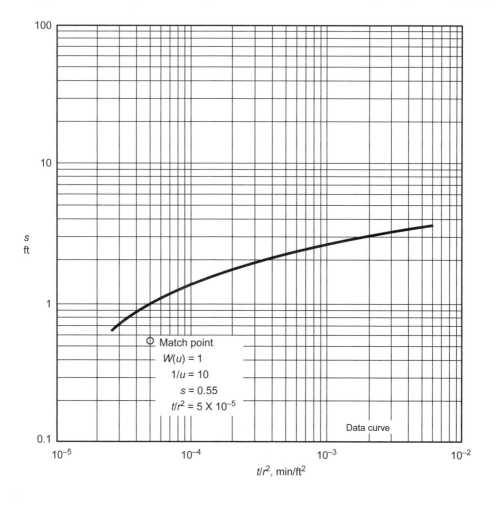

3. Substituting in eq. (4.17) yields

$$T = \frac{Q}{4\pi s} W(u)$$

or

$$T = \frac{(1.11 \text{ ft}^3/\text{s})}{4\pi(0.55 \text{ ft})}(1) = 0.161 \text{ ft}^2/\text{s or } 13,880 \text{ ft}^2/\text{day}$$

4. From eq. (4.18),

$$S = \frac{t}{r^2} 4Tu$$

$$= \left(5 \times 10^{-5} \frac{\text{min}}{\text{ft}^2}\right)\left(4 \times 0.161 \frac{\text{ft}^2}{\text{sec}}\right)\left(\frac{1}{10}\right)\left[\frac{60 \text{ sec}}{1 \text{ min}}\right]$$

$$= 1.93 \times 10^{-4}$$

EXAMPLE 4.10

In a confined aquifer test, the following drawdown data were measured at three observation wells 30 hr after pumping began. The well is pumped at 3000 m³/day. Determine the transmissivity and the storage coefficient.

Observation Well	Distance from Pumped Well (m)	Drawdown (m)
1	150	1.44
2	300	1.17
3	600	0.9

SOLUTION

1. From the data above:

t/r^2 (min/m²)	8×10^{-2}	2×10^{-2}	5×10^{-3}
s (m)	1.44	1.17	0.9

2. The data plot is shown in Figure 4.14. The curve matches with the type curve B. The match point coordinates are:

$$W(u) = 5, \quad \frac{1}{u} = 2.8 \times 10^3, \quad s = 1 \text{ m}, \quad \frac{t}{r^2} = 1 \times 10^{-1} \text{ min/m}^2$$

3. From eq. (4.17),

$$T = \frac{(3000 \text{ m}^3/\text{day})}{4\pi(1 \text{ m})}(5) = 1194 \text{ m}^2/\text{day}$$

Figure 4.14 Data plot for Example 4.10 on drawdown-distance analysis by Theis method.

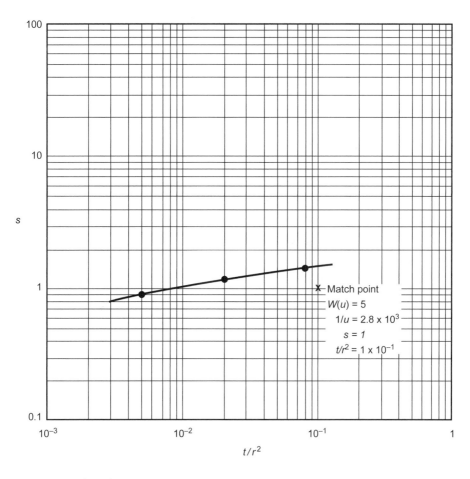

4. From eq. (4.18),

$$S = 4\left(1194\,\frac{m^2}{day}\right)\left(1\times10^{-1}\,\frac{min}{m^2}\right)\left(\frac{1}{2.8\times10^3}\right)\left[\frac{1\,day}{24\times60\,min}\right]$$

$$=1.2\times10^{-4}$$

4.5 UNSTEADY-STATE ANALYSIS OF CONFINED AQUIFERS: COOPER-JACOB METHOD

Cooper and Jacob showed that when $u\ (= r^2S/4T\,t)$ becomes sufficiently small, steady-state conditions tend to develop in the cone of depression and the exponential integral or well function can be closely approximated by only the first two terms of eq. (4.15). Thus eq. (4.13) can be written

$$s = \frac{Q}{4\pi T}(-0.5772 - \ln u) \tag{a}$$

This may be rewritten

$$s = \frac{Q}{4\pi T}\left(\ln 0.562 - \ln \frac{r^2 S}{4Tt} \right) \qquad \text{(b)}$$

or

$$s = \frac{Q}{4\pi T}\ln \frac{2.25Tt}{r^2 S}$$

or

$$s = \frac{2.3Q}{4\pi T}\log \frac{2.25Tt}{r^2 S} \quad [L] \qquad \text{(4.19)}$$

On semilog paper, eq. (4.19) represents a straight line with a slope of $2.3Q/4\pi T$. It is plotted in three different ways.

4.5.1 Drawdown-Time Analysis

The drawdown measurements are made in an observation well at various times; distance r is a constant. A plot is made between drawdown, s (ordinary scale on vertical coordinate), and time, t (log scale on horizontal coordinate), as shown in Figure 4.15. The slope of the line is

$$m = \frac{2.3Q}{4\pi T} \qquad \text{(c)}$$

or

$$\frac{\Delta s}{\log(t_2/t_1)} = \frac{2.3Q}{4\pi T} \qquad \text{(d)}$$

If a change in the drawdown, Δs, is considered for one log cycle, then $\log t_2/t_1 = 1$ and eq. (d) reduces to

$$\Delta s = \frac{2.3Q}{4\pi T} \qquad \text{(e)}$$

or

$$T = \frac{2.3Q}{4\pi \Delta s} \quad [L^2 T^{-1}] \qquad \text{(4.20)}$$

Where the straight line intersects the x-axis, the drawdown is zero and the time is t_0. Substituting these values in eq. (4.19), we have

$$0 = \frac{2.3Q}{4\pi T}\log \frac{2.25Tt_0}{r^2 S} \qquad \text{(f)}$$

or

$$0 = \log \frac{2.25Tt_0}{r^2 S}$$

Figure 4.15 Plot of *s* versus log time.

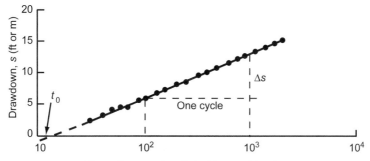

Time after pumping started (log scale), min.

or

$$1 = \frac{2.25Tt_0}{r^2 S}$$

or

$$S = \frac{2.25Tt_0}{r^2} \quad [\text{dimensionless}] \tag{4.21}$$

4.5.2 Drawdown-Distance Analysis

When drawdown measurements are made at a given time in various wells, time is a constant. A plot is made between *s* and distance as shown in Figure 4.16. From a similar consideration as in drawdown-time analysis above

$$T = \frac{2.3Q}{2\pi\Delta s} \quad [\text{L}^2\text{T}^{-1}] \tag{4.22}$$

and

$$S = \frac{2.25Tt}{r_0^2} \quad [\text{dimensionless}] \tag{4.23}$$

where r_0 is the intercept at the *x*-axis.

4.5.3 Measurements in Many Wells at Various Times for Either Drawdown-Time or Drawdown-Distance Analysis

A general procedure is to plot drawdown versus log t/r^2 as shown in Figure 4.17. In this case,

$$T = \frac{2.3Q}{4\pi\Delta s} \quad [\text{L}^2\text{T}^{-1}] \tag{4.24}$$

and

$$S = 2.25T\left(\frac{t}{r^2}\right)_0 \quad [\text{dimensionless}] \tag{4.25}$$

Figure 4.16 Plot of *s* versus log *r*.

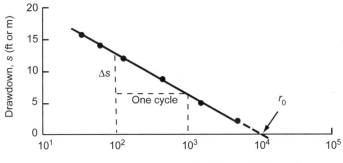

Distance from pumped well, *r* (log scale) (ft or m)

Figure 4.17 Plot of *s* versus t/r^2.

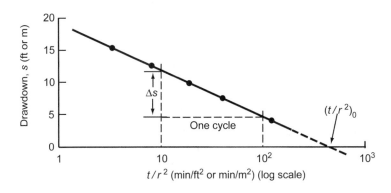

t/r^2 (min/ft^2 or min/m^2) (log scale)

where $(t/r^2)_0$ is the intercept of a straight line on the *x*-axis.

It should be recognized that the derivations by Cooper-Jacob above are based on the assumption of *u* being small, that is,

$$u \leq 0.05$$

or

$$\frac{r^2 S}{4Tt} \leq 0.05 \tag{g}$$

or

$$t \geq \frac{5r^2 S}{T} \quad [\text{T}] \tag{4.26}$$

The data points will begin to fall on a straight line after time, *t*, is sufficiently long to satisfy eq. (4.26).

EXAMPLE 4.11

Solve Example 4.9 by the Cooper-Jacob method.

SOLUTION

1. Drawdown versus time data are plotted in Figure 4.18.
2. For one cycle, $\Delta s = 1.26$ ft, from eq. (4.20)

$$T = \frac{2.3\left(1.11 \text{ ft}^3/\text{s}\right)}{4\pi\left(1.26 \text{ ft}\right)} = 0.161 \text{ ft}^2/\text{s or } 13,880 \text{ ft}^2/\text{day}$$

3. From eq. (4.21),

$$S = 2.25\left(13,880 \frac{\text{ft}^2}{\text{day}}\right)\left(\frac{1}{200^2 \text{ ft}^2}\right)(0.35 \text{ min})\left[\frac{1 \text{ day}}{24 \times 60 \text{ min}}\right]$$

$$= 1.90 \times 10^{-4}$$

4. From eq. (4.26), the Cooper-Jacob formulation is valid when

$$t \geq \frac{5r^2 S}{T}$$

$$t \geq \frac{5(200 \text{ ft})^2 \left(1.90 \times 10^{-4}\right)}{13,880 \text{ ft}^2/\text{day}}$$

$$\geq 0.0027 \text{ day} \quad or \quad 4 \text{ minutes}$$

Measurements after 4 min are valid.

EXAMPLE 4.12

Solve Example 4.10 by the Cooper-Jacob Method.

SOLUTION

1. Drawdown versus distance data are plotted in Figure 4.19.
2. From eq. (4.22), for one cycle, $\Delta s = 0.9$ m; thus

$$T = \frac{2.3(3000)}{2\pi(0.9)} = 1220 \text{ m}^2/\text{day}$$

3. From eq. (4.23),

$$S = 2.25\left(1220 \frac{\text{m}^2}{\text{day}}\right)(30 \text{ hr})\left(\frac{1}{6000^2 \text{ m}^2}\right)\left[\frac{1 \text{ day}}{24 \text{ hr}}\right]$$

$$= 1.0 \times 10^{-4}$$

4. To check whether the Cooper-Jacob formulation is valid to the farthest well, from eq. (4.26),

$$t \geq \frac{5(600)^2 (1 \times 10^{-4})}{1220}$$

$$\geq 0.148 \text{ day} \quad or \quad 3.5 \text{ hr}$$

Since the measurement was made after 30 hr., it is valid.

Figure 4.18 Plot of s versus log t for Example 4.11 on drawdown-time analysis by Cooper-Jacob method.

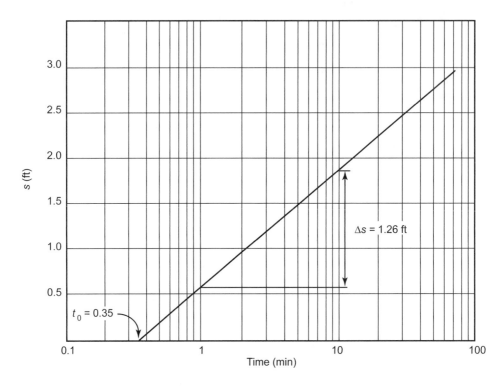

4.6 UNSTEADY-STATE ANALYSIS OF UNCONFINED AQUIFERS

Although the equation of groundwater flow through an unconfined aquifer reduces to the same form as for the confined flow, the solution developed in Section 4.4 is not entirely applicable because of (1) dewatering of the aquifer, (2) vertical flow near the well, and (3) delayed yield due to gravity drainage as mentioned earlier.

If the drawdown is small compared to the depth of the aquifer, the effect of dewatering and vertical flow can be neglected. Also, if pumping continues long enough, the effect of delayed yield becomes negligible. In such situations, the approach of confined aquifer can be applied to water table aquifers as well.

According to Hantush, near a well (in the region less than 0.2 times the depths of the aquifer), the vertical effect is not significant after time period:

$$t > 5b\frac{S_y}{K_z} \quad [\text{T}] \tag{4.27}$$

where

t = time period

b = aquifer thickness

S_y = specific yield

K_z = vertical hydraulic conductivity

Figure 4.19 Plot of s versus log r for Example 4.12 on drawdown-distance analysis by Cooper-Jacob method.

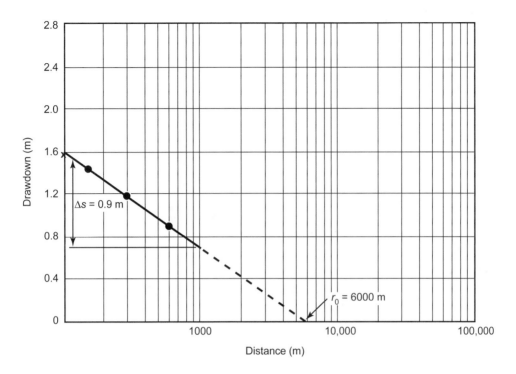

According to Stallman, the delayed yield is pronounced for the period

$$t = 10 S_y \frac{s}{K_z} \quad [\text{T}] \tag{4.28}$$

For a pumping test of a duration shorter than that computed by eqs. (4.27) and (4.28), the type curves developed by Boulton (1963, 1973) and Neuman (1972, 1973, 1975) involving vertical flow and delayed yield should be utilized. These type curves comprise a set of curves between $W(u)$ and $1/u$ drawn for various values of $K_z r^2 / K_h b^2$, where K_z is vertical hydraulic conductivity, K_h is horizontal hydraulic conductivity, r is the distance from the pumping well, and b is the initial saturated thickness of the aquifer. The curves have two distinct segments. In the early part (time), the elastic storativity is responsible for instantaneous release of water to the well similar to a confined aquifer. Later on, the specific yield is responsible for the delayed release of water to the well.

However, if the pumping test can be extended to surpass the time requirements evident from eqs. (4.27) and (4.28), the approach of a confined aquifer can be followed because the late drawdown data follow the Theis curve. As for dewatering (lowering of the saturated thickness), the methods of a confined aquifer can be applied if the drawdown is less than 25% of the initial depth of saturation. The observed (measured) values of drawdown are corrected by the equation

$$s' = s - \frac{s^2}{2b} \quad [\text{L}] \tag{4.29}$$

where

s' = corrected drawdown

s = measured drawdown

b = thickness of aquifer

The value of the specific yield obtained using the method of a confined aquifer is adjusted again as follows:

$$S_y = \frac{(b - \bar{s})S_y'}{b} \quad \text{[dimensionless]} \tag{4.30}$$

where

S_y = adjusted specific yield

S_y' = computed specified yield

$\bar{s}$ = drawdown at the end of pumping at the geometric mean radius of all observation wells

EXAMPLE 4.13

A well fully penetrating an unconfined aquifer of saturated thickness 50 ft is pumped at a rate of 0.8 ft³/sec. The drawdowns as measured in an observation well 30 ft from the pumped well are shown below. Determine the aquifer properties.

Time (min)	20	50	70	110	200	400	800	1200	1700	2000
Drawdown, s (ft)	1.52	1.78	2.0	2.39	2.88	3.42	4.07	4.39	4.72	4.91

SOLUTION

1. Applying corrections to drawdowns by eq. (4.29):

t (min)	20	50	70	110	200	400	800	1200	1700	2000
s' (ft)	1.5	1.75	1.96	2.33	2.80	3.30	3.90	4.20	4.50	4.67

2. By the Cooper-Jacob method:

a. A plot of drawdown and log time is shown in Figure 4.20.

b. For one log time cycle, $\Delta s = 1.85$ ft; $t_0 = 5.7$ min.

c. From eq. (4.20),

$$T = \frac{2.3(0.8 \text{ ft}^3/s)}{4\pi(1.85 \text{ ft})} = 0.079 \text{ ft}^2/\text{sec} \quad \text{or} \quad 6842 \text{ ft}^2/\text{day}$$

d. From eq. (4.21),

$$S_y' = 2.25 \left(6842 \frac{\text{ft}^2}{\text{day}} \right) (5.7 \text{ min}) \left(\frac{1}{30^2 \text{ ft}^2} \right) \left[\frac{1 \text{ day}}{24 \times 60 \text{ min}} \right]$$

$$= 6.8 \times 10^{-2}$$

Figure 4.20 Adjusted drawdown versus log time for Example 4.13 on unconfined aquifer.

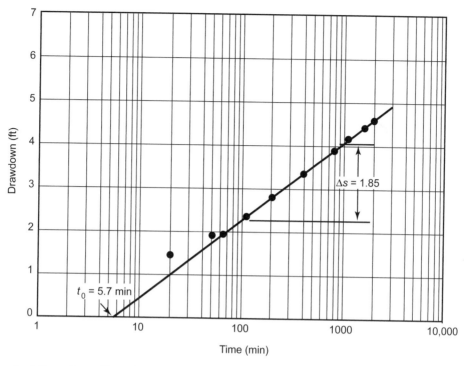

3. Adjusted specific yield, from eq. (4.30):

$$S_y = \left(\frac{50 - 4.67}{50}\right)\left(6.8 \times 10^{-2}\right) = 6.16 \times 10^{-2}$$

4. To check whether the pumping was for a long enough period to neglect vertical flow and delayed yield effects:

   e. $K_z = 6842/50 = 136.84$ ft/day. From eq. (4.27),

   $$t > \frac{5(50 \text{ ft})\left(6.16 \times 10^{-2}\right)}{(136.84 \text{ ft/day})}$$

or

$$t > 0.11 \text{ day} \quad \text{or} \quad 160 \text{ min}$$

The data over 160 minutes are free from a vertical flow effect.

   f. From eq. (4.28),

   $$t = \frac{10\left(6.16 \times 10^{-2}\right)(4.67 \text{ ft})}{(135.84 \text{ ft/day})}$$
   $$= 0.021 \text{ day or 30 min}$$

Thus the delayed yield effect can also be neglected. (The question above could have been solved by the type-curve method as well.)

Section 4.6   Unsteady-State Analysis of Unconfined Aquifers

4.7 SEMICONFINED AQUIFERS: THE THEORY OF LEAKY AQUIFERS

A leaky confined aquifer is underlain by an impervious bed and overlain by a semipervious layer. Above the semiconfining layer is a water-table aquifer, as shown in Figure 4.21. Initially, the artesian (piezometric) level in a semiconfined aquifer and the water table in an unconfined aquifer coincide. Pumping of the semiconfined aquifer lowers its piezometric level, thus creating a head difference between unconfined and confined aquifers, thereby inducing leakage through the semipervious layer. In the theory of leaky aquifers the flow is considered vertical through the semiconfining layer and radially in a horizontal direction in the confined aquifer.

For the semiconfined aquifer, in the equation for groundwater flow a source term, in line with eq. (3.37), is added to reflect leakage from the water-table aquifer into the semiconfined aquifer. This source term is derived applying Darcy's law across a semipervious stratum:

$$q' = K' \frac{H_0 - h}{b'}$$

Thus the equation of an unsteady-state leaky aquifer in one dimension is

$$Kb \frac{\partial^2 h}{\partial r^2} + Kb \frac{1}{r} \frac{\partial h}{\partial r} + K' \frac{H_0 - h}{b'} = S \frac{\partial h}{\partial t} \qquad \text{(a)}$$

In terms of drawdown s, since $s = H_0 - h$,

$$-\frac{\partial^2 s}{\partial r^2} - \frac{1}{r} \frac{\partial s}{\partial r} + \frac{K'}{Kbb'} s = \frac{-S_s}{K} \frac{\partial s}{\partial t} \qquad \text{(b)}$$

or

$$\frac{\partial^2 s}{\partial r^2} + \frac{1}{r} \frac{\partial s}{\partial r} - \frac{s}{B^2} = \frac{S_s}{K} \frac{\partial s}{\partial t} \quad [\text{L}^{-1}] \qquad (4.31)$$

and

$$B = \text{Leakage Factor} = \sqrt{\frac{Kb}{K'/b'}} \quad [\text{L}] \qquad (4.32)$$

Hantush and Jacob in 1954 derived the solutions to eq. (4.31) for steady-state flow when the term $\partial s/\partial t$ was zero, i.e., the heads were stabilized. Later, a more generalized solution of the unsteady flow eq. (4.31) was developed by Hantush and Jacob (1955) to which the steady-state solutions merged. The solution was based on all assumptions made by Theis in Section 4.3, and further added the following two conditions: (1) the hydraulic head in the aquifer above the confining layer remained constant during removal of water from the pumped semiconfined aquifer, and (2) the storage coefficient within the confining layer and, thus, the contribution from storage of the confining layer was negligible. For an infinite leaky aquifer, the solution on the lines of the Theis equation has the form:

$$s = \frac{Q}{4\pi T} W(u, r/B) \quad [\text{L}] \qquad (4.33)$$

where u and B are defined in eq. (4.14) and eq. (4.32), and r is the distance from the pumped well. $W(u, r/B)$ is known as the leaky well function. Hantush (1956) computed the

Figure 4.21 Flow in a semiconfined aquifer.

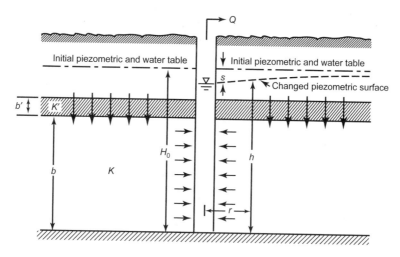

values of $W(u, r/B)$ as listed in Table 4.2. The type curves of $W(u)$ and $1/u$ for selected values of r/B are shown in Figure 4.22. More exhaustive type curves can be prepared from the data in Table 4.2 or can be obtained from other sources (e.g., Walton, 1962).

Of the total discharge Q from the main (semiconfined) aquifer, the rate of flow from storage in the semiconfined aquifer is given by

$$q_s = Q\, e^{-Tt/SB^2} \quad [\mathrm{L^3 T^{-1}}] \tag{4.34}$$

Thus, the rate of leakage from the confining layer is

$$q_L = Q - q_s \quad [\mathrm{L^3 T^{-1}}] \tag{4.35}$$

In 1960, Hantush presented a modified solution to eq. (4.31) wherein the storage capacity of the confining layer was also taken into account, but the first assumption regarding the constant hydraulic head in the top water table aquifer was retained. The solution has the form

$$s = \frac{Q}{4\pi T} H(u, \beta) \quad [\mathrm{L}] \tag{4.36}$$

where β is a function of r/B as well as the ratio of the storage coefficient of the confining bed to the storage coefficient of the semiconfined aquifer. Hantush tabulated $H(u, \beta)$ vs. $1/u$ for various values of β for which reference may be made to Fetter (2000) or any standard hydrogeology textbook.

Neuman and Witherspoon (1969) extended the solution of Hantush by relaxing the first assumption as well. They considered the drawdowns in the top aquifer above the confining layer. Their solutions require calculation of four parameters: r/B and β for both the semiconfined aquifer and the confining layer. The well function has a form $W(u, r/B_{11}, r/B_{12}, \beta_{11}, \beta_{12})$, where subscript 1 and 2 refer to the aquifer and the confining layer. The function is tabulated in many pages. Because of its simplicity, the original solution of Hantush given by eq. (4.33) is very widely used. Its application for steady and unsteady flow conditions is discussed next.

Section 4.7   Semiconfined Aquifers: The Theory of Leaky Aquifers

4.7.1 Steady-State Flow in Leaky Aquifers

Figure 4.22 illustrates that for each r/B value, the function $W(u, r/B)$ becomes asymptotic at a certain value of $1/u$. This is the condition of steady flow. For a steady flow solution, the limiting value of $W(u, r/B)$ can be read for known r/B from Figure 4.22 or Table 4.2. This can be used to solve eq. (4.33). For $r/B < 0.05$, the Hantush solution of steady-state drawdown reduces to

$$s = \frac{Q}{2\pi T} \ln\left(\frac{1.123}{r/B}\right) \quad [\text{L}] \tag{4.37}$$

EXAMPLE 4.14

An aquifer is underlain by an impervious stratum and overlain by a semipervious layer 5 ft thick having a coefficient of permeability of 1×10^{-8} ft/sec. The aquifer has an average thickness of 100 ft and a coefficient of permeability of 1×10^{-3} ft/sec. Groundwater is pumped from the aquifer at a rate of 0.15 ft³/sec through a fully penetrating well of 12 in. diameter. Determine the drawdown **(a)** at 2000 ft from the well, and **(b)** at the face of the well under steady-state conditions.

Figure 4.22 Type curves for leaky aquifer without contribution from confining layer storage.

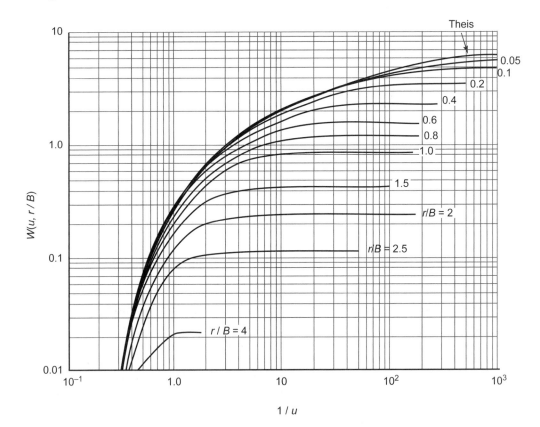

Table 4.2 Values of the Functions $W(u, r/B)$ for Various Values of u

Column headings (after $1/u$ and u) are values of r/B.

$1/u$	u	0	0.002	0.004	0.006	0.008	0.01	0.02	0.04	0.06	0.08	0.1	0.2	0.4	0.6	0.8	1	2	4	6	8
500000	0.000002	12.7	12.7	11.3	10.5	9.89	9.44	8.06	6.67	5.87	5.29	4.85	3.51	2.23	1.55	1.13	0.842	0.228	0.0223	0.0025	0.0003
250000	0.000004	12.1	12.1	11.2	10.5	9.89	9.44														
166700	0.000006	11.6	11.6	11.1	10.4	9.88	9.44														
125000	0.000008	11.3	11.3	10.9	10.4	9.87	9.43														
100000	0.00001	11.0	11.0	10.7	10.3	9.84	9.42														
50000	0.00002	10.2	10.2	10.6	10.2	9.80	9.30	8.06													
25000	0.00004	9.52	9.52	10.1	9.84	9.58	9.01	8.03													
16670	0.00006	9.13	9.13	9.45	9.34	9.19	8.77	7.98													
12500	0.00008	8.84	8.84	9.08	9.00	8.89	8.57	7.91													
10000	0.0001	8.62	8.62	8.81	8.75	8.67	8.40	7.84	6.67	5.87	5.29										
5000	0.0002	7.94	7.94	8.59	8.55	8.48	7.82	7.50	6.62	5.86	5.29										
2500	0.0004	7.24	7.24	7.92	7.90	7.86	7.19	7.01	6.45	5.83	5.29	4.85									
1667	0.0006	6.84	6.84	7.24	7.22	7.21	6.80	6.68	6.27	5.77	5.27	4.85									
1250	0.0008	6.55	6.55	6.84	6.83	6.82	6.52	6.43	6.11	5.69	5.25	4.84									
1000	0.001	6.33	6.33	6.55	6.54	6.53	6.31	6.23	5.97	5.61	5.21	4.83	3.51								
500	0.002	5.64	5.64	6.33	6.32	6.32	5.63	5.59	5.45	5.24	4.98	4.71	3.50								
250	0.004	4.95	4.95	5.64	5.63	5.63	4.94	4.92	4.85	4.74	4.59	4.42	3.48								
167	0.006	4.54	4.54	4.95	4.94	4.94	4.54	4.53	4.48	4.41	4.30	4.18	3.43	2.23							
125	0.008	4.26	4.26	4.54			4.26	4.25	4.21	4.15	4.08	3.98	3.36	2.23							
100	0.01	4.04	4.04	4.26			4.04	4.03	4.00	3.95	3.89	3.81	3.29	2.23	1.55	1.13					
50	0.02	3.35	3.35	4.04			3.35	3.35	3.34	3.31	3.28	3.24	2.95	2.18	1.55	1.13					
25	0.04	2.68	2.68				2.68	2.68	2.67	2.66	2.65	2.63	2.48	2.02	1.52	1.13	0.842				
16.7	0.06	2.30	2.30				2.30	2.29	2.29	2.28	2.27	2.26	2.17	1.85	1.46	1.11	0.839				
12.5	0.08	2.03	2.03					2.03	2.02	2.02	2.01	2.00	1.94	1.69	1.39	1.08	0.832				
10	0.1	1.82	1.82						1.82	1.82	1.81	1.80	1.75	1.56	1.31	1.05	0.819	0.228			
5	0.2	1.22	1.22						1.22	1.22	1.22	1.22	1.19	1.11	0.996	0.857	0.715	0.227			
2.5	0.4	0.702	0.702						0.702	0.702	0.701	0.700	0.693	0.665	0.621	0.565	0.502	0.210			
1.7	0.6	0.454	0.454						0.454	0.454	0.454	0.453	0.450	0.436	0.415	0.387	0.354	0.177	0.0222		
1.3	0.8	0.311	0.311						0.311	0.310	0.310	0.310	0.308	0.301	0.289	0.273	0.254	0.144	0.0218		
1.0	1	0.219	0.219									0.219	0.218	0.213	0.206	0.197	0.185	0.114	0.0207	0.0025	
0.5	2	0.049	0.049										0.049	0.048	0.047	0.046	0.044	0.034	0.011	0.0021	0.0003
0.25	4	0.0038	0.0038											0.0038	0.0037	0.0037	0.0036	0.0031	0.0016	0.0006	0.0002
0.17	6	0.0004	0.0004														0.0004	0.0003	0.0002	0.0001	0
0.13	8	0	0																	0	0

Source: After M. S. Hantush (1956), "Analysis of Data from Pumping Tests in Leaky Aquifers," *Transactions, American Geophysical Union* 37, pp. 702–14.

SOLUTION

$T = bK = (100)(1 \times 10^{-3}) = 0.1 \text{ ft}^2/\text{sec}$

Coefficient of leakage (eq. 3.12),

$$\frac{K'}{b'} = \frac{1 \times 10^{-8}}{5} = 2 \times 10^{-9} \text{ per sec}$$

Leakage factor (eq. 3.14 or eq. 4.32),

$$B = \sqrt{\frac{T}{K'/b'}} = \sqrt{\frac{0.1}{2 \times 10^{-9}}} = 7071 \text{ ft}$$

(a) At 2000 ft from the well

$$\frac{r}{B} = \frac{2000}{7071} = 0.28 > 0.05, \text{ use Figure 4.22 or Table 4.2.}$$

From Table 4.2, for r/B of 0.28, the limiting $W(u, r/B) = 2.9$

$$s = \frac{\left(0.15 \text{ ft}^3/\text{s}\right)}{2\pi\left(0.1 \text{ ft}^2/\text{s}\right)}(2.90) = 0.35 \text{ ft}$$

(b) At the well face

$$\frac{r}{B} = \frac{0.5}{7071} = 7.07 \times 10^{-5} < 0.05, \text{ use eq. (4.37)}$$

From eq. (4.37)

$$s = \frac{\left(0.15 \text{ ft}^3/\text{s}\right)}{2\pi\left(0.1 \text{ ft}^2/\text{s}\right)} \ln\left(\frac{1.123}{7.07 \times 10^{-5}}\right) = 2.31 \text{ ft}$$

4.7.2 Unsteady-State Flow in Leaky Aquifers

Equation (4.33) is solved the following two ways:

1. To determine the drawdown at a specified time at a known distance from the well when the aquifer and confining bed hydraulic characteristics are known:

   From known values, calculate r/B and u. From Figure 4.22 or Table 4.2, read the well function, $W(u, r/B)$. Substitute this in eq. (4.33) to determine the drawdown.

2. To determine the hydraulic characteristics of the aquifer from known time-drawdown data on a leaky aquifer pumping test:

   On log-log graph paper, prepare the plot of the well function type curves of $W(u, r/B)$ and $1/u$ for various values of r/B using the data from Table 4.2, similar to the lines on Figure 4.22. From the pumping data, prepare, on the same size log-log paper, a plot of drawdown versus t/r^2, similar to the Theis method of Section 4.4.1. Overlay the field data curve on the well function type curves. Move it up or down, left or right, keeping the axes parallel to the well function type curves, until an overlap is achieved with respect to one of the r/B curves. Read the matching point coordinates of $W(u, r/B)$, $1/u$,

r/B, t/r^2 and s. These values are substituted in eq. (4.33), eq. (4.18), and eq. (4.32) to ascertain the aquifer and confining bed characteristics.

EXAMPLE 4.15

The following data were recorded in a pumping test on a well confined by a semipervious stratum of 4 m thickness. The well is pumped at a rate of 0.1 m³/min and the drawdown is measured in an observation well 100 m away.

Time, min	5	15	30	45	60	90	120	240
Drawdown, m	0.08	0.3	0.55	0.68	0.74	0.80	0.84	0.88

SOLUTION

1. From the pumping test data

t/r^2, min/m²	5×10^{-4}	15×10^{-3}	3×10^{-3}	4.5×10^{-3}
s, m	0.08	0.3	0.55	0.68
t/r^2	6×10^{-3}	9×10^{-3}	1.2×10^{-2}	2.4×10^{-2}
s	0.74	0.80	0.84	0.88

2. The data are plotted in Figure 4.23. The plot of Figure 4.23 matches the type curve of $r/B = 1$. The coordinates of a match point are:

$s = 0.5$ m, $t/r^2 = 1 \times 10^{-2}$ min/m², $W(u, r/B) = 0.5$, $1/u = 10$, and $r/B = 1$.

Figure 4.23 Field data plot for leaky aquifer of Example 4.15.

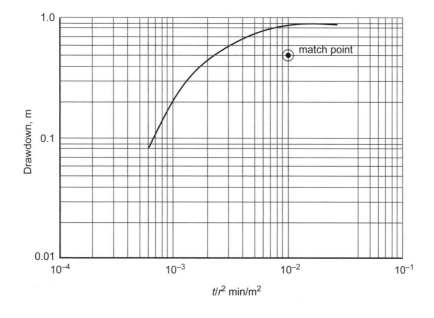

3. Substituting in eq. (4.33)

$$(0.5 \text{ m}) = \frac{(0.1 \text{ m}^3/\text{min})}{4\pi T}(0.5)$$

$$T = 0.008 \text{ m}^2/\text{min}$$

4. From eq. (4.18),

$$S = 4(0.008 \text{ m}^2/\text{min})(1 \times 10^{-2} \text{ min/m}^2)\left(\frac{1}{10}\right)$$

$$S = 3.2 \times 10^{-5}$$

5. $\dfrac{r}{B} = 1$

hence, $B = r = 100 \text{ m}$

$Kb = T = 0.008 \text{ m}^2/\text{min}$

Rearranging eq. (4.32)

$$K' = \frac{Tb'}{B^2}$$

$$= \frac{(0.008 \text{ m}^2/\text{min})(4 \text{ m})}{(100 \text{ m})^2} = 3.2 \times 10^{-6} \text{ m/min}$$

4.8 WELLS NEAR BOUNDARIES: THE THEORY OF IMAGES

An inherent assumption in the equations of groundwater flow in Section 3.11 is that the aquifer is of a really infinite extent. All aquifers are, however, bounded either by recharge boundaries such as streams and lakes or by impermeable boundaries such as buried rocks or tight faults. When wells are located close to such boundaries, they influence the wells, and the formulas based on an infinite aquifer become inapplicable. So that such cases can be analyzed, it is necessary to make the aquifer appear to be of infinite extent. This is achieved through the theory of images. In this method, imaginary wells are introduced in such a manner that the conditions produced by the presence of the boundary(ies) are duplicated utilizing the concepts of an infinite aquifer. Thus the equivalent hydraulic systems are created to which equations developed previously can be applied.

4.8.1 Well Near a Stream

A constant head equivalent to the water level exists along a stream. The cone of depression of a well pumped near a stream should thus terminate at the water surface in the stream. This could be achieved by assuming that an imaginary recharging well is present on the other side of the stream an equal distance opposite the real discharging well. The water is injected into this imaginary well at the same rate as in the real well, so that the increase in head due to the cone of impression of the imaginary well and the decrease in head due to the cone of depression of the real well exactly cancel each other along the line of the stream, as shown in Figure 4.24. The flow net for the system of Figure 4.24 is shown in Figure 4.25. As discussed previously, the graphical solution can be used to analyze such cases.

Figure 4.24 Well near a stream and its equivalent imaginary system in an aquifer of infinite extent.

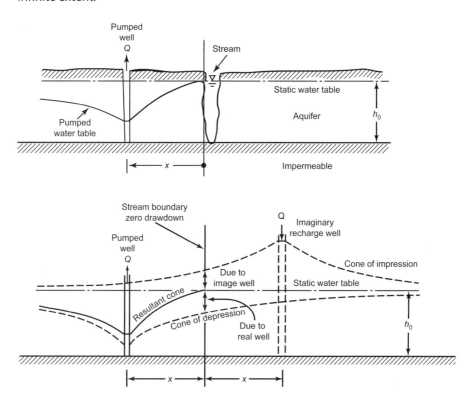

Figure 4.25 Flow net for a discharge well and its imaginary recharge well.

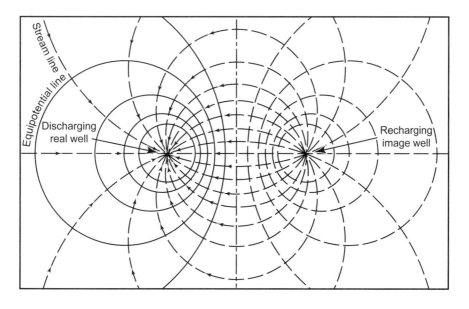

Analytically, if the well is located a distance from the stream, its imaginary recharge well will be as shown in Figure 4.26. Consider any point I having coordinates x and y:

$$r_1 = \sqrt{(a-x)^2 + y^2} \qquad \text{(a)}$$

$$r_2 = \sqrt{(x+a)^2 + y^2} \qquad \text{(b)}$$

For a real well, applying eq. (4.7a) between points O and I, we have

$$s_1 = \frac{Q}{2\pi bK} \ln \frac{a}{r_1} \qquad \text{(c)}$$

where s_1 is the drawdown between O and I. For the imaginary well, applying eq. (4.7a) between O and I, we have

$$s_2 = \frac{-Q}{2\pi bK} \ln \frac{a}{r_2} \qquad \text{(d)}$$

Adding (c) and (d) and substituting r_1 and r_2 give the total drawdown as

$$s = \frac{Q}{4\pi bK} \ln \frac{y^2 + (a+x)^2}{y^2 + (a-x)^2} \quad \text{[L]} \qquad (4.38)$$

where

$a =$ horizontal distance of the well from the stream

$x, y =$ coordinates of the point where drawdown is desired (the head is h)

The origin of the coordinate system is at the intersection of the horizontal line from the well and the vertical stream axis.

The analytical procedure above can be used to derive a relation for an unconfined aquifer. The formula will be the same as eq. (4.38), where b will represent the average saturated thickness of the aquifer. The same procedure can be used to derive a relation for a leaky aquifer by using eq. (4.33) for relations (c) and (d).

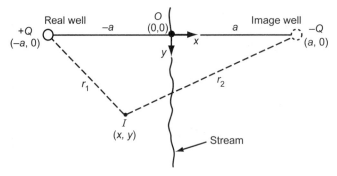

Figure 4.26 Setting up coordinates for a well near a stream with its image.

EXAMPLE 4.16

A 0.5-m well fully penetrates a 30-m-thick confined aquifer of hydraulic conductivity 20 m/day. Due to continuous pumping, if a drawdown of 1 m is registered in the well:

(a) What is the rate of pumping if the well is located 50 m from a stream?

(b) What would the rate of pumping be for the same drawdown if the well was located 5000 m from the stream?

SOLUTION

(a) $a = 50$ m, $x = 50 - 0.25 = 49.75$, and $y = 0$.

From eq. (4.38),

$$s = \frac{Q}{4\pi bK} \ln \frac{y^2 + (a+x)^2}{y^2 + (a-x)^2}$$

or

$$Q = \frac{4\pi bKs}{\ln \dfrac{y^2 + (a+x)^2}{y^2 + (a-x)^2}}$$

$$= \frac{4\pi(30 \text{ m})(20 \text{ m/day})(1 \text{ m})}{\ln[(50 + 49.75)^2 \text{ m}^2/(50 - 49.75)^2 \text{ m}^2]}$$

$$= 629 \text{ m}^3/\text{day}$$

(b)

$$Q = \frac{4\pi(30)(20)(1)}{\ln[(5000 + 4999.75)^2 /(5000 - 4999.75)^2]}$$

$$= 355.6 \text{ m}^3/\text{day}$$

The steeper gradient in the first case contributed to about 80% higher flow.

4.8.2 Well Near an Impermeable Boundary

The desired condition here is that no flow take place across the boundary. If an imaginary discharge well is placed opposite the pumping well an equal distance away from the boundary and both wells pump at the same rate, they will offset each other at the boundary, as shown in Figure 4.27. The flow net for this case is shown in Figure 4.28.

Refer to Figure 4.26. Since both wells are discharging in this case, the sign of eq. (d) is positive. Since there is no stream, a reference distance R is established where there is no drawdown effect (i.e., the head is H_0). This is either the radius of influence or the boundary of an island. Adding eqs. (c) and (d) gives:

$$s = \frac{Q}{2\pi bK} \ln \frac{R^2}{r_1 r_2} \quad [\text{L}] \tag{4.39}$$

where

r_1, r_2 are given by eqs. (a) and (b) of section 4.8.1

$R =$ radius of influence or boundary of the island

Figure 4.27 Well near an impermeable boundary and its equivalent system.

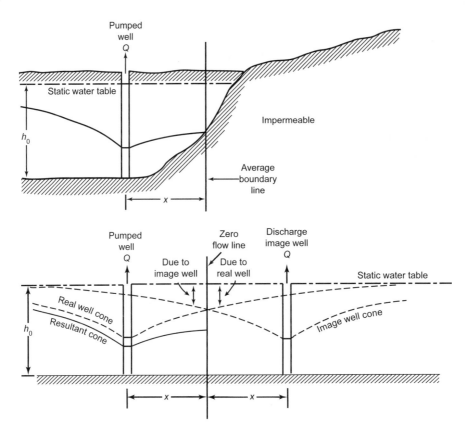

Figure 4.28 Flow net for a discharge well and its imaginary discharge well.

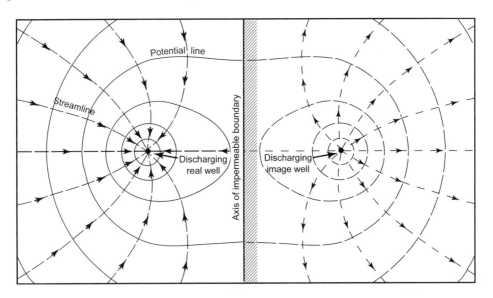

EXAMPLE 4.17

In Example 4.16, if an impermeable boundary is located 50 m from the well having a radius of influence of 1.5 km, what is the rate of pumping?

SOLUTION $r_1 = 0.25$ m and $r_2 = 99.75$ m. From eq. (4.39),

$$Q = \frac{2\pi bKs}{\ln\left(R^2/r_1 r_2\right)}$$

$$= \frac{2\pi(30 \text{ m})(20 \text{ m/day})(1 \text{ m})}{\ln(1500)^2 \text{ m}^2/(0.25 \text{ m})(99.75 \text{ m})]}$$

$$= 330 \text{ m}^3/\text{day}$$

Thus, the flow in the well is less than half of the flow in Example 4.16 for the well near a stream.

4.9 NUMERICAL SOLUTION TO FLOW EQUATIONS

The graphic and analytical solutions, as presented previously, are essentially restricted to simpler flow problems involving simple boundary conditions, mostly in homogeneous, isotropic media. This is due to the inherent problem of obtaining a solution of the governing groundwater flow equation. As it is, no solution exists for the general equation (3.37). Simplifications are introduced to obtain the solutions. Many times, real flow problems cannot be adequately represented. Faced with this situation, scientists and engineers, in dealing with complicated problems, have dispensed with the analytical solution and have attempted to get the desired information directly out of partial differential equations. There are three common techniques of this nature, known as *finite-difference method*, *similarity solution*, and *finite-element method*. The first has a common application in groundwater and is described here briefly.

4.9.1 Finite-Difference Method

In this method a differential equation (of continuous nature) is replaced by a set of algebraic equations called finite difference at discrete points. These algebraic equations are then solved by methods of matrices.

To understand the mechanism of this conversion, consider a plot of head versus horizontal distance as shown in Figure 4.29. The x-axis is divided into intervals of Δx. The differential can be approximated with respect to a subsequent interval known as *forward difference* or over the preceding interval called *backward difference*, as shown in Figure 4.29. Similarly, the *central-difference* approximation will be

$$\left(\frac{\partial h}{\partial x}\right)_n = \frac{h_{n+1} - h_{n-1}}{2\Delta x} \quad \text{[dimensionless]} \tag{4.40}$$

The approximation for the second derivative at n interval will be

$$\left(\frac{\partial^2 h}{\partial x^2}\right)_n = \frac{h_{n+1} - 2h_n + h_{n-1}}{(\Delta x)^2} \quad [L^{-1}] \tag{4.41}$$

The flow equation (3.37), in two dimensions and expressed in terms of transmissivity, has the form

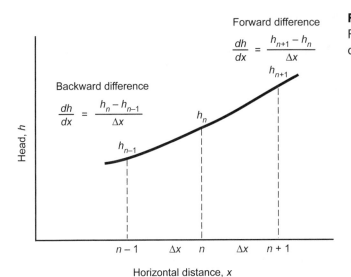

Figure 4.29
Finite difference of a derivative.

Forward difference

$$\frac{dh}{dx} = \frac{h_{n+1} - h_n}{\Delta x}$$

h_{n+1}

Backward difference

$$\frac{dh}{dx} = \frac{h_n - h_{n-1}}{\Delta x}$$

h_n

h_{n-1}

Head, h

$n-1$ Δx n Δx $n+1$

Horizontal distance, x

$$\frac{\partial}{\partial x}\left(T_x \frac{\partial h}{\partial x}\right) + \frac{\partial}{\partial y}\left(T_y \frac{\partial h}{\partial y}\right) - q = S\frac{\partial h}{\partial t} \quad [\mathrm{LT}^{-1}] \tag{4.42}$$

where q is the groundwater withdrawal rate per unit area; for recharge, q will have a negative sign. For homogeneous, isotropic soil, eq. (4.42) reduces to

$$\frac{\partial^2 h}{\partial x^2} + \frac{\partial^2 h}{\partial y^2} = \frac{S}{T}\frac{\partial h}{\partial t} + \frac{q}{T} \quad [\mathrm{L}^{-1}] \tag{4.43}$$

As a first step in obtaining a finite-difference solution, a grid is superimposed on the flow region of interest. Each intersection is called a "node," and as shown in Figure 4.30, indicates coordinates x, y for space dimensions at a particular time level, k (i.e., x_i, y_j, t_k). Thus $h_{i, j, k}$ denotes the head at space dimension x_i, y_j at a time level t_k.

When the space derivatives are approximated at current time level k using an equation of the form (4.41) and the time derivative is approximated with a forward difference, it is known as the *explicit approximation* (method) and contains only one unknown relating to head at advanced level $h_{i,j,k+1}$. When the space derivatives are replaced at an advanced time level $(k + 1)$ and the time derivative is produced by a backward approximation at k (current) level, it is known as *implicit approximation* and contains five unknown values of head of $(k + 1)$ level and only one known value for the current time level. In an alternative procedure, one space derivative is replaced at the advanced time level $(k + 1)$ and the next one is expressed at the current time level. In this way the unknowns are reduced to three, and thus results in a computationally desirable form of tridiagonal matrix. This is known as the *alternating direction implicit (ADI) method*. The method, in this form or with certain alterations, is preferred in finite-difference approximations.

4.9.2 Alternating Direction Implicit Method

Suppose that the first space derivative $\partial^2 h/\partial x^2$ is replaced at the advanced time level $(k+1)$ and the second at the current time level (k). Using difference approximations, we have

Figure 4.30   Grid overlay on an aquifer system.

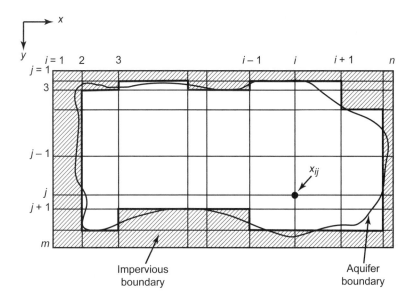

$$\frac{h_{i-1,j,k+1} - 2h_{i,j,k+1} + h_{i+1,j,k+1}}{\Delta x^2} + \frac{h_{i,j-1,k} - 2h_{i,j,k} + h_{i,j+1,k}}{\Delta y^2} =$$

$$\frac{S}{T}\frac{h_{i,j,k+1} - h_{i,j,k}}{\Delta t} + \frac{q_{i,j,k}}{T} \quad [\text{L}^{-1}] \qquad (4.44)$$

There are three unknowns for the $(k + 1)$ time level. Other values at the current time level are known. Collecting the terms together and multiplying both sides by Δx^2 yields

$$h_{i-1,\,j,\,k+1} + B_i h_{i,j,k+1} + h_{i+1,j,k+1} = D_i \quad [\text{L}] \qquad (4.45)$$

where

$$B_i = -\left(2 + \frac{S}{T}\frac{\Delta x^2}{\Delta t}\right)$$

$$D_i = -h_{i,j-1,k}\frac{\Delta x^2}{\Delta y^2} - h_{i,j,k}\left(\frac{S}{T}\frac{\Delta x^2}{\Delta t} - \frac{2\Delta x^2}{\Delta y^2}\right) - h_{i,j+1,k}\frac{\Delta x^2}{\Delta y^2} + \frac{q_{i,j,k}}{T}\Delta x^2$$

Thus B_i and D_i are constants based on known values at the current time level. If eq. (4.45) is written for all nodes along the x-axis with unknown head for that node and two adjacent nodes, there will be n equations for a point j on the y-axis. The values of head at the two outer nodes are known from boundary conditions. There will be $(n-2)$ simultaneous equations to be solved belonging to a tridiagonal coefficient matrix. These equations are for a horizontal line of the grid. Thus, by solving $(n-2)$ equations, it is possible to solve one line at a time parallel to the x-axis (i.e., marching in the direction of Y).

At the next time interval, $\partial^2 h/\partial y^2$ will be expressed at time level $(k+2)$, while $\partial^2 h/\partial x^2$ is retained at the current level of $(k+1)$; thus

$$\frac{h_{i-1,j,k+1} - 2h_{i,j,k+1} + h_{i+1,j,k+1}}{\Delta x^2} + \frac{h_{i,j-1,k+2} - 2h_{i,j,k+2} + h_{i,j+1,k+2}}{\Delta y^2} =$$

$$\frac{S}{T}\frac{h_{i,j,k+2} - h_{i,j,k+1}}{\Delta t} + \frac{q_{i,j,k+1}}{T} \quad [\text{L}^{-1}] \quad (4.46)$$

There are three unknowns for the $(k+2)$ time level. Arranging terms yields

$$h_{i,j-1,k+2} + B_j h_{i,j,k+2} + h_{i,j+1,k+2} = D_j \quad [\text{L}] \quad (4.47)$$

where

$$B_j = -\left(2 + \frac{S}{T}\frac{\Delta y^2}{\Delta t}\right)$$

$$D_j = -h_{i-1,j,k+1}\frac{\Delta y^2}{\Delta x^2} - h_{i,j,k+1}\left(\frac{S}{T}\frac{\Delta y^2}{\Delta t} - \frac{2\Delta y^2}{\Delta x^2}\right) - h_{i+1,j,k+1}\frac{\Delta y^2}{\Delta x^2} + \frac{q_{i,j,k+1}}{T}\Delta y^2$$

Writing eq. (4.47) for all nodes will provide m equations for each value of i on the x-axis (i.e., these belong to a vertical line on the grid). One line at a time can be evaluated by solving $(m-2)$ equations for the interior nodes. For a complete cycle covering eqs. (4.45) and (4.47), two advanced time levels would be required for two-dimensional problems and three advanced levels for three-dimensional problems.

To demonstrate the application, consider a small grid of a five-node row as shown in Figure 4.31. Heads at $i = 1$ and $i = 5$ are known from boundary conditions. Writing eq. (4.45) for node 2 (dropping the time subscript) gives

$$h_{1,j} + B_2 h_{2,j} + h_{3,j} = D_2 \quad (a)$$

known from B.C.

Similarly for nodes 3 and 4, respectively:

$$h_{2,j} + B_3 h_{3,j} + h_{4,j} = D_3 \quad (b)$$

and   known from B.C.

$$h_{3,j} + B_4 h_{4,j} + h_{5,j} = D_4 \quad (c)$$

In matrix form, the equations above are

$$\begin{bmatrix} B_2 & 1 & 0 \\ 1 & B_3 & 1 \\ 0 & 1 & B_4 \end{bmatrix}\begin{bmatrix} h_{2,j} \\ h_{3,j} \\ h_{4,j} \end{bmatrix} = \begin{bmatrix} D_2 \\ D_3 \\ D_4 \end{bmatrix} \quad (4.48)$$

Equation (4.48) is a tridiagonal coefficient matrix with nonzero values on the main and adjacent diagonals only. This can be conveniently converted into an upper triangular matrix to solve for heads. One row at a time, heads can be solved for various values of j. For the next time step, equations can be formed similarly for any column.

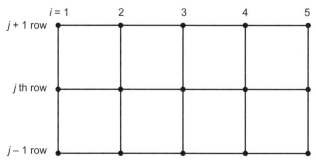

Figure 4.31 Grid of a five-node row and three-node column.

4.9.3 Iterative and Strongly Implicit Methods

In a revised technique known as the *iterative alternate direction implicit (IADI) method*, both the space derivatives and time derivative are replaced by advanced time-level approximations as in the implicit method. However, while solving these equations, one space derivative (in fact, its approximation) is solved one iteration level below compared to the other, on an alternative basis.

In the two-dimensional and three-dimensional model of the USGS (Trescott, 1976; Trescott et al., 1976), a relatively new technique developed by Stone in 1968 has been used in which approximations of both spaces and time are made at an advanced time level. Known as the *strongly implicit procedure* or the *approximate factorization technique,* it is much faster than the alternate direction implicit procedure. In the strongly implicit procedure, the approximate equations are not in tridiagonal form but are reduced to such form through a matrix modification procedure.

4.9.4 Initial and Boundary Conditions

The condition in which an aquifer system exists at the beginning of its operation with respect to a certain dynamic characteristic is required to be defined for the system to be evaluated. Usually, distribution of piezometric heads or water tables are given at initial time $t = 0$.

Boundary conditions are a means to describe the real system appropriately. By proper control of parameters, they are incorporated in finite-difference formulations.

Rivers and lakes are recharge boundaries having constant heads. At these nodes, head may be kept unchanged at initial levels. Some computer models set a very high value for the storage coefficients at such nodes.

Impermeable rocks form no-flow boundaries. These can be treated by considering no hydraulic gradient or the same head as in the preceding node for all such nodes along the boundary. Some models assign zero transmissivities outside the boundary.

If only a part of the flow region is to be considered, it may be important to account for a specified flux into the region. The flux boundary can be treated by assigning recharge or discharge wells to appropriate nodes.

The methods above provide an approximation to the differential equation. When the grid space and time steps are not selected properly, the difference between the approximate and exact solution grows as t increases. The approximate solution is said to be unstable in such cases. The solution is stable in the one-dimensional case when

$$t < \frac{1}{2}\frac{S(\Delta x)^2}{T} \quad [\mathrm{T}] \tag{4.49}$$

EXAMPLE 4.18

Two fully penetrating wells are located in a confined aquifer system as shown in Figure 4.32. Well 1 pumps at a rate of 0.8 cfs and well 2 pumps at a rate of 0.5 cfs. The initial piezometric head is horizontal, in level with the stream. Determine the distribution of head after 1 day of pumping. The aquifer is bound by a stream on one side and impermeable boundaries on two sides. It is extended semi-infinitely on the remaining side, but a width of 2700 ft is represented on the flow system under consideration.

SOLUTION

The grid layout is shown in Figure 4.33. $T = 40 \times 50 = 2000$ ft^2/day. Initial conditions: All $h = 60$ ft at $t = 0$.
Boundary conditions

$$h_{1,1} = h_{1,2} = h_{1,3} = h_{1,4} = 60 \text{ ft (recharge boundary)}$$

$$\left.\begin{array}{l} h_{5,1} = h_{4,1} \\ h_{5,2} = h_{4,2} \\ h_{4,3} = h_{3,3} \\ h_{4,4} = h_{3,4} \end{array}\right\} \text{impermeable boundary}$$

and

$$\left.\begin{array}{l} h_{2,4} = h_{2,3} \\ h_{3,4} = h_{3,3} \\ h_{4,4} = h_{4,3} \end{array}\right\} \text{impermeable boundary}$$

The maximum time step to satisfy the stability requirement:

$$\Delta t < \frac{1}{2}\left[\frac{(0.003)(900)^2}{2000}\right] = 0.61 \text{ day}$$

A time step Δt of 0.5 day is selected.

1. First time step, $\Delta t = 0.5$. Apply eq. (4.45), for one row at a time. In the first time step, heads in row 1($j = 1$) will not be affected since there is no pumping on this row. For row 2, $j = 2$.

Matrix coefficients: From initial conditions, storage coefficient, and transmissivity,

$$B_2 = -\left(2 + \frac{S}{T}\frac{\Delta x^2}{\Delta t}\right) = -\left[2 + \frac{0.003(1000)^2}{2000(0.5)}\right]$$

$$= -5$$

Similarly, $B_3 = B_4 = -5$.

$$D_2 = -60\frac{(1000)^2}{(900)^2} - 60\left[\frac{0.003(1000)^2}{2000(0.5)} - \frac{2(1000)^2}{(900)^2}\right] - 60\frac{(1000)^2}{(900)^2}$$

$$= -180$$

$$D_3 = -180$$

Figure 4.32 (a) Plan of an aquifer system (not to scale); (b) section of a confined aquifer (vertical scale exaggerated).

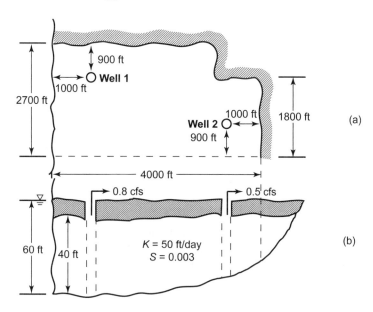

Figure 4.33 Finite-difference grid with grid coordinates.

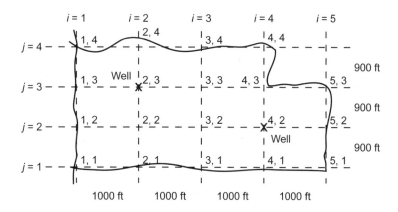

Since

$$q = \frac{0.5}{900(1000)} = 0.556 \times 10^{-6} \text{ ft/sec} \quad or \quad 0.048 \text{ ft/day}$$

hence

$$D_4 = -180 + \frac{0.048(1000)^2}{2000}$$

$$= -156$$

Section 4.9 Numerical Solution to Flow Equations

215

For node 2,

$$h_{1,2,0.5} + B_2 h_{2,2,0.5} + h_{3,2,0.5} = D_2$$

For node 3,

$$h_{2,2,0.5} + B_3 h_{3,2,0.5} + h_{4,2,0.5} = D_3$$

For node 4,

$$h_{3,2,0.5} + B_4 h_{4,2,0.5} + h_{5,2,0.5} = D_4$$

Substituting B and D, initial and boundary conditions, gives

$$60 - 5h_{2,2,0.5} + h_{3,2,0.5} = -180$$
$$h_{2,2,0.5} - 5h_{3,2,0.5} + h_{4,2,0.5} = -180$$
$$h_{3,2,0.5} - 5h_{4,2,0.5} + h_{4,2,0.5} = -156$$

or

$$\begin{bmatrix} -5 & 1 & 0 \\ 1 & -5 & 1 \\ 0 & 1 & -4 \end{bmatrix} \begin{bmatrix} h_{2,2,0.5} \\ h_{3,2,0.5} \\ h_{4,2,0.5} \end{bmatrix} = \begin{bmatrix} -240 \\ -180 \\ -156 \end{bmatrix}$$

Solving by Gaussian elimination

$$h_{2,2,0.5} = 59.7$$
$$h_{3,2,0.5} = 58.7$$
$$h_{4,2,0.5} = 53.7$$

Row 3 is solved by a similar procedure.

Row 4 is an impermeable boundary, hence it will have the same heads as in row 3.

Head distribution obtained after the first time step is shown in Figure 4.34.

2. Second time step, $\Delta t = 0.5$, $t = 1$ day

The head matrix in Figure 4.34 forms the initial heads for the second time step.
Computations are now made column by column. For column 2 ($i = 2$),

$$B_1 = -\left[2 + \frac{0.003(900)^2}{2000(0.5)} \right] = -4.43$$

Similarly, $B_2 = B_3 = -4.43$:

$$D_1 = -60\frac{(900)^2}{(1000)^2} - 60\left[\frac{0.003(900)^2}{2000(0.5)} - \frac{2(900)^2}{(1000)^2} \right] - 60\frac{(900)^2}{(1000)^2}$$

$$= -145.8$$

$$D_2 = -60\frac{(900)^2}{(1000)^2} - 59.7\left[\frac{0.003(900)^2}{2000(0.5)} - \frac{2(900)^2}{(1000)^2} \right] - 58.7\frac{(900)^2}{(1000)^2}$$

$$= -144.5$$

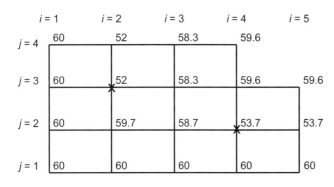

Figure 4.34 Head at the end of $\Delta t = 0.5$ day.

Since

$$q = \frac{0.8}{900(1000)} = 0.889 \times 10^{-6} \text{ ft/sec} \quad \text{or} \quad 0.077 \text{ ft/day}$$

$$D_3 = -60\frac{(900)^2}{(1000)^2} - 52\left[\frac{0.003(900)^2}{2000(0.5)} - \frac{2(900)^2}{(1000)^2}\right] - 58.3\frac{(900)^2}{(1000)^2}$$

$$+ \frac{0.077(900)^2}{2000}$$

$$= -106.8$$

For node 1,*

$$h_{2,0,1} + B_1 h_{2,1,1} + h_{2,2,1} = D_1$$

For node 2,

$$h_{2,1,1} + B_2 h_{2,2,1} + h_{2,3,1} = D_2$$

For node 3,

$$h_{2,2,1} + B_3 h_{2,3,1} + h_{2,4,1} = D_3$$

Substituting B and D, initial and boundary conditions, yields

$$60 - 4.43h_{2,1,1} + h_{2,2,1} = -145.8$$

$$h_{2,1,1} - 4.43h_{2,2,1} + h_{2,3,1} = -144.5$$

$$h_{2,2,1} - 4.43_{2,3,1} + h_{2,3,1} = -106.8$$

or

$$\begin{bmatrix} -4.43 & 1 & 0 \\ 1 & -4.43 & 1 \\ 0 & 1 & -4.43 \end{bmatrix}\begin{bmatrix} h_{2,1,1} \\ h_{2,2,1} \\ h_{2,3,1} \end{bmatrix} = \begin{bmatrix} -205.8 \\ -144.5 \\ -106.8 \end{bmatrix}$$

* Since the aquifer extends beyond the grid boundary in this direction, the head outside the boundary is included here and assigned a value of the initial head of 60 ft.

Section 4.9   Numerical Solution to Flow Equations   **217**

Solving gives

$$h_{2,1,1} = 59.3$$
$$h_{2,2,1} = 56.8$$
$$h_{2,3,1} = 47.7$$

Other columns can be solved similarly. At the end of the second time step, $\Delta t = 0.5$ (i.e., after $t = 1$ day) the head distribution is as shown in Figure 4.35.

4.10 PRODUCTION WELL ANALYSIS

Observation wells serve a very useful purpose in aquifer analysis because of the laminar flow around them. Production wells, on the other hand, contain another component of drawdown in addition to that contributed by the laminar flow as shown in Figure 4.36. It is associated with factors like (1) the increased velocity in the pervious gravel pack surrounding a well that leads to the condition of turbulent flow, (2) the resistance experienced by water as it moves through the well screens and from inside the well to the pump intake, and (3) the miscellaneous losses within the damage zone and filter zone around the well. However, in the absence of observation wells, it is possible to use data from production wells to estimate the hydraulic characteristics of an aquifer. For this purpose, as well as to develop a well field, it is necessary to understand the composition of the drawdown within a production or pumping well.

4.10.1 Well Losses

The total drawdown, s_t in a pumping well is expressed as

$$s_t = s_a + s_w \quad [\text{L}] \tag{4.50}$$

where

s_a = aquifer loss

s_w = well loss $(C Q^n)$

For steady-state flow, the total drawdown is given by

$$s_t = \frac{2.3\,Q}{2\,\pi\,T} \log \frac{r_o}{r_w} + C\,Q^n \quad [\text{L}] \tag{4.51}$$

For unsteady-state flow, the total drawdown is given by

$$s_t = \frac{2.3\,Q}{4\pi\,T} \log \frac{2.25\,T\,t}{r_w^2 S} + C\,Q^n \quad [\text{L}] \tag{4.52}$$

where

C = constant, usually $<0.5\ \text{min}^2/\text{m}^5$ for a properly designed well

n = power of discharge, generally 2

These relations can be expressed in a general form:

$$s_t = B\,Q + C\,Q^n \quad [\text{L}] \tag{4.53}$$

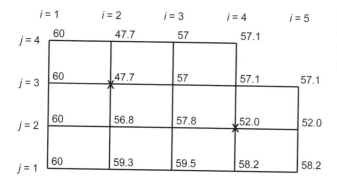

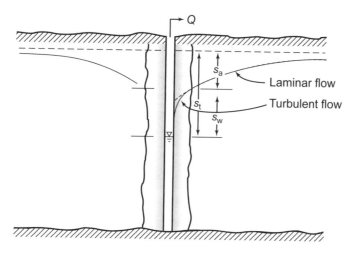

Figure 4.36 Formation loss and well loss in a pumped well.

where

$$B = \text{aquifer loss coefficient} = \frac{2.3}{2\pi T}\log\frac{r_o}{r_w} \text{ or } \frac{2.3}{4\pi T}\log\frac{2.25\,T\,t}{r_w^2\,S}$$

C = well loss coefficient, which is independent of time

To evaluate the coefficients B and C, the drawdown test is performed on a production well for multiple periods with different discharge rates, which is referred to as the step-drawdown test. Jacob (1950), Walton (1962), Bierschenk (1964), and Kasenow (1996b) have proposed methods to analyze data from the test.

4.10.2 Step-Drawdown Test: Bierschenk Solution

Equation (4.53) can be rearranged as follows for $n = 2$:

$$\frac{s_t}{Q} = C\,Q + B \quad [\text{L}^{-2}\text{T}] \tag{4.54}$$

This is a linear equation in s_t/Q vs. Q having a slope of C and a y-intercept of B. During the test, a production well is operated for at least three successive periods (steps) of equal interval. At the beginning of each period the pumping rate is increased and kept constant at that level during that period. B and C are obtained from the arithmetic plot of the data.

Each step of the test is at least one hour in duration and not more than three hours long. The time-drawdown data collected from a production well in the first step of the test can be corrected by subtracting the $C Q^2$ term from the measured drawdowns; the coefficient C being obtained by the above procedure. These data can then be analyzed by the Theis (Section 4.4) or Cooper-Jacob (Section 4.5) method to reasonably estimate aquifer transmissivity and storage coefficient. The corrections to drawdown should not be made after the first step of the test.

EXAMPLE 4.19

The drawdown data collected from a pumping well in Barrington, RI at the end of each of four 100-min steps are given in the first three columns of Table 4.3. Determine the loss coefficients B and C.

Table 4.3   Step-Drawdown Test Data and Computation

(1)	(2)	(3)	(4)
Step	Pumping rate, Q m^3/min	Drawdown, s_t m	s_t/Q min/m^2
1	0.59	0.73	1.24
2	1.07	1.86	1.74
3	1.26	2.46	1.95
4	1.36	2.81	2.07

SOLUTION

1. The s_t/Q ratio for each step is computed in col. 4 of Table 4.3.

2. s_t/Q vs. Q is plotted in Figure 4.37.

   From the graph: intercept, $B = 0.58 \text{ min/m}^2$
   Slope, $C = 1.09 \text{ min}^2/m^5$

4.10.3 Well Efficiency

Well efficiency is related to well loss as follows:

$$E = \frac{s_a}{s_t} \times 100 \quad \text{[dimensionless]} \tag{4.55}$$

or

$$E = \left(1 - \frac{s_w}{s_t}\right) \times 100 \quad \text{[dimensionless]} \tag{4.56}$$

If the well losses are zero, the efficiency is 100%.

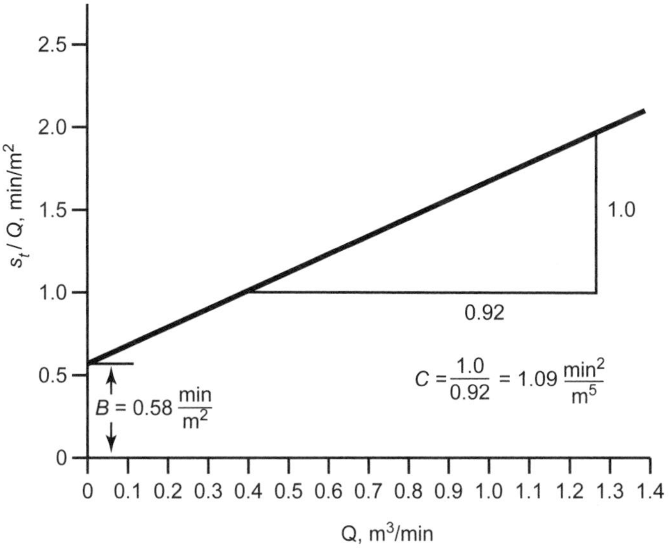
2.5
2.0
1.5
1.0
0.5
0
s_t/Q, min/m^2
1.0
0.92
$C = \frac{1.0}{0.92} = 1.09 \frac{min^2}{m^5}$
$B = 0.58 \frac{min}{m^2}$
0 0.1 0.2 0.3 0.4 0.5 0.6 0.7 0.8 0.9 1.0 1.1 1.2 1.3 1.4
Q, m^3/min

Figure 4.37 Step-drawdown plot of Example 4.19.

4.10.4 Specific Capacity

Specific capacity is defined as the well yield per unit drawdown. Thus

$$\text{specific capacity} = \frac{Q}{s_t} \quad [L^2T^{-1}] \tag{4.57}$$

Depending on steady-state or unsteady-state conditions, s_t will be determined by eq. (4.51) or eq. (4.52), respectively. Specific capacity decreases with pumping rate and time. A reduction of up to 40% in the specific capacity has been observed in 1 year in wells deriving water entirely from storage.

In production well analysis, the specific capacity serves the very useful purpose of estimating the aquifer transmissivity. Equation (4.57), substituted in eq. (4.52) for s_t, provides a means to determine the transmissivity for a known value of the specific capacity based on pumping a well of computed or estimated storage coefficient for a given period of time, usually 12 hours. Many empirical specific capacity equations have been developed to approximate transmissivity.

EXAMPLE 4.20

A fully penetrating 12-in. well in a confined aquifer is pumped at a rate of 400 gpm for 12 hr when a total drawdown of 20 ft is recorded in the well. The transmissivity and storage coefficient for the aquifer are 5000 ft^2/day and 6×10^{-3}, respectively. Determine the **(a)** well losses, **(b)** efficiency of the well, and **(c)** specific capacity of the well.

SOLUTION

$$s_a = \frac{2.3Q}{4\pi T} \log \frac{2.25Tt}{r_w^2 S}$$

$$= \frac{2.3}{4\pi} \left(400\,\frac{\text{gal}}{\text{min}}\right)\left[\frac{1}{5000}\,\frac{\text{day}}{\text{ft}^2}\right]\left[\frac{1\,\text{ft}^3}{7.48\,\text{gal}}\right]\left[\frac{60\times24\,\text{min}}{1\,\text{day}}\right]$$

$$\times \log\left[\frac{2.25}{\left(0.5^2\ \text{ft}^2\right)\left(6\times10^{-3}\right)}\left(5000\,\frac{\text{ft}^2}{\text{day}}\right)(12\,\text{hr})\left[\frac{1\,\text{day}}{24\,\text{hr}}\right]\right]$$

$$= 2.82 \log\left(37.5\times10^5\right)$$

$$= 18.54\ \text{ft}$$

(a) Well loss, $s_w = 20.00 - 18.54 = 1.46$ ft.

(b) Efficiency, $E = \dfrac{s_a}{s_t}\times100 = \dfrac{18.54}{20}\times100 = 92.7\%$

(c) Specific capacity $= \dfrac{Q}{s_t} = \left(400\,\dfrac{\text{gal}}{\text{min}}\right)\left(\dfrac{1}{20\,\text{ft}}\right)\left[\dfrac{1\,\text{ft}^3}{7.48\,\text{gal}}\right]$

$$= 2.67\ \text{ft}^2/\text{min.}$$

4.11 WELL FIELD DESIGN

Well field design consists of determining the location of wells, deciding the number of wells required, and fixing the spacing of these wells. Hydrogeologic characteristics of the aquifer and pumping pattern affect the location of the wells. Two common criteria are that the wells should be located (1) close to the recharging boundaries and parallel to such boundaries, and (2) perpendicular to the impermeable boundary, as far away as possible.

To determine the number of wells that will be needed, the total quantity required to be withdrawn from a well field is divided by the potential yield of a well. The potential yield of a well is obtained in either of the following two ways:

1. The specific capacity of a well, when known, is multiplied by a fraction (one-half to two-thirds) of the available drawdown. The available drawdown is equal to the difference between the static water level and the lowest pumping level that can be imposed on the well, which is normally a meter above the top of the screen.
2. From the Jacob equation (4.23), rearranged below, the value of r_0 is obtained based on the transmissivity and storage coefficient of the flow region and considering a pumping period of 1 year (365 days).

$$r_0^2 = \frac{2.25Tt}{S}\quad [\text{L}^2] \tag{4.58}$$

where r_0 is the distance from the pumping well to zero drawdown.

A semilog plot is prepared by drawing a line connecting point r_0 to a point representing one-half of the available drawdown at the proposed radius of the well, as shown in Figure 4.38.

Figure 4.38 Drawdown-distance analysis for well yield.

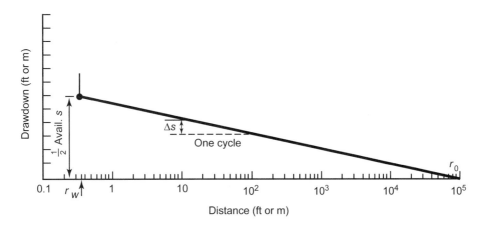

From this figure Δs, corresponding to one log cycle of the distance, is read and used in eq. (4.22) to estimate the pumping rate. This formula can be simplified to:

$$Q = 2.7 T \Delta s \quad [L^3 T^{-1}] \tag{4.59}$$

Only 50% of the available (total) drawdown is assumed to relate to aquifer loss, s_a, and is used to determine the pumping rate; the other 50% is estimated to be lost in well losses, well interferences, boundary effects, and so on. This fraction of the available drawdown can be revised based on the well performance.

The farther apart the wells are, the less their mutual interference, but the greater the cost of interconnecting pipelines and power equipment. Theis gave a formula for spacing that takes economic considerations into account by equating the added cost of pumping due to mutual interference against the capitalized cost of these installations.

Considering well interference alone when wells affect each other's drawdown, the following procedure will indicate the required spacings:

1. If three wells numbered 1, 2, and 3 are located in a straight line, well 2 in the center will be affected by both wells 1 and 3.

2. In the well yield analysis above, one-half of the available drawdown has been allocated to aquifer loss and the other half has been for well losses, well interferences, and boundary effects. The boundary effects are neglected (or determined separately). The well losses, s_w, are determined as $C\,Q^n$ or by well efficiency eq. (4.55). The remaining drawdown can be assigned to well interference.

3. Since two wells interfere, this value of the interference drawdown is divided by 2. Plot the position of this drawdown (effect of each well) on Figure 4.38 and read the distance, which should be the spacing of wells 1 and 3 from well 2.

EXAMPLE 4.21

It is proposed to withdraw a quantity of 2 million gallons of water per day from an aquifer having a transmissivity of 8000 ft^2/day and a storage coefficient of 6×10^{-4}. The maximum permissible drawdown in the aquifer is 40 ft. Wells of 12 in. diameter are to be installed. Determine the number of wells required.

SOLUTION

1. From eq. (4.58),

$$r_0^2 = 2.25\left(8000\frac{\text{ft}^2}{\text{day}}\right)\left(\frac{1}{6\times10^{-4}}\right)(365 \text{ days})$$

$$r_0 = 1.06\times10^5 \text{ ft}$$

2. Available drawdown = 40 ft. One-half of available drawdown = 20 ft. A line connecting one-half available drawdown to r_0 is shown in Figure 4.39. Δs for one distance cycle = 3.8.

3. From eq. (4.59),

$$Q = 2.7T\Delta s$$

$$= 2.7\left(8000 \text{ ft}^2/\text{day}\right)(3.8 \text{ ft}) = 82,080 \text{ ft}^3/\text{day}$$

$$\text{or } 0.95 \text{ cfs or } 6.14\times10^5 \text{ gal/day}$$

4. Number of wells

$$= \frac{\text{withdrawal from well field}}{\text{yield of each well}}$$

$$= \frac{2\times10^6 \text{ gal/d}}{6.14\times10^5 \text{ gal/d}} = 3.26$$

Thus, four wells will be required to withdraw the desired quantity.

Figure 4.39 Plot of the line connecting r_0 to one-half of available drawdown of Example 4.21.

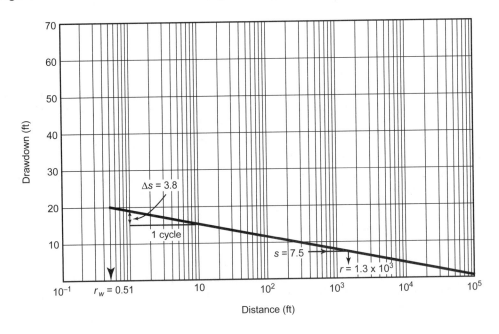

EXAMPLE 4.22

For Example 4.21 determine the spacing among the wells if the well efficiency is 80% and interference is produced only by two outer wells surrounding an interior well.

SOLUTION $s_a = \dfrac{1}{2}$ permissible drawdown $= \dfrac{40}{2} = 20$ ft

From eq. (4.55),

$$E = \frac{s_a}{s_t} \times 100$$

$$s_t = \frac{s_a}{E} \times 100 = \frac{20}{80} \times 100 = 25 \text{ ft}$$

$$s_w = s_t - s_a = 25 - 20 = 5 \text{ ft}$$

Drawdown assigned to well interference $= 20 - 5 = 15$ ft
Drawdown interference per well $= 15/2 = 7.5$ ft
A point corresponding to 7.5 ft drawdown is marked on Figure 4.39. The corresponding distance $= 1.3 \times 10^3$ ft. Thus the well spacing $= 1300$ ft.

PROBLEMS

4.1 Figure P4.1 shows the cross section of a line of sheet piling driven to a depth of 7 m into a homogeneous sandy soil of thickness 15 m. If the soil has a permeability of 25 m/day, determine the **(a)** quantity of seepage under the pile, and **(b)** pore (uplift) pressure at A, B, and C.

Figure P4.1

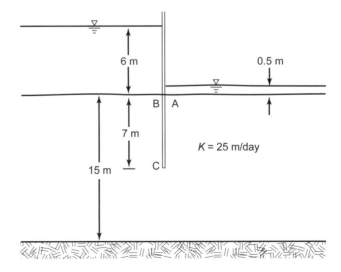

4.2 The concrete dam shown in Fig. P4.2 is founded on isotropic soil of permeability 30 m/day. Determine **(a)** the rate of flow under the foundation, and **(b)** the uplift pressure at various points on the foundation.

Figure P4.2

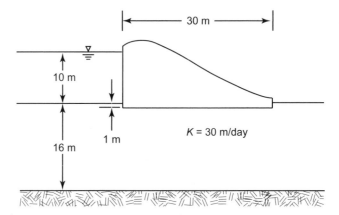

4.3 For the concrete dam with a line of pile at the downstream toe, as shown in Fig. P4.3, determine **(a)** the rate of seepage under the dam, and **(b)** the uplift pressure at various points on the base.

Figure P4.3

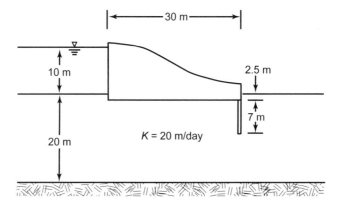

4.4 A masonry dam having a sheetpiling cutoff at the upstream end is located on a pervious stratum having a horizontal hydraulic conductivity of 0.01 ft/min and a vertical hydraulic conductivity of 0.005 ft/min (Fig. P4.4). Determine the **(a)** rate of seepage, and **(b)** uplift force acting at points A, B, and C on the base.

Figure P 4.4

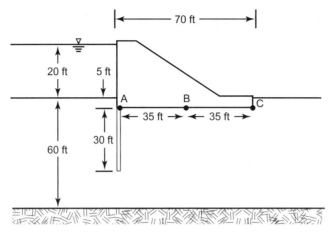

4.5 Two streams are separated by a confined aquifer with an average thickness of 40 ft and hydraulic conductivity of 0.03 ft/sec. The water level in a stream at a higher level is 60 ft. A piezometer located 1000 ft away from this stream in the aquifer has a level of 59.2 ft. Find the rate of flow from one stream to another.

4.6 In an unconfined aquifer having $K = 30$ m/day, two observation wells, 2 km apart, have water levels of 50 and 40 m, respectively, above the base of the aquifer. Determine the rate of flow.

4.7 A confined aquifer underlies an unconfined aquifer as shown in Fig. P4.7. Determine the flow rate from one stream to another.

Figure P 4.7

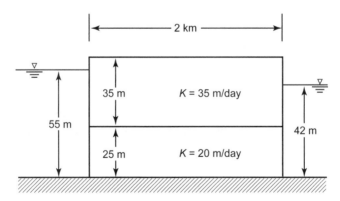

4.8 A well of 0.4 m diameter fully penetrates a 25-m-thick confined aquifer of coefficient of permeability of 12 m/day. The well is located in the center of a circular island of radius 1 km. The water level at the boundary of the island is 80 m. At what rate should the well be pumped so that the water level in the well remains 60 m above the bottom?

4.9 A 12-in. well is drilled through a confined aquifer of average thickness 80 ft. Two observation wells located 50 and 120 ft away register a difference in drawdown of 10 ft. The steady-state pumping rate is 1000 gpm. Determine the transmissivity.

4.10 A 0.3-m-diameter well fully penetrates an unconfined aquifer. The water table is at a height of 70 m from the bottom and the coefficient of permeability is 20 m/day. The well is pumped so that the water level in the well remains steady at 20 m below the original water table. The pumping has no effect on the water table at a distance of 1 km. Determine the rate of flow of the well.

4.11 A fully penetrating unconfined well of 12 in. diameter is pumped at a rate of 1 ft^3/sec. The coefficient of permeability is 750 gal/day per square foot. The drawdown in an observation well located 200 ft away from the pumping well is 10 ft below its original depth of 150 ft. Find the water level in the well.

4.12 For Problem 4.8, find the travel time of groundwater from the boundary of the island to the well. The aquifer has a porosity of 0.25.

4.13 For Problem 4.9, determine the travel time of water from one observation well to another if the porosity is 0.2.

4.14 For Problem 4.11, determine the time of flow from the observation well to the pumping well for the aquifer of porosity 0.2.

4.15 From a well that fully penetrates a confined aquifer, the following drawdowns were observed from an observation well 200 ft from the well being pumped at 500 gpm. Determine the value of transmissivity and storage coefficient by the type-curve method.

Time since pumping (min)	0	1	1.5	2.0	2.5	3	4	5		8
Drawdown (ft)	0	0.66	0.87	0.99	1.11	1.21	1.36	1.49		1.75
Time		10	14	18	24	30	40	50	60	100	150
Drawdown		1.86	2.08	2.20	2.36	2.49	2.65	2.78	2.88	3.16	3.42
Time		180	210	240							
Drawdown		3.51	3.61	3.67							

4.16 From a pumping test on a confined aquifer, the following data were noted on an observation well 50 ft from a well pumped at 300 gpm. Find the transmissivity and storage coefficient by the type-curve method.

Time (min)	30	50	70	90	120	150	200	400	600	900
Drawdown (ft)	6.5	9.0	11	12.4	14.1	15.4	17.0	21.2	23.6	26.0

4.17 From a 2-hour test on a confined aquifer at 400 gpm, the following drawdowns were measured on a number of wells. Determine the transmissivity and storage coefficient by the type-curve method.

Well number:	1	2	3	4	5	6	7
Distance (ft)	29	35	44	60	85	100	163
Drawdown (ft)	14.9	13.8	12.7	11.7	10.1	9.6	7

4.18 During a test on a confined aquifer of 50 m thickness, the following data were noted. The coefficient of permeability was determined from the lab test to be 25 m/day. What were the rate of pumping and the storage coefficient of the aquifer by the type-curve method? $r = 61$ m.

Time (min)	1	3	5	10	18	30	50	100	150	210	240
Drawdown (m)	0.2	0.37	0.45	0.57	0.67	0.76	0.85	0.97	1.05	1.1	1.12

4.19 In a confined aquifer test, the following drawdowns were noted in various observation wells at a time 502 min after pumping commenced at a rate of 300 m³/day.

Well number:	1	2	3	4	5
Distance (m)	4.6	10.4	20	34.5	65.8
Drawdown (m)	1.35	0.9	0.65	0.41	0.20

Determine the transmissivity and storage coefficient by the type-curve method.

4.20 In an unsteady flow, the following data are given: $Q = 0.1$ cfs, $T = 650$ ft²/day, $t = 35$ days, $r = 0.6$ ft, and $S = 6 \times 10^{-3}$. What is the drawdown?

4.21 Solve Problem 4.15 by the Cooper-Jacob method. Check whether this method is valid for this test.

4.22 Solve Problem 4.18 by the Cooper-Jacob method. Check the validity of this approach.

4.23 A well fully penetrating an artesian aquifer was continuously pumped at a rate of 500 gpm. After 200 days, the following drawdowns were noted in three observation wells. Compute the transmissivity and storage coefficient of the aquifer by the Cooper-Jacob method. Also check whether the method is applicable to all wells.

Distance from pumped well (ft)	Drawdown (ft)
100	8.5
1000	5.67
10,000	2.84

4.24 From a pumping test on an unconfined aquifer of 40 ft thickness, the following drawdown measurements were made on an observation well 60 ft from the pumped well. The rate of pumping is 1500 gal/min. Determine the transmissivity and specific yield of the aquifer. Is it justified to treat this the same way as a confined aquifer?

Time (min)	30	60	90	120	300	500	1000	2400
Drawdown (ft)	8	10.6	12.72	13.47	15.83	18.1	23.0	29.0

4.25 An areally infinite aquifer with a transmissivity of 25×10^{-3} m²/s is overlain by a semipervious stratum with a leakage coefficient of 5×10^{-9} per second. A well of 0.5 m diameter is drilled through the aquifer. When the well is pumped at a rate of 0.05 m³/s, determine the drawdown at a distance of 500 m from the well and just outside the well under steady-state conditions.

4.26 A well of 0.4 m diameter fully penetrates a leaky confined aquifer having a hydraulic conductivity of 20 m/day and a thickness of 30 m. The leakage coefficient of the confining layer is 0.05 per day. A drawdown of 1 m is recorded at a distance of 100 m from the well. What is the steady-state rate of pumping of the well?

4.27 A 10-m-thick aquifer is overlain by a 2-m-thick semipervious layer with a hydraulic conductivity of 2×10^{-4} cm/min. The aquifer has a hydraulic conductivity of 1.5×10^{-2} cm/sec and a storage coefficient of 4.4×10^{-4}. If a well in the aquifer is pumped at a rate of 800 m³/day, determine the drawdown at 20 m away from the well after one hour.

4.28 Time-drawdown measurements in an observation well 50 m from a well pumping at 75 gpm in a leaky aquifer are given below. If the confining bed thickness is 12 m, determine the transmissivity and storage coefficient of the main aquifer and the hydraulic conductivity of the confining bed. [*Hint*: The type curve of $r/B = 2$ matches]

Time (min)	3.5	5.0	6.5	7.5	10.0	12.5	15.0	17.5	25	50	125
Drawdown (m)	0.1	0.27	0.4	0.53	0.7	0.85	0.95	1.0	1.1	1.2	1.4

4.29 From a pumping test on a leaky aquifer, the following time-drawdown measurements were recorded in an observation well located 100 ft away from a well pumped at 50 gpm. The confining layer thickness is 15 ft. Determine the transmissivity and storage coefficient of the aquifer and the hydraulic conductivity of the confining layer. [*Hint*: The type curve of $r/B = 1$ matches]

Time (min)	5	10	18	30	40	60	80	100	200	500
Drawdown (ft)	0.4	1	2	3	3.5	4	4.2	4.5	4.7	4.8

4.30 A system is shown in Fig. P4.30. The well is 0.5 m in diameter and fully penetrates a confined aquifer of 50 m thickness and a hydraulic conductivity of 20 m/day. If the pumping rate is 5 m³/min, determine the drawdown (**a**) at point X and (**b**) at the well face.

Figure P4.30

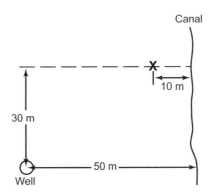

4.31 An unconfined aquifer with $K = 30$ m/day is bounded by a stream on one side as shown in Fig. P4.31. At a distance of 80 m from the stream, a well of 0.4 m diameter is pumped at a rate of 3000 m³/day. What is the drawdown in the well?

Figure P4.31

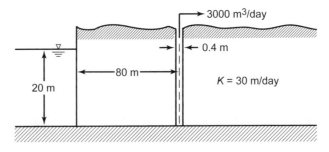

4.32 In Problem 4.30, if an impermeable boundary exists in place of the canal, what is the drawdown at point X? The well has a radius of influence of 2 km.

4.33 For the system shown in Fig. P4.33, determine the drawdown in the well for the pumping rate of 4000 ft^3/day. [*Hint:* $R = 500$ ft.]

Figure P4.33

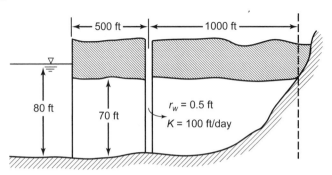

4.34 In the confined aquifer system shown in Fig. P4.34, the pumping rates in well 1 and well 2 are 0.03 m^3/s and 0.02 m^3/s, respectively. The aquifer length extends 2000 m and a width of 1200 m has been included in the flow system being considered. The initial piezometric head is 20 m everywhere. Applying the alternate direction implicit procedure, determine the head distribution after 1 day of pumping.

Figure P4.34   (not to scale)

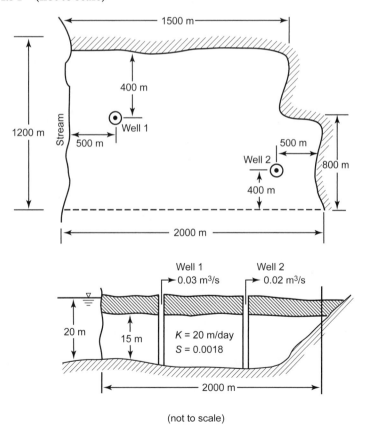

4.35 In a step-drawdown test on a production well, the pumping rates and drawdowns at the end of each one-hour step are shown below. Determine the aquifer and the well loss coefficients.

Step	Rate, m³/min	Drawdown, m
1	0.43	0.86
2	0.76	2.13
3	1.51	7.31
4	1.89	10.64
5	2.27	15.45

4.36 The step-drawdown data for a test in the city of Petersberg, MI are given below. Determine the coefficients in the equation $s = B\,Q + C\,Q^2$.

Step	Discharge, gpm	Drawdown, ft
1	155	2.38
2	283	6.09
3	334	8.08
4	358	9.23

4.37 The water requirement from an aquifer is 3 million gallons per day. The aquifer has a transmissivity of 6000 ft²/day and a storage coefficient of 8×10^{-4}. The available drawdown is 50 ft. It is proposed to install wells of 8 in. diameter. Design the well field for well efficiency of 90%.

4.38 An 8-in.-diameter well pumps a confined aquifer at a rate of 300 gal/min. The transmissivity and storage coefficient for the aquifer are 200 ft²/hour and 0.0003, respectively. If the well has an efficiency of 70%, what is the total drawdown after 12 hours? Also, what is the specific capacity of the well?

4.39 In a water-table aquifer, the rate of pumping is 2000 m³/day. The aquifer has a specific yield of 0.02. The specific capacity after 12 hours of pumping a well of 0.3 m diameter is 150 m²/day. If the well has an efficiency of 70%, determine the transmissivity of the aquifer.

5

Contaminant Transport and Groundwater Monitoring

◆◆◆

5.1 TRANSPORT PROCESSES

5.1.1 Advective Transport

Advective transport is also known as *convective transport*. Advection can be the result of an externally applied pressure difference, known as forced advection, as in the case of pipe flow, or it can be due to gravity or density changes, referred to as free advection. The advective flux, the rate of mass transported with bulk fluid per unit area, is given by

$$N_a = -v_x C \quad [FT/L^3] \tag{5.1}$$

where

N_a = advective mass flux, mass/(area • time)

 (negative sign for transport in direction of decreasing concentration)

v_x = linear water velocity in x-direction

C = concentration of substance, mass/volume

In the case of flow through a porous medium, the velocity term can be characterized by the Darcy equation (3.4).

EXAMPLE 5.1

Groundwater containing a contaminant flows through a cube of sand of 1 m on each side. It enters the left face at el. 99.99 m and leaves from the right face at el. 99.90 m. The hydraulic conductivity is 1×10^{-3} m/s. Assume that the concentration of contaminant is 200 mg/L when it enters the cube. Calculate the mass flux of the contaminant due to advection.

SOLUTION

1. From eq. (3.4),

$$v = K \frac{\Delta h}{L}$$

$$\Delta h = 99.99 - 99.90 = 0.09 \text{ m}$$

$$L = 1 \text{ m}$$

$$v = (1 \times 10^{-3} \text{ m/s}) \left(\frac{0.09 \text{ m}}{1 \text{ m}} \right) = 9 \times 10^{-5} \text{ m/s}$$

2. From eq. (5.1),

$$C = 200 \text{ mg/L or } 200 \text{ g/m}^3$$

$$N_a = -\left(9 \times 10^{-5} \frac{\text{m}}{\text{s}} \right) \left(200 \frac{\text{g}}{\text{m}^3} \right)$$

$$= -0.018 \text{ g/m}^2 \cdot \text{s (in the direction of negative gradient)}$$

5.1.2 Diffusive Transport: Fick's First Law of Diffusion

In Chapter 3, groundwater flow was related to the hydraulic head gradient by Darcy's law. For mass transport, there is a similar law that relates the mass flux to the concentration gradient. This law, known as Fick's First Law of Diffusion, is given by:

$$N = -D \frac{\partial C}{\partial x} \quad [\text{FT}/\text{L}^3] \tag{5.2}$$

where

N = diffusive mass flux, mass/(area $\cdot$ time)

C = concentration of diffusing material (FT^2/L^4)

x = space coordinate in which direction the diffusion is measured

D = diffusion coefficient or diffusivity, L^2/T

The negative sign indicates transport in the direction of decreasing concentration. It is a partial derivative because the concentration is both space- and time-dependent. At steady state, the concentration at any point is constant with respect to time. Accordingly, eq. (5.2) will convert to an ordinary differential form.

Diffusion coefficients in a porous medium are smaller than in a pure liquid medium because solids hinder diffusion. To obtain the diffusion coefficient for a porous medium from a diffusion coefficient of liquid, a constant factor of less than 1 is applied to account for the porous medium structure. The constant factor is a function of both porosity and tortuosity, which is a measure of the circuitous flow path followed by the water molecules. Domenico and Schwartz (1990) have presented several empirical relations for the constant.

EXAMPLE 5.2

Methane concentration in a landfill at a depth of 3 m is 20 mg/L. The atmospheric concentration is 4.87 mg/L. What is the rate of transport of methane from an acre of landfill? The diffusion coefficient is $4.0 \times 10^{-6} \text{ cm}^2/\text{s}$.

SOLUTION

1. $Grad\,(C) = \dfrac{20.0 - 4.87}{3} = 5.043 \dfrac{mg}{L \cdot m} \left[\dfrac{1000\ L}{1\ m^3} \right] = 5043\ mg/m^4$

$$D = 4.0 \times 10^{-6} \dfrac{cm^2}{s} \left[\dfrac{1\ m^2}{10^4\ cm^2} \right] = 4.0 \times 10^{-10}\ m^2/s$$

$$A = 1\ acre \left[\dfrac{4047\ m^2}{1\ acre} \right] = 4047\ m^2$$

2. From eq. (5.2), $\ N = -\left(4.0 \times 10^{-10}\ m^2/s\right)\left(5043\ mg/m^4\right) = -2.017 \times 10^{-6}\ mg/s \cdot m^2$

3. Rate of transport = mass flux $(N) \times$ area of transfer

$$= -\left(2.017 \times 10^{-6}\ mg/s \cdot m^2\right)\left(4047\ m^2\right) = -8.16 \times 10^{-3}\ mg/s$$

(from landfill to atmosphere)

5.1.3 Dispersive Transport

Dispersive transport that originates from spatial and temporal variations of velocity in the flow field can result from several different physical phenomena. Dispersion is described by the mathematics similar to the diffusion process. When substituting a new dispersion coefficient, E, in eq. (5.2), Fick's law for dispersive transport is restated as:

$$N = -E\frac{\partial C}{\partial x} \quad [FT/L^3] \tag{5.3}$$

Based on the characteristics of the flow pattern, dispersion is classified in the following three categories:

1. Taylor dispersion occurs in laminar flow in pipes and narrow channels. It is characterized by the parabolic distribution of velocity. The solute concentrations after a short time become normally distributed longitudinally and transversely. The Taylor dispersion coefficient is inversely proportional to the diffusion coefficient and can be calculated explicitly.

2. Eddy or turbulent dispersion results from velocity fluctuations in advection-dominated flow fields like channels, rivers, streams, and lakes. In turbulent flow, both velocity and concentration fluctuate. These fluctuations are essentially normally distributed in terms of their ensemble averages. In turbulent flow, the dispersion coefficients are formulated so as to get results parallel to the diffusion process. The coefficient weakly relates to the properties of solute and is highly system-specific. Reliable values of the dispersion coefficients must be found from experimental data.

3. Mechanical dispersions are associated with heterogeneity in the flow field such as the flow through a porous medium. Mechanical dispersion is caused by three mechanisms: (1) within individual pore space, molecules travel at different velocities at different points across the pore due to the drag by the pore surface, (2) the difference in pore sizes along the flow paths causes variations of velocity, and (3) the tortuosity, branching, and interfingering of pore channels cause diversions of flow around the obstacles

and the irregular flow paths lead to varied flow velocities. The variation of velocity leads to the distributed front of solute concentration which tends to be generally normally distributed.

Laboratory experiments have shown that the value of the mechanical dispersion coefficient, both longitudinally and transversely, is equal to the velocity of flow times the dispersivity of the medium. Dispersivity has a unit of length and is a characteristic property of the medium. Domenico and Shwartz (1990) describe relations to predict dispersivity.

5.2 Mass Transport Equations

Section 5.1 described the transport processes. These processes are combined with the mass conservation law to develop the framework for the mass transport of contaminants in a system. In words, the law of mass conservation is stated as:

$$\begin{pmatrix} \text{mass inflow} \\ \text{rate} \end{pmatrix} - \begin{pmatrix} \text{mass outflow} \\ \text{rate} \end{pmatrix} = \begin{pmatrix} \text{rate of change of mass} \\ \text{storage within a system} \end{pmatrix} \quad [\text{FT/L}] \qquad (5.4)$$

Consider the parallelepiped element of sides $2dx$, $2dy$, and $2dz$ of Figure 5.1. Let the center of the element be at O, where the concentration of a transport substance is C and the flux (rate of mass transfer) is N. The rate at which the substance enters (inflow) the face ABCD is given by

$$4dydz\left(N_x - \frac{\partial N_x}{\partial x}dx\right)$$

Similarly, the rate of exit (outflow) of the substance through A′B′C′D′ is given by

$$4dydz\left(N_x + \frac{\partial N_x}{\partial x}dx\right)$$

Thus, inflow minus outflow of mass from these two faces is equal to:

$$4dydz\left(N_x - \frac{\partial N_x}{\partial x}dx\right) - 4dydz\left(N_x + \frac{\partial N_x}{\partial x}dx\right)$$

Figure 5.1   An element of volume in a flow field.

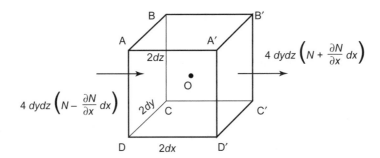

or

$$-8 \; dxdydz \frac{\partial N_x}{\partial x}$$

Similarly from other faces we obtain the net mass of:

$$-8 \; dxdydz \frac{\partial N_y}{\partial y} \text{ and } -8 \; dxdydz \frac{\partial N_z}{\partial z}$$

The rate at which the substance is changing within the element is:

$$(2dx)(2dy)(2dz)\frac{\partial C}{\partial t}$$

Substituting these terms in eq. (5.4), we obtain:

$$8 \; dxdydz \frac{\partial C}{\partial t} = -8 \; dxdydz \frac{\partial N_x}{\partial x} - 8 \; dxdydz \frac{\partial N_y}{\partial y} - 8 \; dxdydz \frac{\partial N_z}{\partial z}$$

or

$$\frac{\partial C}{\partial t} + \frac{\partial N_x}{\partial x} + \frac{\partial N_y}{\partial y} + \frac{\partial N_z}{\partial z} = 0 \quad [FT/L^4] \tag{5.5}$$

For N_x, N_y, N_z, the terms for advective, diffusive, or dispersive transports can be substituted from eqs. (5.1), (5.2), and (5.3), singularly or in combination.

5.2.1 Fick's Second Law of Diffusion

When eq. (5.2) is substituted in eq. (5.5), with a constant diffusion coefficient, it becomes:

$$\frac{\partial C}{\partial t} = D\frac{\partial^2 C}{\partial x^2} + D\frac{\partial^2 C}{\partial y^2} + D\frac{\partial^2 C}{\partial z^2} \quad [FT/L^4] \tag{5.6a}$$

In vector nomenclature, this is expressed as

$$D\nabla^2 C = \frac{\partial C}{\partial t} \quad [FT/L^4] \tag{5.6b}$$

or

$$D \; \text{div (grad } C) = \frac{\partial C}{\partial t} \quad [FT/L^4] \tag{5.6c}$$

In one dimension, this reduces simply to:

$$\frac{\partial C}{\partial t} = D\frac{\partial^2 C}{\partial x^2} \quad [FT/L^4] \tag{5.7}$$

Equation (5.7) is referred to as Fick's Second Law of Diffusion.

5.2.2 The Advection-Diffusion Equation

The combined flux can be described by adding the advective flux of eq. (5.1) with the diffusive flux of eq. (5.2). For the x-direction this is:

$$N_x = -D\frac{\partial C}{\partial x} - v_x C \quad [FT/L^3] \tag{5.8}$$

Similar expressions result for the other two directions. Substituting N_x from eq. (5.8) into eq. (5.5) for the one-dimensional case:

$$\frac{\partial C}{\partial t} = D\frac{\partial^2 C}{\partial x^2} - \frac{\partial}{\partial x}(v_x C) \tag{a}$$

or

$$\frac{\partial C}{\partial t} = D\frac{\partial^2 C}{\partial x^2} - v_x\frac{\partial C}{\partial x} - C\frac{\partial v_x}{\partial x} \quad [\text{FT/L}^4] \tag{5.9}$$

Equation (5.9) is called the advection-diffusion equation. For uniform flow $\partial v_x/\partial x = 0$ and, hence, the last term drops out.

5.2.3 The Advection-Diffusion-Dispersion Equation

In many systems, dispersion is an important process. Logically, dispersion should be reflected in the velocity term since it is a velocity-related phenomenon. However, a simplistic approach is adopted to account for dispersion effects. The coefficient D in eq. (5.9) is substituted with another coefficient that incorporates the combined effects of diffusion and dispersion.

5.2.4 Mass Transport with Reaction

In eq. (5.9), a source or sink term is included depending on whether the substance is being added or removed through chemical or biological processes. The resulting equation, without the last term of eq. (5.9), is:

$$\frac{\partial C}{\partial t} = D\frac{\partial^2 C}{\partial x^2} - v_x\frac{\partial C}{\partial x} \pm r \quad [\text{FT/L}^4] \tag{5.10}$$

where r is the rate of mass produced or consumed by the chemical reaction or by some other means, in the units of mass per unit volume per unit time. Usually, this term is represented by a first order kinetic reaction dependent on the concentration C.

5.3 SOLUTIONS OF THE MASS TRANSPORT EQUATION

A great variety of mass transport problems could be formulated by variations in the mass-transport equation. These can include (1) one, two, or three dimensional approach, (2) homogeneous or nonhomogeneous, isotropic or anisotropic medium, (3) point, line, or plane source, (4) pulse or continuous release, (5) constant or variable flux, (6) constant, time-dependent, or concentration-dependent diffusion and dispersion coefficients, and (7) first-order or heterogeneous reaction kinetics.

The governing equation (5.10) is a second-order partial differential equation that offers only limited possibilities of analytical solutions. In the literature, analytical solutions are available for diffusion coefficients that are constant or time-dependent. For concentration-dependent diffusion coefficients, numerical methods have been proposed. Analytical solutions have two standard forms. Either the solution comprises a series of error functions of related integrals, or it is in the form of a trigonometric series that converges very fast at large time values.

The solutions are based on the pulse or instantaneous release and the continuous release of a contaminant. Three types of sources are considered. A point source indicates that the contaminant of mass M is released at a single point that can diffuse in all three directions. A line source means that the released contaminant is distributed over a line of length l and it can diffuse only in the x-y plane perpendicular to the line. A plane source means that the released contaminant is distributed over a plane of surface area A and it can diffuse in the x-direction perpendicular to the plane. Simple analytical cases have been presented here.

5.3.1 Instantaneous Release from a Plane Source in an Infinite System

A substance having mass M is instantaneously deposited in the y-z plane at $x = 0$ so that it can spread out to an infinite distance positively and negatively in the x-direction. The infinite system condition is rarely encountered in practice. However, this is a basic solution to the mass transport equation that leads to tractable solutions to a variety of complex diffusion problems.

$$\text{Initial conditions are:} \quad C = 0 \text{ as } x \to -\infty \text{ and } x \to +\infty, \ t = 0$$

$$\text{Boundary conditions are:} \quad \frac{dC}{dx} = 0 \text{ as } x \to \infty$$

A general solution of eq. (5.7) by the Laplace transformation is:

$$C = \frac{K}{\sqrt{t}} e^{-\frac{x^2}{4Dt}} \quad [\text{FT}^2/\text{L}^4] \tag{5.11}$$

where K is a constant of integration.
From definition,

$$\frac{M}{A} = \int_{-\infty}^{+\infty} C dx \tag{a}$$

Substituting C from eq. (5.11) into (a)

$$M = A \int_{-\infty}^{+\infty} \frac{K}{\sqrt{t}} e^{-\frac{x^2}{4Dt}} dx \tag{b}$$

The trick to solving this equation is to define a new dimensionless variable

$$\xi = \frac{x}{\sqrt{4Dt}} \tag{c}$$

hence,

$$d\xi = \frac{1}{\sqrt{4Dt}} dx \tag{d}$$

Substituting (c) and (d) into (b)

$$M = 2AK\sqrt{D} \int_{-\infty}^{+\infty} e^{-\xi^2} d\xi \tag{e}$$

or

$$M = 2AK\sqrt{D}\sqrt{\pi} \tag{f}$$

or

$$K = \frac{M}{2A\sqrt{D}\sqrt{\pi}} \qquad \text{(g)}$$

Substituting (g) into eq. (5.11)

$$C = \frac{M}{2A\sqrt{\pi D t}} e^{-\frac{x^2}{4Dt}} \quad [FT^2/L^4] \qquad (5.12)$$

When the advection term is also included in the analysis

$$C = \frac{M}{2A\sqrt{\pi D t}} e^{-\frac{(x-Ut)^2}{4Dt}} \quad [FT^2/L^4] \qquad (5.13)$$

where

M = mass of substance released

U = average velocity in stream

EXAMPLE 5.3

A nonreactive tracer mass of 1 mg is injected instantaneously in the middle of a long tube of cross section 2 cm^2. Plot the distribution along the tube of the tracer concentration at times 1, 5, and 10 min. The dispersion coefficient is 1.8×10^{-2} cm^2/s.

SOLUTION

1. From eq. (5.12),

$$C = \frac{M}{2A\sqrt{\pi D t}} e^{-\frac{x^2}{4Dt}}$$

$$C = \frac{1}{2(2)\sqrt{\pi(1.8\times10^{-2})}\sqrt{60t}} e^{-\frac{x^2}{4(1.8\times10^{-2})(60t)}}; t \text{ taken in min.}$$

or

$$C = \frac{0.136}{\sqrt{t}} e^{-0.23\frac{x^2}{t}}$$

2. The above equation is solved successively for $t = 1$ min, 5 min, and 10 min for various values of x in Table 5.1.

3. The concentration-distance curves are plotted in Figure 5.2.

EXAMPLE 5.4

The tracer of Example 5.3 moves in the column with a velocity of 1×10^{-2} cm/s. Determine the concentration at times 1 min, 5 min, and 10 min.

Figure 5.2 Concentration distribution curves for an instantaneous plane source in an infinite system.

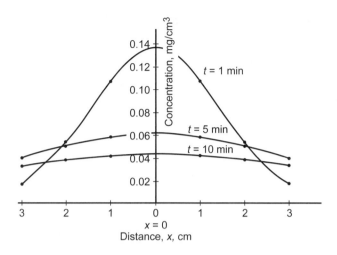

Table 5.1 Distribution of Tracer Concentration in an Infinite System without Advection for Instant Release

x cm	for $t = 1$ min $C = 0.136\,e^{-0.23x^2}$ mg/cm^3	for $t = 5$ min $C = 0.061\,e^{-0.046x^2}$ mg/cm^3	for $t = 10$ min $C = 0.043\,e^{-0.023x^2}$ mg/cm^3
± 0	0.136	0.061	0.043
± 1	0.108	0.058	0.042
± 2	0.054	0.051	0.039
± 3	0.017	0.040	0.035

SOLUTION

1. From eq. (5.13),

$$C = \frac{M}{2A\sqrt{\pi Dt}}\,e^{-\frac{(x-Ut)^2}{4Dt}}$$

or

$$C = \frac{1}{2(2)\sqrt{\pi\left(1.8\times10^{-2}\right)}\sqrt{60t}}\,e^{-\frac{\left[x-\left(1\times10^{-2}\right)(60t)\right]^2}{4\left(1.8\times10^{-2}\right)(60t)}}$$

or

$$C = \frac{0.136}{\sqrt{t}}\,e^{-\frac{0.23(x-0.6t)^2}{t}}$$

Section 5.3   Solutions of the Mass Transport Equation

2. The above equation is solved successively for 1 min, 5 min, and 10 min at various values of x in Table 5.2.

Table 5.2   Distribution of Tracer Concentration in an Infinite System with Advection for Instant Release

x cm	for $t = 1$ min $C = 0.136\,e^{-0.23(x-0.6)^2}$ mg/cm^3	for $t = 5$ min $C = 0.061\,e^{-0.046(x-3)^2}$ mg/cm^3	for $t = 10$ min $C = 0.043\,e^{-0.023(x-6)^2}$ mg/cm^3
± 0	*	*	*
± 1	0.131	*	*
± 2	0.087	*	*
± 3	0.036	0.061	*
± 5	0.002	0.051	*

* For this value of x, Ut is larger than x in eq. (5.13), i.e., the tracer is moved out of this position due to advection.

5.3.2 Instantaneous Release from a Plane Source in a Semi-Infinite System

In this case, a mass M deposited initially in the y-z plane at $x = 0$ can diffuse only in the positive x-direction to an infinite distance. This model can be applied with reasonable accuracy to systems with long lengths, for example a long channel or a pipeline.

By introducing the concept of reflection at the boundary, the basic solution of eq. (5.12) can be extended to solve many problems. For this case, an impervious boundary is inserted at $x = 0$, in which the solution of the negative x-side is reflected back to the $x > 0$ region, and is superimposed on the original distribution of the positive side. Thus, the concentration values of eq. (5.13) get doubled, or

$$C = \frac{M}{A\sqrt{\pi Dt}}\,e^{-\frac{(x-Ut)^2}{4Dt}} \quad [\mathrm{FT}^2/\mathrm{L}^4] \tag{5.14}$$

EXAMPLE 5.5

A laboratory flume filled with sand has a width of 15 cm, water depth of 20 cm, and a flow velocity of 4.5 cm/s. A nonreactive chemical mass of 500 mg enters at one end of the flume across the entire cross section. A concentration of 1.0 µg/cm^3 has been observed at a distance of 11 m from the end where the chemical was introduced, after 4 minutes. What is the dispersion coefficient of the chemical?

SOLUTION

1. $A = 15 \times 20 = 300$ cm^2

   $x = 11 \times 100 = 1100$ cm

   $C = 1\ \mu\text{g/cm}^3$ or 1×10^{-3} mg/cm^3

2. From eq. (5.14),

$$C = \frac{M}{A\sqrt{\pi Dt}} e^{-\frac{(x-Ut)^2}{4Dt}}$$

or

$$1\times 10^{-3} = \frac{500}{(300)\sqrt{\pi (D)(4\times 60)}} e^{-\frac{\left[1100-(4.5)(4\times 60)\right]^2}{4D(60\times 4)}}$$

or

$$1\times 10^{-3} = \frac{1}{16.47\sqrt{D}} e^{-\frac{0.417}{D}}$$

3. Solving by trial and error,

$$D = 7.8 \times 10^{-2} \text{cm}^2/\text{s}$$

5.3.3 Continuous Release from a Plane Source in an Infinite System

In the above two models, all the diffusing substance was released initially. This situation is modified when a substance is discharged continuously in the y-z plane. An example is when a sewer discharges in the middle of a stream.

Initial conditions are: $C = 0$ at $x > 0$, $t = 0$

Boundary conditions are: $C = C_0$ at $x = 0$, $t > 0$

$C = 0$ as $x \to -\infty$, and $x \to +\infty$, $t > 0$

The continuous release is simulated to be composed of an infinite number of instantaneous releases from the plane source. Thus, an infinite number of basic solutions given by eq. (5.12) are superimposed.

With reference to Figure 5.3, consider the diffusing substance in element dx to be a plane source of strength (M/A) of $C_0 dx$. At a distance x from the element, the concentration at O, from eq. (5.12) is

$$\frac{C_0 dx}{2\sqrt{\pi Dt}} e^{-\frac{x^2}{4Dt}}$$

Assume a dimensionless variable $\eta = \frac{x}{2\sqrt{Dt}}$ so that $dx = 2\sqrt{Dt}\, d\eta$

Hence the above relation becomes

$$\frac{C_0}{\sqrt{\pi}} e^{-\eta^2}\, d\eta$$

Figure 5.3 Definition diagram for extended distribution.

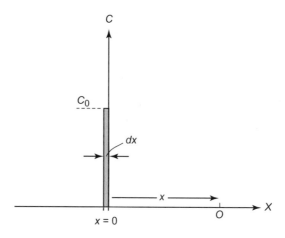

Summing up the successive elements

$$C = \frac{C_0}{\sqrt{\pi}} \int_{\eta}^{\infty} e^{-\eta^2} d\eta \qquad \text{(a)}$$

A standard mathematical function known as the error function, usually written as erf (z), is defined as

$$\text{erf}(z) = \frac{2}{\sqrt{\pi}} \int_{0}^{z} e^{-\eta^2} d\eta$$

The function has the following properties:

$$\text{erf}(-z) = -\text{erf}(z), \ \text{erf}(0) = 0, \ \text{erf}(\infty) = 1, \ 1 - \text{erf}(z) = \text{erfc}(z)$$

and

$$\text{erfc}(-z) = 1 + \text{erf}(z)$$

where erfc (z) is the error function complement. Also,

$$\frac{2}{\sqrt{\pi}} \int_{z}^{\infty} e^{-\eta^2} d\eta = \frac{2}{\sqrt{\pi}} \int_{0}^{\infty} e^{-\eta^2} d\eta - \frac{2}{\sqrt{\pi}} \int_{0}^{z} e^{-\eta^2} d\eta$$

$$= 1 - \text{erf}(z)$$

$$= \text{erfc}(z)$$

Comparing (a) and (b)

$$C = \frac{C_0}{2} \text{erfc}\left(\frac{x}{2\sqrt{Dt}}\right) \quad [\text{FT}^2/\text{L}^4] \qquad (5.15)$$

Including the advection term:

$$C = \frac{1}{2} C_0 \ \text{erfc}\left[\frac{(x - Ut)}{2\sqrt{Dt}}\right] \quad [\text{FT}^2/\text{L}^4] \qquad (5.16)$$

A table for error function, erf (z) is given in Appendix E. To find erfc (z) use the relation erfc $(z) = 1 - \text{erf}(z)$.

EXAMPLE 5.6

A nonreactive tracer of concentration 0.2 mg/L is continuously sent through the middle of a long channel at a velocity of 1×10^{-2} cm/s. Plot the relative concentration (C/C_0) at a distance 30 cm from the release, at 0, 10, 30, 60, 90, and 120 minutes. The coefficient of dispersion is 1.8×10^{-2} cm^2/s.

SOLUTION

1. From eq. (5.16),

$$C = \frac{1}{2} C_0 \, \text{erfc} \left[\frac{x - Ut}{2\sqrt{Dt}} \right]$$

or

$$C = \frac{1}{2}(0.2) \, \text{erfc} \left[\frac{30 - \left(1 \times 10^{-2}\right)(60t)}{2\sqrt{\left(1.8 \times 10^{-2}\right)(60t)}} \right], \text{ taking } t \text{ in min}$$

or

$$C = 0.1 \, \text{erfc} \left[\frac{30 - 0.6t}{2.078\sqrt{t}} \right]$$

2. The above relation is solved for various values of t in Table 5.3 and plotted in Figure 5.4.

Table 5.3   Distribution of Tracer Concentration in an Infinite System with Advection for Continuous Release

t min	$z = \dfrac{30 - 0.6t}{2.078\sqrt{t}}$	erfc(z)	$C = 0.1$ erfc(z)	$\dfrac{C}{C_0}$
0	∞	0*	0	0
10	3.65	0	0	0
30	1.05	0.138	0.014	0.07
60	−0.37	1.399**	0.14	0.7
90	−1.22	1.916	0.19	0.95
120	−1.85	1.991	0.20	1.00

* erfc (z) = 1− erf(z) from Appendix E.
** erfc (−z) = 1+ erf(z) from Appendix E.

5.3.4 Continuous Release from a Plane Source in a Semi-Infinite System

For a continuous release in the y-z plane which can diffuse only in the positive x-direction, such as the discharge at the end of a channel, the diffusing material of the negative x domain is simply added back into the positive x domain as though reflected back at the boundary $x = 0$. Thus, the concentration given by eq. (5.16) is doubled.

Figure 5.4 Variation in concentration with time at a specified distance for continuous release.

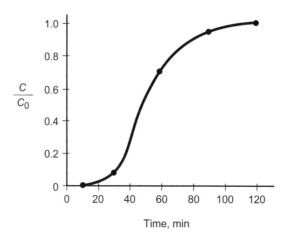

$$C = C_0 \ \mathrm{erfc}\left[\frac{x - Ut}{2\sqrt{Dt}}\right] \quad [FT/L^4] \tag{5.17}$$

5.4 FATE OF CONTAMINANTS

Advection, dispersion, and diffusion, as described in Section 5.1, constitute the physical processes by which contaminants are distributed through an ecosystem. Simultaneously, certain chemical and biological processes also take place that remove or transform the contaminants. Chemically, the contaminants might be sorbed onto organic carbon in water or air, sorbed onto mineral grains in an aquifer, undergo chemical precipitation or oxidation-reduction type reactions, or degrade biologically. Volatile compounds could undergo volatization and radioactive compounds could decay.

The rate at which a chemical process progresses is stated by the rate law, according to which the reaction rate is equal to a constant times the product of the concentrations of the reactants raised to some power. When this power is zero, i.e., the rate is independent of the reactants' concentrations, the reaction is said to be of zero order. It is a first-order reaction when the rate directly depends on the concentration (raised to power one) of one reactant, i.e., $r = KC$, where K is the rate constant. It is a second- or higher-order reaction when the rate depends on the concentration raised to the second power or more of one reactant or when it depends on the concentrations of two different reactants each raised at least to the first power. First-order reactions commonly are adopted. In some cases, they oversimplify a process, but higher-order processes are difficult to measure and apply analytically. The biodegradation process has reaction aspects similar to chemical processes. Biodegradation in a natural system can be modeled as a first-order reaction.

To quantify chemical and biological processes, a general approach consists of including a sink or a source term in the mass transport equation. The sorption process is represented in the advection-dispersion eq. (5.9) by multiplying the first term of the equation $\partial C/\partial t$ by $R\partial C/\partial t$, where R is the retardation factor. The effect of sorption is to reduce (retard) the velocity of movement of the contaminants.

For all transformation processes involving a first-order kinetic reaction like hydrolysis, volatilization, radioactive decay, and biodegradation, the mathematical relation is described through eq. (5.10) by substituting $r = KC$, where K denotes the overall first-order rate coefficient composed of the first-order rate coefficients of individual processes. Bear (1979) presented the solution for mass transport with transformation processes for continuous release from a plane source.

5.5 AQUEOUS PHASE OR SOLUBLE CONTAMINANTS

Once a contaminant reaches an aquifer, it moves through groundwater forming a plume of contamination. Two states of contaminants exist: soluble contaminants that dissolve to a large extent in groundwater, and insoluble or immiscible fluids, like oil, that do not appreciably mix with water. The movement of the first group of substances is controlled by the advection, diffusion, dispersion, and reaction processes depicted by the mass transport equation (5.10). The solutions of Section 5.3 are, therefore, applicable to soluble or aqueous phase contaminants.

In the context of groundwater flow, the velocity term in eqs. (5.13), (5.14), (5.16), and (5.17) pertains to seepage velocity through a porous medium, as given by Darcy's equation (3.5) rewritten below:

$$U = \frac{K}{\eta}\frac{dh}{dx} \quad [\text{L/T}] \tag{5.18}$$

where

K = hydraulic conductivity in the direction of flow

η = porosity

dh/dx = hydraulic gradient

U = advection velocity through porous medium

As stated in Section 5.1.3, flow through a porous medium experiences mechanical dispersion in addition to diffusion. To account for this effect, the coefficient of the mass transport equation is replaced with a combined coefficient recognized as the hydrodynamic dispersion coefficient, as follows:

$$D_L = D_d + D_m \quad [\text{L}^2/\text{T}] \tag{5.19}$$

where

D_L = longitudinal hydrodynamic dispersion coefficient

D_d = effective molecular diffusion coefficient

D_m = mechanical dispersion coefficient

The effective molecular diffusion coefficient D_d equals τD_0, where D_0 is the molecular diffusion in water and τ is tortuosity of the medium, which accounts for the hindrance to flow through a porous medium. For granular porous media τ is typically in the range of 0.6 to 0.7. The mechanical dispersion coefficient D_m equals αU, where U is the seepage velocity and α is known as the dispersivity parameter.

Substituting the expressions for D_d and D_m, the hydrodynamic dispersion coefficient is

$$D_L = \tau D_0 + \alpha U \quad [\text{L}^2/\text{T}] \tag{5.20}$$

A similar relation is given for the transverse hydrodynamic dispersion coefficient. However, field studies have indicated that the transverse dispersion in porous media is very small and in many situations close to zero.

Perkins and Johnston (1963) plotted the relative hydrodynamic coefficient (ratio of D_L to D_0) against the Peclet Number, defined as Ud / D_0, where U is the seepage velocity and d is the average grain diameter, as shown in Figure 5.5. Up to the Peclet Number of 0.02, the dispersion coefficient is a constant given by the diffusion term τD_0. This is the case for groundwater flow through clays and fine-grained deposits. When the Peclet Number is larger than 6, the dispersion coefficient is given by αU and the diffusion contribution is negligible. In between is the transition zone where both terms are relevant.

Hydrodynamic dispersion coefficients are estimated by laboratory methods that utilize columns packed with porous media representing the formations and by field methods that involve on-site testing. There are three types of field testing:

1. Natural gradient tracer tests in which a tracer is injected into the ground via a well. The plume that develops under the existing water table gradient is measured by means of small amounts of water withdrawn from down-gradient observation wells. For the plume of known concentrations, the mass transport equation is solved with the dispersion coefficient as the unknown.

2. Well tests that involve injecting a tracer into a well as a pulse or on a continuous basis and pumping out the same or another well to collect the concentration vs. time data for the analysis.

3. Model calibration that starts with an estimate of the dispersion coefficient. The value is then varied during successive model runs until the computer model yields a reasonable match with the observed contaminant plume.

5.5.1 Uncertainties of Dispersion Coefficients for Porous Media

It has been observed that the measured values of the dispersion coefficient are scale dependent. Laboratory tests performed on small-scale models provide much smaller values (the dispersion parameter α of 0.0001 to 0.01 m) as compared to field tests by the natural gradient and well test methods (α of 0.01 to 15 m), which in turn have lower values than model calibration tests (α of 3 to 100 m). The longitudinal dispersion coefficient increases with increased transport distances in a complex manner. It is a consensus among contaminant hydrologists that the macroscopic dispersions that occur in the field are not the pore-scale processes described through diffusion and mechanical dispersions.

Macrodispersion is the result of aquifer heterogeneity. Even in an aquifer described to be homogeneous, the hydraulic conductivity is known to vary by over two orders of magnitude and the porosity up to 60%. From Darcy's law, it is obvious that both of these properties affect the velocity of flow. Since dispersion depends on velocity variations, hydraulic conductivity and porosity play an important role in dispersion phenomena. The former is more important since it varies over a much larger range.

Based on the scale effect, the dispersion coefficient will increase with the transport distance. Eventually the flow path will become long enough to incorporate all possible variations of hydraulic conductivity and porosity and the value of the dispersion then will reach an optimum value. It has been found that the dispersion follows Fick's law only at the laboratory scale or for very long flow paths under optimum macrodispersion conditions.

The application of the advection-dispersion equation to field studies requires the use of an unrealistically high longitudinal dispersion coefficient. The coefficient obtained from

Figure 5.5 Hydrodynamic dispersion coefficient (from Perkins and Johnston, *A Review of Diffusion and Dispersion in Porous Media*, © 1963 Society of Petroleum Engineers).

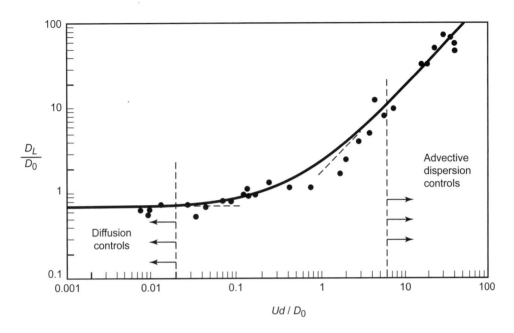

the field is a fitted parameter and does not represent the intrinsic property of the aquifer. Dispersion is a phenomenon that often is used as a mathematical convenience to correct for ignorance about the heterogeneous nature of the aquifer and for poor understanding of the processes occurring within the aquifer (USEPA, 1989).

EXAMPLE 5.7

A contaminant continuously discharges into a sandy aquifer that has a hydraulic conductivity of 12 m/day, porosity of 0.6, and average grain size of 1 mm. The molecular diffusion coefficient of the contaminant in water is 1.25×10^{-5} cm^2/s. The water table gradient is 0.005. How far will 10% of the initial concentration reach in one-half hour?

SOLUTION

1. From eq. (5.18),

$$U = \frac{12}{0.6}(0.005) = 0.1\frac{m}{d} \text{ or } 6.94 \times 10^{-3} \text{ cm/min}$$

$$\frac{Ud}{D_0} = \left(6.94 \times 10^{-3}\frac{cm}{min}\right)(1 \text{ mm})\left(\frac{1}{1.25 \times 10^{-5}}\frac{s}{cm^2}\right)\left[\frac{1 \text{ cm}}{10 \text{ mm}}\right]\left[\frac{1 \text{ min}}{60 \text{ s}}\right]$$

$$= 0.93$$

2. From Fig. 5.5,

$$\frac{D_L}{D_0} = 2.5$$

$$D_L = 2.5(1.25 \times 10^{-5}) = 3.13 \times 10^{-5} \text{ cm}^2/\text{s}$$

3. From eq. (5.16),

$$\frac{C}{C_0} = \frac{1}{2} \text{ erfc}\left[\frac{x - Ut}{2\sqrt{Dt}}\right]$$

or

$$\frac{0.1}{1} = \frac{1}{2} \text{ erfc}\left[\frac{x \text{ cm} - \left(6.94 \times 10^{-3} \text{ cm/min}\right)(0.5 \times 60 \text{ min})}{2\sqrt{\left(3.13 \times 10^{-5} \text{ cm}^2/\text{s}\right)(0.5 \times 60 \times 60 \text{ s})}}\right]$$

or

$$0.1 = 0.5 \text{ erfc}\left(\frac{x - 0.21}{0.475}\right)$$

or

$$\text{erfc}\left(\frac{x - 0.21}{0.475}\right) = 0.2$$

or

$$\text{erf}\left(\frac{x - 0.21}{0.475}\right) = 1 - 0.2 = 0.8$$

4. From Appendix E, for $z = 0.9$ erf(z) has a value of 0.8

5. Hence,

$$\left(\frac{x - 0.21}{0.475}\right) = 0.9$$

$$\text{or } x = 0.9(0.475) + 0.21 = 0.64 \text{ cm}$$

5.6 IMMISCIBLE OR NONAQUEOUS PHASE LIQUIDS (NAPL)

Liquids that do not mix with water move as a separate phase distinct from water flow. Only to the extent that the liquids dissolve in water is their flow controlled by the principles discussed in the previous section. Some immiscible liquids, such as gasoline and diesel fuel, may be lighter than water; these are known as *light nonaqueous phase liquids* (LNAPL). Others that are denser than water, such as chlorinated hydrocarbons, are called *dense nonaqueous phase liquids* (DNAPL). When a liquid makes contact with another substance—whether solid, another immiscible liquid, or gas—a force of attraction acts on the interface as demonstrated by the formation of a curved surface. This phenomenon, related to molecular attraction among substances, is called *surface tension*. When two liquids compete for a single contact surface for interfacial tension, one dominates and coats the solid surface as a *wetting fluid*. The other acts as the nonwetting fluid. In a water-oil system, water is the wetting fluid unless the surface is already coated with oil before water makes contact.

Consider a pore space that is saturated with a wetting fluid, and a nonwetting fluid slowly starts displacing the wetting fluid. At a certain stage, this draining process stops and no more wetting fluid will be displaced. The content of the wetting fluid at this stage is referred to as the *residual wetting saturation*.

Now consider that the wetting fluid is introduced again to displace the nonwetting fluid. When this process stops, some nonwetting fluid still remains in the pores. This amount is referred to as the *residual nonwetting saturation*. These phenomena are important in understanding the flow of immiscible fluids through porous media.

Water will not flow in a pore space until its content exceeds the residual wetting fluid saturation. Similarly, a nonwetting fluid will not begin to flow until the residual nonwetting fluid saturation is exceeded. In other words, in a two-phase water-oil system, if the water content is less than the residual saturation, the oil can flow but the water will be held within the pores. Similarly, when the oil concentration is less than the residual nonwetting saturation, water can flow but oil will be held within the pores. When their contents exceed the residual saturations, both immiscible fluids will flow through the pore space in distinct phases at different rates. In a steady-state condition (when there is a continuous flow at a uniform rate), a part of the pore space is filled with one fluid and the remainder with the other fluid. Water is the wetting phase; it tends to line the edges of the pores and cover the sand grains. Oil is the nonwetting phase; it moves through the central portion of the pores.

5.6.1 Two-Phase Flow through a Porous Medium

Since neither water nor nonaqueous phase liquid occupies the entire pore space, the permeability or hydraulic conductivity of the medium with respect to each is less than when the pore space is fully filled by one single phase. When two phases (liquids) are present in a system, Darcy's law is used to compute the flow rate of each phase:

$$Q_i = \frac{k_{ri}k\,\gamma_i}{\mu_i}\frac{dh_i}{dx}A \quad [\mathrm{L}^3/\mathrm{T}] \tag{5.21}$$

where

$$Q_i = \text{rate of flow of fluid (phase) } i$$
$$k_{ri} = \text{relative permeability of fluid } i \text{ in the presence of another fluid}$$
$$k = \text{intrinsic permeability of medium}$$
$$\gamma_i = \text{unit weight of fluid } i$$
$$\mu_i = \text{dynamic viscosity of fluid } i$$
$$dh_i/dx = \text{gradient due to fluid } i$$
$$A = \text{cross sectional area of flow}$$

The term $k\gamma_i/\mu_i$ denotes K_i, the hydraulic conductivity (for saturated flow) of fluid i.

EXAMPLE 5.8

Figure 5.6 represents the groundwater flow pattern for an aquifer system. The contaminant source consists of water and carbon tetrachloride (CTC)—a nonaqueous phase liquid at 15 °C. The intrinsic permeability of the medium is 1.12×10^{-11} m^2 and porosity is 0.4. The relative permeability of water is 0.08 and the relative permeability of CTC is 0.23. Determine the rate of flow of water and CTC through the aquifer.

SOLUTION

1. For water at 15 °C, $\rho = 0.999 \dfrac{\text{g}}{\text{cm}^3}$, $\mu = 1.14 \times 10^{-2} \dfrac{\text{g}}{\text{cm} \cdot \text{s}}$

$$K_w = k\frac{\rho_w g}{\mu}$$

$$= \left(1.12 \times 10^{-11} \, \text{m}^2\right)\left(0.999 \frac{\text{g}}{\text{cm}^3}\right)\left(980 \frac{\text{cm}}{\text{s}^2}\right)\left(\frac{1}{1.14 \times 10^{-2}} \frac{\text{cm s}}{\text{g}}\right)\left[\frac{100 \, \text{cm}}{1 \, \text{m}}\right]$$

$$= 96.18 \times 10^{-6} \text{ m/s or 8.3 m/d}$$

2. For CTC at 15 °C, $\rho = 1.593 \dfrac{\text{g}}{\text{cm}^3}$, $\mu = 1.03 \times 10^{-2} \dfrac{\text{g}}{\text{cm} \cdot \text{s}}$

$$K_{CTC} = k\frac{\rho_{CTC} g}{\mu_{CTC}}$$

$$= \left(1.12 \times 10^{-11}\right)\left(1.593\right)\left(980\right)\left(\frac{1}{1.03 \times 10^{-2}}\right)\left[\frac{100}{1}\right]$$

$$= 169.76 \times 10^{-6} \text{ m/s or 14.67 m/d}$$

3. From Fig. 5.6, $\dfrac{dh}{dx} = \dfrac{100 - 95}{2000} = 0.0025$

4. From eq. (5.21) for water,

$$Q_w = k_{rw} K_w \frac{dh_w}{dx} A$$

or

$$Q_w = \left(0.08\right)\left(8.3 \frac{\text{m}}{\text{d}}\right)\left(0.0025\right)\left(10 \times 1 \, \text{m}^2\right)$$

$$= 0.017 \text{ m}^3/\text{day per m width}$$

5. For CTC, $Q_{CTC} = \left(0.23\right)\left(14.67\right)\left(0.0025\right)\left(10 \times 1\right)$

$$= 0.084 \text{ m}^3/\text{day per m width}$$

Figure 5.6 Hydrologic profile of system in Example 5.8.

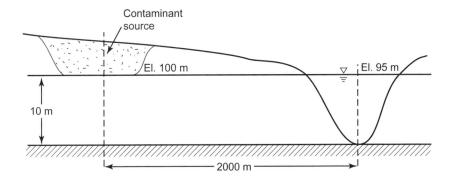

5.6.2 Transport of NAPL

When a nonaqueous phase liquid (NAPL) is spilled at the land surface, it travels through the larger pore openings as a nonwetting fluid vertically downward through the vadose or aeration zone, displacing the air. A view of the distribution of NAPL in soil is shown in Figure 5.7 for light nonaqueous phase liquids, and Figure 5.8 for dense nonaqueous phase liquids. As the NAPL moves forward, a fraction remains behind throughout the thickness of the vadose zone as residual oil. In moving down, the NAPL may displace some water in the vadose zone, causing a water layer to move in advance of the NAPL front. Once the capillary zone is reached, NAPL accumulates. Eventually, the capillary fringe is squeezed out and an oil table forms on top of the water table.

If it is an LNAPL, a core of the LNAPL will remain, slightly depressing the water table by its own weight as shown in Figure 5.7. LNAPL can migrate at the top of the water table following its slope. If the water table drops, the pool of LNAPL also drops. When the water table rises again, part of the LNAPL is pushed upward but a portion at the residual saturation remains below the new water table. Thus, variations in water table cause spreading of the LNAPL to a greater thickness.

Conversely, a DNAPL (Figure 5.8) will continue to move downward below the water table. For DNAPL to migrate downward, the water in the pores must be expelled. To achieve this, the DNAPL must have sufficient height so that its weight can displace the pore water. The critical height of the DNAPL can be determined from the principle of interfacial tension and capillary pressure between water and DNAPL. For well-sorted, well-rounded grains, the critical height can be expressed by:

$$h_c = \frac{16.5\sigma \cos\theta}{d(\gamma_0 - \gamma_w)} \quad [\text{L}] \tag{5.22}$$

where

h_c = critical height of DNAPL

σ = surface tension between fluids (water and oil)

θ = wetting angle

d = diameter of grains

γ_w = unit weight of water

γ_0 = unit weight of DNAPL

Figure 5.7 Typical behavior of light nonaqueous phase liquid (LNAPL) underground (from Gupta, 2004).

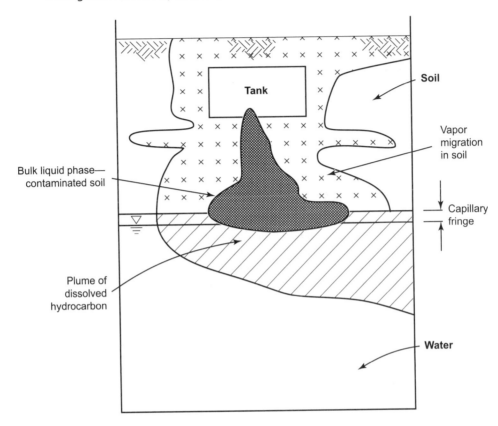

Eq. (5.22) indicates that the critical height is inversely proportional to both the grain size and the density of DNAPL. Thus, smaller grains and less dense fluid will require more height to overcome the capillary pressure.

EXAMPLE 5.9

The average grain size of a saturated porous medium is 0.5 mm. Determine the critical height of carbon tetrachloride to migrate down through the medium. Temperature is 20 °C. The interfacial (surface) tension between fluids is 46×10^{-3} N/m and the angle of contact is zero.

SOLUTION

1. At 20 °C, $\gamma_w = 9.8$ kN/m^3

$$\gamma_{CTC} = 15.64 \text{ kN/m}^3$$

Figure 5.8 Typical behavior of dense nonaqueous phase liquid (DNAPL) underground (from Gupta, 2004).

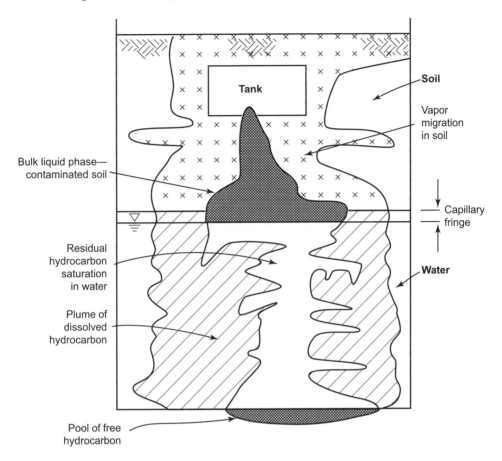

2. From eq. (5.22)

$$h_c = 16.5\left(46\times10^{-3}\,\frac{N}{m}\right)(\cos 0°)\left(\frac{1}{0.5\ mm}\right)\left(\frac{1}{15.64-9.80}\,\frac{m^3}{kN}\right)$$

$$\left[\frac{1000\ mm}{1\ m}\right]\left[\frac{1}{1000}\,\frac{kN}{N}\right]=0.26\ m$$

If an adequate amount of DNAPL is present, it will continue to migrate down. A layer of the DNAPL will be formed at the aquifer bottom. In the column of water above, the pores will contain residual saturation of DNAPL. As the DNAPL string moves down, the flowing groundwater tends to displace it laterally to some extent. On reaching the bottom, the DNAPL will move laterally down the bottom slope even though the groundwater flow may be in the other direction.

The residual saturation NAPL (both LNAPL and DNAPL) in the vadose zone and groundwater can partition into vapor phase through pores. The degree of partitioning depends upon the relative volatility and solubility of NAPL.

If the amount of NAPL spilled is small, it will flow until residual saturation is reached to some depth within the vadose zone. The infiltrating water that will dissolve a small amount of the NAPL from the vadose zone can still cause groundwater contamination.

5.7 SALINE WATER INTRUSION

The intrusion of saline water is a common groundwater pollution problem. This occurs due to invasion of seawater in coastal aquifers, seepage of saline wastes from the surface, and upward movement of saline waters of geologic origin in other aquifers. The first category has been recognized very widely.

5.7.1 Freshwater and Saltwater Interface

Figure 5.9 shows a cross section of a coastal aquifer. If at any point on the interface the pressure from the top of the fresh water balances the pressure of saline water from the bottom, then

$$\rho_s g Z = \rho_f g (Z + h)$$

or

$$Z = \frac{\rho_f}{\rho_s - \rho_f} h \quad [\text{L}] \tag{5.23}$$

where

ρ_s = density of saline water

ρ_f = density of fresh water

Z = depth of the interface at any point

h = water table or piezometric head above seawater level at any point

Equation 5.23 is recognized as the Ghyben-Herberg relation. The movement of water has been ignored in this relation.

For typical seawater, $\rho_s = 1.025$ and $\rho_f = 1.0$; then

$$Z = 40h \tag{a}$$

It had been observed in the past by many investigators that salt water occurred at a depth below sea level of 40 times the height of the fresh water as given by eq. (a). If h is not positive (i.e., the piezometric or water table head is not above sea level), sea water will advance directly inland.

5.7.2 Upconing of Saline Water

In a situation where a saline water layer underlies a freshwater zone and a well penetrating only the freshwater portion is pumped, a local rise of the interface of saline and fresh water occurs as shown in Figure 5.10. This phenomenon is known as *upconing*.

Figure 5.9 Fresh and saline water balance in an unconfined coastal aquifer.

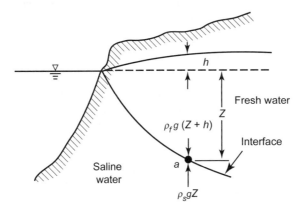

Figure 5.10 Upconing of salt water under a pumping well.

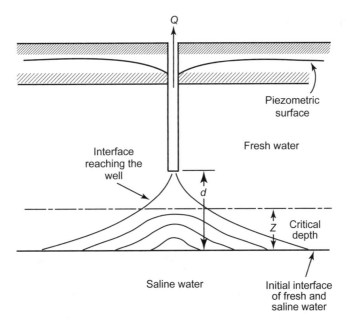

The rise of upconing under steady-state conditions is given by

$$Z = \frac{Q\rho_f}{2\pi d K \left(\rho_s - \rho_f\right)} \quad [\text{L}] \tag{5.24}$$

where

K = hydraulic conductivity

d = depth of the initial freshwater-saline water interface below the bottom of the well

When this rise becomes critical (i.e., $Z/d = 0.3$ to 0.5), salt water reaches the well, contaminating the supply. Thus the maximum discharge to keep the rise below the critical limit can be given by substituting $Z = 0.5d$ in eq. (5.24):

$$Q_{max} = \pi d^2 K \frac{\rho_s - \rho_f}{\rho_f} \quad [L^3 T^{-1}] \tag{5.25}$$

In a real situation, a zone of brackish water occurs between salt water and fresh water. Even with a low rate of pumping, some saline water reaches the well. However, the upconing effect can be minimized by increasing d (i.e., separating wells from the saltwater layer as far as possible) and by decreasing Q, the flow rate.

EXAMPLE 5.10

In a deep aquifer, fresh water extends to a depth of 120 ft, below which there is a deposit of salt water of specific weight 64 lb/ft^3. A water well is drilled through the freshwater zone to a depth of 80 ft. Determine the maximum rate at which this well can be pumped without drawing saline water from the well if the aquifer permeability is 1000 ft/day.

SOLUTION   From eq. (5.25),

$$Q_{max} = \pi d^2 K \frac{\rho_s - \rho_f}{\rho_f}$$

or

$$Q_{max} = \pi d^2 K \frac{\gamma_s - \gamma_f}{\gamma_f}$$

$$= \pi \left(40^2 \ \text{ft}^2 \right) \left(1000 \frac{\text{ft}}{\text{day}} \right) \left(\frac{64.0 - 62.4}{62.4} \frac{\text{lbs/ft}^3}{\text{lbs/ft}^3} \right)$$

$$= 128,800 \ \text{ft}^3/\text{day or } 1.49 \text{ cfs}$$

5.8 ANALYSIS OF GROUNDWATER MONITORING DATA

5.8.1 Statistical Evaluation of Groundwater Monitoring Data

The U.S. Environmental Protection Agency (EPA) under the Resource Conservation and Recovery Act (RCRA) of 1976 promulgated regulations in October 1988 that required groundwater monitoring at permitted hazardous waste land disposal facilities. The regulations enacted in October 1991 mandated groundwater monitoring at municipal solid waste disposal facilities as well.

Facility owners and operators are required to sample groundwater at regular intervals to obtain at least four samples from each well semi-annually. They are also required to apply a statistical procedure to determine whether or not constituents from the facility are contaminating groundwater. The pollutant concentrations vary in space and time in a random manner. It is difficult to predict the next value of a random variable, but a probability can be attached to various possible values. Hence, statistical procedures are preferred for groundwater monitoring. Background or upgradient monitoring wells are located upgradient of the waste disposal facility, geographically removed from the potential source of contamination. These provide historical or background measurements of various constituents for comparison. The background data aim to gauge average levels of naturally occurring constituents or to confirm the absence of some other specified constituents. A series of compliance point monitoring wells are constructed downgradient of the waste disposal facility from which the samples are collected on a periodic basis. Statistical comparisons are made of the compliance data to the background data or the compliance data to fixed regulatory standards, e.g., maximum contaminant levels (MCLs).

Statistical tests are required for each monitored constituent in each compliance well. The regulations prohibit the *pooling* of constituents and the *pooling* of wells except for background samples. The background wells should be sampled as often as is feasible; when background wells cannot be sampled frequently, the pooling of data from multiple background wells might be considered to increase overall background sample size. A minimum of eight to ten background samples is highly desirable. The EPA has prescribed certain specific statistical procedures and sampling methods that meet the minimum performance standards for a low probability of the following two occurrences: (1) indicating contamination when it is not present, i.e., identifying contamination at a *clean well*, or a *false positive* or *type I error* (denoted by α), and (2) failing to detect contamination that actually is present, or a *false negative* or *type II error* (denoted by β).

All statistical methods assume that the observed data follow some kind of a pattern which can be described through an appropriate mathematical formula. Many probability distribution models have been formulated with their specific forms of mathematical equations. The relation to which a set of observed data best fits is known as its *probability distribution model*, and the related mathematical formula is its *probability distribution function*. With reference to streamflows, the important probability distributions and their density functions are described in Section 8.9.

5.8.2 Statistical Measures of Sample Data

Two basic properties of distribution functions of probability models are: (1) the average behavior of the data (mean or median); and (2) the variability of the data from measurement to measurement (standard deviation, interquartile range (IQR), variance). Another important property of some models is the skewness, which is a measure of the asymmetry of the data distribution about its mean or median.

The mean value of a data set is computed by:

$$\overline{X} = \frac{1}{n}\sum_{i=1}^{n} x_i \tag{5.26}$$

where

$\overline{X}$ = mean observed (historical) concentration

n = total numbers (values) of samples

$\mu = (E(\overline{X})$ i.e., $\overline{X}$ is expected to be equal to the population mean μ as n tends to infinity)

x_i = ith number of observed concentration

The sample estimate of the variance or standard deviation, S, is given by

$$S^2 = \frac{1}{n-1}\sum_{i=1}^{n}(x_i - \bar{X})^2 \quad * \tag{5.27}$$

The sample coefficient of skewness, g, is given by

$$g = \frac{n\sum_{i=1}^{n}(x_i - \bar{X})^3}{(n-1)(n-2)S^3} \quad [\text{dimensionless}] \tag{5.28}$$

EXAMPLE 5.11

In four monitoring wells, the following concentrations of lead were measured on a monthly basis (see Table 5.4). Compute the statistical characteristics of the combined sample based on logarithmic values of the data.

Table 5.4   Lead Concentration in Monitoring Wells, ppb

Month	Well 1	Well 2	Well 3	Well 4
1	88	28.5	58.5	4.75
2	2.2	122.3	226.5	2000
3	393	496.5	40.5	128.4
4	88	21	32.1	15
5	7.0	96.6	867	955.5

SOLUTION

1. Natural logarithms of measurements are given in column 2 of Table 5.5.

2. From eq. (5.26): mean concentration

$$\bar{X} = \frac{1}{20}(86.655) = 4.333 \text{ ppb}$$

3. From eq. (5.27): variance

$$S^2 = \frac{1}{(20-1)}(63.415) = 3.338$$
$$S = 1.827 \text{ ppb}$$

4. From eq. (5.28): skewness coefficient

$$g = \frac{20(-9.688)}{(19)(18)(1.827)^3} = -0.093$$

---

* Units are the concentration units squared.

Table 5.5   Computation of the Statistical Parameters

		(1)	(2) In conc x_i ppb	(3) $\left(x_i - \bar{X}\right)$	(4) $\left(x_i - \bar{X}\right)^2$	(5) $\left(x_i - \bar{X}\right)^3$
	Sample					
Well 1	Month 1		4.477	0.144	0.021	0.003
	Month 2		0.79	−3.543	12.55	−44.465
	Month 3		5.974	1.641	2.693	4.419
	Month 4		4.477	0.144	0.021	0.00
	Month 5		1.946	−2.387	5.698	−13.601
Well 2	Month 1		3.35	−0.983	0.966	−0.95
	Month 2		4.806	0.473	0.224	0.106
	Month 3		6.208	1.875	3.516	6.593
	Month 4		3.045	−1.288	1.659	−2.137
	Month 5		4.571	0.238	0.057	0.014
Well 3	Month 1		4.069	−0.264	0.07	−0.018
	Month 2		5.423	1.09	1.188	1.295
	Month 3		3.701	−0.632	0.399	−0.252
	Month 4		3.469	−0.864	0.746	−0.645
	Month 5		6.765	2.432	5.915	14.39
Well 4	Month 1		1.558	−2.775	7.701	−21.37
	Month 2		7.601	3.268	10.68	34.902
	Month 3		4.855	0.522	0.272	0.142
	Month 4		2.708	−1.625	2.641	−4.292
	Month 5		6.862	2.529	6.396	16.175
	Σ		86.655		63.415	−9.688

5.9   CHECKING DATA FITNESS FOR STATISTICAL PROCEDURES

A model has to be chosen that matches with the distribution of data for proper interpretation of the results. All statistical procedures assume independence of data in order to represent true variability of random variables. In the context of statistical procedures, there is another important test that seeks to determine whether the different groups of data have approximately the same variance. It is called the test for homogeneity of variance. Accordingly, the first step in a statistical analysis comprises:

1. Checking for the assumed probability distribution of the data.
2. Ensuring that the data are statistically independent.
3. Testing for homogeneity of variance.

The EPA's experience, which is shared by the U.S. Geological Survey (USGS), is that the lognormal distribution is generally an appropriate statistical model for groundwater data. Accordingly, the following framework is recommended for checking the groundwater monitoring data:

1. The underlying model is assumed to follow the lognormal distribution. All data are first logged (the natural logarithm of each observation is taken) and checked for lognormality.

2. If the statistical tests reject the lognormality model, the same tests are performed on the data in original form to check for normality.

3. If the data are approximately lognormal or normal, one of the parametric procedures is used to analyze the sampled data.

4. If the sample data are grossly non-lognormal or non-normal, there are two options to follow:

   a. Find a transformation (square root, reciprocal, cube root, etc.) that leads to normality or choose another distribution (Gamma, Weibull, Beta, etc.) that fits adequately to the data and then apply a parametric procedure, or alternatively,

   b. Use one of the non-parametric procedures based on ranks of the data and not on their values. The non-parametric procedures do not require any specific distribution.

5. Further, there might be a number of *non-detect* (ND) concentrations in many samples due to constraints imposed by method detection limits (MDL) and practical quantification limits (PQL). If there are too many non-detects, a non-parametric procedure should be used because of the difficulty of verifying the lognormality/normality of the distribution.

5.10 Tests for Lognormality/Normality

The test procedures are similar for both lognormality and normality. The logged values of the data are used in the former case and the original measurements in the latter. Tests for lognormality or normality should be run separately on background wells and compliance wells. EPA's *Interim Final Guidance to Statistical Analysis* (April 1989) outlined three methods for checking lognormality: (1) Probability Plots (p-plots), (2) Coefficient of Variation (CV), and (3) Chi-square test (χ^2-test)*. The addendum to the *Interim Final Guidance* (July 1992) did not recommend the Chi-square test, substituted the Coefficient of Skewness for the CV test, and suggested that although the p-plot is an excellent visual and qualitative indicator of normal distribution, it should be supplemented with a numerical test. Three numerical methods have been suggested, namely the Shapiro-Wilk test (for less than 50 data points), the Shapiro-Francia test (for more than 50 data points), and the Probability Plot Correlation Coefficient. These numerical tests compute a distinct test statistic whose high value suggests normality of distribution. The Shapiro-Wilk test is considered to be one of the best available tests of normality. The test is described below, along with the p-plot.

5.10.1 Probability Plot and the Shapiro-Wilk Test

The p-plot presents a clear, visual picture of the distribution of data; if a symmetrical model is rejected it indicates whether a left skewed or right skewed model should be considered or whether the departures from the assumed model are in the center or in the tail parts of the data. To prepare a plot, the data are ordered sequentially and plotted on graph paper specifically scaled for the probability model being fitted. The procedure to construct a probability graph paper is described in Section 8.6.1. Several types of probability paper are commercially available, e.g., normal (or lognormal), type I extreme or Gumbel distribution, and type II extreme or Weibull distribution papers.

The testing procedure consists of the following steps; the first four steps pertain to p-plots and the remaining to the Shapiro-Wilk test.

---

* χ is the Greek lowercase letter Chi (pronounced "kai") and χ^2 is Chi-square.

1. Take the natural logarithm of each observation. Arrange the observations in ascending order from lowest to highest.

2. Compute the cumulated probability as $p = i/(n + 1)$, where i is the order or rank number of the observation and n is the total number of observations.

3. On normal probability paper, plot the probability versus the log of observations.

4. If the plot shows an apparent curvature or a bend, then the data do not belong to lognormal distribution. Further steps are not necessary. For a linear or approximately linear plot continue with the remaining steps.

5. Compute the differences between the corresponding extreme observations from beginning and end, i.e., $x_{(n-i+1)} - x_i$.

6. Compute k as the greatest integer (rounded value) less than or equal to $n/2$: for example, if $n = 19$ then $k = 9$.

7. Look up the coefficients $a_{(n-i+1)}$ from Appendix F. Multiply the differences of step 5 by the respective coefficient. Add the first k products to obtain Σb_i.

8. Calculate the Shapiro-Wilk test statistic, W, by eq. (5.29):

$$W = \frac{1}{n-1}\left(\frac{\Sigma b_i}{S}\right)^2 \quad \text{[dimensionless]} \tag{5.29}$$

9. Compare the computed W to the critical value for sample size n at the 5% significance level from Table 5.6. A computed W value higher than the critical table value indicates evidence of lognormality. A 5% level of significance means that there is a 5% chance that a model will be rejected when it should have been accepted.

Table 5.6   Critical Values of the W-Test at 5% Significance Level

Sample Size	Critical W-value
3	0.767
5	0.762
10	0.842
15	0.881
20	0.905
25	0.918
30	0.927
35	0.934
40	0.940
45	0.945
50	0.947

EXAMPLE 5.12

Check whether the data collected in Example 5.11 follow a lognormal distribution.

SOLUTION

1. Natural logarithm of the data are arranged from lowest to highest value in column 2 of Table 5.7. Alternatively, logarithmic normal paper can be used to plot the data directly.

2. Cumulated probabilities are computed in column 3 by $100i/(n + 1)$, where i is rank and n is the sample size, $n = 20$.

3. The plot of ln concentration (col. 2) versus cumulated probabilities (col. 3) is shown in Figure 5.11. It is nearly a straight line fit.

4. From Example 5.11, $S = 1.827$ ppb

Table 5.7 Computations for P-Plot and the Shapiro-Wilk Test

(1) Rank i	(2) ln concentration x_i	(3) probability $p=100i/n + 1$	(4)[a] x_{n-i+1} $= x_{21-i}$	(5)[b] $x_{n-i+1} - x_i$	(6)[c] a_{n-i+1}	(7)[d] b_i
1	0.79	4.8	7.601	6.811	0.4734	3.224
2	1.558	9.5	6.862	5.304	0.3211	1.703
3	1.946	14.3	6.765	4.819	0.2565	1.236
4	2.708	19.0	6.208	3.500	0.2085	0.730
5	3.045	23.8	5.974	2.929	0.1686	0.494
6	3.350	28.6	5.423	2.073	0.1334	0.277
7	3.469	33.3	4.855	1.386	0.1013	0.140
8	3.701	38.1	4.806	1.105	0.0711	0.079
9	4.069	42.9	4.571	0.502	0.0422	0.021
10.5	4.477	50.0	4.477	0	0.014	0
10.5	4.477	50.0	4.477			
12	4.571	57.1	4.069			
13	4.806	61.9	3.701			
14	4.855	66.7	3.469			
15	5.423	71.4	3.350			
16	5.974	76.2	3.045			
17	6.208	81.0	2.708			
18	6.765	85.7	1.946			
19	6.862	90.5	1.558			
20	7.601	95.2	0.79			

$$\Sigma b_i = 7.904$$

[a] Col. 2 values in the reverse order starting with the last value first
[b] Col. 4 – Col. 2
[c] From Appendix F
[d] Col. 5 × Col. 6

Figure 5.11 Probability (p) plot of Example 5.12.

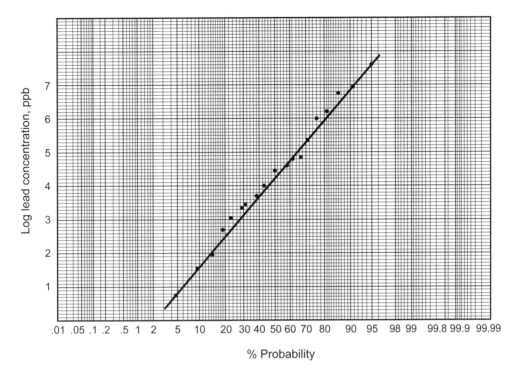

5. From eq. (5.29),

$$W = \frac{1}{(20-1)}\left(\frac{7.904}{1.827}\right)^2 = 0.985$$

6. From Table 5.6, the $W_{critical} = 0.905$. Since the computed $W > 0.905$, the validity of lognormal distribution holds.

5.10.2 Coefficient of Variation and Coefficient of Skewness

The coefficient of variation (CV) is defined as the ratio of standard deviation to mean of a data set. The original RCRA guideline had recommended rejecting normality if the CV was not larger than 1. However, it is not considered a reliable indicator because even when the true CV of a series is between 0.5 and 1, the sample could yield a CV of greater than 1.

The coefficient of skewness can be computed by eq. (5.28). Normal data are expected to have a zero skewness coefficient. However, a sample skewness up to 1, in terms of absolute value, is accepted to assume that the data are roughly normally distributed.

EXAMPLE 5.13

For the logged data of Example 5.11, use the skewness coefficient to test for lognormality of the data.

SOLUTION

1. From Example 5.11, skewness coefficient, $g = -0.093$, or $|g| = 0.093$
2. Since $g = 0.093 < 1$, the assumption of lognormality cannot be rejected.

5.11 TESTING FOR STATISTICAL INDEPENDENCE

The serial correlation or autocorrelation is a measure of the extent to which the measurements are interdependent. Standard correlation is calculated between two sets of data and autocorrelation is computed on only one set of data. The k-lag serial correlation (autocorrelation) coefficient, in which the effect extends by k time units, is given by[*]

$$r_K = \frac{\sum_{i=1}^{n-K} (x_i - \bar{X})(x_{i+K} - \bar{X})}{(n-k)S^2} \quad \text{[dimensionless]} \tag{5.30}$$

The one-lag coefficient, in which the current measurement is affected only by the previous measurement, can be obtained by substituting $k = 1$ in eq. (5.30). Groundwater samples from a well can be treated as being statistically independent if the one-lag serial correlation coefficient is close to zero. However, the number of samples obtained from each well is often too small to run autocorrelation or any other statistical analysis. EPA recommends that the sampling plan be developed so that the samples of groundwater are physically independent and, thus, statistically independent. The plan needs to allow enough time between sampling and ensure that sampling is done on different volumes. In a series of measurements, individual values should fluctuate around the median in an unpredictable manner; too many consecutive values above or below the median are indicative of statistical dependence. A formal statistical procedure has been developed based on this concept.

Streamflows and other time series also make applications of the serial correlation or the autocorrelation, as discussed in sections 7.19 through 7.21. Since hydrologic time series show statistical dependence, a high serial correlation coefficient is desirable in streamflow techniques as opposed to groundwater monitoring.

EXAMPLE 5.14

Test for the independence of data on the natural log of the concentrations of nickel, measured in a well on a monthly basis.

month	1	2	3	4	5	6	7	8	9
ln nickel, ppb	3.26	4.56	3.22	1.73	2.35	3.52	4.64	2.11	0.9
month	10	11	12	13	14	15	16	17	18
ln nickel	5.48	3.56	1.84	2.93	4.02	2.64	5.01	3.34	2.11

SOLUTION

1. Computations for statistical parameters are arranged in Table 5.8.
2. From eq. (5.26),

$$\bar{X} = \frac{57.22}{18} = 3.18 \text{ ppb}$$

[*] This equation is slightly biased. It is, however, widely used for simplicity.

3. From eq. (5.27),

$$S^2 = \frac{26.10}{(18-1)} = 1.535$$

$$S = 1.24 \text{ pbb}$$

4. From eq. (5.30),

$$r_1 = \frac{\sum_1^{n-1}(x_i - \overline{X})(x_{i+1} - \overline{X})}{(n-1)S^2}$$

$$= \frac{-3.66}{(18-1)(1.535)}$$

$$r_1 = -0.14, \text{ low; hence the data in the set are independent}$$

Table 5.8 Computation of Statistical Parameters for Data of Example 5.14

(1)	(2)	(3)	(4)	(5)[a]	(6)
Month	ln conc.	$(x_i - \overline{X})$	$(x_i - \overline{X})^2$	$(x_{i+1} - \overline{X})$	$(x_i - \overline{X})(x_{i+1} - \overline{X})$
i	x_i				
1	3.26	0.08	0.01	1.38	0.11
2	4.56	1.38	1.90	0.04	0.06
3	3.22	0.04	0.00	−1.45	−0.06
4	1.73	−1.45	2.10	−0.83	1.20
5	2.35	−0.83	0.69	0.34	−0.28
6	3.52	0.34	0.12	1.46	0.50
7	4.64	1.46	2.13	−1.07	−1.56
8	2.11	−1.07	1.14	−2.28	2.44
9	0.90	−2.28	5.20	2.30	−5.24
10	5.48	2.30	5.29	0.38	0.87
11	3.56	0.38	0.14	−1.34	−0.51
12	1.84	−1.34	1.80	−0.25	0.34
13	2.93	−0.25	0.06	0.84	−0.21
14	4.02	0.84	0.71	−0.54	−0.45
15	2.64	−0.54	0.29	1.83	−0.99
16	5.01	1.83	3.35	0.16	0.29
17	3.34	0.16	0.03	−1.07	−0.17
18	2.11	−1.07	1.14		
Σ	57.22		26.10		−3.66

[a] Value from the succeeding month of Col. 3.

5.12 Checking for Equality of Variances across Well Groups

A very important assumption for the analysis of variance (ANOVA) test of groundwater statistical analysis, as stated in Section 5.9, is that the samples collected from different wells have approximately the same variance; mild differences in variance are acceptable. The effect becomes noticeable when the largest and smallest group variances differ by a ratio of about 4 and becomes quite severe when the ratio exceeds 10. The homogeneity (equality) of variances is tested by the Fisher (F-) distribution. It is described in Section 8.5.2 in the context of peak-discharge or flood flow data.

5.13 Statistical Procedures for Groundwater Monitoring

There are three ways to monitor groundwater wells:

1. *Detection monitoring:* Compliance (downgradient) wells data are compared with background (upgradient) wells data.

2. *Compliance monitoring:* Compliance wells data are compared to a Groundwater Protection Standard (GWPS). The GWPS is either fixed as an Alternate Contaminant Level (ACL) from the average of the background data or it is a risk-based MCL or ACL limit.

3. *Intrawell comparison:* Changes within a single compliance well are observed over a period of time.

EPA has recommended many standard statistical procedures. The suitability of a procedure depends on the distribution of data and type of monitoring. When the data are lognormally or normally distributed, a parametric procedure is recommended. For nonlognormal or non-normal data, a non-parametric or rank-based procedure is to be used. The standard procedures are summarized in Table 5.9. There are two forms of statistical inference: hypothesis testing and estimation. The t-test and ANOVA are hypothesis testing methods where the null hypothesis is that the mean concentration of pollutants in the background wells is the same as the mean concentration in the compliance wells. The intervals methods are estimation tests that construct intervals and make inferences from the data falling within or outside the intervals.

5.14 Strategies for Procedure Selection

Type I error, or the false positive, is the probability that a test will falsely indicate contamination when it is not present. The aim should be to try to lower the type I error. Unfortunately, in general, attempts to lower the false positive rate lower the statistical power of a test to detect real contamination at the well; if the power drops too much, real contamination will not be identified when it occurs. EPA mandates that the type I error for any individual comparison be at least 1% and that the network-wide false positive rate across all wells and all constituents tested per period should be kept to approximately 5%.

ANOVA procedures allow inclusion of multiple gradient wells in a single test, but it is not designed to test multiple contaminants simultaneously. The overall false positive rate may be 5% per contaminant, leading to a high network-wide false positive rate. ANOVA combines all downgradient wells simultaneously so it will not indicate which wells are contaminated without further testing. Furthermore, the power of ANOVA depends upon at least 3 to 4 samples and preferably 6 to 8 per well having independent data.

Table 5.9 Types of Statistical Procedures

Distributing Data	Type of Test	Type of Monitoring
Lognormal or normal	**Parametric**	
	1. t-test including CABF t-test	Detection monitoring for only two sets of data
	2. Analysis of variance (ANOVA)	Detection monitoring for at least three sets (groups) of data
	3. Interval Estimation-Confidence Interval	Compliance monitoring
	4. Interval Estimation-Tolerance Interval	Detection monitoring
	5. Interval Estimation-Prediction Interval	Detection monitoring and intrawell comparison
	6. Control Charts	Intrawell comparisons
Non-lognormal or non-normal or data with many non-detects	**Non-parametric** 7. Wilcoxon Rank-Sum or Mann-Whitney U-Test	Detection monitoring for two sets of data
	8. Kruskal-Wallis or Nonparametric ANOVA	Detection monitoring for at least three sets of data
	9. Non-parametric Intervals	As items 3, 4, and 5 above, for non-parametric cases

Parametric and non-parametric ANOVA both assume that the variances across well groups are approximately equal. This condition must be statistically satisfied before the application of ANOVA procedures. Other methods like tolerance and prediction intervals allow statistical testing of each sample as it is collected.

The t-tests are similar to ANOVA procedures but they can be run on only two groups. On multiple wells or contaminants, a series of t-tests have to be made. A high number of test comparisons will enhance the false positive rate. EPA recommends that the number of statistical comparisons should be reduced to the lowest extent possible. Alternative strategies to ANOVA (and t-tests) are advisable for these reasons.

5.15 CONFIDENCE INTERVAL TECHNIQUE

Two numbers within which a statistical parameter may be considered to lie is called the *interval* estimate of that parameter. A confidence interval is the most common interval in statistics. It is designed to contain some specified statistical parameter within a given level of certainty or *confidence*, denoted as $(1-\alpha)$. A 95% confidence level for the mean indicates that with 95% surety, the true mean concentration is contained between the indicated interval limits. Mean concentration in a well (or group of wells) is the usual statistical parameter of confidence interval. However, when comparing MCL or ACL fixed on a risk basis, the upper 95th percentile of concentration in a well is the statistical parameter used since the MCL sets an upper limit on the acceptable contamination. A confidence interval is used in two situations.

1. Confidence interval based on mean parameter: In compliance monitoring when down-gradient (compliance) samples are being compared to the Groundwater Protection Standard (GWPS) based on the average of the background samples.

2. Confidence interval based on 95th percentile parameter: In compliance monitoring for MCL or risk-based ACL.

5.15.1 Confidence Interval Containing Mean of Compliance Data

1. Compute sample mean ($\bar{X}$) and standard deviation (S) of the compliance wells data (or logged data).

2. The lower confidence limit is given by

$$CI_L = \bar{X} - t_{(1-\alpha)(n-1)} \frac{S}{\sqrt{n}} \quad [FT^2/L^4] \tag{5.31}$$

where α is type I error or significance level and $(1 - \alpha)$ is the confidence level, $t_{(1-\alpha)(n-1)}$ is the confidence coefficient based on the t-distribution for which $(1 - \alpha)$ is the confidence level and $(n - 1)$ degree of freedom, as given in Appendix G.

On lognormal data the confidence limits for percentile, the tolerance limits, and the prediction limits are calculated on the logarithms of the data and then converted back to the scale of the original data by inversed-log or exponentiated values. However, on the logged data the confidence limits for the mean and standard deviation do not translate to the original scale.

Hence, for lognormal distribution, the lower confidence limit is given by:

$$CI_L = \bar{X} + 0.5\,S^2 + \frac{S\,H_\alpha}{\sqrt{n-1}} \quad [FT^2/L^4] \tag{5.32}$$

where H_α is the coefficient for lognormal mean given in Appendix I.

3. Compare the GWPS to the lower limit of the confidence interval. If the GWPS is less than the lower limit, there is evidence of contamination.

5.15.2 Confidence Interval Containing 95th Percentile of Compliance Data

1. Compute sample mean ($\bar{X}$) and standard deviation (S) of the compliance wells data (or logged data).

2. Find the confidence coefficient, k, from Appendix J. The coefficient is read for k (percentile, 1–confidence level, # of data). For the 95th percentile at 95% confidence, it corresponds to k ($p = 0.95$, $\alpha = 0.05$, n).

3. Compute the lower limit of the confidence interval by

$$CI_L = \bar{X} + k\,S \quad [FT^2/L^4] \tag{5.33}$$

The plus sign is because the upper 95th percentile is larger than the mean.

4. If the risk-based ACL/MCL or its logged value is less than the lower limit, there is evidence of contamination.

EXAMPLE 5.15

The logged data in Example 5.11 pertain to four compliance wells. The average concentrations of lead (raw values) observed in three background wells are 33, 25, and 40 ppb. Determine, with 95% confidence, whether the compliance wells are contaminated.

SOLUTION

1. Before application of the confidence interval, the data should be checked for normality. From Example 5.12, the data are lognormally distributed; hence, eq. (5.32) should be used.

2. For logged data $\bar{X} = 4.333$ ppb, $S = 1.827$ ppb   (from Example 5.11)

3. At 95% confidence value and $n = 20$, from Appendix I, one-sided $H_\alpha = -2.12$.

4. From eq. (5.32),

$$\begin{aligned} CI_L &= \bar{X} + 0.5\, S^2 + \frac{S\, H_\alpha}{\sqrt{n-1}} \\ &= 4.333 + 0.5(1.827)^2 + \frac{(1.827)(-2.12)}{\sqrt{20-1}} \\ &= 5.113 \text{ ppb} \end{aligned}$$

5. Background wells concentrations

Well #	conc., ppb	ln conc.
1	33	3.497
2	25	3.219
3	40	3.689
GWPS (mean)		$\overline{3.468}$ ppb

6. Since GWPS $<$ CI_L, there is evidence of contamination.

EXAMPLE 5.16

Consider each compliance well sample of Example 5.11 separately. Compute the 95% confidence limit for well #4 and determine whether there is evidence of contamination. The logged GWPS value is 3.47 ppb.

SOLUTION

1. The statistical parameters for well #4 are computed below:

Month	ln conc. x_i, ppb	$(x_i - \bar{X})^2$
1	1.558	10.00
2	7.601	8.300
3	4.855	0.018
4	2.708	4.048
5	6.862	4.588
Total	$\overline{23.584}$	$\overline{26.954}$

$$\bar{X} = 23.584/5 = 4.72 \text{ ppb}$$
$$S^2 = 26.954/4 = 6.738$$
$$S = 2.60 \text{ ppb}$$

2. At a 95% confidence level, $n = 5$, and $S = 2.6$, $H_\alpha = -2.03$

3. From eq. (5.32),

$$CI_L = 4.72 + 0.5(2.6)^2 + \frac{(2.6)(-2.03)}{\sqrt{4}}$$

$$= 5.46 \text{ ppb}$$

4. Since GWPS $< CI_L$, there is evidence of contamination.

EXAMPLE 5.17

EPA has fixed the MCL of 50 ppb for lead. Determine if the compliance wells in Example 5.15 have evidence of contamination at 95th percentile and 99% confidence level.

SOLUTION

1. From Example 5.11, $\bar{X} = 4.333$ ppb, $S = 1.827$ pbb
2. For the 95th percentile, at 99% confidence and n of 20, $k = 1.008$ (from Appendix J)
3. From eq. (5.33),

$$CI_L = 4.333 + 1.008\,(1.827)$$

$$= 6.175 \text{ ppb}$$

4. Since MCL $< CI_L$, there is evidence of contamination.

5.16 TOLERANCE INTERVAL TECHNIQUE

As stated earlier, a confidence interval defines the limits within which the population mean (or any other statistical parameter) lies with a specified degree of certainty. Distinguished from this, a tolerance interval provides limits to contain a specific percentage of the population with a known degree of certainty. There are two percentages associated with a tolerance interval. One of these refers to the percentage of the population that the interval contains, known as the *coverage*, and the second specifies the confidence level associated with that percentage, known as the *tolerance interval*. Interpretation of a 95% tolerance interval with 90% coverage means that with 95% surety 90% of the population measurements will fall within the limit. Tolerance intervals are generally wider than confidence intervals.

Tolerance intervals are used in detection monitoring when the interval is computed from background wells data and compared to individual compliance wells. Since a tolerance interval covers all but a small percentage of population measurements, the compliance value should not exceed the upper tolerance limit when testing small to medium sample sizes. When the sample size is larger, 1 in every 20 samples might fail without evidence of contamination.

Although tolerance intervals are also used in compliance monitoring by constructing the interval on compliance data and comparing with a MCL or ACL, this produces more false positives. Hence, it is better to use the confidence interval in compliance monitoring.

5.16.1 Computing Tolerance Intervals

1. Compute sample mean and standard deviation from the background samples (or their logged values for lognormal distribution).

2. For a 95% tolerance interval and 95% coverage, obtain the upper limit tolerance coefficient, k, from Appendix J. The higher coefficient that corresponds to k is read ($p = 0.95$, $(1 - \alpha) = 0.95$, n).

3. Set the upper tolerance limit to

$$\text{TL}_U = \bar{X} + kS \quad [\text{FT}^2/\text{L}^4] \tag{5.34}$$

4. Compare each compliance sample to the upper tolerance limit, TL_U. The compliance value exceeding the TL_U has evidence of contamination.

EXAMPLE 5.18

Chrysene concentration levels in a background well were observed as follows. In a compliance well, a concentration of 118 ppb was measured. Determine whether there is any evidence of contamination. The data are lognormally distributed.

Month	1	2	3	4	5	6
Chrysene, ppb	21.67	43.12	8.58	14.08	11.22	7.92

SOLUTION

1. The data are stated to be lognormally distributed, otherwise they have to be checked for distribution.

2. The logged values of the data are arranged and mean and standard deviation computed.

Month	ln conc. x_i, ppb	$(x_i - \bar{X})^2$
1	3.076	0.151
2	3.764	1.160
3	2.149	0.289
4	2.645	0.002
5	2.418	0.072
6	2.069	0.382
Σ	16.121	2.056

$$\bar{X} = 16.121/6 = 2.687 \text{ ppb}$$
$$S^2 = 2.056/5 = 0.411$$
$$S = 0.641 \text{ ppb}$$

3. At 95% coverage and 95% tolerance for n of 6, k (0.95, 0.95, 6) = 3.708 from Appendix J

4. From eq. (5.34), $\text{TL}_U = 2.687 + 3.708(0.641) = 5.064$ ppb

5. Compliance well ln conc. = ln 118 = 4.77 ppb

6. Since the compliance well ln conc. < TL_U, there is no evidence of contamination.

5.17 PREDICTION INTERVAL TECHNIQUE

A confidence interval sets a range within which the mean value of a normally distributed population resides. The tolerance interval defines limits that contain data representing a specified proportion of the population. On the other hand, a prediction interval estimates the future values for a set of data. Prediction intervals generally have wider limits than comparable confidence intervals but are shorter than tolerance intervals. Prediction intervals are used for the following two kinds of monitoring:

1. For detection monitoring that contains one or more future observations. The prediction interval is constructed from the background data and the compliance well data are compared to the upper prediction limit, similar to the tolerance interval.

2. For intrawell comparison on an uncontaminated well. The prediction interval is constructed based on the past data, to compare a specified number of future observations from the same well. There is evidence of a recent contamination at the well if any one or more of the future samples fall above the upper prediction limit.

Tolerance intervals and prediction intervals, when applied to detection monitoring, are both constructed from the background data. Given n background measurements and a desired confidence level, a tolerance interval covers a very high percentage and, thus, misses only a small percentage of the samples from the compliance wells. A prediction interval, on the other hand, ensures that the next k future samples will fall below the upper limit. A prediction limit may be thought to be a 100% coverage tolerance limit for the next k future samples.

The number of future observations k for which the prediction interval is used must be specified in advance. Thus, the interval has to be reconstructed on a periodic basis. The number of future observations k is considered the following two ways:

1. It is the sum of all individual observations in all sampling periods.

2. The mean is taken of samples in each period and k represents the number of sampling periods.

5.17.1 Computing Prediction Intervals

1. Calculate sample mean and standard deviation of the observed data (on logged data for lognormal distribution).

2. Calculate the prediction factor c_p, as

$$c_p = t_{(1-\alpha/k)(n-1)}\sqrt{\frac{1}{m}+\frac{1}{n}} \quad \text{[dimensionless]} \quad (5.35)$$

where

$$n = \text{number of the background data or the past data for an intrawell comparison}$$

$$m = \text{number of observations in a sample for which mean is taken; when mean is not taken, m is equal to 1}$$

$$k = \text{number of future samples to be collected}$$

$$(1-\alpha) = \text{confidence level or prediction probability}$$

$$t_{(1-\alpha/k)(n-1)} = \text{t-value at } (1-\alpha/k) \text{ level for } (n-1) \text{ degrees of freedom from Appendix G. Refer to Gibbons (1994) for a detailed table.}$$

3. Compute the upper prediction limit.

$$PI_U = \bar{X} + c_p S \quad [ML^{-3}] \tag{5.36}$$

4. If any one or more of k compliance samples exceeds the upper prediction limit, there is significant evidence of contamination.

EXAMPLE 5.19

Consider the background data for chrysene concentration given in example 5.18. The samples from the compliance well recorded concentrations of 50 and 118 ppb. Determine whether there is evidence of contamination at the 95% confidence prediction limit.

SOLUTION

1. First the normality or lognormality of the data has to be established. The data are stated to be lognormally distributed.

2. From example 5.18, $\bar{X} = 2.687$ ppb, $S = 0.641$ ppb

3. $k = 2$, $n = 6$, $\alpha = 0.05$, $\left(1 - \dfrac{\alpha}{k}\right) = \left(1 - \dfrac{0.05}{2}\right) = 0.975$

   $t_{(0.975,5)} = 2.57$ (from Appendix G)

4. $c_p = 2.57\sqrt{\dfrac{1}{1} + \dfrac{1}{6}} = 2.776$

5. From eq. (5.36), $PI_U = 2.687 + 2.776(0.641) = 4.47$ ppb

6. Compliance well concentration (natural logged values): sample #1 = 3.91 ppb, sample #2 = 4.77 ppb.

7. Since sample #2 exceeds the PI_U, there is evidence of contamination.

Note that for the same concentration, there was no evidence of contamination according to the tolerance interval in Example 5.18.

5.18 NON-PARAMETRIC INTERVALS

The application of non-parametric tolerance and prediction intervals is considered when the data are not normally or lognormally distributed or when the data have over 15% non-detects. The non-parametric tolerance limit is taken to be the maximum value of the background data. Each compliance sample is compared to this value for evidence of contamination. The non-parametric tolerance interval is shown to follow the Beta Distribution. EPA (1992) provides a table to indicate the minimum coverage levels at the 95% tolerance interval for various sample sizes. For a desired level of coverage, the number of background samples required can be determined from this table. For nonparametric intervals, more background samples are needed as compared to the parametric setting.

The non-parametric prediction interval is also taken to be the maximum value of the background data, similar to the non-parametric tolerance limit against which the future k samples are compared. EPA (1992) provides a table that indicates the confidence level associated with various chances of n (# of background data) and k (# of future samples to be compared). For a desired confidence level, the number of background samples required can be determined from this table for a specified number of future samples.

5.19 CONTROL CHARTS

A control chart plots time (on the x-axis) versus a concentration parameter (on the y-axis) for a well or a group of wells. It provides a visual tracking of contamination for a well over a period of time. As such, the chart is an alternative method to prediction intervals in the case of intrawell comparison, or comparison of a compliance well (or group of wells) to historically monitored background wells in detection monitoring. The method expects that the requirements with respect to the normality (lognormality) distribution and independence of data are both fulfilled.

5.19.1 The Combined Shewart-Cusum Control Chart Procedure

1. Initial sample data comprised of at least 8 independent samples are collected from an intrawell or background well in order to establish the baseline parameters with respect to mean and variance. None of these data are plotted. Update the baseline when more data become available without any contamination.

2. As future samples from an intrawell or a compliance well become available, the baseline parameters are used to standardize the data. At each sampling period, a standardized mean is computed:

$$Z_i = \frac{\sqrt{n_i}\left(X_i - \mu\right)}{\sigma} \quad \text{[dimensionless]} \tag{5.37}$$

where

   i = sampling period index

   n = # of samples collected in period i

   $\bar{X}_i$ = mean of samples in period i

   μ = baseline mean

   σ = baseline standard deviation

3. Also compute a cumulative summation (Cusum), S_i by:

$$S_i = \max\{0, (Z_i - K + S_{i-1})\} \quad \text{[dimensionless]} \tag{5.38}$$

where

$$\max \left\{ 0, \left(Z_i - K + S_i - 1 \right) \right\} = \text{the maximum of 0 and } \left(Z_i - K + S_{i-1} \right)$$
$$K = \text{a control chart parameter, EPA (1989) has suggested } K = 1$$

4. Plot Z_i and S_i each against time on the control chart.
5. Select the threshold values of the Shewart-Cusum parameter h, and the upper Shewart limit, SCL. Plot them on the control chart. EPA (1989) adopted the value of $h = 5$ and SCL = 4.5 for groundwater monitoring.
6. The chart is declared out of control in one of the following two ways signifying evidence of contamination:
   a. when the standardized mean (Z_i) at any time becomes too large to cross the SCL, or
   b. when the Cusum (S_i) becomes too large to cross the h line.

EXAMPLE 5.20

For the following nickel concentrations observed on two wells, construct a control chart. A prior monitoring of background wells provided mean and standard deviation of 32.4 ppb and 30 ppb, respectively. The data are normally distributed.

Month	Sample 1	Sample 2
1	18.4	27.1
2	49.3	33.4
3	21.0	21.7
4	18.8	37.8
5	44.6	38.9
6	119.2	77.0

SOLUTION

1. Standardized mean for each period from eq. (5.37) and Cusum from eq. (5.38) are arranged in the following table.
2. There is no evidence of contamination because neither Z_i nor S_i exceeds the SCL or h value at any time.

Month, i	Mean of sample	Z_i	$Z_i - K$	S_i
1	22.75	-0.45^a	-1.45	0^b
2	41.35	0.42	-0.58	0
3	21.35	-0.52	-1.52	0
4	28.3	-0.19	-1.19	0
5	41.75	0.44	-0.56	0
6	98.10	3.10	2.10	2.10

[a] From eq. (5.37), $Z_i = \dfrac{\sqrt{2}\,(22.75 - 32.4)}{30} = -0.45$

[b] From eq. (5.38), $S_i = \max\{0, (-0.45 - 1 + 0)\} = 0$, S_0 being zero

PROBLEMS

5.1 Contamination by advection develops in a groundwater system as shown in Figure P5.1. Estimate the contaminant flux per second to the stream.

Figure P5.1

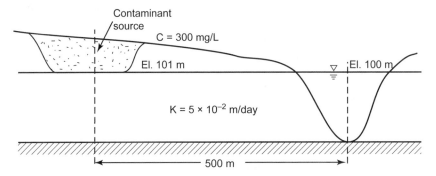

5.2 Two bulbs are connected by a capillary tube of 2 mm diameter and 1 m length. One bulb contains 100 mg/L concentration of carbon dioxide and the other is filled with air. The system is at a constant temperature and pressure. Determine how fast the carbon dioxide will mix initially. The diffusion coefficient is 0.15 cm^2/s.

5.3 There is a leak of gasoline from a gas station to a nearby residence. A gasoline concentration of 4×10^{-6} g/cm^3 has been measured 3 m below the dirt floor of the basement of the residence. What is the flux of gasoline vapor into the house? The diffusion coefficient of gasoline vapor is 1.1×10^{-2} cm^2/s. Assume zero concentration at the basement floor level.

5.4 An experiment is conducted with instantaneous input of 2 mg mass of a nonreactive tracer at the center of a long column of 1 cm radius. The dispersion coefficient is 35 cm^2/hr. Plot the concentration-distance curves at elapsed times of 1, 5, and 10 min.

5.5 In Problem 5.4, if the velocity of the tracer through the column is 1×10^{-2} cm/s, prepare the concentration-distance curves and compare with Problem 5.4.

5.6 Solve Problem 5.5 if the tracer is placed at one end of the column.

5.7 An experiment is conducted with the continuous input of a nonreactive tracer of 0.5 mg/L concentration at the center of a long column. The velocity of the tracer through the column is 30 cm/hr. Plot the relative concentration after 0, 10, 30, 60, 90, 120, and 150 min at 40 cm distance from the release. The dispersion coefficient is 9×10^{-3} cm^2/s.

5.8 A nonreactive chemical species is continuously sent through the middle of a long column at a certain velocity. The C/C_o ratios of 0.42 and 0.60 were observed 25 cm away from the release plane after 45 min and 54.2 min, respectively. Determine the velocity and the dispersion coefficient for the chemical.

5.9 A nonreactive contaminant of 100 mg/L concentration is disposed of continuously in the center of a long narrow trench that fully penetrates an aquifer. The advection velocity through the aquifer is 3×10^{-4} cm/s. Calculate the contaminant concentration after 30 days at 10-meter intervals from the source. The dispersion coefficient is 0.1 cm^2/s. Plot the concentration-distance curve.

5.10 Hydrogen sulfide flows through an aquifer consisting of silty sand. The aquifer porosity is 0.5, hydraulic conductivity is 5 m/day and the average grain size is 0.07 mm. The water table drops by 10 ft in 1 mile. Determine the hydrodynamic dispersion coefficient and whether the dispersion is diffusion-dominated or advection-dominated. The diffusion coefficient D_0 of hydrogen sulfide in water is 1.41×10^{-5} cm^2/s.

5.11 A contaminant continuously leaks into an aquifer. The aquifer properties are: hydraulic conductivity of 50 m/day, porosity of 0.5, and average grain size of 0.5 mm. The water table gradient is 0.008. The diffusion coefficient of the contaminant in water is 6×10^{-5} cm^2/s. (**a**) How far will the plume of the 0.1 relative concentration (10% of the original concentration) travel in the aquifer in one day? (**b**) What will the relative concentration be 50 cm away from the source in one day?

5.12 In a test on a sand column of 3 ft height, a water line is opened to fully saturate the sand column. A carbon tetrachloride (CTC) line is then also opened to inject CTC into the column. At steady-state, the relative permeability of water is 0.2 and of CTC is 0.05. The head of water over the sand column is 10 ft and the head of CTC is 12 ft. Determine the flow rate of the two phases through the column.

5.13 In Example 5.8 of two-phase flow, if the flow of water has to be maintained at 0.05 m^3/day per m width, what will be the relative permeability of water?

5.14 Determine the critical height of a glycerin column to migrate downward through a saturated sandy porous medium of average 1 mm grain size at 15 °C. Surface tension of glycerin with air is 63.41×10^{-3} N/m and of water with air is 72.75×10^{-3} N/m. (Assume a zero wetting angle.) [*Hint*: Difference of two surface tensions will be the surface tension between water and glycerine.]

5.15 A 2-ft-thick layer of a DNAPL of density 1.2 g/cm^3 can migrate through water in a sand bed of 0.1 mm average size. Determine how much the thickness of the DNAPL layer has to build up with similar capillary characteristics but 1.08 g/cm^3 density to migrate through a bed of fine sand of 0.5 mm diameter.

5.16 In a deep saline water aquifer, there is a zone of fresh water of 100 ft depth. A well is drilled through a depth of 80 ft to pump out the water at a rate of 0.5 cfs. The aquifer permeability is 100 ft/hr. Determine the upconing rise of the saline water under steady state (assume a saltwater unit weight of 64 lb/ft^3).

5.17 In Problem 5.16, will the pump draw the saline water? If not, what will be the maximum rate of pumping without pumping the saline water out of the well?

5.18 The following copper concentrations were observed in three monitoring wells. Check whether the observations follow a normal distribution by the p-plot and the Shapiro-Wilk method.

Copper Concentration (ppb)

Month	Well 1	Well 2	Well 3
1	8.2	18.4	10.4
2	11.6	21.8	12.8
3	22.6	29	22.4
4	14	32.2	23
5	14.6	43	20.2
6	16.4	35.2	19.4

5.19 Determine whether the data set in Problem 5.18 follows a lognormal distribution by the p-plot and the Shapiro-Wilk method.

5.20 The concentrations of arsenic observed in six monitoring wells in monthly samples are given below. Check whether these observations follow a normal distribution by the p-plot and the Shapiro-Wilk method.

Arsenic Concentration (ppm)

Month	Well 1	Well 2	Well 3	Well 4	Well 5	Well 6
1	23	3.26	3.26	14.4	28.5	0.46
2	6.66	1.2	165.8	19	1.65	11.14
3	63.8	14.4	9.5	35.4	0.82	5.7
4	8.02	92.4	4.5	3.26	63.8	2.1

5.21 Check whether the data in Problem 5.20 follow a lognormal distribution by the p-plot and the Shapiro-Wilk method.

5.22 Cadmium concentrations in ppb in four monitoring wells are given. Determine whether the distribution of the concentrations follows a normal distribution by the p-plot and the Shapiro-Wilk method.

Cadmium Concentration (ppb)

Month	Well 1	Well 2	Well 3	Well 4
1	4.07	2.94	3.66	1.13
2	0	4.4	5.02	6.85
3	5.57	5.8	3.3	4.45
4	4.03	2.64	3.06	2.3
5	2.16	4.17	6.36	6.46

5.23 The monthwise concentrations of copper observed in a well are listed. Verify the independence of the data set.

Month	1	2	3	4	5	6	7	8	9	10
Copper, ppb	5.33	7.54	14.69	9.1	10.66	22.88	11.96	6.76	14.17	18.85

Month	11	12	13	14	15	16	17	18	19	20
Copper, ppb	20.93	9.49	27.95	6.76	8.32	14.56	12.61	14.95	5.33	18.05

5.24 Toluene concentrations in a well are given below. Check for the statistical independence of the data set.

Month	1	2	3	4	5	6	7	8	9	10
Toluene, ppb	17.5	14.5	6.4	4.0	5.0	12.5	8.0	11.2	13.7	25.0

Month	11	12	13	14	15	16
Toluene, ppb	19.0	7.8	20.2	18.2	20.1	15.3

5.25 Arsenic concentrations in six compliance wells are as given in Problem 5.20. Three background wells recorded average concentrations of 5, 4.8, and 6.2 ppm, respectively. Determine with 95% confidence whether or not the compliance wells are contaminated. The data are lognormally distributed.

5.26 Consider each compliance well of Problem 5.20 separately. Determine whether well #1 and well #6 have evidence of contamination at a 95% confidence level. The background wells' concentrations are the same as in Problem 5.25. Data for each well are lognormally distributed.

5.27 The EPA has established a MCL of 50 ppm for arsenic. Determine whether well #1 and well #6 of Problem 5.20 have evidence of contamination at a 99% confidence level for the 95th percentile. Data are lognormally distributed.

5.28 In Problem 5.22, determine whether well #3 has any evidence of contamination at a 99% confidence level for the 95th percentile. The EPA has established a MCL of 5 ppb for cadmium. Data are normally distributed.

5.29 Chromium concentrations in ppb in a background well are noted below. Compute the upper tolerance limit for 95% coverage at a 95% tolerance limit. A compliance well has a concentration of 100 ppb. Is there any evidence of contamination when the data are normally distributed?

Month	1	2	3	4	5	6
Chromium, ppb	11.22	17.71	6.27	74.8	53.79	3.11

5.30 The following data were recorded at two background wells for Chrysene concentrations. Two compliance wells had concentrations of 10 ppb and 30 ppb respectively. Determine whether these wells have evidence of contamination for 95% coverage and 95% tolerance limit. Both background wells' data considered together are normally distributed.

Chrysene (ppb)

Month	Well 1	Well 2
1	4.2	5.2
2	5.8	6.4
3	11.3	11.2
4	7	11.5
5	7.3	10.1
6	8.2	9.7

5.31 The table below contains the lognormally distributed data for two background wells and one compliance well for arsenic concentrations in ppm. Determine whether the compliance wells have any evidence of contamination for 95% coverage and a 99% tolerance interval. Both background wells' data are considered together.

Background wells (ppm)

Month	#1	#2	Compliance well (ppm)
1	29.48	74.8	33.55
2	19.47	53.79	26.5
3	35.09	33.11	55.74
4	24.42	41.91	61.7
5	30.1	17.71	80.2

5.32 Investigate Problem 5.29 by prediction interval at a 95% confidence prediction level.

5.33 Investigate Problem 5.30 by prediction interval at a 95% confidence prediction level.

5.34 In an intrawell comparison, the past observations of benzene concentrations used to construct the prediction interval are listed below. Four new samples collected from the same well for comparison in a test are also listed. Determine whether there is evidence of recent contamination at a 90% confidence prediction level. Data are normally distributed.

Past Data		New Data	
Month	Conc. ppb	Month	Conc. ppb
1	12.6	9	47.6
2	38	10	70.3
3	32.5	11	85
4	43.4	12	51.9
5	49.2		
6	21		
7	58.1		
8	27.2		

5.35 Construct the control chart for the data presented in Problem 5.34. The past data relate to the prior monitoring for computing of the baseline parameters. Determine whether there is evidence of recent contamination of the well.

5.36 The following concentrations of toluene were observed in a compliance well. Based on the prior monitoring of background wells, the sample mean and standard deviation were determined to be 60 ppb and 12 ppb, respectively. Construct the control chart and test for evidence of contamination of the compliance well. Data are normally distributed.

Quarter (2005)	Toluene conc. ppb	Quarter (2006)	Toluene conc. ppb
1	60	1	84
2	48	2	72
3	70	3	120
4	50	4	90

6

Measurement of Surface Water Flow

◆◆◆

6.1 DETERMINATION OF STREAMFLOW

The quantity of water flowing in a stream, its distribution in space, and its variability with time are required information in order to plan any surface water supply project or to design a hydraulic structure. The most direct and desirable method is to measure the quantity of flow per unit time, referred to as the *streamflow* or *discharge*. For this purpose, a stream gaging station is set up. Since long-term flow records are needed for the planning of a project because of the high variability of flow, a network of stream gaging stations is designed from which data are continuously collected for use at any time in the future. For measurements in small creeks and open channels, hydraulic instruments such as weirs, notches, and flumes are convenient means for computing the discharge. During periods of flooding, it is not always possible to make direct measurements due to such problems as inaccessibility of the site, damage to the measuring structure, and short duration of peak. In such a situation, indirect methods are used by making measurements of certain data after the flood.

It is not possible to gage every site where flow data are desired. If time permits, a temporary gage can be installed to collect direct information. However, when the project formulation has to proceed without delay, three alternatives are available. In order of preference, they are: extend information from nearby gaging sites, estimate streamflows from precipitation data, and use generalized information or the empirical approach. The procedure to assess the streamflow can be summarized as follows:

I. Measurement of streamflow
    A. Direct measurement or stream gaging
        1. Measurement by current meter method
        2. Measurement by floats
        3. Tracer-dilution technique
        4. Ultrasonic method
        5. Electromagnetic method
    B. Measurement through hydraulic devices
        1. Weirs and notches
        2. Orifices
        3. Flumes

C. Indirect measurement of peak flows
    1. Slope-area method
    2. Contracted-opening method
    3. Flow-over-structure method
II. Estimation of streamflow
    A. Application of precipitation data
        1. Precipitation-runoff relation
        2. Hydrograph analyses
        3. Empirical formulas
    B. Extension of gage-sites data
    C. Generation of synthetic flows
    D. Use of generalized data, charts, tables, and empirical approach

In this chapter we discuss direct measurement of streamflow. Measurements through hydraulic devices and by indirect methods are discussed in Chapter 10. The estimation procedure of streamflow is described in Chapter 7.

6.2 STREAM GAGING

Stream gaging or hydrometry is a procedure for measuring the water stage (level) and discharge at a gaging station with the objective of obtaining a continuous record of stage and discharge at the station. For this purpose, equipment is installed at the stream site that enables continuous or regular observation of water stage and frequent measurement of discharge, as well as optional recording of any other hydrologic parameter, such as sediment load. A number of such stations in a basin form a hydrologic network that provides information on the water resources of the basin. The network of continuous-record stations, known as the *basin network*, is often augmented by an *auxiliary network* of partial-record stations: for example, to provide data on peak discharge only.

The systematic records of streamflows as published by the U.S. Geological Survey for each gaging site from year to year involve the measurement and computational steps outlined below.

1. Measurement of water level (stage) on a continuous or daily basis

2. Measurement of discharge from time to time

3. Establishing a relation between stage and discharge

4. Conversion of the measured daily stage into discharge using the relation of step 3

5. Presentation and publication of measured and computed data

The measurement of discharge in step 2 is performed using a current meter, floats, tracer dilution, or the ultrasonic or electromagnetic method, although the current meter is most commonly used.

6.3 STAGE MEASUREMENT

The stage, also known as gage height, is the height of the water surface in a stream above a fixed datum. The datum can be a recognized reference level, such as mean sea level, or an arbitrary level chosen for convenience. Two or three reference marks of known gage height are established on stable structures to maintain a permanent datum.

There are two broad categories of gages: nonrecording gages and recording gages. Nonrecording gages are manually observed at fixed hours. These comprise the staff gage, wire weight gage, float-tape gage, and crest-state gage. The simplest is the staff gage, which consists of porcelain-enameled iron sheet sections (gaged plates) of 4 in. (0.1 m) width and 3.4 ft (1 m) length graduated every 0.02 ft (1 cm).

Automatic or self-recording gages provide a continuous record of stage. These have advantages over nonrecording gages. Their features are discussed below.

6.4 COMPONENTS OF AUTOMATIC GAGES

There are three components of automatic recording gages. These pertain to sensing of the water stage, method of its recording, and storage of the data. Various types of each of these components and their combination in a recording system (gage) are shown in Figure 6.1. A brief description follows.

6.5 SENSOR DEVICES

6.5.1 Float Sensor

This consists of a float attached to one end of a cable that passes over a pulley and is counter-weighted at the other end. The float follows the rise and fall of the water level rotating the pulley. Through a system of other pulleys, this moves a pen up and down, recording the stage.

The float is installed inside a stilling well, which protects the float and dampens the water surface fluctuations. Stilling wells can be made of bricks, concrete blocks, concrete, concrete pipe, or steel pipe. They are placed directly in the stream or on a bank in the vicinity of the stream. In the latter case, an intake pipe connects the stream to the well as shown in Figure 6.2. The bottom of the well is at least 1 ft below the minimum stage and the top is above the 100-year flood level. The dimension of the well is usually 4 ft in diameter or 4 ft by 4 ft in size. For intake the most common size is a 2-in.-diameter pipe.

Three errors associated with float gages are float-lag error, line-shift error, and counterweight submergence error. These are discussed by Rantz and others (1982a).

6.5.2 Bubble-Gage Sensor

This is a pressure-actuated system in which an orifice at the end of a length of tubing is installed underneath the water surface at the location of the gage datum. The water level is directly proportional to the pressure at the orifice. A gas, usually nitrogen, is passed through the tube to bubble freely into the stream through the orifice. The gas pressure is equal to the head on the orifice or the gage height. A servomanometer or a bellows system converts the pressure to the shaft movement for stage recording. For this system the orifice is installed directly inside a stream; the tubing runs along the stream bank and the recorder can be installed away from the stream as shown in Figure 6.3.

6.5.3 Pressure Transducers for Stage Sensing

These devices convert water pressure into electrical signals that can be recorded at the gage site or at a remote location. A transducer has two components: a force-summing-up device that responds to pressure changes and an output device to convert force-summing-up device signals to electrical signals.

Figure 6.1   Components of self-recording gages.

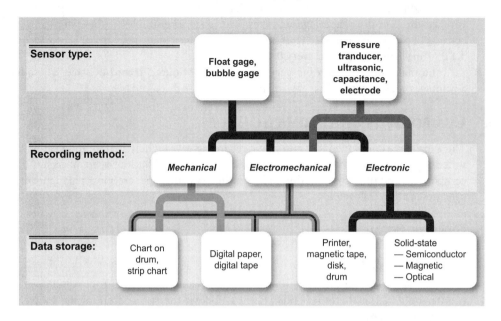

Figure 6.2   Stilling well for a float-type recorder (from Herschy, 1985a).

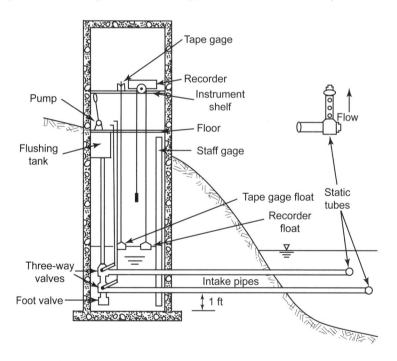

Figure 6.3 Bubble-gage installation (from Herschy, 1985a).

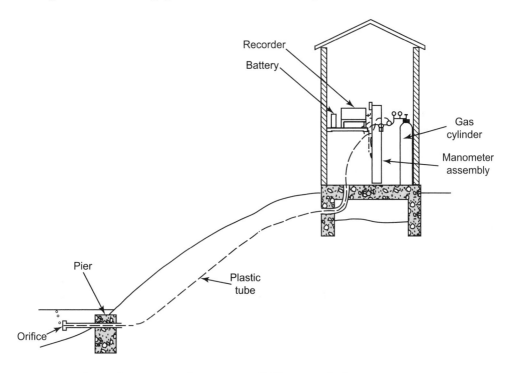

6.5.4 Capacitance, Resistance, and Ultrasonic Sensors

These have not been widely used in stream gaging. The capacitance sensor operates by sensing the change in capacitance between an insulated wire electrode mounted vertically for the full depth of the stream and the surrounding water. The change in water level varies the capacitance in a controlled manner which is sensed electronically.

Since water is an electrical conductor, the resistance sensor operates by detecting the change in the electrical resistance of a vertically mounted sensing element as the water surface moves over it.

In an ultrasonic sensor, an acoustic pulse is directed toward the water surface by the sensor. This pulse is reflected from the surface back to the sensor. The time of travel from transmission to reception is measured and from the velocity of sound, the distance is computed.

6.6 RECORDING MECHANISMS

6.6.1 Mechanical Devices

This type of device consists of a time element and a water-level element. The time element, controlled by a clock, moves the chart or tape at a predetermined fixed rate. The water-level element, actuated by a float or bubble gage, moves the pen stylus or the punch block.

6.6.2 Electromechanical Devices

In such a device, the time element is controlled by an electrically driven clock. Where an AC power source is available at the site, it can be used to power the device.

6.6.3 Electronic Data-Acquisition Systems

Modern devices have been developed that are capable of accepting and storing data entirely electronically. Such devices are able to:

1. Receive a signal from a sensor
2. Convert that signal to compatibility with the recording format
3. Record corresponding time information
4. Record other site identification data
5. Transfer the data to an acquisition system

A recorder may function as a single-site unit, or the operations of many recorders may be controlled by a central processing unit. These devices represent a significant advancement in stage recording technology.

6.7 DATA STORAGE

6.7.1 Graphic (Analog) Record on a Drum

A chart is mounted on a drum, which is installed inside a box either horizontally or vertically. The rotation of the drum and the movement of the stylus pen provide a continuous trace of the water stage with respect to time.

6.7.2 Graphic Record on a Strip Chart

A roll of paper (chart) moves horizontally from one small drum to another at a preadjusted fixed rate. Over the chart, the stylus pen makes a continuous record of the gage height, similar to that on the drum chart.

6.7.3 Digital Record on Paper Tape

The paper tape slowly moves vertically from one small drum to another. At preselected time intervals, the tape is punched for the gage height transmitted to the instrument by rotation of the input shaft, which drives two code disks. Electronic translators are used to read the punch tape onto a tape suitable for input into a computer.

6.7.4 Magnetic Tape

Signals from transducers, capacitance, resistance, and ultrasonic sensors can be recorded in digital format on a magnetic tape. A replay system then reconstructs the data from the magnetic tape records. It is often necessary to translate the recorded tape data into a computer tape format.

6.7.5 Computer Data Storage

In an electronic data-acquisition system, records are transferred to the data memory, which may be either an integral part of the device or may be detachable for data extraction at a processing center. The data memory is capable of accepting large numbers of records, which enables broad-based data gathering, including recording of air and water temperatures and other site information, as well as correction and checking of the data.

6.8 REMOTE TRANSMISSION OF STAGE DATA: TELEMETERING SYSTEM

Stream gaging stations are usually visited by hydrologic field parties at regular intervals to service the stations and to conduct hydrologic measurements. The record of the stage data is collected at the time of such visits. However, when stage information is needed at frequent intervals, or it is not practical to visit the gage site for a considerable time, a telemetering system is used for remote transmitting of the data. There are two types of telemetering system.

6.8.1 Continuous Transmission of Data

Position-motor system. This system, for a distance up to 15 miles, employs a pair of self-synchronizing motors. One motor, on the transmitter side, is actuated by a float or bubble-gage. The other, on the receiving unit, follows the rotary motion of the first one, to which it is electrically connected by a transmission line.

Impulse system. This system, operating over a greater distance than a position-motor system, utilizes telephone lines. An impulse sender at the gaging site is actuated by a float or bubble-gage and sends electrical pulses over the line to the receiver, to be recorded as digital signals.

6.8.2 Intermittent Transmission at Predetermined Intervals

Telemark system. This system codes the instantaneous stage and signals it via a telephone circuit, by radio, or through cellular communication on VHF, UHF, and microwave frequencies. The distance is unlimited in this case. The device consists of a positioning element that is actuated by a sensor and a signaling element that makes contact across the signaling drums positioned in correspondence with the stage.

Resistance system. This system, for distances up to 40 miles, consists of two potentiometers and a microammeter null indicator. One potentiometer, at the gage site, is actuated by the sensor. The other potentiometer, at the observation site, is adjusted for a null balance and the gage height is read from a dial coupled to the potentiometer.

Satellite data-collection system. From stream gaging stations, stage data are transmitted using inexpensive battery-operated radios, known as data-collection platforms (DCPs), to a communication satellite which then retransmits the data to a ground receiving center. Data for transmission are obtained either directly from a digital recorder or through a memory device. The two basic types of satellites used for data transmission are (1) polar-orbiting and (2) geostationary. Polar-orbiting satellites have a low nominal altitude of 870 km and each orbit takes 101 minutes to complete. Thus, every day these satellites make 14 orbits and because of the earth's rotation, their orbit paths shift westward so that the entire earth is covered in 18 days. They provide coverage everywhere on the globe. Several geostationary-type satellites are located at a relatively high altitude of about 35,000 km. Each satellite orbit synchronizes with the rotation of the earth. Thus, these satellites appear to be stationary over a fixed point on the earth; i.e., they observe the same area of the globe continuously.

6.9 THEORY OF DISCHARGE MEASUREMENT

Discharge or streamflow is the volume rate of water flow in a stream, expressed as cubic feet per second or cubic meters per second. It is a product of the area of cross section and the velocity of flow. A natural stream channel can have an irregular shape and thus a standard formula cannot be used to compute the area. Similarly, there is no fixed velocity; it varies in both width and depth in a stream section. Thus the discharge can be given by

$$Q = \int_A v dA \quad [\mathrm{L^3 T^{-1}}] \tag{6.1a}$$

Except for the tracer-dilution technique, in which the equation of the mass rate of flow is used, the other discharge methods are based on eq. (6.1). The current meter and float methods perform algebraic integration (summation). The stream section is divided into a number of subsections. The depth and velocity measurements are arranged to determine an average velocity for each subsection. The discharge is computed by

$$Q = \sum av \quad [\mathrm{L^3 T^{-1}}] \tag{6.1b}$$

where

a = individual subsection area

v = mean velocity of flow in the subsection

Measurements using ultrasonic and electromagnetic methods give the average velocity for the entire stream section. Then the discharge is obtained by multiplying the average velocity by the entire area of cross section.

6.10 MEASUREMENT BY CURRENT METER

The use of the current meter is a common method of discharge measurement. The current meter consists of a cup- or propeller-type rotor. The number of revolutions of the rotor in a given period is directly proportional to the velocity of water. The relation between revolutions per second, n, and velocity of flow, v, is of a straight-line form ($v = a + bn$). Values of a and b are established from the calibration of the meter by the manufacturer and are known as the meter rating. For convenience, the rating data are produced in a table form.

The current meter is divided into two broad categories of vertical-axis meter and horizontal-axis meter, depending on the direction of the rotor shaft. The vertical-axis rotors are mounted with cups or vanes that rotate with the current. The horizontal-axis rotors have a propeller-type attachment. In both cases, each revolution of the rotor completes a circuit through a battery connection that produces an audible click in a headphone or moves a digital counter. A stopwatch is used to measure the time over which revolutions are counted. Both types are available in a standard size and a miniature size for use in very small depths.

The vertical-axis cup meter, known as the Price current meter after its inventor, is most common in the United States. A complete assembly of suspension cable, headphone, battery unit, and sounding weight is shown in Figure 6.4

The U.S. Geological Survey has developed an optical current meter. This is a stroboscopic device used to measure surface velocities at the time of floods without immersing the instrument.

Figure 6.4 Assembly of a type AA current meter (courtesy of Geophysical Instrument and Supply Co.).

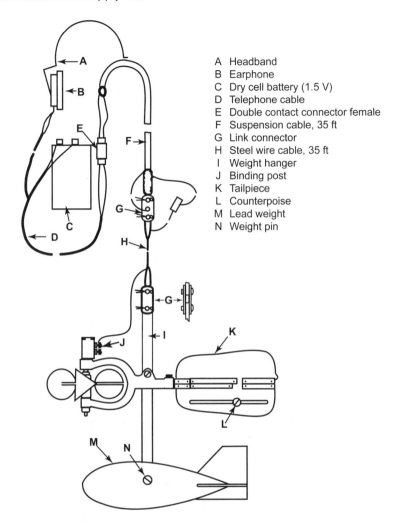

A Headband
B Earphone
C Dry cell battery (1.5 V)
D Telephone cable
E Double contact connector female
F Suspension cable, 35 ft
G Link connector
H Steel wire cable, 35 ft
I Weight hanger
J Binding post
K Tailpiece
L Counterpoise
M Lead weight
N Weight pin

6.10.1 Procedures of Current Meter Measurement

Measurements by current meter are classified as follows in terms of the procedure used to cross a stream during the measurement.

- By wading
- From a bridge
- From a cableway
- By boat
- Over ice cover

In the wading procedure, measurements are made by entering the stream. The method is thus applicable to shallow depths up to 4 ft and velocities of less than 3 to 4 ft/sec.

In the bridge measurement, either a handline or a sounding reel supported by a bridge board or a portable crane is used to suspend the meter and the sounding weight. The size of the sounding weight should be greater than the maximum product of velocity and depth in the cross section.

Cableway measurement is superior to bridge measurement because there is no obstruction of the flow passage, but it involves more initial and operational expenses. The sounding reel carrying the meter and the weight is attached to the cable car.

In deep rivers, where no cableways or suitable bridges are available, the measurement is made by boat. A tag line is first stretched across the section. The tag line serves the dual purpose of holding the boat in position during the measurement and measuring of the width of the river.

For measurement under an ice cover, the most desirable section is just upstream from a riffle because the ice cover is thickest there. At least 20 holes are cut across the section using an ice drill. The effective depth of the water is the total depth minus the depth of the ice cover. A meter with vanes is preferred because the vanes do not become filled with slush ice.

6.11 VELOCITY DISTRIBUTION IN A STREAM SECTION

The velocity in a stream section is not uniformly distributed, due to the presence of the free surface and friction along the stream wall. It varies both across the width and along the depth. Figure 6.5 indicates the general pattern of velocity distribution in a stream channel. The maximum velocity usually occurs below the free surface near the center of the channel section. The velocity decreases toward the banks. Also, the closer to the banks, the deeper the point of highest velocity in a vertical section. Factors that affect the velocity distribution are the shape of the section, the roughness of the channel, and the presence of bends. The surface wind has very little effect. A spiral type of motion has been observed in laboratory investigations. In natural rivers, the spiral motion is usually so weak that its effect is practically eliminated by the channel friction (Chow, 1959).

The problem of the horizontal variation of velocity is resolved by dividing the width of the river into a number of segments while performing the velocity measurements. The vertical variation, however, has to be considered at each segment. The vertical velocity distribution is based on the concept of the *boundary layer* theory because analogies have been found between turbulent boundary-layer flow and turbulent pipe and channel flows. A boundary layer is a region next to the boundary of an object in which the fluid velocity is diminished because of the shear resistance created by the boundary. The turbulent channel flow can be visualized as a turbulent layer that has become as thick as the depth of flow.

6.12 BOUNDARY LAYER THEORY

On a flat plate placed inside a fluid along the direction of flow, a laminar boundary layer develops which grows in the downstream direction and becomes a turbulent boundary layer. The turbulent boundary layer has three zones: (1) laminar sublayer, (2) zone of logarithmic velocity distribution, and (3) zone of velocity defect law, as illustrated in Figure 6.6. The logarithmic velocity distribution law has the widest application since it covers a major portion of the boundary layer, overlaps into the velocity defect law, and because flows broadly conform to this law. This law, derived from Prandtl's mixing-length theory, has the following form for a flat plate:

Figure 6.5 Typical velocity distribution in a stream channel.

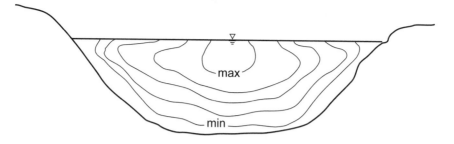

Figure 6.6 Velocity distribution in turbulent boundary layer on a flat plate.

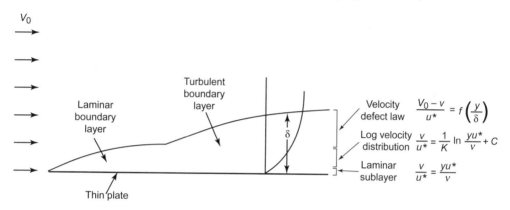

$$\frac{v}{\sqrt{\tau_0/\rho}} = \frac{1}{K} \ln \frac{y\sqrt{\tau_0/\rho}}{v} + C \quad \text{[dimensionless]} \tag{6.2}$$

where

v = velocity at any distance y from the plate

τ_0 = shear stress at the plate (wall)

ρ, v = density and kinematic viscosity of fluid

$u^* = \sqrt{\tau_0/\rho}$ = shear velocity

K = von Kármán universal turbulent constant, ≈ 0.4

C = a constant

Nikuradse's experiments and numerous other tests on rough pipes confirmed the validity of the logarithmic velocity law.

For pipes, the von Kármán logarithmic velocity distribution has the following form, which is a direct derivation from eq. (6.2) after substitution of the boundary condition of $v = V_{max}$ at $y = r_0$, thus ascertaining constant C.

$$\frac{v - V_{max}}{\sqrt{\tau_0/\rho}} = \frac{1}{K} \ln \frac{y}{r_0} \quad \text{[dimensionless]} \tag{6.3}$$

where r_0 represents the radius of the pipe.

Section 6.12 Boundary Layer Theory

For open channels, by substituting $\tau_0 = gds$ and $r_0 = d$ and manipulating eq. (6.3), Vanoni (1941) obtained the following relation:

$$v = \bar{V} + \frac{1}{K}\sqrt{gds}\left(1 + \ln\frac{y}{d}\right) \quad [\text{LT}^{-1}] \tag{6.4}$$

where

$\bar{V}$ = average velocity

d = depth of flow

s = slope of channel

Analyses have shown that the power law equation of the following form provides a result similar to the logarithmic distribution law that conforms to the experimental data very closely in the boundary layer, as well as pipe and channel flows. Also, it is more convenient to apply. An extensive study by Dickinson (1967) led to the consensus that the distribution in streams fits the parabolic curve given by the power law.

$$v = V_0\left(\frac{y}{a}\right)^{1/m} \quad [\text{LT}^{-1}] \tag{6.5}$$

where

v = velocity at a distance y from the bed

V_0 = a known velocity at a distance a from the bed

m = a constant that varies from 6 to 10 depending on the Reynolds number

(Daily and Harleman, 1966); usually, m equals 7

6.13 MEAN VERTICAL VELOCITY

The mean velocity in a vertical plane is

$$\bar{V} = \frac{1}{d}\int_0^d v\,dy \quad [\text{LT}^{-1}] \tag{6.6}$$

Substituting eq. (6.5),

$$\bar{V} = \frac{1}{d}\int_0^d V_0\left(\frac{y}{a}\right)^{1/m} dy \tag{a}$$

or

$$\bar{V} = \frac{1}{d}\frac{am}{m+1}V_0\left[\left(\frac{y}{a}\right)^{1+1/m}\right]_0^d \tag{b}$$

or

$$\bar{V} = \frac{m}{m+1}V_0\left(\frac{d}{a}\right)^{1/m} \quad [\text{LT}^{-1}] \tag{6.7}$$

Suppose that the mean velocity occurs at a distance Z from the bottom. Making $v = \bar{V}$ in eq. (6.5) by substituting $y = Z$, and equating to eq. (6.7), we obtain

$$V_0 \left(\frac{Z}{a} \right)^{1/m} = \frac{m}{m+1} V_0 \left(\frac{d}{a} \right)^{1/m} \tag{c}$$

or

$$Z = \left(\frac{m}{m+1} \right)^m d \quad [\text{L}] \tag{6.8}$$

For values of m between 6 and 10, eq. (6.8) provides Z to be approximately equal to $0.4d$ (i.e., the average velocity occurs at 0.6 depth below the surface).

Also, if $V_{0.2}$ is the velocity at $y = 0.2d$ and $V_{0.8}$ is the velocity at $y = 0.8d$,

$$V_m = \frac{1}{2} \left(V_{0.2} + V_{0.8} \right) \tag{d}$$

From eq. (6.5),

$$V_m = \frac{1}{2} \left[V_0 \left(\frac{0.2d}{a} \right)^{1/m} + V_0 \left(\frac{0.8d}{a} \right)^{1/m} \right] \tag{e}$$

For $m = 7$,

$$V_m = 0.88 V_0 \left(\frac{d}{a} \right)^{1/7} \quad [\text{LT}^{-1}] \tag{6.9}$$

For $m = 7$, eq. (6.7) gives $\bar{V} = 0.88 V_0 (d/a)^{1/7}$, which is equal to V_m of eq. (6.9). Thus the mean of 0.2-depth and 0.8-depth velocities is equal to the average velocity.

6.14 MEASUREMENT OF VELOCITY

The current meter or any other instrument measures velocity at a point, whereas the mean value of velocity in a vertical is required to evaluate the discharge. The mean velocity in a vertical is obtained from the point velocity measurements by one of the following methods:

- Two-point method
- Six-tenths-depth method
- Vertical-velocity curve method
- Integrated measurement method
- Three-point method
- Five-point method
- Six-point method
- Two-tenths-depth method
- Subsurface-velocity method
- Surface-velocity method

The first two methods are common. As proved in the preceding section, the velocity at 0.6 depth from the surface or mean of 0.2 and 0.8 depths is the average velocity by the logarithmic distribution and power laws. A field study by Savini and Bodhaine (1971) indicated that the average velocity determined by both one-point and two-point methods differed from 10-point measurement by 0.7%. The two-point method is slightly better, but it is not used where the depth is less than 2.5 ft. An indication as to whether the two-point method is adequate is derived from two conditions; the 0.2-depth velocity should be greater than the 0.8-depth velocity, and the 0.2-depth velocity should be less than twice the 0.8-depth velocity.

Although there is a striking similarity between observed velocity distribution and the logarithmic (and power) law, the actual distribution in an open channel is not strictly logarithmic. According to the logarithmic (and power) law, the maximum velocity should occur at the surface, which is not the actual case. In natural streams there is further deviation from the theoretical distribution. For this reason, methods 3 through 7, which involve observations at a larger number of points, are used. Methods 8 through 10 are employed in special circumstances.

In the vertical-velocity curve method, a number of velocity observations are made at points well distributed between the water surface and the streambed at each vertical. A plot is made between observed velocities and observation depths as a ratio of total depth. A graphic integration is carried out by measuring the area between the curve and the ordinate axis. The mean velocity is obtained from dividing the area by the length of the ordinate axis. Arithmetic integration (summation) is also a very convenient way to obtain the mean velocity. These procedures on the velocity curve method are demonstrated in Examples 6.1 through 6.4.

In the integrated measurement method the current meter is lowered to the bed and raised to the surface at a uniform rate. The measurement of velocity thus obtained represents the mean velocity for the section. The vertical-axis current meter is not used in this method.

The three-point method combines the two-point and six-tenths-depth methods. Five- and six-point methods make five and six observations, respectively, evenly distributed throughout the depth. In the two-tenths-depth method the velocity is observed at 0.2 of the depth below the surface and a coefficient is applied to the velocity observed. The USGS studies determined a coefficient of 0.87. In the subsurface velocity method, observations are made at some arbitrary distance below the surface when it is not possible to obtain the depths with reliability at very high flow conditions. The coefficients are necessary to convert these to the mean velocity. To determine the coefficients, depths of measurement, as compared to total depth, are estimated after the stage has receded. In conditions of very high flow (i.e., floods), the surface velocity method is preferred over subsurface velocity if an optical current meter is available. A coefficient between 0.85 and 0.90 is used to compute the mean velocity. For smoother sections a value toward the upper limit of 0.9 is applied.

EXAMPLE 6.1

The water velocity in a stream channel has a distribution across a vertical section given by $v = 2(4-y)^{1/7}$, where v is the velocity in ft/s and y is the distance from the water surface. The water depth at this section is 4 ft. (a) Determine the mean velocity across the vertical section. (b) Determine the discharge in the channel if the vertical section represents the average condition for the channel of rectangular shape of 50 ft width.

SOLUTION

(a) Mean velocity across the vertical section,

$$\bar{V} = \frac{1}{4} \int_0^4 v \, dy \text{ (since } y \text{ from top)}$$

$$= \frac{1}{4} \int_0^4 2(4-y)^{1/7} \, dy$$

$$= \frac{1}{4}(2)\left(\frac{7}{8}\right)\left[(4-y)^{8/7}\right]_0^4$$

$$= 2.13 \text{ ft/s}$$

(b) Discharge in the channel,

$$Q = A\bar{V} = (50)(4)(2.13)$$

$$= 426 \text{ cfs}$$

EXAMPLE 6.2

The vertical-velocity distribution in a 4-ft-deep channel is given by $v = 2y^{1/2}$, where y is the distance from the bottom. Determine the mean velocity by graphic integration.

SOLUTION The velocities for various depths are computed below and plotted in Figure 6.7

Height from bottom (ft)	Ratio of depth/total depth	Velocity, $v = 2y^{1/2}$ (ft/sec)
1	0.25	2
2	0.50	2.83
3	0.75	3.46
4	1.0	4.0

x scale   10 divisions $= 1$ ft/sec

y scale   10 divisions $= 0.2$

100 squares* $= 0.2$ ft/sec

1 square $= 0.002$ ft/sec

Area covered by the curve $= 1340$ squares

or $\bar{V} = 1340 \times 0.002$

$= 2.68$ ft/sec

EXAMPLE 6.3

Solve Example 6.2 by algebraic summation.

---

* A small square formed by one division on x and y scales.

Section 6.14   Measurement of Velocity

Figure 6.7 Plot of velocity versus depth for Example 6.2.

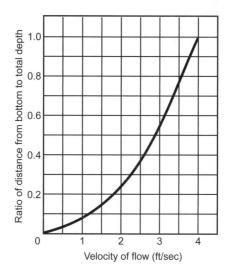

SOLUTION

Either from the velocity equation of Example 6.2 or from the plot in Figure 6.7, the velocities corresponding to different values of depths are as follows:

(1) Height (ft)	(2) Velocity (ft/s)	(3) Mean V[a] (ft/s)	(4) Area of Curve[b] (ft²/s)
0	0		
		0.71	0.36
0.5	1.41		
		1.70	0.85
1.0	2		
		2.22	1.11
1.5	2.45		
		2.64	1.32
2.0	2.83		
		3.0	1.50
2.5	3.16		
		3.31	1.66
3.0	3.46		
		3.60	1.80
3.5	3.74		
		3.87	1.94
4.0	4.0		
Total			10.54

[a] Average of two successive values of col. 2
[b] col. 3 × (difference of two successive values in col. 1)

$$\bar{V} = \frac{1}{(4 \text{ ft})}\left(10.54 \text{ ft}^2/\text{s}\right) = 2.64 \text{ ft/sec}$$

EXAMPLE 6.4

The following point-velocity observations were made in a vertical section of a stream channel. Determine the mean velocity by various methods and compare their results. The total depth of flow is 4 m.

Ratio of observation to total depth	Velocity (m/s)
0.05	0.36
0.2	0.35
0.4	0.34
0.6	0.32
0.8	0.28
0.95	0.20

SOLUTION

1. Vertical-velocity curve method:

   Depth versus velocity data are plotted in Figure 6.8.

   x-scale   10 divisions $= 0.04$ m/s

   y-scale   10 divisions $= 0.2$

   $\qquad\qquad$ 100 squares $= 0.008$ m/s

   $\qquad\qquad$ 1 square $= 8 \times 10^{-5}$ m/s

   $\qquad$ Area under the plot $= 1420$ squares $\left(8 \times 10^{-5}\right) = 0.114$ m/s

   $\qquad\qquad$ Mean velocity $=$ datum $+ 0.114$

   $$\overline{V} = 0.2 + 0.114 = 0.314 \text{ m/s}$$

2. Two-point method:

   $$\overline{V} = 0.5(V_{0.2} + V_{0.8}) = 0.5(0.35 + 0.28) = 0.315 \text{ m/s}$$

3. Six-tenths-depth method:

   $$\overline{V} = V_{0.6} = 0.32 \text{ m/s}$$

4. Three-point method:

   $$\overline{V} = 0.25(V_{0.2} + 2V_{0.6} + V_{0.8})$$
   $$= 0.25\left[0.35 + 2(0.32) + 0.28\right] = 0.318 \text{ m/s}$$

5. Two-tenths-depth method:

   $$V_{0.2} = 0.35$$
   $$\overline{V} = (\text{coefficient})V_{0.2} = (0.87)(0.35) = 0.305 \text{ m/s}$$

6. Surface-velocity method:

   $$\overline{V} = (\text{coefficient})\, V_{\text{surface}} = (0.85)(0.36) = 0.306 \text{ m/s}$$

Figure 6.8 Vertical-velocity profile for Example 6.4.

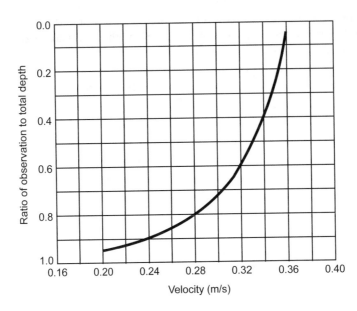

7. Five-point method:

$$\bar{V} = 0.1\left(V_{surf} + 3V_{0.2} + 3V_{0.6} + 2V_{0.8} + V_{bed}\right)$$
$$= 0.1\left[0.36 + 3(0.35) + 3(0.32) + 2(0.28) + 0.20\right] = 0.313 \text{ m/s}$$

8. Six-point method:

$$\bar{V} = 0.1\left(V_{surf} + 2V_{0.2} + 2V_{0.4} + 2V_{0.6} + 2V_{0.8} + V_{bed}\right)$$
$$= 0.1\left[0.36 + 2(0.35) + 2(0.34) + 2(0.32) + 2(0.28) + 0.20\right]$$
$$= 0.314 \text{ m/s}$$

Summary of Results

Method	Mean Velocity (m/s)	Deviation	Error (%)
1. Vertical-velocity curve	0.314	0	0
2. Two-point	0.315	0.001	0.3
3. Six-tenths	0.320	0.006	1.9
4. Three-point	0.318	0.004	1.3
5. Two-tenths-depth	0.305	0.009	2.9
6. Surface-velocity	0.306	0.008	2.5
7. Five-point	0.313	0.001	0.3
8. Six-point	0.314	0	0

Thus, the vertical-velocity curve, six-point, five-point, and two-point methods provide better results in this case.

6.15 MEASUREMENT OF DEPTH (SOUNDING)

Along with velocity, measurement of depth at vertical sections is required to compute discharge. The following four methods are used for depth measurement:

1. Wading rod
2. Sounding weight suspended by a hand line
3. Sounding weight suspended by a reel line
4. Sonic sounder

6.15.1 Wading Rod

This is a graduated steel rod of hexagonal or round shape, with a diameter of 1/2 in. The rod is placed in the stream so that the base plate rests on the streambed and the depth of water is read on the graduated rod. The current meter can be set at a desired position of 0.2, 0.6, or 0.8 depth.

6.15.2 Weight with a Hand Line

When it is not possible to use a wading rod due to deep or swift water, a sounding weight is suspended below the current meter. The assembly is attached to a cable and is used from a bridge, boat, or cableway to perform measurements. The weights are streamlined to a bomb shape.

6.15.3 Weight with a Reel Line

For high-water measurements requiring heavier weights, a sounding reel is used. It has a drum for winding the sounding cable; a crank and ratchet assembly for lowering, raising, and holding the current meter and weight assembly; and a depth indicator. Often, air and wet-line corrections have to be applied to the sounding cable measurements as discussed below.

6.15.4 Sonic Sounder

Based on the principle of echo sounding, a sonic sounder provides a continuous strip-chart record of the depth of the stream. The portable sounder works on a 6- or 12-V storage battery. Its transducer releases pulses of ultrasonic energy at fixed intervals. The instrument measures the time taken by these pulses of energy to travel to the streambed, to be reflected, and to return to the transducer. With a known velocity of sound in water, the instrument computes and records the depth.

6.15.5 Air Correction for Depth

The position that a sounding line will take is shown in Figure 6.9. The air correction is de and from trigonometry given by

$$de = \left(\frac{1 - \cos \theta}{\cos \theta} \right) ab \quad [\text{L}] \tag{6.10}$$

The air corrections from eq. (6.10) as a percent of vertical depth ab are given in Table 6.1 for various values of θ.

Figure 6.9 Deflection of current meter cable in deep, swift water.

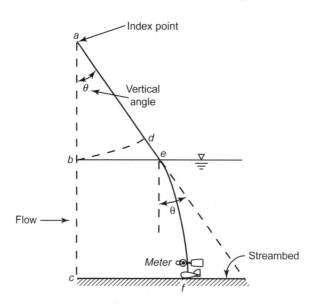

Table 6.1 Air Correction

Vertical Angle (deg)	Correction (%)	Vertical Angle (deg)	Correction (%)
4	0.24	18	5.15
6	0.55	20	6.42
8	0.98	22	7.85
10	1.54	24	9.46
12	2.23	26	11.26
14	3.06	28	13.26
16	4.03	30	15.47

6.15.6 Wet-Line Correction for Depth

Below the water surface, the tangent at any point of the cable is equal to the total horizontal force (of the current) divided by the total vertical force (of the sounding weight) at that point. This provides the value of the angle that the cable makes at any point below the water surface. These angles for incremental depths are computed and then using the relation of eq. (6.10), corrections are computed. The summation of these provides total wet-line correction, which has been tabulated as a function of wet-line depth *ef* in Table 6.2 for various values of θ.

The following procedure is followed to apply this correction:

1. Depth, *aef*, is measured by the sounding line.

2. Measure the vertical distance, *ab*, by taking the reading when the weight is placed at the water surface. Determine the air correction from Table 6.1 for *ab*.

Table 6.2   Wet-Line Correction

Vertical Angle (deg)	Correction (%)	Vertical Angle (deg)	Correction (%)
4	0.06	18	1.64
6	0.16	20	2.04
8	0.32	22	2.48
10	0.50	24	2.96
12	0.72	26	3.50
14	0.98	28	4.08
16	1.28	30	4.72

3. Wet-line depth, $ef = aef - (ab + \text{air correction})$. Determine wet-line correction from Table 6.2 for ef.

4. The corrected bc is computed to be

$$bc = aef - ab - (\text{air correction} + \text{wet-line correction}) \quad [\text{L}] \quad (6.11)$$

5. To position the current meter at 0.2 depth: For 0.2 depth, the wet-line curvature is disregarded.

$$\begin{pmatrix} \text{vertical distance} \\ \text{to 0.2 depth} \end{pmatrix} = ab + 0.2bc + \begin{pmatrix} \text{distance from the} \\ \text{bottom of weight} \\ \text{to current meter} \end{pmatrix} \quad [\text{L}] \quad (6.12a)$$

$$(\text{corrected 0.2 depth}) = \begin{pmatrix} \text{vertical distance} \\ \text{to 0.2 depth} \end{pmatrix} + \begin{pmatrix} \text{air correction for} \\ \text{vertical distance} \\ \text{to 0.2 depth} \end{pmatrix} \quad [\text{L}] \quad (6.12b)$$

6. To position the meter at 0.6- or 0.8-depth:

$$\begin{pmatrix} \text{corrected} \\ \text{0.8 or 0.6 depth} \end{pmatrix} = aef - 0.2 \text{ or } 0.4 \begin{pmatrix} bc \\ \text{from} \\ \text{eq. (6.11)} \end{pmatrix} + \begin{matrix} \text{wet-line} \\ \text{correction} \end{matrix} + \begin{pmatrix} \text{distance from} \\ \text{weight to the} \\ \text{current meter} \end{pmatrix} \quad [\text{L}] \quad (6.13)$$

Equations (6.12b) and (6.13) are used for placing the current meter at 0.2 and 0.8 depths in cable-suspended measurements even when air and wet-line corrections are not involved. In such cases the correction terms are treated as being equal to zero.

EXAMPLE 6.5

In gaging a deep, swift stream through a cableway, the total depth of the sound line was found to be 25.2 ft. The depth from the guide pulley to the surface was measured to be 10.3 ft. A protractor measured the vertical angle of 24°. The weight hanger separates the current meter from the weight by 1 ft. Determine the **(a)** true depth of the water, **(b)** position for 0.2 depth, and **(c)** 0.8 depth of the current meter.

SOLUTION   Refer to Figure 6.9.

(a) $aef = 25.2$ ft

$ab = 10.3$ ft

From Table 6.1, for 24°,

$$\% \text{ air correction} = 9.46$$

$$\text{air correction} = \frac{9.46}{100}(10.3) = 0.974 \text{ ft}$$

$$\text{Wet-line depth, } ef = aef - (ab + \text{air correction})$$

$$= 25.2 - (10.3 + 0.974) = 13.93 \text{ ft}$$

From Table 6.2, for 24°,

$$\% \text{ wet-line correction} = 2.96$$

$$\text{wet-line correction} = \frac{2.96}{100}(13.93) = 0.412 \text{ ft}$$

From eq. (6.11), corrected depth,

$$bc = aef - ab - (\text{air correction} + \text{wet-line correction})$$

$$= 25.2 - 10.3 - (0.974 + 0.412) = 13.51 \text{ ft}$$

(b) Position of 0.2 depth:

From eq. (6.12a),

$$\text{vertical distance} = 10.3 + 0.2(13.51) + 1 = 14.00$$

$$\text{air correction for 14-ft depth} = \frac{9.46}{100} \times 14.0 = 1.32$$

From eq. (6.12b), 0.2 depth $= 14.00 + 1.32 = 15.32$ ft

(c) Position of 0.8 depth:

From eq. (6.13), 0.8 depth $= 25.2 - 0.2(13.51 + 0.412) + 1 = 23.42$ ft

---

EXAMPLE 6.6

A stream is gaged using a hand line from a bridge. The total depth of the sound line from the rail of the bridge is measured to be 8.25 m. The depth up to the water surface is 4.4 m. If the distance from the center of the current meter to the bottom of the weight is 0.3 m, determine the position where the current meter is to be placed for 0.2 depth and 0.8 depth, respectively.

SOLUTION With the hand line, no air and wet-line corrections are involved. Depth of water, $bc = 8.25 - 4.4 = 3.85$ m.

From eq. (6.12a),   0.2 depth $= ab + 0.2bc + \text{distance from meter to weight}$

$$= 4.4 + 0.2(3.85) + 0.3 = 5.47 \text{ m}$$

From eq. (6.13),   0.8 depth $= aef - 0.2bc + \text{distance of meter to weight}$

$$= 8.25 - 0.2(3.85) + 0.3 = 7.78 \text{ m}$$

6.16 COMPUTATION OF DISCHARGE

Measurements of velocity and depth, made at a number of locations across a stream chan-nel, are used to compute discharge by summing up the product of mean velocity and area of cross section of the segment between successive locations. Usually, between 20 and 30 verti-cals of equidistant or variable spacings are used to divide a stream width. These spacings should be arranged so that no segment contains more than 10% of the total flow. Depend-ing on the procedure used to obtain the multiplication of velocity and area of various ele-ments constituting the channel section, methods are known as midsection, mean-section, velocity-depth integration, and velocity contour methods of discharge computation. The first two methods are arithmetic summation procedures and the last two are graphic meth-ods. Midsection is a preferred method.

6.16.1 Midsection Method

In this method it is assumed that the velocity at each vertical represents a mean velocity for a section that extends half the distance into the preceding and following segments, as shown in Figure 6.10.

$$\text{Area for subsection } 3 = \frac{W_2 + W_3}{2} d_3 \tag{a}$$

$$\text{Discharge through subsection } 3 = \bar{V}_3 \frac{W_2 + W_3}{2} d_3 \tag{b}$$

$$\text{Discharge through subsection } x = \bar{V}_x \frac{W_{x-1} + W_x}{2} d_x \quad [\text{L}^3\text{T}^{-1}] \tag{6.14}$$

Figure 6.10   Subsection in the midsection method.

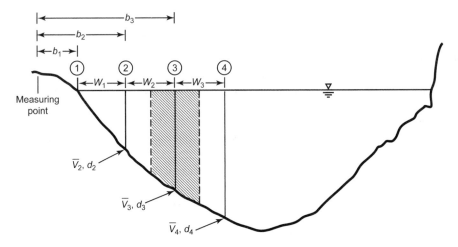

1, 2, 3, ... Stations
$b_1, b_2, b_3, ...$ Distance from the initial point to the station
(observation verticals)
$d_1, d_2, d_3, ...$ Depth of water at the observation verticals
$W_1, W_2, W_3, ...$ Width between successive verticals

When the cross section is such that there is a depth at the edge of the water as at the last vertical in Figure 6.10, the velocity is estimated as a certain percentage (between 65 and 90%) of the adjacent vertical because it is not possible to measure velocity by the current meter. At the beginning section, using eq. (6.14), W_0 will have no significance and should be dropped. Example 6.7 provides the data in the format as they are recorded in the field book while taking measurements.

EXAMPLE 6.7

Compute the discharge by midsection method for the following measurement data. The current-meter rating is given by $v = 0.1 + 2.2N$, where v is velocity in ft/sec and N is the number of revolutions per second.

Distance from Initial Point (ft)	Depth (ft)	Observed Depth	Revolutions	Time (sec)
10	1			
12	3.5	0.2	35	50
		0.8	22	50
14	5.2	0.2	40	60
		0.8	30	55
17	6.3	0.2	45	60
		0.8	30	55
19	4.4	0.2	33	45
		0.8	30	50
21	2.2	0.6	22	50
23	0.8	0.6	10	45
25	0			

SOLUTION   The computations are shown in Table 6.3.

Width (col. 8) is the successive difference of col. 1.

Effective width (col. 9) is the average of preceding and following widths in col. 8.

Area (col. 10) = col. 2 × col. 9.

Discharge (col. 11) = col. 7 × col. 10.

Note that the conditions of a minimum 20 verticals and less than 10% flow in any subsection are violated to reduce computation. Also for the first vertical, having a depth of 1 ft, a velocity of 0.65 times the adjacent velocity has been taken. For the last vertical, this value is taken as zero since there is no water depth.

Table 6.3 Computation of Discharge by Midsection Method (Example 6.7)

(1)	(2)	(3)	(4)	(5)	(6)	(7)	(8)	(9)	(10)	(11)
					Velocity (ft/sec)					
Distance from Initial Point	Depth (ft)	Observed Depth	Revolu- tions	Time (sec)	At Points	Mean in Section	Width (ft)	Effective Width (ft)	Area[b] (ft²)	Discharge[c] (ft³/sec)
10	1					0.88[a]		1	1.0	0.88
							2			
12	3.5	0.2	35	50	1.64	1.36	2	2	7.0	9.52
		0.8	22	50	1.07					
							2			
14	5.2	0.2	40	60	1.57	1.44		2.5	13.0	18.72
		0.8	30	55	1.30					
							3			
17	6.3	0.2	45	60	1.75	1.53		2.5	15.75	24.10
		0.8	30	55	1.30					
							2			
19	4.4	0.2	33	45	1.71	1.57		2	8.8	13.82
		0.8	30	50	1.42					
							2			
21	2.2	0.6	22	50	1.07	1.07		2	4.4	4.71
							2			
23	0.8	0.6	10	45	0.59	0.59		2	1.6	0.94
							2			
25	0					0		1	0	0
Total									51.55	72.69

[a] 0.65×1.36 of subsequent vertical = 0.88
[b] col. 2 × col. 9
[c] col. 7 × col. 10

6.16.2 Mean-Section Method

The segment area (subsection) extends from vertical to vertical as shown in Figure 6.11.

$$\text{Area for subsection } 3-4 = \frac{d_3 + d_4}{2} W_3 \tag{c}$$

$$\text{Discharge through subsection } 3-4 = \left(\frac{\bar{V}_3 + \bar{V}_4}{2}\right)\left(\frac{d_3 + d_4}{2}\right) W_3 \tag{d}$$

$$\text{Discharge through subsection } x \text{ and } x+1 = \left(\frac{\bar{V}_x + \bar{V}_{x+1}}{2}\right)\left(\frac{d_x + d_{x+1}}{2}\right) W_x \quad [\text{L}^3\text{T}^{-1}] \tag{6.15}$$

Figure 6.11 Subsection in the mean-section method.

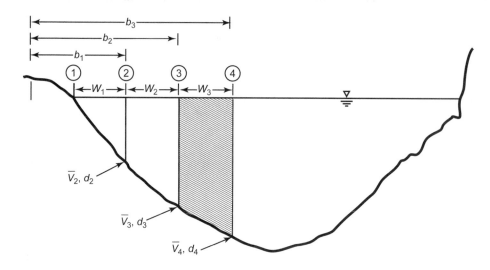

EXAMPLE 6.8

Solve Example 6.7 by the mean-section method.

SOLUTION Computations are arranged in Table 6.4.
Average velocity (col. 8) is the average of mean velocities of two verticals in col. 7.
Average depth (col. 9) is the average of depths of two verticals in col. 2.
Width (col. 10) is the successive difference of col. 1.
Area (col. 11) = col. 9 × col. 10.
Discharge = col. 8 × col. 11.

6.16.3 Velocity-Depth Integration Method

This is a graphic method in which velocity measurements in each vertical should preferably be performed at a number of depths to plot the vertical-velocity curve for each vertical.
The procedure is as follows:

1. Draw the vertical-velocity curve for each vertical and determine the area under this curve
   that will represent (velocity depth) at each vertical. (If the mean velocity in a vertical has
   been determined by any other method, it can be multiplied by the vertical depth.)
2. Plot (velocity depth) values at the location of respective verticals across the stream cross
   section as shown in Figure 6.12. Draw a smooth curve through these points. The area
   enclosed by this curve will provide the discharge.

EXAMPLE 6.9

The velocity-depth values for various verticals of a stream cross section as obtained from
vertical-velocity curve analyses are indicated in Figure 6.13. Determine the discharge of the
stream by the velocity-depth integration method.

Table 6.4 Computation of Discharge by Mean-Section Method (Example 6.8)

(1)	(2)	(3)	(4)	(5)	(6)	(7)	(8)	(9)	(10)	(11)	(12)
				Time	Velocity (ft/sec)						
Distance from Initial Point (ft)	Depth (ft)	Observed Depth	Revolutions	(sec)	At Points	Mean in Vertical	Average Velocity for Subsection (ft/sec)	Average Depth for Subsection (ft)	Width (ft)	Area (ft²)	Discharge (ft³/sec)
10	1					0.88[a]					
							1.12	2.25	2	4.5	5.04
12	3.5	0.2	35	50	1.64	1.36					
		0.8	22	50	1.07						
							1.40	4.35	2	8.7	12.18
14	5.2	0.2	40	60	1.57	1.44					
		0.8	30	55	1.30						
							1.49	5.75	3	17.25	25.70
17	6.3	0.2	45	60	1.75	1.53					
		0.8	30	55	1.30						
							1.55	5.35	2	10.70	16.59
19	4.4	0.2	33	45	1.71	1.57					
		0.8	30	50	1.42						
							1.32	3.3	2	6.60	8.71
21	2.2	0.6	22	50	1.07	1.07					
							0.83	1.5	2	3.00	2.49
23	0.8	0.6	10	45	0.59	0.59					
							0.30	0.40	2	0.80	0.24
25	0				0	0					
Total										51.55	70.95

[a] 0.65 × 1.36 of subsequent vertical

Figure 6.12 Graphic procedure of velocity-depth integration (subsections are enlarged).

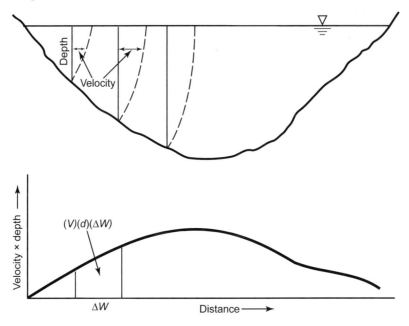

Figure 6.13 Results of vertical curve analysis for a stream cross section.

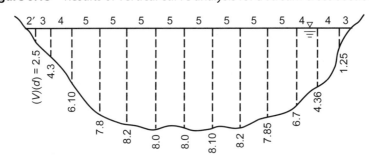

SOLUTION   The velocity-depth versus distance has been plotted in Figure 6.14.

Scale factor   x-scale   10 divisions = 10 ft

y-scale   10 divisions = 2 ft^3/sec

100 squares = 20 ft^3/sec

1 square = 0.2 ft^3/sec

Area under the curve = 1900 squares

Discharge = 1900 × 0.2 = 380 ft^3/sec

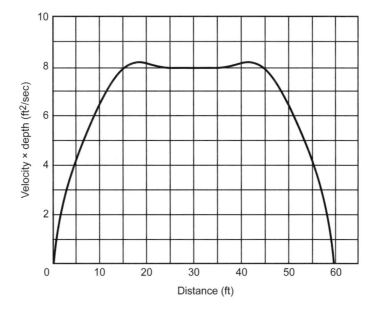

6.16.4 Velocity-Contour Method

This is also a graphical method. Since velocity contours have to be drawn, velocity measurements at a number of points in each vertical are required for the application of this method. The procedure is as follows:

1. Draw the river cross section to a convenient scale. On each vertical, write the point velocity measurements. Connect points of equal velocity to draw the velocity contours (isovels), as shown in Figure 6.15.

2. Starting from the highest value, determine the areas between successive velocity contours. Plot the lower limit of contour velocity against cumulated area up to that velocity contour, and extend it back to the y-axis as shown in Figure 6.16. The area enclosed by the curve represents the total discharge.

EXAMPLE 6.10

The velocity distribution (contours) in ft/s in a river cross section has the pattern shown in Figure 6.15. The areas measured by a planimeter between successive contour lines are tabulated below. Determine the discharge at the site.

Contours (ft)	Area between Contours (ft^2)	Cumulated Area (ft^2)
>2.5–2.5	5.0	5.0
2.5–2.0	20.5	25.5
2.0–1.5	25.2	50.7
1.5–1.0	16.3	67.0
1.0–0.5	10.8	77.8
0.5–0	4.2	82.0

SOLUTION The lower limit of contour velocity and the corresponding cumulated area are plotted in Figure 6.17, extending to the vertical axis.

$$\text{Scale Factor} \quad x \text{ scale} \quad 10 \text{ divisions} = 10 \text{ ft}^2$$
$$y \text{ scale} \quad 10 \text{ divisions} = 0.5 \text{ ft/sec}$$
$$100 \text{ squares} = 5 \text{ ft}^3/\text{sec}$$
$$1 \text{ square} = 0.05 \text{ ft}^3/\text{sec}$$
$$\text{Area under the curve} = 2691 \text{ squares}$$
$$\text{Discharge} = 2691(0.05) = 134.6 \text{ ft}^3/\text{sec}$$

6.17 MEASUREMENT OF DISCHARGE BY FLOATS

During high-flow conditions, when excessive velocities, depths, or floating drifts prohibit the use of a current meter, the float method is used to measure discharge. It is essentially observing the time required for a float to transverse a known length, so that

$$V = \frac{L}{t} \quad [\text{LT}^{-1}] \tag{6.16}$$

The three types of floats used are (1) surface floats, (2) subsurface floats with a submerged canister, and (3) rod floats.

Two cross sections that are about four or five times the stream width apart are selected along a reach of a straight channel. A number of floats are distributed uniformly across the stream width, and their position with reference to distance from the bank is posted. The floats are introduced a short distance upstream from the upper cross section so that they take up the speed of the current when they reach the upper cross section. They are introduced from a bridge or cableway, or tossed in from the bank. A boat is used for wide rivers.

The velocity of each float derived from eq. (6.16) is converted to the mean velocity in the vertical by a reduction coefficient, which is commonly 0.85 for surface floats and 1.0 for subsurface and rod floats.

At the time of measurement, water surface elevation is referenced to stakes along the bank at each cross section and also at one or more intermediate sites. At a later date these sections are surveyed to derive an average stream cross section to compute the area.

The computation of discharge is similar to that of current meter computation. The discharge in each subsection of the average cross section is obtained by multiplying the area of subsection by the mean velocity for that subsection.

When it is impractical to obtain proper float movement across the entire width or when floats tend to move toward the center of the flow, an unadjusted discharge is obtained based on the mean of surface velocity. A coefficient is applied to determine the stream discharge.

6.18 MEASUREMENT OF DISCHARGE BY TRACER DILUTION

This method is used when conditions are not favorable for current meter measurement. In rock-strewn shallow streams or rough channels carrying highly turbulent flow, the dilution technique provides an effective method of flow measurement. In this method, a tracer

Figure 6.15 Velocity distribution in a channel cross section.

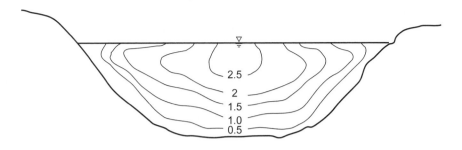

Figure 6.16 Plot of velocity versus area to compute discharge.

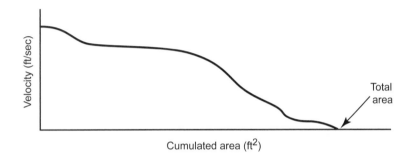

Figure 6.17 Velocity-area plot for Example 6.10.

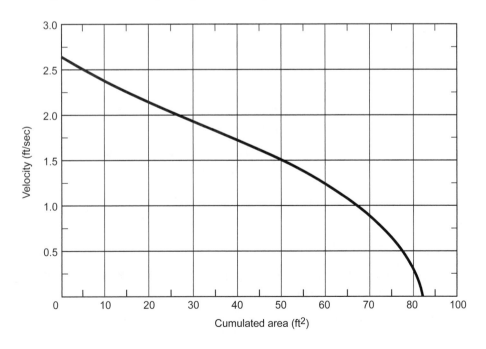

solution of known concentration, in known quantity, is injected into the stream to be diluted by the discharge of the stream. At a cross section downstream from the injection site, water is sampled to ascertain the concentration of the tracer solution. Applying the principle of conservation of mass, the discharge is determined directly.

There are two basic injection techniques, several sampling techniques, and a large number of tracers of chemical, fluorescent dye, and radioactive types. The injection techniques differ in terms of period of injection of the tracer solution into the stream. The two methods are described in the following sections.

6.18.1 Constant Injection Rate Method

A solution of concentration C_1 of a selected tracer is injected at a constant rate q at a beginning section. At a second section downstream, the concentration of the tracer is measured for a sufficient period of time at a number of points to ensure proper mixing and attaining of constant dilution. This has been shown in Figure 6.18. Assuming no concentration of the same tracer in natural flow:

$$\text{Mass rate at station } 1 = qC_1$$

$$\text{Mass rate at station } 2 = (Q + q)C_2$$

Equating two rates gives

$$qC_1 = (Q + q)C_2$$

or

$$Q = \frac{C_1 - C_2}{C_2}q$$

In general, C_1 is much greater than C_2; then

$$Q = \frac{C_1}{C_2}q \quad [\text{L}^3\text{T}^{-1}] \tag{6.17}$$

where

q = rate of injection of tracer solution, ft^3/sec, m^3/s, or liters/s

Q = discharge of the stream, ft^3/sec, m^3/s, or liters/s

C_1 = concentration of tracer as injected, lb/ft^3, kg/m^3, or mg/liter

C_2 = diluted concentration downstream, lb/ft^3, kg/m^3, or mg/liter

$\frac{C_1}{C_2} = N$ = dilution ratio

EXAMPLE 6.11

In tracer dilution measurement by the constant injection rate method, the following observations are made. Determine the streamflow.

1. Rate of injection of tracer = 2.5×10^{-4} ft^3/sec.

2. Concentration of injected solution = 1.8 g/liter.

Figure 6.18 Constant rate injection of tracer solution.

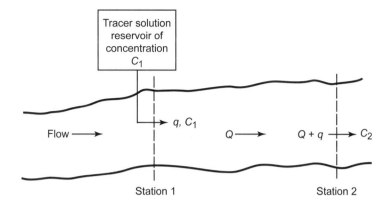

3. Concentration at three sampling points evenly spaced:

$$At\ point\ A,\ C_2 = 5.42\ \mu g/liter\ (5.42 \times 10^{-6}\ g/liter)$$
$$At\ point\ B,\ C_2 = 5.35\ \mu g/liter$$
$$At\ point\ C,\ C_2 = 5.50\ \mu g/liter$$

SOLUTION   Because the three points are spaced evenly, each is given equal weight in computing mean C_2:

$$C_2 = \frac{5.42 + 5.35 + 5.50}{3} = 5.42\ \mu g/liter \quad or \quad 5.42 \times 10^{-6}\ g/liter$$

From eq. (6.17),

$$Q = \frac{C_1}{C_2} q$$

$$= \frac{1.8}{5.42 \times 10^{-6}} \left(2.5 \times 10^{-4}\right) = 83\ ft^3/sec$$

6.18.2 Sudden Injection Method

A volume V of a tracer solution of concentration C_1 is injected over a short period at the beginning of the section (station 1). At a second section downstream (station 2), the concentration of the tracer, C_i, is determined from time to time for a period sufficiently long enough to ensure that all of the tracer has passed through the second section.

$$Mass\ of\ tracer\ at\ station\ 1 = VC_1$$

$$Mass\ of\ tracer\ at\ station\ 2 = Q \int_0^\infty C_i\,dt$$

$$= Q \sum_{i=1}^{N} \frac{C_i \left(t_{i+1} - t_{i-1}\right)}{2}$$

where N is the number of samples collected over a sufficiently long period of time. Equating yields

$$Q = \frac{VC_1}{\displaystyle\sum_{i=1}^{N} C_i \left(t_{i+1} - t_{i-1} \right)/2} \quad [\mathrm{L}^3\mathrm{T}^{-1}] \tag{6.18}$$

where

V = volume of injected solution

C_1 = concentration of injected tracer

C_i = measured tracer concentration downstream at a given time in sample i

i = sequence number of sample

N = total number of samples

t_i = time when sample C_i is obtained

EXAMPLE 6.12

Gaging by the sudden injection method of the tracer dilution technique provided the following data. Calculate the discharge.

1. Volume of tracer injected = 3.12 liters.
2. Concentration of injected tracer = 1.75 g/liter.
3. Concentration of tracer in samples collected at various times:

Time (min)	Concentration (µg/liter)
0	0.0
2	5.0
3	12.1
4	11.5
5	9.5
6	5.2
7	1.8
8	1.0
10	0.3
15	0.0

Sample	Concentration C_i (μg/liter)	t_i (min)	$(t_{i+1}-t_{i-1})/2^a$ (min)	$C_i(t_{i+1}-t_{i-1})/2$ μg · min/liter
1	0	0	1	0
2	5	2	1.5	7.5
3	12.1	3	1	12.1
4	11.5	4	1	11.5
5	9.5	5	1	9.5
6	5.2	6	1	5.2
7	1.8	7	1	1.8
8	1.0	8	1.5	1.5
9	0.3	10	3.5	1.1
10	0	15	2.5	0
Total				50.2

[a] For example, for sample 3, $(t_4-t_2)/2 = (4-2)/2 = 1$.

From eq. (6.18),

$$Q = \frac{VC_1}{\displaystyle\sum_{i=1}^{N} C_i\left(t_{i+1}-t_{i-1}\right)/2}$$

$$= \frac{3.12(1.75)}{50.2 \times 10^{-6}} = 108{,}765 \text{ liters/min} \quad \text{or} \quad 1.81 \text{ m}^3/\text{s}$$

6.19 Discharge Measurement by the Ultrasonic (Acoustic) Method

This method has been applied selectively to measure discharge of rivers, canals, penstocks, conduits, and tunnels. The method uses two transducers, two receivers, and a digital processor. The block diagram is shown in Figure 6.19. The transducers are mounted on each bank in an oblique direction, as shown in the figure. Sound pulses sent by A are received by B and, in the opposite direction, pulses transmitted by B are received by A. Sound waves traveling downstream have a higher velocity than those traveling upstream, due to the stream velocity component parallel to the flight (acoustic) path. Since the stream velocity is much less than the sound velocity in water, the difference in upstream and downstream time is very small and needs to be recorded precisely.

The travel time downstream is $t_{BA} = L/(C + v_p)$ and that upstream is $t_{AB} = L/(C - v_p)$, C being the acoustic velocity in water. Since $v_p = v\cos\theta$, from the difference of time or difference of frequency (l/time) the following is derived, for the average velocity:

$$\bar{V} = \frac{L\,\Delta t}{2t_{AB}t_{BA}\,\cos\theta} \quad [LT^{-1}] \tag{6.19}$$

When multiplied by the average depth of flow and channel width, this leads to the following discharge equations:

1. Travel-time difference method:

$$Q = \frac{L^2 \bar{d} \, \Delta t \, \tan\theta}{2t_{AB}t_{BA}} \quad [\mathrm{L^3T^{-1}}] \tag{6.20}$$

2. Frequency difference method:

$$Q = \frac{L^2}{2}\left(\frac{1}{t_{BA}} - \frac{1}{t_{AB}}\right)\bar{d} \, \tan\theta \quad [\mathrm{L^3T^{-1}}] \tag{6.21}$$

where

L = flight (path) length

θ = angle of acoustic path to direction of flow

$\bar{d}$ = average depth of flow along AB

t_{AB} = travel time from A to B

t_{BA} = travel time from B to A

$\Delta t = t_{AB} - t_{BA}$

In a single-path system, measurement is made at one depth only with a pair of transducers set at 0.6 of the most frequently occurring depth. More common, however, is the multipath system, in which several pairs of transducers are installed at various water depths. The average velocity of each path when multiplied by spacing between paths (transducers), and the length of each, provides the total discharge. Holmes et al. (1973) have presented the Gaussian quadrature method of computation of discharge with three or more paths of measurement of velocity.

6.20 DISCHARGE MEASUREMENT BY THE ELECTROMAGNETIC METHOD

This method is capable of measuring flow in weedy rivers and rivers with moving beds. However, it requires on-site calibration with a current meter. According to the theory of electromagnetic induction, the flowing water in a river cuts the vertical component of the earth's magnetic field and an electromotive force (emf) is induced in the water. This emf, sensed by electrodes at each bank, is directly proportional to the velocity of flow. The emf due to the earth's magnetic field is, however, very feeble. To generate a measurable potential in the electrodes, a magnetic field is generated by means of coils buried in the river across its width.

Based on empirical tests, the following equation has been developed:

$$Q = K\left(\frac{E}{I}h\frac{r_w}{r_b}\right) \quad [\mathrm{L^3T^{-1}}] \tag{6.22}$$

Figure 6.19 Ultrasonic method of discharge measurement (from Holmes, Whirlow and Wright, 1973).

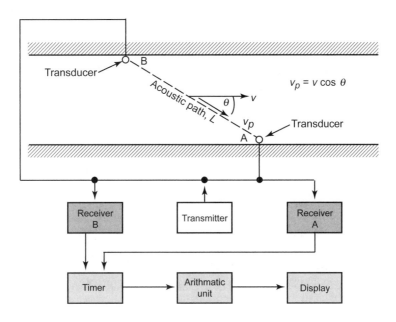

where

K = a constant

E = voltage at electrodes, mV

I = coil current, A

h = depth of flow, m or ft

r_w = water resistivity, $\Omega \cdot$ m or Ω-ft

r_b = bed resistance, Ω

The value of the constant K is evaluated by calibration using current-meter measurements.

6.21 MEASUREMENTS THROUGH HYDRAULIC DEVICES

In relatively shallow rivers, small creeks, open channels, and pipes and closed conduits, certain devices can be installed to measure the discharge. These devices (weirs, orifices, and flumes) include permanently constructed structures across the river and portable devices. The measurements these provide are based on the energy principle. The hydraulic structures for flow measurements are described in Section 10.2.

6.22 DISCHARGE RATING

The discharge rating depicts the relation between stage and discharge for a gaging station. It is applied to a stream's records of stage to convert them into discharge. The rating for a site is established by performing periodic field measurements of discharge and stage. Measured discharge is plotted against concurrent stage to define a rating curve for the site. At least 10 to 12 points covering the range of low to high flows are needed to determine the stage and discharge relation, and periodic measurements are needed thereafter to check the validity of the relation. Certain physical characteristics at the gaging section or in the channel bed, known as *station controls*, stabilize the stage and discharge relation. When an appropriate control is missing at a gaging site, the rating or stage-discharge relation will change (shift) from time to time. When different types of controls become operative at different stages, different relations will hold from stage to stage, which is usually the case.

The discharge rating may be simple if there is a direct relation between stage and discharge. It may, however, be complex if any other parameter is also needed to define the stage-discharge relation. Usually, there are three types of ratings.

1. Simple stage-discharge, or two-parameter discharge relation
2. Slope-stage-discharge, or three-parameter discharge relation with slope
3. Velocity index-stage-discharge, or three-parameter discharge relation with velocity index

6.22.1 Controls for Stage-Discharge

As stated above, controls tend to make a stage-discharge curve stable. There are two types of controls: section control and channel control. Section controls comprising physical features at a particular section, such as a riffle, rock ledge, weir, or spillway, can be natural or engineered. Channel controls include all features, such as size, shape, slope, roughness, alignment, constriction, or expansion in a reach of channel downstream of the gage, that provide rigidity and stability to the bed and banks of the stream. If a control is effective for the entire range of low to high flows, it is known as a complete control. More commonly, however, control is partial. Section control is often effective only at low stages and is submerged by channel control at medium or high stages unless it is a high dam. Channel control is generally effective at high stages, but the reach of the channel acting as the control may lengthen with increasing stage, inducing new features that may affect the stage-discharge relation.

6.23 SIMPLE STAGE-DISCHARGE RELATION

The rating curve or stage-discharge relation for each gaging site has its own features based on the control characteristics for the station. A plot of a series of discharge measurements made at medium and high stages will indicate whether a simple stage-discharge relation applies, because in an unsteady flow situation of complex relation, plotted points will have a scattered pattern. For a simple stage-discharge relation, the curve has a parabolic form, given by

$$Q = A(h \pm a)^n \quad [\mathrm{L^3 T^{-1}}] \tag{6.23}$$

where

Q = discharge

h = gage height

a = stage reading at zero flow (datum correction)

A, n = constants

Traditionally, discharge measurements are plotted on the horizontal scale (abscissa) and the gage height on the vertical scale (ordinate).* A curve is fitted by eye to the plotted

---

* This is unusual. In eq. (6.23), Q is a dependent variable and, as such, should normally be plotted on the y-scale. The arrangement on a rating curve is, however, reversed. The slope n is accordingly computed as the ratio of horizontal to vertical distance.

Figure 6.20 Simple stage-discharge relation.

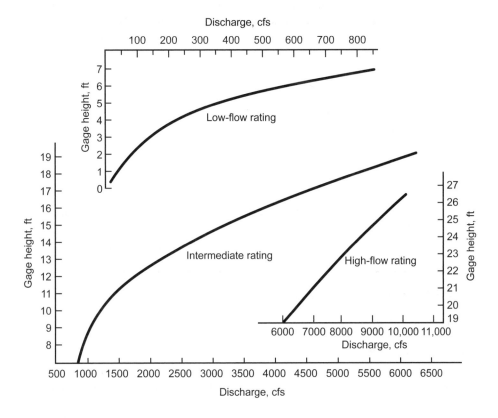

points. A plot on rectangular-coordinate paper is shown in Figure 6.20. Two different controls coming into effect at different stages produce a compound curve formed of two different parabolic curves.

6.23.1 Logarithmic Rating Curve

Taking the logarithm of eq. (6.23) will transform it to a straight line as follows:

$$\log Q = n \log (h \pm a) + \log A \tag{6.24}$$

A plot of Q and $(h \pm a)$ on log-log paper will produce a straight line. A straight-line plot is preferred because (1) it can be extended or extrapolated, (2) it can be described by a simple mathematical equation, and (3) by noting changes in the slope of the line, the ranges in stage for which the individual controls are effective can be identified. However, often an additional plot of the low-flow data on rectangular-coordinate paper is prepared so that the point of zero flow may be plotted. For this purpose, logarithmic rating-curve sheets have been designed with a rectangular-coordinate scale in one corner. Sometimes, when the stage-discharge equation changes too frequently with stage, the logarithmic method may not be suitable and a parabolic curve on rectangular-coordinate paper is used.

The log plot is between Q and $(h \pm a)$, not gage height. If the control is a section control of regular shape, the value of datum control, a, is the distance between zero gage height and the lowest point of the control. However, for a channel control or section control of irregular shape, the value of *a* is a mathematical constant, to maintain the concept of a logarithmic linear relation. It thus is necessary to ascertain the value of the datum control to prepare the plot.

The logarithmic rating relation is seldom one straight line throughout the entire range of stage for the gaging stations. It is usually necessary to fit two or more lines, each corresponding to the range over which a particular control is applicable.

6.24 DETERMINING THE STAGE OF ZERO FLOW

When discharge (*x*-scale) and stage (*y*-scale) are plotted on log-log paper, the shape of the plot determines the type of equation, as follows:

Type of plot	Type of equation	Remark
Straight line	$Q = Ch^n$	Stage of zero flow coinciding with zero gage height
Concave up [Figure 6.21(a)]	$Q = C(h-a)^n$	Stage of zero flow above zero gage height
Concave down [Figure 6.21(b)]	$Q = C(h+a)^n$	Stage of zero flow below zero gage height

The following two methods are used to determine the stage of zero flow, *a*.

6.24.1 Trial-and-Error Procedure

1. Prepare a log-log plot of gage height and discharge, shown as curve *ef* in Figure 6.22.
2. If the plot is concave upward, add the trial value of *a* to the chosen scale (i.e., slide the scale downward) and make a plot, as shown by *gh*. (If it is concave downward, subtract the trial value of *a*.)
3. Continue adding (or subtracting) trial values of *a* for the concave upward (or concave downward) curve until a value is found that results in a straight-line plot as *ij*. For a concave upward plot, the relation is $Q = C(h-a)^n$.

6.24.2 Arithmetic Procedure

1. Select two well-separated points on the gage-discharge plot and read values Q_1, Q_2, and h_1, h_2. Compute Q_3 as follows:

$$Q_3 = \sqrt{Q_1 Q_2}$$

2. From the plot, read h_3 corresponding to Q_3.
3. According to the straight-line property on the log plot,

$$a = \frac{h_1 h_2 - h_3^2}{h_1 + h_2 + 2h_3} \quad [\text{L}] \qquad (6.25)$$

When *a* is positive, the relation is $Q = C(h-a)^n$.

Figure 6.21 Type of curves and zero flow correction.

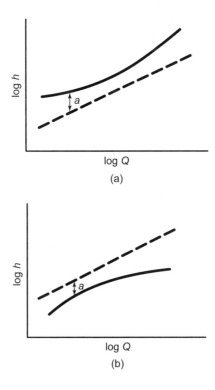

(a)

(b)

Figure 6.22 Trial-and-error procedure to determine the stage of zero flow.

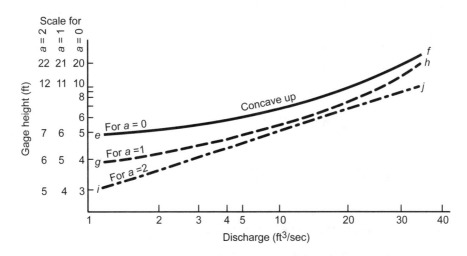

EXAMPLE 6.13

Discharge measurements and the corresponding stages observed at a stream gaging station are listed below. Determine the datum correction (stage of zero flow) by **(a)** the trial-and-error procedure, and **(b)** the arithmetic method.

Discharge (ft³/sec)	Stage (ft)	Discharge (ft³/sec)	Stage (ft)
0.40	0.85	12.23	3.40
0.6	0.98	18.31	4.27
1.26	1.30	19.84	4.62
1.61	1.46	25.10	5.26
3.77	2.02	35.24	6.46
6.78	2.62	45.75	7.44
8.98	2.94	58.12	8.67
9.73	3.06	72.02	10.09

SOLUTION   The data are plotted on log-log graph paper in Figure 6.23. The curve has two segments. At point A, corresponding to a gage height of 3.06 ft, the high-stage control became operative. There will be two stage-discharge relations, one below point A and the other for higher stages. This example solves the relation for the lower stage up to 3.06 ft. A similar procedure will apply for the other curve.

(a) Trial-and-error procedure

    1. A log-log plot of discharge and stage for the lower range is reproduced in Figure 6.24, as shown by curve ab.

    2. Since this curve is slightly concave upward, the trial value of $a = 0.1$ ft is added to the stage scale (y-scale shifted down by 0.1) and the data are plotted again as cd, which is a straight line. Hence $a = 0.1$.

(b) Arithmetic procedure

    1. Select $Q_1 = 0.40$ and $Q_2 = 8.98$ ft³/sec, corresponding $h_1 = 0.85$ ft and $h_2 = 2.94$ ft.

    2. $Q_3 = \sqrt{Q_1 Q_2} = \sqrt{0.4(8.98)} = 1.90$ ft³/sec

       $h_3 = 1.56$ ft (from Fig. 6.23)

    3. $a = \dfrac{h_1 h_2 - h_3^2}{h_1 + h_2 - 2h_3} = \dfrac{0.85(2.94) - (1.56)^2}{0.85 + 2.94 - 2(1.56)} = 0.098$ ft

6.25 EQUATION OF STAGE-DISCHARGE CURVE

It is desirable to express the stage-discharge relation in mathematical form by determining the equation so that it can be used directly for discharge conversion or in preparation of a conversion (rating) table. Once a straight-line form of the stage and discharge has been obtained after ascertaining the value of the stage of zero flow, the equation of this line defining the stage-discharge relation can conveniently be determined graphically or by

Figure 6.23   Log-log plot of stage and discharge data for Example 6.13.

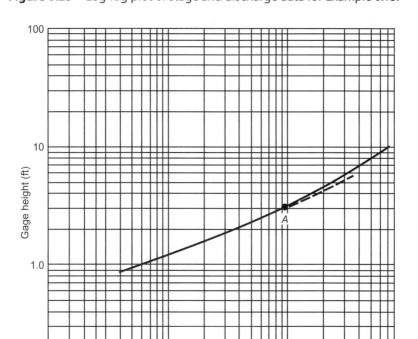

linear regression analysis. When the rating curve is composed of more than one straight-line segment, the equation for each is determined separately.

6.25.1 Graphic Procedure to Determine Rating Equation

In log Q versus log $(h \pm a)$ plot of the equation log $Q = n$ log $(h \pm a) + $ log A (eq. 6.24), n is the slope of the line. Since in a rating curve, the dependent variable Q is plotted on the x-axis, the slope is taken as the ratio of the horizontal distance to the vertical distance. Further, since both the x- and y-scales are logarithmic (fixed), the slope can be determined by measuring horizontal and vertical projections of the line using a ruler/scale marked in inches or millimeters.

When $(h \pm a) = 1$, log $(h \pm a) = 0$ and $Q = A$. Thus, the value of A is obtained by reading Q corresponding to $(h \pm a)$ equal to 1. If the scale does not contain $(h \pm a) = 1$, any value of Q and corresponding $(h \pm a)$ are read and substituted in the rating equation (6.24) to obtain A.

6.25.2 Linear Regression Analysis to Determine Rating Equation

Regression analysis is the procedure to establish a curve that fits a given set of data. Simple regression involves two variables, one dependent and one independent, as opposed to multiple regression involving several independent variables. If the equation of the curve relates to a straight line, it is known as *linear regression*. The best-fitting curve through the set of

Figure 6.24 Trial-and-error procedure for determining the stage of zero flow for Example 6.13.

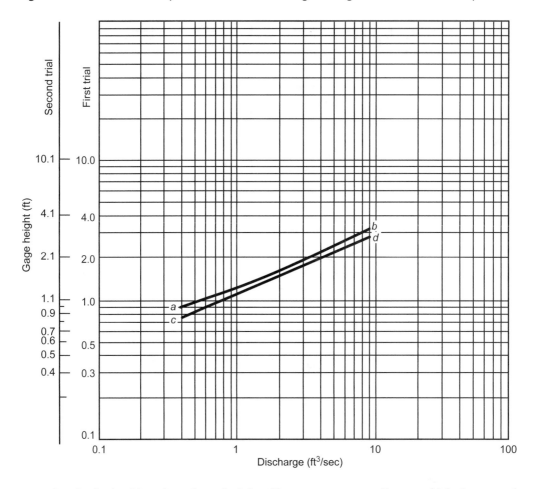

data is obtained based on the principle of least squares, according to which the sum of the squares of deviations (differences) of the measured value of the dependent variable in the data set from the value estimated from the fitted curve should be minimal. Whether the derived equation represents the relationship adequately is indicated by the correlation coefficient. Its value of 1 represents a perfect relation, and 0 indicates no relation among variables.

For a straight-line equation,

$$y = mx + C \quad [L] \tag{6.26}$$

the least-squares line has properties given by

$$\sum y = CN + m\sum x \quad [L]$$
$$\sum xy = C\sum x + m\sum x^2 \quad [L^2] \tag{6.27}$$

In the previous equations, known as *normal equations* for the least-squares line, N represents the total number of observations between variables x and y.

Equations (6.27), when solved, provide the following values for m and C:

$$C = \frac{(\Sigma y)(\Sigma x^2) - (\Sigma x)(\Sigma xy)}{N(\Sigma x^2) - (\Sigma x)^2} \quad [\text{L}] \tag{6.28}$$

$$m = \frac{N(\Sigma xy) - (\Sigma x)(\Sigma y)}{N(\Sigma x^2) - (\Sigma x)^2} \quad [\text{dimensionless}] \tag{6.29}$$

The standard deviation, standard error, and correlation coefficient are related as follows:

$$S_y^2 = \frac{\Sigma y^2 - (\Sigma y)^2 / N}{N - 1} \quad [\text{L}^2] \tag{6.30}$$

$$S_{yx}^2 = \frac{\Sigma y^2 - C\Sigma y - m\Sigma xy}{N - 2} \quad [\text{L}^2] \tag{6.31}$$

$$r = \left(1 - \frac{S_{yx}^2}{S_y^2}\right)^{1/2} \quad [\text{dimensionless}] \tag{6.32}$$

When using the equations above on the rating equation (6.24), the variable y is denoted by log Q and variable x by log $(h \pm a)$. Thus, first, the logarithmic values of Q and $(h \pm a)$ have to be determined to be used in eq. (6.28) and eq. (6.29) to ascertain constant parameters C and m. The process is explained in Table 6.5 (Example 6.14).

EXAMPLE 6.14

Find the equation of the rating curve in Example 6.13 by **(a)** a graphic procedure, and **(b)** regression analysis.

SOLUTION   From Example 6.13, $a = 0.1$. Hence the equation of the line is log $Q = n$ log $(h - 0.1) + $ log A.

(a) Graphic procedure. A log-log plot of Q and $(h - 0.1)$ for the discharge (Q) and stage (h) data in Example 6.13 is given in Figure 6.25. From the graph,

$$n = \text{slope} = \frac{\text{horizontal distance}}{\text{vertical distance}} = \frac{46.5 \text{ mm}}{20 \text{ mm}} = 2.325$$

For $(h - 0.1) = 1$, $Q = 0.8$; therefore, $A = Q = 0.8$ or log $A = -0.097$. Hence the equation of the curve

$$\log Q = -0.097 + 2.325 \log (h - 0.1)$$

or

$$\log Q = \log 0.8 + 2.325 \log (h - 0.1)$$

or

$$Q = 0.8(h - 0.1)^{2.325}$$

(b) Regression analysis. Refer to Table 6.5.

Figure 6.25 Plot of log Q and log (h ± a) to determine the rating equation for Example 6.14.

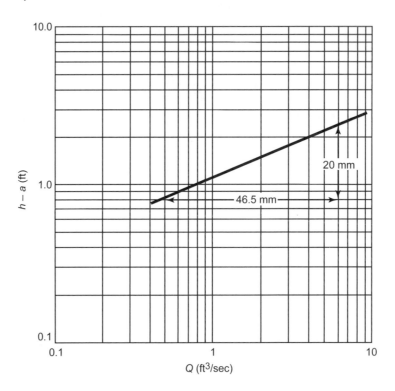

Table 6.5   Regression Analysis for the Rating Equation[a]

(1)	(2)	(3)	(4)	(5)	(6)	(7)	(8)	(9)
No.	Q from data	h from data	h–0.1	log Q = y	log (h–0.1) = x	x^2	y^2	xy
1	0.4	0.85	0.75	−0.40	−0.125	0.016	0.16	0.050
2	0.61	0.98	0.88	−0.215	−0.056	0.003	0.046	0.012
3	1.26	1.30	1.20	0.100	0.079	0.006	0.01	0.008
4	1.61	1.46	1.36	0.207	0.134	0.018	0.043	0.028
5	3.77	2.02	1.92	0.576	0.283	0.080	0.332	0.163
6	6.78	2.62	2.52	0.831	0.401	0.161	0.691	0.333
7	8.98	2.94	2.84	0.953	0.453	0.205	0.908	0.432
Σ				2.052	1.169	0.489	2.19	1.026

[a] $N = 7$; log Q (col. 5) = logarithm of col. 2; log (h − 0.1) (col. 6) = logarithm of col. 4; x^2 (col. 7) = square of col. 6; xy (col. 9) = col. 5 × col. 6; y^2 (col. 8) = square of col. 5.

From eq. (6.28),

$$\log A \text{ or } C = \frac{(\Sigma y)(\Sigma x^2) - (\Sigma x)(\Sigma xy)}{N(\Sigma x^2) - (\Sigma x)^2}$$

$$= \frac{2.052(0.489) - 1.169(1.026)}{7(0.489) - (1.169)^2}$$

$$= -0.095$$

From eq. (6.29),

$$m = \frac{N(\Sigma xy) - (\Sigma x)(\Sigma y)}{N(\Sigma x^2) - (\Sigma x)^2}$$

$$= \frac{7(1.026) - 1.169(2.052)}{7(0.489) - (1.169)^2}$$

$$= 2.326$$

Hence the equation is

$$\log Q = -0.095 + 2.326 \log (h - 0.1)$$

or

$$\log Q = \log 0.8 + 2.326 \log (h - 0.1)$$

or

$$Q = 0.8(h - 0.1)^{2.326}$$

Next, we check the adequacy of the relationship.

From eq. (6.30),

$$S_y^2 = \frac{(\Sigma y^2) - (\Sigma y)^2 / N}{N - 1}$$

$$= \frac{2.19 - (2.052)^2 / 7}{6}$$

$$= 0.265$$

From eq. (6.31),

$$S_{yx}^2 = \frac{\Sigma y^2 - C\Sigma y - m\Sigma xy}{N - 2}$$

$$= \frac{2.19 - (-0.095)(2.052) - 2.326(1.026)}{5}$$

$$= 0$$

From eq. (6.32),

$$r = \left(1 - \frac{0}{0.265}\right)^{1/2} = 1.0$$

That is, there is a perfect correlation.

6.26 Slope-Stage-Discharge Relation

When variable backwater* conditions exist in a stream due to channel constriction, artificial structures, downstream tributaries, or natural flood waves, the discharge is not merely a function of stage but is also affected by the slope of the water surface (energy gradient). The slope or fall is used as a third parameter in such cases. In addition to a main gage, known as a *base gage*, an *auxiliary gage* is installed downstream of the base gage and simultaneous gage readings are made along with the discharge measurements at both gages. The procedure described here to establish the rating is known as the Unit Fall Method.

According to both Chezy and Manning, the discharge is directly proportional to the square root of the slope. Following this relation, we obtain

$$\frac{Q}{Q_r} = \left(\frac{F}{F_r}\right)^n \quad \text{[dimensionless]} \tag{6.33}$$

where

Q = measured discharge of a stream for a given base gage height

Q_r = discharge from the rating curve corresponding to the same base gage height

F = measured fall between base and auxiliary gages

F_r = fall between base and auxiliary gages corresponding to rating curve discharge, Q_r

n = close to 0.5, between 0.4 and 0.6

For $F_r = 1$,

$$Q_r = \frac{Q}{F^n} \quad [\text{L}^3\text{T}^{-1}] \tag{6.34}$$

First the rating curves are prepared as follows:

1. Adopting $n = 0.5$ in eq. (6.34), determine Q_r from the measured discharge values Q and fall values F.
2. Plot Q_r against the corresponding base gage stage. Fit a curve to the plotted points.
3. Repeat steps 1 and 2 using $n = 0.4$ and subsequently, 0.45, 0.55, and 0.6.
4. From the five curves with different values of n, select the one that best fits the points plotted.
5. Compute ratios Q/Q_r for various measured values of discharge and corresponding best fit computed Q_r values.
6. Make another plot between Q/Q_r and corresponding fall measurements, F.

Two curves from steps 4 and 6, as shown in Figure 6.26, are used together to convert the stage values to discharges. For an observed base gage height, determine Q_r using the first curve. From the second curve, for the fall observed, read Q/Q_r, and substituting the determined value of Q_r, find Q.

---

* A constant backwater, such as that caused by section control, does not affect a simple stage-discharge relation.

Figure 6.26 Slope-stage-discharge curves.

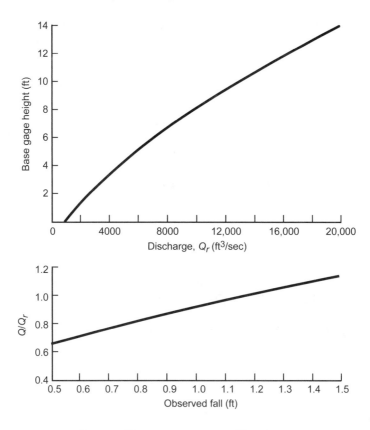

6.27 CONVERTING STAGE RECORDS INTO DISCHARGE

For a water year, the daily record of discharge is computed from the record of stage using the discharge rating for the gaging station. The process, performed manually or by computer, involves the following steps.

6.27.1 Station Analysis

As a preliminary to computing the discharge record, a study is performed of the data collected at each station. This includes (1) a review of the stage data and determination of the datum corrections, if any, to be applied to observed stages; (2) appraisal of the accuracy of the discharge measurements and computations; (3) analysis of the discharge rating to determine whether the last rating used is applicable for part or all of the year, or whether departures are more than 5% and thus require a new rating curve; and (4) preparation of tables from the rating curves on the standard rating-table form that lists discharge at intervals of 0.1 ft (1 cm) of stage or checking the range and validity of formulas when mathematical expressions are to be used.

6.27.2 Preparation of Gage-Height Record

For a nonrecording gage, the first step is to check and reconcile observers' readings with hydrographers' readings made on field visits. The datum corrections, if any, are applied next. For gradually changing stages, the mean daily gage heights are computed as an average of two observed readings for each day. For rapid changes, a stage hydrograph is sketched through plotted points of gage heights using the graphic stage record from a nearby recording gage station. Peaks are determined from high-water marks or crest-stage readings.

For a recording station, the corrections that need to be made to a stage record are (1) time correction due to a slow or fast clock, which is prorated by straight-line interpolation and applied by changing the positions of the midnight lines for the affected days; (2) gage-height corrections, based on differences in readings of the recorder pen and inside staff gage at inspection times prorated with time, and also reversal errors that occur when the pen reverses direction on reaching the top and bottom of the chart; and (3) datum correction, if required.

6.27.3 Determination of Daily Mean Gage Height

This is usually a template that is placed over a 24-hour segment of the recorder chart and adjusted so that areas above and below a line on the template approximately balance each other. For the chart from a recording gage, this is the uncorrected gage height, which is corrected for gage-height correction and datum corrections. From the sketched hydrograph for a nonrecording gage, corrected daily mean gage heights are obtained. When there is a large variation in stage during a day, the day is subdivided into smaller increments and the mean gage height for each time increment is determined.

6.27.4 Applying Rating Tables to the Gage Heights

A form containing columns for daily mean gage height and discharge for the 12-month period, and also a place to record monthly and annual summaries, is used. Discharges are determined by applying the appropriate rating tables to the gage heights.

6.28 PRESENTATION AND PUBLICATION OF DATA

A summary of daily mean discharges for a gaging site for a water year is prepared from computations of the discharge records. The USGS publishes an annual report that includes a sheet each for daily mean discharges of each gaging station. In addition, it contains a map showing the location of gaging stations, introductory text, reservoir records, tabulation of discharge records for partial record stations, and water quality data. Groundwater data are also included in the annual report.

6.29 FUTURE TRENDS IN STREAMFLOW MEASUREMENTS

Knowledge is growing very rapidly in the fields of electronics, acoustics, optics, and nuclear energy. Based on advances in these fields, many instruments capable of more accurate streamflow measurements are either under development or are in the experimental stage. Water stage observation and discharge measurement are two basic procedures in stream gaging.

Digital recorders are gaining preference for water stage recording, to take advantage of computer computation and processing and automatic graph plotting. A level recording device based on an acoustic technique has been developed to record stage data in a form that is directly accessible to computer usage.

Use of a telemetry system, including a satellite data collection system, is limited today but is expected to grow in the future. The trend is toward multidata recording systems, where different kinds of data are recorded on the same plot, and also toward multistation data collection at a centralized platform for remote transmission to receiving centers.

Discharge measurement involves width, depth, and velocity elements. For depth measurements, the echo sounder, which is already in widespread use, is likely to be in even greater use. Electronic instruments that permit precise location of the position of observations are available now. Also, sophisticated instruments for measuring the bed profile with sufficient resolution have been developed (e.g., Hydrodist, Ralog, and Hi-fix). The current meter is still a popular method of discharge measurement. A number of new types of velocity meters are in the experimental stage. These include the acoustic velocity meter, Doppler velocity meter, electromagnetic velocity meter, optical current meter, nuclear current meter, and the eddy shedding current meter.

In the dilution method, radio element generators, such as the indium-tin generator, open up the possibility for continuous discharge measurements. The ultrasonic method has proved successful, but the electromagnetic method has not made a significant impact due to its complexity, unsatisfactory accuracy, and high cost.

In the search for new methods of discharge measurement, one that has already been established is the "moving boat" method, which uses a sonic counter to record the geometry of the cross section and a continuously operating current meter sensing the combined stream and boat velocities during a traverse, which is made without stopping.

With regard to indirect measurement through structures, electronic measurement and recording of the head of water constitute notable advances.

Automated observation platforms for the acquisition of observations from a large number of stations and the use of aircraft and earth satellites for acquisition and relay of data to receiving centers have promising prospects.

PROBLEMS

6.1 The velocity distribution in a pipe is given by the following Prandtl type relation

$$\frac{v}{u^*} = 5.75 \ \log\frac{u^* y}{v} + 5.5$$

where $u^* = \sqrt{\tau_0/\rho}$

The axial velocity at 100 mm from the wall in a 200-mm-radius pipe is 1.2 m/s. Determine the maximum velocity at the center and the shear stress at the wall at 20 °C.

6.2 The axial velocities measured at 25 mm and 75 mm across from the inner wall of a 200-mm-diameter pipe conveying water at 20 °C are 0.8 m/s and 1 m/s, respectively. Calculate the shear stress at the wall and the maximum flow velocity. If the maximum velocity is twice the mean velocity, determine the discharge. [Apply eq. (6.3)]

6.3 The velocities at 0.25 m and 0.75 m from the bed of a 1.5-m-deep channel conveying water at 20 °C are 0.8 m/s and 1 m/s, respectively. Determine the mean velocity and slope of the channel. If the channel cross section is rectangular with a width of 10 m, determine the discharge. [Apply eq. (6.4)].

6.4 The equation for vertical velocity distribution (in ft/sec) in a stream was found to be $v = 1.9y^{1/7}$, where y is the distance from the bottom. If the depth of the flow is 3.5 ft, what is the mean velocity of flow by the direct integration method? What is the discharge per foot width?

6.5 Solve Problem 6.4 by the graphic integration method.

6.6 Solve Problem 6.4 by the arithmetic summation method.

6.7 The equation for the vertical velocity distribution with respect to the water surface is $v = 3.8(4 - y)^{1/2}$. For a depth of flow of 4 ft, what is the average velocity of flow?

6.8 The equation for a fully developed turbulent flow has the logarithmic form

$$v = 4.45 + 2.62 \log \frac{y}{d}$$

in which y is the distance from the bed. For the depth of flow, d, of 5 ft, determine the mean velocity of flow. (The equation applies a small distance above the bed, beyond the laminar sublayer.)

6.9 The following observations were recorded for point velocities in a 15-ft-deep vertical section of a stream. Determine the mean velocity in the section using different methods. Determine the percent error in each method compared to the velocity-curve method.

Ratio of Observation to Total Depth	Velocity (ft/sec)
0.04	2.35
0.2	2.30
0.4	2.20
0.6	2.00
0.8	1.75
0.9	1.50

6.10 In a gaging measurement of a deep swift stream from a bridge, the total depth of the sound line was measured to be 7.55 m. The depth from the guide pulley to the surface was 3.0 m. The angle of sound line from the vertical was 20°. The distance from the centerline of the current meter to the bottom of the weight was 0.3 m. Determine **(a)** the true depth of the water, and **(b)** the positions of 0.2 depth and 0.8 depth.

6.11 In a hand-line measurement of a stream, the total depth from the pulley was measured to be 25.2 ft. The depth to the water surface from the pulley was 10.5 ft. The meter was suspended by a hanger having 1-ft distance from the centerline of the meter to the bottom of the weight. Determine the position of the hand line to place the current meter at 0.2 depth and 0.8 depth.

6.12 A river cross section has been divided into 10 sections. The area of each section and the mean velocity observed are indicated below. Determine the rate of flow in the river.

Section:	1	2	3	4	5	6	7	8	9	10
Area (m²)	2.2	4.5	6.2	6.8	7.2	5.5	4.5	3.2	2.1	1.0
Velocity (m/min)	3.5	3.8	4.5	4.8	4.2	4.0	3.8	3.7	3.6	3.2

6.13 A river cross-section profile is shown in Fig. P6.13. The depths measured at various verticals are indicated together with the point measurements of velocity in each vertical at 0.2 and 0.8 depths or 0.6 depth. Compute the discharge at the cross section by the midsection method.

Figure P6.13

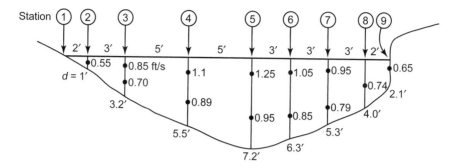

6.14 Field observations for discharge measurement at a site are recorded below. Determine the streamflow by the midsection method. The current-meter rating is $v = 0.03 + 0.25N$, where v is velocity in m/s and N is the number of revolutions per second.

Distance from Initial Point (m)	Depth (m)	Meter Position	Revolutions	Time (sec)
4	0		0	
6	0.4	0.6	45	59
8	0.85	0.6	52	61
10	1.58	0.2	58	62
		0.8	46	61
12	1.71	0.2	65	62
		0.8	51	63
14	1.17	0.2	51	62
		0.8	39	60
16	0.81	0.6	41	63
17	0.48	—	—	—

6.15 Solve Problem 6.13 by the mean-section method.

6.16 Solve Problem 6.14 by the mean-section method.

6.17 Solve Problem 6.13 by the velocity-depth integration method.

6.18 Velocity measurement data were plotted in the form of vertical-velocity curves for various verticals of a stream cross section. Analyses of these data yielded the velocity-distance values shown in Fig. P6.18. Determine the discharge of the stream.

Figure P6.18

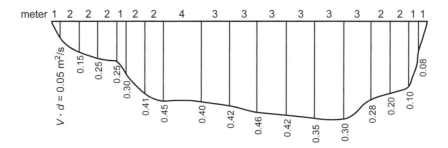

6.19 To determine discharge by the velocity-contour method, velocity measurements at a number of points on each vertical were made and isovels were drawn. The areas computed between successive isovels are listed below. Determine the streamflow.

Isovel (m/s)	Area between Isovels (m²)
>0.25	0.8
0.25–0.20	2.2
0.20–0.15	3.42
0.15–0.10	4.16
0.10–0.05	1.62
0.05–0	1.7

6.20 In measuring discharge by the tracer-dilution technique using the constant injection rate method, a tracer of concentration 6.23 g/liter was injected at a rate of 8 mL/s. The samples collected at a downstream cross section from four points evenly distributed across the width registered concentrations of 0.048, 0.05, 0.045, and 0.052 mg/liter, respectively. Calculate the stream discharge.

6.21 Discharge measurement by the sudden tracer injection method provided the data listed below. Volume of dye injected $= 10.85 \times 10^{-3} m^3$. Concentration of dye injected $= 3.55$ g/ liter. Samples were collected at various times. Determine the stream discharge.

Sample	Time (min)	Concentration (µg/liter)
1	0	0
2	2	0.6
3	4	4.5
4	6	10.2
5	7	10.8
6	9	8.0
7	12	3.2
8	15	1.0
9	20	0.0

6.22 Results from a sudden injection of tracer into a stream to measure streamflow by the dilution technique provided the following data: Weight (mass) of tracer = 1800 g. Sampling interval = 30 sec. Number of samples = 25. Average concentration of 25 samples = 7.55 mg/liter. Calculate the discharge. [*Hint*: Mass injected = VC_1 = 1800 g; mass at downstream station = $N\bar{C}\,\Delta tQ = 25(7.55\times10^{-3})(30)Q$].

6.23 For the upper range of stages in Example 6.13 when high-water control became effective, determine the datum correction (stage of zero flow) by **(a)** the trial-and-error procedure, and **(b)** the arithmetic procedure. The discharge stage data are reproduced below.

Discharge (ft³/sec)	Stage (ft)	Discharge (ft³/sec)	Stage (ft)
9.73	3.06	35.24	6.46
12.23	3.40	45.75	7.44
18.31	4.27	58.12	8.67
19.84	4.62	72.02	10.09
25.10	5.26		

6.24 On Zor Creek at Gbanka, Liberia (West Africa), discharge measurements are recorded along with corresponding gage heights. Determine the datum correction (stage of zero flow) by **(a)** the trial-and-error method, and **(b)** the arithmetic method.

Gage Height (cm)	Discharge (m³/s)	Gage Height (cm)	Discharge (m³/s)
30	0.02	80	2.35
40	0.16	100	4.4
50	0.40	120	7.5
70	1.55	150	13.5

6.25 The following rating table was developed from field measurements on the St. John River at Baila, Liberia (West Africa). Determine the stage of zero flow by **(a)** the trial-and-error method, and **(b)** the arithmetic method.

Gage Height (ft)	Discharge (cfs)	Gage Height (ft)	Discharge (cfs)
1.0	70	7.0	9,500
2.0	430	8.0	13,500
3.0	1,200	9.0	17,500
4.0	2,350	10.0	22,500
5.0	4,250	11.0	27,500
6.0	6,500	11.5	30,000

6.26 Find the equation for the rating curve of Problem 6.23 by **(a)** the graphic procedure, and **(b)** regression analysis.

6.27 Determine the rating equation for the stage-discharge data in Problem 6.24 by **(a)** the graphic procedure, and **(b)** regression analysis.

6.28 Determine the rating equation for the stage-discharge data in Problem 6.25.

6.29 The following observations relate to discharge and stage at the base station and stage at an auxiliary gage 2000 m downstream. Develop the slope-stage-discharge relationship. What is the estimated discharge when the base and auxiliary stages are 9.0 m and 8.25 m, respectively?

Discharge (m³/s)	Stage at Base Gage (m)	Stage at Auxiliary Gage (m)
34	1.012	0.951
206	2.206	1.279
78	1.359	1.155
165	1.963	1.347
164	1.755	1.054
200	2.139	1.331
995	7.638	4.986
780	7.108	5.188
1415	10.558	7.678
445	4.026	2.429
760	6.105	3.923
580	4.907	2.990

6.30 An auxiliary gage was used downstream of a base gage in a river to provide correction for the backwater effect. The following data were noted. The value of exponent n for all observations was 0.5.

Base Gage (m above datum)	Auxiliary Gage (m above datum)	Discharge (m³/s)
80.0	79.5	142
85.5	84.8	260
90.5	89.5	500

If the base gage and auxiliary readings are 92.5 m and 91.6 m, estimate the discharge.

7

Estimation of Surface Water Flow

◆◆

Part A: Hydrograph Analysis

7.1 RUNOFF AND STREAMFLOW

The term *runoff* is used for water that is on the run or in a flowing state, in contrast to water held in storage or evaporated into the atmosphere. Since such flow conditions take place in various stages of a hydrological cycle, there are various types of runoff, as shown in Figure 7.1. In this figure, boxes indicate storage units. *Surface runoff* or *overland runoff* is that part of the runoff that travels over the surface of the ground to reach a stream channel and through the channel to the basin outlet. (To be precise, the surface runoff also includes the precipitation directly falling over the channel reach, while the overland runoff excludes the channel precipitation.) Surface runoff appears relatively quickly as streamflow.

Subsurface runoff is that part of the runoff that travels under the ground to reach a stream channel and ultimately the basin outlet. It consists of two parts. One part moves laterally through the upper soil horizons within the unsaturated zone or through the shallow perched saturated zone toward the stream channel. This is known as the *subsurface stormflow, interflow,* or *throughflow.* Another part infiltrates deeper to the saturated zone to form the *groundwater flow.* This flow discharges into the stream channel as the base runoff or *baseflow.* The groundwater flow into a stream is due to the past infiltrated precipitation. It is, accordingly, referred to as *delayed runoff.* The interflow or subsurface stormflow has an intermediate travel time to the stream between the surface and base flows. When the response time is short, the interflow is sometimes considered as a contribution to the surface flow. Total runoff, comprising surface runoff and subsurface runoff at the downstream end of any reach of a stream channel, forms the *streamflow.* Thus the runoff is that part of the precipitation that eventually appears as streamflow. Excluding the baseflow contribution, the balance is the *direct streamflow* or *direct runoff.*

The runoff, particularly the overland runoff, is measured in terms of the depth of the water. The streamflow, on the other hand, is measured as a volume rate of flow in cubic feet per second or cubic meters per second. The volume is an equating term in two units of measurement; that is, when a runoff depth is multiplied by the contributing area, it provides the same volume as the direct streamflow multiplied by the time for which the flow has contributed the runoff. The baseflow, accounted for separately because of its different nature and area of contribution, is discussed subsequently. Various methods of assessing runoff resulting from rainfall or snowmelt in terms of depth of water over a watershed are

Figure 7.1 Forms of runoff in the hydrologic cycle.

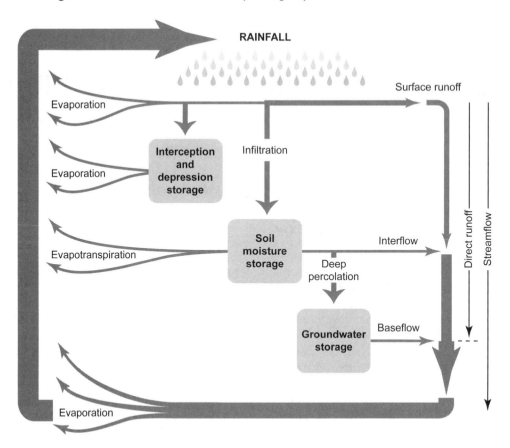

described in sections 2.13 through 2.17. This chapter deals with assessment of streamflow, the form in which runoff appears at the outlet of a watershed.

A plot of streamflow (discharge) against time at any section of a stream channel is known as a *hydrograph*. The runoff volume is equal to the excess rainfall from a precipitation storm over a drainage area, and results in an equivalent volume increase in the flow of the receiving stream as represented in the area under the hydrograph. There are two general approaches to prepare hydrographs from runoff data. The unit hydrograph approach applies data based on actual streamflow and runoff measurements; it is a field-based streamflow-runoff correlation. The routing procedure, on the other hand, applies the theory of hydraulics of water flow to ascertain flow conditions from section to section. This chapter discusses the former. The kinematic technique of hydrograph formulation, a relatively simple approach to routing procedure, is described in Chapter 9.

7.2 MECHANISM OF RUNOFF GENERATION

The baseflow of a stream is contributed by the groundwater discharge, shown as path 2 in Figure 7.2. The runoff process that leads to the direct streamflow is, however, not that straightforward. Despite considerable research in this field, the mechanism of runoff is not fully resolved. There are three widely accepted theories.

Figure 7.2 Paths of runoff (after Dunne, 1982).

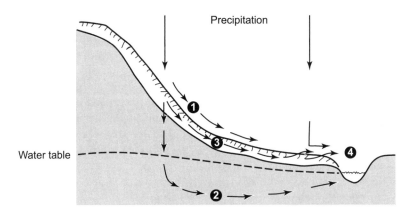

The classic concept of Horton (1933) holds that any soil surface has a certain maximum rate of water absorbance, known as the *infiltration capacity*. This capacity is high at the onset of rainfall and then declines rapidly, to achieve a constant rate. If rainfall intensity at any time during a storm exceeds the infiltration capacity of the soil, water accumulates on the surface, fills small depressions, and runs downslope as overland flow. According to this theory, the major contribution to direct streamflow is from the overland runoff, recognized as the *Horton overland flow*, shown as path 1 in Figure 7.2. Practically the entire basin area contributes to this overland flow. Horton's runoff concept serves as a base for the unit hydrograph technique (Section 7.8) and the infiltration curve technique (Section 2.13). Horton's theory has applicability in arid and semiarid landscapes and cultivated fields, paved areas, construction sites, and rural roads of humid regions that lack a dense vegetation cover and well-aggregated topsoil.

For forests and densely vegetated humid regions, Hewlett and Hibbert (1967), Kirkby and Chorley (1967), and others suggested the theory of subsurface stormflow (throughflow). According to this, a densely vegetated humid region has the capacity to absorb all except the rarest, most intense storms. A major part of this absorbed water moves laterally through the shallow soil horizon in the zone of aeration, shown as path 3 in Figure 7.2. This process of transmission effectively contributes to the streamflow. The flow is confined to intergranular pores, root holes, worm holes, and structural openings. It travels more slowly than Horton overland flow, but some of it arrives quickly enough to produce floods. Freeze (1972b) and many others hold the opinion that this is a viable mechanism but it cannot provide a very large contribution to the total quantity of direct streamflow.

The third type of storm runoff for humid regions is based on the concept of saturation overland flow (Musgrave and Holton, 1964). Rainfall causes a thin layer of soil on some parts of a basin to saturate upward from some restricting boundary to the ground surface, especially in zones of shallow, wet, or less permeable soil. Then the rainfall cannot infiltrate further in the saturated soil and runs over as the saturation overland flow. Some water moving through the topsoil also appears as the return flow, shown as path 4 in Figure 7.2. Thus direct precipitation on the saturated soil with or without return flow contributes to the streamflow. This process occurs frequently on the footslopes of hills, bottoms of valleys, swamps, and shallow soils. It expands outward from the stream channels as shown in Figure 7.3. Unlike Horton's concept of the entire area contributing to runoff, the flow at any moment is contributed by the saturated area, which is only a small percentage of the total basin area. This source area expands and shrinks. The source area increases at the beginning and decreases at the end of a rainstorm. Accordingly, the process is referred to as the *variable source area concept* (Hewlett and Hibbert, 1967) or the *dynamic watershed concept* (Tennessee Valley Authority; see Dunne, 1982). Studies on hillslope hydrology are based on this theory.

The general consensus is that in densely vegetated humid regions, streamflow is mostly generated by a combination of subsurface stormflow (throughflow) and saturation overflow. The relative contributions by each mechanism depend on soil and topographic conditions. Figure 7.4 summarizes the conditions affecting the runoff process.

In the literature, another descriptor, *partial source area*, has been used for a runoff process (Betson, 1964; Dunne and Black, 1970). This concept is based on Horton's theory of predominantly overland flow. The contributing area is, however, considered only a portion of the basin, which is taken as a relatively fixed source area, as against the variable one in the saturation flow concept.

Figure 7.3   Expansion of source area.

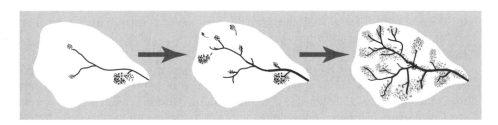

Figure 7.4   Conditions controlling the runoff mechanism (after Dunne, 1982).

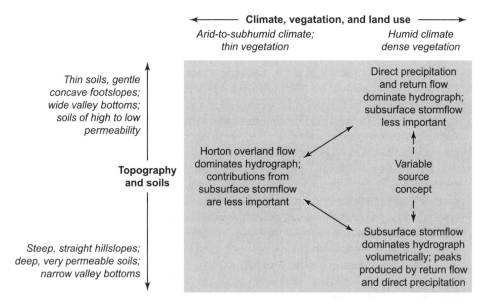

7.3 TECHNIQUES OF STREAMFLOW ESTIMATION

Measurements of streamflow from a hydrological stream-gaging network, as described in Chapter 6, are the main and best source of surface water flow data. However, no national data collection program anywhere in the world collects sufficient data to satisfy all the design and decision-making needs in any one catchment (Fleming, 1975). The World Meteorological Organization recommended (1976) that when the data are inadequate, project activity should begin with installation of a hydrological gaging network. Project planning and design necessitate at least one decade of hydrological data, and it is often not feasible to postpone implementation of a project for this duration after setting up a network— hence the need for streamflow estimation. The data to be estimated relate to natural flow conditions comprising average annual flows and distribution of the flow during a year over days, months, or seasons, and peak discharges and minimum discharges for various durations.* We discuss natural streamflows in this chapter and describe extreme flow conditions in Chapter 8.

There are four approaches to the estimation of streamflows.

7.3.1 Hydrograph Analysis

Since streamflow represents precipitation returning to a stream, a comparison of a precipitation storm in a basin and the resulting streamflow hydrograph at the outlet of the basin provides a site-specific rainfall-runoff model to convert all precipitation storms to streamflow hydrographs. Based on this concept, the unit hydrograph method and kinematic hydrograph method have been developed.

7.3.2 Correlation with Meteorological Data

Hydrological and meteorological processes are both natural and interrelated. The precipitation and streamflow data collected can be considered as statistical samples derived from an indefinite natural series of meteorological and hydrological events. The standard statistical techniques and probability theory, including cross-correlation or regression analysis and frequency analysis, can be applied to these data. In multiple linear† regression analysis, the drainage basin characteristics and other meteorological parameters are also included in the correlation beside the precipitation. Since the meteorological data series is often longer, it is used to extend the short hydrological data series.

7.3.3 Correlation with Hydrological Data at another Site

Many similarities have been noticed in the natural streamflow of various streams over a fairly large region. The data generated at one gage site thus have a transferable value for a neighboring site on the same or a different stream. This may involve a simple extrapolation or interpolation of information gathered at two stations on the same stream or cross-correlation by regression analysis or duration analysis between two different streams.

7.3.4 Sequential Data Generation

In the techniques just described, the data were reconstituted for the past period up to the present time by using either meteorological data or hydrological information from another

---

* Additional hydrological data include sediment transport and ice phenomena.

† In the case of nonlinear relationships, a linearization is usually possible by an appropriate transformation, such as a logarithmic transformation.

site. When the data have to be extended into the future or when there are no parallel data to provide a relation, the synthetic data are produced based on a time series that includes a random component. Further, the estimation of a large number of values by the correlation technique produces a series that has a low variance. It is, therefore, necessary to use a technique that incorporates a random component. The data generated have statistical characteristics similar to the short-duration series used in the generation process. This involves a *stochastic process*.

Any of these four techniques is capable of estimating streamflows, but depending on the available data and need for their extension, a specific technique is chosen because of its suitability and convenience. There are five situations in which streamflows have to be estimated. The information that exists in each of these cases and applicable techniques are indicated in Table 7.1. The correlation technique, besides estimating streamflows in situations when they are lacking, is also used to extend a record by using another record, which is more than 25% longer, to improve the data because the error introduced by the correlation is usually less than the sampling error of the short-duration record. Also, extending and filling in data are necessary for regional studies, in which every record should be adjusted to the same length.

Table 7.1   Data Situation and Estimation Techniques

Case	Available Data	Technique
Gaged site		
1. Assessing streamflow data from precipitation	Precipitation data for the site	Hydrograph analysis
2. Augmenting streamflow data	1. Short-term streamflow data and long-term precipitation data for the site	Rainfall-runoff relation
	2. Short-term streamflow data for the site and long-term streamflow data for another site	1. Correlation of stream-gaging stations 2. Comparison of flow duration curves
3. Estimating gaps in streamflow data	(Same as item 2)	
4. Generation of data	Short-term streamflow data	Synthetic flow generation
Ungaged site		
5. Assessing streamflow data	1. Streamflow data at one or two neighboring sites on the same river	Interpolation or extrapolation
	2. Overall precipitation and other meteorological data	Hydrologic cycle model
	3. Overall precipitation and soil data	NRCS method
	4. Drainage basin characteristics	Generalized regional relation
	5. Channel geometry	Generalized regional relation

7.4 HYDROLOGICAL PROCESSES IN STREAMFLOW ESTIMATION

Descriptive hydrology presents the theory of water distribution in a subjective manner, which by means of *quantitative hydrology* is expressed in terms of numbers, either measured or calculated. The functional relationship between the numbers in a quantitative representation is termed *mathematical hydrology*. In this context the process of representing a phenomenon mathematically is known as the *mathematical model*.

Mathematical hydrology is further divided into *physical* or *deterministic hydrology, statistical hydrology, probabilistic hydrology, stochastic hydrology*, and *systems hydrology*. Deterministic hydrology is subdivided into empirical and conceptual procedures.

The empirical method yields an output for a given set of inputs without considering the relationships of the parameters involved in the process being considered. It considers only the extremes or treats discrete time periods. The hydrograph analysis for inadequate data and filling in gaps and the use of generalized relations for the ungaged sites constitute the empirical approach.

In the conceptual approach, the various processes and their interrelationships are identified, although sometimes the empirical relationship is also used. The meteorological-runoff relationship is initially approached by the conceptual method.

The correlation and regression techniques for inadequate data and data with gaps are the statistical methods that establish functional relationships between the two sets of data. The relation is assessed in statistical terms by the standard deviation, correlation coefficient, and significance tests.

Probabilistic methods involve the concept of frequency or probability. Peak flow and minimum flow estimates, which ignore the sequence of events and treat the data as time independent, are based on the probabilistic approach.

The stochastic process involves the generation of a synthetic long-duration data series from a limited sample of data. This method treats the sequence of events as time dependent and incorporates a random component to represent the natural phenomenon.

Thus, all hydrological processes are utilized in the estimation of streamflows, except for the systems approach, which is essentially an optimization process used in planning.

7.5 HYDROGRAPH ANALYSIS FOR ESTIMATION OF STREAMFLOW

As defined earlier, a hydrograph is a graphic representation of river discharge with respect to time. The strip chart from a water-level recorder provides a stage hydrograph, which is transformed by the application of a rating curve into a discharge hydrograph of a stream. A streamflow hydrograph is a result of the runoff processes, as discussed in Section 7.2, comprising overland flow, interflow, and baseflow that are generated by precipitation storms. A hydrograph resulting from a precipitation storm is known as a *storm hydrograph*. The streamflow hydrograph is thus the cumulative effect of storm hydrographs. The shape of a single storm hydrograph has a typical pattern, as shown in Figure 7.5.

There is a rising limb, the shape of which is characterized by the basin properties and the duration, intensity, and uniformity of the rain. The crest segment includes the part of the hydrograph from the inflection point on the rising limb to an inflection point on the recession limb. It contains the peak flow rate. The peak represents the arrival of flow at the outlet from all parts of the basin. For short-duration rain that does not last long enough so that the entire area contributes, the peak represents the flow from that portion of the basin receiving the highest concentration of runoff. The end of the crest segment, known as the

Figure 7.5 Simple storm hydrograph.

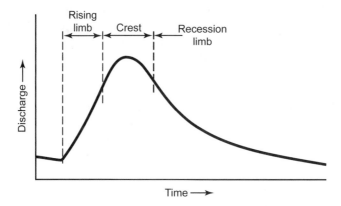

inflection point, marks the time when direct runoff from the overland flow (excess rainfall) into the stream outlet ceases. The recession limb thus indicates the storage contribution from detention storage (depth of water built up over the land surface), interflow, and groundwater flow. The recession curve is independent of the characteristics of the rainstorm. It can be considered as a rate of discharge resulting from the draining-off process. If there is no added inflow, the time variation of discharge due to the draining-off process, and hence the equation of the recession curve, can be given by

$$Q_t = Q_0 K^t \quad [\text{L}^3\text{T}^{-1}] \tag{7.1}$$

where

Q_t = discharge t time units after Q_0

Q_0 = initial discharge at the start of recession $(t = 0)$

K = recession constant

This equation, on a semilogarithm plot of log Q_t and time, represents a straight line. The plot is usually a curve which can be approximated by three straight lines of decreasing slope or increasing values of K, representing three different types of storage contribution to the recession: surface detention, interflow, and groundwater. The slope of each of the three lines provides a value of the recession constant for each type of storage contribution.

7.6 DIRECT RUNOFF HYDROGRAPH AND BASEFLOW HYDROGRAPH

It is convenient to consider the total flow to be divided into two parts: the storm, or direct, runoff and the baseflow. (Interflow is included with the direct runoff; sometimes it is treated separately and the hydrograph is separated into three components.) A stream carries baseflow during most of the year when there are no storms in the basin. This comes from the groundwater. Groundwater accretion resulting from any storm is released over an extended period; thus a particular storm contributing to direct runoff is not directly concerned with the baseflow. The precipitation excess of a storm (i.e., net rainfall after all abstractions) constitutes the direct runoff. The arrival of direct runoff at the stream outlet is the starting point for the direct runoff hydrograph (DRH). As time elapses, progressively distant areas add to the outlet flow until a peak flow is attained. If the rainfall continues beyond this period and maintains a constant intensity for a long period of time, a state of equilibrium will be reached and the constant peak flow should continue. The condition of equilibrium is, however, seldom attained because even in extended rainfall, variations in intensity occur throughout its duration. After the peak, the DRH begins to descend. A point of inflection comes when the overland flow to the outlet ceases and the storage contribution from surface detention and interflow begins. When the contribution pertains to baseflow only, it indicates the end of the DRH.

The first step in the hydrograph analysis is to separate the baseflow and direct runoff hydrograph. When multiple storms occur it is sometimes necessary to separate the overlapping parts of consecutive direct runoff hydrographs.

7.7 HYDROGRAPH SEPARATION

There are two common approaches for separating the baseflow from the direct runoff. The first approach relates to the use of the recession curve equation (7.1), since the last part of the curve is for the baseflow. This approach makes it possible to separate interflow also, if desired. The second approach is of an arbitrary nature. There are many techniques under the second approach.

7.7.1 Separation by Recession Curve Approach

When the data from a stream-gaging station are available for rainless periods reflecting baseflow only, several time intervals of equal value are selected and considered as a unit time. Flow at the beginning of each interval is analogous to Q_0, and at the end of each interval is analogous to Q_1. The values of Q_0 versus Q_1 corresponding to various selected intervals are plotted on ordinary graph (grid) paper. A straight line is fitted graphically to these points. The slope of the line is $Q_1/Q_0 = K$. Once the equation is known, beginning with Q at the starting point of the direct runoff hydrograph and with the selected time as the unit time, the groundwater recession curve is plotted from eq. (7.1).

Commonly, however, the separation has to be done without a dry period streamflow record. The hydrograph is plotted on semilogarithmic paper with log of discharge values on the y-axis and corresponding time on the x-axis. Since the logarithmic form of eq. (7.1) is a straight line, the tail part of the hydrograph is extended back from F (Figure 7.6) by a straight line under the hydrograph to point O directly under the inflection point E on the recession limb and then point O is joined to point B at the beginning of the direct runoff. Point B is selected arbitrarily where runoff begins. The area below the line BOF approximates the groundwater flow. The ordinates above the line are plotted as the DRH.

EXAMPLE 7.1

The daily streamflow data for the Merrimack River, Massachusetts, at a site having a drainage area of 6500 km2 are given in Table 7.2. Separate the baseflow from the direct runoff hydrograph (DRH) by the recession curve method. Determine the equivalent depth of the direct runoff.

SOLUTION

1. The semilogarithm plot of log Q versus t is shown in Figure 7.6.

2. The last straight-line segment is extended backward to O under the inflection point and then joined to B.

Figure 7.6 Baseflow separation by the recession curve approach.

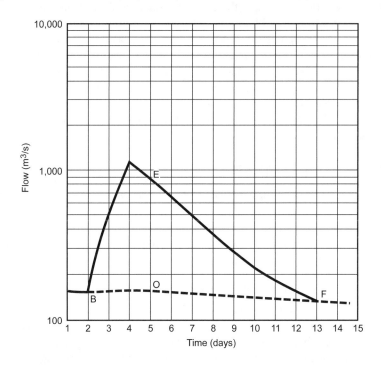

Table 7.2 Daily Discharge of Merrimack River

Time (days)	Flow (m³/s)	Time (days)	Flow (m³/s)
1	168	9	280
2	160	10	220
3	500	11	185
4	1130	12	160
5	860	13	140
6	650	14	138
7	500	15	135
8	380		

3. The ordinates of the DRH (residual ordinates) are listed in Table 7.3 and the area under the DRH is computed, which signifies the runoff volume.

$$\text{Volume of runoff} = \left(3342\frac{\text{m}^3}{\text{s}}\text{day}\right)\left[\frac{24 \times 60 \times 60 \text{ s}}{1 \text{ day}}\right]$$

$$= 288.75 \times 10^6 \text{ m}^3$$

$$\text{Runoff depth} = \frac{\text{runoff volume}}{\text{drainage area}} = \frac{288.75 \times 10^6 \text{ m}^3}{\left(6500 \text{ km}^2\right)\left[1 \times 10^6 \text{m}^2/\text{km}^2\right]}$$

$$= 0.0444 \text{ m or } 44.4 \text{ mm}$$

Table 7.3 Computation of Direct Runoff Volume

Time (days)	Direct Runoff (m³/s)	Average Runoff (m³/s)	Duration (days)	Runoff × Time (m³ · day/s)
1	0			
		0	1.0	0
2	0			
		170	1.0	170
3	340			
		655	1.0	655
4	970			
		835	1.0	835
5	700			
		596	1.0	596
6	492			
		419.5	1.0	419.5
7	347			
		288	1.0	288
8	229			
		179.5	1.0	179.5
9	130			
		101.5	1.0	101.5
10	73			
		57.5	1.0	57.5
11	42			
		30.5	1.0	30.5
12	19			
		9.5	1.0	9.5
13	0			
Total				3342

7.7.2 Separation by Arbitrary Approach

Since the method of Section 7.7.1 is approximate and there is no clear basis for making a precise distinction between direct and baseflow after they have been intermixed in a stream, many other approximate procedures are commonly used. The inaccuracies involved in the separation of the baseflow are usually not important.

Method 1. Refer to Figure 7.7. The simplest method is to join the beginning of the direct runoff (point A) to the end of direct runoff (point B) by a straight line. If point B is not well defined, draw a horizontal line from point A.

In the U.S. Army Corps of Engineers HEC-1 model, the point of baseflow recession on the falling limb can be specified as a ratio to peak flow, i.e., the recession point is when the falling limb discharge drops to 0.1 of the peak discharge. The Corps (ASCE, 1997) has prepared a plot of the discharge at the beginning of baseflow recession versus basin drainage area based on observations on the Upper Hudson and Mohawk rivers in New York.

Method 2. Extend the recession curve before the storm to point C beneath the peak. Connect point C to point D by a straight line. Point D on the hydrograph represents N days after the peak, given by the formula

$$N = aA^{0.2} \quad \text{[unbalanced]} \tag{7.2}$$

Figure 7.7 Methods of baseflow separation.

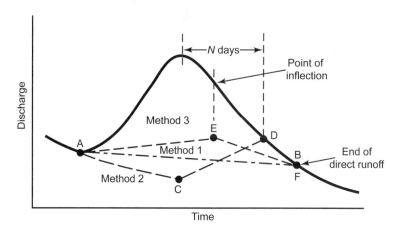

where

N = time, days

A = drainage area

a = 0.8 when A is in square kilometers

or 1.0 when A is in square miles

The value of N is best determined by an inspection so that the time base is neither too long nor too short.

Method 3. Extend the recession curve backward to point E below the inflection point. Connect A to E by a straight line or an arbitrary shape. If a recession curve is fitted to the hydrograph, the point of departure F of the computed curve from the actual curve marks the end of direct runoff. Otherwise, the endpoint B is arbitrarily selected.

EXAMPLE 7.2

For the data of Example 7.1, separate the baseflow by the methods of Section 7.7.2.

SOLUTION

1. The hydrograph has been plotted on Figure 7.8.

2. *Method 1.* Join point A, the beginning of direct runoff, to point B, the end of direct runoff. Both points are selected by judgment.

3. *Method 2.* Extend the recession curve before the storm up to point C below the peak. Join point C to point D, computed from eq. (7.2), as follows:

$$N = 0.8A^{0.2}$$
$$= 0.8(6500)^{0.2} = 4.6 \text{ days} \approx 5 \text{ days}$$

4. *Method 3.* Extend the recession curve backward to point E. Join point E to A.

5. The ordinates of the DRH (residual ordinates) by three methods are indicated in Table 7.4.

Figure 7.8 Baseflow separation for Example 7.2.

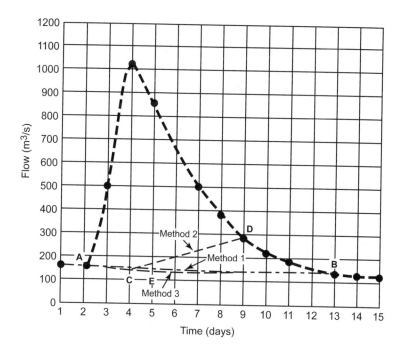

Table 7.4   Ordinates of the DRH by Different Methods

Time (days)	Direct Runoff (m³/s)		
	Method 1	Method 2	Method 3
1	0	0	0
2	0	0	0
3	340	347	343
4	972	980	977
5	705	685	710
6	497	455	500
7	349	270	352
8	230	125	234
9	132	0	136
10	74	0	78
11	42	0	44
12	19	0	20
13	0	0	0

7.8 UNIT HYDROGRAPH AND INSTANTANEOUS UNIT HYDROGRAPH

7.8.1 Time Parameters

Time base of a hydrograph. There are three time parameters related to the direct runoff hydrograph and unit hydrograph. The time base of a hydrograph is considered to be the time from the beginning to the end of the direct or unit hydrograph, shown as T in Figure 7.9a.

Lag time. The lag time or basin lag, t_p, is a basic time parameter that is defined as the difference in time between the center of mass of rainfall excess and center of mass of runoff (or peak rate of flow), as indicated in Figure 7.9a. Many other time intervals between rainfall and its hydrograph are also referred to as the time lag. Singh (1988) has listed various other definitions of the time lag.

Time of concentration. The time of concentration, t_c, is defined in two ways. In terms of physical characteristics of a watershed, which is more important in peak flow assessment, it is defined as the travel time of a water particle from the hydraulically most remote point in the basin to the outflow location. This is explained in Section 13.10.5. Based on rainfall and hydrograph characteristics, it is taken as the time from the end of the rainfall excess to the point of inflection on the falling limb of the DRH or the unit hydrograph that signifies the end of direct rainfall inflow into the stream and the start of the detention storage contribution, as indicated in Figure 7.9a.

7.8.2 Unit Hydrograph

In 1932, L. K. Sherman introduced the concept of the unit hydrograph, a very important contribution commonly used to transform rainfall to streamflow. The *unit hydrograph* is defined as a hydrograph of direct runoff (excluding the baseflow) observed at the downstream limit of a basin due to one unit of rainfall excess (precipitation) falling for a unit time t_r (Figure 7.9a). The unit of excess precipitation is taken as 1 in. or 1 mm (though it also can be defined to result from 1 cm of precipitation excess). The unit of time for precipitation may be 1 day or less, but must be less than the time of concentration explained above. Sherman (1942) suggested the following:

Basin Area (mi²)	Unit of Time (Duration) of Precipitation Excess (hr)
Over 1000	12 to 24 (preferably 12)
100–1000	6, 8, or 12
20	2
Small areas	One-third to one-fourth of the time of concentration

The effect of a small difference in storm (rainfall) duration is not significant; a tolerance of ± 25% in duration is acceptable (Linsley et al., 1982).

The following characteristics of a hydrograph form the basis of the unit hydrograph concept:

1. A hydrograph reflects all of the combined physical characteristics of the drainage basin (shape, size, slope, soils) and that of the causative storm (pattern, intensity, duration).

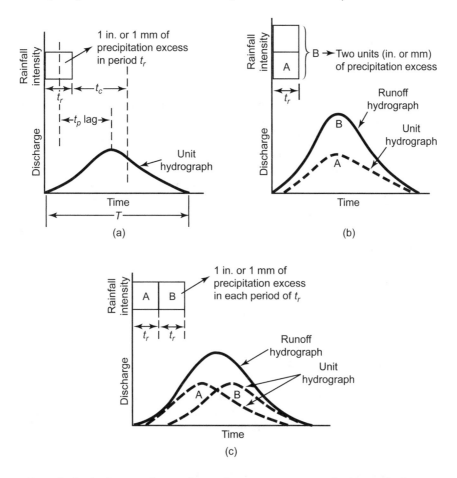

Figure 7.9 Principles of the unit hydrograph: (a) unit hydrograph; (b) runoff hydrograph for two units of precipitation of duration t_r; (c) runoff hydrograph from unit precipitation for two consecutive periods of duration t_r.

2. Since the basin features do not change from storm to storm, hydrographs from storms of similar duration and pattern are considered to have a similar shape and time base. The theory of superposition applies and linearity of the relation is assumed. Thus, if 2 in. of precipitation excess of a specified unit time occurs, the resulting hydrograph will have the same shape as the hydrograph from 1 in. of precipitation excess of the same unit duration except that all of the ordinates will be twice as large, as shown in Figure 7.9b.

Similarly, when 1 in. of precipitation excess occurs in each of the two consecutive unit time durations, the resulting hydrograph will be the sum of two 1-in. hydrographs, with the second hydrograph beginning 1 time unit later, as shown in Figure 7.9c.

3. Variations in storm characteristics do, however, have a significant effect on the shape of hydrographs. This includes (a) rainfall duration, (b) intensity, and (c) areal distribution in the basin.

   If the rainfall time duration is increased, the time base of the unit hydrograph will be lengthened. Since a unit hydrograph contains 1 unit of runoff by definition, the peak will be lowered for an increased duration. There is a technique, discussed later, to develop unit hydrographs for storms of longer duration from a unit hydrograph of short duration.

The effect of rainfall intensity on the hydrograph is related to basin size. On large basins, the changes in storm intensity have to last for a considerable time to cause a noticeable effect on the hydrograph, whereas in very small basins, short duration heavy bursts of rain will produce marked peaks in the hydrographs.

As regards the areal pattern, precipitation concentrated in the lower part of the basin will produce a hydrograph of rapid rise, sharp peak, and rapid recession. On the other hand, the precipitation occurring in the upper part will result in slower rise, low peak, and slower recession. Thus, only one unit hydrograph is not sufficient for a basin. Theoretically, a separate unit hydrograph is necessary for each possible duration of rainfall, but a few unit hydrographs of short durations are adequate, since a tolerance of $\pm 25\%$ is permissible and techniques are available for adjustment of the duration. Also, the unit hydrograph concept is applicable to relatively small basins where the difference in areal distribution of the rainfall does not significantly affect the hydrograph. This concept should not be used for basins much larger than 2000 mi^2 (5000 km^2) unless reduced accuracy is acceptable (Linsley et al., 1982).

When only daily rainfall records are available (hydrograph unit duration is 24 hours), there is a lower limit of 1000 mi^2 for the area below which the unit hydrograph theory should not be applied (Gray, 1973).

According to Sherman (1942), the unit hydrograph method does not apply to runoffs originating from snow or ice. The Hydrologic Engineering Center of the U.S. Army Corps of Engineers, however, makes use of the unit hydrograph technique in snowmelt excess as well.

7.8.3 Distribution Graph

In 1935, M. Bernard suggested a dimensionless form of the unit hydrograph, in which each ordinate of the hydrograph is obtained by dividing the runoff volume at a particular time by the total runoff volume under the hydrograph, thus representing the relative fraction or percentage values at different times. This is known as the *distribution graph*. The area under the graph is 100% and the graph holds true for all storms of the same duration in the basin, regardless of their intensity.

7.8.4 Instantaneous Unit Hydrograph

In 1943, C. O. Clark used the concept of an instantaneous unit hydrograph in routing analysis. As the duration of precipitation excess approaches zero, an *instantaneous unit hydrograph* (IUH) results. It is a hydrograph that results from 1 unit of precipitation excess applied instantly (in an infinitesimally small time) over a basin. This is a fictitious situation used for the purpose of analysis. The IUH is indicative of the basin storage characteristics, since the rainfall-duration effects are eliminated. Further, since storms of different durations produce varying hydrograph shapes, the IUH, which is independent of duration, is unique for a basin. However, this is true in the linear theory on which the concept of unit hydrograph is based. Several studies since 1960 indicate that the IUH varies from storm to storm, and a nonlinearity exists in the hydrograph properties. But the simplified concept is still very useful. For determination of an IUH, the data from a basin on a storm and the corresponding hydrograph are required. The IUH can then be used to compute a unit hydrograph (UH).

7.9 DERIVATION OF UNIT HYDROGRAPH

The unit hydrograph has a very important application in the estimation of natural stream-flows and peak flows from rainfall records.

There are two approaches to deriving a unit hydrograph. In the first approach, the unit hydrograph is derived directly from a storm hydrograph recorded in the basin. This is an inverse problem, since the basic use of the unit hydrograph is to construct storm or streamflow hydrographs. The second approach is to make use of the instantaneous unit hydrograph concept.

7.9.1 Derivation by the Inverse Procedure

It is necessary to have data on a rainfall storm and the corresponding runoff (streamflow). A hydrograph resulting from an isolated, intense, short-duration storm of nearly uniform distribution in space and time is most satisfactory. If such a well-defined single-peaked hydrograph is unavailable, the unit hydrograph is derived from a complex hydrograph.

The duration (unit time) of the unit hydrograph will be the same as the duration of the storm that had produced the storm hydrograph. It is, however, possible to adjust the duration of the unit hydrograph by the technique of superposition described subsequently.

The procedure to derive a unit hydrograph is as follows:

1. The hydrograph associated with a storm is plotted. The baseflow is separated by the technique of Section 7.7, thus obtaining the direct runoff hydrograph (DRH).

2. The area under the DRH that represents the volume of surface runoff is computed. This volume of runoff is converted to a depth, P_n, of the net storm over the basin by the equation

$$P_n = \frac{KV}{A} \quad [L] \tag{7.3}$$

where

P_n = runoff depth of the storm

K = conversion factor, as given in Table 7.5

V = volume under the hydrograph

A = drainage area of the basin

3. Each of the ordinates of the DRH is divided by P_n. The result is a unit hydrograph of duration equal to the duration of the storm.

Table 7.5   Factor to Convert Runoff Volume to Depth

Unit of Runoff Ordinate	Unit of Time Base	Unit of Volume	Unit of Area	Unit of Depth	K
cfs	day	cfs · day	mi^2	in.	3.27×10^{-2}
cfs	hour	cfs · hr	mi^2	in.	1.55×10^{-3}
m^3/s	day	m^3/s · day	km^2	mm	86.4
m^3/s	hour	m^3/s · hr	km^2	mm	3.6

EXAMPLE 7.3

The hydrograph of Example 7.1 was produced by a storm of 12-hour duration considered to have uniform intensity over the basin. Determine the unit hydrograph.

SOLUTION

1. The data are tabulated in Table 7.6. The baseflow separated by the recession curve technique in Example 7.1 is given in column 3 and the direct runoff in column 4 of the table. The runoff volume, from Example 7.1, is 3342 $m^3/s \cdot$ day.

2. From eq. (7.3),

$$P_n = \frac{86.4(3342)}{6500} = 44.4 \text{ mm}$$

3. Each ordinate of the DRH (column 4) is divided by 44.4. The resultant unit hydrograph of 12-hour duration is given in column 5. This has been plotted in Figure 7.10.

7.9.2 Derivation by the IUH Technique

The unit hydrographs derived from storms of different durations by the inverse procedure will have different shapes. The instantaneous unit hydrograph eliminates the duration effect and defines a unique unit hydrograph for the basin (in the linear theory). Of the many procedures for preparing an IUH, the method attributed to Clark (lag and route technique), who was the first to use the IUH concept, has been described here.

There are two steps in the procedure: (1) preparation of an IUH from the field data of a storm and the corresponding runoff, and (2) conversion of the IUH to a unit hydrograph.

In developing the IUH, Clark (1943) conceived a fictitious linear reservoir located at the outlet of the stream such that the storage is proportional to the outflow [i.e., $S = KO$, where K is an attenuation constant, equivalent to the recession constant of eq. (7.1)]. Since the difference between inflow, I, and outflow, O, is the rate of change of storage,

$$\frac{I_1 + I_2}{2} - \frac{O_1 + O_2}{2} = \frac{\Delta S}{\Delta t} \tag{a}$$

or

$$\frac{I_1 + I_2}{2} - \frac{O_1 + O_2}{2} = \frac{K(O_2 - O_1)}{\Delta t} \quad \text{(since } S = K\,O\text{)} \tag{b}$$

or

$$O_2 = C_1 I_2 + C_2 O_1 \quad \text{(treating } I_1 = I_2\text{)} \tag{c}$$

In general terms,

$$O_i = C\,I_i + (1 - C)O_{i-1} \quad [\text{L}^3\text{T}^{-1}] \tag{7.4}$$

where

O_i = outflow at the end of period i

I_i = inflow at the end of period i

$$C = \frac{2\Delta t}{2K + \Delta t} \quad [\text{dimensionless}] \tag{7.5}$$

Table 7.6 Computation of Unit Hydrograph

(1) Time (days)	(2) Total Runoff (m^3/s)	(3) Baseflow (m^3/s)	(4) Direct Runoff (m^3/s)	(5) Unit Hydrograph (m^3/s) per mm
1	168	168	0	0
2	160	160	0	0
3	500	160	340	7.66
4	1130	160	970	21.85
5	860	160	700	15.77
6	650	158	492	11.08
7	500	153	347	7.82
8	380	151	229	5.16
9	280	150	130	2.93
10	220	147	73	1.64
11	185	143	42	0.95
12	160	141	19	0.43
13	140	140	0	0

Figure 7.10 Derivation of unit hydrograph from a storm hydrograph.

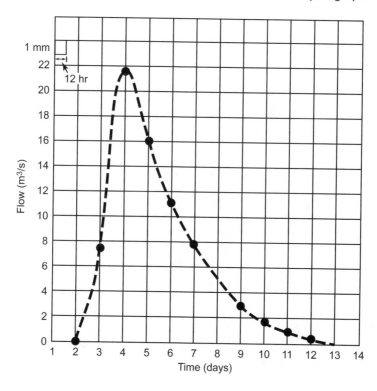

Clark routed the inflow comprising the runoff over the basin divided into several areas (subbasins) and represented by a time-area curve. The outflow from eq. (7.4) thus provided the IUH.

The following input data are needed for application of the Clark method:

Time of concentration, t_c. This information is obtained from the runoff hydrograph of a storm. It is estimated as the time from the end of the excess rainfall to the inflection point on the recession limb of the hydrograph. If field data are not available, t_c is estimated empirically.

Attenuation constant, K. This parameter accounts for the effect of storage in the channel on the hydrograph. It is also estimated from the runoff hydrograph by dividing the flow at the point of inflection of the direct runoff hydrograph by the rate of change of discharge at the same time (slope of the hydrograph at the inflection point), as shown in Figure 7.11.

Time-area relation. On the map of the basin, isochrones defining points with equal travel times to the outlet are marked, as shown in Figure 7.12a. The area between each pair of isochrones is measured. A curve of travel time versus area (between each pair or cumulated) is drawn, known as the time-area relation. From this curve, the values of areas $a_1, a_2, a_3, \ldots$ that are Δt apart are read, where Δt is the computational interval that must be equal to or less than the duration of the rainfall excess. The areas are in units of volume for one in. or mm depth (in.-mi^2 or mm-km^2). These are converted to discharge units, representing the inflows, by the relation

$$I_i = \frac{Fa_i}{\Delta t} \quad [L^3 T^{-1}] \tag{7.6}$$

where

I_i = ordinate in discharge unit cfs or m^3/s at the end of period i of inflow

(time-area curve)

a_i = ordinate of time-area relation at the end of period i, in-mi^2 or mm-km^2

Δt = time period of computational interval in hours

F = conversion factor:

= 645 to convert in.-mi^2 per hr to cfs

= 0.278 to convert mm-km^2 per hr to m^3/s

The inflows I determined for various times by eq. (7.6) are substituted in eq. (7.4) to determine corresponding outflows that indicate the IUH ordinates.

To derive the unit hydrograph of the same unit duration Δt as the time step of computations of the IUH, two IUHs spaced an interval of Δt apart are averaged as follows:

$$Q_i = 0.5 (O_i + O_{i-1}) \quad [L^3 T^{-1}] \tag{7.7}$$

where Q_i is the ordinate of unit hydrograph at time i.

EXAMPLE 7.4

The watershed of the St. Paul River at the Walker Bridge is shown in Figure 7.12a, on which isochrones have been marked. A hydrograph observed at the Walker Bridge site due to a storm of 2-hour duration is also shown in Figure 7.12b. Determine **(a)** the instantaneous unit hydrograph, and **(b)** the 2-hour duration unit hydrograph.

Figure 7.11 Determination of the attenuation constant.

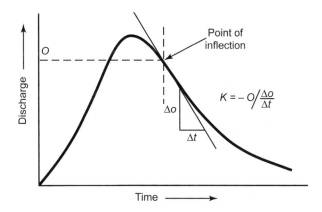

SOLUTION

1. Time of concentration (from the hydrograph) = 10 hr.

2. Attenuation constant, K (from the hydrograph)

$$K = \frac{\text{ordinate of DRH at inflection}}{\text{slope of DRH at inflection}} = \frac{420 \text{ m}^3/\text{s}}{80 \text{ m}^3/\text{s} \cdot \text{hr}} = 5.25 \text{ hr}$$

3. Derivation of time-area curve (from isochronal diagram):

Map Area Number	Area (km²)	Cumulated Area (km²)	Travel Time (hr)	Travel Time (%)
1	10	10	1	10
2	25	35	2	20
3	50	85	3	30
4	55	140	4	40
5	70	210	5	50
6	65	275	6	60
7	80	355	7	70
8	100	455	8	80
9	110	565	9	90
10	30	595	10	100

4. Plot the percent time versus cumulated area as shown in Figure 7.13. Areas are expressed in units of volume corresponding to a unit depth of runoff over the area.

5. From Figure 7.13, areas are read at points that are one computational interval, Δt, apart (equal to the desired unit hydrograph duration; in this example 2 hr) and recorded in column 3 of Table 7.7.

6. From eq. (7.6),

$$I_i = \frac{0.278}{2} a_i = 0.139 a_i$$

By this formula, a_i is converted to I_i in column 5 of the table.

Figure 7.12 (a) Isochronal map of the St. Paul Basin; (b) 2-hr hydrograph of the St. Paul River at the Walker Bridge.

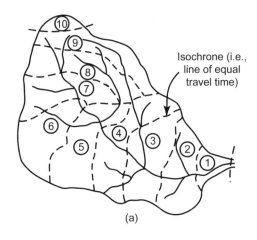

(a)

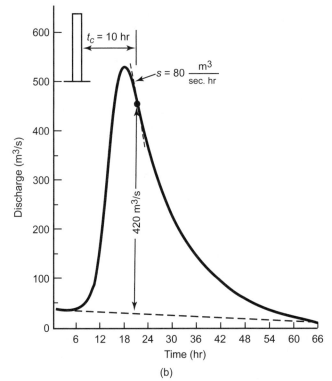

(b)

7. From eqs. (7.5) and (7.4),

$$C = \frac{2\Delta t}{2K + \Delta t} = \frac{2(2)}{2(5.25) + 2} = 0.32$$

$$O_i = CI_i + (1 - C)O_{i-1} = 0.32I_i + 0.68O_{i-1}$$

Route the inflow (column 5) to outflow (column 6) using the relation above.

Figure 7.13 Time-area relation for St. Paul Basin.

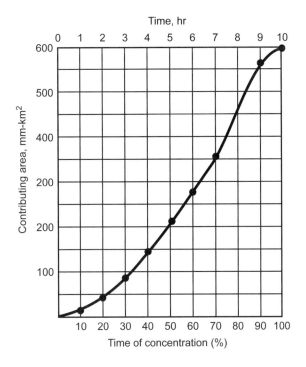

8. From eq. (7.7), $Q_i = 0.5\,(O_{i-1} + O_i)$

Thus, average the ordinate of the IUH with another ordinate of the IUH one interval before to obtain the unit hydrograph in column 7.

7.10 CHANGING THE UNIT HYDROGRAPH DURATION

A unit hydrograph represents 1 inch (or 1 mm) of direct runoff from the rainfall of a specified period of time. Rainfall of different durations will produce different shapes of the unit hydrograph. The longer duration of the rainfall will lengthen the time base and lower the peak, and vice versa, since a unit hydrograph by definition contains 1 unit of direct runoff. There are two common techniques by which a unit hydrograph can be adjusted from one duration to another.

7.10.1 Lagging Method

This is restricted to cases when a duration has to be converted to a longer duration which is a multiple of the original duration. If a unit hydrograph of duration t_r is added to another identical unit hydrograph lagged by t_r, the resulting hydrograph represents the hydrograph of 2 units of storm occurring in $2t_r$ time, as shown in Figure 7.14. If the ordinates of this hydrograph are divided by 2, a unit hydrograph will result. In general terms,

$$\text{UH of } nt_r \text{ duration} = \frac{\text{sum of } n, \text{ UH of } t_r \text{ duration each lagged by } t_r \text{ time}}{n} \quad [\text{L}^3\text{T}^{-1}] \tag{7.8}$$

Table 7.7   Unit Hydrograph Computation by the Clark Method

(1)	(2)	(3)	(4)	(5)	(6)[a]	(7)[b]
			Inflow (from Fig. 7.13)			2-hr
		Cumulated	Incremental Area, a_i	I (m^3/s)	IUH, O_i	Unit Hydrograph
No.	Time (hr)	Area	(mm-km^2)	(eq. 7.6)	(m^3/s)	Q_i (m^3/s)
1	0	0	0	0	0	0
2	2	35	35	4.9	1.57	0.79
3	4	140	105	14.6	5.74	3.66
4	6	275	135	18.8	9.92	7.83
5	8	455	180	25.0	14.75	12.33
6	10	595	140	19.5	16.27	15.51
7	12		0	0	11.06	13.67
8	14				7.52	9.29
9	16				5.11	6.32
10	18				3.47	4.29
11	20				2.36	2.92
12	22				1.60	1.98
13	24				1.09	1.35
14	26				0.74	0.92
15	28				0.50	0.62
16	30				0.34	0.42
17	32				0.23	0.29
18	34				0.16	0.20

[a] At any time step: col. (6) = $C \times$ col. (5) + (1 − C) × col. (6) of previous step.
[b] Col. 7 = 0.5 [col. 6 + col. 6 of previous step].

EXAMPLE 7.5

The following unit hydrograph results from a 2-hour storm. Determine the hourly ordinates of a 6-hour unit hydrograph.

Time (hr)	0	1	2	3	4	5	6
Q (m^3/s)	0	1.42	8.50	11.30	5.66	1.45	0

SOLUTION   See Table 7.8.

$$t_r = 2 \text{ hr}$$

$$n = \frac{6\text{-hr duration}}{2\text{-hr duration}} = 3$$

Figure 7.14 Lagging procedure to convert unit hydrograph duration.

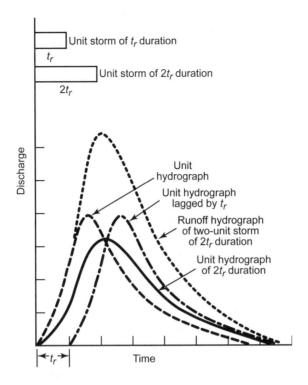

Table 7.8 Conversion of Unit Hydrograph Duration by Lagging

(1)	(2)	(3)	(4)	(5)	(6)	(7)
		Three 2-hr Hydrographs Each Lagged by 2 hr				6-hr Unit
	2-hr Unit Hydrograph					Hydrograph (m^3/s)
Time (hr)	(m^3/s)	$1 \times UH$	$1 \times UH$	$1 \times UH$	Total	(col. 6/3.0)
0	0	0			0	0
1	1.42	1.42			1.42	0.47
2	8.50	8.50	0		8.50	2.83
3	11.30	11.30	1.42		12.72	4.24
4	5.66	5.66	8.50	0	14.16	4.72
5	1.45	1.45	11.30	1.42	14.17	4.72
6	0	0	5.66	8.50	14.16	4.72
7			1.45	11.30	12.75	4.25
8			0	5.66	5.66	1.89
9				1.45	1.45	0.48
10				0	0	0

7.10.2 S-Curve Method

By the summation-curve method, a unit hydrograph can be converted to any other duration of shorter or longer time period. The S-curve results when the unit rate of excess rainfall continues indefinitely. The S-curve is thus constructed by adding together a series of t_r-duration unit hydrographs each lagged the unit duration, t_r, with respect to the preceding one, as shown in Figure 7.15. The curve assumes the S-shape and its ordinates in its equilibrium condition acquire a constant outflow rate equivalent to the rainfall excess. If T is the time base of the unit hydrograph, the summation of T/t_r unit hydrographs will produce the S-curve. It is not necessary, however, to add so many unit hydrographs to prepare an S-curve. A simplified procedure is illustrated in Table 7.9. In column 3, the values of column 4, corresponding to the previous t_r duration, have been repeated. Column 4, which represents the S-curve, is the sum of columns 2 and 3.

The next three columns of the table demonstrate the conversion process. In column 5, the t_r-duration S-curve (values of column 4) are lagged by the desired revised duration, t'_r. Column 6 indicates the difference of S-curve ordinates, that is, column 4 minus column 5. The ordinates of the t'_r-duration hydrograph are computed by multiplying the S-curve differences (col. 6) by the ratio t_r/t'_r.

EXAMPLE 7.6

Solve Example 7.5 by the S-curve method.

SOLUTION   Computations are shown in Table 7.10.

7.11 FORMULATION OF SYNTHETIC UNIT HYDROGRAPH

For sites where hydrologic records are not available to derive a unit hydrograph, the unit graph is synthesized from the physical characteristics of the watershed. Several methods (Snyder, 1938; Commons, 1942; Williams, 1945; Mitchell, 1948; Taylor and Schwarz, 1952; U.S. SCS, 1957; Hickok et al., 1959; Bender and Roberson, 1961; Gray, 1961) have been developed, some to serve special purposes, such as flood estimation. Two of these common methods are described below.

7.11.1 Snyder's Method

The four parameters—lag time, peak flow, time base, and standard duration—of rainfall excess for the unit hydrograph have been related to the physical geometry of the basin by the following relations:

$$t_p = C_t \left(LL_C \right)^{0.3} \quad \text{[unbalanced]} \tag{7.9}$$

$$Q_p = \frac{C_p A}{t_p} \quad \text{[unbalanced]} \tag{7.10}$$

$$T = 3 + \frac{t_p}{8} \quad \text{[T]} \tag{7.11}$$

$$t_D = \frac{t_p}{5.5} \quad \text{[T]} \tag{7.12}$$

Figure 7.15 Illustration of the S-curve.

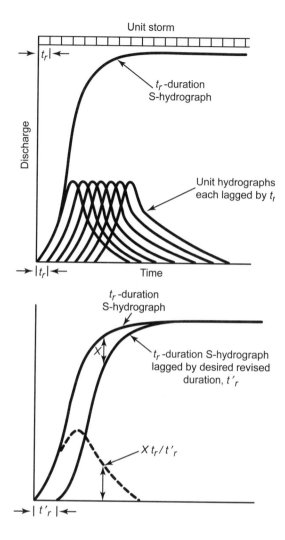

When the duration of rainfall excess, t_r, is other than the standard duration, t_D, the following adjustments in lag time and peak discharge are made:

$$t_{pR} = t_p + 0.25 \, (t_r - t_D) \quad [\text{T}] \tag{7.13}$$

$$Q_{pR} = Q_p \frac{t_p}{t_{pR}} \quad [\text{L}^3\text{T}^{-1}] \tag{7.14}$$

Section 7.11   Formulation of Synthetic Unit Hydrograph

Table 7.9 Computation of S-Curve and Conversion of Unit Hydrograph Duration

(1)	(2)	(3)	(4)	(5)	(6)	(7)
	Computation of S-Curve from Unit Hydrograph			Conversion of t_r-Duration Unit Hydrograph to t'_r-Duration UH		
Time	t_r-Duration Unit Hydrograph	S-Curve Additions	t_r-Duration S-Curve (col. 2 + col. 3)	t_r-Duration S-Curve Lagged by t'_r	S-Curve Difference (col. 4 − col. 5)	t'_r-Duration UH (col. 6 × t_r/t'_r)
duration of rainfall excess = t_r	a	0	a'			
	b	0	b'			
	c	0	c'	a'		
	d	0	d'	b'		
	e	0	e'	c'		
	f	a'	f'	d'		
	g	b'	g'	e'		
	h	c'	h'	f'		
	i	d'	i'	etc.		
	j	e'	j'			
	k	f'	k'			
	l	g'	l'			
	m	h'	m'			
	n	i'	n'			
	o	j'	o'			
etc.						

Table 7.10 Computation of 2-Hour S-Curve and 6-Hour Unit Hydrograph

(1)	(2)	(3)	(4)	(5)	(6)	(7)
	2-hr Unit Hydrograph	S-Curve		2-hr S-Curve Lagged by	S-Curve	6-hr Unit Hydrograph
Time (hr)	(m³/s)	Addition	2-hr S-Curve	6-hr	Difference	(col. 6 × 2/6)
0	0	+0	0		0	0
1	1.42	+0	1.42		1.42	0.47
2	8.50	+0	8.50		8.50	2.83
3	11.30	+1.42	12.72		12.72	4.24
4	5.66	+8.50	14.16		14.16	4.72
5	1.45	+12.72	14.17		14.17	4.72
6	0	14.16	14.16	0	14.16	4.72
7		14.17	14.17	1.42	12.75	4.25
8		14.16	14.16	8.50	5.66	1.89
9		14.17	14.17	12.72	1.45	0.48
10		14.16	14.16	14.16	0	0
11		14.17	14.17	14.17	0	0

where

t_D = standard duration of rainfall excess, hours

t_r = duration of rainfall excess other than standard duration adopted in the study, hours

t_p = lag time from midpoint of rainfall excess duration, t_D, to peak of the unit hydrograph, hours

t_{pR} = lag time from midpoint of duration, t_r, to the peak of the unit hydrograph, hours

T = time base of unit hydrograph, days

Q_p = peak flow for standard duration, t_D

Q_{pR} = peak flow for duration, t_r

L_C = stream mileage from the outlet to a point opposite the basin centroid

L = stream mileage from the outlet to the upstream limits of the basin

A = drainage area, mi² or km²

C_t = coefficient representing slope of the basin;

varies from 1.8 to 2.2 for distance in miles, or from 1.4 to 1.7 for distance in kilometers; Taylor and Schwarz state that C_t equals $0.6/\sqrt{S}$ for distance in miles, S being the basin slopes

C_p = coefficient indicating the storage capacity;

varies from 360 to 440 for English units, and from 0.15 to 0.19 for metric units

If the ungaged basin and the gaged basin are located in close proximity to each other within a region, the coefficients C_t and C_p are computed from the data of the gaged basin.

The coefficients so obtained are used in the preceding equations to construct the unit hydrograph for the ungaged basin. Otherwise, generalized values are used for the coefficients.

A unit hydrograph is sketched, from the lag time, peak discharge, and time base computed from eqs. (7.9) through (7.14), to represent a unit runoff amount (area under the graph). Equation (7.11) usually gives long base length for small to medium basins. The following Corps of Engineers formulas give additional assistance in plotting time width, W_{50}, in hours, at the discharge point equal to 50% of the peak discharge, and the width, W_{75}, in hours, at the discharge point equal to 75% of the peak flow.

$$W_{50} = \frac{770 A^{1.08}}{Q_{pR}^{1.08}} \quad \text{(English units)} \quad \text{[unbalanced]} \tag{7.15a}$$

or

$$W_{50} = \frac{0.23 A^{1.08}}{Q_{pR}^{1.08}} \quad \text{(metric units)} \quad \text{[unbalanced]} \tag{7.15b}$$

and

$$W_{75} = \frac{440 A^{1.08}}{Q_{pR}^{1.08}} \quad \text{(English units)} \quad \text{[unbalanced]} \tag{7.16a}$$

or

$$W_{75} = \frac{0.13 A^{1.08}}{Q_{pR}^{1.08}} \quad \text{(metric units)} \quad \text{[unbalanced]} \tag{7.16b}$$

In eqs. (7.15a) and (7.16a), A is in mi^2 and Q in cfs, and in eqs. (7.15b) and (7.16b), A is in km^2 and Q in m^3/s.

As a rule of thumb, the widths W_{50} and W_{75} are proportioned each side of the unit hydrograph peak in the ratio 1:2, with the short side on the left of the synthetic unit hydrograph.

EXAMPLE 7.7

For a basin of 500 km^2 having $L = 25$ km and $L_C = 10$ km, derive the 4-hour unit hydrograph. Assume that $C_t = 1.6$ and $C_p = 0.16$.

SOLUTION

1. $t_r = 4$ hr (given). From eq. (7.9),

$$t_p = 1.6(25 \times 10)^{0.3} = 8.38 \text{ hr}$$

2. From eq. (7.10),

$$Q_p = \frac{0.16(500)}{8.38} = 9.55 \text{ m}^3/\text{s}$$

3. From eq. (7.11),

$$T = 3 + \frac{8.38}{8} = 4.05 \text{ days or } 97 \text{ hr}$$

4. From eq. (7.12),

$$t_D = \frac{8.38}{5.5} = 1.5 \text{ hr}$$

5. From eq. (7.13),

$$t_{pR} = 8.38 + 0.25(4 - 1.5) = 9 \text{ hr}$$

6. From eq. (7.14),

$$Q_{pR} = \frac{9.55(8.38)}{9.0} = 8.89 \text{ m}^3/\text{s}$$

7. Time from beginning to peak,

$$P_r = \frac{t_r}{2} + t_{pR} = 2 + 9 = 11 \text{ hr}$$

8. From eq. (7.15b),

$$W_{50} = \frac{0.23(500)^{1.08}}{(8.89)^{1.08}} = 18 \text{ hr}$$

9. From eq. (7.16b),

$$W_{75} = \frac{0.13(500)^{1.08}}{(8.89)^{1.08}} = 10 \text{ hr}$$

The unit hydrograph has been sketched in Figure 7.16.

7.11.2 Natural Resources Conservation Service (NRCS) Method

The NRCS employs an average dimensionless hydrograph developed from an analysis of a large number of unit hydrographs from field data of various-sized basins in different geographic locations.

This dimensionless hydrograph has its ordinate values of discharge expressed as the dimensionless ratio with the peak discharge and its abscissa values of time interval as the dimensionless ratio with the period of rise (time from beginning to the peak flow). The ratios for the NRCS dimensionless unit hydrograph are given in Table 7.11.

The unit hydrograph ordinates for different time periods can be obtained from Table 7.11. However, to use this table, the values of P_r and Q_p are required, which are computed as follows:

$$Q_p = \frac{484A}{P_r} \quad \text{(English units)} \quad \text{[unbalanced]} \tag{7.17a}$$

or

$$Q_p = \frac{0.208A}{P_r} \quad \text{(metric units)} \quad \text{[unbalanced]} \tag{7.17b}$$

$$P_r = \frac{t_r}{2} + t_p \quad \text{[T]} \tag{7.18}$$

The time lag, t_p, is computed by eq. (7.9) or by a regional empirical relation, or by the NRCS equation involving the NRCS curve number.

Figure 7.16 Synthetic unit hydrograph by Snyder's method.

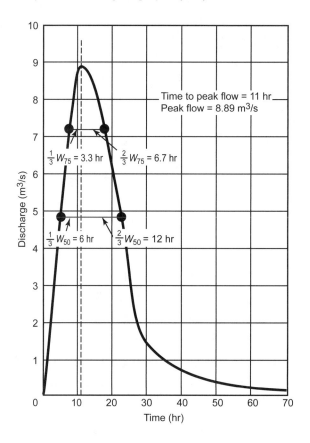

EXAMPLE 7.8

Solve Example 7.7 by the NRCS method.

SOLUTION

1. t_p = 8.38 hr, from eq. (7.9) computed in Example 7.7.

2. From eq. (7.18), P_r = 4/2 + 8.38 = 10.38 hr ≈ 10.5 hr.

3. From eq. (7.17b),

$$Q_p = \frac{0.208(500)}{10.5} = 9.90 \text{ m}^3/\text{s}$$

4. Using Table 7.11, the hydrograph ordinates are given in Table 7.12.

7.12 ESTIMATION OF STREAMFLOW FROM UNIT HYDROGRAPH

A direct application of the unit hydrograph is to synthesize the storm runoff (DRH) and the streamflow (by adding the baseflow) from a series of rainfall events of varying intensity. If the rainfall records are available on a daily basis, the resulting hydrograph is of daily

Table 7.11   Ratios for the NRCS Dimensionless Unit Hydrograph

Time Ratio, t/P_r	Hydrograph Discharge Ratio, (Q/Q_p)
0	0
0.1	0.030
0.2	0.100
0.3	0.190
0.4	0.310
0.5	0.470
0.6	0.660
0.7	0.820
0.8	0.930
0.9	0.990
1.0	1.000
1.1	0.990
1.2	0.930
1.3	0.860
1.4	0.780
1.5	0.680
1.6	0.560
1.8	0.390
2.0	0.280
2.2	0.207
2.4	0.147
2.6	0.107
2.8	0.077
3.0	0.055
3.5	0.025
4.0	0.011
4.5	0.005
5.0	0.000

Source: NRCS (formerly Soil Conservation Service), 1972.

streamflow. When the rainfall record belongs to a heavy storm, the hydrograph produced is for the flood flow, as discussed in Chapter 8.

Consider that a storm consists of a series of rainfall excesses $i_1, i_2, \ldots, i_n$, each of duration d. To formulate the storm hydrograph, a unit hydrograph of d duration (unit time) will be required for the basin. The ordinates of the d-duration unit hydrograph will be multiplied by i_1. Shifting the base by time d, the unit hydrograph ordinates will be multiplied by i_2, and so on, covering all excesses. Each of these hydrographs represents the DRH for individual rainfall excesses, and their sum is the DRH for the entire storm. The estimated baseflow added to this will provide the streamflow (storm) hydrograph.

Table 7.12   Synthetic Unit Hydrograph by NRCS Method

(1) t/P_r	(2)[a] t (hr)	(3) Q/Q_p (from Table 7.11)	(4)[b] Q (m³/s)
0	0	0	0
0.2	2.1	0.100	0.99
0.5	5.25	0.470	4.65
0.8	8.4	0.930	9.21
1.0	10.5	1.00	9.90
1.5	15.75	0.680	6.73
2.0	21.0	0.280	2.77
3.0	31.5	0.055	0.54
4.0	42.0	0.011	0.11
5.0	52.5	0.000	0.00

[a] Col. 2 = col. 1 × P_r

[b] Col. 4 = col. 3 × Q_p

EXAMPLE 7.9

Given below is a hydrograph that resulted from an isolated 2-hour duration storm of 1.5 in. rainfall excess. Determine the streamflow hydrograph from the storm sequence indicated. Assume that the losses amount to 60% of the precipitation.

Hydrograph data:

Time (hr)	0	1	2	3	4	5	6	7	8	9	10
Flow (cfs)	151	146	268	562	660	630	510	370	250	190	150

Storm sequence data:

Time units = 2 hr	Unit 1	Unit 2	Unit 3	Unit 4
Precipitation (in.)	1	2.75	4.5	2.25

SOLUTION

1. The baseflow is separated from the hydrograph by the technique of Section 7.7. From the DRH thus obtained, the unit hydrograph is derived using the procedure of Section 7.9.1. The values are shown in columns 1 and 2, respectively, of Table 7.13.

2. The storm (rainfall) excess sequence is obtained, excluding 60% losses from the precipitation amounts given in the problem. These will be 0.4 in., 1.1 in., 1.8 in., and 0.9 in. for unit 1 through unit 4.

3. The ordinates of the UH are multiplied by the successive values of the rainfall excess, each lagged by the effective duration as shown in columns 3, 4, 5, and 6.

4. The total of columns 3 through 6 results in the DRH. Adding baseflow to this provides the streamflow hydrograph.

Table 7.13 Streamflow Hydrograph from Unit Hydrograph

(1)	(2)	(3)	(4)	(5)	(6)	(7)	(8)	(9)
				DRH Ordinates				
Time (hr)	2-hr UH (cfs)	Unit 1 $0.4 \times$ UH	Unit 2 $1.1 \times$ UH	Unit 3 $1.8 \times$ UH	Unit 4 $0.9 \times$ UH	Total (cfs)	Baseflow (cfs)	Streamflow Hydrograph (cfs)
0	0	0				0	151	151
1	0	0				0	146	146
2	81	32	0			32	146	178
3	277	111	0			111	147	258
4	342	137	89	0		226	147	373
5	321	128	305	0		433	148	581
6	241	96	376	146	0	618	148	766
7	147	59	353	499	0	911	149	1060
8	67	27	265	616	73	981	149	1130
9	27	11	162	578	249	1000	150	1150
10	0	0	74	434	308	816	150	966
11			30	265	289	584	150	734
12			0	121	217	338	150	488
13				49	132	181	150	331
14				0	61	61	150	211
15					24	24	150	174
16					0	0	150	150

EXAMPLE 7.10

During the month of September, the weighted average values of the daily rainfall recorded at various stations in the Housatonic basin above New Milford (CT) are as given below. A representative 24-hour duration unit hydrograph* for the basin is also indicated, along with the baseflow observed at the New Milford site. A study indicates that the infiltration and other losses constitute 85% of the precipitation. Compute the streamflow for the month.

Date	5	8	15	21
Rainfall (in.)	1.21	0.48	3.22	1.02

Time (days)	0	1	2	3	4	5
Unit hydrograph (cfs)	0	1000	2950	125,000	5600	1750
Baseflow (cfs)	120	120	120	125	130	135

* This is obtained from a 1-day segregated rainfall and the corresponding record of streamflow. C. W. Sherman indicated that the actual duration of a 1-day rainstorm is 10 to 13 hours.

Time (days)	6	7	8	9	10	11	12	
Unit hydrograph (cfs)	1000	620	500	350	270	250	0	
Baseflow (cfs)	140	125	120	120	120	120	120	120 for all other days

Solution   Computations are shown in Table 7.14.

Part B: Correlation Technique

7.13 Augmenting Short Streamflows and Filling in Missing Data

The most common techniques for extending data of short duration are (1) precipitation runoff relation, (2) correlation of two sets of streamflow data, and (3) comparison of flow-duration curves. The same methods are applicable for filling in the gaps of a stream-gaging station record. The application of the multiple correlation technique by utilizing meteorological and drainage basin data has been made to develop the generalized relations for the ungaged sites, as described in Part D. Commonly, the technique of rainfall-runoff relation is used to estimate annual flows for missing years, and correlation of the stream-gaging station records is used to extend short-term monthly records.

7.14 Stationary and Homogeneous Check of Data

For the standard statistical techniques to be applied, the long-duration data series should satisfy the conditions of stationariness and homogeneity. When a series is divided into several segments and a statistical parameter such as the mean value is used to characterize the data of each segment, the expected value of the statistical parameter is practically the same for each segment in a stationary series.

The temporal homogeneity check is performed to detect any sudden changes or inconsistency in the data at any time in the period of record. The double-mass curve analysis (Section 2.6.2) is often used for this purpose. The spatial homogeneity is checked to ensure that the data are representative of hydrologically and meteorologically similar areas. The homogeneity of the mean of data is tested by Student's distribution and the homogeneity of the standard deviation (variance) by the Fisher distribution. These tests are described in Section 8.5.

7.15 Precipitation-Runoff Relation for Estimation of Streamflow

The runoff in this context signifies the streamflow. The rainfall and runoff, being parts of the same hydrologic cycle, are intimately correlated. The runoff relates to the flow passing through a section in a stream and it reflects the cumulative effect of the precipitation falling anywhere in the area representing the catchment of that section. An areal average of the precipitation from the measured values at the rain gages within the catchment is determined by Thiessen's polygon or Isohyetal method (Section 2.7). The average monthly or annual precipitation is related to the corresponding short-duration record of monthly or annual runoff (streamflow). The variability of streamflow is not reflected in the relation. Further, the precipitation in the catchment area takes time to reach the point of flow, depending on the

Table 7.14 Estimation of Daily Streamflow by Unit Hydrograph

(1)	(2)	(3)	(4)	(5)	(6)	(7)	(8) DRH (cfs)	(9)	(10)	(11)	(12)
Date	Rainfall (in.)	Losses[a] (in.)	Rainfall Excess (in.)	UH (cfs)	Sept. 5	Sept. 8	Sept. 15	Sept. 21	Total	Baseflow (cfs)	Streamflow (cfs)
1				1,000					0	120	120
2				2,950					0	120	120
3				12,500					0	125	125
4				5,600					0	130	130
5	1.21	1.03	0.18	1,750	180				180	135	315
6				1,000	531				531	140	671
7				620	2,250				2,250	125	2,375
8	0.48	0.41	0.07	500	1,008	70			1,078	120	1,198
9				350	315	207			522	120	642
10				270	180	875			1,055	120	1,175
11				250	112	392			504	120	624
12				0	90	123			213	120	333
13					63	70			133	120	253
14					49	43			92	120	212
15	3.22	2.74	0.48		45	35	480		560	120	680
16					0	25	1,416		1,441	120	1,561
17						19	6,000		6,019	120	6,139
18						18	2,688		2,706	120	2,826
19						0	840		840	120	960
20							480		480	120	600
21	1.02	0.87	0.15				298	150	448	120	568
22							240	443	683	120	803
23							168	1,875	2,043	120	2,163
24							130	840	970	120	1,090
25							120	263	383	120	503
26							0	150	150	120	270
27								93	93	120	213
28								75	75	120	195
29								53	53	120	173
30								41	41	120	161

[a] 85% of Col. 2

distance. Hence the runoff at a stream point may in part be a result of the precipitation that had occurred in the catchment in the past. In relation to runoff, precipitation should include all carryover portions that contribute to a current level of runoff. The antecedent precipitation index (API) is a weighted average of current and antecedent precipitation that is "effective" in correlating with runoff. It is determined by trial and error.

In a monthly correlation study, the precipitation carryover effect extends many months in the past. The procedure is as follows. For each individual month, the correlations are attempted by assuming a relation of the type:

$$Q = b_0 + b_1 X_1 + b_2 X_2 \quad \text{[unbalanced]} \tag{7.19}$$

where

$$Q = \text{mean monthly flow for the month in question}$$
$$X_1, X_2 = \text{values of monthly precipitation for the current month}$$
$$\text{and series of previous months in different combinations}$$
$$b_0, b_1, b_2 = \text{constants obtained from the multiple regression}$$

For example,

$$X_1 = P_i + P_{i-1} + P_{i-2} \quad \text{[L]} \tag{7.20a}$$

$$X_2 = P_{i-3} + P_{i-4} + P_{i-5} \quad \text{[L]} \tag{7.20b}$$

where

$$P_i = \text{average precipitation for the month in question}$$
$$P_{i-1}, P_{i-2}, \ldots = \text{precipitation in previous months}$$

The relationship yielding the best value of the correlation coefficient is adopted for an individual month. The separate relations are derived for each month. The trial nature of the problem, with many possible combinations, makes it desirable to solve this by computer. An element of complexity is introduced in case of a significant groundwater contribution to the streamflow when the precipitation effect may extend years in the past.

The approach becomes less tedious in correlating annual series because the antecedent precipitation index consists of the current and the previous year or, at most, a few additional years. The antecedent precipitation index in an annual correlation study is ascertained by a trial-and-error procedure applying rank analysis. The two steps involved are (1) determining the antecedent precipitation index by rank analysis, and (2) deriving a precipitation-runoff equation by simple regression analysis.

7.15.1 Rank Analysis for Antecedent Precipitation Index (API)

In annual series, the API is that portion of the current year's precipitation and the proportion of the preceding years' precipitation that furnishes the current year's streamflow:

$$P_e = aP_0 + bP_1 + cP_2 + \cdots \quad \text{[L]} \tag{7.21}$$

where

$$P_e = \text{annual API}$$
$$P_0 = \text{current year's precipitation}$$
$$P_1, P_2, \ldots = \text{previous years' precipitation}$$
$$a, b, c, \ldots = \text{coefficient that must add to unity}$$

The trial should start with the two years' precipitation, P_0 and P_1, and should be extended beyond if good correlation is not achieved with these two years. The procedure is explained in the following example.

EXAMPLE 7.11

The Warren River in Warren, Rhode Island, has an annual streamflow record from 1979 to 2005 as shown in Table 7.15. The average annual precipitation values computed from the stations in the drainage basin are also listed. Both the streamflow and precipitation data are checked by the double-mass technique and found to be consistent. Determine the antecedent precipitation index for the basin.

Table 7.15   Annual Flow of the Warren River, Rhode Island

Year:	1979	1980	1981	1982	1983	1984	1985	1986	1987
Streamflow (m^3/s)	82.33	97.68	97.68	31.14	108.72	54.64	50.45	66.53	70.22
Precipitation (m)	0.82	0.79	0.82	0.44	0.91	0.57	0.59	0.67	0.68

Year:	1988	1989	1990	1991	1992	1993	1994	1995	1996
Streamflow (m^3/s)	77.86	89.14	64.84	104.76	31.14	29.73	77.58	58.04	—
Precipitation (m)	0.73	0.78	0.64	0.87	0.46	0.50	0.74	0.61	0.49

Year:	1997	1998	1999	2000	2001	2002	2003	2004	2005
Streamflow (m^3/s)	—	—	—	—	—	—	—	—	—
Precipitation (m)	0.73	0.44	0.46	0.76	0.61	0.49	0.45	0.52	0.60

SOLUTION

1. Refer to Table 7.16.

2. Annual runoff is arranged in column 2 and precipitation in column 4 of the table.

3. The yearly runoff from second row onward is assigned a rank number beginning with the highest runoff as number 1, as shown in column 3. When two values are identical, they are both assigned the average of the two sequence numbers they would have if they were slightly different from each other, as for years 1980 and 1981, and 1982 and 1992.

4. The yearly precipitation has also been assigned a rank number in column 5. The difference in rank between the precipitation and the runoff is squared and noted in column 6. The sum of column 6 is obtained.

5. A formula for antecedent precipitation index is assumed. In the first trial a relation of P_e $= 0.9P_0 + 0.1P_1$ has been assumed and the values based on this formula have been computed in col. 7. The computed precipitation values are again assigned rank numbers from highest to lowest in col. 8, and the difference of the rank has been squared as before. The sum of the squares of the difference in rank, i.e., 5.0 is compared with that for the previous case, i.e., 11.0. A lower value means that the trial relation is better than the previous case.

6. Different trial relations are assumed, gradually raising the proportion of the previous year precipitation, until a lowest point in the sum of the squares is reached. Beyond that

point, the sum of the squares increases with raised proportion of the preceding year's precipitation. For this example the best relation is

$$P_e = 0.8P_0 + 0.2P_1 \quad [\text{L}] \tag{7.22}$$

7.15.2 Correlation of Antecedent Precipitation Index and Runoff by Regression Analysis

The equation of the relation between antecedent precipitation index and runoff is a straight line of the form

$$Q = C + mP_e \quad [\text{unbalanced}] \tag{7.23}$$

where

Q = runoff (streamflow)

C, m = constants representing abstractions

P_e = antecedent precipitation index

The yearly values of antecedent precipitation index computed in Section 7.15.1 are plotted against the annual runoff and a straight line is drawn to average the pattern of plotted points. In many instances it is difficult to draw a line through the shotgun pattern of the points. A line of least-squares fit is drawn by simple regression analysis as described in Chapter 6.

For a straight-line relation, $y = mx + C$, the following set of equations (reproduced from Chapter 6) provide the values of constants C and m and other statistical parameters.

$$C = \frac{\left(\sum y\right)\left(\sum x^2\right) - \left(\sum x\right)\left(\sum xy\right)}{N\left(\sum x^2\right) - \left(\sum x\right)^2} \quad [\text{L}] \tag{6.28}$$

$$m = \frac{N\left(\sum xy\right) - \left(\sum x\right)\left(\sum y\right)}{N\left(\sum x^2\right) - \left(\sum x\right)^2} \quad [\text{dimensionless}] \tag{6.29}$$

Standard variance:

$$S_y^2 = \frac{\sum y^2 - \left(\sum y\right)^2 / N}{N - 1} \quad [\text{L}^2] \tag{6.30}$$

Standard error:

$$S_{yx}^2 = \frac{\sum y^2 - C\sum y - m\sum xy}{N - 2} \quad [\text{L}^2] \tag{6.31}$$

Correlation coefficient:

$$r = \left(1 - \frac{S_{yx}^2}{S_y^2}\right)^{1/2} \quad [\text{dimensionless}] \tag{6.32}$$

where N is the number of pairs of data.

Table 7.16 Computation of Antecedent Precipitation Index

(1)	(2)	(3)	(4)	(5)	(6)	(7)	(8)	(9)	(10)	(11)	(12)	(13)	(14)	(15)
	Runoff		Observed Precipitation			First Assumption			Second Assumption			Third Assumption		
Year	Yearly (m³/s)	Rank	Average Yearly (m)	Rank	Col. $(5-3)^2$	$0.9P_0+0.1P_1$	Rank	Col. $(8-3)^2$	$0.8P_0+0.2P_1$	Rank	Col. $(11-3)^2$	$0.7P_0+0.3P_1$	Rank	Col. $(14-3)^2$
1979	82.33		0.82											
1980	97.68	3.5	0.79	4	0.25	0.79	4	0.25	0.80	4	0.25	0.80	2.5	1.0
1981	97.68	3.5	0.82	3	0.25	0.82	3	0.25	0.81	3	0.25	0.81	1	6.25
1982	31.14	14.5	0.44	16	2.25	0.48	16	2.25	0.52	15	0.25	0.55	15	0.25
1983	108.72	1	0.91	1	0	0.86	1	0	0.82	1.5	0.25	0.77	4.5	12.25
1984	54.64	12	0.57	13	1	0.60	12	0	0.64	11.5	0.25	0.67	9.5	6.25
1985	50.45	13	0.59	12	1	0.59	13	0	0.59	13	0.0	0.58	13.5	0.25
1986	66.53	9	0.67	9	0	0.66	9	0	0.65	10	1.0	0.65	11.5	6.25
1987	70.22	8	0.68	8	0	0.68	8	0	0.68	8	0.0	0.68	7.5	0.25
1988	77.86	6	0.73	7	1	0.73	6	0	0.72	6	0	0.72	6	0
1989	89.19	5	0.78	5	0	0.78	5	0	0.77	5	0	0.77	4.5	0.25
1990	64.84	10	0.64	10	0	0.65	10	0	0.67	9	1.0	0.68	7.5	6.25
1991	104.76	2	0.87	2	0	0.85	2	0	0.82	1.5	0.25	0.80	2.5	0.25
1992	31.14	14.5	0.46	15	0.25	0.50	14.5	0	0.54	14	0.25	0.58	13.5	1.0
1993	29.73	16	0.50	14	4	0.50	14.5	2.25	0.49	16	0	0.49	16	0
1994	77.58	7	0.74	6	1	0.72	7	0	0.69	7	0	0.67	9.5	6.25
1995	58.04	11	0.61	11	0	0.62	11	0	0.64	11.5	0.25	0.65	11.5	0.25
Total					11.00			5.0			4.0			47.0

EXAMPLE 7.12

For Example 7.11, determine the precipitation-runoff relation by **(a)** graphic procedure, **(b)** by regression analysis, and **(c)** using this relation, extend the streamflow record through 2005.

SOLUTION

(a) Precipitation-runoff relation by graphic procedure

1. Refer to Table 7.16. Runoff data are given in column 2 and the best antecedent precipitation index data in column 10 of the table.

2. These data are plotted in Figure 7.17, which is a straight line having a slope of 241:1.

3. From Fig. 7.17, for $y = 88$, $x = 0.75$

    or $88 = 241(0.75) + C$

    thus $C = -92.75$

(b) Regression analysis is arranged in Table 7.17.

4. From eq. (6.28),

$$C = \frac{(\Sigma y)(\Sigma x^2) - (\Sigma x)(\Sigma xy)}{N(\Sigma x^2) - (\Sigma x)^2}$$

$$= \frac{(1110.2)(7.53) - (10.85)(794.47)}{16(7.53) - (10.85)^2}$$

$$= -94.34$$

5. From eq. (6.29),

$$m = \frac{N(\Sigma xy) - (\Sigma x)(\Sigma y)}{N(\Sigma x^2) - (\Sigma x)^2}$$

$$= \frac{16(794.47) - (10.85)(1110.2)}{16(7.53) - (10.85)^2}$$

$$= 241.42$$

6. From eq. (6.30),

$$S_y^2 = \frac{\Sigma y^2 - (\Sigma y)^2 / N}{N - 1}$$

$$= \frac{(87.18 \times 10^3) - (1110.2)^2 / 16}{16 - 1}$$

$$= 676.4$$

7. From eq. (6.31),

$$S_{yx}^2 = \frac{\Sigma y^2 - C \Sigma y - m \Sigma xy}{N - 2}$$

$$= \frac{(87.18 \times 10^3) - (-94.34)(1110.2) - (241.42)(794.47)}{16 - 2}$$

$$= 8.24$$

Figure 7.17 Precipitation-runoff relation of Example 7.12.

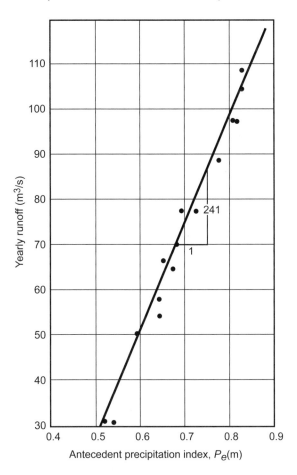

8. From eq. (6.32),

$$r = \left(1 - \frac{8.24}{676.4}\right)^{1/2}$$

$$= 0.99 \text{ (good correlation)}$$

9. Thus the correlation equation is

$$Q = 241.42\, P_e - 94.34 \qquad (7.24)$$

(c) Extension of streamflow record

10. Using eq. (7.22), the average annual precipitation is converted to the antecedent precipitation index as per column 2 of Table 7.18.

11. By eq. (7.24), the streamflow is ascertained from the effective precipitation as shown in column 3 of Table 7.18.

Table 7.17   Regression Analysis for Precipitation-Runoff Correlation

No.	$Q = y$	$P_e = x$	x^2	xy	y^2 $(\times 10^3)$
1	97.68	0.80	0.64	78.14	9.54
2	97.68	0.81	0.66	79.12	9.54
3	31.14	0.52	0.27	16.19	0.97
4	108.72	0.82	0.67	89.15	11.82
5	54.64	0.64	0.41	34.97	2.99
6	50.45	0.59	0.35	29.77	2.54
7	66.53	0.65	0.42	43.24	4.43
8	70.22	0.68	0.46	47.75	4.93
9	77.86	0.72	0.52	56.06	6.06
10	89.19	0.77	0.59	68.68	7.95
11	64.84	0.67	0.45	43.44	4.20
12	104.76	0.82	0.67	85.90	10.97
13	31.14	0.54	0.29	16.82	0.97
14	29.73	0.49	0.24	14.57	0.88
15	77.58	0.69	0.48	53.53	6.02
16	58.04	0.64	0.41	37.14	3.37
Σ	1110.2	10.85	7.53	794.47	87.18

Table 7.18   Computation of Runoff from Precipitation

(1)	(2)	(3)
Year	Antecedent precipitation index (m)	Streamflow (m³/s)
1996	0.51	28.78
1997	0.68	69.82
1998	0.50	26.37
1999	0.46	16.71
2000	0.70	74.65
2001	0.64	60.17
2002	0.51	28.78
2003	0.46	16.71
2004	0.51	28.78
2005	0.58	45.68

7.16 CORRELATION OF GAGING-STATION RECORDS FOR ESTIMATION OF STREAMFLOW

The technique of correlation, which is a process of establishing a mutual relation between a variable and related variable(s), has been used in the rainfall-runoff relationship. Another convenient application is to correlate directly stream-gaging records between two or more stations, one with short-term data and others with long-term records. The relation thus established, based on the concurrent data corresponding to the short period of the record, is used to obtain correlative estimates of discharges for the short-duration station from the records of the long-duration station outside the common period.

The runoff depends on the climate and the basin characteristics. The complex interrelations of climate and drainage basin characteristics are integrated in the flow of the stream, and their aggregate effect is measured directly at the stream-gaging station (Searcy, 1960). Evaluating the effect of various drainage and climatic factors in order to predict the runoff is a difficult task. The direct comparison of their result (i.e., streamflows) is more convenient. However, there has to be a common basis for this comparison, which is provided by the similarity of the climate and certain common characteristics of the basins being compared. Thus a correlation between a mountain stream and a desert stream may not be satisfactory. The degree of reliability of the relation will decrease as the distance between two gaging stations increases. When two basins differ greatly in size (by more than 10 times), they can be poorly related.

7.16.1 Simple Correlation

Although the relations between the flows can be based on concurrent daily, weekly, monthly, or annual discharges, monthly flow is a commonly used duration. The relation between two stream-gaging records is usually expressed by two curves, one for the high flow and one for the low flow, and occasionally by a third curve at extremely low flows. Very seldom is only one curve adequate to represent the relation. It is thus desirable as a first step to plot the discharge of one station against the corresponding discharge at the other station to identify the position of changed relations (shift in the curve). The logarithmic plot (converting discharges to their log values or using log-log paper) is convenient, as it tends to transform common curvilinear relations to straight lines and to normalize the streamflow data.

The fitting of the curve (line) through the scatter of the data is performed either graphically or numerically. In the graphic procedure (Searcy, 1960), the plot is first divided into a number of slices of equal width using the vertical lines, and the median of the points in each slice is determined in both x and y directions. Then, using the horizontal lines, the plot is divided into a number of equally spaced slices and the median of the points in each slice is determined again in both x and y directions. The straight line(s) are drawn through the average of the median points, giving reduced weight to the endpoints.

In the numerical procedure, a linear regression for a short sequence (one station) and a long sequence (another station) of streamflows is used to lengthen the short sequence. If the observed events (streamflows) for the long and short sequences are represented as

$$x_1, x_2, \ldots, x_{N_1}, x_{N_1+1}, \ldots, x_{N_1+N_2}$$

$$y_1, y_2, \ldots, y_{N_1}$$

then the estimate of y_i beyond N_1 is obtained from the following regression of y on x, using the known values of x_i,

$$y_i = mx_i + C \quad [\text{L}^3\text{T}^{-1}] \tag{7.25}$$

where y_i and x_i are short- and long-term sequences, respectively. The values of C and m are calculated from eqs. (6.28) and (6.29).

The estimated values of y_i from eq. (7.25) tend to yield a smaller variance than the real observations. To "preserve" the variance that is inherent in the observed values, a random component is added to the regression estimates (Matalas and Jacobs, 1964). This component, referred to as *noise*, is normally distributed with zero mean and variance proportional to the variance of the short sequence. The noise is shown to have no effect on the reliability of the estimate of the mean, but improves the reliability of the variance for the lengthened sequence. Including the noise term, the equation for streamflow estimates is given by

$$y_i = mx_i + C + \sqrt{1 - r^2}\, S_y e_i \quad [\text{L}^3\text{T}^{-1}] \tag{7.26}$$

where

r = product-moment correlation coefficient given by eq. (6.32)

S_y = standard deviation of y from short sequence by eq. (6.30)

e_i = random normal variable with zero mean and unit variance (Table 7.23)

The strength (goodness) of the regression analysis is measured by the product-moment correlation coefficient, r. Table 7.19 gives the minimum values of r for the means. When the computed r by eq. (6.32) is equal to or larger than the value from the table for the length of the short sequence, the lengthened sequence, by including estimated values, provides a better estimate of the mean than does the short sequence alone. Matalas and Jacobs (1964) suggested similar tables of minimum r for the variance with and without the noise component. If the correlation coefficient exceeds 0.8, the noise component need not be added to get a reliable estimate of the variance. Thus eq. (7.25) can be used to estimate streamflows, since the use of pseudorandom numbers is not too appealing, with every investigator deriving different values of streamflows (although within the limits expected by chance). The procedure is exactly similar to Section 7.15.2 as explained in Example 7.12. Streamflow data, y_i, from the station with short sequence are equivalent to Q values, and the longer streamflow sequence, x_i, from the other station is equivalent to P_e in the example.

Table 7.19   Minimum Values of the Correlation Coefficient for the Mean

Period of short record	10	15	20	25	30	35
Minimum r	0.35	0.28	0.24	0.21	0.19	0.17

Source: Matalas and Jacobs (1964).

7.17 ESTIMATION OF STREAMFLOW BY COMPARISON OF FLOW-DURATION CURVES

A flow-duration curve is a plot between discharge and percent of time that discharge is equaled or exceeded in a stream. The analysis to prepare a flow-duration curve is described in detail in Section 7.27.2. The shape of the curve may change depending on the length of the streamflow record. This characteristic is used to extend the data at a site with a short-term record, provided that long-term data are available at another site having similar hydrologic conditions. The following procedure can be applied:

1. Construct the flow-duration curve for the site in question (site A) based on the short-term data.

2. For an adjacent site B, construct two flow-duration curves, one based on the long-term data and one based on the data for the period concurrent to the short-term record of the first site.

3. Compare the flow-duration curves based on the short-term record for site A and site B, determine the ratios of the site A curve to the site B curve at different percents of exceedance. Apply these ratios to the long-term flow-duration curve of site B to produce an adjusted curve representing the long period of record at site A.

4. For a recorded discharge on site B corresponding to which discharge at site A is not available, from the long-term flow-duration curve of site B, read the percent time the value is exceeded from the abscissa. For the same percent exceedance, ascertain the discharge at site A from the adjusted curve of site A. If the curves of site A and site B are drawn on the same graph paper, this conversion can be obtained directly.

EXAMPLE 7.13

Stream-gaging station A has a short-term streamflow record of 7 years, from 1967 to 1973, as compared with stream-gaging station B, which has an up-to-date record from 1967 through 2005 of 39 years. The mean monthly flow-duration curves prepared from the data of 1967 through 1973 for both stations have been summarized in Table 7.20. The ordinates of the duration curve for station B based on the entire period of record are also indicated. Construct the flow-duration curve for site A representing the long period of record. If a flow of 2000 cfs was recorded in May 2005 for station B, what was the corresponding flow at site A?

Table 7.20   Mean Monthly Flow-Duration Curves

Percent Equaled or Exceeded	Streamflow Based on Short-Term Record (cfs)		Streamflow Based on Long-Term Record for Station B (cfs)
	Station A	Station B	
10	1500	2880	2520
20	1280	2500	2380
30	1160	2200	2250
40	1080	1960	2100
50	1000	1760	1960
60	960	1600	1800
70	900	1460	1660
80	850	1300	1520
90	800	1160	1380
95	760	1080	1300

Table 7.21   Correlation of Long-Term Flow-Duration Curves

(1)	(2)	(3)	(4)	(5)	(6)
	Short-Term Record (cfs)			Long-Term Record (cfs)	
Percent Equaled or Exceeded	Station A	Station B	Ratio $\frac{A}{B}$	Station B	Station A (col. 5 × col. 4)
10	1500	2880	0.52	2520	1310
20	1280	2500	0.51	2380	1214
30	1160	2200	0.53	2250	1193
40	1080	1960	0.55	2100	1155
50	1000	1760	0.57	1960	1117
60	960	1600	0.60	1800	1080
70	900	1460	0.62	1660	1029
80	850	1300	0.65	1520	988
90	800	1160	0.69	1380	952
95	760	1080	0.70	1300	910

SOLUTION   Refer to Table 7.21.

1. Column 4 determines the ratio of station A to station B for various percents of exceedance based on a short-term record.

2. This ratio is applied to the data of station B in column 5 to derive the adjusted flow of station A in column 6 for a long-term period.

3. The flow-duration curves for stations A and B, based on the data of columns 5 and 6, are plotted in Figure 7.18.

4. Corresponding to a discharge of 2000 cfs for station B, the percent exceedance is 47%. For the same exceedance level of 47%, the discharge for station A is read from Figure 7.18. Thus $Q = 1120$ cfs.

Part C: Synthetic Technique

7.18 GENERATION OF SYNTHETIC STREAMFLOWS

A very common constraint encountered in the context of water resources planning is inadequacy of streamflow records. The available streamflows, known from historical records, are often quite short, generally less than 25 years in length. These do not cover even the economic life of a project of 50 to 100 years. The same flow sequence is not likely to be repeated within the span of 100 years. A system designed on the basis of the historical record faces a chance of being inadequate for the unknown flow sequence that the system might experience. Further, the historical record comprising a single short series does not cover a sequence of low flows as well as high flows. The reliability of a system has to be evaluated under these conditions, which is not possible with the historical records alone. In a statistical sense, the historical record is a sample out of a population of natural streamflow process. If this process is considered stationary, that is, the statistical properties of a

Figure 7.18   Long-term flow-duration curves.

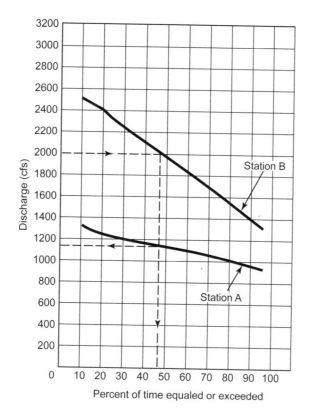

series do not change with time, then many series representing such samples can be formulated that will be statistically similar to the historical record. This is the basis of the synthetic technique. Thus the generated flows are neither historical flows nor a prediction of future flows but are representative of likely flows in a stream.

7.19 HYDROLOGIC TIME SERIES AND STOCHASTIC PROCESS

A sequence of values of any variable represented with respect to time is known as a *time series*. The time can be a discrete value, a time interval, or a continuous function. The hydrologic data of streamflows, precipitation, groundwater or lake levels, water temperatures, or oxygen concentration fall under the category of time series. These data can be deterministic, random, or a combination of the two. Streamflow, being a natural phenomenon, has a random component. However, it is not fully random, since it has been observed that a low flow tends to follow another low flow, and a high flow has a similar pattern. The word *stochastic* is used to denote the randomness in statistics, but in hydrology it refers to a partial random sequence as well. Thus the streamflow data represent a time series involving a stochastic process.

The analysis of a time series, in the time domain, is performed by a parameter known as the serial correlation coefficient or the autocorrelation coefficient. This parameter indicates the dependence in successive values of a time series. This coefficient is determined not only for successive values (elements) of a time series but for elements that are various time intervals apart, known as the lag period. A graph of the autocorrelation coefficient against the lag period is known as the correlogram. If a correlogram shows zero or nearly zero values for all lag periods, the process is purely random. A value close to 1 will suggest a dominating deterministic process.

The analysis of a time series in the frequency domain is done by the spectral density that identifies the cyclic nature or periodicity in the series. The density indicates the cycle in the deterministic data. In a purely random process it oscillates randomly. The purpose of streamflow synthesis, however, is not to analyze a time series but to generate the data based on the series. This does not require the decomposition of the time series by the analysis above but an understanding of its statistical properties to reproduce series of similar statistical characteristics. Various stochastic processes used for generating the hydrologic data are discussed in the following section.

7.20 MARKOV PROCESS OR AUTOREGRESSIVE (AR) MODEL

The Markov process considers that the value of an event (i.e., streamflow) at one time is correlated with the value of the event at an earlier period (i.e., a serial or autocorrelation exists in the time series). In a first-order Markov process, this correlation exists in two successive values of the event. The first-order Markov model, which constitutes the classic approach in synthetic hydrology, states that the value of a variable x in one time period is dependent on the value of x in the preceding time period plus a random component. Thus the synthetic flows for a stream represent a sequence of numbers, each of which consists of two parts:

$$x_i = d_i + e_i \quad [\text{L}^3\text{T}^{-1}] \qquad (7.27)$$

where

x_i = flow at ith time (ith number of a time series)

d_i = deterministic part at ith time

e_i = random part at ith time

The values of x_i are tied up with the historical data by ensuring that they belong to the same frequency distribution and possess similar statistical properties (mean, deviation, skewness) as the historical series.

The general Markov procedure of data synthesis comprises: (1) determining statistical parameters from the analysis of the historical record, (2) identifying the frequency distribution of the historical data, (3) generating random numbers of the same distribution and statistical characteristics, and (4) constituting the deterministic part considering the persistence (influence of previous flows) and combining with the random part. The various forms and combinations of deterministic and random components are recognized as different models. The simplest of these is a single season (annual) flow model of lag 1, which assumes that the magnitude of the current flow is significantly correlated with the previous flow value only. Multiple-season models divide the yearly flow into seasons or months.

Multilag models have a long memory; that is, they consider the influence on the current flow to extend beyond the previous year alone. When the flows at the various locations are strongly correlated, multisite models are needed that involve comparatively complicated mathematics and often include simplifying assumptions.

7.20.1 Statistical Parameters of Historical Data

The four parameters that are important in a synthetic study are (1) mean flow, (2) standard deviation, (3) coefficient of skewness, and (4) correlation coefficient. These parameters, described in Chapter 5, are reproduced here.

$$\overline{X} = \frac{1}{n}\sum_{i=1}^{n} x_i \tag{5.26}$$

where

$\overline{X}$ = mean observed (historical) flow

n = total numbers (values) of flow

$\mu = E(\overline{X})$ (i.e., $\overline{X}$ is expected to be equal to the population mean μ as

n tends to infinity)

x_i = ith number of observed flow

The sample estimate of the variance or standard deviation, S, which is a measure of the variability of the data, is given by

$$S^2 = \frac{1}{n-1}\sum_{i=1}^{n}\left(x_i - \overline{X}\right)^2 \tag{5.27}$$

The sample coefficient of skewness, g, which is a measure of the lack of symmetry, is given by

$$g = \frac{n\sum_{i=1}^{n}\left(x_i - \overline{X}\right)^3}{(n-1)(n-2)S^3} \quad \text{[dimensionless]} \tag{5.28}$$

The serial correlation coefficient is a measure of the extent to which a flow at any time is affected by the flow at another time. The K-lag coefficient, in which the effect extends by K time units, is given by

$$r_K = \frac{\sum_{i=1}^{n-K}\left(x_i - \overline{X}\right)\left(x_{i+K} - \overline{X}\right)}{(n-K)S^2} \quad \text{[dimensionless]} \tag{5.30}$$

The one-lag serial coefficient, in which the current flow is affected only by the previous flow, can be obtained by substituting $K = 1$ in eq. (5.30). The additional lags should be included as long as they produce a model that explains more about the pattern of flows than does one with fewer lags (Fiering and Jackson, 1971).

7.20.2 Identifying the Distribution of Streamflow Data

The historic data are expected to follow a theoretical frequency distribution. The theory of frequency distributions is described in Chapter 8. In streamflow generation, the distributions of interest are normal, lognormal, and gamma families. The bell-shaped normal distribution is most extensively used in statistical applications because the sum of variables derived from any distribution tends to be distributed normally according to the central limit theorem. The procedure described in Section 5.10 in the context of groundwater well samples can be used to test for normality or lognormality of streamflow data, too. In a simple test for normality, the historic data are arranged in ascending order. For each value x_i, the percent probability is computed by $100i/(n+1)$ where i is the rank of value x_i and n is the number of historic values. If the plot is a straight line, the distribution is normal. Also, the coefficient of skewness should be close to zero, since the normal distribution has no skewness.

If the data do not fit a normal distribution, the second distribution tried is lognormal distribution. To test for lognormality, either the logarithms of the flows versus probability are plotted on normal arithmetic probability paper or the flows are directly plotted on log probability paper against the probability percentages computed as discussed above. A straight-line plot indicates the lognormal distribution is correct. Although the skewness computed from the logarithms of the values should be close to zero, the lognormal distribution for the original data is positively skewed, which is a characteristic of many hydrologic variables. Since small changes in low values produce comparatively large changes in their logarithmic values, this distribution is especially preferred in low-flow studies.

When the historical records of flows or logarithms of flows show appreciable skewness, the gamma distribution is used, which has a distinctly positive skewness. This distribution should, however, be excluded when multiple lags exist (i.e., when a flow is affected by many previous flows). Often, the historical data do not clearly fit any of these distributions. The choice has to be made based on the purpose of the project, economics, and any other considerations.

EXAMPLE 7.14

The annual flows for the Lamprey River near Newmarket, New Hampshire, are given in column 2 of Table 7.22. Compute the statistical parameters for this flow sequence.

SOLUTION

1. From eq. (5.26), mean flow:

$$\bar{X} = \frac{1}{21}\sum_{i=1}^{21} x_i = \frac{1}{21}\sum \text{col. 2} = \frac{13{,}711}{21} = 653 \text{ cfs}$$

2. From eq. (5.27), variance:

$$S^2 = \frac{1}{21-1}\sum_{i=1}^{21}\left(x_i - \bar{X}\right)^2 = \frac{1}{20}\sum \text{col. 4} = \frac{696{,}690}{20} = 34{,}834.5$$

$$S = 186.6 \text{ cfs}$$

3. From eq. (5.28), coefficient of skewness:

$$g = \frac{21(\sum \text{col. 5})}{20(19)(186.6)^3} = \frac{21\left(11{,}661\times 10^4\right)}{20(19)(186.6)^3} = 1$$

4. From eq. (5.30), lag 1 serial correlation coefficient: The value in column 6 is obtained by multiplying the number in column 3 by its successive value:

$$r_1 = \frac{\sum_{1}^{21-1}\left(x_i - \bar{X}\right)\left(x_{i+1} - \bar{X}\right)}{(20)S^2}$$

$$= \frac{\sum \text{col.6}}{(20)S^2} = \frac{2555.6\times 10^2}{20(186.6)^2} = 0.37$$

Table 7.22 Annual Flows and Computation of the Statistical Parameters for the Lamprey River, New Jersey

(1)	(2)	(3)	(4)	(5)	(6)	(7)
i	Annual Flow, x_i (cfs)	$x_i - \bar{X}$	$(x_i - \bar{X})^2$	$(x_i - \bar{X})^3$ ($\times 10^4$)	$(x_i - \bar{X})(x_{i+1} - \bar{X})$ ($\times 10^2$)	$(x_i - \bar{X})(x_{i+2} - \bar{X})$ ($\times 10^2$)
1	472	−181	32,761	−593	432.6	155.7
2	414	−239	57,121	−1,365	205.5	614.2
3	567	−86	7,396	−63.6	221	9.5
4	396	−257	66,049	−1,697.5	28.3	447.2
5	642	−11	121	−0.1	19.1	17.3
6	479	−174	30,276	−526.8	273.2	212.3
7	496	−157	24,649	−387	191.5	−249.6
8	531	−122	14,884	−181.6	−194	−297.7
9	812	159	25,281	402	388	−256.0
10	897	244	59,536	1,452.7	−392.8	−65.9
11	492	−161	25,921	−417.3	43.5	77.3
12	626	−27	729	−2.0	13	−32.7
13	605	−48	2,304	−11.0	−58	−249.1
14	774	121	14,641	177.2	628	119.8
15	1,172	519	269,361	13,980	513.8	980.9
16	752	99	9,801	97	187	16.8
17	842	189	35,721	675	32.1	139.9
18	670	17	289	0.5	12.6	17.3
19	727	74	5,476	40.5	75.5	−46.6
20	755	102	10,404	106	−64.3	
21	590	−63	3,969	−25		
Σ	13,711		696,690	$11,661 \times 10^4$	2555.6×10^2	1610×10^2

5. Lag 2 serial correlation coefficient: The value in column 7 is obtained by multiplying the column 3 value by the value two time intervals following, in the same column.

$$r_2 = \frac{\sum\limits_{i=1}^{21-2} (x_i - \bar{X})(x_{i+2} - \bar{X})}{(n-2)S^2}$$

$$= \frac{\sum \text{col. }7}{19(186.6)^2} = \frac{1610 \times 10^2}{19(186.6)^2} = 0.24$$

Hence lag 1 is better related.

7.20.3 Generating Random Numbers

The source of random numbers in a simulation study is either a computer-based pseudorandom-number generator or random number tables. The tables provide a set of random numbers drawn from the uniform distribution or from the normal distribution. A set of normally distributed random numbers are given in Table 7.23. The random number should belong to the same distribution to which the historical record belongs for the generated flow to have similar characteristics. It is a simple matter to convert a sequence of uniformly distributed random numbers to normal random numbers if only a uniform distribution table exists. According to the central limit theorem, the numbers formed by summing up many (say, 12) consecutive uniformly distributed numbers will be approximately normally distributed. The normally distributed random numbers for normal or lognormal flow sequences are thus conveniently available. For flows that are distributed as gamma variates, use the Wilson and Hilferty (Fiering and Jackson, 1971, p. 53) formula that incorporates the skewness coefficient and the serial correlation coefficient.

Further, the normal random numbers given in Table 7.23 have a zero mean and a standard deviation of one, known as the *standard normal deviates*. A transformed random variable with zero mean and variance of random variables of S_e^2 can be given by

$$n_i = S_e t_i \quad \text{[dimensionless]} \tag{7.28}$$

where

t_i = standard normal variate (with mean zero and variance of one from Table 7.23)

n_i = transformed random variable with zero mean and

variance of random numbers of S_e^2

S_e = standard deviation of random numbers

7.20.4 Deterministic and Random Components

Streamflows show persistence as reflected in their flow pattern. The Markovian process or autoregressive (AR) model considers that this persistence is indicated through the serial correlation within the sequence. Thus in the p-order model the effect runs through p terms, and the autoregressive model, AR(p), takes the form

$$a_i = \phi_{p,1}\, a_{i-1} + \phi_{p,2}\, a_{i-2} + \phi_{p,3}\, a_{i-3} + \ldots + \phi_{p,p}\, a_{i-p} + n_i \quad \text{[dimensionless]} \tag{7.29}$$

Table 7.23 Standard Normal Random Sampling Deviates

	0	1	2	3	4	5	6	7	8	9
0	-0.523	0.611	-0.359	-0.393	0.084	-0.931	-0.027	0.798	1.672	-1.077
1	-1.536	-0.454	0.071	-2.129	1.525	0.261	2.319	0.972	0.767	-2.849
2	-0.121	0.968	-1.943	0.581	-0.711	-0.060	-0.482	-0.746	-0.747	1.254
3	-0.542	-0.807	0.168	0.839	-0.756	-0.453	-1.912	0.766	-0.890	0.205
4	0.131	-0.859	-1.096	-0.785	0.310	1.314	-0.231	0.029	1.819	-1.602
5	-0.234	0.551	0.743	-0.900	0.435	-2.999	0.212	0.869	-0.716	-0.410
6	-1.010	1.347	0.230	0.009	-1.495	2.145	-1.033	0.729	0.309	0.920
7	0.273	-0.885	-0.016	0.775	-1.740	0.353	-1.519	0.958	-0.448	2.185
8	-0.102	-1.111	-0.585	1.461	-0.307	1.489	-0.196	0.506	-0.662	-1.175
9	0.368	-0.710	0.407	0.066	-0.617	-0.580	0.107	-2.247	1.616	-1.060
10	-1.762	1.382	1.142	-2.056	-0.400	-1.701	-0.914	-1.000	-0.172	0.903
11	0.306	-0.607	-0.324	1.171	1.016	-1.829	1.723	-0.513	-0.657	2.011
12	-0.465	-1.214	-0.174	0.894	0.245	-0.987	-1.155	0.592	-0.411	-0.109
13	-0.004	-0.029	-0.633	0.004	-0.603	1.104	-0.655	1.191	0.938	-0.805
14	0.593	0.252	-0.541	0.318	1.268	1.972	0.875	-1.030	-1.175	0.445
15	0.233	0.430	-0.331	-1.272	-0.289	-0.060	-0.754	0.789	0.546	0.687
16	0.571	-0.215	-1.090	0.610	-0.810	-0.364	-1.282	0.010	0.586	0.926
17	0.370	0.976	1.017	1.106	0.441	-2.376	0.793	0.016	-0.704	0.146
18	-0.009	-1.285	-0.346	-0.323	0.609	-0.373	0.078	-1.034	0.153	0.997
19	0.416	-0.131	0.668	0.662	-1.835	1.646	0.197	0.131	0.783	0.076

Source: Fiering and Jackson (1971).

where

$$a_i = i\text{th variable of the sequence (stochastic}$$
$$\text{component) of zero mean and unit variance}$$
$$n_i = \text{random number at } i\text{th time}$$
$$\phi_{p,1}, \phi_{p,2}, \ldots = \text{autoregressive parameters or weights}$$

The stochastic component can be related to the flow at any time by

$$x_i = \bar{X} + Sa_i \quad [\text{L}^3\text{T}^{-1}] \tag{7.30}$$

where $\bar{X}$ and S are the mean and variance of the historical flow sequence.

The first-order autoregressive model AR(1), commonly known as a Markov model, reduces to the form

$$a_i = \phi_{1,1} a_{i-1} + n_i \quad [\text{dimensionless}] \tag{7.31}$$

When the conditions of mean $a_i = 0$, variance $a_i = 1$, and expectation $E(n_i a_{i-1}) = 0$ are included for the sequence a, the following relations are derived:

$$\phi_{1,1} = r_1 \quad [\text{dimensionless}] \tag{7.32}$$

where r_1 is the lag 1 autocorrelation coefficient and

$$S_e^2 = 1 - r_1^2 \quad [\text{dimensionless}] \tag{7.33}$$

Section 7.20   Markov Process or Autoregressive (AR) Model

7.20.5 Formulating the Markov Model

When equations (7.30), (7.31), (7.28), (7.32), and (7.33) are combined, we arrive at the following Markov model for annual flow comprising deterministic and random parts.

$$x_i = \underbrace{\bar{X} + r_1\left(x_{i-1} - \bar{X}\right)}_{\text{deterministic}} + \underbrace{S\sqrt{\left(1 - r_1^2\right)}t_i}_{\text{random}} \quad [\text{L}^3\text{T}^{-1}] \tag{7.34}$$

where

x_i = streamflow at ith time

$\bar{X}$ = mean of recorded flows

r_1 = lag 1 serial or autocorrelation coefficient

S = standard deviation of recorded flows

t_i = random variate from an appropriate distribution
with a mean of zero and variance of unity

i = ith position in series from 1 to N years

The procedure to generate a series of flows is as follows:

1. Determine the mean flow, variance, coefficient of skewness, and lag 1 serial correlation coefficient from the historical record and identify the distribution.

2. Pick up the random numbers from a generator or a table. To use the table, close your eyes and place a pencil point at any number on any page. The random numbers are taken consecutively across the row succeeding the number selected.

3. Starting with x_{i-1} equal to $\bar{X}$, determine x_i, which becomes x_{i-1} to compute the next value. This is a cyclic computation.

4. If the generated flow is negative, use this to generate the next flow and then discard it from the series.

5. To neglect the effect of the starting condition, discard the first 25 to 50 generated flows in the series.

6. If the distribution is lognormal, use the logarithm of the values and finally convert back to flows. Other procedures for generating lognormal flows are given by Beard (1965) and Matalas (1967). The gamma distribution needs conversion of random variates, as previously stated.

A model on the same lines for monthly flows, developed by Thomas and Fiering, has the following form (Maass et al., 1962, p. 467):

$$q_{i,j} = \bar{q}_j + b_j\left(q_{i-1,j-1} - \bar{q}_{j-1}\right) + t_{i,j}S_j\left(1 - r_j\right)^{1/2} \quad [\text{L}^3\text{T}^{-1}] \tag{7.35}$$

where

$i =$ month in series, measured from the beginning

$j =$ month in year, $j = 1,2,\ldots,12$ for January to December

$q_{i,j} =$ flow in ith month from the beginning, for jth month of the year

$q_{i-1,\,j-1} =$ immediate previous month

$\bar{q}_j =$ mean of flows of jth month (12 values)

$b_j =$ regression coefficient of flows of jth month and flows of

$(j-1)$th month is equal to $r_j S_j / S_j - 1$ (12 values)

$r_j =$ correlation coefficient between flows of jth month and

$(j-1)$th month (12 values)

$S_j =$ standard deviation for jth month (12 values)

$t_{i,\,j} =$ random normal deviate of zero mean and unit standard deviation

Models of monthly flows are handled expediently by a computer. The Hydrologic Engineering Center of the U.S. Army Corps of Engineers has prepared a very versatile model, HEC-4, that is capable of analyzing monthly streamflows at 10 stations simultaneously to produce a sequence of hypothetical flows with maximum, minimum, and average values for each month.

EXAMPLE 7.15

Generate a synthetic annual series similar to the historical record of Example 7.14. Assume the normal distribution.

SOLUTION

1. From Example 7.14, $\bar{X} = 653$ cfs, $S = 186.6$ cfs, and $r_1 = 0.37$.
2. Use Table 7.23 for standard normal deviates, t_i.
3. Using eq. (7.34), values for x_i are computed in Table 7.24, starting with $x_{i-1} = \bar{X}$, in column 2.

7.21 AUTOREGRESSIVE-MOVING AVERAGE (ARMA) MODEL

In the autoregressive models of Section 7.20, the flows generated depend on the preassigned number of past flow values and a random variate. There is another category, known as the moving average (MA) model, that considers a stochastic component (streamflow event) to be a constituent of a number of random variates, with the current variate assigned a weight of unity and random variates generated at antecedent times multiplied by the assigned factors. Such models are generally inappropriate for direct application to hydrology. A moving average model is, however, combined with an autoregressive model to produce a mixed model known as an ARMA model, which appropriately represents the effect of linear aquifer, linear reservoir, and independent rainfall pattern. One of the first applications of ARMA models in hydrography was made by Carlson et al. (1970).

Table 7.24 Computation of Annual Flow by One-Lag Markov Model

(1)	(2)	(3)	(4)	(5)	(6)	(7)	(8)
				$\bar{X}+r_1\left(x_{i-1}-\bar{X}\right)$ Deterministic	t_i	$t_i S\sqrt{1-r_1^2}$ Random	
i	x_{i-1}	$x_{i-1}-\bar{X}$	$r_1(x_{i-1}-\bar{X})$	Component	(Table 7.23)	Component	x_i
1	653	0.0	0.0	653	0.131[a]	22.7	675.7[b]
2	675.7[c]	22.7	8.4	661.4	−0.859	−148.9	512.5
3	512.5	−140.5	−52.0	601.0	−1.096	−190.0	411.0
4	411.0	−242.0	−89.5	563.5	−0.785	−136.1	427.4
5	427.4	−225.6	−83.5	569.5	0.310	53.7	623.2
6	623.2	−29.8	−11.0	642.0	1.314	227.8	869.8
7	869.8	216.8	80.2	733.2	−0.231	−40.0	693.2
8	693.2	40.2	14.9	667.9	0.029	5.0	672.9
9	672.9	19.9	7.4	660.4	1.819	315.3	975.7
10	975.7	322.7	119.4	772.4	−1.602	−277.7	494.7
11	494.7	−158.3	−58.6	594.4	−0.234	−40.6	553.8
12							
.							
.							
.							

[a] picked randomly
[b] col. 5 + col. 7
[c] col. 8 of previous step

ARMA (p, q) consists of two polynomials of order p and q, respectively, as follows:

$$a_i = \phi_{p,1}\, a_{i-1} + \phi_{p,2}\, a_{i-2} + \dots + \phi_{p,p}\, a_{i-p} + n_i - \theta_{q,1}\, n_{i-1} - \theta_{q,2}\, n_{i-2} - \cdots$$
$$- \theta_{q,q}\, n_{i-q} \quad \text{[dimensionless]} \tag{7.36}$$

where $\theta_{q,1}, \theta_{q,2}, \dots$ are the random variate weights.

Estimation of the parameters ϕ and θ is not a straightforward procedure. In the class of mixed models, the simplest is the ARMA (1, 1) model given by

$$a_i = \phi_{1,1}\, a_{i-1} + n_i - \theta_{1,1}\, n_{i-1} \quad \text{[dimensionless]} \tag{7.37a}$$

or

$$a_i = \phi_{1,1}\, a_{i-1} + S_e(t_i - \theta_{1,1}\, t_{i-1}) \tag{7.37b}$$

The a_i values are converted to flows, x_i by eq. (7.30).

From the known statistical conditions of the series in eq. (7.37), the following relations are obtained:

$$\phi_{1,1} = \frac{r_2}{r_1} \quad \text{[dimensionless]} \tag{7.38}$$

$$S_e^2 = \frac{1-\phi_{1,1}^2}{1+\theta_{1,1}^2 - 2\phi_{1,1}\theta_{1,1}} \quad \text{[dimensionless]} \tag{7.39}$$

and

$$r_1 = \frac{\left(1 - \phi_{1,1}\theta_{1,1}\right)\left(\phi_{1,1} - \theta_{1,1}\right)}{1 + \theta_{1,1}^2 - 2\phi_{1,1}\theta_{1,1}} \quad \text{[dimensionless]} \tag{7.40}$$

where r_1 and r_2 are lag 1 and lag 2 serial correlation coefficients.

An initial estimate of $\phi_{1,1}$ is obtained from eq. (7.38); the initial estimate of $\theta_{1,1}$ is obtained from eq. (7.40) by substituting $\phi_{1,1}$ and computed r_1. Final estimates of $\phi_{1,1}$ and $\theta_{1,1}$ are made by a least-squares fitting procedure (Kottegoda, 1980, p. 128).

If the value of parameter $\phi_{1,1}$ is too close to its limits of -1 and $+1$, this indicates non-stationary behavior of the historical hydrologic sequence. The nonstationariness is accounted for by means of a dth-order difference operator, which represents successive difference of d terms of stochastic variables (i.e., a_i values). This is known as an autoregressive-integrated moving average ARIMA (p, d, q) model. Box and Jenkins (1976) consider that the parameters p, d, and q need not be greater than 2 for practical purposes. By changing the form of variable a_i, an ARIMA (p, d, q) model can be transformed to an ARMA (p, q) type (Kottegoda, 1980, p. 129).

The ARMA $(1, 1)$ model should be considered when an AR (1) type does not fit the data. Parameters $(p + q)$ should be minimum. For example, ARMA $(1, 1)$ is preferable to AR (3) if it is found that both fit an observed sequence.

EXAMPLE 7.16

Solve Example 7.15 by the ARMA $(1, 1)$ model. Use the following values for standard normal deviates, t_i: 0.131, –0.859, –1.096 from Table 7.23.

SOLUTION   From Example 7.14, $\bar{X} = 653$ cfs, $S = 186.6$ cfs, $r_1 = 0.37$, and $r_2 = 0.24$.

From eq. (7.38), $\phi_{1,1} = 0.24/0.37 = 0.65$

From eq. (7.40), $\theta_{1,1} = 0.33$

From eq. (7.39), $S_e^2 = 0.85$, $S_e = 0.92$

The values of t_i are obtained from Table 7.23 (given in this problem). In each step, t_{i-1} and a_{i-1} are substituted from the previous step.

From eq. (7.37b),

$$a_1 = \left(0.65 \times 0\right) + 0.92\left(0.131 - 0.33 \times 0\right) = 0.121$$
$$a_2 = \left(0.65 \times 0.121\right) + 0.92\left(-0.859 - 0.33 \times 0.131\right) = -0.75$$
$$a_3 = \left(0.65 \times 0.75\right) + 0.92\left[-1.096 - \left(0.33 \times -0.859\right)\right] = -1.24$$
$$\vdots$$

From eq. (7.30),

$$x_1 = \bar{X} + Sa_i = 653 + (186.6)(0.121) = 676 \text{ cfs}$$
$$x_2 = 653 + (186.6)(-0.75) = 513 \text{ cfs}$$
$$x_3 = 653 + (186.6)(-1.240) = 422 \text{ cfs}$$
$$\vdots$$

7.22 Disaggregation Model

In this type of model, the flows of a higher level (aggregate flows) are created by the sequential technique of Sections 7.20 and 7.21, which are distributed at a lower level by the disaggregation procedure. A disaggregation model divides annual flows into seasonal or monthly flows and the aggregate basin flows (monthly or annual) into flows at individual sites. Since introduction of the disaggregation approach by Valencia and Schaake (1973), many models of this type have been proposed. Some of these are by Mejia and Rousselle (1976), Lane (1979), Salas et al. (1980), and Stedinger and Vogel (1984). The advantage of a disaggregation model over a sequential model (AR or ARMA) results from (1) the sets of parameters required in the monthly sequential model are too large, and (2) relevant statistics of higher levels of (annual) flows are not necessarily preserved in lower-level (monthly) flows.

The disaggregation model takes the following mathematical form in matrix notation:

$$\boldsymbol{Y} = \bar{\boldsymbol{A}}\boldsymbol{X} + \bar{\boldsymbol{B}}\boldsymbol{V} \quad [\mathrm{L^3T^{-1}}] \tag{7.41}$$

where

$\boldsymbol{Y} = (n \times 1)$ vector of disaggregated variables (i.e., monthly flows)

$\bar{\boldsymbol{A}} = (n \times m)$ coefficient matrix

$\boldsymbol{X} = (m \times 1)$ vector of aggregated variables (i.e., annual flows)

$\bar{\boldsymbol{B}} = (n \times n)$ coefficient matrix

$\boldsymbol{V} = (n \times 1)$ vector of random standard normal deviates

It is assumed that the sample (historical) values of variables X and Y have been adjusted to have zero mean by subtracting their average values. Variable V, being the standard normal deviate, has zero mean, too.

Given generated values of X at m sites for a particular year (or for m years at one site), the seasonal monthly flows Y are generated at m sites for a particular year (or for m years at one site) by eq. (7.41), provided that the coefficient matrices $\bar{\boldsymbol{A}}$ and $\bar{\boldsymbol{B}}$ are estimated from historical data.

The criteria used in parameters (coefficients) estimation are that the expected means, variances, and covariances of generated data are equal to the historical means, variances, and covariances. By transposing the matrices and taking expected values of samples (historical data), the following relations for estimation of $\bar{\boldsymbol{A}}$ and $\bar{\boldsymbol{B}}$ are obtained (Valencia and Schaake, 1973):

$$\bar{\boldsymbol{A}} = \mathbf{E}\left[\boldsymbol{YX}^{\mathrm{T}}\right]\mathbf{E}\left[\boldsymbol{XX}^{\mathrm{T}}\right]^{-1} \quad [\text{dimensionless}] \tag{7.42}$$

and

$$\bar{\boldsymbol{B}}\bar{\boldsymbol{B}}^{\mathrm{T}} = \mathbf{E}\left[\boldsymbol{YY}^{\mathrm{T}}\right] - \mathbf{E}\left[\boldsymbol{YX}^{\mathrm{T}}\right]\mathbf{E}\left[\boldsymbol{XX}^{\mathrm{T}}\right]^{-1}\mathbf{E}\left[\boldsymbol{XY}^{\mathrm{T}}\right] \quad [\text{dimensionless}] \tag{7.43}$$

where $\mathbf{E}[\boldsymbol{XX}^{\mathrm{T}}]$, $\mathbf{E}[\boldsymbol{YX}^{\mathrm{T}}]$, etc., are covariance matrices; equivalent to $\bar{\mathbf{S}}_{\mathrm{XX}}$, $\bar{\mathbf{S}}_{\mathrm{YX}}$, etc. and $\boldsymbol{X}^{\mathrm{T}}$ and $\boldsymbol{Y}^{\mathrm{T}}$ are transpose of $\boldsymbol{X}$ and $\boldsymbol{Y}$.

It is not necessary that $\bar{\boldsymbol{B}}$ have a solution. For a solution, $\bar{\boldsymbol{B}}\bar{\boldsymbol{B}}^{\mathrm{T}}$ should be positive semidefinite, which is determined by the theory of Gramian matrix.

7.23 Autorun Model

Among the flow generation schemes, a comparatively recent methodology relates to run analysis, which has a distribution-free behavior (is independent of the probability distribution of the data). A *run* is defined as a succession of the same kinds of observations, preceded and succeeded by at least a single observation of a different kind. Thus a run is made up of a wet period (water surpluses) and a drought period (water deficits) with respect to a truncation level, x_0, often taken as the median value of the observations, as shown in Figure 7.19.

The models in previous sections are based on the sequential properties of the historical data. An autorun model reproduces the runs (i.e., preserves the lengths of wet and dry periods) as observed in the historical data. Two basic parameters of autorun analysis are (1) the autorun coefficient, and (2) the run length, such as positive run length, n_p, and negative run length, n_n, in Figure 7.19.

Along the lines of the autocorrelation coefficient of the Markovian process, Sen (1976) defined the *autorun coefficient* as a conditional probability of an observation being greater than the truncation level, given that the observation preceding the k lag is greater than the truncation level. The following relationship between the autocorrelation coefficient of Markovian nature and the autorun coefficient has been derived by Sen (1978):

$$r_k = \sin \pi [(r_0)_k - 0.5] \quad \text{[dimensionless]} \tag{7.44}$$

where

$$r_k = k\text{-lag autocorrelation coefficient}$$
$$(r_0)_k = k\text{-lag autorun coefficient}$$

A value of $(r_0)_k = 0.5$ means that the data are independent of each other, and $(r_0)_k = 1$ indicates perfectly correlated data.

The run length is a basic parameter to indicate the property of runs. It also serves as a test parameter. If the average positive or negative run length calculated from the available data is equal to 2, the historical sequence is independent; otherwise, it is dependent. In dependent hydrologic series, high values of streamflow tend to follow high values, and low values tend to follow low values.

Figure 7.19 Runs in a hydrologic series (Sen, 1978).

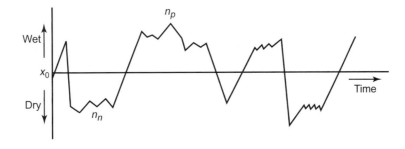

The average positive (wet) and negative (dry) run lengths at any truncation level are calculated as follows:

$$\bar{n}_p = \frac{1}{m_p} \sum_{i=1}^{m_p} (n_p)_i \quad [\text{L}] \tag{7.45}$$

and

$$\bar{n}_n = \frac{1}{m_n} \sum_{i=1}^{m_n} (n_n)_i \quad [\text{L}] \tag{7.46}$$

where

m_p, m_n = numbers of wet and dry periods

$(n_p)_i, (n_n)_i = i$th wet and dry period length in the historical record

For a stationary process the run lengths are distributed according to a geometric distribution. The geometric PDF (probability distribution function) parameters $\bar{r}_p$ of wet periods and $\bar{r}_n$ of dry periods have been obtained by Sen (1985) as

$$\bar{r}_p = \frac{\bar{n}_p - 1}{\bar{n}_p} \quad [\text{dimensionless}] \tag{7.47}$$

$$\bar{r}_n = \frac{\bar{n}_n - 1}{\bar{n}_n} \quad [\text{dimensionless}] \tag{7.48}$$

The data generation mechanism first constructs alternately wet and dry periods and then within these periods determines the flow values of surplus and deficit magnitudes.

Geometrically distributed wet and dry periods are constructed with parameters $\bar{r}_p$ and $\bar{r}_n$ through

$$y = 1 + \frac{\log \varepsilon}{\log *} \tag{7.49}$$

where

y = geometrically distributed random variable indicating a wet or dry period

ε = uniformly distributed random variable between 0 and 1 (from a table)

$*$ = assumes the value of $\bar{r}_p$ and $\bar{r}_n$ alternately

The initial wet or dry period can be selected according to the final period of the historical sequence.

Once the alternate sequence of wet and dry periods is determined, the flow values are selected randomly from the appropriate PDF corresponding to the historical sequence. The values of random variables below the threshold (lower than truncation level) are disregarded in generating surpluses, and vice versa.

Part D: Estimation of Flow at an Ungaged Site

7.24 INTERPOLATION OR EXTRAPOLATION OF DATA

Often a site for which streamflow data are needed is not gaged, but a gaging station exists on the same river upstream or downstream, or both upstream and downstream. The interpolation of the record provides a means for estimating the flow in such cases.

Consider that the gaging-station record is available at a site X having a drainage area A_x, and an estimate has to be made for another site, Y, on the same river with a drainage area of A_y. The flow will be distributed in direct proportion to the drainage area; that is,

$$Q_y = \frac{Q_x}{A_x} A_y \quad [\text{L}^3\text{T}^{-1}] \tag{7.50}$$

A better estimate is made when records at two gaging sites exist, preferably one upstream and one downstream of the ungaged site because between a gaged site and the ungaged site there might be some changes in the drainage pattern, such as the meeting of a tributary or extraction of water. Figure 7.20 shows that Bowie Creek meets the Leaf River between station A and station B. The variation in flow between the two gaged sites is adjusted either on the basis of the drainage area, as under

$$\frac{Q_y}{A_y} = \frac{Q_x}{A_x} + \left(\frac{\bar{Q}_z}{A_z} - \frac{\bar{Q}_x}{A_x}\right)\frac{A_y - A_x}{A_z - A_x} \quad [\text{LT}^{-1}] \tag{7.51}$$

where

Q_x = flow at gaged site X of drainage area A_x

Q_y = flow at ungaged site Y of drainage area A_y

Q_z = flow at gaged site Z of drainage area A_z

$\bar{Q}_x, \bar{Q}_z$ = average of the entire record at X and Z

or on the basis of the distance between the sites, as follows:

$$\frac{Q_y}{A_y} = \frac{Q_x}{A_x} + \left(\frac{\bar{Q}_z}{A_z} - \frac{\bar{Q}_x}{A_x}\right)\frac{L_y}{L_z} \quad [\text{LT}^{-1}] \tag{7.52}$$

where

L_z = distance between stations X and Z

L_y = distance between stations X and Y

EXAMPLE 7.17

The monthly mean discharge of the Leaf River near Collins, Mississippi, which has a drainage area of 752 mi², and the corresponding discharge of the same river at Hattiesburg, Mississippi, which has a drainage area of 1760 mi², are given in Table 7.25. The distance between the two stations is 35 miles. Estimate the discharge for a site near Laurel that has a drainage area of 1035 mi² and is also located about 10 miles from the site near Collins.

Figure 7.20 Section of Southern Mississippi Basin.

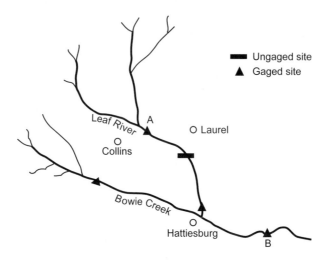

SOLUTION

Average flow near Collins, $\bar{Q}_x = 2099$ cfs

Average flow at Hattiesburg, $\bar{Q}_z = 5065$ cfs

1. Adjustment based on the drainage area:

$$\left(\frac{\bar{Q}_z}{A_z} - \frac{\bar{Q}_x}{A_x}\right)\frac{A_y - A_x}{A_z - A_x} = \left(\frac{5065}{1760} - \frac{2099}{752}\right)\left(\frac{1035 - 752}{1760 - 752}\right)$$

$$= (2.88 - 2.79)\left(\frac{283}{1008}\right)$$

$$= 0.025$$

2. From eq. (7.51),

$$\frac{Q_y}{1035} = \frac{Q_x}{752} + 0.025 \quad [LT^{-1}] \qquad (7.53)$$

Q_y values are computed from eq. (7.53) for various Q_x of Table 7.25 as shown in column 2 of Table 7.26.

3. Adjustment according to the distance:

$$\left(\frac{Q_z}{A_z} - \frac{Q_x}{A_x}\right)\frac{L_y}{L_z} = (2.88 - 2.79)\left(\frac{10}{35}\right)$$

$$= 0.026$$

4. From eq. (7.52),

$$\frac{Q_y}{1035} = \frac{Q_x}{752} + 0.026 \quad [LT^{-1}] \qquad (7.54)$$

Q_y values computed from eq. (7.54) are shown in column 3 of Table 7.26.

Table 7.25 Monthly Mean Discharge of Leaf River, Mississippi

Month	Leaf River near Collins	Leaf River at Hattiesburg
October	312	1046
November	3651	6985
December	2524	6740
January	2803	5401
February	3592	8900
March	3329	7792
April	2944	7428
May	2029	4565
June	1003	2788
July	1237	3727
August	906	2796
September	858	2611
Average	2099	5065

Table 7.26 Estimated Discharge of the Leaf River near Laurel

(1) Month	(2) Discharge Based on the Drainage Area (cfs)	(3) Discharge Based on the Distance (cfs)
October	455	456
November	5051	5052
December	3499	3500
January	3883	3884
February	4970	4971
March	4608	4609
April	4078	4079
May	2818	2819
June	1406	1408
July	1728	1729
August	1273	1274
September	1207	1208

7.25 STREAMFLOW FROM DRAINAGE-BASIN CHARACTERISTICS

The U.S. Geological Survey conducted a statistical multiple regression analysis to derive generalized relations for the natural streamflows in four regions: eastern, central, southern, and western United States (Thomas and Benson, 1970). Long-term streamflow records were used in the regression and a large number of topographic and climatic indices were included to define the drainage-basin characteristics. The regression considered various categories of flows, such as low flows, high flows, mean monthly flows, and annual flows. The multiple regression analyses defined the relation between each category of flow and the drainage basin characteristics. It was concluded that the flow can be defined more accurately in the humid eastern and southern regions than in the more arid western and central regions. Also, the mean flow can be more accurately defined than the high flows, and the low flows are very poorly defined.

The regression relation has the following form in which the constant, a, and coefficients, b_1, b_2, . . . , have different values for different regions and different categories of flows (i.e., high, low, mean, etc.).

$$Q = aA^{b_1}S^{b_2}L^{b_3}S_t^{b_4}E^{b_5}I_{24,2}^{b_6}P^{b_7}S_n^{b_8}F^{b_9}S_i^{b_{10}}t_1^{b_{11}}t_7^{b_{12}}E_v^{b_{13}}A_a^{b_{14}} \quad \text{[unbalanced]} \quad (7.55)$$

where

Q = discharge, cfs

A = drainage area, mi^2

S = channel slope, ft/mi

L = channel length, miles

S_t = percent of total drainage area occupied by lakes, swamps, ponds

E = mean elevation of the basin, thousands of feet above sea level

$I_{24,2}$ = maximum 24-hour precipitation expected to be exceeded once
   every 2 years, in.

P = mean annual precipitation, in.

S_n = mean annual snowfall, in.

F = forest cover (i.e., percent of total area forested)

S_i = soil index for infiltration, in.

t_1 = mean of minimum January temperatures, °F

t_7 = mean of minimum July temperatures, °F

E_v = annual evaporation, in.

A_a = alluvial area in the basin, mi^2

For each category of flow characteristics, all of the indices above need not be included. The indices most highly related to streamflow are drainage basin size and mean annual precipitation. The standard error of estimate for the mean annual flow ranged from 8.6% for the eastern region to 33% for the western region.

EXAMPLE 7.18

The following relation was derived by the U.S. Geological Survey (Thomas and Benson, 1970) for the annual mean flow of the eastern region. Determine the annual mean flow for the Shetucket basin near Willimantic, Connecticut, which has a drainage area of 226 mi^2, a slope of 1.2 ft/mi, and an annual precipitation of 25 in. including 10 in. of snowfall.

$$Q_A = \left(2.89 \times 10^{-4}\right)A^{1.06}S^{0.1}P^{1.87}S_n^{0.18} \quad \text{[unbalanced]} \quad (7.56)$$

SOLUTION From eq. (7.56),

$$Q_A = \left(2.89 \times 10^{-4}\right)(226)^{1.06}(1.2)^{0.1}(25)^{1.87}(10)^{0.18}$$

$$= 57.32 \text{ cfs}$$

7.26 The Hydraulic Geometry of Stream Channels

A natural river system tends to attain an approximate equilibrium between the channel form and the quantity of water and sediment it must transport. This equilibrium exists for all discharges up to the bankfull stage. The hydraulic characteristics of stream channels—depth, width, and velocity—are related to discharge in the form of simple power functions, expressed by

$$w = aQ^b \quad [\text{L}] \tag{7.57}$$

$$d = cQ^f \quad [\text{L}] \tag{7.58}$$

$$v = kQ^m \quad [\text{L/T}] \tag{7.59}$$

where

w = water surface width

d = mean depth

v = mean velocity

a, c, k = the scale factors (numerical constants) for a channel

b, f, m = the exponent constants for a channel

These relationships are described by the term *hydraulic geometry*. The hydraulic geometry characteristic of rivers also applies to stable or regime canals that do not scour or aggrade. This has been discussed in Section 11.11 with regard to the design of regime channels.

Since the classic work of Leopold and Maddock (1953), voluminous studies have been made on the hydraulic geometry of stream channels. The scale factors a, c, and k in hydraulic geometry relations are highly variable from site to site, but the exponents b, f, and m show a degree of consistency and seem independent of the sites and type of river channels. However, there is no consensus on the reasonable values of the exponents among researchers.

According to hydraulic principle, $Q = (A)(v)$ or $Q = (w)(d)(v)$ for a natural, wide stream. Substituting the above three relations

$$Q = a\,c\,k\,Q^{b+f+m}$$

which means that the scale factors and exponents should satisfy:

$$(a)(c)(k) = 1 \quad \text{and} \quad b + f + m = 1 \tag{7.60}$$

Studies conducted on natural river systems seem to satisfy these conditions, in spite of a wide variation in the values of the exponents among various studies. If a river system could be calibrated for the reasonable values of the constants, hydraulic geometry is a powerful tool in assessing streamflows and channel characteristics.

EXAMPLE 7.19

From field investigations by the USGS, the following data have been obtained for Cheyenne River near Hot Springs, South Dakota. Determine the scale factors and the exponents for the stream channel. At a high flow of 2000 cfs, what are the channel characteristics?

Width, ft	Depth, ft	Velocity, ft/s	Discharge, cfs
107	1.1	2.18	250
150	1.3	2.62	500

SOLUTION

1. Equations (7.57), (7.58), and (7.59) plotted on log-log paper with the above data are straight lines, as shown in Figure 7.21.

2. Width-discharge,

$$\text{slope } b = \frac{\log 150 - \log 107}{\log 500 - \log 250} = 0.49$$

substituting a known value

$$a = \frac{w}{Q^b} = \frac{150}{500^{0.49}} = 7.14$$

Thus, the equation is $w = 7.14\ Q^{0.49}$

3. Depth-discharge,

$$\text{slope } f = \frac{\log 1.3 - \log 1.1}{\log 500 - \log 250} = 0.24$$

$$\text{factor } c = \frac{d}{Q^f} = \frac{1.3}{500^{0.24}} = 0.29$$

Thus, the equation is $d = 0.29\ Q^{0.24}$

4. Velocity-discharge,

$$\text{slope } m = \frac{\log 2.62 - \log 2.18}{\log 500 - \log 250} = 0.265$$

$$\text{factor } k = \frac{v}{Q^m} = \frac{2.62}{500^{0.265}} = 0.50$$

Thus the equation is $v = 0.50\ Q^{0.265}$

5. For $Q = 2000$ cfs,

$$w = 7.14(2000)^{0.49} = 296 \text{ ft}$$

$$d = 0.29(2000)^{0.24} = 1.8 \text{ ft}$$

$$v = 0.50(2000)^{0.265} = 3.75 \text{ ft/s}$$

7.27 VARIABILITY OF STREAMFLOW

The previous sections provide an indication that streamflow is not constant in space or with respect to time, which makes it necessary to measure or estimate the data at a specific location for a long duration of time. From the available record at a site for a certain period, the trends in streamflows can be detected. Three devices used to study the variability of streamflow that have direct application to resource planning are: (1) frequency curve, (2) mass curve, and (3) duration curve. The frequency curve is discussed in the context of peak flows in Chapter 8. The mass curve and duration curve are described here.

Figure 7.21   Hydraulic geometry of the Cheyenne River.

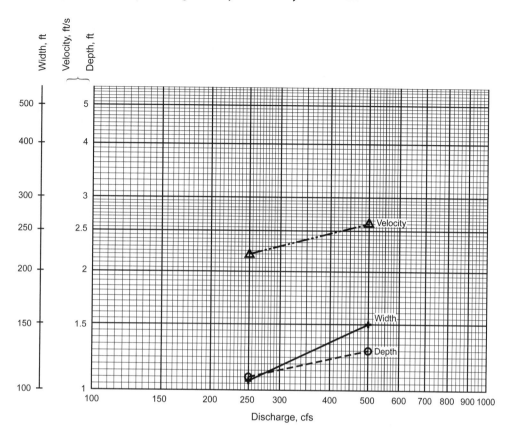

7.27.1 Flow-Mass Curve

A graph of the cumulated values of a hydrologic quantity as the ordinate, against time or date as the abscissa, is known as a *mass curve*. When the streamflow is taken as the hydrologic quantity, it is known as a flow-mass curve. The flow-mass curve is a summation of the hydrograph that represents the area under the hydrograph from one time to another. Mathematically, a flow-mass curve can be expressed as

$$V = \sum_{t=t_1}^{t_2} Q_t \, \Delta t \quad [\text{L}^3] \tag{7.61}$$

where

V = volume of streamflow

Q_t = discharge as a function of time or the hydrograph ordinate at time $t = (t_1 + t_2)/2$

Δt = period between time t_1 and t_2

This is described in Chapter 10 in the context of reservoir storage requirements.

7.27.2 Flow-Duration Curve

To ascertain how often flow of a given magnitude occurred during the period of record, a flow-duration curve is prepared. From the available data, the discharge is plotted as the ordinate against the percent of time that discharge is exceeded on the abscissa, as shown in Figure 7.22. This is referred to as a *complete series analysis*. Two other series, the annual series and the partial duration series, are described in the context of frequency analysis in Chapter 8. In a statistical sense, a duration curve represents the cumulation of the frequency distribution curve. The duration curve can be of daily flows, mean monthly flows, or mean annual flows. The curve of mean annual flows is likely to be different from the monthly or daily flow curves. If the river flows fluctuate from month to month, but the total flow every year is nearly the same, the monthly duration curve will be similar to Figure 7.22, but the mean annual duration curve will be nearly a horizontal line.

The procedure used to prepare a duration curve is as follows, which has been demonstrated in Example 7.20.

1. The total range of discharge is divided into a number of classes. For instance, the discharge ranging from 0 to 10,000 cfs is divided into 20 classes of 500 cfs each.

2. The entire record is scanned day by day for daily flows, month by month for monthly flows, and year by year for yearly flows curves.

3. A mark is made in the appropriate class for each item in the record. The plot of the number of items in each class against the discharge value of the class will represent the frequency distribution, as shown in Figure 7.23a.

4. The number of items in each class are cumulated starting with the highest flow. The percent of the accumulated number of items of each class with respect to total items of all classes is determined.

5. The average discharge value of each class is plotted on the vertical scale against the percent (of time) determined in item 4, as the abscissa, as shown in Figure 7.23b.

Figure 7.22   Flow-duration curve.

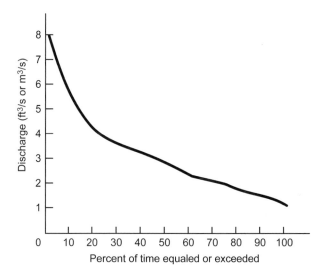

Figure 7.23 Frequency curve and flow-duration curve.

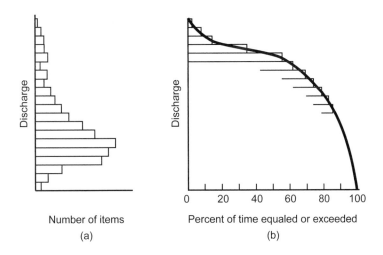

Number of items

(a)

Percent of time equaled or exceeded

(b)

Sometimes the ordinates of the flow-duration curve are plotted in dimensionless form, in terms of ratios, to the overall average discharge rather than actual discharges. When no flow data are available, a synthetic flow duration can be constructed by a procedure described by Chow (1964, pp. 14–43).

The flow-duration curve is a very important tool for appraising the flow values of various dependabilities, and thus is indispensable for water resources study. The slope of the curve indicates streamflow characteristics. A flat-sloped curve suggests large natural storage in a stream and a steep slope indicates the flashy nature of a stream. The flow corresponding to 50% exceedance is called the median flow. This is usually different from the mean flow, which can be obtained by (1) computing the average of the data in the record, (2) dividing the duration curve in two equal parts by a horizontal line, or (3) using the following area summation equation:

$$Q_{mean} = 0.025\left(\tilde{Q}_0 + \tilde{Q}_{100}\right) + 0.05\left(Q_5 + Q_{95}\right) + 0.075\left(Q_{90} + Q_{10}\right)$$
$$+ 0.10\left(Q_{20} + Q_{30} + Q_{40} + Q_{50} + Q_{60} + Q_{70} + Q_{80}\right) \quad [\text{L}^3\text{T}^{-1}] \quad (7.62)$$

where

$$Q_{mean} = \text{mean discharge}$$
$$Q_5, Q_{10}, \ldots = \text{discharge corresponding to 5\%, 10\%, etc., exceedance levels}$$
$$\tilde{Q}_0, \tilde{Q}_{100} = \text{discharge nearly 0 and 100\% of time (any discharge of less}$$
$$\text{than 5\%, and more than 95\%, respectively)}$$

The firm power of a run-of-river (without storage) power plant is estimated on the basis of the flow available at 90% of the time from an unregulated flow-duration curve. The minimum flow of a stream can be improved by creating reservoir storage, which will modify the duration curve to a flatter slope. This will enhance the firm power potential of a hydroelectric plant. The ordinate and abscissa of the curve can be replaced to represent kilowatts and hours, respectively, so that the area under the curve will provide the annual energy output of a plant in kilowatt-hours.

EXAMPLE 7.20

Prepare the flow-duration curve from the mean monthly flow data given in Table 7.27. Determine the percent of time that a monthly flow of 650 cfs is equaled or exceeded. Also determine the median and mean monthly flows.

Table 7.27   Mean Monthly Flow in cfs for Example 7.20

Water Year	Oct.	Nov.	Dec.	Jan.	Feb.	Mar.	Apr.	May	June	July	Aug.	Sept.
2000	468	710	1462	841	899	735	971	1538	1247	839	576	481
2001	1114	1140	800	739	581	499	690	1592	2652	1088	691	467
2002	466	643	649	432	581	762	1121	2280	1762	1128	666	473
2003	414	960	769	655	694	924	892	1570	2660	1874	862	552
2004	818	1448	1536	727	1313	624	1169	1504	1484	1129	638	422
2005	556	544	501	330	642	428	970	1407	1249	1004	625	433
2006	341	227	243	1043	978	416	665	1149	1302	1404	715	500

SOLUTION

1. The discharge, in the range of 0 to 3000 cfs, has been divided into 15 classes of 200 cfs each, as shown in column 1 of Table 7.28.
2. The data are scanned and each item is noted in the class group to which it belongs. The total in each class is shown in column 2.
3. Column 3 accumulates the number of items of column 2, starting from the bottom.
4. The items accumulated are shown as percents in column 4.
5. The plot of average value in each class in column 1 against column 4 is given in Figure 7.24.
6. From the plot, a flow of 650 cfs is equaled or exceeded 75% of the time.
7. The flow corresponding to 50% (i.e., Q_{median}) = 900 cfs.
8. Mean flow, from eq. (7.62),

$$Q_{mean} = 0.025(2700 + 300) + 0.05(1860 + 500) + 0.075(1580 + 520)$$
$$+ \ 0.10(1320 + 1160 + 1020 + 900 + 810 + 700 + 620)$$
$$= 75 + 118 + 158 + 653$$
$$= 1004 \text{ cfs}$$

7.28 COMPUTER MODELS FOR SURFACE WATER HYDROLOGY

A large number of computer models have been developed for hydrology and hydraulic engineering. The main developers of these models are the Hydrologic Engineering Center (HEC) and other departments of the U.S. Army Corps of Engineers, U.S. Geological Survey (USGS), Natural Resources Conservation Service (NRCS), National Weather Service (NWS), U.S. Environmental Protection Agency (EPA), and many universities and private enterprises. The hydrologic phenomena that have been modeled extensively relate to (1) watershed hydrology, (2) statistical hydrology, (3) river hydraulics, and (4) reservoir planning and analysis.

Table 7.28   Computation of Flow-Duration Curve

(1) Class (Flow Range) (cfs)	(2) Number of Items	(3) Cumulated Number of Items	(4) Percent of Time
0–200	0	84	100
201–400	4	84	100
401–600	20	80	95.2
601–800	20	60	71.4
801–1000	11	40	47.6
1001–1200	10	29	34.5
1201–1400	4	19	22.6
1401–1600	10	15	17.9
1601–1800	1	5	6.0
1801–2000	1	4	4.8
2001–2200	0	3	3.6
2201–2400	1	3	3.6
2401–2600	0	2	2.4
2601–2800	2	2	2.4
2801–3000	0	0	0

Figure 7.24   Flow-duration curve for Example 7.20.

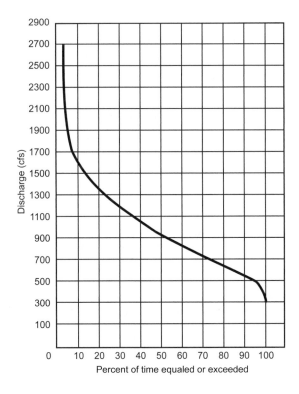

Watershed hydrology models primarily include the components of rainfall-runoff correlation, overland runoff study, and routing process. Many models combine quality aspects also.

The HEC has developed models of all hydrologic categories and continually strives to upgrade them. Its HEC-1 and HEC-2 (upgraded version HEC-RAS) models are universally recognized and very extensively used. The USGS, with over two dozen models, is also a significant contributor to streamflow modeling of watershed hydrology, statistical hydrology, and river hydraulics. The NRCS has formulated models of watershed hydrology, statistical application, and river hydraulics. These models make application of the NRCS (SCS) Curve Number. The NWS has incorporated the element of hydrologic forecasting in its watershed and river hydraulics models. The Department of Agriculture covers the transportation of nutrients and sediments in its watershed models, and the EPA simulates the movement of pollutants with the flow.

Important models developed by various government agencies are commercially distributed by private vendors. Private vendors also have developed user-friendly counterparts to some of the well-known models. Some university models are also popular, for example, Pennsylvania State University's river hydraulics model, PSUPRO, and Cornell University's statistical model, MAX.

Models relating to river hydraulics and reservoirs have special relevance to the study of floods. These are described in Chapter 8. The watershed and statistical models related to streamflows are summarized here. These models are applicable to flood analysis as well when used with maximum storm rainfall.

Watershed hydrology models commonly fall in two groups: event-oriented models and continuous simulation models. The first group of models considers the effect of a single storm. The storm can be of any degree of complexity, and of any duration. Continuous simulation models, on the other hand, provide a long-term simulation involving multiple storm events and keep track of the soil moisture conditions between storms. Two very popular event-oriented models are the HEC-HMS of the Hydrologic Engineering Center and the Technical Release (TR)-20 program of the NRCS. HEC-HMS, last released in December 2005, is a successor to the HEC-1 program. HEC-1 was a comprehensive hydrograph-simulation program built around a stream network model that segmented a basin into a number of subbasins. For each subbasin, the input was a precipitation hyetograph. Precipitation excesses were computed by subtracting the infiltration and detention losses. The resulting precipitation excesses were routed by the unit hydrograph or kinematic techniques to the outlet of the subbasin, providing a runoff hydrograph. Runoff hydrographs from subbasins were routed through river reaches and combined at control points. Reservoir routing functioned similarly, wherein upstream inflows were routed using the storage routing method of Chapter 8.

HEC-HMS not only provides the unit hydrograph, hydrologic, and reservoir routing options similar to those in HEC-1, but also provides linear-distributed runoff transformations that can be applied to gridded (e.g., radar) rainfall data, continuous multi-layered soil moisture simulation, and a versatile parameter optimization option.

TR-20, last released in 2005, can be used to analyze multiple storms (rainfalls by frequency) within one model run. Direct runoff is computed from watershed land areas resulting from synthetic or natural rain events. The runoff is routed through channels and/or impoundments to the watershed outlet. The model assists in the hydrologic evaluation of flood events for use in the analysis of water resource projects.

Important continuous simulation models and their salient features are listed in Table 7.29. The water balance component is a core feature of all these models. Evapotranspiration is either computed directly or inferred from other hydrologic information. The

Table 7.29 Continuous Simulation Watershed Models

Model	Developing agency	Latest release	Features
Gridded Surface Subsurface Hydrologic Analysis (GSSHA)	USACE	2005	Distributed parameters, simulates streamflows generated from various types of runoff processes along with groundwater discharge, includes various in-stream hydraulic structures.
Watershed Modeling System	SSG	2006	Distributed parameters, comprehensive graphical modeling for all phases of watershed hydrology and hydraulics, supports models of HEC and NRCS.
National Weather Service River Forecasting System (NWSRFS)	NWS	1997	Lumped parameters, contains three systems: (1) calibration system to calibrate model from historic data and conversion of point to areal values, (2) operational forecasting system to simulate streamflows, and (3) extended streamflow prediction for probabilistic forecasts of hydrologic variables.
Streamflow Synthesis and Reservoir Regulation (SSARR)	USACE	1991	Lumped parameter, computes runoff from rainfall and/or snowmelt, simulation is based on defining the specific rainfall-runoff relationship applicable to the watershed, input of parameters required to describe the relationship.
Simulation for Water Resources in Rural Basins (SWRRB)	USDA	1995	Distributed parameter, for large basins, simulation covers elements of the hydrologic cycle, ponds and reservoir storage, sedimentation, crop growth, nutrient yields, and pesticide fate for up to 10 subbasins.

USDA   U.S. Department of Agriculture
NWS   National Weather Service
USGS   U.S. Geological Survey
EPA   U.S. Environmental Protection Agency
USACE   U.S. Army Corps of Engineers
SSG   Scientific Software Group

(continued)

Table 7.29 Continuous Simulation Watershed Models (Continued)

Model	Developing agency	Latest release	Features
Distributed Routing Rainfall-Runoff Model (DR3M)	USGS	1996	Distributed parameter, for small (<10 mi^2) urban basins, simulates runoff from rainfall (no snow); infiltration accounted by the Green and Ampt method; kinematic wave routing for overland and channel flow; modified Puls method for reservoir routing.
Precipitation-Runoff Modeling System (PRMS)	USGS	1998	Distributed parameter, simulates basin response to normal and extreme rainfall and snowmelt to evaluate streamflows and sediment yields; can be used to simulate daily flows or storm hydrographs.
Hydrological Simulation Program-FORTRAN (HSPF)	EPA	1997	Distributed parameter model of watershed hydrology and water quality that allows integrated simulation of runoff and soil contamination processes and reaction processes for ascertaining flow rates, sediment loads, nutrient and pesticide concentrations, and chemical concentrations.
Stormwater Management Model (SWMM)	EPA	1994	Distributed parameter, single event, or continuous simulation of catchments with storm sewers or combined sewers and natural drains for prediction of flows and pollutant concentrations; statistical analysis performed on precipitation data.

USDA   U.S. Department of Agriculture
NWS   National Weather Service
USGS   U.S. Geological Survey
EPA   U.S. Environmental Protection Agency
USACE   U.S. Army Corps of Engineers
SSG   Scientific Software Group

runoff element is either based on the empirical rainfall-runoff approach or accounts for the infiltration component separately. The soil moisture conditions are considered throughout the simulation period. The channel routing and storage routing procedures are included. Some models include special features such as sediment transport, sediment yield, groundwater flow, reaction processes, snow accumulation and melting routines, pesticide runoff, and weather components.

Related to statistical hydrology of natural streamflows, the HEC has formulated the Multi-Linear Regression Program (MLRP) and the HEC-4 Monthly Streamflow Simulation models. The MLRP utilizes regression analysis to determine relationships among rainfall-runoff parameters and basin characteristics. The program automatically deletes the least significant variables after each iteration. HEC-4 performs analysis of monthly streamflows at a number of interrelated stations to determine statistical characteristics like mean, standard deviation, and skew coefficient. Up to ten stations can be analyzed simultaneously. By correlating two stations, the missing streamflows for a station are filled in from the data of the other station. Using historic data, the model also can generate a sequence of hypothetical streamflows for a station with the same statistical characteristics as the historic data and include a random component. The computer codes, documents, and manuals for all the programs are available from the developers in microcomputer versions.

PROBLEMS

7.1 From the following hourly streamflow record due to a storm, separate the baseflow by the recession curve technique. The drainage area is 30 acres. Also determine the runoff depth.

Time (hr)	Flow (cfs)	Time (hr)	Flow (cfs)
1	30	9	45
2	29.4	10	31.5
3	66	11	22.5
4	155	12	18
5	190	13	16
6	140	14	14.5
7	100	15	13
8	63		

7.2 Tabulated on the following page are the ordinates of a hydrograph at a section of a stream having a drainage area of 250 km^2. Prepare the direct runoff hydrograph (DRH) after separating the baseflow using the recession curve technique. Determine the equivalent depth of runoff.

Time (hr)	Flow (m³/s)	Time (hr)	Flow (m³/s)
0	1.37	50	8.8
5	1.25	55	6.8
10	1.12	60	5.50
15	5.00	65	4.10
20	12.00	70	2.75
25	15.60	75	2.00
30	17.15	80	1.20
35	14.40	85	0.65
40	12.50	90	0.60
45	10.70	95	0.56

7.3 The daily streamflow data for Fall River, Massachusetts at a site having a drainage area of 650 km² are given in the table. Separate the baseflow from the runoff hydrograph by the recession curve method.

Time (days)	Flow (m³/s)	Time (days)	Flow (m³s)
1	16	8	50
2	15	9	28
3	26	10	21
4	50	11	17
5	86	12	14
6	113	13	13
7	90	14	12

7.4 Solve Problem 7.3 by the methods of the arbitrary approach.

7.5 Solve Problem 7.1 by the methods of the arbitrary approach.

7.6 Solve Problem 7.2 by the methods of the arbitrary approach.

7.7 If the hydrograph of Problem 7.1 resulted from an isolated storm of 2-hour duration, determine the unit hydrograph by the inverse procedure.

7.8 The hydrograph of Problem 7.2 was produced by an isolated storm of 5-hour duration. Derive the unit hydrograph by the inverse procedure.

7.9 Given below are the measured streamflows in cfs from a storm of 6-hour duration on a stream having a drainage area of 185 mi². Derive the unit hydrograph by the inverse procedure. Assume a constant baseflow of 550 cfs.

Hour	Day 1	Day 2	Day 3	Day 4
Midnight	550	5000	1900	550
6 a.m.	600	4000	1400	
Noon	9000	3000	1000	
6 p.m.	6600	2500	750	

7.10 In a basin of 80 mi^2, the following data were observed from a hydrograph and from an isochronal map of the watershed. Determine **(a)** the instantaneous hydrograph and **(b)** the 1-hour unit hydrograph, using the Clark method. Time of concentration = 8 hr; attenuation constant = 7.7 hr.

Area	Basin Area (mi^2)	Travel Time (hr)
A_1	3.3	1
A_2	9.3	2
A_3	19.4	3
A_4	14.3	4
A_5	9.4	5
A_6	10.8	6
A_7	10.0	7
A_8	1.5	8

7.11 The isochronal map of a basin is shown in Figure P7.11. Assume that the hydrograph given in Problem 7.2 results from a storm of 5-hour duration in the basin on which the ordinate corresponding to a time scale of 40 hours represents the inflection point. Prepare **(a)** the time-area curve for the basin, **(b)** the instantaneous unit hydrograph, and **(c)** the 5-hour-duration unit hydrograph.

Figure P7.11

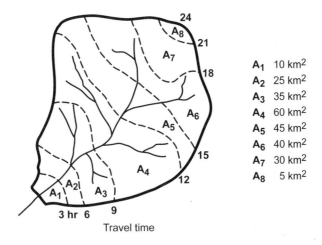

A_1	10 km^2
A_2	25 km^2
A_3	35 km^2
A_4	60 km^2
A_5	45 km^2
A_6	40 km^2
A_7	30 km^2
A_8	5 km^2

Travel time

7.12 Given below is a 2-hour unit hydrograph. Derive an 8-hour unit hydrograph by the lagging method.

Time (hr)	0	1	2	3	4	5	6	7	8
Q (cfs)	0	100	200	400	300	200	100	50	0

7.13 The ordinates of a 4-hour unit hydrograph for a 42-mi^2 basin are given in the following table. Find the 8-hour unit hydrograph by the lagging method.

Time (hr)	Q (cfs)	Time (hr)	Q (cfs)
0	0	11	1350
1	200	12	1100
2	1250	13	900
3	2200	14	700
4	3000	15	550
5	3500	16	400
6	3000	17	300
7	2600	18	200
8	2300	19	100
9	1900	20	50
10	1600	21	0

7.14 From the 2-hour unit hydrograph of Problem 7.12, determine the 8-hour unit hydrograph by the S-curve method.

7.15 From the 4-hour unit hydrograph of Problem 7.13, determine the 2-hour and 6-hour unit hydrographs by the S-curve method.

7.16 A basin of 140 mi^2 has a total stream length of 20 miles up to the upstream boundary of the basin and has a distance of 8 miles along the stream opposite the point of the basin centroid. Determine the 3-hour unit hydrograph using Snyder's method. Assume that $C_t = 2.0$ and $C_p = 400$.

7.17 For a basin of 400 km^2 having $L = 28$ km and $L_c = 12$ km, determine the 2-hour unit hydrograph using Snyder's method. $C_t = 1.5$ and $C_p = 0.20$.

7.18 Solve Problem 7.16 by the NRCS (SCS) method.

7.19 Solve Problem 7.17 by the NRCS (SCS) method.

7.20 Given a 500-km^2 basin with a lag of 10 hours, derive the 4-hour unit hydrograph using the NRCS (SCS) method.

7.21 For the following unit hydrograph and storm pattern, determine the composite direct runoff hydrograph. The unit hydrograph is a triangle as follows:

$$\text{Base length} = 4 \text{ time units}$$

$$\text{Time to peak} = 1 \text{ time unit*}$$

$$\text{Peak flow} = \tfrac{1}{2} \text{ rainfall unit*}$$

Storm pattern (time units)	1	2	3	4
Rainfall excess (rainfall units)	1.5	0	2.5	0.8

7.22 The baseflow in a stream and the 3-hour unit hydrograph for the basin are given below. Determine the total flow hydrograph for a storm of the pattern indicated.

---

* The "time unit" is a generic unit of time (e.g., hour) and the "rainfall unit" for rainfall (e.g., inch).

Estimation of Surface Water Flow   Chapter 7

Figure P7.22

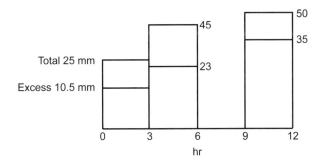

Time (hr)	Unit Hydrograph (m³/s)	Baseflow (m³/s)
1200	0	10
1500	4.7	10
1800	7.5	11
2100	5.7	11
2400	4.3	11
0300	3.1	12
0600	2.4	12
0900	1.4	12
1200	0.8	13
1500	0.2	13
1800	0	13
		(same baseflow continues)

7.23 The following hydrograph resulted from an isolated 5-hour storm of 7.3 mm rainfall excess. Determine the streamflow hydrograph resulting from the storm sequence indicated.

Figure P7.23

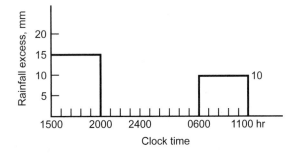

Time (hr)	Flow (m³/s)
0	1.37
10	1.12
20	12.0
30	17.15
40	12.5
50	8.8
60	5.5
70	2.75
80	1.20
90	0.60

7.24 The rainfall excesses derived from the weighted-average daily rainfall in a basin are indicated below. The 24-hour unit hydrograph from the observed data in the basin is also given. Estimate the daily streamflow. A constant baseflow of 20 m³/s has been estimated.

Date, July 2006	6	14	18
Rainfall excess (mm)	28	5.2	48.5

Unit Hydrograph:

Time (days)	0	1	2	3	4	5	6
Flow (m³/s)	0	1.5	11.6	7.7	5.0	2.3	0

7.25 The yearly streamflow record of a stream and the corresponding average annual precipitation data from the basin are listed below. Assuming that these data have already been tested for consistency, determine the antecedent precipitation index that contributes to the runoff.

Year:	1980	1981	1982	1983	1984	1985	1986	1987
Streamflow (m³/s)	1.00	1.25	0.50	0.25	3.0	1.26	0.90	2.45
Precipitation (mm)	600	740	500	450	870	640	770	730

Year:	1988	1989	1990	1991	1992	1993	1994	1995	1996
Streamflow (m³/s)	1.55	1.65	1.00	1.10	2.00	0.60	2.70	1.20	1.90
Precipitation (mm)	680	670	590	570	900	420	830	620	820

Year:	1997	1998	1999	2000	2001	2002	2003	2004	2005
Streamflow (m³/s)	0.4	1.15	0.50	—	—	—	—	—	—
Precipitation (mm)	480	610	760	560	450	730	490	500	620

7.26 (Adapted from Searcy and Hardison, 1960.) The following annual runoff and precipitation data relate to the Colorado River near the Grand Canyon, Arizona. Determine the antecedent precipitation index that furnishes the annual runoff to the river.

Year:		1920	1921	1922	1923	1924	1925	1926	1927
Yearly runoff (million acre-ft)		—	21.27	17.84	17.05	13.01	11.74	14.42	17.26
Precipitation (in.)		16.68	16.52	14.16	16.30	11.56	15.15	13.78	20.36

Year:	1928	1929	1930	1931	1932	1933	1934	1935	1936
Runoff (million acre-ft)	15.63	19.43	13.42	6.74	15.97	10.01	4.66	10.22	12.32
Precipitation (in.)	12.49	18.43	14.00	9.94	13.85	12.12	10.09	13.12	12.72

Year:	1937	1938	1939	1940	1941	1942	1943	1944	1945	1946
Runoff (million acre-ft)	12.41	15.63	9.62	7.44	16.94	17.26	11.43	13.53	11.87	9.09
Precipitation (in.)	12.47	13.99	11.14	12.31	16.94	13.65	14.63	12.51	14.00	12.39

7.27 Determine the precipitation-runoff relation for Problem 7.25. Use this relation to extend the streamflow data up to 2005.

7.28 Determine the precipitation-runoff relation for Problem 7.26.

7.29 A stream-gaging station X with the short-term record of 10 years has to be extended with the help of an adjacent station Y having a long period of record of 50 years. The flow-duration data based on the short period of record for station X and the corresponding duration for station Y are given here along with the flow-duration values based on the entire period of record for station Y. Estimate the flow at station X corresponding to a discharge of 680 cfs at station Y.

Percent of Time Equaled or Exceeded	Short-Term Flow (cfs)		Long-Term Flow at Station Y (cfs)
	Station X	Station Y	
10	2300	920	820
20	1920	700	720
30	1620	580	620
40	1380	500	550
50	1180	420	480
60	1020	380	400
70	880	320	350
80	720	270	300
90	580	220	250
95	500	180	230

7.30 Average annual streamflows for Saint John River at Baila from 1989–2001 and for Saint Paul River at Walker Bridge from 1989–2005 in Liberia, West Africa are given below. Perform the correlation analysis to **(a)** check how well the two sites are correlated, **(b)** ascertain the correlation equation between the two sites, and **(c)** extend the streamflow for Saint John River from 2002 to 2005.

Year	1989	1990	1991	1992	1993	1994	1995	1996	1997	1998
Saint John River (cfs)	4134	3696	3623	2596	4793	5741	4702	4976	5327	5177
Saint Paul River (cfs)	8923	8388	8073	7573	10,756	12,818	9840	10,653	10,934	10,575

Year	1999	2000	2001	2002	2003	2004	2005
Saint John	6371	4560	4237	—	—	—	—
Saint Paul	12,809	9554	8761	9203	12,123	7273	11,950

7.31 From streamflow records at two gaging sites, one at Little River near Hanover, CT, and one at Quinebaug River near West Thompson, CT, the following flows were observed in the month of December. Quinebaug River has a shorter period of record. Perform the correlation analysis to **(a)** check how well the December month data are correlated, **(b)** find the correlation equation, and **(c)** fill in the streamflows for previous December months at Quinebaug River.

Year	1976	1977	1978	1979	1980	1981	1982	1983	1984	1985
Little River (cfs)	31	36	120	89	15	28	44	29	74	73
Quinebaug River (cfs)	—	—	—	—	—	—	—	120	376	211

Year	1986	1987	1988	1989	1990	1991	1992	1993	1994	1995	1996
Little River	88	44	61	177	148	80	103	22	113	58	45
Quinebaug River	480	152	430	844	776	531	435	118	587	213	278

Year	1997	1998	1999	2000	2001	2002	2003	2004	2005
Little River	35	72	42	125	27	56	133	42	49
Quinebaug River	156	453	197	619	153	317	620	206	230

7.32 The annual flows for the Delaware River at Montague, NJ, are given below. Determine the statistical parameters— mean, standard deviation, skew coefficient, and serial coefficient— and identify the frequency distribution of flow sequence.

Year	1990	1991	1992	1993	1994	1995	1996	1997	1998	1999
Streamflow (cfs)	5816	6900	6900	2200	7680	3860	3564	4700	4960	5500

Year	2000	2001	2002	2003	2004	2005	2006
Streamflow (cfs)	6300	4580	7400	2200	2100	5480	4100

7.33 The annual flow record of Shetucket River near Willimantic, CT, is given below. Determine the statistical parameters of this site.

Year	1986	1987	1988	1989	1990	1991	1992	1993	1994	1995
Flow (m^3/s)	19.35	17.1	23.22	16.2	26.28	20.25	19.62	21.78	33.3	36.72

Year	1996	1997	1998	1999	2000	2001	2002	2003	2004	2005
Flow (m³/s)	20.16	25.65	24.75	31.68	47.88	30.78	34.47	27.45	29.7	24.12

7.34 Generate a synthetic annual series using the AR(1) model for the data of Problem 7.32. For random number values, use Table 7.23 starting with column 0 of row 0.

7.35 Using the historical record of Problem 7.33, generate a sequence of annual flows for the Shetucket River by the AR(1) model. Assume a normal distribution. For random numbers, use Table 7.23 starting at column heading 1 of row 3.

7.36 Annual flow data of the Batten Kill River near Middle Falls, New York, provide a mean and standard deviation of 40.7 m³/s and 9.58 m³/s, respectively. The first and second serial correlation coefficients are computed to be 0.56 and 0.45, respectively. Estimate an initial set of parameters of an ARMA (1, 1) model. Generate three additional values of streamflows using the following standard normal deviates: $1.123, -0.821, -0.342$.

7.37 Solve Problem 7.35 by the ARMA(1, 1) model.

7.38 The mean monthly discharges of the White River at Greenwater, Washington, with a drainage area of 220 mi², are given below. Estimate the discharge of the same river at Putnam, which has a drainage area of 75 mi².

Month	Oct.	Nov.	Dec.	Jan.	Feb.	Mar.	Apr.	May	June	July	Aug.	Sept.
Discharge (cfs)	1150	1200	900	760	591	511	702	1602	2680	1095	705	510

7.39 The mean monthly discharges of the Naugatuck River at Seymour, CT, with a drainage area of 880 mi², are given below along with the discharges on the same river at Rosenfield, which has a drainage area of 289 mi². The distance between the two sites is 45 mi. Determine the monthly discharge at site X, of drainage area 475 mi², which is located 16 miles from the Seymour site. Estimate both according to distance and drainage area.

Month:	Oct.	Nov.	Dec.	Jan.	Feb.	Mar.
Discharge at Seymour (cfs)	302	353	1140	730	1468	1510
Discharge at Rosenfield (cfs)	107	127	438	270	505	542

Month:	Apr.	May	June	July	Aug.	Sept.
Discharge at Seymour (cfs)	1040	900	340	261	268	223
Discharge at Rosenfield (cfs)	375	295	125	102	113	79

7.40 The regression relation developed by the U.S. Geological Survey for the mean January monthly flow in cfs in the eastern region is given below. Determine the mean January monthly flow for the Shetucket basin near Willimantic, CT, which has a drainage area of 226 mi², a slope of 1.2 ft/mi, and an annual precipitation of 25 in., including 10 in. of snowfall. Assume that 25% of the drainage area is occupied by the water body (lakes, swamps, etc.).

$$Q_1 = \left(4.61 \times 10^{-5}\right) A^{1.03} S_t^{0.27} P^{2.30} S_n^{0.42}$$

7.41 From the USGS record, the following data have been obtained for the Rio Grande at San Acacia in New Mexico. Determine the scale factors a, c, and k and the exponents b, f, and m

for the river section. What are the channel width, depth, and flow velocity corresponding to a flow of 3000 cfs?

Width, ft	Depth, ft	Velocity, ft/s	Discharge, ft³/s
140	1.08	3.28	500
182	1.42	3.82	1000

7.42 Along the main trunk of the Missouri and lower Mississippi River, the following data were recorded from upstream to downstream for a constant frequency of flow. Determine the scale factors and the exponents for the basin. At Memphis, Tennessee, the depth observed was 15.24 m; what are the discharge and velocity of flow?

Flow, m³/s	Width, m	Depth, m	Velocity, m/s
849	340	3.15	1.0
2830	480	6.10	1.27
6900	619	9.95	1.52

7.43 The mean monthly discharge data (cfs) for a river are given below. Prepare the flow-duration curve (perform the complete series analysis) for the stream. Determine the percent of time that a flow of 100 cfs is equaled or exceeded. Also determine the mean monthly and median monthly flows.

Year	Oct.	Nov.	Dec.	Jan.	Feb.	Mar.	Apr.	May	June	July	Aug.	Sept.
2000	77.2	163	533	263	245	270	341	408	255	111	58.1	47.1
2001	181	504	320	225	177	135	241	605	747	184	92.3	72.1
2002	86.3	156	159	102	171	282	433	833	425	160	67.3	45.3
2003	64.7	180	226	203	240	286	291	575	900	371	102	57.2
2004	94.0	328	400	235	327	175	411	528	319	114	57.5	43.1
2005	69.5	97.2	118	64.4	158	118	323	391	218	119	55.5	38.3
2006	32.1	32.9	35.0	276	297	124	206	364	357	191	71.9	47.7

7.44 The data reconstituted for mean monthly flow (m³/s) for the Pee Dee River at Peedee, South Carolina, are given here. Prepare the flow-duration curve for the stream. Determine whether the stream carries 150 m³/s of flow 90% of the time. Also determine the mean monthly flow.

Year	Oct.	Nov.	Dec.	Jan.	Feb.	Mar.	Apr.	May	June	July	Aug.	Sept.
1976	169.8	228.9	170.4	162.0	159.0	180.3	471.0	492.0	309.0	216.3	166.5	158.1
1977	153.0	177.9	141.3	147.6	106.2	217.5	309.0	246.9	145.5	126.6	137.4	118.8
1978	104.4	168.3	147.6	137.7	143.7	172.2	263.1	254.1	186.0	306.0	155.1	171.6
1979	171.0	173.4	175.8	177.3	172.8	183.3	597.0	729.0	303.0	207.6	197.4	147.0
1980	130.2	146.7	145.2	151.5	133.5	204.0	801.0	513.0	429.0	477.0	279.9	366.0
1981	441.0	351.0	261.0	230.7	226.8	199.5	657.0	282.0	260.1	459.0	291.0	175.8
1982	136.8	150.6	155.7	178.2	187.5	309.0	501.0	399.0	498.0	522.0	262.5	201.0
1983	168.3	160.2	190.8	171.0	200.1	205.5	546.0	552.0	414.0	219.6	183.0	273.3
1984	369.0	253.8	205.5	198.6	183.0	211.5	762.0	324.0	318.0	146.1	168.3	114.0
1985	179.1	189.0	172.2	177.0	164.4	149.1	810.0	324.0	265.2	348.0	250.2	184.5
1986	146.4	159.3	160.8	148.2	138.3	209.1	390.0	279.0	195.0	124.2	108.9	156.9
1987	158.1	219.9	179.4	162.0	153.9	185.4	369.0	197.7	186.0	396.0	141.0	175.5
1988	143.1	206.9	144.6	140.1	129.3	165.0	330.0	315.0	170.1	144.9	234.0	426.0
1989	465.0	366.0	234.6	246.9	213.0	187.2	726.0	1119.0	405.0	250.2	216.3	242.4
1990	241.5	306.0	204.6	157.8	147.6	231.0	474.0	558.0	247.5	182.4	161.4	146.7
1991	173.7	202.8	189.3	161.7	157.2	207.6	504.0	504.0	321.0	177.9	156.3	221.4
1992	195.9	164.1	154.2	140.4	131.4	168.3	263.1	360.0	297.3	130.8	121.8	114.0
1993	102.6	106.8	107.1	111.6	101.4	125.4	210.6	393.0	189.6	126.3	170.7	195.3
1994	166.8	209.4	162.6	147.9	132.3	152.7	519.0	1083.0	250.8	155.1	136.8	168.6
1995	202.5	229.8	261.6	235.5	199.8	393.0	471.0	351.0	339.0	133.8	189.0	130.5
1996	187.8	191.1	177.6	179.4	192.0	195.6	885.0	399.0	405.0	300.0	198.9	154.8
1997	258.0	324.0	189.9	151.5	174.0	217.8	402.0	324.0	504.0	423.0	234.3	465.0
1998	281.4	270.6	248.1	269.7	270.0	253.2	687.0	471.0	354.0	261.3	153.3	127.8
1999	195.9	206.7	174.6	188.1	153.0	176.4	342.0	327.0	471.0	170.7	139.8	147.3
2000	192.3	366.0	292.5	217.2	194.1	268.5	852.0	465.0	279.9	202.5	153.0	141.0
2001	277.2	264.6	240.6	197.7	170.1	205.5	582.0	762.0	265.8	204.6	324.0	273.0
2002	402.0	414.0	231.6	255.9	221.4	678.0	720.0	321.0	333.0	184.8	259.2	233.4
2003	200.7	247.5	198.9	188.4	196.2	220.2	426.0	321.0	333.0	184.8	259.2	233.4
2004	176.4	318.0	226.5	198.9	198.0	205.2	579.0	627.0	348.0	164.4	124.5	189.6
2005	130.2	207.6	258.9	202.5	207.6	290.7	894.0	441.0	237.3	123.6	110.7	90.9

Computation of Extreme Flows

8.1 COMPUTATION METHODS

Floods and droughts are extreme hydrological events. Both are defined differently by various agencies throughout the world. In general qualitative terms, they refer to periods of unusually high and low water supplies. Any hydraulic structure in a river system, such as a dam, spillway, channel, road drainage, or railway drainage, has to accommodate floods and droughts related to that system. Floods must be considered when determining the capacities of these hydraulic structures. On the other hand, the depth of a navigable channel, the water supply during a dry period, and the quantity of flow below a regulatory structure are concerns that accompany drought conditions. When the streamflow and/or precipitation records are available, these form the basis for estimating flood and drought flows. These records, however, are not long enough to provide the extreme values directly. Extrapolation is carried out by a statistical process or by the physical analysis of critical hydrometeorological events. Where streamflow data are not available, analysts apply empirical and other indirect methods.

Floods may arise from extreme rainstorms, the rapid melting of extensive snow deposits, or a combination of the two. Where records of streamflows are not available, the flood flow estimate is made from the data on extreme rainfall. Methods of flood flow computation and the procedures used in the study of droughts are summarized in Figure 8.1.

8.2 THE CONCEPT OF PROBABILITY IN HYDROLOGY

Since the magnitude of the flows recorded in the past will be repeated, a specific flood value will be equaled or exceeded (equaled or less in low-flow analysis) in a period of time. The actual time between exceedances is called the *recurrence interval*. Statistical analysis of hydrological events considers the average elapsed time between occurrences of an event (i.e., flow of a certain magnitude or greater). This average recurrence interval for a certain event is also known as the *return period* of that event. The chance of a flood occurring with a return period *T*, in a unit time, is $1/T$, called the *probability of occurrence*. Because the period is usually measured in years and the probability is expressed in percent, it is referred to as the percent probability of annual exceedance.

A flood discharge is a continuous variable that can acquire any value between two numbers. An individual observation or value of a variable, in this case flood flow, is known as a *variate*. An array of variates, constituting a time series, represents a sample from the population of peak discharges recorded in the past and to be observed in the future at the study site. The continuous series can be reduced to a discrete form by grouping the data

Figure 8.1   Methods for flood flow computation and procedures used in drought analysis.

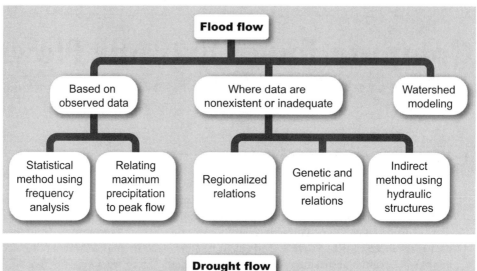

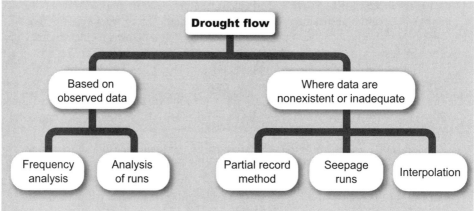

into a number of classes of equal discharge interval, each class representing a discrete variate. The number of items in a class (number of occurrences of a variate) within the entire database is called its *frequency*. A complete description of the frequency of all classes (variates), such as a plot of the number of items in each class against the respective class interval as shown in Figure 8.2, is called a *frequency distribution*. When the number of items in a class is divided by the total number of items in all classes, the result is the probability of that class or variate, as defined earlier; that is,

$$p = \frac{n_i}{N} \quad \text{[dimensionless]} \tag{8.1}$$

where

$\qquad p$ = probability of occurrence of flood flow of class i (variate i)

$\qquad n_i$ = number of items in the ith class

$\qquad N$ = total number of items in a series

Figure 8.2   Frequency distribution curve.

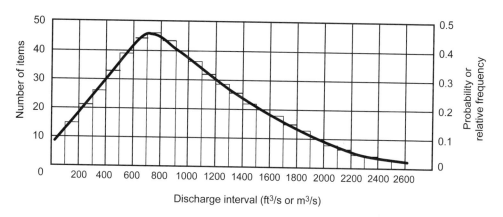

The distribution of the probabilities of all classes (instead of their frequencies) is known as a *probability distribution.* The ordinates of the frequency distribution and the probability distribution are proportional to each other.

In the case of a continuous random variable, when a variate, x, takes a continuous value, the probability becomes a continuous function, p_x, called the *probability density function* (PDF). Statisticians have demonstrated that the distribution of a large number of natural phenomena, including hydrologic data, can be expressed by certain general mathematical equations. These are recognized to be theoretical probability distribution functions. There are many different types of probability distributions. Some of these, such as binomial, geometric, and Poisson distributions, consider the discrete process, while others, such as uniform, normal, gamma, beta, Pearson, and extreme value distributions, are for the continuous random process, as described subsequently. These mathematical functions (equations) are very convenient for analysis because of their known solutions.

If the frequencies or probabilities, as shown in Figure 8.2, are successively summed up (accumulated) starting from the highest value, a curve of type (a) in Figure 8.3 results. This is known as the *cumulative frequency or probability, P,* which indicates the probability that a variable has a value equal to or greater than a certain assigned value. This probability is designated as $P(X \geq x)$. When the probabilities of a variate are summed up starting with the lowest value, a curve of type (b) (Figure 8.3) is obtained. This curve indicates the cumulative probability, $P(X \leq x)$, that a variable has a value equal to or less than a certain assigned value. In the case of a continuous variable, the summation for cumulative probability can be expressed by an integration of the aforementioned probability density function (PDF), as follows:

$$P(X \leq x) = \int_{-\infty}^{x} p_x dx \quad \text{[dimensionless]} \tag{8.2}$$

This integration (area under the PDF) is called the *cumulative distribution function* (CDF). The total area must be equal to unity.

In eq. (8.1), the number of possible values for a continuous variable, and hence N, approaches infinity. Thus the probability that a variable will have an exact value, x, has no meaning [zero probability from eq. (8.1)] for a continuous function. Therefore, the probability of occurrence is expressed for a variate having a value greater than x or less than x.

Figure 8.3 Cumulative probability curve.

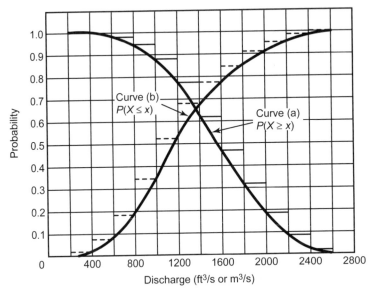

Curve (a): Cumulative frequency or probability of equal, or greater than, $P(X \geq x)$
Curve (b): Probability of equal, or less than, $P(X \leq x)$

The probability estimates for continuous variables are, accordingly, related to areas under the PDF (i.e., CDF rather than the ordinates of the PDF). For the theoretical distribution functions, tables are available for areas under the curves in standardized units. On graph paper specially constructed for a specific distribution, CDF plots as a straight line.

8.3 DESIGN FLOOD FOR HYDRAULIC STRUCTURES

The hydrologic design of a project is based on an optimum peak flood discharge. There are two approaches to estimating the optimum design discharge. The major hydraulic structures for flood control and other structures where a high degree of protection is required due to the danger to human lives and extensive property damage, such as the spillway of an earthen dam, are designed on the basis of probable maximum flood and standard project flood associated with a critical combination of meteorological and hydrological conditions. No attempt is made to associate the design discharge with any specific probability of exceedance. The precipitation maximization method described subsequently is used for this purpose.

In the case of storage capacity of reservoirs, spillway design for concrete structures in remote areas, carrying capacity of channels and culverts, channel improvement schemes, and storm sewer systems, if the design capacity is exceeded, some damage will result, but not of a catastrophic nature. The optimum design flood of these structures is based on a certain probability of exceedance or return period. A frequency analysis is required in such cases. The design probability of flood discharge is determined from a consideration of (1) acceptable level of risk, (2) economic factors, and (3) standard practice.

8.3.1 Risk Basis for Design Flood

A structure designed for any level of peak discharge bears a certain risk of being overflowed in its lifetime. Consider that the return period of 10,000 cfs discharge is 100 years. This is referred to as a *100-year exceedance* flood. This means that the probability of exceeding 10,000 cfs in any one year is $1/100 = 0.01$, or the exceedance probability, P, is 0.01 or 1%. The exceedance probability is defined with respect to a single trial (year). If the probability of exceeding 10,000 cfs in a total period of 100 years is desired, the answer is not straightforward. A 100-year flood does not mean that it will definitely be exceeded exactly once in every 100 consecutive years, but it implies that in a very large number of occurrences it is the average return period. The ASCE (1996) recommends that to avoid confusion and improve understanding, the reporting of return periods should be avoided. It is clearer to refer to exceedance probability.

Flood flows follow Bernoulli's process, according to which the probability of an event occurring is independent of time and independent of the past history of occurrences or nonoccurrences. In such a case, at any time, an event may either occur with probability P or not occur with probability $Q(= 1 - P)$. The probability of one event in three years is $PQQ + QPQ + QQP$, which is equal to $3PQ^2$. Thus the probability of k events in n years is equal to the number of ways of arranging k values of P among n items. This is indicated in terms of the exceedance probability and is referred to as the binomial probability distribution. It considers a discrete time scale.

$$f_x\{\text{exactly } k \text{ events in } n \text{ years}\} = C_k^n P^k (1-P)^{n-k} \quad \text{[dimensionless]} \qquad (8.3)$$

where

$$P = \text{exceedance probability of an event in any one year}$$

$$f_x = \text{probability of } k \text{ events (exceedances) in } n \text{ years}$$

$$C_k^n = \frac{n!}{k!(n-k)!}$$

In hydrologic study it is usually not important to know the probability that an event (e.g., flood) will exceed exactly k times, but to ascertain the probability that an event will occur once or more in n years. Thus

$$f_x\{1 \text{ or more event in } n \text{ years}\} = 1 - f_x\{\text{zero event in } n \text{ years}\}$$

From eq. (8.3),

$$f_x\{1 \text{ or more flood in } n \text{ years}\} = 1 - C_0^n P^0 (1-P)^{n-0}$$

or

$$f_x\{\text{at least one flood in } n \text{ years}\} = 1 - (1 - P)^n \quad \text{[dimensionless]} \qquad (8.4)$$

Equation (8.4) gives the probability, f_x, of a structure overtopping at least once or the risk level, in n years, associated with a flood of any exceedance probability P (return period, $T = 1/P$). Alternatively, for a project life, n, and an acceptable risk level, R (where $R = f_x \times 100$), the exceedance probability (P) and hence the return period ($1/P$) of the design flood can be computed from eq. (8.4). The values are shown in Table 8.1 for various acceptable risk levels and a project life of 25, 50, and 100 years. When only a 1% chance (risk level) of a structure being overtopped in 50 years can be taken, it should be designed for a 5260-year-return-period flood (Table 8.1).

Table 8.1   Return Period, 1/P, For Various Risk Levels [eq. (8.4)]

Acceptable Level of Risk, R (%)	Project Life, n (years)		
	25	50	100
		Return Period	
1	2440	5260	9950
25	87	175	345
50	37	72	145
75	18	37	72
99	6	11	27

EXAMPLE 8.1

A culvert has been designed for a 50-year exceedance interval. What is the probability that exactly one flood of the design capacity will occur in the 100-year lifetime of the structure?

SOLUTION $n = 100$, $k = 1$. Exceedance probability, $P = \dfrac{1}{50} = 0.02$. From eq. (8.3),

$$f_x\{1 \text{ event in 100 years}\} = C_1^{100} P^1 (1-P)^{100-1}$$

$$= \frac{100!}{1!(100-1)!}(0.02)^1 (1-0.02)^{99}$$

$$= 0.27$$

EXAMPLE 8.2

In Example 8.1, what is the probability that the culvert will experience the design flood one or more times (at least once) in its lifetime?

SOLUTION   From eq. (8.4),

$$f_x\{\text{at least once in 100 years}\} = 1 - (1-0.02)^{100}$$

$$= 0.87$$

EXAMPLE 8.3

The spillway of a dam has a service life of 75 years. A risk of 5% for the failure of the structure (exceeding of the flood capacity) has been considered acceptable. For what return period should the spillway capacity be designed?

SOLUTION

$$f_x = \frac{5}{100} = 0.05$$

$$n = 75$$

From eq. (8.4),

$$f_x = 1 - (1 - P)^n$$

$$0.05 = 1 - (1 - P)^{75}$$

$$P = 0.000684$$

$$T = \frac{1}{P} = 1460 \text{ years}$$

8.3.2 Economic Basis for Design Flood

From economic considerations, the optimum design discharge is the peak flow rate corresponding to a return period whose use in the project design will minimize the average annual cost of the project. The average annual cost involves the following:

1. Annual cost allocated from the total cost of construction of a structure, apportioned over the economic life of the structure

2. Annual operation and maintenance cost of the structure

3. Annual flood damages in money terms with the proposed structure in position

A flood-frequency curve (a plot of flood magnitude versus exceedance probability) is necessary for this analysis, which is described in Section 8.4. Various development levels for the project in terms of design discharge are considered. For each alternative of project development, the costs of the first two items listed above are computed by standard procedures of engineering economics. The last item is computed by the following steps:

1. Prepare a flood-frequency curve of peak discharge (Q) versus exceedance probability, (P).

2. For a selected alternative, collect data on flood damage from the field study for flood levels higher than the selected alternative; that is, monetary flood damages (J) for various flood stages (H) or flood discharges (Q). If it is in terms of flood stages, convert these stages into corresponding discharges, using the stage-discharge relation for the site of study.

3. Combining steps 1 and 2, prepare a damage-frequency relation or curve of damage (J) versus exceedance probability (P) for floods onwards of the selected alternative.

4. Determine the area under the damage-frequency curve graphically or arithmetically to obtain the annual flood damage associated with the selected alternative.

5. Repeat steps 1 through 4 for each alternative.

When the annual construction cost and the annual cost of operation and maintenance of the selected alternative are added to the annual damage cost computed above, the total annual cost is obtained. Perform a similar analysis with all alternative designs. Plot the computed data as shown in Figure 8.4. The optimum design discharge is the point at which the total cost is minimum. Determine the exceedance probability for this discharge from the flood-frequency curve.

8.3.3 Standard Practice for Design Exceedance Probabilities

The extensive analysis described above is justified for major projects. Moreover, the parameters used in the analysis are often not well defined. Consequently, it has become a practice

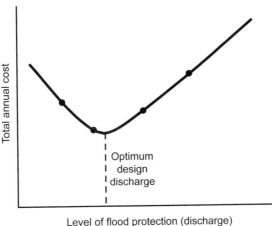

to adopt a standard design exceedance probability based on (1) the type of structure, (2) the importance of the structure, and (3) the development of the area subject to flooding.

Large and flood-vulnerable hydraulic structures are designed for a recurrence interval of 1000 years or more. A common frequency level is 100 years, for which small to medium-sized hydraulic structures, navigable waterways, and river ports are designed. Minor structures, the culverts on highways, and railway bridges are designed for a frequency of 10 to 50 years. The storm drainage in residential areas is designed using flood exceedance of 2.5 to 10 years.

8.4 STATISTICAL METHODS

There are four applications of statistical methods in hydrologic study. First, they are used in monitoring groundwater data (Section 5.8). Second, regression analysis is used to determine the rating equation (Section 6.25.2) and to extend the short-duration record (Section 7.13). Third, statistical parameters are computed in synthetic streamflow generation (7.18). Fourth, statistical methods are used in flood-frequency studies to prepare a curve that indicates the magnitude of floods of various probabilities of occurrence. Once the selection of the design exceedance probability (or return period) has been made by the procedures of the preceding section, the peak discharge corresponding to that probability or return interval on the flood-frequency curve becomes the design discharge. The procedure of analysis and the reliability of a flood-frequency curve depend on the type and quality of observed flood flow series on which the curve is based. No amount of statistical sophistication can improve the quality of the data.

8.5 TYPE AND QUALITY OF DATA

When a set contains all available data observed over a certain period of time, it constitutes a *complete-duration series*. An application of such a series is considered in the flow-duration analysis (Section 7.27.2). All of these data, however, have no significance in flood or drought estimation, which is governed by extreme flows. Accordingly, from a complete-duration series, two types of data are selected; annual series and partial or partial-duration series. The *annual series* includes the largest or smallest values recorded each year (or equal time intervals apart). The *partial series* contains all the data that have a magnitude greater than a certain base value irrespective of their year or duration of occurrence. If the base value selected is such that the number of values in the series is equal to the number of years of the record, the series is called an *annual exceedance partial-duration series*. The relationship between the probabilities of the annual exceedance partial-duration series and the annual series has been investigated by Langbein (1949) and Chow (1950).

Where two types of flood peaks occur each year, such as spring snowmelt floods and winter rainstorm floods, or hurricane and nonhurricane floods, they are known as mixed population data that require special treatment, as discussed subsequently.

For a reliable computation of a flood-frequency curve, the peak-discharge data should meet the following requirements.

8.5.1 Stationariness of Data

The characteristics of stationary data do not change with time. As described in Section 7.14, stationariness can be checked by dividing a long flood flow series into a number of segments (subsets). The statistical parameters of mean, standard deviation, and coefficient of skew should be comparable for each subset. For data of short length, it is not feasible to perform this test, but their adequacy can be checked as per Sections 8.5.3 and 8.5.4.

8.5.2 Homogeneity of Data

Homogeneity is an indicator that all data of a series belong to the same population. The homogeneity check may be performed qualitatively by studying factors that have a disturbing effect on it. The quantitative analysis is made by the statistics theory known as *test of hypothesis*. According to this, it is hypothesized (assumed) that the data follow a certain distribution. An acceptable probability, or risk level, of making the wrong decision from the hypothesis is specified; this is known as the *level of significance*.

In flood flow computations the significance level is usually assumed to be 1, 2, or 5%. As an example, a 5% level of significance means that there is about a 5% chance that a hypothesis will be rejected when it should have been accepted. This represents an area at the extreme end of the probability distribution curve and is referred to as the *critical region*, as shown in Figure 8.5. Using the equation of the hypothesized distribution and the statistical parameters derived from the sample, the value of the standard variate, called the *test statistic*, is computed. This is compared to the theoretical value of the variate obtained from a standard distribution table for the specified level of significance. If the computed value is more than the tabular or theoretical value, it belongs to the critical region or in the region of rejection of the hypothesis. The hypothesis is accepted if the computed value is lower than the tabular value.

The distributions suitable for small samples of less than 30 values, as is usually the case with flood flow series, are Student's distribution and the chi-square distribution. Both of these assume that the population, from which a sample has been derived, is normally distributed. Often the logarithmic values of peak discharges are used in analysis since the flood flows are normally distributed in that form.

To test the hypothesis that two samples of sizes N_X and N_Y come from the same population, the standard variate or test statistic of Student's distribution has the following form:

$$t = \frac{|\bar{Y} - \bar{X}|}{\sqrt{N_X S_X^2 + N_Y S_Y^2}} \sqrt{\frac{N_X N_Y \nu}{N_X + N_Y}} \quad \text{[dimensionless]} \qquad (8.5)$$

Figure 8.5 Critical region in hypothesis testing.

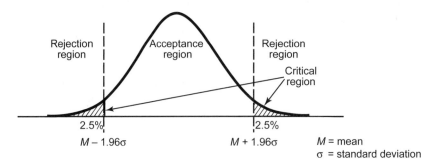

where

$$\overline{X}, \overline{Y} = \text{mean values of two samples}$$
$$S_X, S_Y = \text{standard deviations of two samples}$$
$$v = \text{degrees of freedom} = N_X + N_Y - 2$$

The theoretical standard variates for Student's distribution are given in Appendix G.

Equation (8.5) tests the homogeneity of the mean of the series. In its formulation it was considered that the two series are homogeneous with respect to the standard deviations. Hence, before performing the test above by Student's criterion, the homogeneity for the standard deviation is tested by the Fisher distribution, which is an extension of the chi-square distribution. The statistic, or variate, of χ^2 [χ is the Greek lowercase letter chi (pronounced "kai")] is given by $\chi^2 = NS^2/\sigma^2$, where σ (Greek lowercase letter sigma) is the standard deviation of the population, of which a sample of size N has the standard deviation S. The distribution has a degree of freedom, $v = N - 1$. For two samples, one of chi-square variate with $v = m$ degrees of freedom and the other of chi-square variate with $v = n$ degrees of freedom, the Fisher distribution is

$$F = \frac{\chi_m^2}{\chi_n^2}$$

or

$$F = \frac{S_X^2}{S_Y^2} \quad \text{[dimensionless]} \tag{8.6}$$

The cumulative F distribution with $v_1 = m$ and $v_2 = n$ degrees of freedom (m and n are considered as the numerator and denominator degrees of freedom, respectively) are given in Appendix H. As in the case of Student's criterion, when the computed F value is less than the theoretical value from the standardized statistical table at a specified level of significance, the hypothesis cannot be rejected (is accepted).

EXAMPLE 8.4

The annual peak discharges of the Merrimack River, below Manchester, New Hampshire, are given in Table 8.2. The discharges in parentheses have been contributed by snowmelts

436   **Computation of Extreme Flows   Chapter 8**

and the others by rainstorms. Determine whether the two types of peak discharges are part of a single population of flood peaks. Adopt a 5% level of significance.

SOLUTION

1. The sequence above is divided into two series, one for rainstorm discharges and the other for snowmelt discharges.

2. The statistical parameters of mean and standard deviation for the two series, as computed by eqs. (5.26) and (5.27), are listed in Table 8.3.

3. Since the homogeneity of standard deviations is a prerequisite for Student's criterion for the mean, the Fisher distribution is tested first. From eq. (8.6),

$$F(\text{computed}) = \frac{S_X^2}{S_Y^2} = \frac{(14,570)^2}{(12,248)^2} = 1.42$$

4. Using $(N_X - 1) = 17$ degrees freedom for S_X (numerator) and $(N_Y - 1) = 21$ degrees of freedom for S_Y (denominator), F (theoretical), from the table in Appendix H at 5% significance level,[*] is $F = 2.15$.

Table 8.2   Annual Peak Discharge Data for Merrimack River below Manchester, NH

Date	Discharge (cfs)	Date	Discharge (cfs)	Date	Discharge (cfs)
12/16/54	18,805	3/21/68	32,200	2/27/81	36,708
4/18/56	(29,604)	3/28/69	(27,105)	6/8/82	(19,700)
3/1/57	6,768	4/13/70	(23,006)	3/20/83	45,700
4/27/58	(27,903)	4/5/71	15,103	4/7/84	(32,000)
4/5/59	33,006	3/25/72	20,304	3/14/85	(14,208)
4/6/60	44,705	4/4/73	(30,305)	1/29/86	19,307
4/17/61	(18,602)	12/18/73	14,308	4/7/87	(75,706)
4/2/62	(33,203)	4/5/75	(15,405)	3/29/88	(13,905)
12/8/62	23,908	11/15/75	22,403	11/4/88	17,403
11/9/63	12,406	3/15/77	50,006	4/3/90	15,000
2/26/65	7,887	1/29/78	(16,106)	8/22/91	17,208
3/27/66	(10,705)	3/9/79	(29,704)	3/25/92	(13,105)
4/4/67	24,805	4/12/80	16,004	4/1/93	34,005
				3/26/94	(22,106)

Table 8.3   Statistical Parameters for Snowmelt and Rainstorm Series

	Snowmelt (X)	Rainstorm (Y)
1. Number of floods	18	22
2. Mean, $\bar{X}$ (cfs)	25,132	23,998
3. Standard deviation, S (cfs)	14,570	12,248

[*] For 5% or 0.05 significance, α in the table $= 1 - 0.05 = 0.95$.

5. Since F (computed) $< F$ (theoretical), the hypothesis cannot be rejected. The data are homogeneous with respect to the standard deviations.

6. After the Fisher criterion is satisfied, Student's criterion is, from eq. (8.5),

$$t(\text{computed}) = \frac{|\bar{Y} - \bar{X}|}{\sqrt{N_X S_X^2 + N_Y S_Y^2}} \sqrt{\frac{N_X N_Y v}{N_X + N_Y}}$$

$$v = N_X + N_Y - 2 = 18 + 22 - 2 = 38$$

$$t(\text{computed}) = \frac{|23{,}998 - 25{,}132|}{\sqrt{18(14{,}570)^2 + 22(12{,}248)^2}} \sqrt{\frac{18(22)(38)}{18 + 22}}$$

$$= 0.26$$

7. From the table in Appendix G for the 5% level and 38 degrees of freedom, t (theoretical) = 1.68.

8. Since t (computed) $< t$ (theoretical), there is a homogeneity of means. Hence the data belong to a single homogeneous population.

8.5.3 Consistency of Data

The record of peak stages and discharges should be as complete as possible. The peak stage and discharge, when an instrument has failed, should be determined from high-water marks in the vicinity of the gaging station. If there is a reason for doubting that the data for the entire period are not related to the same datum, the streamflow data of a particular site should be checked for consistency and accuracy by comparison with several surrounding gaging sites by the mass curve analysis (Section 2.6.2). The inconsistencies should be reconciled and erroneous data should be recomputed or excluded. Natural flood flows are the basic data required for hydrologic design. Where regulated flows are available, the effect of regulation should be corrected by the flood-routing techniques (Chapter 9) to obtain natural discharge values.

8.5.4 Adequacy of Data

The length of record is an important factor since a short-duration record may not be representative of the true nature of peak flows at a site. The computed statistical parameters will thus not be reliable. A minimum period of record of 25 years has been considered desirable for the statistical analysis of peak flows. For a shorter observation period, the adequacy of the record may be evaluated by analysis with respect to a long-term base gaging station in the region (Sokolov et al., 1976).

Consider that the gaging station at the study site has a short record of S years. Also, that a base gaging station in the homogeneous region has a long-term record of L years that includes the period of S years. For the base gaging station, the statistical parameters (mean, $\bar{X}$; standard deviation, S_X; and coefficient of skewness, g) are computed for two sets of records of period S and L years, respectively. If the ratios of parameters for short and long length, $\bar{X}_S / \bar{X}_L, (S_X)_S / (S_X)_L$, and g_S/g_L, do not depart from unity by more than 15%, the short-term record of the study site is considered adequate. If these ratios show more than a 15% difference, the short-period record is extended by correlation analysis (Section 7.16) with the long-term base gage station.

8.6 Methods of Flood-Frequency Analysis

There are three methods to prepare a flood-frequency curve from the array of flood flow data: (1) graphical method, (2) empirical method, and (3) analytical method. The last of these has a wider application. The first two methods make a plot on probability graph paper as described below.

8.6.1 Probability Graph Paper

The plot of peak-discharge magnitude against probability of exceedance (cumulative distribution function) is a curve on arithmetic paper. The purpose of probability paper is to linearize this plot, so that extrapolation of the data, as often needed, is simplified. The equation of the cumulative distribution function (CDF) has to be transformed to the form $Y = mX + C$ to plot as a straight line, where X is a function of exceedance probability and statistical parameters and Y indicates the peak flow. On graph paper, the linear transformation can be achieved by the distortion of the probability scale (abscissa).

Since there are different equations of the CDF, separate graph paper has to be constructed for each theoretical distribution. Further, since the transformed function includes the statistical parameters, it is feasible to construct the paper for a distribution that is defined by two parameters only. Hence, probability paper has been designed for normal distribution, lognormal distribution (the ordinate is in log scale), type I extreme value or Gumbel distribution (two parameters and a fixed skew of 1.14), and type III extreme value or Weibull distribution, which is essentially a logarithmically transformed type I distribution.

The frequency factors, K, related to various distributions have a linear form as described in Section 8.10. These factors for normal, Pearson type III, and extreme value distributions are listed in Tables 8.6 through 8.8. If the K values from a distribution table are plotted to a scale on the abscissa of Cartesian (regular grid) paper and labeled by their respective probability level, the resulting graph will be a probability graph paper for that distribution. The theoretical distributions will plot as a straight line on respective papers. The natural flow data do not necessarily follow any exact theoretical distribution.

8.7 Graphical Method

In this method, the array of flood flows is divided into a number of class intervals of equal range in discharge. The number of occurrences of flood flows in each class interval is noted. The number of occurrences in each class interval is cumulated, starting with the highest value. The percentage of the accumulated number of items or occurrences of each class with respect to the total occurrences of all classes is determined. The computed percent is then plotted against the lower discharge limit of each class on probability paper. The plotting paper commonly used in the graphic method is the lognormal probability graph.

This procedure can be applied only when the array consists of a very large number of flood events. The method has been described in detail in connection with the flow-duration curve or complete series analysis (Section 7.27.2). Typically, however, the record of peak-flow data is not extensive enough to support this method. As such, the other two methods are more often used.

8.8 EMPIRICAL METHOD

This is also a graphic procedure. In this method, however, the plotting position of the magnitude of a flood is determined by an empirical formula. If an array of n flood flow values is arranged in descending order of magnitude starting with the highest discharge, when n approaches infinity, a discharge ranking m in order of magnitude will have an exceedance probability as follows:

$$P_m = \left(\frac{m}{n}\right)_{n \to \infty} \quad \text{[dimensionless]} \tag{8.7}$$

Equation (8.7), when applied to a smaller sequence, however, will assign a probability of 100% to the lowest-valued flood of the sequence having a rank of n, which means that there is no possibility of a flow of less than that value. This is obviously erroneous. To remove the bias in plotting positions at two extreme ends, analysts have suggested many empirical formulas that are special cases of the following general formula:

$$P_m = \frac{m-a}{n+b}(100) \quad \text{[dimensionless]} \tag{8.8}$$

where a and b are constants. Several of these formulas are summarized by Adamowski (1981). All formulas give practically the same results in the middle range of discharge but produce different positions near the upper and lower tails of the distribution.

The following formula proposed by Weibull in 1939 is widely used:

$$P_m = \frac{m}{n+1}(100) \quad \text{[dimensionless]} \tag{8.9}$$

For an annual series, the return period or recurrence interval T is the inverse of P_m.

The steps in the empirical procedure are summarized below.

1. Rank the data from the largest to the smallest values. If two or more observations have the same value, count each of them separately and assign the average rank.

2. Calculate the plotting position from eq. (8.9).

3. Do not omit any years during the period of record since it will have a biasing effect. If any data are missing, make their estimates (Section 7.13). The data could be excluded when the cause of the interruption in data is known to be independent of the flow condition.

4. Often, data on one or more historical flood events that occurred prior to the period of record may be known. The plotting positions for the historical events and other peak flows higher than the historical events are determined based on the total period from the time of the first historical event to the end of the flood flow record. The plotting positions of other peak flows are based on the period of record of the data only. For example, suppose flood flows are available for a period of 40 years, from 1967 to 2006. The magnitude of a historical peak flow that occurred in 1908 has been observed which is the highest of all recorded flows. The plotting positions for historical flow will be based on 99 years (1908 to 2006), while that of other floods will be based on 40 years of record. The modification of the plotting positions due to historical floods has been discussed by Dalrymple (1960, pp. 16 – 18).

5. Select the type of probability paper to be used; lognormal graph paper is common.

6. Plot the magnitude of flood on the ordinate and the corresponding plotting position on the abscissa, representing the probability of exceedance as one side of the scale and the return period on the other side.

Frequently, one or two extreme events may plot far off from other points as "outliers." This occurs because an extreme event in the recorded data may actually represent a much higher return period than the period of record. There is a probability of at least one occurrence of a T-year (higher) event in a n-year (smaller) period of record from eq. (8.4). If possible, the return period of these extreme events might be investigated based on available historical or regional information. A test for detection of high and low outliers and their treatment are described by the U.S. Water Resources Council (1981, pp. 17–18).

The extrapolation of the data for longer return periods should be done very cautiously because the probability distribution is very sensitive in the tail part of the curve.

EXAMPLE 8.5

The maximum annual instantaneous flows of a river are given in Table 8.4. From the historic record, a peak flow of 30,000 cfs was noted in 1938. **(a)** Prepare a flood-frequency curve, **(b)** determine the probability of flow of 20,000 cfs, **(c)** determine the magnitude of flow corresponding to an exceedance probability of 0.5, and **(d)** determine the magnitude of flow of a return period of 100 years.

Table 8.4   Annual Peak Flows of the River in Example 8.5

(1)	(2)	(3)	(4) Plotting Position (%)	(1)	(2)	(3)	(4) Plotting Position (%)
Year	Flow (cfs)	Rank		Year	Flow (cfs)	Rank	
1982	14,400	5	20	1994	6,240	17	68
1983	6,720	16	64	1995	22,700	1	4
1984	13,390	7	28	1996	11,140	10	40
1985	15,360	4	16	1997	4,560	21	84
1986	8,856	13	52	1998	5,376	19	76
1987	5,136	20	80	1999	12,480	9	36
1988	6,770	15	60	2000	19,200	3	12
1989	9,600	12	48	2001	12,984	8	32
1990	980	24	96	2002	5,450	18	72
1991	4,030	22	88	2003	13,440	6	24
1992	10,440	11	44	2004	22,680	2	8
1993	3,100	23	92	2005	8,400	14	56

SOLUTION

(a) Flood-frequency curve

1. The rank of each flow value, starting with the highest flood, is indicated in Table 8.4. Many designers prefer to make a separate table by arranging values in descending order.

2. For historic flow, the base period is 68 years (the period between 1938 and 2005); hence the plotting position.

$$P = \frac{1}{69}(100) = 1.45\%$$

3. For each rank in Table 8.4, the plotting position is computed from eq. (8.9) for $n = 24$.

4. Peak discharge versus plotting position (exceedance probability) is plotted on log-normal paper in Figure 8.6.

5. The following values are read directly from the graph.

(b) Probability of flow of 20,000 cfs = 8% or 0.08

(c) Flow of 0.5 (or 50%) probability = 9500 cfs

(d) For $T = 100$ years, $P = \frac{1}{100} (100) = 1.0\%$

Flow for 1.0% probability = 28,000 cfs

8.9 ANALYTICAL METHOD

This method makes use of the theoretical probability distribution functions. The plot of the cumulative density function (CDF) of any distribution is a frequency curve by definition. The CDF equations contain statistical parameters that are computed from recorded data series. The requirement, however, is that the theoretical distribution should represent characteristics similar to those demonstrated by the natural recorded data. There are many distributions; four commonly applied for fitting the hydrologic sequences are described below. The features of these distributions, comprising PDF, range, mean, and variance, are summarized in Table 8.5. Of these, the log-Pearson type III has been widely adopted as the distribution of choice in flood flow analysis. The extreme value type III distribution is preferred in the study of low flows (droughts).

8.9.1 Normal (Gaussian) Distribution

This is a bell-shaped frequency function symmetrical about the mean value. It has very wide applications, although, due to its range from $-\infty$ to $+\infty$, it does not fit well to hydrologic sequences that do not have negative values. The distribution has two parameters, the mean, μ, and the standard deviation, σ ($\bar{X}$ and S for sample data), as shown in Table 8.5. It can be transformed to a single-parameter function using a standard variate, z, in terms of μ and σ by defining $z = (x - \mu)/\sigma$. The standard variate z is normally distributed with zero mean and variance of one.

8.9.2 Lognormal Distribution

This is an extension of the normal distribution wherein the logarithmic values of a sequence are considered to be normally distributed. The PDF and all other properties of the normal distribution are applicable to this distribution when the data are converted to logarithmic form, $y = \ln x$, as indicated in Table 8.5. It is a two-parameter, bell-shaped, symmetrical distribution in this form. In terms of an untransformed variate, x, it is a three-parameter (skewed) distribution having a range from 0 to ∞.

Figure 8.6 Flood-frequency curve for Example 8.5.

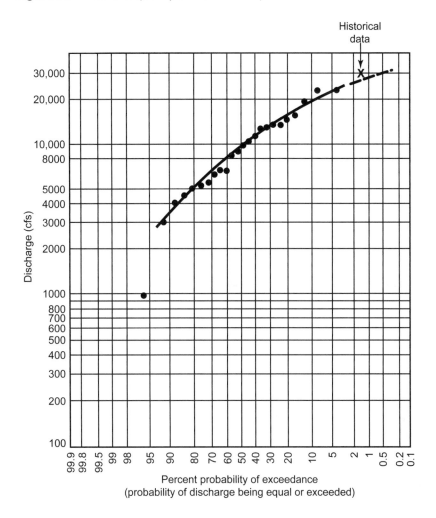

This distribution suits hydrologic data and has the advantage of a link with the normal distribution.

8.9.3 Extreme Value Distribution

Consider n data series with m observations in each series. A largest or smallest (extreme) value is obtained out of m observations in each series. There will be n such extreme values. The probability distribution of these extreme values depends on the sample size, m, and the parent distribution of the series. Frechet, in 1927, and Fisher and Tippett, in 1928, found that the distribution of extreme values approaches an asymptotic form as m is increased indefinitely. The type of the asymptotic form is dependent on the parent distribution from which the extreme values were obtained. Three types of asymptotic distributions have been developed based on different parent distributions.

The type I extreme value distribution, also known as the Gumbel distribution, results from the exponential-type parent distribution. The parent distribution is unbounded (has

Table 8.5   Properties of Common Distributions

Distribution	Probability Density Function (PDF), $p(X)$	Cumulative Density Function (CDF), $P(X \le x)$	Range	Mean μ or $\overline{X}$	Standard Deviation σ or S
1. Normal	$\dfrac{1}{\sqrt{2\pi}\,\sigma}e^{-(x-\mu)^2/2\sigma^2}$ or $\dfrac{1}{\sqrt{2\pi}}e^{-z^2/2}$ where $z=\dfrac{x-\mu}{\sigma}$	$\displaystyle\int_{-\infty}^{x}\dfrac{1}{\sqrt{2\pi}\,\sigma}e^{-(x-\mu)^2/2\sigma^2}\,dx$	$-\infty \le x \le \infty$	μ	σ
2. Lognormal $y=\ln x$	$\dfrac{1}{\sqrt{2\pi}\,\sigma_y}e^{-(y-\mu_y)^2/2\sigma_y^2}$	$\displaystyle\int_{-\infty}^{y}\dfrac{1}{\sqrt{2\pi}\,\sigma_y}e^{-(y-\mu_y)^2/2\sigma_y^2}\,dy$	$-\infty \le y \le \infty$ $0 \le x \le \infty$	μ_y	σ_y
3. Extreme value Type I, $y=(x-\beta)/\alpha$	$\dfrac{1}{\alpha}e^{-y-e^{-y}}$	$e^{-e^{-y}}$	$-\infty \le x \le \infty$	$\beta + 0.577\alpha$	1.283α
Type III	$\alpha\,x^{\alpha-1}\beta^{-\alpha}e^{-(x/\beta)\alpha}$	$1-e^{-(x/\beta)\alpha}$	$x \ge 0$	$\beta\Gamma(1+1/\alpha)$	$\beta[\Gamma(1+2/\alpha)-\Gamma^2(1+1/\alpha)]^{1/2}$
4. Log-Pearson Type III $y=\ln x$	$p_0\left(1+y/\alpha\right)^c e^{-cy/\alpha}$	$\displaystyle\int_{-\infty}^{y}p_0\left(1+y/\alpha\right)^c e^{-cy/\alpha}\,dy$ (known as incomplete gamma function)	$-\infty \le y \le \infty$ $0 \le x \le \infty$	$(c+1)\dfrac{\alpha}{c}$	$\sqrt{c+1}\,\dfrac{\alpha}{c}$
	where p_0 = prob. at the mode $=\dfrac{N}{\alpha\,e^c\,\Gamma(c+1)}c^{c+1}$				

Γ is the gamma function; $\Gamma(n) = (n-1)!$.

α and β are evaluated from relations under the columns of mean and standard deviation.

c and α are evaluated from relations under the columns of mean and standard deviation.

no limit) in the direction of the extreme value. The density functions of the type I distribution, which are in terms of parameters α and β, are given in Table 8.5. Using the mean and the standard deviation of the flood flow series, α and β can be evaluated from the relations given under mean and standard deviation in Table 8.5. The distribution has a constant skew coefficient of ± 1.14. Gumbel used the type I distribution first in an analysis of floods in 1941. He argued that the daily discharge of each year constituted a sufficiently large sample, with $m = 365$, from which an extreme value of flood flow was picked up.

The type II distribution originates from the Cauchy-type distribution of the parent distribution, but it has little application in hydrologic events.

The type III or Weibull distribution also arises from an exponential-type parent distribution, but the parent distribution is limited in the direction of the extreme value (e.g., the low flows are bounded by zero on the left). The density functions are given in Table 8.5. This distribution is essentially a logarithmically transformed type I distribution. Gumbel applied this for low-flow analyses.

8.9.4 Log-Pearson Type III (Gamma-Type) Distribution

Karl Pearson proposed a general equation for a distribution that fits many distributions— including normal, beta, and gamma distributions—by choosing appropriate values for its parameters. A form of the Pearson function, similar to the gamma distribution, is known as the Pearson type III distribution. It is a distribution in three parameters with a limited range in the left direction, unbounded to the right, and has a large skew. Since the flood flow series commonly indicate considerable skew, this is used as the distribution of flood peaks. The distribution is usually fitted to the logarithms of flood values because this results in lesser skewness. The density functions of the distribution in terms of c and α are given in Table 8.5. The values of c and α are evaluated from the relations given under the columns of mean and standard deviation in the table. The log-Pearson type III distribution has been adopted as a standard by U.S. federal agencies for flood analyses.

8.9.5 Probability Distribution of Extreme Flow Data

Section 5.10 and Section 7.20.2 discussed, in the context of groundwater monitoring and streamflow generation, respectively, the procedures for checking whether the sample data belong to normal, lognormal, or gamma (similar to log-Pearson type III) distribution. The same procedures are applicable to extreme flows as well. However, the log-Pearson type III is a commonly used distribution in flood studies and the logarithmic form of extreme value type I (which is essentially type III in nonlogarithmic form) is common in studies of droughts. The recorded data should be plotted with the computed data for an assessment of the distribution used. The distribution should be changed if two sets of data do not compare reasonably.

8.10 APPROACH TO ANALYTICAL METHOD

The CDF indicated in Table 8.5 for various probability distributions is of the type for which a direct integration is generally not possible. However, integration tables have been developed for various distributions from approximate or numerical analyses. These tables are used to obtain the cumulative (exceedance) probability for a desired magnitude of flow corresponding to the statistical parameters from the sample data.

A simplified approach was suggested by Chow (1951). He suggested that most frequency functions applicable to hydrologic sequences, including the four distributions previously discussed, can be resolved to the linearized form

$$X = \bar{X} + KS \quad [\text{L}^3\text{T}^{-1}] \tag{8.10}$$

where

X = flood of a specified probability

$\bar{X}$ = mean of the sample (observed data)

S = standard deviation of the sample

K = frequency factor

8.10.1 Use of Frequency Factors

The frequency factor is a property of a specific probability distribution at a specified probability level. For a given distribution, the relationship has been developed between the frequency factor and the corresponding return interval, known as the K-T relationship. For various distributions, these are expressed in mathematical terms, by tables or by curves called K-T curves.

The procedure used for analysis is as follows:

1. Compute the statistical parameters (mean, standard deviation, and skewness coefficient, if necessary) from the flood flow series.

2. Use the K-T relationship (commonly in the form of a table) for the proposed distribution.

3. For a given return interval, determine the corresponding frequency factor from the K-T relation of step 2.

4. Compute the magnitude of flood by eq. (8.10).

5. Repeat steps 3 and 4 for various exceedance probabilities (return intervals) and make a frequency plot. The normal distribution is plotted on normal probability paper; the lognormal and log-Pearson III distributions are plotted on lognormal paper and the extreme value distribution is graphed on Gumbel extreme probability paper.

The frequency curve computed by the above procedure is based on the sample statistics. It is biased in relation to average future expectation because of uncertainty as to the true mean and standard deviation. The effect of this bias can be eliminated by an adjustment called the *expected probability adjustment* that accounts for the actual sample size. This is demonstrated in Section 8.11.

The K-T relations for normal, log-Pearson type III, and extreme value type I distributions are given in Table 8.6 through Table 8.8. The values of K in Table 8.6 are equivalent to the normal standard variate z as defined in Section 8.9.1 and available from a normal probability distribution table.

8.10.2 Generalized Skew Coefficient

In the log-Pearson type III distribution, the frequency factor K is dependent on the skew coefficient, g. It is difficult to obtain accurate skew estimates from sample data that usually comprise less than 100 events. The Interagency Advisory Committee on Water Data (IACWD, 1982) Guide recommends use of a weighted generalized skew coefficient. From

Table 8.6   Frequency Factor for Normal Distribution

Exceedance Probability	Return Period	K	Exceedance Probability	Return Period	K
0.0001	10,000	3.719	0.450	2.22	0.126
0.0005	2,000	3.291	0.500	2.00	0.000
0.001	1,000	3.090	0.550	1.82	−0.126
0.002	500	2.88	0.600	1.67	−0.253
0.003	333	2.76	0.650	1.54	−0.385
0.004	250	2.65	0.700	1.43	−0.524
0.005	200	2.576	0.750	1.33	−0.674
0.010	100	2.326	0.800	1.25	−0.842
0.025	40	1.960	0.850	1.18	−1.036
0.050	20	1.645	0.900	1.11	−1.282
0.100	10	1.282	0.950	1.053	−1.645
0.150	6.67	1.036	0.975	1.026	−1.960
0.200	5.00	0.842	0.990	1.010	−2.326
0.250	4.00	0.674	0.995	1.005	−2.576
0.300	3.33	0.524	0.999	1.001	−3.090
0.350	2.86	0.385	0.9995	1.0005	−3.291
0.400	2.50	0.253	0.9999	1.0001	−3.719

the data of all nearby stations within a 100-mile radius, a regionalized skew coefficient for the specific site is estimated by the methods suggested in the guide. In the absence of detailed records, the regionalized skew coefficient could be read from the map in Figure 8.7 adapted from the guide. The weighted generalized skew is obtained by the weighted average of the sample (station) skewness and the regional (map) skewness, as follows:

$$g = Wg_s + (1 - W)g_m \quad \text{[dimensionless]} \tag{8.11}$$

where

g = generalized skew coefficient

W = weighted factor

g_s = sample skew coefficient

g_m = map (regional) skew coefficient

Tung and Mays (1981) and the IACWD (1982) suggested assigning weights in accordance with the variance of the sample skew and the variance of the regional skew. Accordingly,

$$W = \frac{V(g_m)}{V(g_s) + V(g_m)} \quad \text{[dimensionless]} \tag{8.12}$$

where $V(\ )$ stands for the mean square error of the variable indicated within the parentheses in the equation.

Table 8.7 Frequency Factors for Log-Pearson Type III Distribution

Skew Coefficient, g	Probability							
	0.99	0.80	0.50	0.20	0.10	0.04	0.02	0.01
	Return Period							
	1.0101	1.2500	2	5	10	25	50	100
3.0	−0.667	−0.636	−0.396	0.420	1.180	2.278	3.152	4.051
2.8	−0.714	−0.666	−0.384	0.460	1.210	2.275	3.114	3.973
2.6	−0.769	−0.696	−0.368	0.499	1.238	2.267	3.071	3.889
2.4	−0.832	−0.725	−0.351	0.537	1.262	2.256	3.023	3.800
2.2	−0.905	−0.752	−0.330	0.574	1.284	2.240	2.970	3.705
2.0	−0.990	−0.777	−0.307	0.609	1.302	2.219	2.912	3.605
1.8	−1.087	−0.799	−0.282	0.643	1.318	2.193	2.848	3.499
1.6	−1.197	−0.817	−0.254	0.675	1.329	2.163	2.780	3.388
1.4	−1.318	−0.832	−0.225	0.705	1.337	2.128	2.706	3.271
1.2	−1.449	−0.844	−0.195	0.732	1.340	2.087	2.626	3.149
1.0	−1.588	−0.852	−0.164	0.758	1.340	2.043	2.542	3.022
0.8	−1.733	−0.856	−0.132	0.780	1.336	1.993	2.453	2.891
0.6	−1.880	−0.857	−0.099	0.800	1.328	1.939	2.359	2.755
0.4	−2.029	−0.855	−0.066	0.816	1.317	1.880	2.261	2.615
0.2	−2.178	−0.850	−0.033	0.830	1.301	1.818	2.159	2.472
0	−2.326	−0.842	0	0.842	1.282	1.751	2.054	2.326
−0.2	−2.472	−0.830	0.033	0.850	1.258	1.680	1.945	2.178
−0.4	−2.615	−0.816	0.066	0.855	1.231	1.606	1.834	2.029
−0.6	−2.755	−0.800	0.099	0.857	1.200	1.528	1.720	1.880
−0.8	−2.891	−0.780	0.132	0.856	1.166	1.448	1.606	1.733
−1.0	−3.022	−0.758	0.164	0.852	1.128	1.366	1.492	1.588
−1.2	−3.149	−0.732	0.195	0.844	1.086	1.282	1.379	1.449
−1.4	−3.271	−0.705	0.225	0.832	1.041	1.198	1.270	1.318
−1.6	−3.388	−0.675	0.254	0.817	0.994	1.116	1.166	1.197
−1.8	−3.499	−0.643	0.282	0.799	0.945	1.035	1.069	1.087
−2.0	−3.605	−0.609	0.307	0.777	0.895	0.959	0.980	0.990
−2.2	−3.705	−0.574	0.330	0.752	0.844	0.888	0.900	0.905
−2.4	−3.800	−0.537	0.351	0.725	0.795	0.823	0.830	0.832
−2.6	−3.889	−0.499	0.368	0.696	0.747	0.764	0.768	0.769
−2.8	−3.973	−0.460	0.384	0.666	0.702	0.712	0.714	0.714
−3.0	−4.051	−0.420	0.396	0.636	0.660	0.666	0.666	0.667

Table 8.8   Frequency Factors for Extreme Value Type I Distribution

	Probability								
	0.2	0.1	0.067	0.05	0.04	0.02	0.0133	0.01	0.001
Sample					Return Period				
Size, *n*	5	10	15	20	25	50	75	100	1000
15	0.967	1.703	2.117	2.410	2.632	3.321	3.721	4.005	6.265
20	0.919	1.625	2.023	2.302	2.517	3.179	3.563	3.836	6.006
25	0.888	1.575	1.963	2.235	2.444	3.088	3.463	3.729	5.842
30	0.866	1.541	1.922	2.188	2.393	3.026	3.393	3.653	5.727
35	0.851	1.516	1.891	2.152	2.354	2.979	3.341	3.598	
40	0.838	1.495	1.866	2.126	2.326	2.943	3.301	3.554	5.576
45	0.829	1.478	1.847	2.104	2.303	2.913	3.268	3.520	
50	0.820	1.466	1.831	2.086	2.283	2.889	3.241	3.491	5.478
55	0.813	1.455	1.818	2.071	2.267	2.869	3.219	3.467	
60	0.807	1.446	1.806	2.059	2.253	2.852	3.200	3.446	
65	0.801	1.437	1.796	2.048	2.241	2.837	3.183	3.429	
70	0.797	1.430	1.788	2.038	2.230	2.824	3.169	3.413	5.359
75	0.792	1.423	1.780	2.029	2.220	2.812	3.155	3.400	
80	0.788	1.417	1.773	2.020	2.212	2.802	3.145	3.387	
85	0.785	1.413	1.767	2.013	2.205	2.793	3.135	3.376	
90	0.782	1.409	1.762	2.007	2.198	2.785	3.125	3.367	
95	0.780	1.405	1.757	2.002	2.193	2.777	3.116	3.357	
100	0.779	1.401	1.752	1.998	2.187	2.770	3.109	3.349	5.261
∞ [a]	0.719	1.305	1.635	1.866	2.044	2.592	2.911	3.137	4.936

[a]Additional data for $n = \infty$:

Probability	K
0.3	0.354
0.4	0.0737
0.5	−0.164
0.6	−0.383
0.8	−0.821
0.9	−1.100

Figure 8.7 Map skew coefficients of logarithmic annual maximum streamflow (Interagency Advisory Committee on Water Data, 1982).

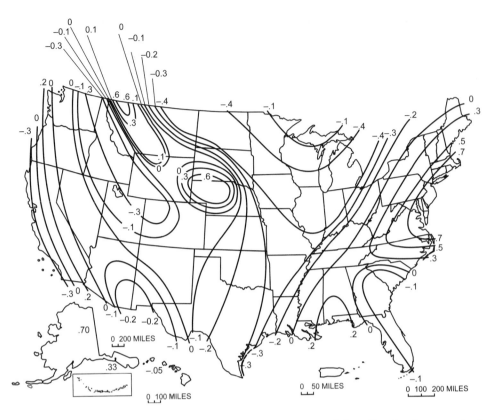

The value of $V(g_m)$ has to be estimated from the regional data. For the skew map in the guide of the IACWD, $V(g_m)$ is estimated to be 0.3025.

Tung and Mays (1981) have discussed various methods to compute $V(g_s)$. The IACWD suggests the following approximate equation:

$$V(g_s) = 10^{A-B \log N/10} \qquad (8.13)$$

where

N = record length in years

$A = -0.33 + 0.08|g_s|$ for $|g_s| \le 0.9$

$A = -0.52 + 0.30|g_s|$ for $|g_s| \ge 0.9$

$B = 0.94 - 0.26|g_s|$ for $|g_s| \le 1.5$

$B = 0.55$ for $|g_s| > 1.5$

$|g_s|$ = absolute value of the station skew, g_s

Equation (8.11) gives improper weight to the regional skew if the regional and sample skews differ by more than 0.5. In such a situation, the weight should be determined by studying the flood-producing characteristics of the watershed.

In Example 8.6, an application of the log-Pearson type III distribution has been made. The procedure is the same for other distributions except that different K-T tables have to be used.

EXAMPLE 8.6

On the flood flow sequence of Example 8.5, perform the frequency analysis by the theoretical method, adopting the log-Pearson type III distribution. The generalized skew from the regional map is -0.7.

SOLUTION

1. The data are converted to log form by $y = \log x$.
2. The statistical parameters computed by eqs. (5.26) through (5.28) are as follows:
    a. Mean, $\bar{X} = 3.920$ cfs
    b. Standard deviation, $S = 0.308$ cfs
    c. Coefficient of skewness, $g = -1.1$
3. $V(g_m) = 0.3025$ and for $N = 24$:

    $A = -0.52 + 0.30(1.1) = -0.19$

    $B = 0.94 - 0.26(1.1) = 0.654$

    From eq. (8.13),

    $$V(g_s) = 10^{-0.19-0.654 \log 24/10} = 0.364$$

    From eq. (8.12),

    $$W = \frac{0.3025}{0.3025 + 0.364} = 0.45$$

    From eq. (8.11),

    $$g = 0.45(-1.1) + 0.55(-0.7) = -0.88 \quad \text{or} \quad -0.9$$

4. For various percent exceedance probabilities, peak flows are computed in Table 8.9. The values of K (frequency factor) in column 2 are obtained from Table 8.7.
5. The data are plotted on lognormal paper in Figure 8.8.
6. For $T = 100$ or $P = 1\%$, flow = 27,000 cfs

8.11 CONFIDENCE LIMITS AND PROBABILITY ADJUSTMENTS

Frequency curve by the analytical method is only an estimate of the population curve. The uncertainty (error) of the estimated peak flow for a given probability is a function of the errors in estimating the mean and standard deviation for a known skew coefficient. Confidence limits provide a measure of uncertainty of the estimated peak flow. Beard (1962) proposed a method for constructing the error limit curves above and below a theoretically fitted frequency curve to form a reliability band. Table 8.10 provides the factors by which the standard deviation of flood series is multiplied to get the error limits. For a 5% error curve, the flood values from the fitted frequency curve are added to the computed error limits for the corresponding exceedance probabilities. For a 95% curve, the error limits are

Table 8.9 Computation of Peak Flows of Different Exceedance Probability

(1) Percent Probability	(2) K	(3) $X = \overline{X} + KS$	(4) $Q = \log^{-1}X\,(\text{cfs})$
1	1.66	4.431	27,000
4	1.407	4.353	22,540
10	1.147	4.273	18,750
20	0.854	4.183	15,240
50	0.148	3.966	9,250
80	−0.769	3.683	4,820
90	−1.339	3.507	3,210
99	−2.957	3.009	1,020

Table 8.10 Error Limits for Frequency Curve

Years of Record, N	Percent Exceedance Frequency (at 5% Level of Significance)[a]						
	0.1	1	10	50	90	99	99.9
5	4.41	3.41	2.12	0.95	0.76	1.00	1.22
10	2.11	1.65	1.07	0.58	0.57	0.76	0.94
15	1.52	1.19	0.79	0.46	0.48	0.65	0.80
20	1.23	0.97	0.64	0.39	0.42	0.58	0.71
30	0.93	0.74	0.50	0.31	0.35	0.49	0.60
40	0.77	0.61	0.42	0.27	0.31	0.43	0.53
50	0.67	0.54	0.36	0.24	0.28	0.39	0.49
70	0.55	0.44	0.30	0.20	0.24	0.34	0.42
100	0.45	0.36	0.25	0.17	0.21	0.29	0.37
	99.9	99	90	50	10	1	0.1
	Percent Exceedance Frequency (at 95% Level of Significance)[a]						

[a] Chance of true value being greater than the value represented by the error curve.
Source: Beard (1962).

subtracted from the flood values at the same exceedance probabilities. There is a 90% probability that the true value lies between the 5% and 95% curves.

The probabilities computed from the theoretical distribution are of an infinite population. As given in Table 8.11, Beard (1962) proposed an adjustment to these probabilities to reflect the limited size of the sample. The table is based on the normal distribution but can be applied approximately to the log-Pearson type III distribution with small skew coefficients between −0.5 to +0.5.

EXAMPLE 8.7

Plot the 5% and 95% error limit curves (upper and lower confidence limits) on the flood-frequency curve of Example 8.6. Also determine the probabilities adjusted for the limited sample size.

Figure 8.8   Log-Pearson type III frequency curve and a reliability band.

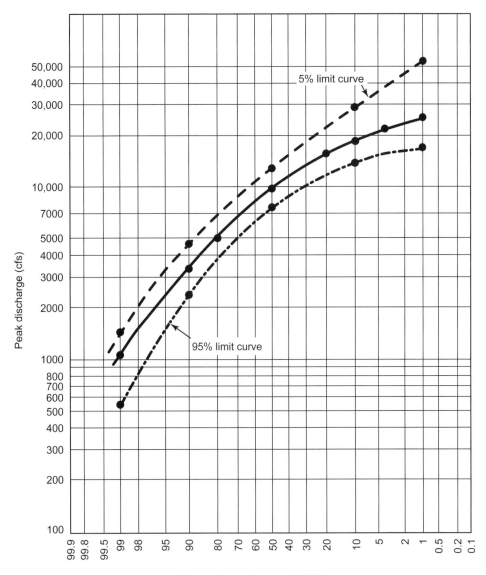

Solution

1. The log-Pearson type III curve of Example 8.6 is shown in Figure 8.8.
2. Error limit and probability adjustment computations are given in Table 8.12.
3. For a 5% limit curve, the values of row (e) are plotted against row (a) from Table 8.12 as shown in Figure 8.8 and for a 95% curve, the values of row (i) versus row (a) plotted on the same graph.

Section 8.11   Confidence Limits and Probability Adjustments

Table 8.11 P_n Versus $P\infty$ for Normal Distribution (Percent)[a] for Expected Probability Adjustment

	$P\infty$						
N–1	50	30	10	5	1	0.1	0.01
	Adjusted Probability, P_n						
5	50.0	32.5	14.6	9.4	4.2	1.79	0.92
10	50.0	31.5	12.5	7.3	2.5	0.72	0.25
15	50.0	31.1	11.7	6.6	1.96	0.45	0.13
20	50.0	30.8	11.3	6.2	1.7	0.34	0.084
25	50.0	30.7	11.0	5.9	1.55	0.28	0.06
30	50.0	30.6	10.8	5.8	1.45	0.24	0.046
40	50.0	30.4	10.6	5.6	1.33	0.20	0.034
60	50.0	30.3	10.4	5.4	1.22	0.16	0.025
120	50.0	30.2	10.2	5.2	1.11	0.13	0.017
∞	50.0	30.0	10.0	5.0	1.0	0.10	0.01

[a] Values for probability > 50 by subtraction from 100 [i.e., $P_{90} = (100 - P_{10})$].

Source: Beard (1962).

Table 8.12 Error Limits and Probability Adjustments

(a) P_α (%) (select)	0.1	1	10	50	90	99	99.9
(b) 5% level (from Table 8.10) ($N = 24$)	1.11	0.88	0.58	0.36	0.39	0.54	0.67
(c) Error limit, [(b) × S]	0.342	0.271	0.179	0.111	0.120	0.166	0.206
(d) Log value [X^a + (c)]		4.702	4.452	4.077	3.627	3.175	
(e) Curve value, [$\log^{-1}$ (d)] (cfs)		50,350	28,310	11,940	4240	1496	
(f) 95% level (from Table 8.10)	0.67	0.54	0.39	0.36	0.58	0.88	1.11
(g) Error limit, [(f) × S]	0.206	0.166	0.120	0.111	0.179	0.271	0.342
(h) Log value, [X^a – (g)]		4.265	4.153	3.855	3.328	2.738	
(i) Curve value, [$\log^{-1}$ (h)] (cfs)		18,410	14,220	7160	2130	550	
(j) P_N (for N – 1 = 23) (from Table 8.11)	0.29	1.58	11.0	50.0	89.0	98.42	99.71

S = Standard deviation of flood sample

[a] = Column 3 of Table 8.9

8.12 Special Cases of Flood-Frequency Analysis

8.12.1 Combined-Population (Composite) Frequency Analysis

When the homogeneity test of Section 8.5.2 indicates nonhomogeneousness of the data, the series consists of events caused by different types of hydrologic phenomena, such as rainstorm floods and snowmelt floods or hurricane and nonhurricane events. A composite flood-frequency curve is derived for such cases. The composite probability of events having individual probabilities of $P_1, P_2, P_3, \ldots, P_n$ is given by

$$P_c = 1 - (1 - P_1)(1 - P_2) \ldots (1 - P_n) \quad \text{[dimensionless]} \tag{8.14}$$

When only two types of events are involved in the observed peak discharges, eq. (8.14) reduces to

$$P_c = P_1 + P_2 - (P_1 P_2) \quad \text{[dimensionless]} \tag{8.15}$$

where

P_1 = exceedance probability of one type of

phenomenon (i.e., rainstorm peak discharge)(as decimal)

P_2 = exceedance probability of other nonhomogeneous type of

phenomenon (i.e., snowmelt peak discharge)(as decimal)

Equation (8.15) means that a given discharge may occur as an annual maximum in the form of either a rainstorm flood or a snowmelt flood $(P_1 + P_2)$ but not as both a rainstorm and a snowmelt flood [i.e., $(P_1 P_2)$].

The annual peak discharge series is separated into subseries, each of which is homogeneous (i.e., rainstorm flood series and snowmelt flood series). Each of these subseries is analyzed separately, either by empirical (Section 8.8) or analytical (Section 8.9) procedures, and separate flood-frequency curves are derived. About 10 discharge values are selected and the exceedance probability of each of these discharges is obtained from each of the flood-frequency curves. The composite probability, P_c, for each of the discharges is obtained by eq. (8.14) or (8.15). The composite flood-frequency curve is plotted based on P_c values.

EXAMPLE 8.8

From the annual peak flow record of a gaging site, the flood-frequency curves of nonhomogeneous rainstorm events and snowmelt events are derived separately, as shown in Figure 8.9. Prepare a composite flood-frequency curve for the flood flow sequence that includes both annual rainstorm and annual snowmelt discharges.

Figure 8.9 Flood-frequency curves of snow and rain events and the composite curve.

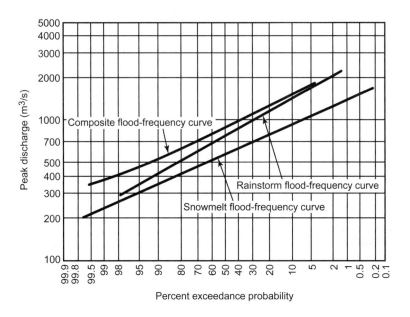

1. For selected discharges, the exceedance probabilities for rainstorm and snowmelt events are read from Figure 8.9, as shown in columns 2 and 3 of Table 8.13.

2. The composite exceedance probabilities of column 4 are computed by eq. (8.15).

3. The composite curve is plotted on Figure 8.9.

Table 8.13   Computation of Composite Probabilities

Q (m³/s)	$P_{rainstorm}$ (%)	$P_{snowmelt}$ (%)	P_c [from eq. (8.15)] (%)
350	95	90	99.5
500	81	63	93.0
600	70	44	83.2
700	57	29	69.5
800	46	18	55.7
900	36	11	43.0
1000	28	7	33.0
1200	17	2.5	19.1
1400	10	0.9	10.8
1600	6	0.3	6.3

8.12.2 Frequency Analysis of Partial-Duration Series

A partial-duration series comprises all peak discharges greater than some arbitrary base discharge. The introduction of more than one discharge per year causes the problem of computation of probabilities of annual exceedances. As such, the partial series is not widely used in flood studies. When used, the empirical method is applied for analysis, and the frequency plot is made on semilogarithm paper because probability paper is generally designed for plotting hydrologic data of annual and complete series (Chow, 1964). The logarithmic scale is used for exceedance probability or return period.

When there are several peak discharges per year, the average number of flood peaks per year ($\bar{m}$) that exceed any given discharge, Q, is given by

$$\bar{m} = \frac{N_Q}{N} \quad [\text{dimensionless}] \qquad (8.16)$$

where

$\bar{m}$ = number of flood peaks per year that exceed Q

N_Q = total number of peak discharges greater than the given discharge Q

N = total number of years of record

For each value of Q considered, N_Q, and hence $\bar{m}$, will be different. The annual exceedance probability (P_m) of the given discharge Q has been related to $\bar{m}$ by the following (Sokolov et al., 1976):

$$P_m = 1 - e^{-\bar{m}} \quad [\text{dimensionless}] \qquad (8.17)$$

For various values of Q, the values of $\bar{m}$ are computed by eq. (8.16) and then, by eq. (8.17), annual exceedance probabilities, P_m, are computed to make a frequency plot.

Equation (8.17) indicates that for a value of $\bar{m} < 0.1$(recurrence interval of greater than 10 years), P_m is practically equal to $\bar{m}$, i.e., both annual and partial series have almost the same probabilities.

8.12.3 Frequency Analysis of Flood Volume

Our analysis until now has considered instantaneous maximum flows (flood peaks). In a study for reservoir design, it is important to know the frequency of flood volumes comprising maximum one-day discharge (runoff), maximum 2-day (consecutive) discharge, and maximum 3-day, 7-day, . . . 90-day or any other period discharge in cfs-days. Such information is obtained from the daily discharge record at a gaging site. After flood-volume data are tabulated (the U.S. Geological Survey uses a special form for this purpose), the frequency analysis is performed exactly in a manner similar to that used for the flood peaks. The same procedures are applied to prepare stage curves, rainfall curves, and curves for other hydrologic factors as well.

8.12.4 Regional Frequency Analysis

The preceding discussion of frequency analysis relates to the analysis at an individual site. The flood flow data at a single site involve a larger sampling error than a group of stations. Moreover, it is only rarely that the flood-frequency information is required exactly at the gage site. It is more often desired anywhere in a region, including ungaged locations. The USGS accordingly recommends a study of frequency analysis on a regional basis. However, the region should be homogeneous with respect to flood-producing characteristics. A test has been developed by Langbein to define a homogeneous region (Dalrymple, 1960, p. 38).

There are many ways to perform regional studies. The procedure of the USGS, described below, is very widely used (Dalrymple, 1960).

1. List maximum annual floods for all gaging stations in the region having a record of 5 years or more.
2. Select the longest period of record at any station as the base period.
3. Adjust all records to the base period by a correlation study.
4. Perform a test for homogeneity and exclude the stations that are nonhomogeneous.
5. For each station, arrange the data in descending order of magnitude.
6. Compute the plotting position and recurrence intervals.
7. Plot the frequency curves, one for each station.
8. For each station, determine the mean annual flood represented by the discharge corresponding to a 2.33-year recurrence interval on the frequency curve.
9. Compute the ratios of floods of different recurrence intervals to the mean annual flood of each station.
10. Tabulate ratios of all stations in one table and select median ratios for various recurrence intervals.
11. Draw a regional frequency curve by plotting the median flood ratios against the corresponding recurrence intervals.
12. Plot the mean annual flood of each station against its drainage area.

13. For the known drainage area of any place in the region, the curve of item 12 will provide the mean annual flow. From the curve of item 11, the ratios of peak flow to mean can be converted to flood flows for various recurrence intervals.

The Hydrologic Engineering Center of the U.S. Army Corps of Engineers relates the mean of annual maximum flows with the drainage basin characteristics—such as drainage area, slope, surface storage, stream length, and infiltration characteristics—through a multiple linear regression analysis of all stations in the region (HEC, 1975). A similar regression equation is derived for the standard deviation of annual maximum flows. These equations have the following form:

$$\bar{X} = a\,A^{b}L^{c}\left(1+I\right)^{d} \quad \text{[unbalanced]} \tag{8.18}$$

and

$$S = e\,A^{f}G^{g}L^{h} \quad \text{[unbalanced]} \tag{8.19}$$

where

$\bar{X}$ = mean of the logarithms of annual series peak flood events

S = standard deviation of the logarithms of annual series peak flood events

A, L, I, G = physical basin characteristics

a, b, c, d, e, f, g, h = regression coefficients

From the known basin characteristics and the regression coefficients obtained through a multiple regression analysis or from a generalized map of regression constants, the values of $\bar{X}$ and S are obtained by the above equations. For computed $\bar{X}$ and S, a frequency curve is prepared by eq. (8.10) using a log-Pearson type III distribution for K values.

In another method, the flood peaks of various return periods in the region are directly related to basin characteristics by multiple regression analysis. The USGS has developed regional equations for various areas of the country. An example of this type of equation is the following:

$$Q_{100} = 19.7\,A^{0.88}P^{0.84}H^{-0.33} \quad \text{[unbalanced]} \tag{8.20}$$

where

Q_{100} = 100-year flood in cfs

A = drainage area in mi^2

P = mean annual precipitation in inches

H = average main channel elevation in thousands of feet

A different relation exists for every return period for the same region. The standard error of estimate associated with each equation should be identified, which indicates the possible range of error in the value of flow predicted by the equation.

8.13 COMPUTATION OF PEAK FLOW FROM PRECIPITATION

As mentioned earlier, in situations where a risk to human life is involved, the structures are not designed with respect to any specific frequency but for the worst expected conditions.

Snyder (1964) suggested that the spillway of a major dam, having a storage capacity larger than 50,000 acre-ft, which involves considerable risk to life and excessive damage potential, should be designed for the probable maximum flood (PMF). For intermediate dams of 1000 to 50,000 acre-ft capacity, where there is a possibility of loss of life, the spillway should be designed for the Standard Project Flood (SPF). The *probable maximum flood* is defined as the most severe flood considered reasonably possible in a region. The *standard project flood* excludes extremely rare storm conditions and thus is the most severe flood considered reasonably characteristic of the specific region. The peak discharge of an SPF is about 40 to 60% of that of a PMF for the same basin.

The floods result, directly or indirectly, from precipitation. The PMF is produced by the *probable maximum precipitation* (PMP), which was defined by the American Meteorological Society in 1959 as the theoretically greatest depth of precipitation for a given duration that is physically possible over a particular drainage area at a certain time of the year. The spatial and temporal distribution of the probable maximum precipitation, determined on the basis of maximization of the factors that operate to produce a maximum storm, leads to the development of the *probable maximum storm* (PMS).

A PMS thus developed can be used with a precipitation-runoff relation (Section 7.15) or a unit hydrograph (Section 7.12) or a simulation model to compute a PMF hydrograph.

8.14 ESTIMATION OF PMP

There are two common approaches for determining PMP: rational estimation and statistical estimation. The rational approach has the following basis:

$$\left(\frac{\text{precipitation}}{\text{moisture}}\right)_{\text{max}} \times \left(\frac{\text{moisture}}{\text{supply}}\right)_{\text{max}} = \text{PMP} \quad [\text{L}] \qquad (8.21)$$

The first term, relating to the ratio of the observed rainfall value to the atmospheric moisture at the time of the actual storm, is available from the record of actual storms. The optimum moisture supply for the second term is obtained from the meteorological tables of the *effective precipitable water* based on the maximum persisting dew point for the basin. The maximization is achieved by considering all major storms in an area and transposed storms from the homogeneous region. For each of these storms the precipitation depth, computed by applying eq. (8.21), is plotted against the duration of the storm in a drainage area. The enveloping curve is the depth-duration curve for the given area. Such curves are prepared for different sizes of the drainage area in the study basin, thus providing depth-duration-area curves.

The statistical approach considers the linearized frequency distribution of the form of eq. (8.10), as follows:

$$X_m = \bar{X}_m + K_m S_m \quad [\text{L}] \qquad (8.22)$$

where

X_m = maximum observed rainfall (i.e., PMP of a given duration)

$\bar{X}_m$ = mean of a series of maximum annual rainfalls of a specified duration from a drainage area

K_m = standard variate for probable maximum precipitation; maximization of K value

S_m = standard deviation of annual maxima series

The value of K depends on the type of distribution and the recurrence interval. For an extreme value type I distribution, which often adequately describes the rainfall extremes, K has a value of 3.5 for a 100-year rainfall (for the data size of 50). The value of K for the maximum probable precipitation is, obviously, higher than that of 100-year rain. It can be ascertained from the enveloping curve of extreme historic storms. Hershfield (1961) analyzed 24-hour rainfalls from 2600 stations, with a total of 95,000 station-years of data, and determined an enveloping K_m value of 15. From eq. (8.22), for the selected K_m value, the depth-area-duration curves can be prepared by considering storms of different durations from various drainage areas.

The generalized diagrams of PMP based on the rational approach and historic record have been prepared by the National Weather Service (formerly, U.S. Weather Bureau) for the entire country divided in two parts: east of the 105th meridian and west of the 105th meridian. For east of the 105th meridian, the monthwise variations of PMP for storm areas from 10 to 1000 mi^2 and durations of 6, 12, 24, and 48 hours were prepared in 1956 (U.S. Weather Bureau, 1956), and contained in Hydrometeorological Report 35, commonly known as HMR 35. The all-season values for east of the 105th meridian of PMP have since been revised and extended to 20,000 mi^2 and for durations of 6 to 72 hours in HMR 51 (NWS, 1978). The seasonal variations for only the 10 mi^2 storm area have been revised again in HMR 52 and HMR 53 (1980). Figure 8.10 shows the all-season envelope probable maximum 24-hour precipitation for a 200-mi^2 storm area for east of the 105th meridian. The PMP of a different duration and for a different-sized storm area can be obtained by applying the conversion factor of Table 8.14, corresponding to the relevant region indicated in Figure 8.10. Thus, Table 8.14 provides the generalized depth-area-duration relations of the PMP. The depth-area-duration analysis is described in Section 2.9.

To determine PMP for an area other than one listed in Table 8.14, the conversion factor of the table can be interpolated from a semilog plot of the factor (ordinary scale) versus areal size (log scale). Similarly, the value for an intermediate duration can be ascertained from a Cartesian or semilog plot of the factor versus duration. For both interpolations together, the areal interpolation is done first for different durations and then the durational interpolation is carried out.

For regions west of the 105th meridian, all-season values of PMP for areas to 400 mi^2 and duration to 24 hours were presented in Technical Paper 38 (U.S. Weather Bureau, 1960), and have been updated in other hydrometeorological reports. The specific manuals of the National Weather Service from which PMP data for west of the 105th meridian can be obtained are listed in Figure 8.10.

8.15 DEVELOPMENT OF PMS

The magnitude of PMP varies with duration. The duration and the temporal (time varying) and spatial (isohyetal) distribution of the PMP estimated in the preceding section are necessary for determination of the probable maximum flood.

8.15.1 Critical Duration

The duration of the PMP that causes the largest flood at the site of interest is the critical duration for that drainage basin. It is determined by routing various hydrographs resulting from PMP of various durations and selecting the one causing maximum flood discharge. It is short for a small basin and long for a large basin. In general, it should never be less than the time of concentration.

Figure 8.10 All-season PMP (in.) for 200 mi², 24 hr.

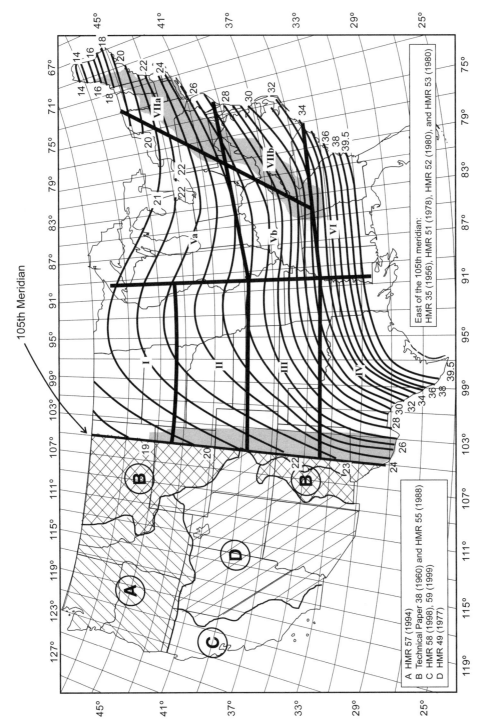

105th Meridian

East of the 105th meridian:
HMR 35 (1956), HMR 51 (1978), HMR 52 (1980), and HMR 53 (1980).

A HMR 57 (1994)
B Technical Paper 38 (1960) and HMR 55 (1988)
C HMR 58 (1998), 59 (1999)
D HMR 49 (1977)

Table 8.14 Depth-Area-Duration Relation of Maximum Probable Precipitation[a]

Storm Area mi²	km²	Duration (hr)	I	II	III	IV	Va	Vb	VI	VIIa	VIIb
							Regions				
10	26	6	1.00	1.09	1.03	0.93	1.04	1.01	0.90	1.04	1.00
		12	1.20	1.29	1.22	1.10	1.26	1.18	1.07	1.21	1.16
		24	1.28	1.38	1.31	1.25	1.34	1.31	1.25	1.34	1.33
		48	1.38	1.50	1.45	1.40	1.50	1.45	1.40	1.50	1.45
		72	1.47	1.60	1.55	1.50	1.52	1.53	1.50	1.52	1.53
200	518	6	0.75	0.78	0.74	0.66	0.76	0.72	0.67	0.73	0.68
		12	0.90	0.93	0.87	0.82	0.93	0.86	0.81	0.87	0.85
		24	1.00	1.00	1.00	1.00	1.00	1.00	1.00	1.00	1.00
		48	1.10	1.12	1.14	1.16	1.13	1.14	1.16	1.17	1.15
		72	1.15	1.20	1.20	1.22	1.22	1.23	1.23	1.23	1.25
1,000	2,590	6	0.57	0.56	0.54	0.50	0.56	0.52	0.50	0.52	0.52
		12	0.67	0.71	0.66	0.63	0.70	0.69	0.63	0.63	0.66
		24	0.77	0.80	0.79	0.79	0.80	0.79	0.83	0.80	0.80
		48	0.85	0.90	0.92	0.93	0.90	0.92	0.94	0.93	0.93
		72	0.96	0.97	0.98	1.00	0.97	0.98	1.04	0.98	0.98
5,000	12,950	6	0.36	0.36	0.31	0.28	0.36	0.31	0.28	0.33	0.31
		12	0.45	0.47	0.43	0.39	0.48	0.43	0.40	0.45	0.43
		24	0.52	0.54	0.54	0.55	0.54	0.54	0.55	0.56	0.56
		48	0.63	0.67	0.68	0.65	0.67	0.65	0.68	0.70	0.69
		72	0.70	0.74	0.76	0.76	0.74	0.76	0.78	0.74	0.76
10,000	25,900	6	0.26	0.27	0.23	0.21	0.28	0.23	0.22	0.28	0.23
		12	0.36	0.37	0.33	0.30	0.38	0.35	0.32	0.37	0.35
		24	0.42	0.45	0.43	0.43	0.47	0.44	0.45	0.47	0.45
		48	0.50	0.58	0.54	0.55	0.58	0.57	0.58	0.60	0.60
		72	0.60	0.62	0.64	0.65	0.66	0.64	0.67	0.67	0.65
20,000	51,800	6	0.18	0.20	0.17	0.16	0.20	0.17	0.16	0.20	0.16
		12	0.27	0.28	0.25	0.23	0.30	0.28	0.25	0.33	0.28
		24	0.35	0.36	0.35	0.32	0.38	0.36	0.36	0.40	0.37
		48	0.45	0.47	0.45	0.45	0.48	0.47	0.48	0.50	0.49
		72	0.50	0.55	0.55	0.55	0.56	0.55	0.56	0.57	0.55

[a] Factors derived by the author from the figures in National Weather Service (1978) to be applied to 24-hour values on 200-mi² area of Figure 8.10.

8.15.2 Temporal Distribution

The National Weather Service (1982) considered 6-hour increments for sequencing a PMP. Based on the examination of 28 storm samples, the NWS recommended the following:

1. Arrange the individual 6-hour increments such that they decrease progressively to either side of the greatest 6-hour increment. This implies that the lowest 6-hour increment will be either at the beginning or the end of the sequence.

2. Place the four greatest 6-hour increments at any position in the sequence, except within the first 24-hour period of the storm sequence.

   When it is necessary to consider values for a duration of less than 6 hours, the Hydrologic Engineering Center recommends the breakup of 6 hours according to the percentages given in Table 8.15.

   In another procedure, the increments of precipitation are first aligned to match the ordinates of the unit hydrograph. The sequence of precipitation increments is then reversed to form the design hyetograph.

Table 8.15 Hyetograph of 6-Hour PMP According to the HEC (1979)

Sequence (hr)	Percent of 6-hr depth
First hour	10
Second hour	12
Third hour	15
Fourth hour	38
Fifth hour	14
Sixth hour	11

EXAMPLE 8.9

For a storm in an 890-mi^2 area over the watershed of the Ponaganset River near Foster, Rhode Island, determine the 6-hour PMP from the generalized chart. Prepare the hyetograph for this storm.

SOLUTION

1. 24-hour PMP for 200 mi^2 (Fig. 8.10) = 24 in.

2. Conversion factor for 6-hour duration and different area:

Area (mi^2)	200	1000	5000
Factor (Table 8.14)	0.73	0.52	0.33

   These values are plotted on semilog paper and for 890 mi^2 the factor is 0.54, from the plot.

3. 6-hour PMP = 24(0.54) = 12.96 in.

4. Distribution of PMP (based on Table 8.15):

	Percent	Rainfall (in.)
First hour	10	1.30
Second hour	12	1.56
Third hour	15	1.94
Fourth hour	38	4.92
Fifth hour	14	1.81
Sixth hour	11	1.43

8.15.3 Spatial Distribution

For distributing the area-averaged PMP over a drainage basin, the important considerations are (1) the shape of the isohyets, (2) the number of isohyets, (3) the magnitude of isohyets, and (4) the orientation of isohyetal pattern.

For drainage areas of less than approximately 100 mi^2, a uniform depth of precipitation over the entire drainage basin can be assumed (HEC, 1979). For larger areas, the procedure for the determination of the PMS has been developed in HMR 52 (NWS, 1982). A standard elliptical storm pattern has been adopted with a major-axis to minor-axis ratio of 2.5:1 for the distribution of 6-hour increments of precipitation for the entire region east of the 105th meridian. This pattern is shown in Figure 8.11. It contains 14 isohyets for areas up to 6500 mi^2 and 19 isohyets for coverage of an area of 60,000 mi^2. Since the area of an ellipse is given by πab, where a and b are semimajor (one-half major) and semiminor axes, and that area covered by each isohyet is known, as given in Figure 8.11, each isohyet can be drawn to a selected scale. It is necessary to draw only as many of the isohyets as needed to cover the drainage area.

HMR 52 has provided a map in the form of orientation-contour lines for east of the 105th meridian. The map indicates the angle of orientation of the isohyetal pattern at a place to be most conducive to a PMP. If the longitudinal axis of the storm pattern is oblique to this recommended orientation by 40° or more, a reduction factor as given in HMR 52 is applied, depending on the storm pattern area and the orientation difference.

The value of PMP is the average precipitation depth over the specified storm area for a given duration. When a standard isohyetal pattern is superimposed, there will be greater precipitation depth near the center of the pattern and lesser depth near the edges. HMR 52 provides a nomograph from which the isohyetal adjustment factors for each isohyet (contour) could be read for a given storm area. When this factor corresponding to an isohyet is multiplied by the PMP value for the storm area, it represents the precipitation depth over the entire area covered by that isohyet. The distribution of precipitation depths and the average depth over the watershed due to a PMP can be determined by considering successive isohyets of the storm pattern as shown in Example 8.10 below.

EXAMPLE 8.10

Given the PMP of Example 8.9, determine the spatial distribution and the average depth of precipitation over the watershed. The drainage area is 400 mi^2. For a storm area of 890 mi^2, the isohyetal adjustment factors obtained from the HMR 52 for each isohyet are shown in col. 3 of Table 8.16.

Figure 8.11 Standard isohyetal pattern of PMP.

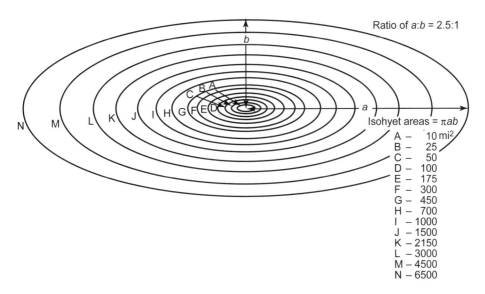

SOLUTION   Refer to Table 8.16.

1. From Example 8.9, for 890-mi^2 storm area, 6-hour PMP = 12.96 in.

2. Within a 400-mi^2 watershed, the first 6 isohyetal contours from A to F are fully contained. For isohyet G, the contributory area is the portion of the isohyetal area that can be contained within the watershed.

3. The col. 3 values are multiplied by 12.96 in. to get isohyetal precipitation depth in col. 4.

4. Other computations are explained by the footnotes of Table 8.16.

5. Average depth over watershed = 5752.5/400 = 14.38 in.

8.15.4 Design Storm

It is economically prohibitive to design a structure for the PMF except for very large spillways with excessive damage potential as described earlier. Usually, the design flood is the standard project flood (SPF). The largest storm that may be reasonably expected at the site is estimated from the standard project storm (SPS). The SPS is derived from a detailed analysis of the severe storms at a site, which includes selecting an isohyetal pattern that is considered reasonably characteristic of the region. Both PMS and SPS require considerable effort in preparation and, as such, are applied to major structures.

8.16 TRANSFORMATION OF DESIGN STORM TO FLOOD FLOW HYDROGRAPH

The unit hydrograph method is widely used to develop design flood hydrographs. The first step is to estimate the rainfall losses. The losses per unit duration of the rainfall are subtracted from each increment of total precipitation to give increments of rainfall excess.

Table 8.16 Computation of Spatial Distribution due to PMP

(1)	(2)	(3)	(4)	(5)	(6)	(7)
Isohyet contour	Area covered, mi^2	Isohyetal adjustment factor	Isohyetal[a] precipitation depth, in.	Average[b] depth between isohyets, p	Incremental[c] watershed area, ΔA	$p\Delta A$[d]
A	10	1.45	18.79	18.79	10	187.9
				18.15	15	272.3
B	25	1.35	17.50			
				16.98	25	424.5
C	50	1.27	16.46			
				15.88	50	794.0
D	100	1.18	15.29			
				14.78	75	1108.5
E	175	1.10	14.26			
				13.61	125	1701.3
F	300	1.00	12.96			
				12.64	100	1264.0
G	400	0.95	12.31			
Σ						5752.5

[a] PMP multiplied by col. 3
[b] average of successive values of col. 4
[c] difference of col. 2 values
[d] col. 5 × col. 6

The computation of losses by infiltration capacity curve and index methods is discussed in Sections 2.13 through 2.16.

The unit hydrograph (UHG) is derived from the rainfall-runoff data of a basin where such data are available; otherwise, it is formulated using the synthetic procedure from regional relations, as described in Section 7.11. However, an uncertainty arises from the fact that the design storm and resulting flood are invariably of greater magnitude than the storms and corresponding flows used to derive the UHG. Studies by the U.S. Army Corps of Engineers (1959) showed that the peak ordinate of a UHG derived from data for a major flood is 25 to 50% greater than the peak from data for a minor flood. Therefore, modification of the UHG is made when the derived UHG does not represent conditions similar to those during the design storm. The modification procedure is as follows.

1. Read the widths of the UHG at ordinates equal to 50% and 75% of the peak discharge, referred to as W_{50} and W_{75}.
2. Plot the widths against the peak discharge on Figure 8.12, which is based on the study of a large number of drainage basins. Draw lines A-A' and B-B' through the plotted points parallel to curves W_{50} and W_{75}.
3. Increase the peak discharge of the UHG by 25 to 50%, depending on judgment.

4. Draw a horizontal line for the revised peak discharge, Q_u of item 3 to cross lines A-A' and B-B' on Figure 8.12.

5. At the intersection points read revised W_{50} and W_{75} on the x-axis.

6. Sketch the modified UHG on the same time base as the original UHG through the revised plotted points Q_u, W_{50}, and W_{75}. The W_{50} and W_{75} widths are plotted on each side of Q_u in the same ratio as existed on the original UHG.

7. The minor adjustment on the recession curve of the revised UHG can be made by (a) preparing an S-curve from the plotted modified UHG, (b) smoothing the S-curve, (c) deriving the UHG from it, (d) adjusting the UHG ordinates, (e) drawing the S-curve again, and repeating the process until the most logical forms of both the UHG and S-curve are obtained. This is usually not done.

Once the UHG is derived, it is applied to the rainfall sequence of the design storm and the baseflow is added to produce the discharge hydrograph (Section 7.12). The application is shown in Example 7.9.

8.17 PEAK SNOWMELT DISCHARGE

In some regions the contribution from snowmelt is vital. The maximum floods in many areas are either due to a combination of snowmelt and rainfall runoff or from snowmelt events

Figure 8.12 Width and peak discharge relation of the unit hydrograph (from Sokolov et al., 1976).

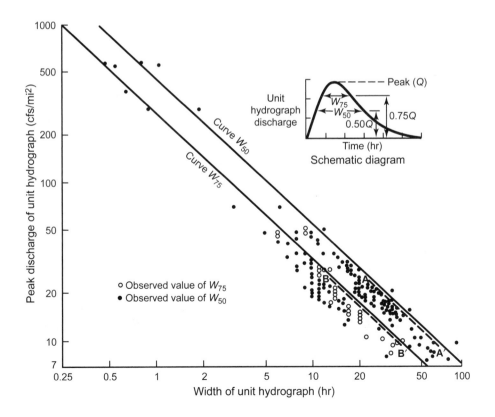

alone. In the former Soviet Union, empirical equations similar to rainstorm floods (Section 8.18) are applied to compute the direct snowmelt peak discharge. In the United States, the intensity (rate) of snowmelts in terms of depth per unit time is computed from equations that have been derived by combining the heat balance equation with empirical factors. These equations have been developed separately for plains and mountainous regions during rainfall and rain-free periods, as described in Section 2.17. The duration of a given intensity is the period for which the set of conditions, as applied in the equations of snowmelt, existed.

After the depth of melt has been estimated for the portion covered by the snowpack, it can be treated like the rainfall to ascertain the flow; that is, losses are deducted and melt excess is converted into a streamflow hydrograph by the application of the unit hydrograph or routing technique. The losses are treated in two ways. For a rain-on-snow event, the amount of water that is delayed very long in reaching a stream is considered lost. For a primarily snowmelt event, the losses consist of evapotranspiration, deep percolation, and permanent retention of water in the snowpack. The rates of melt excess are small but continue for a long period (are approximately continuous). Special long-tailed unit hydrographs or S-hydrographs are used for snowmelt streamflow computation.

8.18 FLOOD FLOW COMPUTATION BY GENETIC AND EMPIRICAL EQUATIONS

When the hydrologic data are insufficient to use the preceding methods, genetic and empirical equations are used. The genetic equations are based on physical considerations that embody the theoretical concepts of the runoff generation (Sokolov et al., 1976). On the other hand, empirical relations are not concerned with mechanisms of runoff generation but bring out the resultant effect of the relevant factors. For small basins of less than 20 mi^2, a genetic equation known as the rational method is commonly used. It is applied to the drainage analysis on a small watershed. The method is described in Section 13.10. For rivers having a drainage area larger than 20 mi^2, empirical relations are used. The parameters of the empirical equations are determined from the regional analysis. A large number of equations exist to determine the peak discharge from basin and climate parameters. Gray (1973) has presented about 50 empirical relations drawn from throughout the world. One commonly used relation is described below.

8.18.1 Myers-Jarvis Enveloping Curves

This relation has the simple form

$$Q = pA^{1/2} \quad \text{[unbalanced]} \tag{8.23}$$

where

Q = peak discharge, cfs

p = Myers rating, as given in Table 8.17

A = drainage area, mi^2

On a log-log graph, eq. (8.23) plots a straight line with a slope of 1/2 (1 vertical to 2 horizontal) and an intercept of p. From the regional data, the peak discharge and the drainage area are plotted. A line as an envelope through the upper points is drawn with a slope of 1/2. Known as the Myers curve, this provides a value of p and an estimate of flood peaks that could occur anywhere in the region. When the regional data are not available to plot the curve, the value of p is ascertained guided by Table 8.17.

Table 8.17 Myers Rating for Maximum Flood Discharge

Region	Approximate Range of Drainage Area (mi²)	Myers Rating
North Atlantic and Middle Atlantic slope basins	10–100	7,000
	500–80,000	5,500
South Atlantic slope and eastern Gulf of Mexico	10–10,000	6,000
Ohio River basin	10–900	7,000
	1,000–5,000	6,000
	5,000–200,000	4,300
Eastern Great Lakes and eastern St. Lawrence River basins	10–60	6,000
	70–1,000	3,500
Upper Mississippi River and western Great Lakes basins	10–150	4,000
	200–800	2,700
	900–15,000	1,800
Lower Missouri and western Mississippi River tributaries	10–5,000	7,500
	6,000–15,000	4,000
Western Gulf of Mexico basins	50–3,000	11,000
	1,000–9,000	10,000
	10,000	5,000
Pacific slope basins in California	10–3,000	6,000

EXAMPLE 8.11

The Miami River near Hamilton, Ohio, has a drainage area of 3630 mi². Determine the peak flood discharge.

SOLUTION

1. From Table 8.17, for a drainage area of 3000 mi² in the Ohio River basin, $p = 6000$.
2. Substituting in eq. (8.23), $Q = 6000(3630)^{1/2} = 361{,}500$ cfs
3. An actual peak flow of 352,000 cfs was recorded on the Miami River at Hamilton on March 16, 1913.

8.19 MEASUREMENT OF PEAK DISCHARGE BY INDIRECT METHODS

In indirect methods, the hydraulic equations of flow applicable to different hydraulic systems are used. These relations, derived from the energy equation and continuity equation, are in terms of the head or its variation. The high-water marks left by peak floods provide this information. To apply the method, analysts select a location of a particular structure for which the specified hydraulic equation is applicable. The hydraulic structures to measure peak flows are discussed in Section 10.8.

8.20 COMPUTATION OF LOW FLOW

Low streamflows are significant because drought conditions upset the entire ecological balance; more specifically, low flows are important in terms of the adequacy of a stream to receive wastes, to supply municipal and industrial water requirements, to meet supplemental irrigation, or to maintain aquatic life. Of the following procedures of low-flow computation, the first two relate to gaged sites and the others to ungaged locations.

1. *Frequency analysis.* This analysis is similar to peak-flow analysis (Sections 8.6 through 8.10).

2. *Analysis of runs.* The deficit in flow with respect to the baseflow is counted along with the duration (of the drought).

3. *Partial-record method.* A few baseflow measurements at a site are related to the concurrent discharges at a neighboring station for which a low-flow frequency curve is available.

4. *Seepage runs.* During a period of baseflow, discharges at intervals along a channel reach are measured to identify the loss or gain in the flow along a river channel.

5. *Interpolation.* By plotting the low-flow characteristics at gaged sites against the channel distance, the flows at intermediate points are interpolated.

Frequency analysis is a common procedure in which an array of annual low flows is formulated from the records of daily discharge of a gaging station. The sequence can be prepared for the lowest daily discharge in each year or the lowest mean discharge for 2, 3, 7, 10, or more consecutive days. The frequency curves can be prepared from the extracted values of low flows for different periods of days. An empirical or analytical procedure, similar to flood-flow analysis, can be used.

8.20.1 Low-Flow Frequency Analysis by the Empirical Method

1. Arrange the annual values in ascending order beginning with the smallest as number 1 (in peak flows these are arranged in descending order).

2. Compute the plotting position using eq. (8.9), that is, $P_m = m(100)/(n+1)$; the return period, T, is the inverse of P_m.

3. Plot each value on graph paper [lognormal or log-extreme value type I (Gumbel)]. Draw a smooth curve.

EXAMPLE 8.12

The data for the annual 7-day minimum average flows at the Walkill River near Walden, New York, are given in Table 8.18. **(a)** Prepare the low-flow frequency curve. **(b)** Determine **(1)** the probability of the 7-day low flow to be less or equal to 2.5 ft³/s, **(2)** the return period of a flow of 3.5 ft³/s, and **(3)** 10-year 7-day low flow.

Table 8.18 Annual 7-Day Minimum Average Flow of Walkill River, Walden, New York

(1)	(2)	(3)	(4)	(5)	(6)
					Return period, $T = \dfrac{1}{P_m}$
Year	7-day Low-flow (cfs)	Log Flow (cfs)	Rank	Plotting Position, P_m %	Years
1976	2.74	0.438	21	67.74	1.48
1977	1.83	0.262	9	29.03	3.44
1978	2.33	0.367	13	41.93	2.38
1979	2.97	0.473	26	83.87	1.19
1980	1.76	0.246	6	19.35	5.17
1981	1.93	0.286	10	32.25	3.10
1982	2.66	0.425	19	61.29	1.63
1983	4.00	0.602	30	96.77	1.03
1984	2.93	0.467	25	80.65	1.24
1985	2.48	0.394	16	51.61	1.94
1986	2.83	0.452	23	74.19	1.35
1987	1.56	0.193	2	6.45	15.50
1988	1.71	0.233	4	12.90	7.75
1989	1.57	0.196	3	9.68	10.33
1990	2.39	0.378	14	45.16	2.21
1991	2.63	0.420	18	58.06	1.72
1992	3.13	0.496	27	87.10	1.15
1993	2.09	0.320	12	38.71	2.58
1994	2.80	0.447	22	70.97	1.41
1995	3.53	0.548	29	93.55	1.07
1996	1.46	0.164	1	3.23	30.96
1997	2.70	0.431	20	64.52	1.55
1998	2.53	0.403	17	54.84	1.82
1999	1.75	0.243	5	16.13	6.20
2000	1.81	0.258	8	25.81	3.87
2001	1.99	0.299	11	35.48	2.82
2002	3.33	0.522	28	90.32	1.11
2003	2.90	0.462	24	77.42	1.29
2004	2.41	0.382	15	48.39	2.07
2005	1.79	0.253	7	22.58	4.43

SOLUTION

(a) **1.** The rank is shown in column 4 of Table 8.18. It is advisable to arrange values in ascending order in a separate table.

2. The plotting position is computed by eq. (8.9) in column 5 of the table.

3. The log of the flow values (col. 3) and the recurrence interval (col. 6) are plotted on extreme value (Gumbel) paper in Figure 8.13 [logs of the values are plotted since the logarithmic transform of type I (Gumbel) is essentially type III (Weibull) distribution, which is more suitable for low-flow sequence].

(b) **4.** From Figure 8.13:

(1) For $Q = 2.5$, log $Q = 0.4$, $T = 1.65$, $P = 1/T = 0.61$

(2) For $Q = 3.5$, log $Q = 0.54$, $T = 1.1$ years

(3) For $T = 10$ years, log $Q = 0.2$, $Q = \log^{-1} 0.2 = 1.6$ ft^3/s

8.20.2 Low-Flow Frequency Analysis by Analytical Method

The analytical approach discussed in the context of flood flows (Section 8.9) is applicable to low flows. The linear equation (8.10), reproduced below, is used in theoretical analysis.

$$X = \bar{X} + KS \quad [L^3 T^{-1}] \tag{8.10}$$

The distributions suitable for low flows are lognormal, log-Pearson type III, and extreme value type III distributions, with a preference for the latter. The frequency factors, K, are given in Tables 8.6 through 8.8. The K factors of the extreme value type I distribution in Table 8.8 are used with the logarithmic flow values to fit to the extreme value type III distribution.

The frequency factors have been tabulated for use with the peak flows that indicate the probability of exceedance. However, the low-flow frequency curve indicates the probability of flow being equal to or less than the value indicated. The probability value of the table should thus be converted by subtracting from 1 (from 100 in the case of percent).

EXAMPLE 8.13

Solve Example 8.12 by the analytical method using log-extreme value type I distribution.

SOLUTION

1. The flow data are converted to log form.

2. The statistical parameters computed by eqs. (5.26) and (5.27) are as follows: mean, $\bar{X} = 0.369$ cfs and standard deviation, $S = 0.116$ cfs.

3. For various probability levels, the flows are computed in Table 8.19. The values of K are obtained from Table 8.8.

4. The frequency curve has been plotted on Figure 8.13.

5. From Figure 8.13:

(1) For $Q = 2.5$, log $Q = 0.4$, $T = 1.5$, $P = 1/T = 0.67$

(2) For $Q = 3.5$, log $Q = 0.54$, $T = 1.12$ years

(3) For $T = 10$ years, log $Q = 0.245$, $Q = \log^{-1} 0.245 = 1.76$ ft^3/s

---

Figure 8.13 Frequency curve of 7-day annual minimum flow of Walkill River, NY.

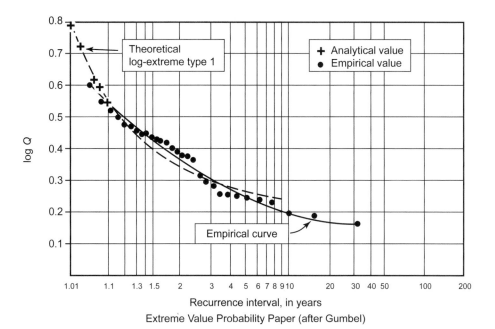

Extreme Value Probability Paper (after Gumbel)

Table 8.19   Computation of Low Flow of Various Probabilities

(1)	(2)	(3)	(4)	(5)	(6)
$P(X>)$ Table 8.8 Value	$P(X<)$ $1-P(X>)$	$T = \dfrac{1}{P(X<)}$	K	$X = \bar{X} + KS$	$Q = \log^{-1} X$
0.9[a]	0.1	10	−1.10	0.241	1.743
0.8[a]	0.2	5	−0.821	0.273	1.875
0.4[a]	0.6	1.67	0.0737	0.378	2.388
0.2	0.8	1.25	0.866	0.469	2.944
0.10	0.90	1.11	1.541	0.548	3.532
0.05	0.95	1.053	2.188	0.623	4.198
0.02	0.98	1.02	3.026	0.720	5.248
0.01	0.99	1.010	3.653	0.793	6.209
0.001	0.999	1.001	5.727	1.033	10.789

[a] Based on sample size $n = \infty$.

8.20.3 Regionalization and Application to Ungaged Sites

The low-flow characteristics of streams in adjacent basins vary significantly. Geology-lithology and rock structure influence low flows. It has not been possible to describe quantitatively the effect of geology on low flows by certain indexes. Hence the regionalization of low-flow characteristics has met with only limited success. Attempts have been made to regionalize by multiple regression on several basin characteristics, but the standard errors of regressions have been too large. The regional relations are developed only for limited conditions where all significant basin characteristics except drainage area in a region have extremely limited range. The lack of general regionalization does not permit estimates at ungaged sites by the extension of frequency relations.

8.21 Computer Models for Flood Analysis

Section 7.29 discusses watershed simulation models. In the context of extreme precipitation, those models lead to simulation of flood flows. However, the models described here are those that directly deal with flooding of basins. Such models belong to the last three categories outlined in Section 7.29. They can be rearranged as follows based on the nature of tasks they perform.

8.21.1 Statistical Flood Frequency Models

Flood frequency models perform frequency computations on historic flood series to make predictions of probability-based flood magnitudes. Important models of this category are: (1) the Hydrologic Engineering Center's Flood Frequency Analysis Model (HEC.FFA)— last released in 1995—and (2) the U.S. Geological Survey's model PEAKFQ—last released in 2005. Both of these are based on "Guidelines for Determining Flood Flow Frequency," Bulletin 17B of the Interagency Advisory Committee on Water Data (1982) that uses the log-Pearson Type III distribution.

8.21.2 Steady-State Flood Hydraulics Models

Steady-state flood hydraulics models compute depths and velocities or the water surface profile along a river reach as a flood wave moves downstream under the condition of steady-state nonuniform flow. Important models of this group are: (1) HEC River Analysis System (HEC RAS)—regularly updated, with last release in 2006—which is an improvement of the HEC-2 model with graphic display, data storage and management capability, and now a GIS interface. It is capable of handling an entire network of channels, all kinds of steady flows, and all types of obstructions in flood plains. (2) U.S. Geological Survey and Federal Highway Administration, Water Surface Profile Model (WSPRO), last released in 1998, is easy to use for highway design, flood plain mapping, flood insurance studies, and stage-discharge relations. It can analyze all types of flows through bridges, culverts, roadway crossings with multiple openings, and embankment overflows.

8.21.3 Unsteady-State Flood Hydraulics Models

Unsteady-state flood hydraulics models are used when unsteady conditions occur in channels. The dynamic wave method is used for analysis wherein both the continuity and momentum equations, known as the St. Venant equations, are solved simultaneously. Since this represents a more realistic condition, many models of this type have been developed. The most important of these are summarized below.

USGS Models. The USGS has four dynamic flow models. The Branch-Network Dynamic Flow Model (BRANCH), last released in 1997, is used to simulate unsteady (or steady) flow in a single channel or network of channels. BRANCH is particularly suitable for complex geometric configuration of channels. Surface water and groundwater interaction can be simulated by the coupled MODBRANCH model.

The Full Equations Model (FEQ), last released in 1999, simulates flow in a wide range of stream configurations by solving the full dynamic wave equations for channels, including the control structures.

The Finite-Element Surface-Water Modeling System (FESWMS), last released in 1995, is a program for two-dimensional flow in the horizontal plane. Shallow rivers, flood plains, estuaries, and coastal seas are suitable for the model. It is used to analyze flow at bridge crossings where complicated hydraulic conditions exist.

The Four-Point (FOURPT) model, last released in 1997, simulates flow in large, complex interconnected networks of open channels and through hydraulic structures. The options include kinematic, diffusion, and dynamic equations.

NWS Models. The NWS released a revised Flood Wave Model (FLDWAV) in 2000 that replaces DWOPER/NETWORK and DAMBRK models. It includes the best capabilities of both models. FLDWAV can simulate unsteady-state flows occurring in a system of interconnected rivers. Any river may contain any type of structure that breaches when failure conditions are reached. Flow may be of any type.

The Dynamic Wave Operation Model (DWOPER/NETWORK), last released in 1989, simulated unsteady flow of a single or a network of channels or pipes. The Dam-Break Flood Forecasting Model (DAMBRK) simulated the breaching of a dam and routed the subsequent flood wave or flood hydrograph through a single downstream channel. Downstream structures such as bridges, levees, and dams were considered.

HEC Models. The HEC Unsteady-Flow Network Model (UNET), revised in 2006, simulates unsteady flow through a full network of open channels. Hydraulic structures such as levees, spillways, weirs, bridges, culverts, and dams can be included. RAS2UNET utility is available to convert HEC-RAS files to UNET.

8.21.4 Reservoir Regulation Models

Reservoir regulation models simulate the operation of a reservoir or a system of reservoirs for flood control or other objectives like water supply, irrigation, and hydropower. It may be recalled that routing through uncontrolled reservoirs is part of the watershed models described in Section 7.29. The main regulation model is the HEC-5, Simulation of Flood Control and Conservation Systems, last released in 1991. This model simulates the sequential operation of a series of reservoirs connected by channel reaches in a branched network configuration. Flood damage and other economic computations are made. The HEC has also developed other models for damage analysis and benefit-cost analysis of projects.

PROBLEMS

8.1 A spillway is designed for a 100-year exceedance flood. What is the probability of the design flood occurring exactly twice in a 50-year period?

8.2 What is the probability that five 50-year floods will occur in a 100-year period?

8.3 What is the probability of the 100-year design flood occurring one or more times in 50 years?

8.4 Determine the probability of one or more floods of 50-year severity occurring in 100 years.

8.5 If a designer accepts a 5% chance of a flood control levee being overtopped in 25 years, what return period of flood should be used in the design?

8.6 For a flood control project having a service life of 100 years, only a 1% risk of overtopping can be taken. Determine the design exceedance level (required return period) of the flood.

8.7 A reinforced-concrete culvert has to be designed for a roadway. The flood-frequency data at the proposed site are given in the table below. The flood damages associated with various magnitudes of floods are also indicated in the table. A culvert of 30 in. diameter will pass a discharge of 40 cfs. Determine the total annual cost of this culvert including the cost of damage due to higher than design-level floods. Assume that the following relationships hold for annual direct cost and cost of operation and maintenance: (1) Annual direct cost $= 950 + 70D$ (D is diameter in inches) and (2) Annual O&M cost = 30% of direct cost.

Discharge (cfs)	Annual Exceedance Probability	Estimated Damage (dollars)
20	0.46	0
24	0.36	600
32	0.22	4,000
40	0.14	10,000
60	0.06	22,500
74	0.03	75,000
108	0.01	110,000
200	0.001	150,000

8.8 Four alternative designs are proposed for the culvert of Problem 8.7. The cost of each alternative is analyzed per the procedure adopted in Problem 8.7 and the total cost of each alternative is indicated below. The level of flood protection provided by each alternative is also given below. Determine the optimum design flood and its exceedance probability.

Culvert Size (in.)	Level of Protection (cfs)	Total Annual Cost (dollars)
27	30	10,100
30	40	9,750
40	70	10,250
50	100	11,800

8.9 The annual peak discharges of the Little Madawaska River near Fort Fairfield, Maine, are shown in the following table. The data in parentheses relate to snowmelt floods. Determine whether the peak discharges due to rainstorms are homogeneous to the peak discharges due to snowmelts at the 5% significance level.

Date	Discharge (m³/s)	Date	Discharge (m³/s)	Date	Discharge (m³/s)
4/5/65	(223)	4/12/79	(84)	2/23/93	(269)
3/25/66	(194)	3/30/80	(200)	7/18/94	356
2/24/67	(266)	2/28/81	(103)	6/18/95	(323)
12/20/68	302	7/13/82	178	3/25/96	(123)
3/11/69	(129)	3/15/83	(205)	2/28/97	(70)
8/15/70	215	9/17/84	215	4/3/98	(350)
4/4/71	(103)	12/23/85	197	1/5/99	140
7/1/72	253	4/20/86	(135)	3/25/00	(97)
3/3/73	(157)	3/25/87	(178)	4/1/01	(185)
5/7/74	80	8/8/88	215	4/15/02	(130)
4/3/75	(401)	4/1/89	(297)	7/25/03	171
7/27/76	241	11/26/90	190	10/20/04	192
9/20/77	347	7/20/91	195	8/5/05	109
8/20/78	464	3/31/92	(215)	7/7/06	97

8.10 The annual maximum flow data for two different periods of record for Genesee River at Rochester, NY, are given below. Determine whether there is homogeneity in the sequence of data of two periods, at the 2.5% level of significance.

Year	Discharge (m³/s)	Year	Discharge (m³/s)
1948	21,600	1993	20,300
1949	16,400	1994	23,300
1950	33,100	1995	13,800
1951	20,200	1996	28,200
1952	17,700	1997	19,400
1953	17,100	1998	18,200
1954	17,500	1999	16,300
1955	19,100	2000	18,000
1956	24,300	2001	20,000
1957	17,000	2002	17,800
1958	14,900	2003	18,400
1959	17,700	2004	11,500
1960	25,800	2005	21,000
		2006	14,900

8.11 Prepare the flood-frequency curve by the empirical method for the annual peak-discharge data of Problem 8.9 on lognormal probability paper. Determine the exceedance probability of a flood of 100 m³/s. What is the magnitude of flood of a 200-year return period?

8.12 The maximum annual instantaneous flows from a river are listed below. Plot the flood frequency curve on lognormal probability paper by the empirical method. Determine the flood having a probability of exceedance of 0.5. Also determine the flood having a return period of 10 years.

Water Year	Flow (m³/s)	Water Year	Flow (m³/s)	Water Year	Flow (m³/s)
1980	224	1989	460	1998	223
1981	305	1990	1330	1999	218
1982	210	1991	340	2000	185
1983	333	1992	447	2001	256
1984	500	1993	298	2002	288
1985	98	1994	397	2003	483
1986	288	1995	195	2004	790
1987	370	1996	690	2005	500
1988	212	1997	410		

8.13 The following is a record of peak discharge data from a gaging station. From the city records, a historic flood of 10,000 cfs was noted for the year 1926. Plot the flood frequency curve on lognormal paper by the empirical method. Compute the flow of the 200-year return period.

Year	Peak Discharge (cfs)	Year	Peak Discharge (cfs)	Year	Peak Discharge (cfs)
1976	7420	1986	3560	1996	7345
1977	5290	1987	4680	1997	7420
1978	7880	1988	3560	1998	6944
1979	8800	1989	5550	1999	7250
1980	7590	1990	5770	2000	4290
1981	5550	1991	8260	2001	5385
1982	6175	1992	3430	2002	5984
1983	4170	1993	8060	2003	3040
1984	6215	1994	6855	2004	2064
1985	5010	1995	4370	2005	5150

8.14 The peak-flow data on an annual basis from Cedar River near Austin, MN, are listed in the following table. Plot the flood-frequency curve on lognormal probability paper. Determine the (**a**) magnitude of a flood having a return period of 100 years (probability of 1%), and (**b**) probability of a flow of 100 cfs.

Year	Peak-flow cfs	Year	Peak-flow cfs	Year	Peak-flow cfs	Year	Peak-flow cfs
1956	7750	1969	979	1982	3880	1994	8690
1957	5440	1970	4940	1983	2110	1995	3410
1958	3580	1971	4260	1984	8270	1996	2190
1959	5260	1972	9400	1985	5740	1997	4440
1960	4000	1973	9530	1986	4140	1998	4070
1961	8800	1974	2330	1987	3910	1999	1400
1962	7070	1975	990	1988	1300	2000	3290
1963	7520	1976	9400	1989	12400	2001	7580
1964	6990	1977	3740	1990	4720	2002	4640
1965	5570	1978	3250	1991	3250	2003	3190
1966	2710	1979	2920	1992	4810	2004	10800
1967	2190	1980	3830	1993	3060	2005	2500
1968	2250	1981	1430				

8.15 On the flood sequence of Problem 8.14, perform a frequency analysis by the theoretical method adopting the lognormal distribution.

8.16 Perform a flood-frequency analysis of the annual maximum flow data of Problem 8.9 by the theoretical method, adopting the log-Pearson type III distribution. The regional skew coefficient from the map is 0.0.

8.17 On the flood-flow sequence of Problem 8.12, perform a frequency analysis by the theoretical method, adopting the log-Pearson type III distribution. The regional skew coefficient from the map is 0.50.

8.18 Using the peak-discharge data of Problem 8.13, compute the flood-frequency curve based on the log-Pearson type III distribution. The regional skew coefficient is 0.10.

8.19 Solve Problem 8.16 by the extreme value type I distribution.

8.20 Solve Problem 8.17 by the lognormal distribution.

8.21 Solve Problem 8.18 by the normal distribution.

8.22 Construct a reliability band of 0.05 and 0.95 error limits on the log-Pearson type III frequency curve of Problem 8.16.

8.23 On the frequency curve of Problem 8.18, plot the 5% and 95% error limit curves. Also compute the probability adjusted for the size of the flood flow series.

8.24 The observed peak discharges at a gaging site consist of hurricane and nonhurricane events for which separate flood-frequency curves have been computed. From these curves the following exceedance probabilities have been noted for two types of floods, corresponding to the discharge levels indicated. Prepare a composite flood-frequency curve for the combined series. What is the return period of a peak flow of 40,000 cfs?

Q (cfs)	P_h (for hurricane)	P_n (for nonhurricane)
26,500	0.048	0.002
25,000	0.051	0.009
20,000	0.064	0.153
17,500	0.072	0.308
15,000	0.084	0.52
12,500	0.098	0.727
10,000	0.119	0.883

8.25 A basin in the central coastal region of California has a drainage area, $A = 250$ mi^2; stream length, $L = 80$ mi; infiltration index, $I = 0.6$; and surface storage, $G = 16,000$ acre-ft. The regression coefficients for eq. (8.18) and (8.19) are: a = 0.26, b = 0.15, c = 0.3, d = 1.2, e = 0.05, f = 0.1, g = 0.018, and h = 0.2. The weighted regional skewness coefficient is −0.9. Determine the flood-frequency curve for the region (use log-Pearson type III, Table 8.7). Determine the magnitude of 100-year flood.

8.26 For the basin in Problem 8.25, annual precipitation is 16 in. and altitude index, *H*, is 1.1 thousand ft. Determine the 100-year flood by eq. (8.20).

8.27 Using the generalized map, determine the 18-hour PMP for a watershed of 320 mi^2 located in Boston, MA. Construct a hyetograph, assuming that the rainfall in each 3-hour duration is according to HEC distribution (Table 8.15).

8.28 For the Shetucket River at Willimantic, CT (latitude 41.5 N, longitude 72 E), from Figure 8.10 and Table 8.14 read the PMP values for various storm areas for 6-hour and 12-hour durations. Develop the depth-area-duration curve of the type shown in Figure 2.8.

8.29 For a storm area of 1000-mi^2 on the Delaware River Basin at Callicoon, PA (latitude 41.45 N, longitude 75 E), determine the 6-hour PMP. Also determine the average depth of precipitation on the watershed if its area is about 1800 mi^2. The isohyetal adjustment factors for 1000-mi^2 are given below. Assume the orientation recommended by HMR 52.

Isohyet	A	B	C	D	E	F	G	H	I	J	K
Factor	1.5	1.4	1.3	1.22	1.12	1.04	0.96	0.88	0.82	0.6	0.42

8.30 The ordinates of a 4-hour unit hydrograph for a basin of 100 mi^2, which has resulted from an ordinary storm, are given below. Prepare the modified unit hydrograph that can be used for computing the design flood hydrograph. Assume that the peak of the major storm UHG is 130% of the ordinary storm.

Time (hr)	1	2	3	4	5	6	7	8
Flow (cfs)	475	3000	5240	7140	8300	7260	6190	5360

Time (hr)	9	10	11	12	13	14	15	16
Flow (cfs)	4520	3800	3200	2600	2140	1670	1300	950

Time (hr)	17	18	19	20
Flow (cfs)	710	475	240	120

8.31 A peak flow of 329,000 cfs was recorded on August 20, 1955, in the Delaware River near Trenton, NJ, having a drainage area of 6780 mi². If the peak flow was estimated by the Myers method with a rating of 4000, what is the error of estimation?

8.32 A peak discharge of 1,110,000 cfs was recorded on the Ohio River near Louisville on January 26, 1937. The river has a drainage area of 91,170 mi². Estimate the peak-flow rate by the Myers method and the percent error of estimation. For the Myers rating, refer to Table 8.17.

8.33 The following data were observed during floods in a river basin. Determine the Myers rating. What peak is expected for a drainage area of 20,000 square miles?

Drainage Area, mi²	Peak flow, ft³/s
220	400
500	680
1500	1100
3000	1500
5400	2500
10,000	3100
14,000	4000

8.34 The data for 7-day low flows of the English River near Altona, NY, from the annual series are given here. Prepare the low-flow frequency curve on extreme value type III (log Gumbel) paper. Determine (a) the probability of a 7-day low flow of 7.5 m³/s, and (b) the return period for a flow of 4.5 m³/s.

Year	7-day Low Flow (m³/s)	Year	7-Day Low Flow (m³/s)
1972	4.95	1989	4.15
1973	4.90	1990	6.00
1974	5.80	1991	5.80
1975	4.95	1992	4.90
1976	6.20	1993	7.25
1977	6.25	1994	10.10
1978	4.85	1995	6.65
1979	4.00	1996	5.55
1980	4.15	1997	5.05
1981	5.55	1998	3.60
1982	5.60	1999	4.00
1983	6.35	2000	3.55
1984	4.85	2001	4.20
1985	3.15	2002	2.80
1986	6.40	2003	5.40
1987	6.30	2004	6.15
1988	4.65	2005	6.10

8.35 The annual 7-day minimum average flows for the Buffalo River in Tennessee are given below. Plot the low-flow frequency curve on log Gumbel paper. Determine **(a)** the 10-year 7-day low flows, and **(b)** the probability of a 160 cfs flow.

Year	Flow (cfs)	Year	Flow (cfs)
1972	207.8	1989	128.5
1973	209.0	1990	141.2
1974	196.6	1991	156.2
1975	151.4	1992	187.6
1976	117.2	1993	195.3
1977	141.1	1994	184.0
1978	184.0	1995	144.9
1979	206.6	1996	153.7
1980	262.1	1997	157.5
1981	284.5	1998	187.8
1982	252.0	1999	211.2
1983	151.0	2000	121.0
1984	184.0	2001	138.6
1985	229.3	2002	200.3
1986	214.2	2003	163.8
1987	165.1	2004	182.7
1988	122.2	2005	124.7

8.36 Solve Problem 8.34 by the analytical method using log-extreme value I distribution.

8.37 Solve Problem 8.35 by the analytical method using log-extreme value I distribution.

9

Hydraulics: Principles and Applications

◆◆

9.1 HYDRODYNAMIC EQUATIONS OF FLOW

There are three basic hydrodynamic equations that can describe all types of flow problems in hydraulics. These are the continuity equation, the energy equation, and the momentum equation. The combined application of the continuity and energy equations or the continuity and momentum equations is common to hydraulic problems. The combination of the continuity equation and the momentum equation is known as the *Saint-Venant equations* after the researcher who first used them together in 1871 to formulate the theory of one-dimensional analysis of unsteady-state flow.

The momentum equation accounts for all forces in a hydraulic system. It is distinct from the energy equation since energy is a scalar quantity, whereas momentum is a vector quantity. In certain problems, both lead to similar results. In other problems, only one of them can be used conveniently. The energy equation is simpler to use and has a wider application. However, in problems involving unknown energy losses such as a hydraulic jump, the momentum equation has a direct application.

9.2 THE CONTINUITY EQUATION

The continuity equation originates from the law of conservation of mass. To develop a continuous form, consider an element of the fluid volume as shown in Figure 9.1. In the figure, q_L is the lateral inflow rate per unit length of channel. Q and A are initial discharge and cross-sectional area. All variables are functions of time and space.

$$\text{Total inflow} = Q + q_L \Delta s \qquad (a)$$

$$\text{Total outflow} = Q + \frac{\partial Q}{\partial s} \Delta s \qquad (b)$$

The rate of change in volume stored within the element is equal to the change in cross-sectional area multiplied by the length of section, i.e.,

$$\frac{\partial A}{\partial t} \Delta s$$

According to the conservation law:

$$\text{Input} - \text{Output} = \text{Rate of change in volume} \qquad (c)$$

483

Figure 9.1 Control volume for the continuity equation.

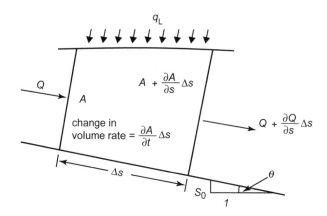

Substituting the inflow and outflow quantities (a) and (b) in (c) and dividing by Δs,

$$\frac{\partial A}{\partial t} + \frac{\partial Q}{\partial s} = q_L \quad [L^2/T] \tag{9.1}$$

Since $Q = Av$, eq. (9.1) becomes

$$\frac{\partial A}{\partial t} + A\frac{\partial v}{\partial s} + v\frac{\partial A}{\partial s} = q_L \quad [L^2/T] \tag{9.2}$$

This is the continuity equation in *differential form* for incompressible flow. For a very wide channel or for overland flow, area A can be substituted by depth y or by pressure head h for a pressurized system. If the flow is steady ($\partial A/\partial t = 0$) and where water does not run in or out along the course of flow ($q_L = 0$), from eq. (9.1),

$$\frac{dQ}{ds} = 0 \quad \text{or} \quad Q = \text{constant}$$

Thus, a discrete form of the most widely used continuity equation is

$$Q = v_1 A_1 = v_2 A_2 = \cdots \quad [L^3 T^{-1}] \tag{9.3}$$

where

$$Q = \text{rate of flow or discharge}$$
$$v_1, v_2,\ldots = \text{velocity of flow at section 1, 2,...}$$
$$A_1, A_2,\ldots = \text{cross-sectional area of section 1, 2,...}$$

9.3 THE ENERGY EQUATION

According to Newton's second law of motion, the net force acting on a fluid element of Figure 9.2 is equal to its mass times its acceleration ($F = ma$). The forces acting on the element include the net pressure force, gravity force, and any other internal force such as the friction acting along the surface of contact.

Figure 9.2 Control volume for the energy and momentum equations.

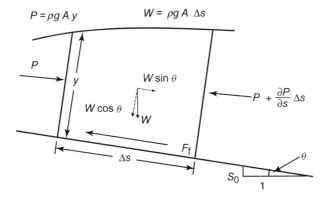

1. The term for acceleration in the s direction is given by

$$a = \frac{\partial v}{\partial t} = \frac{\partial s}{\partial t} \cdot \frac{\partial v}{\partial s} = v\frac{\partial v}{\partial s} = \frac{1}{2}\frac{\partial v^2}{\partial s}$$

Thus,

$$ma = \frac{1}{2}\frac{W}{g}\frac{\partial v^2}{\partial s}$$

2. From Figure 9.2, the net pressure force is

$$-\frac{\partial P}{\partial s}\Delta s = -\frac{\partial p}{\partial s}\Delta A \Delta s = -\frac{W}{\gamma}\frac{\partial p}{\partial s}$$

3. The gravitational force is given by the element weight component in the direction of flow, i.e.

$$W \sin \theta \quad \text{or} \quad -W\frac{\partial Z}{\partial s} \quad (-\text{ for drop in elevation})$$

4. Other internal forces acting within the element are expressed similarly in terms of the slope multiplied by the weight of element, i. e., $W\dfrac{\partial H}{\partial s}$

Equating the *ma* term of item 1 with the forces from items 2, 3, and 4

$$\frac{1}{2}\frac{W}{g}\frac{\partial v^2}{\partial s} = -\frac{W}{\gamma}\frac{\partial p}{\partial s} - W\frac{\partial Z}{\partial s} - W\frac{\partial H}{\partial s}$$

This can be simplified to

$$\frac{1}{2g}dv^2 + \frac{1}{\gamma}dp + dZ + dH = 0 \quad [\text{L}] \tag{9.4}$$

This is the *differential form* of the energy equation. This can be integrated to give

$$\frac{v^2}{2g} + \frac{p}{\gamma} + Z + H = \text{constant} \quad [\text{L}] \tag{9.5}$$

Section 9.3   The Energy Equation

Including the internal loss of energy between the two sections, $h_f = (H_2 - H_1)$, the common form of the energy equation is

$$Z_1 + \frac{p_1}{\gamma} + \frac{v_1^2}{2g} = Z_2 + \frac{p_2}{\gamma} + \frac{v_2^2}{2g} + h_f \quad [L] \qquad (9.6)$$

In Bernoulli's equation, the internal energy term h_f is ignored.

For channel flow, since $p = \gamma y$, the pressure terms in p_2/γ and p_1/γ are replaced by depths y_2 and y_1, respectively. Various forms of energy in open channel flow are shown in Figure 9.3.

EXAMPLE 9.1

Determine the discharge per unit width of a broad-crested weir in a rectangular channel as shown in Figure 9.4. Disregard the losses.

SOLUTION

1. Applying the energy equation at point 1 and at point 2 in the center of the weir, and disregarding the velocity of approach at point 1, we have

$$Z_1 + y_1 + \frac{v_1^2}{2g} = Z_2 + y_2 + \frac{v_2^2}{2g}$$

$$10 + 0 + 0 = 4 + 3 + \frac{v^2}{2g}$$

$$\text{or } v = 13.90 \text{ ft/sec}$$

2. From the continuity equation,

$$q = Av$$

$$= (3 \text{ ft} \times 1 \text{ ft})(13.90 \text{ ft/s}) = 42 \text{ cfs/ft width}$$

(The coefficient of discharge, C_d, has not been included in the computation.)

9.4 THE MOMENTUM EQUATION

The resultant force on a fluid element is also equal to the change of rate of momentum. In Figure 9.2 the forces acting on the element are the net pressure force and gravity force, and the friction force acts along the surface of contact between the water and channel bottom. When there is a lateral inflow, the momentum transfer due to lateral inflow should be included as well. The equation is developed as follows:

1. For a fluid of mass m and velocity v, the change of rate of momentum is

$$m\frac{dv}{dt}$$

The velocity term can be expressed

$$\frac{dv}{dt} = \frac{\partial v}{\partial t} + v\frac{\partial v}{\partial s}$$

Figure 9.3   Various forms of energy in open channel flow.

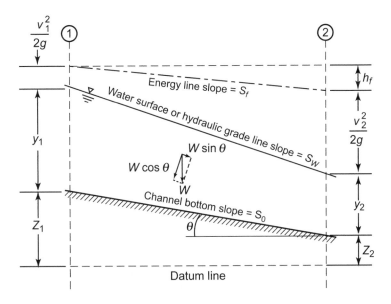

Figure 9.4   Application of the energy principle on a broad-crested weir.

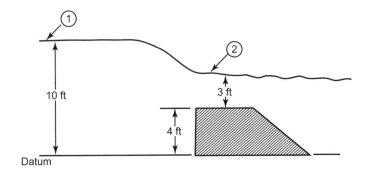

Thus, the rate of change of momentum consists of two parts:

a. The change of momentum with respect to time, referred to as the temporal or local acceleration, given by $m\partial v/\partial t$, and

b. The spatial change of momentum, referred to as the convection acceleration, given by $mv\partial v/\partial s$, where m is the mass of element $= \rho\Delta A\Delta s$

2. From Figure 9.2, the net pressure force is

$$-\frac{\partial P}{\partial s}\Delta s \quad \text{or} \quad -W\frac{\partial y}{\partial s}$$

since $P = \gamma y\Delta A$ and $W = \gamma\Delta A\Delta s$

3. The gravitational force is given by the element weight component in the direction of flow, i.e., $W\sin\theta$ or WS_0, S_0 being the channel bottom slope.

4. The friction force along the bottom is equal to the friction slope multiplied by the weight of the element, i.e., WS_f.

Equating the momentum terms of item 1(a) and (b) above with the external forces from items 2, 3, and 4:

$$\rho \Delta A \Delta s \left[\frac{\partial v}{\partial t} + v \frac{\partial v}{\partial s} \right] = -\frac{\partial P}{\partial s} \Delta s \quad + \quad WS_0 \quad - \quad WS_f \quad [F] \qquad (9.7)$$

$$\left(\text{inertia term} \right) = \left(\begin{matrix} \text{pressure} \\ \text{term} \end{matrix} \right) + \left(\begin{matrix} \text{gravity} \\ \text{term} \end{matrix} \right) - \left(\begin{matrix} \text{friction} \\ \text{term} \end{matrix} \right)$$

Equation (9.7) is the momentum equation. The term WS_f measures the force exerted by the water. It is not related to internal losses. In many applications this can be omitted.

For steady-state flow, $\partial v / \partial t$ is zero. This reduces eq. (9.7), in a commonly used discrete form, to

$$\left(\rho A v_n v_s \right)_2 - \left(\rho A v_n v_s \right)_1 = \Sigma F_s \quad [F] \qquad (9.8)$$

where the subscript 2 relates to the outflow section and the subscript 1 refers to the inflow section. The subscript s denotes the direction such as x, y, or z, to which the momentum equation is applied, while v_n is velocity normal to the flow and v_s is the velocity component in the s-direction. The outflow velocity vector v_s denotes the positive direction with reference to which signs are assigned to all terms. The term ΣF_s denotes all applied forces, including pressure forces at sections 1 and 2, $P_1 - P_2$, the weight component in s direction, $W \sin \theta$, and other applied forces.

EXAMPLE 9.2

Solve Example 9.1 by the momentum principle. Assume no friction force.

SOLUTION　Refer to Figure 9.5.

1. Considering the horizontal bed, the weight component in the direction of flow is zero, and in the positive direction, left to right.

2. Pressure forces acting are:

   a. Upstream pressure at section 1, taking 1-foot width, b

   $$P_1 = \frac{1}{2} \gamma b y_1^2 = \frac{1}{2} \gamma (1)(10)^2 = 50\gamma$$

   b. Pressure at section 2,

   $$P_2 = \frac{1}{2} \gamma b y_2^2 = \frac{1}{2} \gamma (1)(3)^2 = 4.5\gamma$$

   c. Water pressure on the weir face at section 2,

   $$P_3 = b \text{ (area of pressure trapezoid)}$$

   $$= (1)(4) \frac{[6+10]\gamma}{2}$$

   $$= 32\gamma$$

Figure 9.5 Application of the momentum principle.

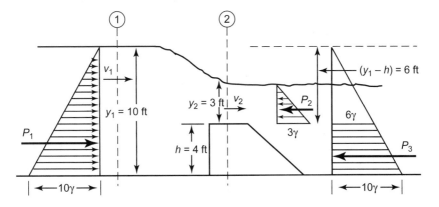

3. Applying the momentum equation (9.8) between sections 1 and 2:

$$\frac{\gamma}{g}(3)(1)v_2^2 - \frac{\gamma}{g}(10)(1)v_1^2 = (50\gamma - 4.5\gamma - 32\gamma)$$

4. From the continuity equation,

$$q = A_1 v_1 \quad \text{or} \quad v_1 = \frac{q}{(1)(10)} = \frac{q}{10}$$

and

$$q = A_2 v_2 \quad \text{or} \quad v_2 = \frac{q}{(1)(3)} = \frac{q}{3}$$

5. Substituting in step 3 yields

$$\frac{1}{g}q^2\left(\frac{1}{3} - \frac{1}{10}\right) = 13.5$$

$$q = 43 \text{ cfs/ft width}$$

A slight difference in the answer compared to Example 9.1 is due to the assumptions specifically disregarding the velocity of approach in Example 9.1.

9.4.1 Application of the Momentum Equation

Equation (9.7) describes complete flow in an open channel. The following different types of flow can be represented by this equation:

1. If inertia and pressure terms are neglected in eq. (9.7), then $S_0 = S_f$, which means flow is steady and uniform. The equation of motion can, then, be described by the uniform flow formula of Chezy (eq. 11.8) or Manning (eq. 11.9). The analysis of this equation, along with the continuity equation, is known as *kinematic wave analysis*. The term "kinematics" refers to the branch of mechanics that treats the motion of a body in the abstract without reference to the action of forces on it. This is the case with the uniform flow formula of open channel flow.

2. If only inertia terms in eq. (9.7) are disregarded, then $(dP/ds)\Delta s = W(S_0 - S_f)$ or $\partial y/\partial x = S_0 - S_f$. This is steady nonuniform flow. This equation can be combined with the continuity equation into a single equation of the form of the classic advection-diffusion eq. (5.9). Accordingly, analysis with this assumption is known as *diffusion* or *non-inertial wave analysis.* The dispersion acts to laterally spread a hydrograph as it travels down a river and the diffusion is responsible for attenuation or subsidence of the hydrograph crest.

3. When all terms of eq. (9.7) are considered, it represents unsteady nonuniform flow. Equations (9.2) and (9.7) are known as the dynamic wave equations. The dynamic wave solutions are quite complicated. The above two approximations have accordingly found a wide application in hydrology and the simplest of these, the kinematic wave theory, is now a well-accepted approach to hydrologic analysis.

9.5 KINEMATIC WAVE THEORY

Kinematic theory uses the continuity eq. (9.1) and a simplified momentum equation expressed by the uniform channel flow formula of Chezy (11.8) or Manning (11.9). As far as momentum is concerned, the flow is assumed to be steady. The unsteadiness of flow is maintained through the continuity equation. The theory has limited validity. Manning's equation 11.10 can be rewritten as:

$$Q = \frac{K}{n} \frac{A^{5/3} S_0^{1/2}}{P_w^{2/3}} \quad [L^3/T] \tag{9.9}$$

where K is 1.486 for FPS units and 1 in metric units, and P_w is wetted perimeter.

For a shallow water depth in overland flow or flow in a very wide channel, the wetted perimeter is practically independent of the flow area. Equation (9.9) can be expressed in a general form:

$$Q = \alpha A^m \quad [L^3/T] \tag{9.10}$$

where $\alpha = \dfrac{K}{n} \dfrac{S_0^{1.2}}{P_w^{2/3}}$ and $m = 5/3$ for the Manning case.

Differentiating eq. (9.10)

$$\partial Q = \alpha m A^{m-1} \partial A \tag{a}$$

or

$$\partial Q = \alpha m \left(\frac{A^m}{A} \right) \partial A \tag{b}$$

In eq. (b), substituting Q for αA^m from eq. (9.10)

$$\partial Q = m \left(\frac{Q}{A} \right) \partial A = mv \partial A = c \partial A \tag{c}$$

where v is the flow velocity and c is wave speed (celerity). Since $m > 1$, the celerity is greater than the flow velocity.

When ∂A from eq. (c) is substituted in terms of ∂Q in eq. (9.1):

$$\frac{\partial Q}{\partial s} + \frac{1}{c}\frac{\partial Q}{\partial t} = q_L \quad [\text{L}^2\text{T}] \tag{9.11}$$

This is the kinematic flow equation for discharge.

When ∂Q from eq. (c) is substituted in terms of ∂A in eq. (9.1):

$$\frac{\partial A}{\partial t} + c\frac{\partial A}{\partial s} = q_L \quad [\text{L}^2/\text{T}] \tag{9.12}$$

This is the equation for kinematic flow in terms of flow area.

9.5.1 Methods of Solving the Kinematic Equations

The kinematic equations (9.1) and (9.10) or the combined eq. (9.12) are solved by the finite-difference, finite-element, or characteristics methods. The finite-difference or finite element techniques commonly consider the full dynamic form of the momentum eq. (9.7). For the kinematic eq. (9.12), the method of characteristics is a convenient mode of solution. The concept of characteristics is essentially a graphic representation. A *characteristic* is a propagation path followed by some entity. A rectangular grid of roads could be considered as two families of propagation paths along which vehicles propagate. In a mechanical system, the directions along which a force is propagated are its characteristics. In a hydraulic system, the propagation paths of a water wave are its characteristics. The propagation paths could be represented by lines in terms of x and y coordinates. In many cases, the propagation process has to be considered at different times. The characteristics, then, have physical distances and times as coordinates.

The concept of characteristics is of great significance in applied mathematics. Through its use it is possible to reduce a partial differential equation to one or the other form of the standard ordinary differential equations, known as the *characteristic equations* or *auxiliary equations*. The solutions of these auxiliary ordinary differential equations are then expressed by one or two families of curves in an *x-y* or *x-t* plane. These curves are said to be the characteristics of the partial differential equation.

To solve a linear first-order partial differential equation, the approach of the French mathematician d'Alembert (1717–1783), which was later followed by Monge (1809) and extended by Riemann (1860), involves converting the partial differential to the following form so as to derive its total differential.

$$F = \frac{\partial f}{\partial t} + c\frac{\partial f}{\partial x} \tag{a}$$

The kinematic eq. (9.12) is already of this type. It can be solved easily by the following procedure:

In eq. (9.12), let us associate c with $\dfrac{ds}{dt}$ by defining

$$\frac{ds}{dt} = c \quad [\text{L}/\text{T}] \tag{9.13}$$

Then eq. (9.12) becomes

$$\frac{\partial A}{\partial t} + \frac{ds}{dt}\frac{\partial A}{\partial s} = q_L \tag{b}$$

Since $\dfrac{dA}{dt} = \dfrac{\partial A}{\partial t} + \dfrac{ds}{dt} \cdot \dfrac{\partial A}{\partial s}$, the left side of eq. (b) is the total differential, i.e.,

$$\frac{dA}{dt} = q_L \quad [\text{L}^2/\text{T}] \tag{9.14}$$

The solution consists of the following steps:

1. Find directions by eq. (9.13) which can be associated with the characteristic directions.
2. Solve eq. (9.14) for function A for given initial and boundary conditions.

EXAMPLE 9.3

Find the characteristic curve and the solution of the following equation. Initially, at $t = 0$, $x = 1$ m, and $u = 2$ m/s.

$$\frac{\partial u}{\partial t} + 0.2x\frac{\partial u}{\partial x} = 3$$

SOLUTION

1. From eq. (9.13), $dx/dt = 0.2x$
2. Integrating, $\ln x = 0.2t + \text{Const.}$
3. Substituting the initial conditions, $\ln 1 = 0.2(0) + \text{Const.}$, or Const. $= 0$
4. Hence, $x = e^{0.2t}$ $\quad$ [L] $\hspace{3cm}$ (9.15)

t, sec (select)	x, m (from eq. 9.15)
0	1
1	1.22
2	1.49
5	2.72

This is plotted as the characteristic curve in Figure 9.6

5. From eq. (9.14), $du/dt = 3$
6. Integrating, $u = 3t + \text{Const.}$
7. Substituting initial condition, $2 = 3(0) + \text{Const.}$ or Const. $= 2$
8. Hence, the solution is $u = 3t + 2$

9.6 FORMULATION OF HYDROGRAPH BY THE KINEMATIC THEORY

Conceptually, the solution of eq. (9.14) gives the flow area at different times at the end of a specified section. Since discharge is related to flow area by eq. (9.10), the above solution also provides the discharge at different times at a section that represents the hydrograph at that section. When the kinematic theory is applied to overland flow on a watershed in which q_L represents the input rate of rainfall on the watershed, the resultant solution is the hydrograph at the basin outlet due to the rainfall. This is an elegant way to formulate a

Figure 9.6 The characteristic curve for the equation in Example 9.3.

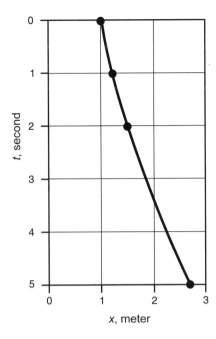

runoff hydrograph from a hyetograph of the rainfall excess. The explicit analytical solutions can be obtained for simple conditions on rectangular plane catchments.

For sheet flow on a wide plane, $A = by$, $Q = bq$, and since the depth is quite small compared to the width, $R = y$ and $P = b$, where b is the width of the plane, y is the water depth, and q is the flow per unit width. The lateral inflow rate, $q_L = i_e b$, where i_e is the excess rainfall intensity (after subtracting the loss rate). For unit width ($b = 1$), eqs. (9.12), (9.10), (9.13), and (9.14) can be rewritten in the x-direction, by substituting the above expressions, as follows:

$$\frac{\partial y}{\partial t} + c\frac{\partial y}{\partial x} = i_e \qquad [\text{L/T}] \tag{9.16}$$

$$q = \alpha y^m \qquad [\text{L}^2/\text{T}] \tag{9.17}$$

$$\frac{dx}{dt} = c = \alpha m y^{m-1} \quad [\text{LT}] \tag{9.18}$$

(since $c = mv$ or mq/y)

and

$$\frac{dy}{dt} = i_e \quad [\text{L/ T}] \tag{9.19}$$

where α, m, and K are defined in eqs. (9.10) and (9.9).

9.6.1 Solution for Rising Hydrograph

Suppose the following initial conditions are set: At $t = 0$ (at the beginning), $y = 0$ (the catchment surface is dry) and $x = 0$ (the wave characteristics emanate from the upslope end of the plane). Integration of eq. (9.19) with the initial conditions yields

$$y = i_e t \quad [\text{L}] \tag{9.20}$$

Substituting in eq. (9.17),

$$q = \alpha(i_e t)^m \quad [\text{L}^2/\text{T}] \tag{9.21}$$

Equations (9.20) and (9.21) are for the water depth and the discharge along each characteristic as that characteristic moves from origin toward the downstream end of the plane. The location of the characteristic at any point in time or the downslope position of depth y, after a given time t, is obtained from eq. (9.18). In eq. (9.18), substituting $y = i_e t$ and integrating with the initial conditions, gives

$$x = \alpha i_e^{m-1} t^m \quad [\text{L}] \tag{9.22}$$

A characteristic path for the indefinite duration of rainfall is shown in Figure 9.7. Since the built-in time unit in Manning's and Chezy's equations is seconds, it will be dimensionally consistent to use seconds as the unit for time.

EXAMPLE 9.4

Determine the rising hydrograph for a rainfall rate of 3 cm/hr on a 50 m × 50 m parking lot sloped at 1%. $n = 0.02$

SOLUTION

1. $i_e = 3$ cm/hr or 8.33×10^{-6} m/s

2. Per meter width, $\alpha = \dfrac{1}{n} S_0^{1/2} = \dfrac{(0.01)^{1/2}}{0.02} = 5$

3. From eq. (9.21): $q = 5(8.33 \times 10^{-6} t)^{5/3}$, where t is in sec.

4. Using the above equation, the computations are arranged in Table 9.1.

Table 9.1   Rising Hydrograph Ordinates

Time		Discharge
sec	min	m³/s × 10⁻⁵ per m
0	0	0
60	1	1.57
120	2	5.00
180	3	9.83
300	5	22.96

9.6.2 Time of Concentration

It will be seen from Figure 9.7 that for a uniform rainfall intensity, the depth profile remains constant at a given downslope position regardless of how long the rainfall continues. Once the characteristic has reached the downstream end of the plane, the optimum depth will be reached and will remain constant as long as rainfall persists. This is the equilibrium state in which the rate of outflow equals the rainfall rate. The time required for this to happen is the equilibrium time or the time of concentration, t_c. The concentration time can be obtained from eq. (9.22) for the condition that at $x = L$, $t = t_c$. Thus,

$$t_c = \left(\frac{L}{\alpha i_e^{m-1}} \right)^{1/m} \quad [T] \tag{9.23}$$

For Manning-kinematic flow, $m = 5/3$, hence

$$t_c = \left(\frac{L}{\alpha i_e^{0.667}} \right)^{0.6} \quad [T] \tag{9.24}$$

EXAMPLE 9.5

Estimate the time of concentration for Example 9.4.

SOLUTION

1. From Example 9.4, $\alpha = 5$, $i_e = 8.33 \times 10^{-6}$ m/s, and $L = 50$ m.

2. From eq. (9.24),

$$t_c = \left[\frac{50}{(5)(8.33 \times 10^{-6})^{0.667}} \right]^{0.6}$$

$$= 429 \text{ sec or } 7.2 \text{ min}$$

Figure 9.7 Kinematic wave characteristic for rainfall on a plane catchment.

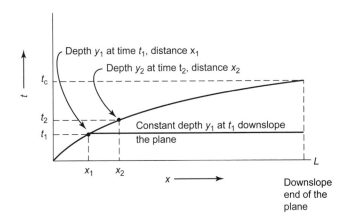

9.6.3 Receding Hydrograph

After the rainfall stops, $dy/dt = 0$ from eq. (9.19). Thereafter, the accumulation from upstream travels down to the exit at a rate (velocity) given by eq. (9.18). The integration of eq. (9.18) beyond the rainfall duration, t_d, leads to the following implicit relation of the receding limb of the hydrograph:

$$L - \frac{q}{i_e} - m\alpha^{1/m} q^{m-1/m} \left(t - t_d \right) = 0 \qquad (9.25)$$

The above relation involves the following three cases:

1. When the rainfall duration, t_d, equals time of concentration, t_c, the recession hydrograph will start at t_c. Up to t_c, eq. (9.21) is used for the rising limb, thereafter the recession eq. (9.25) is applied by substituting t_c for t_d.

2. When $t_d > t_c$, the constant equilibrium depth of flow will continue from time t_c through t_d. Up to time t_c, use the rising limb eq. (9.21), continue the same flow till t_d, and thereafter apply eq. (9.25).

3. When $t_d < t_c$, the discharge will rise by eq. (9.21) up to $q = \alpha(i_e t_d)^m$, thereafter it will remain constant from t_d through t_k given by the following equation:

$$t_k = t_d + \frac{t_d}{m} \left[\left(\frac{t_c}{t_d} \right)^m - 1 \right] \quad [\text{T}] \qquad (9.26)$$

After the time t_k, the recession will proceed by eq. (9.25).

EXAMPLE 9.6

For Example 9.4, the rainfall duration is 5 min. Determine the equation of the recession hydrograph. Plot the complete hydrograph.

SOLUTION

1. From Example 9.5, $t_c = 7.2$ min, since $t_d < t_c$, it is case 3 above.

2. From eq. (9.26),

$$t_k = t_d + \frac{t_d}{m} \left[\left(\frac{t_c}{t_d} \right)^m - 1 \right]$$

$$= 5 + \frac{5}{1.667} \left[\left(\frac{7.2}{5} \right)^{1.667} - 1 \right]$$

$$= 7.51 \text{ min or } 450 \text{ sec}$$

3. The discharge will rise as in Figure 9.1 up to $t_d = 5$ min. It will remain constant from 5 min to 7.51 min at 22.96×10^{-5} m³/s per meter width corresponding to the 5 min value.

4. After 7.51 min, from eq. (9.25),

$$L - \frac{q}{i_e} - m\alpha^{1/m} q^{m-1/m} \left(t - t_d \right) = 0$$

or

$$50 - \frac{q}{8.33 \times 10^{-6}} - 1.667(5)^{1/1.667} q^{0.667/1.667} (t-300) = 0$$

or

$$50 - 1.2 \times 10^5 q - 4.38 q^{0.4} (t - 300) = 0 \quad [L] \tag{9.27}$$

5. Equation (9.27) is solved graphically. For selected values of q, t has been computed by equation (9.27) as given in the inset on Figure 9.8 and plotted in the figure.

6. From Figure 9.8, the receding hydrograph ordinates at various times are read as given in Table 9.2.

7. The complete hydrograph is plotted in Figure 9.9 from values in Tables 9.1 and 9.2.

9.6.4 Validity of the Kinematic Theory of Hydrographs

The kinematic theory was pioneered by Lighthill and Whitham (1955) for the runoff process. Since then it has been used increasingly in hydrologic applications. However, as discussed in Section 9.5, the theory develops from an approximate form of the momentum equation. The conditions under which the kinematic approximation holds in overland flow are investigated by Woolhiser and Liggett (1967), Overton and Meadows (1976), Morris and Woolhiser (1980), and Vieira (1983). The kinematic solution offers a close approximation to the full dynamic equation when the kinematic flow number (defined by $K = S_0 L / y_0 Fr^2$, where Froude number, $Fr = v_0 / g y_0$) has a large value (the subscript 0 denoting the uniform flow condition). The following criterion has been developed:

$$KFr^2 = \frac{\alpha^{0.6} S_0 L^{0.4}}{i_e^{0.6}} \quad [\text{dimensionless}] * \tag{9.28}$$

When $KFr^2 > 5$ the kinematic theory can be applied. As apparent from eq. (9.28), the higher values are produced on steep, long planes with low rainfall rates and lower roughness. Most overland flow problems satisfy kinematic flow conditions.

9.7 ROUTING PROCESS

Flow routing is a process whereby an outflow hydrograph is determined at a point in a stream, reservoir, or lake resulting from a known inflow hydrograph at an upstream point. The inflow and outflow hydrographs may represent daily or monthly streamflows or flood flows resulting from a short duration storm. Routing serves the useful purpose of (1) deriving the hydrographs from rainfall distributions, (2) estimating the water yield at a specified point, (3) developing design elevations of flood walls and levees, (4) studying the effect of a reservoir on the modification of a flood peak, (5) determining the size of a spillway, and (6) other flow related objectives, e.g., compliance with water allocations, dilution of regulated wastes, and forecasting of floods.

There are two approaches to flow routing:

1. In hydraulic routing, the flow is described through a set of hydrodynamic differential equations of unsteady-state flow and simultaneous solutions of those equations lead to determination of the outflow hydrograph. Hydraulic routing is based on the principles

---

* A manipulation of the Morris and Woolhiser (1980) relation done by the author in terms of the physical and hydraulic characteristics.

Figure 9.8 Solution of implicit kinematic equation.

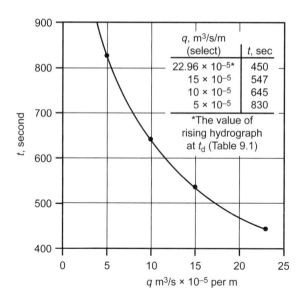

q, m³/s/m (select)	t, sec
22.96 × 10⁻⁵*	450
15 × 10⁻⁵	547
10 × 10⁻⁵	645
5 × 10⁻⁵	830

*The value of rising hydrograph at t_d (Table 9.1)

Figure 9.9 Kinematic hydrograph shape.

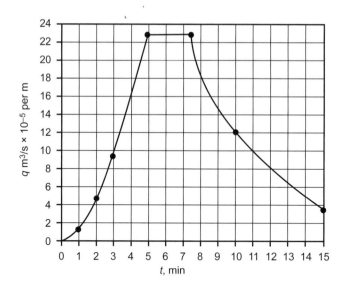

Table 9.2 Receding Hydrograph Ordinates

Time		Discharge
sec	min	m³/s × 10⁻⁵ per m
450	7.5	22.96
600	10	12.0
900	15	3.5

of hydraulics in which flow is computed as a function of time at several locations along a hydraulic system. It involves complexities of varying degrees.

2. Hydrologic routing procedures, on the other hand, do not solve the hydrodynamic equations but approximate solutions of those equations without their direct use. Hydrologic routing is a simple but lumped flow process that provides the direct result at the outlet. The hydrologic routing process is shown in Figure 9.10.

9.8 HYDRAULIC ROUTING

Depending on the number of terms included in the solution of the hydrodynamic equations, the hydraulic routing procedures are called *kinematic routing, diffusion routing*, or *dynamic wave routing*. The simplest of the three methods, the kinematic approach, has been successfully applied to overland flow and streamflow routing problems. Application of the kinematic wave theory is expanding very rapidly to encompass virtually all components of the hydrologic cycle.

The routing process can be envisioned as a movement of shallow water waves over land surface, in rivers and channels, or through lakes and reservoirs. Wave phenomena can be described by the hydrodynamic equations. The continuity equation of waves has two unknowns: the velocity of propagation and area (or depth) of flow. A second equation is therefore required to describe the phenomenon fully. Where energy losses are insignificant or when such losses can be assessed accurately, the energy equation is very convenient for the second equation. However, routing problems consider a wide range of inflows into the routing reach that involve considerable losses of energy that are difficult to quantify. Also, whenever abrupt waves are formed, substantial energy losses occur. As such, it is desirable to use the momentum equation, which deals with the forces acting in a section.

9.9 STREAMFLOW ROUTING BY THE KINEMATIC THEORY

Kinematic eq. (9.11) or eq. (9.12) can be applied to route the flow through a stream channel. If there is no lateral inflow into a channel reach, then q_L is zero. By eq. (9.14), dA/dt (and accordingly dQ/dt) is zero, and through its integration, A or Q is a constant. The analytical solutions are tractable to such simple cases only. For practical streamflow problems, numerical methods are used to solve the kinematic equations. By itself, a kinematic equation is nondiffusive, i.e., the hydrograph crest does not attenuate or subside by the kinematic equation. However, numerical methods of solution introduce some amount of numerical diffusion, i.e., the error resulting from the numerical representation of the kinematic equation constitutes the numerical diffusion. This is desirable since diffusion is present in most natural unsteady channel flows.

Figure 9.11 shows a finite-difference rectangular grid for a cell in the *x-t* plane. The derivatives $\partial Q/\partial t$ and $\partial Q/\partial x$ at point P are expressed in terms of the four adjacent values of Q on the nodes. The variable c, the celerity, also can be expressed in terms of space and time by the four values on the nodes. In the linear form of the equation, c is assumed constant. In the figure, a and b are the weighting factors in space and time directions. The derivatives in the numerical forms are expressed as

$$\frac{\partial Q}{\partial x} = \frac{\Delta Q}{\Delta x} = \frac{1}{\Delta x}\left[(1-b)\left(Q_{x+1}^{t+1}-Q_x^{t+1}\right)+b\left(Q_{x+1}^t-Q_x^t\right)\right] \quad [L^2/T] \qquad (9.29)$$

and

$$\frac{\partial Q}{\partial t} = \frac{\Delta Q}{\Delta t} = \frac{1}{\Delta t}\left[(1-a)\left(Q_{x+1}^{t+1}-Q_{x+1}^t\right)+a\left(Q_x^{t+1}-Q_x^t\right)\right] \quad [L^2/T] \qquad (9.30)$$

Figure 9.10 Flood routing process.

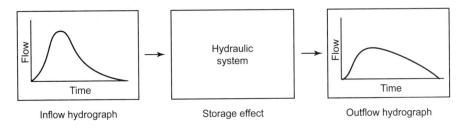

Inflow hydrograph Storage effect Outflow hydrograph

Figure 9.11 Finite-difference grid for a cell.

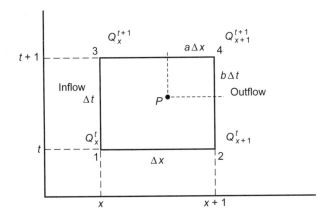

Usually, Q_{x+1}^{t+1} at point 4 is unknown and the values at points 1, 2, and 3 are known. The solution marches forward in either the x or t direction. A desired level of numerical diffusion is achieved by an appropriate choice of a and b values.

9.9.1 Muskingum-Cunge Kinematic Routing Method

By taking celerity c to be an average constant value for the reach, centering the time derivation ($b = 0.5$) and substituting eqs. (9.29) and (9.30) in eq. (9.11), the kinematic equation can be written in the finite-difference form as follows:

$$\frac{1}{2\Delta x}\left[Q_{x+1}^{t} - Q_{x}^{t} + Q_{x+1}^{t+1} - Q_{x}^{t+1}\right] + \frac{1}{c\Delta t}\left[(1-a)\left(Q_{x+1}^{t+1} - Q_{x+1}^{t}\right)\right.$$
$$\left. + a\left(Q_{x}^{t+1} - Q_{x}^{t}\right)\right] = 0 \quad [\mathrm{L^2/T}] \tag{9.31}$$

Solving for Q_{x+1}^{t+1}

$$Q_{x+1}^{t+1} = C_0 Q_x^{t+1} + C_1 Q_x^{t} + C_2 Q_{x+1}^{t} \quad [\mathrm{L^3/T}] \tag{9.32}$$

where

$$K = \Delta x / c \quad [\mathrm{T}] \tag{9.33}$$

$$C_0 = \frac{0.5\Delta t - aK}{K(1-a) + 0.5\Delta t} \quad \text{[dimensionless]} \tag{9.34}$$

$$C_1 = \frac{0.5\Delta t + aK}{K(1-a) + 0.5\Delta t} \quad \text{[dimensionless]} \tag{9.35}$$

$$C_2 = \frac{K(1-a) - 0.5\Delta t}{K(1-a) + 0.5\Delta t} \quad \text{[dimensionless]} \tag{9.36}$$

and

$$C_1 + C_2 + C_3 = 1 \tag{9.37}$$

The subscript x denotes inflows into a reach and $(x+1)$ outflows out of a reach. The subscript t indicates a reference time and $(t + 1)$ refers to the next time interval. Equation (9.32) can be expressed in a simple form

$$O_2 = C_0 I_2 + C_1 I_1 + C_2 O_1 \quad [\text{L}^3/\text{T}] \tag{9.38}$$

The negative values of C_1 and C_2 do not adversely affect the accuracy, but a negative value of C_0 should be avoided.

Cunge (1969) demonstrated that the numerical diffusion might be matched with the hydraulic diffusion, if the parameter, a, is selected as follows:

$$a = \frac{1}{2}\left(1 - \frac{Q}{BS_0 c\Delta x}\right) \quad \text{[dimensionless]} \tag{9.39}$$

where

Q = representative channel discharge

B = width of channel

S_0 = bed slope

For the value of a given by equation (9.39), equations (9.32) through (9.37) constitute an approximation to the diffusion equation.

The routing is performed with either (1) a constant value of c during the entire analysis, or (2) variable values of c when the parameters a; K; and C_0, C_1, and C_2 are calculated at each time step. For an adequate resolution of the outflow hydrograph, the values of Δt and Δx should not be very large. Nominally, the time to peak of an inflow hydrograph is divided into a minimum of five time increments, Δt. If the total reach length is too large it can be divided into several distance increments, Δx; the outflow from each is treated as the inflow to the next reach.

EXAMPLE 9.7

Apply the Muskingum-Cunge method to route the inflow hydrograph indicated in columns 1 and 2 of Table 9.3. Peak flow is 680 cfs. The area of cross section of the river at the peak flow is 95 ft^2 and the width is 20 ft. The channel has a bottom slope of 0.0001 and the reach length is 25 mi. Select $\Delta t = 12$ hours.

SOLUTION

$$\Delta x = 25 \text{ mi or } 132,000 \text{ ft}$$

$$v = \frac{Q_p}{A} = \frac{680}{95} = 7.16 \text{ ft/s}$$

$$c = mv = \frac{5}{3}(7.16) = 11.93 \text{ ft/s}$$

From eq. (9.39):

$$a = \frac{1}{2}\left[1 - \frac{680}{(20)(.0001)(11.93)(132,000)}\right] = 0.39$$

From eq. (9.33):

$$K = \frac{\Delta x}{c} = \frac{132,000}{11.93} = 11,064.5 \text{ s or } 3.07 \text{ hr}$$

From eqs. (9.34) through (9.36):

$$C_0 = \frac{0.5(12) - (0.39)(3.07)}{3.07(1 - 0.39) + 0.5(12)} = \frac{4.80}{7.87} = 0.61$$

$$C_1 = \frac{0.5(12) + (0.39)(3.07)}{3.07(1 - 0.39) + 0.5(12)} = \frac{7.20}{4.87} = 0.91$$

$$C_2 = \frac{3.07(1 - 0.39) - (0.5)(12)}{3.07(1 - 0.39) + 0.5(12)} = -\frac{4.13}{7.87} = -0.52$$

Equation (9.38) is solved successively in Table 9.3.

Table 9.3 Computations by the Muskingum-Cunge Method

(1) Time hr.	(2) Inflow cfs	(3) $C_0 I_2$[a]	(4) $C_1 I_1$[b]	(5) $C_2 O_1$[c]	(6) O_2[d]
12	100	—	—	—	100[e]
24	300	183	91	−52	222
36	680	415	273	−115	573
48	500	305	619	−298	626
60	400	244	455	−326	373
72	310	189	364	−194	359
84	230	140	282	−187	235
96	180	110	209	−122	197
108	100	61	164	−102	123
120	50	31	91	−64	58

[a] C_0 multiplied by the value of col. 2.
[b] C_1 multiplied by the value of col. 2 from the preceding time step.
[c] C_2 multiplied by the value of col. 6 from the preceding time step.
[d] col. 3 + col. 4 + col. 5.
[e] Assume the same inflow and outflow initially.

9.9.2 Validity of the Kinematic Theory of Routing

Henderson (1963) compared theoretical results with actual flood hydrographs and concluded that the kinematic theory is applicable to steep channels. In steep channels, the Froude number is very high and, thus, the momentum equation becomes $S_0 = S_f$.

Ponce et al. (1978) developed a criterion based on a dimensionless expression which has to be larger than a specified value for kinematic flow. The expression comprises the products of the channel slope, average velocity, duration of the flood wave or time-of-rise of the inflow hydrograph, and division by the average flow depth. The kinematic theory applies if the flood waves are of long duration or travel on a channel of steep slope with shallow depth.

Hager and Hager (1985) argued that the kinematic approximation holds for thoroughly subcritical flows of very shallow depth. They presented the following simple criteria, in metric units, for any plane or channel reach.

$$\eta_1 = \frac{S_0^{1/2}}{n} < 3 \quad \text{(metric units)} \quad [L^{1/3}/T] \tag{9.40}$$

and

$$\eta_2 = \frac{i\,S_0^{1/2}}{n^3 g^2} < 0.07 \quad \text{(metric units)} \tag{9.41}$$

where

S_0 = slope of channel or plane

n = Manning's coefficient (Table 11.4)

i = lateral inflow per unit width in m/s

g = gravitational constant (9.81 m/s^2)

9.10 Hydrologic Routing

In hydrologic routing, the three elements of the routing process are connected by the basic continuity equation

$$I\Delta t \qquad + \qquad \Delta S \qquad = \qquad O\Delta t \quad [L^3] \tag{9.42}$$

$$\begin{pmatrix} \text{inflow volume in} \\ \text{time, } \Delta t \end{pmatrix} + \begin{pmatrix} \text{change in volume of} \\ \text{water stored by the} \\ \text{hydraulic system} \\ \text{during time, } \Delta t \end{pmatrix} = \begin{pmatrix} \text{outflow} \\ \text{volume} \\ \text{in time,} \\ \Delta t \end{pmatrix}$$

where I and O are the rates of inflow and outflow, respectively.

The hydraulic system in Figure 9.10 can be represented by a reservoir or by a stream-channel section, and the routing process is, accordingly, classified into two broad types: reservoir routing and channel, or streamflow, routing.

In eq. (9.42), I is a known input, O has to be determined, but S also is an unknown parameter. To solve the equation, either both O and S have to be related to a common unknown parameter or S has to be defined in terms of O. The former approach is applied to reservoir routing and the latter is adopted in streamflow routing.

Conceptually, eq. (9.42), expressed in differential form, can be integrated to provide the outflow as a function of time. Generally, however, the terms in the equation have a form that is not amenable to direct solution. The numerical solution is preferred. In terms of numerical approximation, eq. (9.42) can be written as

$$\frac{I_1 + I_2}{2} + \frac{S_1 - S_2}{\Delta t} = \frac{O_1 + O_2}{2} \quad [L^3 T^{-1}] \tag{9.43}$$

where subscript 1 is at the beginning and subscript 2 at the end of the routing interval, Δt. The procedure used to solve eq. (9.43) is described separately for reservoir and streamflow routings.

9.11 STREAMFLOW ROUTING BY THE HYDROLOGIC METHOD: MUSKINGUM METHOD

The approach of defining the storage and the outflow in terms of the stage is not applicable to streamflow routing because of the varied flow conditions in a river channel. The problem of storage and outflow, being two unknowns in eq. (9.42), is resolved by considering storage as being related to outflow (as well as to inflow). The storage in a channel reach under varied flow conditions consists of two parts, as shown in Figure 9.12: the prism storage comprises the water below the line drawn parallel to the channel bottom, and the wedge storage is the water between this line and the actual water surface shown by the hatched portion. As seen from Figure 9.12, the wedge storage increases the storage volume during rising stages and reduces the volume in the falling stages for the same outflow. Thus, in a plot of storage versus outflow, a loop is observed due to the effect of the wedge storage. In simpler methods, the wedge storage is neglected and the channel storage is indicated in terms of the outflow only. (This is the case with reservoir routing, also.) To incorporate the effect of wedge storage, the inflow is also included as a parameter in the relation of storage. The inflow storage is related to the inflow rate and the outflow storage to the outflow rate, as follows:

$$S_I = K I^n \quad [L^3] \tag{9.44}$$

$$S_O = K O^n \quad [L^3] \tag{9.45}$$

where the subscripts $_I$ and $_O$ refer to inflow and outflow, n is an exponent, and K is a storage constant. If x is a weight factor to account for the relative effect of inflow and outflow on storage, then

$$S = xS_I + (1 - x)S_O \quad [L^3] \tag{9.46}$$

Substituting eqs. (9.44) and (9.45) into eq. (9.46) yields

$$S = K [x I^n + (1 - x) O^n] \quad [L^3] \tag{9.47}$$

Many applications of eq. (9.47) have been made. In a common case, the exponent n is taken to be unity, which results in

$$S = K [xI + (1 - x)O] \quad [L^3] \tag{9.48}$$

Substituting eq. (9.48) into eq. (9.43) gives us

$$\frac{I_1 + I_2}{2} + \frac{K\left[xI_1 + (1-x)O_1\right] - K\left[xI_2 + (1-x)O_2\right]}{\Delta t} = \frac{O_1 + O_2}{2} \quad [L^3 T^{-1}] \tag{9.49}$$

Figure 9.12 Components of channel storage.

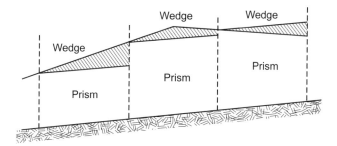

Arranging the terms yields

$$O_2 = C_0 I_2 + C_1 I_1 + C_2 O_1 \quad [\text{L}^3\text{T}^{-1}] \tag{9.50}$$

$$C_0 = \frac{0.5\Delta t - Kx}{K(1-x) + 0.5\Delta t} \quad [\text{dimensionless}] \tag{9.51}$$

$$C_1 = \frac{0.5\Delta t + Kx}{K(1-x) + 0.5\Delta t} \quad [\text{dimensionless}] \tag{9.52}$$

$$C_2 = \frac{K(1-x) - 0.5\Delta t}{K(1-x) + 0.5\Delta t} \quad [\text{dimensionless}] \tag{9.53}$$

Equation (9.50) is identical to equation (9.38) and equations (9.51) through (9.53) are similar to equations (9.34) through (9.36). The basic difference in the two methods is that in the Muskingum method the routing parameters K and x (or a) are calibrated using historic streamflow observations at the site. They bear no physical significance to channel properties. On the other hand, in the kinematic Muskingum-Cunge method, the parameters can be computed from the channel characteristics and the measured flow, as they relate to channel hydraulics.

In fact, the hydrologic routing method of this section is the older method that was used by McCarthy in a study of the Muskingum Conservancy District Flood Control Project in 1934–1935.

To use eqs. (9.51) through (9.53), the values of K and x have to be established. This is done on the basis of actual observed inflow and outflow hydrographs, as described below.

9.11.1 Determination of Routing Constants

1. To determine the routing constants K and x, plot the actual observed inflow and outflow hydrographs as shown in Figure 9.13(a).
2. Divide the time scale into a number of time intervals. At the end of each time interval read the inflow and outflow hydrograph ordinates and their differences. These are shown in col. 1 through 4 of Table 9.4 and plotted in Figure 9.13(b).
3. Determine the area under the curve in Figure 9.13(b) to the end of successive time intervals as shown in Figure 9.13(c). This has been computed arithmetically in col. 5 and col. 6 of the table. The peak value represents the maximum accumulated storage.

Figure 9.13 Computation of storage from hydrographs.

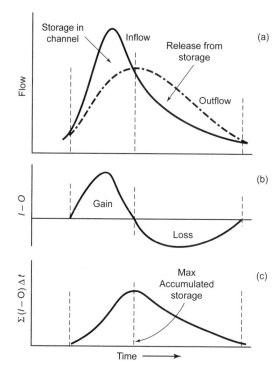

4. Compute the weighted discharges $[xI + (1 - x)O]$ for the end of successive time interval I and O values, using different selected values of x in col. 7, col. 8, and col. 9.

5. The values of accumulated storage at the end of successive time intervals, represented by the ordinates of Figure 9.13(c) or col. 6 of Table 9.4, are graphed against the weighted discharges $[xI + (1 - x)O]$ for different values of x as shown in Figure 9.14.

6. Since eq. (9.48) is a linear relation, the correct value of x is that which gives a straight line plot or the narrowest loop. The value of K is obtained automatically as the slope of the line.

The value of x ranges from 0 to 0.5 with a value of 0.25 as an average for river reaches. Analysis of many flood waves indicates that the time required for the center of mass of the flood wave to pass from the upstream end to the downstream end of a reach is equal to the factor K. The time between the upstream and downstream peaks is approximately equal to K.

EXAMPLE 9.8

The inflow and outflow hydrograph for a river reach are as shown in columns 2 and 3 of Table 9.4. Determine K and x for the reach.

SOLUTION

1. The storage and the weighted discharge for values of x of 0.1, 0.2, and 0.3 have been computed in Table 9.4.

2. The storage versus weighted discharge for various x is plotted in Figure 9.14.

Hydraulics: Principles and Applications   Chapter 9

Figure 9.14 Determination of K and x in Muskingum method.

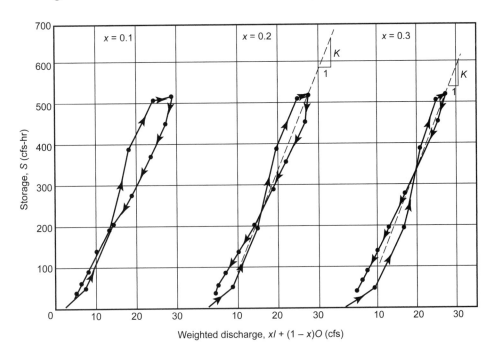

Weighted discharge, $xI + (1 - x)O$ (cfs)

3. When x is increased from 0.2 to 0.3, the loop is reversed from top to bottom. A value of x between 0.2 and 0.3 (say, 0.25) will provide the best linear relation.

4. From Figure 9.14, for $x = 0.2$, $K = 24$ hr (slope of line); for $x = 0.3$, $K = 24.2$ hr. The factor K is practically the same for $x = 0.2$ and 0.3.

9.11.2 Application of the Muskingum Method

Once the values of constants K and x have been determined, the routing parameters C_0, C_1, and C_2 are computed from eqs. (9.51) through (9.53). The sum of these three parameters is equal to 1. The time interval, Δt, is selected so that all three parameters have positive values. The routing operation is simply a solution of eq. (9.50). The O_2 value of one routing period is used as O_1 for the succeeding period. Initially, outflow is equal to inflow.

EXAMPLE 9.9

For the inflow hydrograph indicated in columns 1 and 2 of Table 9.5, perform the routing through a river reach when $K = 20$ hours and $x = 0.25$.

Table 9.4 Determination of Storage and Weighted Discharge

(1)	(2)	(3)	(4)	(5)	(6)	(7)	(8)	(9)
						\multicolumn — $xI + (1 - x)O$, cfs		
Time	Inflow, I	Outflow, O	$I - O$	S $\Delta t = 12$ hr	Cumulated S			
(hr)	(cfs)	(cfs)	(cfs)	(cfs-hr)[a]	(cfs-hr)	$x = 0.1$	$x = 0.2$	$x = 0.3$
0	0	0	0					
				1.8	1.8	2.03[b]	2.06	2.09
12	2.3	2.0	0.3					
				46.8	48.6	7.75	8.50	9.25
24	14.5	7.0	7.5					
				145.2	193.8	13.37	15.04	16.71
36	28.4	11.7	16.7					
				192.0	385.8	18.03	19.56	21.09
48	31.8	16.5	15.3					
				126.0	511.8	24.57	25.14	25.71
60	29.7	24.0	5.7					
				11.4	523.2	28.72	28.34	27.96
72	25.3	29.1	−3.8					
				−70.8	452.4	27.60	26.80	26.00
84	20.4	28.4	−8.0					
				−93.0	359.4	23.05	22.30	21.55
96	16.3	23.8	−7.5					
				−85.8	273.6	18.72	18.04	17.36
108	12.6	19.4	−6.8					
				−76.8	196.8	14.70	14.10	13.50
120	9.3	15.3	−6.0					
				−63.0	133.8	10.75	10.30	9.85
132	6.7	11.2	−4.5					
				−46.2	87.6	7.88	7.56	7.24
144	5.0	8.2	−3.2					
				−33.0	54.6	6.17	5.94	5.71
156	4.1	6.4	−2.3					
				−23.4	31.2	5.04	4.88	4.72
168	3.6	5.2	−1.6					

[a] $\frac{1}{2}$ (sum of the value in column 4 with its previous value) $\times \Delta t$.

[b] $0.1(2.3) + (1 - 0.1)(2.0)$

Hydraulics: Principles and Applications   Chapter 9

SOLUTION Consider $\Delta t = 12$ hours.

$$C_0 = \frac{0.5\Delta t - Kx}{K(1-x)+0.5\Delta t} = \frac{0.5(12) - 20(0.25)}{20(1-0.25)+0.5(12)} = 0.05$$

$$C_1 = \frac{0.5\Delta t + Kx}{K(1-x)+0.5\Delta t} = \frac{0.5(12) + 20(0.25)}{20(1-0.25)+0.5(12)} = 0.52$$

$$C_2 = \frac{K(1-x)-0.5\Delta t}{K(1-x)+0.5\Delta t} = \frac{20(1-0.25)-0.5(12)}{20(1-0.25)+0.5(12)} = 0.43$$

Refer to Table 9.5.

Table 9.5 Computation by the Muskingum Method

(1) Time (hr)	(2) Inflow (cfs)	(3) $C_0 I_2$	(4) $C_1 I_1$ [a]	(5) $C_2 O_1$ [b]	(6) O_2
12	100	—	—	—	100 (O_1 for next step)
24	300	15	52	43	110
36	680	34	156	47.3	237.3
48	500	25	353.6	102	480.6
60	400	20	260	206.7	486.7
72	310	15.5	208	209.3	432.8
84	230	11.5	161.2	186.1	358.8
96	180	9	119.6	154.3	282.9
108	100	5	93.6	121.6	220.2
120	50	2.5	52	94.7	149.2

[a] C_1 multiplied by the value of column 2 from the preceding step.

[b] C_2 multiplied by the value of column 6 from the preceding step.

9.12 RESERVOIR ROUTING BY THE HYDROLOGIC METHOD: THE PULS METHOD

In the case of a reservoir, the volume of storage can be expressed as a function of water surface elevation by planimetering the reservoir surface area from the topographic map for successive elevations and multiplying the average area by the water depth. A typical relation is shown by curve (a) in Figure 9.15.

Also, the outflow of water through the reservoir (besides the controlled releases through sluices, turbines, etc.) depends on the depth of flow over the spillway and thus on the depth of water in the reservoir. A spillway rating curve for the relation between discharge and water surface elevation can be prepared as shown by curve (b) of Figure 9.15.

Since the outflow and the storage are both functions of water surface elevation or stage, the above continuity equation becomes a relation between the known inflow and the unknown water stage, from which the stage can be computed as a function of time. These stages can readily be converted to outflows from the spillway rating curve. For this purpose, eq. (9.43), in numerical form, is rearranged as follows:

$$\left(I_1 + I_2\right) + \left(\frac{2S_1}{\Delta t} - O_1\right) = \left(\frac{2S_2}{\Delta t} + O_2\right) \quad [\mathrm{L^3 T^{-1}}] \tag{9.54}$$

Figure 9.15   Reservoir routing curves.

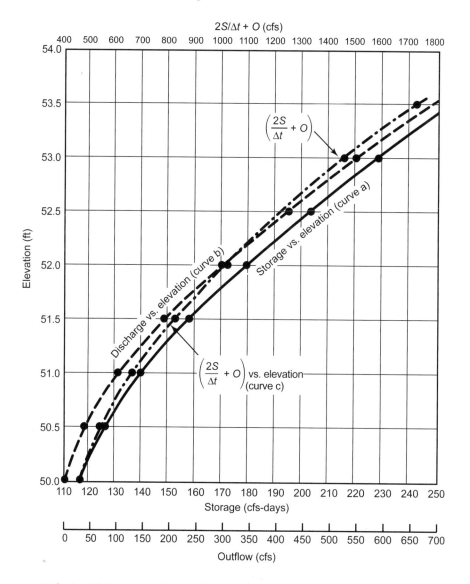

At the initial time, $t = 0$ (start of the routing just before flood arrives), $I_1 = I_2 = O_1$ and S_1 corresponds to the storage at the spillway crest elevation. The left side of equation (9.54) has known quantities that yield a value of $(2S_2/\Delta t + O_2)$, but still does not yield O_2 and S_2 separately. For computational expediency, by combining curves (a) and (b) of Figure 9.15, another curve of the relation between $(2S/\Delta t + O)$ and surface elevation, or alternatively, $(2S/\Delta t + O)$ versus O, is constructed on the same paper for a selected value of Δt, as shown by curve (c) in Figure 9.15.

Using curve (c), for known $(2S_2/\Delta t + O_2)$, the elevation will be obtained which will provide S_2 and O_2 directly from curves (a) and (b), respectively. These values will be used as initial values on the left side of eq. (9.54) for the next time step of the routing period.[*] The computation is repeated for succeeding routing periods.

As a slight modification of the procedure above, only two curves, S versus O and $(2S/\Delta t + O)$ versus O, are constructed. From these curves it is possible to split $(2S/\Delta t + O)$ into O and S.

EXAMPLE 9.10

Route the inflow hydrograph indicated below through a reservoir. The storage data (water surface elevation versus storage volume) for the reservoir are given below. The spillway discharge is $Q = 3LH^{3/2}$. The crest height of the spillway is 50 ft and the length of the spillway is 35 ft.

Inflow hydrograph:

Time (days)	0	0.5	1.0	1.5	2.0	2.5	3.0	3.5	4.0
Flow (cfs)	0	70	185	360	480	300	165	80	0

Storage data:

Elevation (ft)	50	50.5	51.0	51.5	52.0	52.5	53.0	53.5
Storage (acre-ft)	231	247	277	313	353	400	452	509

SOLUTION

1. The above storage data are listed in col. 3 and the discharge data computed from $Q = 3(35)H^{3/2}$ are listed in column 5 of Table 9.6.

2. $(2S/\Delta t + O)$ has been calculated in column 6 for $\Delta t = 0.5$ day, from the values of S and O in columns 4 and 5, respectively. These have been plotted in Figure 9.15.

3. The routing computations are performed in Table 9.7 and explained below.

   Column 3: Addition of two successive values of column 2.

   Columns 4 and 5: Obtained from the storage and the discharge curves (Figure 9.15), entering from the $(2S/\Delta t + O)$ curve corresponding to the value of column 7 in the previous line.

   Column 6: Obtained from the values in columns 4 and 5.

   Column 7: Equal to left side of eq. (9.54), column 3 + column 6.

   Column 8: From Figure 9.15, for the value in column 7.

4. The inflow hydrograph (column 2) and the outflow hydrograph (column 4) are plotted in Figure 9.16.

9.13 HYDRAULIC TRANSIENTS

The interim unsteady stage when a flow changes from one steady-state condition to another steady-state condition is known as the *transient state* of flow. In conduits and open

---

[*] A variation of this procedure uses a $(2S/\Delta t + O)$ versus O curve. From known $(2S/\Delta t + O)$, the value of O becomes available from this curve for the next step. The subtraction of twice O provides $(2S/\Delta t - O)$ directly for application in eq. (9.54).

Table 9.6   Storage, Discharge, and $(2S/\Delta t + O)$ Data

(1)	(2)	(3)	(4)	(5)	(6)
Water Surface Elevation (ft)	Head, H= (col. 1–crest level)	Storage, S (from given data) Acre-ft	Storage, S (from given data) cfs-day	Outflow, O (from formula) (cfs)	$(2S/\Delta t + O)$ (cfs)
50	0	231	116.4	0	465.6
50.5	0.5	247	124.5	37.1	535.1
51.0	1.0	277	139.6	105.0	663.4
51.5	1.5	313	157.8	192.9	824.1
52.0	2.0	353	177.9	297.0	1008.6
52.5	2.5	400	201.6	415.0	1221.4
53.0	3.0	452	227.8	545.6	1456.8
53.5	3.5	509	256.5	687.5	1713.5

Table 9.7   Reservoir Flood Routing Computation

(1)	(2)	(3)	(4)	(5)	(6)	(7)	(8)
			LHS	LHS		RHS	
Time (days)	Inflow, I (cfs)	$I_1 + I_2$ (cfs)	Outflow, O (cfs)	Storage, S (cfs-day)	$\dfrac{2S_1}{\Delta t} - O_1$ (cfs)	$\dfrac{2S_2}{\Delta t} + O_2$	Water Elevation (ft)
Before flood arrives:	0	0	0	116.4	465.6	465.6	50.00
Inflow hydrograph:							
0	0	70	0	116.4	465.6	535.6	50.50
0.5	70	255	37	124.5	461	716	51.20
1.0	185	545	140	146.0	444	989	51.95
1.5	360	840	285	176.0	419	1259	52.60
2.0	480	780	450	210	390	1170	52.40
2.5	300	465	380	197	408	873	51.60
3.0	165	245	215	163	437	682	51.05
3.5	80	80	115	142	453	533	50.45
4.0	0	0	30	124	466	466	50.0
4.5	0	0	0	116.4	465.6	465.6	50.0

channels, such conditions occur when the flow is decelerated or accelerated due to sudden closing or opening of the control valves, starting or stopping of the pumps, rejecting or accepting of the load by a hydraulic turbine, quick lifting and shutting of reservoir gates or similar situations of sudden increased or decreased inflows. The variations in velocity result in a change of momentum. The fluid is subject to an impulse force equivalent to the rate of change of the momentum according to the momentum principle. An appreciable

Figure 9.16 Inflow hydrograph and routed outflow hydrograph.

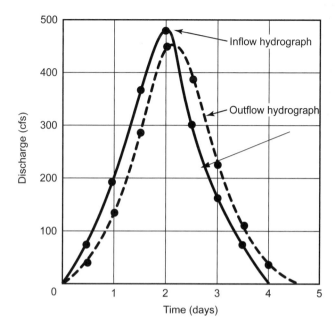

increase of pressure occurs within a short time due to this impulse force. The system design should be adequate to withstand both the normal static pressure and the maximum rise in pressure due to hydraulic transients.

The dynamic form of the momentum equation (9.7) describes the transient state of flow because the unsteady condition prevails. Along with the momentum equation, the continuity (conservation of mass) equation is used to fully describe the transient flow phenomenon.

9.14 TRANSIENT PHENOMENON IN PIPES

The pressure fluctuation caused by the impulse force is known as the *water* (or oil) *hammer* because of the hammering noise it produces. In simplified form, after dropping the convective acceleration term, which is very small compared to the other terms, and expressing in terms of pressure head, the two equations (9.2) and (9.7) as applied to a horizontal pipe are as follows:

Continuity equation, expressing A in terms of h and since $h = c^2/g$:

$$\frac{\partial h}{\partial t} + \frac{c^2}{g}\frac{\partial v}{\partial x} = 0 \quad [\mathrm{LT^{-1}}] \tag{9.55}$$

Momentum equation, since $P = \gamma h \Delta A$, dropping the gravity term for horizontal pipe, and expressing the friction term as $fv|v|/2Dg$

$$\frac{\partial h}{\partial x} + \frac{1}{g}\frac{\partial v}{\partial t} + \frac{fv|v|}{2Dg} = 0 \quad [\text{dimensionless}] \tag{9.56}$$

where

h = pressure head above a specified datum

c = speed of propogation of pressure wave in a specified fluid and specific conduit material, is equal to $\sqrt{gy}$ or $\sqrt{gh}$

v = flow velocity; $|v|$ is absolute value of velocity

f = friction factor

D = conduit diameter

The speed of the pressure wave is related to the medium of flow (fluid), conduit properties, and the type of anchoring. A general expression is as follows (Halliwell, 1963):

$$c = \sqrt{\frac{K}{\rho\left[1+(K/E)\psi\right]}} \quad [\text{LT}^{-1}] \tag{9.57}$$

where

K = bulk modulus of elasticity of fluid

ρ = density of fluid

E = modulus of elasticity of conduit material

ψ = dimensionless parameter that depends on the elastic properties of conduit and the type of anchoring, as follows

1. For rigid conduits:

$$\psi = 0 \quad [\text{dimensionless}] \tag{9.58}$$

2. For a thin-walled conduit without expansion joints, anchored throughout its length:

$$\psi = \frac{D}{e}\left(1-v^2\right) \quad [\text{dimensionless}] \tag{9.59}$$

where

D = conduit diameter

e = conduit wall thickness

v = Poisson's ratio

3. For a thin-walled conduit without expansion joints, anchored at the upper end:

$$\psi = \frac{D}{e}(1-0.5v) \quad [\text{dimensionless}] \tag{9.60}$$

4. For a thin-walled conduit with frequent expansion joints:

$$\psi = \frac{D}{e} \quad [\text{dimensionless}] \tag{9.61}$$

An important factor that affects the pressure rise is the time of closure (stoppage) of flow, t_c. For a very slow closure, when $t_c > 20L/c$, where L is the conduit length, the conduit acts as a rigid body and the fluid behaves as an incompressible mass. The entire column of fluid is subjected to a uniform deceleration. The rigid column theory applies and the transient or surge pressure can be computed from the momentum principle [eq. (9.56)] applied to the rigid column of fluid.

As closing time, t_c, decreases, the inertia force increases. A point is reached when the force (pressure) is sufficient to cause the fluid to compress and the pipe material to expand. Under this condition, the transient process is radically changed. The kinetic energy of flow is converted into elastic energy. The elastic wave theory applies, in which the shock or pressure wave travels through the fluid. The continuity and momentum equations [eqs. (9.55) and (9.56)] are solved simultaneously to compute the pressure fluctuations. There are two cases of this category: (1) when the time of closure, t_c, is not more than $2L/c$ – this is known as rapid closure, and (2) when t_c is greater than $2L/c$, but less than $20L/c$ – this is known as slow closure. These are discussed subsequently.

9.14.1 Mechanism of Pressure-Wave Propagation

Consider the piping system shown in Figure 9.17a, in which steady-state flow is taking place from a reservoir at a velocity, V_0, and head, H_0, with no friction losses. When the valve at the downstream end of the pipe is instantly closed, the following sequence of events takes place.

1. Immediately following the valve closure, the fluid in proximity of the valve is brought to rest. This causes a local pressure increase. Due to the increased pressure, the liquid is compressed and the pipe walls expand. This provides a little extra volume in which the liquid enters to come to a stop. An instant later, the next section immediately upstream follows the same process. Thus a wave of increased pressure propagates upstream toward the reservoir. The process that starts in Figure 9.17b completes in Figure 9.17c. When the wave reaches the reservoir, the entire pipe is expanded; the liquid column is compressed and comes to a complete stop. The time taken by the wave to reach the reservoir is L/c.

2. At the end of step 1, the pressure in the pipe is much higher than the pressure in the reservoir. Hence the halted water begins to flow from the pipe into the reservoir. This relieves the pressure in the pipe. The liquid column decompresses and the pipe material contracts. The process starts at the reservoir end of the pipe [Figure 9.17d] and continues toward the valve [Figure 9.17e]. The time in this step is L/c, and the total time taken by the wave to return to the valve is $2L/c$, which is known as the water-hammer period.

3. The wave of backwater motion cannot go past the closed valve. The sudden stoppage (inertia) of this moving fluid at the valve causes the pressure to drop below the normal level. This sends a wave of negative pressure again upstream toward the reservoir. The process starts at Figure 9.17f and ends at 9.17g.

4. At the end of step 3, since the pressure at the reservoir end of the pipe is negative (less than that in the reservoir), liquid starts flowing from the reservoir into the pipe. The process starts at Figure 9.17h and ends at 9.17i. At the end of this step, the conditions are identical to the beginning of step 1 (i.e., one cycle is completed in a period of $4L/c$).

This process could continue indefinitely but the friction losses in the system diminish the pressure successively, until the waves finally die out.

9.14.2 Very Slow Closure of the Valve

For $t_c > 20L/c$, the fluid and pipe material act as solid media. There is little or no pressure wave propagation, as illustrated in Section 9.14.1. The pressure surge acts due to uniform

Figure 9.17 Propagation of pressure wave due to instantaneous closure.

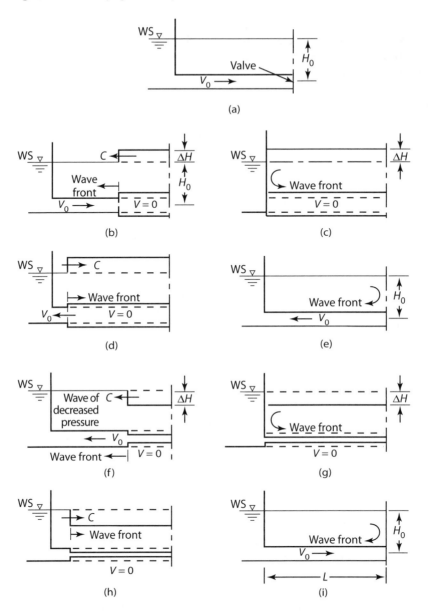

deceleration of the column of water in the pipe, which can be ascertained from the momentum (or energy) principle. Expressing the terms in the lumped finite forms in eq. (9.56), the maximum pressure difference is given by

$$\Delta h_m = \frac{L\,V_0}{g\,t_c} \quad [\text{L}] \tag{9.62}$$

where

Δh_m = maximum pressure rise

L = length of pipeline

V_0 = initial velocity of flow

t_c = time of closure (velocity changes from V_0 to 0 in time t_c)

EXAMPLE 9.11

From a reservoir, water at a rate of 0.744 m³/s flows through a 1000-m-long horizontal cast-iron pipeline discharging into the atmosphere through a valve at the downstream end. The pipe is 500 mm in diameter and 15 mm thick. The valve is fully closed in 20 seconds. Calculate the transient pressure at the valve and at 400 m upstream of the valve. The pipe is without expansion joints and anchored throughout. K = 2.2 GPa, E (cast-iron) = 150 GPa, v = 0.25. For water, ρ = 1000 kg/m³.

SOLUTION

1. For steady-state flow when the valve is open:

$$Q = A_0 \, V_0 = 0.744 \text{ m}^3/\text{s}$$

or

$$\frac{\pi}{4}(0.5)^2 V_0 = 0.744$$

or

$$V_0 = 3.79 \text{ m/s}$$

2. Speed of wave, c:

$$\psi = \frac{D}{e}\left(1-v^2\right) = \frac{500}{15}\left[1-(0.25)^2\right] = 31.3$$

$$c = \sqrt{\frac{2.2 \times 10^9}{1000\left[1+\left(2.2 \times 10^9 / 150 \times 10^9\right) \times 31.3\right]}} = 1230 \text{ m/s}$$

3. Wave travel time:

$$\frac{L}{c} = \frac{1000}{1230} = 0.8 \text{ sec}$$

Since a t_c of 20 seconds > $20L/c$, this is a case of very slow valve closing.

4. At the valve when the valve is fully closed:

$$\Delta h_m = \frac{LV_0}{gt_c} = \frac{1000(3.79)}{9.81(20)} = 19.3 \text{ m}$$

5. At 400 m upstream from the valve when the valve is closed,

$$\Delta h_m = \frac{600(3.79)}{9.81(20)} = 11.6 \text{ m}$$

9.14.3 Rapid Closure of the Valve

When the time of closure, t_c, is not more than $2L/c$, the return pressure wave finds the valve closed. The maximum transient pressure in such a case is the same as for the instantaneous closing of the valve. Analytical solution of this elastic wave phenomenon is obtained by solving the simultaneous differential equations (9.55) and (9.56). For the two successive time periods $i-1$ and i, the following general relation results:

$$h_i + h_{i-1} - 2h_0 = \frac{c}{g}\left(V_{i-1} - V_i\right) \quad [L] \tag{9.63}$$

For rapid and instantaneous valve closure, $i-1$ denotes the initial condition at time 0 and i denotes the next period when the valve is closed; thus $V_i = 0$. Substituting in eq. (9.63) yields

$$\Delta h_m = \frac{cV_0}{g} \quad [L] \tag{9.64}$$

where $\Delta h_m = h_1 - h_0$ = maximum transient head.

If a pipeline is operating at partial opening and complete closure is attained starting from this opening, the maximum pressure rise can still be given by eq. (9.64), when V_0 will correspond to the velocity at partial opening.

EXAMPLE 9.12

In Example 9.11, the valve is now closed in 1.5 seconds. Determine the maximum transient pressure at the valve.

SOLUTION

1. Since $t_c < 2L/c$, this is a case of rapid closure.

2. From eq. (9.64),

$$\Delta h_m = \frac{1230(3.79)}{9.81} = 475.2 \text{ m}$$

9.14.4 Slow Closure of the Valve

When the time of closure exceeds $2L/c$ (but is less than $20L/c$, so that the elastic theory is applicable), the reflection or negative wave returning at the valve partially compensates the pressure rise and the transient pressure is not as high as for the rapid closure. There are many analytical methods (besides the numerical procedures) to ascertain the pressure variation with time. The classic procedure of Allievi (1925) is described here. Using the equation of discharge through a valve (sluice), we have

$$V_0 = C_d \frac{A_0}{A_p}\sqrt{2gh} \tag{a}$$

$$\frac{V_i}{V_0} = \frac{C_{d,i}}{C_{d,0}}\frac{A_i}{A_0}\sqrt{\frac{h_i}{h_0}} \tag{b}$$

where

A_0, A_i = areas of valve initially and at time i

A_p = area of pipe

If C_d is a constant, $\eta_i = A_i/A_0$ and $\xi_i = \sqrt{h_i/h_0}$. Then $V_i = \eta_i \xi_i V_0$ and $h_i = h_0 \xi_i^2$. Substituting in eq. (9.63) gives

$$h_0 \xi_i^2 + h_0 \xi_{i-1}^2 - 2h_0 = \frac{cV_0}{g}\left(\eta_{i-1}\xi_{i-1} - \eta_i \xi_i\right) \tag{c}$$

Denoting $N = cV_0/2gh_0$, we obtain

$$\xi_i^2 + \xi_{i-1}^2 - 2 = 2N\left(\eta_{i-1}\xi_{i-1} - \eta_i \xi_i\right) \quad \text{[dimensionless]} \tag{9.65}$$

Equation (9.65) is solved at different time intervals, $i = 1, 2, 3, \ldots$, to calculate the head variation. In practice, the slower rate of closing is selected wherever possible to prevent the excessive rise in pressure associated with the instant closing.

EXAMPLE 9.13

In Example 9.11, the valve is fully closed in 8 sec. The area of the valve opening varies as follows. Calculate the variation of pressure head at the valve during closure. Disregard the friction losses. $C_d = 0.6$.

Time (s)	0	1.6	3.2	4.8	6.4	8.0
Valve area (m)^2	0.06	0.048	0.036	0.024	0.012	0

SOLUTION

1. From Example 9.11, $L/c = 0.8$ second, time interval $= 2L/c = 1.6$ seconds, $V_0 = 3.79$ m/s.

$$A_p = \frac{\pi}{4}(.5)^2 = 0.196 \text{ m}^2$$

2. $Q = A_p V_0 = C_d A_0 \sqrt{2gh_0}$. Hence,

$$h_0 = \left(\frac{A_p V_0}{C_d A_0 \sqrt{2g}}\right)^2 = \left[\frac{0.196(3.79)}{0.6(0.06)\sqrt{2(9.81)}}\right]^2 = 21.7 \text{ m}$$

3. $N = \dfrac{cV_0}{2gh_0} = \dfrac{1230(3.79)}{2(9.81)(21.7)} = 10.95$

4. At $i = 1$ ($t = 1.6$ seconds), from eq. (9.65),

$$\xi_1^2 + \xi_0^2 - 2 = 2N\left(\eta_0 \xi_0 - \eta_1 \xi_1\right)$$

$$\xi_1^2 + 1 - 2 = 2(10.95)\left[1(1) - \left(\frac{0.048}{0.06}\right)\xi_1\right]$$

or

$$\xi_1^2 + 17.52\xi_1 - 22.90 = 0$$

or

$$\xi_1 = 1.22,$$
$$h_1 = \xi_1^2 h_0 = (1.22)^2 (21.7) = 32.3 \text{ m}$$

and

$$V_1 = \eta_1 \xi_1 V_0 = \left(\frac{0.048}{0.060}\right)(1.22)(3.79) = 3.70 \text{ m/s}$$

5. At $i = 2$ ($t = 3.2$ sec),

$$\xi_2^2 + (1.22)^2 - 2 = 2(10.95)\left[\left(\frac{0.048}{0.060}\right)(1.22) - \left(\frac{0.036}{0.06}\right)\xi_2\right]$$

$$\xi_2^2 + 13.14\xi_2 - 21.89 = 0$$

or

$$\xi_2 = 1.50$$

Hence

$$h_2 = (1.50)^2 (21.7) = 48.83 \text{ m}$$

and

$$V_2 = \eta_2 \xi_2 V_0 = 0.6(1.50)(3.79) = 3.41 \text{ m/s}$$

6. The other steps are completed in Table 9.8.

9.15 TRANSIENT PHENOMENON IN OPEN CHANNELS

A surge occurs when there is a sudden change of depth during lifting or closing of a gate or sudden change in flow conditions under gravity. The surge associated with an increase in depth is called a *positive surge* and the surge associated with a decreasing depth is a *negative surge*. A positive or negative surge can move in an upstream direction or a downstream direction. Thus there are four types of surges.

The top part of a surge moves faster than the lower part. In a positive surge where depth increases, this causes the wave to roll as it moves. This is a stable situation. In a negative surge, however, the fact that the top part moves faster than the lower part causes a stretching of the wave as shown in Figure 9.18. This is an unstable condition.

9.15.1 Type I: Positive Surge Moving Downstream

This occurs when an upstream gate is suddenly lifted, as shown in Figure 9.19. In all types of surges, the subscript 1 refers to the section representing the initial condition, prior to surge, and the subscript 2 refers to the section representing the condition after the surge. Vw is the *absolute velocity of surge*. The relative velocity of surge ($Vw - V$) is known as the *wave celerity*.

Table 9.8 Computation of Transient Pressure by Allievi Method

(1)	(2)	(3)	(4)	(5)[a]	(6)[b]
	$t = 2L/c$	$\eta_i = \dfrac{A_i}{A_0}$	ξ_i	$h_i = \xi_i^2 h_0$	$V_i = \eta_i \xi_i V_0$
i	(s)		[eq. (9.65)]	(m)	(m/s)
0	0	1	1	21.7	3.79
1	1.6	0.8	1.22	32.3	3.70
2	3.2	0.6	1.50	48.83	3.41
3	4.8	0.4	1.84	73.47	2.79
4	6.4	0.2	2.23	107.91	1.69
5	8.0	0	2.61	147.82	0

[a] (col. 4)^2 × 21.7
[b] (col. 3) × (col. 4) × 3.79

Figure 9.18 Stretching of a negative surge.

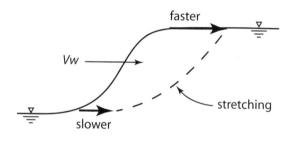

Figure 9.19 Positive surge moving downstream.

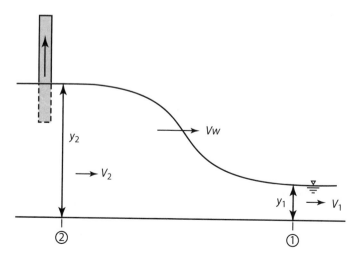

The imposition of the surge velocity, Vw, makes the flow unsteady. If Vw is subtracted, then the flow could be considered steady as shown in Figure 9.20. The continuity and momentum equations per unit channel width for a controlled volume (see Figure 9.21) can be written as follows according to eq. (9.3) and eq. (9.8):

Continuity equation

$$y_1\left(Vw-V_1\right)=y_2\left(Vw-V_2\right) \quad [L^2T^{-1}] \tag{9.66}$$

Momentum equation

$$\rho y_1\left(Vw-V_1\right)\left[\left(Vw-V_2\right)-\left(Vw-V_1\right)\right]=\gamma\frac{y_1^2}{2}-\gamma\frac{y_2^2}{2} \quad [FL^{-1}] \tag{9.67}$$

Substituting V_2 from eq. (9.66) into eq. (9.67) and solving the equation for positive surge:

$$\frac{\left(Vw-V_1\right)^2}{gy_1}=\frac{1}{2}\frac{y_2}{y_1}\left(1+\frac{y_2}{y_1}\right) \quad [\text{dimensionless}] \tag{9.68}$$

Figure 9.20 Equivalent steady-state condition for a positive surge.

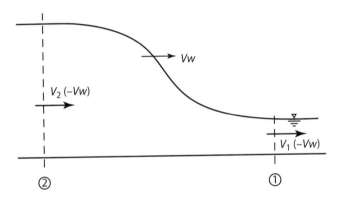

Figure 9.21 Controlled volume section of an equivalent steady surge.

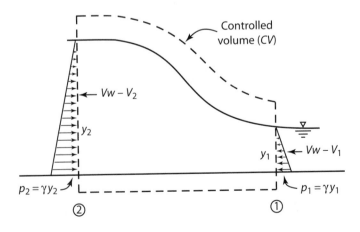

9.15.2 Type II: Positive Surge Moving Upstream

This occurs when a downstream gate is suddenly closed, as shown in Figure 9.22. An equivalent steady flow could be conceived by adding the surge velocity Vw. Along the lines of eqs. (9.66) and (9.68), the relations will be

Continuity equation

$$y_1(Vw+V_1) = y_2(Vw+V_2) \quad [L^2T^{-1}] \tag{9.69}$$

Momentum equation

$$\frac{(Vw+V_1)^2}{gy_1} = \frac{1}{2}\frac{y_2}{y_1}\left(1+\frac{y_2}{y_1}\right) \quad [\text{dimensionless}] \tag{9.70}$$

EXAMPLE 9.14

The depth and velocity of flow in a rectangular channel are 0.9 m and 1.5 m/s, respectively. If a gate at the downstream end of a channel is suddenly closed, what will be the height and absolute velocity of the surge?

SOLUTION

1. This is a type II: positive surge moving upstream.

$$y_1 = 0.9 \text{ m}, \ V_1 = 1.5 \text{ m/s}, \ V_2 = 0$$

2. From eq. (9.69),

$$0.9\,(Vw+1.5) = y_2 Vw$$

or

$$Vw = \frac{1.35}{(y_2-0.9)} \tag{a}$$

3. From eq. (9.70),

$$\frac{(Vw+1.5)^2}{(9.81)(0.9)} = \frac{1}{2}\frac{y_2}{(0.9)}\left(1+\frac{y_2}{0.9}\right) \tag{b}$$

4. Substituting for Vw in (b) from (a) and simplifying

$$\sqrt{\frac{y_2}{0.9+y_2}} = 1.556(y_2-0.9) \tag{c}$$

5. Solving (c) by trial and error, $y_2 = 1.4$ m

6. From eq. (a), $Vw = 2.7$ m/s

9.15.3 Type III: Negative Surge Moving Downstream

This occurs when an upstream gate is suddenly closed, as shown in Figure 9.23. The resulting unstable wave is considered to be composed of a series of small wavelets of celerity $\sqrt{gy}$ superimposed on the steady flow. Each step of the velocity and depth is changed by ΔV and Δy respectively. Thus, $V_1 = V$ and $y_1 = y$, and $V_2 = V - \Delta V$ and $y_2 = y - \Delta y$.

Figure 9.22   Positive surge moving upstream.

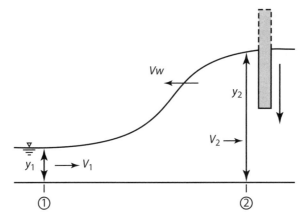

Figure 9.23   Negative surge moving downstream.

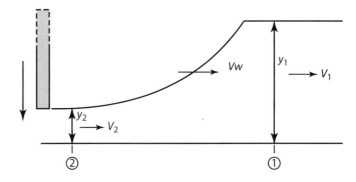

Substituting these in the continuity equation (9.66) and simplifying by disregarding the product of $\Delta V \Delta y$:

$$\frac{\Delta V}{\Delta y} = \frac{(Vw - V)}{y} \qquad \text{(a)}$$

Since $c = (Vw - V) = \sqrt{gy}$

$$\frac{\Delta V}{\Delta y} = \sqrt{\frac{g}{y}} \qquad \text{(b)}$$

Integrating (b),

$$V = 2\sqrt{gy} + C$$

Substituting the boundary conditions that at $y = y_1$, $V = V_1$, thus $C = V_1 - 2\sqrt{gy_1}$

$$V = V_1 + 2\sqrt{gy} - 2\sqrt{gy_1} \quad [\text{LT}^{-1}] \qquad (9.71)$$

Since $Vw = x/t$, the position of the surge is given by

$$x = Vw\, t$$

Hydraulics: Principles and Applications   Chapter 9

or

$$x = \left(\sqrt{gy} + V\right)t \qquad \text{(c)}$$

Substituting V from eq. (9.71) into (c),

$$x = \left(V_1 + 3\sqrt{gy} - 2\sqrt{gy_1}\right)t \quad [L] \qquad (9.72)$$

9.15.4 Type IV: Negative Surge Moving Upstream

This occurs when a downstream gate is suddenly lifted, as shown in Figure 9.24. In this case,

$$Vw + V = \sqrt{gy}$$

and

$$\frac{\Delta V}{\Delta y} = -\sqrt{\frac{g}{y}}$$

By operations similar to Section 9.15.3, the following relations result

$$V = V_1 + 2\sqrt{gy_1} - 2\sqrt{gy} \quad [LT^{-1}] \qquad (9.73)$$

and

$$x = \left(V_1 - 3\sqrt{gy} + 2\sqrt{gy_1}\right)t \quad [L] \qquad (9.74)$$

EXAMPLE 9.15

A gate in a rectangular channel carrying a discharge of 8 m³/s per meter width at a depth of 2.0 m is lowered upstream to reduce discharge by 50%. Calculate the height of the surge and the velocity of flow after the surge.

SOLUTION

1. This is a type III negative surge moving downstream. $y_1 = 2$ m, $V_1 = q/y_1 = 8/2 = 4$ m/s. Since q reduces by 50%, $V_2 = 4/y_2$

2. From eq. (9.71),

$$V_2 = V_1 + 2\sqrt{gy_2} - 2\sqrt{gy_1}$$

$$\frac{4}{y_2} = 4 + 2\sqrt{9.81(y_2)} - 2\sqrt{9.81(2)}$$

   or

$$\frac{4}{y_2} = -4.86 + 2\sqrt{9.81y_2}$$

3. By trial and error, $y_2 = 1.465$ m

4. $V_2 = 4/1.465 = 2.73$ m/s

5. Height of surge $= y_1 - y_2 = 2.0 - 1.465 = 0.535$ m

Figure 9.24 Negative surge moving upstream.

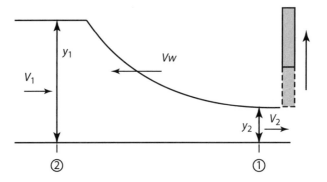

PROBLEMS

9.1 Water flows steadily from tank A to tank B as shown. Determine the water depth in tank A.

Figure P9.1

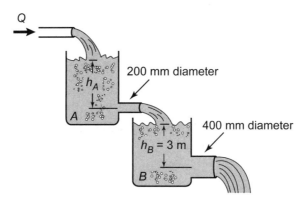

9.2 In a branching pipe shown, the flow rate in pipe 2 is 40% of pipe 1. Determine the **(a)** size of pipe 2, **(b)** velocity in pipes 1 and 3, and **(c)** pressure in pipes 2 and 3.

Figure P9.2

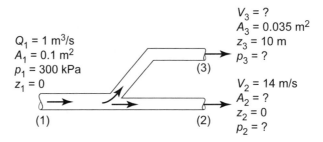

9.3 A venturimeter is introduced in a 200-mm-diameter horizontal pipe carrying water under a pressure of 150 kN/m². The throat diameter is 100 mm and the pressure at the throat is − 50 kN/m². Determine the flow rate using the energy principle. Disregard losses.

9.4 Solve Problem 9.3 by the momentum principle. A force of 2.6 kN is exerted by water at the throat acting to the left.

9.5 For the 50-ft-wide rectangular channel section shown in Fig. P9.5, determine the depth of flow and the velocity at section 2 using the energy principle. Disregard the energy losses.

Figure P9.5

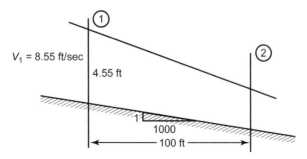

9.6 Solve Problem 9.5 by the momentum principle.

9.7 A rectangular channel section increases in width from 40 ft to 50 ft in a length of 100 ft. The channel slope is 0.1%. If the discharge and the velocity of flow at section 1 are 2950 cfs and 9.8 ft/sec, respectively, determine the depths of flow at sections 1 and 2 and the velocity at section 2. Use the energy principle. Disregard the losses.

9.8 Solve Problem 9.7 by the momentum principle.

9.9 A horizontal bend in a pipeline conveying 1 m^3/s of water gradually reduces from 600 mm diameter at section 1 to 300 mm at the open end (see Figure P9.9). It deflects the flow through an angle of 60°. At the larger section the pressure is 170 kN/m^2. Determine the magnitude and direction of the force exerted.

Figure P9.9

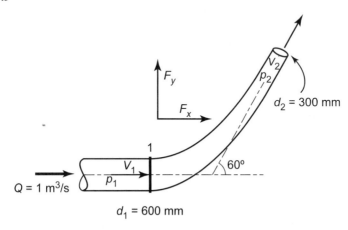

9.10 A vertical jet of water leaves a nozzle at a speed of 10 m/s and a diameter of 20 mm. It suspends a plate having a weight of 15 N as shown. What is the vertical distance h? (Hint: Three unknowns are the vertical velocity at impact, area of jet at impact, and height h. Apply the continuity eq., the momentum eq., and the energy eq. in that order.)

Figure P9.10

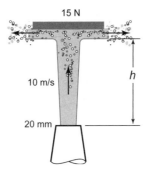

9.11 Find the characteristic curve and the solution of the following equation. Initially, $t = 0$ sec, $x = 0$ ft, and $v = 1$ ft/s.

$$\frac{\partial v}{\partial t} + 0.15(1-x)\frac{\partial v}{\partial x} = 0.2v$$

9.12 In a canal the celerity c is given by $x\sqrt{0.5t}$, where x is in ft, t is in seconds, and c is in ft/sec. If the lateral inflow rate is 201 ft^3/min/ft, determine the characteristic curve and the equation for the area of flow. At $t = 0$ sec, the flow area is 500 ft^2 at 10 ft from the beginning. [Use eq. (9.12)]

9.13 The rainfall intensity on a 50-m-long plane of slope 0.005 is 2.5 cm/hr. Draw the characteristic path (x-t diagram) to the end of the plane. Indicate the flow depths in mm at time 1 min and 5 min and determine their position downslope (x-distances). Also determine the equivalent depth at the end of the plane. $n = 0.013$.

9.14 Determine the rising hydrograph for Problem 9.13 for a rainfall rate of 2.5 cm/hr on a 50-meter-long plane having a slope of 0.005 and $n = 0.013$.

9.15 Precipitation is falling at a rate of 1.75 in./hr on a 320-ft-long plane for a duration of 10 min. The slope of the plane is 1% and $n = 0.023$. Determine the time of concentration and the ordinates of the rising and falling hydrograph limbs. Plot the hydrograph.

9.16 For a rainfall rate of 2.5 cm/hr on a parking lot 50 m long and sloped at 1%, determine the complete hydrograph. The rainfall duration is 12 min and $n = 0.023$.

9.17 Rainfall intensity on a plane is 3.5 cm/hr for a duration of 8 min. The plane of 60 m × 60 m has a slope of 0.005. Draw the complete hydrograph. $n = 0.03$.

9.18 Check whether the kinematic wave theory can be applied to Problem 9.15 for formulating the hydrograph.

9.19 Determine whether the kinematic wave theory is valid for Problem 9.16 to formulate the hydrograph.

9.20 Use the Muskingum-Cunge method to route the following hydrograph:

Time (hr)	12	24	36	48	60	72	84	96
Flow (cfs)	300	450	750	825	740	600	400	270

The flood channel characteristics are: peak flow = 825 cfs, area of channel cross section = 120 ft^2, width = 30 ft, channel slope = 0.00015, reach length = 50 mi.

9.21 Use the Muskingum-Cunge method to route the following hydrograph:

Time	Inflow (cfs)	Time	Inflow (cfs)
6 A.M.	150	Midnight	300
Noon	180	6 A.M.	200
6 P.M.	300	Noon	150
Midnight	750	6 P.M.	100
6 A.M.	850	Midnight	100
Noon	550	6 A.M.	80
6 P.M.	400	Noon	50

The channel and flow characteristics are: channel bed slope = 0.0001, reach length = 30 mi, peak flow = 1500 cfs, area of cross section of channel = 270 ft^2 and width = 20 ft.

9.22 By the Muskingum-Cunge method, route the following hydrograph.

Time, hr.	0	1	2	3	4	5	6	7	8	9	10
Flow, m^3/s	15	27	75	160	220	200	160	88	50	25	12

Peak flow rate is 220 m^3/s, channel cross sectional area is 183 m^2, and width is 35 m. The bed slope is 0.001 and reach length is 6 km.

9.23 A 10-m-wide stream channel has a bed slope of 1%. The lateral inflow into the stream due to excess rain is 150 mm/hr. Determine whether the kinematic approximation is applicable to the channel. $n = 0.05$.

9.24 Given the following hydrographs at the upstream and downstream ends of a river reach, determine the Muskingum routing constants.

Date	Time (hr)	Inflow (cfs)	Outflow (cfs)	Date	Time (hr)	Inflow (cfs)	Outflow (cfs)
1	6	0	0	3	6	25	48.5
	12	75	32.3		12	25	37.7
	18	170	66.3		18	25	31.8
	24	125	107.8		24	25	28.7
2	6	100	112.3				
	12	77.5	103.3				
	18	57.5	88.2				
	24	25.0	69.3				

9.25 From the following hydrographs, compute the Muskingum routing constants.

Time	Inflow (m^3/s)	Outflow (m^3/s)	Time	Inflow (m^3/s)	Outflow (m^3/s)
Midnight	17.3	3.1	Noon	12.6	21.9
Noon	28.8	7.5	Midnight	9.8	17.0
Midnight	35.9	16.3	Noon	7.7	13.6
Noon	37.0	26.8	Midnight	6.2	10.9
Midnight	30.6	33.7	Noon	5.0	9.0
Noon	22.0	34.0	Midnight	4.2	7.5
Midnight	16.4	28.0	Noon	3.6	6.2

9.26 The routing constants for a reach of a river have been found to be 24 hours and $x = 0.2$. Route the inflow hydrograph of Problem 9.20 through the reach of the river by the Muskingum method.

9.27 Determine the hydrograph at the downstream section if a storm produced the hydrograph at the upstream section given in Problem 9.21. The Muskingum constants are $K = 8$ hours and $x = 0.15$ for the reach.

9.28 Route the flood hydrograph indicated below through a reservoir. The storage (elevation versus volume) data obtained from the reservoir survey also are given. The spillway has the following characteristics:

1. Flow $= 3LH^{3/2}$

2. Length $= 70$ ft

3. Crest height $= 60$ ft

Inflow hydrograph:

Time (hr)	0	0.4	0.8	1.2	1.6	2.0	2.4	2.8	3.2	3.6
Flow (cfs)	0	600	2100	2500	1600	950	550	300	80	0

Storage data:

Elevation (ft)	60	61	62	63	64	65
Storage (acre-ft)	300	330	360	395	430	470

9.29 The reservoir storage data and the spillway rating data are given below. Route the following flood hydrograph through the reservoir.

Inflow hydrograph:

Time (hr)	0	0.5	1.0	1.5	2.0	2.50	3.0	3.50	4.0
Flow (cfs)	0	20	70	160	280	330	140	100	40

Storage and discharge data:

Elevation (ft)	15	16	17	18	19	20
Storage in 1000 ft³	180	252	414	655	990	1350
Outflow (cfs)	0	15	55	105	175	240
(1 cfs-hr $= 3600$ ft³)						

9.30 A steel pipeline 1600 m long discharges 150 liters per second of water from a reservoir to the atmosphere through a valve at the downstream end. The pipeline is 250 mm in diameter with a wall thickness of 10 mm. The valve is fully closed in 30 seconds. Determine the maximum transient pressure at the valve and 500 m upstream of the valve. The pipeline is without expansion joints and anchored at the upstream end only. $K = 2.1$ GPa, $E = 210$ GPa, $v = 0.3$.

9.31 A pipe system operates under a head of 50 ft at the valve to carry a discharge of 3.6 cfs. The cast-iron pipe has a length of 4000 ft, a diameter of 9 in., and a wall thickness of 0.375 in. If the valve at the downstream end is fully closed in 20 seconds, determine the maximum transient pressure at the center of the pipe. The pipe has frequent expansion joints. $K = 320$ ksi, $E = 20 \times 10^3$ ksi, and $v = 0.25$.

9.32 In Problem 9.30, the valve is fully closed in 2 seconds. Determine the maximum pressure rise at the valve.

9.33 In Problem 9.31, the valve is fully closed in 1.9 seconds. Determine the maximum transient pressure at the valve.

9.34 A steel pipe 1625 m long discharges water through a valve under a head of 10 m. The pipe has a diameter of 300 mm. The speed of the pressure wave is 1300 m/s. The area of the valve opening varies as shown in the table. Determine the variation of head at the valve if it is fully closed in 20 seconds. Disregard the friction losses. $C_d = 0.6$.

Time (s)	0.0	2.5	5.0	7.5	10.0	12.5	15.0	17.5	20.0
Valve area (m²)	0.03	0.0234	0.0165	0.011	0.008	0.006	0.004	0.002	0

9.35 In Problem 9.31, if the valve is fully closed in 15.2 seconds, calculate the pressure variation at the valve. The area of the valve opening reduces linearly with time from initially open to fully closed in 15.2 sec. $C_d = 0.65$.

9.36 For a positive surge traveling downstream in a horizontal rectangular channel, fill in the following blanks:

Number	y_1 (m)	V_1 (m/s)	y_2 (m)	V_2 (m/s)	Vw (m/s)
a	2	1.5	4		
b	1	1.75			5.5
c		3.0	1.5		−3.0
d	0.3		0.6		3.5

9.37 A rectangular channel carries a discharge of 1.5 m³/s per meter width at a depth of 0.75 m. If a sudden operation of the gate at an upstream section causes the discharge to increase by 33%, estimate the height and absolute velocity of the positive surge in the channel.

9.38 A sluice gate in a horizontal channel is suddenly partially closed to reduce the flow by 60 percent. If the initial velocity and depth are 6.1 m/s and 3 meters, respectively, determine the height of the negative surge and the water surface profile one second after the gate operation.

9.39 A negative wave of height 0.75 m is produced in a rectangular channel due to the sudden lifting of a gate downstream. The initial depth upstream of the gate is 3.0 m. and initial flow per unit width is 5 m³/s. Determine the discharge per unit width and the height of the negative wave at the gate 4 seconds after the gate is opened.

9.40 A gate is completely closed on a rectangular channel. The water depth is 10 m on upstream side of the gate (the initial velocity is zero). The gate is suddenly removed (fails). Draw the profile of the surge and the final velocity of flow at the gate. [This sudden release is called a *dam break* problem.]

Problems 531

10

Hydraulic Principles of Structures

▲▲

10.1 HYDRAULIC STRUCTURES

There are a large variety of hydraulic structures to serve the many purposes for which water resources are used. A classification based on the function performed by the structure is given in Table 10.1. This lists only common structures. There are other specialized hydraulic structures, such as hydrofoils, offshore structures, and hydrodynamic transmissions. This chapter presents the hydraulic principles of common structures belonging to the first three categories of Table 10.1. A hydraulic structure in an integral part of a river system; the structure impacts the basin hydrology and the hydrology controls the structure's design.

10.2 FLOW-MEASURING STRUCTURES

The following devices are commonly used for flow measurements. These measurements are based on the energy principle. Localized losses involved due to inertia and viscous effects are included in the form of a coefficient of discharge, which is preferably ascertained experimentally.

- Orifices and mouthpieces
- Weirs and notches
- Flumes
- Pipe-flow measuring devices

10.3 ORIFICES AND MOUTHPIECES

An orifice is a hole or an opening in a barrier placed in a stream through which water discharges under pressure. An orifice also can be made in the side or bottom of a tank or vessel or in a plate placed between the flanges of a pipeline to measure flow through these structures. Orifices are classified according to size (small and large), shape (circular, rectangular, triangular), and the shape of the upstream edge (sharp edged or round cornered). Some orifices contain a mouthpiece, which is a cylindrical extension of an orifice. An orifice may discharge free or may be submerged under a downstream level.

533

Table 10.1   Classification of Hydraulic Structures

Type	Purpose	Structure
Flow measurement structures	To determine discharge	Weirs, orifices, flumes
Storage structures	To store water	Dams, tanks
Flow control structures	To regulate the quantity and pass excess flow	Spillways, outlets, gates, valves
Diversion structures	To divert the main course of water	Coffer dams, weirs, canal head-works, intake works
Conveyance structures	To guide flow from one location to another	Open channels, pressure conduits, pipes, canals, sewers
Collection structures	To collect water for disposal	Drain inlets, infiltration galleries, wells
Energy dissipation structures	To prevent erosion and structural damage	Stilling basins, surge tanks, check dams
Shore protection structures	To protect the banks	Dikes, groins, jetties, rivetments, breakwaters, seawalls
River training and waterway stabilization structures	To maintain a river channel and water transportation	Levees, cutoffs, locks, piers, culverts
Sediments and quality control structures	To control or remove sediments and other pollutants	Racks, screens, traps, sedimentation tanks, filters, sluiceways
Hydraulic machines	To convert energy from one form to other	Pumps, turbines, rams

10.3.1 Flow through a Small Orifice

When the area of an orifice is sufficiently small with respect to the size of the container, the velocity of flow can be considered constant throughout the orifice. For the orifice section shown in Figure 10.1, apply Bernoulli's theorem at points 1 and 2 with the datum at the center of the orifice.

$$0 + \frac{v_1^2}{2g} + h = 0 + \frac{v_2^2}{2g} + 0 \tag{a}$$

The approach velocity, v_1, is very small compared to v_2 and can be disregarded. Hence

$$v_2 = \sqrt{2gh} \tag{b}$$

The actual velocity is slightly less, due to the viscous shear effect between water and orifice edge. Hence, including a coefficient of velocity, we have

$$v_2 = C_v \sqrt{2gh} \tag{c}$$

Figure 10.1   Stream jet through an orifice.

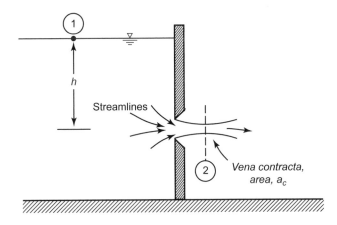

The size of the jet is narrowest at a distance of about one-half the orifice diameter. At the narrowest section, the *vena contracta*, the streamlines are parallel and perpendicular to the orifice. At the *vena contracta*, discharge

$$Q = a_c C_v \sqrt{2gh} \qquad \text{(d)}$$

In terms of the orifice area,

$$Q = C_c C_v A \sqrt{2gh} \qquad \text{(e)}$$

where C_c is the ratio of the area of jet at the *vena contracta* to the area of the orifice, known as the coefficient of contraction. The two coefficients are combined into a single coefficient of discharge, C_d. Thus

$$Q = C_d A \sqrt{2gh} \quad [L^3 T^{-1}] \qquad (10.1)$$

10.3.2 Flow through a Large Orifice

When the head over the orifice is less than five times the size (diameter or height of opening) of the orifice, it is a large orifice for which eq. (10.1) is not true because the streamlines of the jet are not normal to the orifice plane and the velocity is not constant throughout the orifice. Instead, it acts like a weir under pressure, with the water level always above the top edge of the weir on the upstream side.

In the rectangular orifice under the low head shown in Figure 10.2 the velocity of flow through an elemental strip at a depth of h from the free surface is $\sqrt{2gh}$, and the discharge is

$$dQ = (B\, dh)\sqrt{2gh} \qquad \text{(a)}$$

For the total discharge, integrating between the limits of H_1 and H_2 and introducing a coefficient,

$$Q = \frac{2}{3} C_d \sqrt{2g} B \left(H_1^{3/2} - H_2^{3/2} \right) \quad [L^3 T^{-1}] \qquad (10.2)$$

Figure 10.2 Large rectangular orifice.

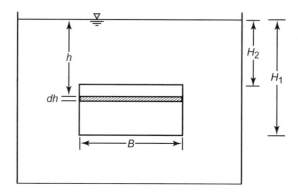

For a circular or any other shape of orifice, the area term in eq. (a) is expressed in terms of h and then the equation is integrated. If the velocity of approach cannot be disregarded, the velocity head, $v^2/2g$, should be added in both H_1 and H_2 of eq. (10.2).

10.3.3 Mouthpieces

A mouthpiece has a length of two to three times the jet diameter. It is used to increase the coefficient of contraction, C_c. There are four types of mouthpieces.

Cylindrical external mouthpiece. The vena contracta occurs at a distance of half the orifice diameter from the outlet of the orifice. In an external mouthpiece of a length 2.5 times the diameter of the orifice, the vena contracta will occur within the tube. The tube will be full when discharge takes place, which makes C_c equal to 1. But then there are turbulence losses that will reduce the coefficient of velocity.

Convergent mouthpiece. The losses due to sudden enlargement can be avoided with a convergent mouthpiece. The discharge coefficient could even be equal to 1.

Divergent or Bell mouthpiece. This type of mouthpiece is convergent until the vena contracta and then diverges. The coefficient is close to 1.0. For diverging conical tubes (without convergence), a coefficient of discharge of more than 1 has been reported. Brater and King (1996) have provided a detailed discussion on orifice coefficients.

Reentrant or Borda mouthpiece. This mouthpiece projects inside the container. When running free it has a low coefficient of contraction. However, if the mouthpiece is running full at the outlet, it increases the coefficient.

The typical values of the coefficients of contraction, velocity, and discharge are given in Table 10.2.

EXAMPLE 10.1

In a stream of 5 ft width and 3 ft depth, a plate is placed that has a rectangular orifice 3 ft in length and 1.2 ft in height. The upper edge of the orifice is 9 in. below the water surface. Determine the orifice discharge **(a)** treating it as a small orifice and **(b)** using the large orifice approach. $C_d = 0.6$.

Table 10.2 Typical Orifice and Mouthpiece Coefficients

	Rounded orifice	Sharp-edged orifice	External mouthpiece	Convergent mouthpiece	Divergent mouthpiece	Borda (Reentrant) mouthpiece	
						Free	Full
C_c	1.0	0.61	1.0	1.0	1.0	0.51	1.0
C_v	0.98	0.98	0.8	1.0	1.0	0.98	0.75
C_d	0.98	0.60	0.8	1.0	1.0	0.50	0.75

SOLUTION

(a) **1.** Water depth to orifice center = 1.35 ft.

  2. $A = 3 \times 1.2 = 3.6 \text{ ft}^2$

  3. From eq. (10.1),

$$Q = 0.6(3.6)\sqrt{2(32.2)(1.35)}$$
$$= 20.14 \text{ cfs}$$

(b) **4.** Disregard the velocity of approach:

  5. $H_2 = 0.75 \text{ ft}, H_1 = 0.75 + 1.2 = 1.95 \text{ ft.}$

  6. From eq. (10.2),

$$Q = \frac{2}{3}(0.6)\sqrt{2(32.2)}(3)\left(1.95^{3/2} - 0.75^{3/2}\right)$$
$$= 19.96 \text{ cfs}$$

  7. Velocity of approach $= \dfrac{19.96}{(5)(3)} = 1.33 \text{ fps.}$

  8. Velocity head $= \dfrac{(1.33)^2}{2(32.2)} = 0.03 \text{ ft.}$

  9. Including the velocity head, we have

$$Q = \frac{2}{3}(0.6)\sqrt{2(32.2)}(3)\left(1.98^{3/2} - 0.78^{3/2}\right)$$
$$= 20.20 \text{ cfs}$$

EXAMPLE 10.2

Discharge from an orifice of 75 mm diameter is 0.02 m³/s under a constant head of 3 m. An external mouthpiece of the same diameter is installed that raises the coefficient of contraction from 0.63 to 1.0. The coefficient of velocity is not known and remains unchanged. Determine discharge from the mouthpiece.

For the orifice:

1. $A = \dfrac{\pi}{4}(0.075)^2 = 0.0044 \text{ m}^2$

2. From eq. (10.1), $0.02 = C_d(0.0044)\sqrt{2(9.81)(3)}$

$$C_d = 0.59$$

3. $C_v = \dfrac{0.59}{0.63} = 0.937$

For the mouthpiece:

4. $C_d = C_v C_c = (0.937)(1) = 0.937$

5. From eq. (10.1), $Q = 0.937(0.0044)\sqrt{2(9.81)(3)}$

$$= 0.0316 \text{ m}^3/\text{s}$$

10.3.4 Time to Empty

In the case of a tank or vessel, if the water level is not kept constant by an inflow, the level will drop due to discharge from the orifice. The rate of flow through the orifice will vary with the change in head. Consider that at any instant the head over the orifice is h, and in time dt it falls by dh. If the volume of water leaving the tank is equated to the volume of flow through the orifice, then

$$-A_t \, dh = C_d A \sqrt{2gh} \, dt \quad [\text{L}^3] \tag{10.3}$$

By expressing the water surface area in the tank, A_t, by a suitable formula for a specified shape and by integrating between two levels, the time needed to lower the water surface can be determined. Simultaneously with orifice discharge, if an inflow at a constant rate of Q_i takes place into the vessel, the term $Q_i dt$ should be subtracted from the right side of eq. (10.3).

EXAMPLE 10.3

A vessel has the shape of a cone as shown in Figure 10.3. The orifice at the bottom has a diameter of 100 mm. How long will it take the cone to become one-half empty from its full depth? $C_d = 0.6$.

SOLUTION

1. Area of orifice $= \dfrac{\pi}{4}(0.1)^2 = 0.0079 \text{ m}^2$

2. At any instant when the head over the orifice is h, by the similarity of the triangles:

$$\frac{d}{h} = \frac{3}{3.5}$$

$$d = 0.857h$$

Figure 10.3 Emptying of a conical vessel.

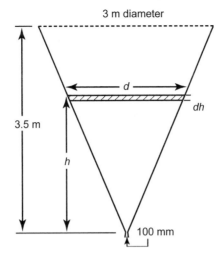

3. Water surface area, $A_t = \dfrac{\pi}{4}d^2 = \dfrac{\pi}{4}(0.857h)^2 = 0.577h^2$

4. Substituting in eq. (10.3) and integrating between 3.5 m and 1.75 m (one-half full)

$$-\int_{3.5}^{1.75} 0.577 \frac{h^2}{\sqrt{h}}dh = \int_0^t 0.6(0.0079)\sqrt{2(9.81)}dt$$

or $t = 207.4$ s or 3.46 min

10.4 WEIRS AND NOTCHES

A weir may be defined as a regular obstruction across a channel section over which flow takes place. It may be a vertical flat plate with a sharpened upper edge; then it is known as a *sharp-crested* weir or notch. It may have a solid broad section of concrete or other material; then it is known as a *broad-crested* weir. Weirs are classified according to their shapes [i.e., rectangular, triangular (V-notch), trapezoidal (Cipolletti), and parabolic]. The rectangular weir is the most popular. A rectangular section that spans the full width of the channel is known as a *suppressed* weir. If the width of the weir section is less than the width of the channel, it is a *weir with end contraction*. Where the downstream water level is lower than the crest, the weir is said to have a *free discharge*. If the downstream level is higher than the crest level, it is known as a *drowned* or *submerged* weir.

10.4.1 Flow over Sharp-Crested Weirs

The water flowing over a sharp-crested weir under free-discharge conditions falls away from the downstream face of the weir. This forms a nappe, as shown in Figure 10.4. Air is trapped between the lower nappe surface and the downstream face of the weir. Thus the underside of the jet or lower nappe is exposed to the atmospheric pressure. If means of restoring this air are not provided, the entrapped air will be carried away by the flowing water, creating a negative pressure. This can increase the discharge as much as 25% but can damage the structure. In a contracted weir, air is restored naturally from the sides, and in a suppressed weir, through installed ventilation pipes.

10.4.2 Rectangular Sharp-Crested Suppressed Weir

Apply the energy equation at points 1 and 2 on the streamline AA′ in Figure 10.5. Crest as a datum:

$$H + \frac{v_1^2}{2g} = h + \frac{v_2^2}{2g} \tag{a}$$

At point 1, H is the sum of the datum and pressure head on the streamline. At point 2, the pressure is assumed to be atmospheric (after Weisbach). The flow over the weir forms an overspilling jet whose under and top sides are exposed to the atmosphere. When the approach velocity, v_1, is disregarded,

$$v_2 = \sqrt{2g(H-h)} \tag{b}$$

The flow through the elemental strip of area Ldh is

$$dQ = Ldh\sqrt{2g(H-h)} \tag{c}$$

After introducing a discharge coefficient to account for the inertia and shear losses, the total discharge is

$$Q = C_d\sqrt{2g}L\int_0^H \sqrt{H-h}\,dh \tag{d}$$

or

$$Q = \frac{2}{3}C_d\sqrt{2g}LH^{3/2} \quad [\text{L}^3\text{T}^{-1}] \tag{10.4}$$

10.4.3 Coefficient of Discharge of Sharp-Crested Weirs

The coefficient of discharge should preferably be determined experimentally. Many empirical formulas have been proposed to determine the coefficient of discharge. Based on the experiments performed at Georgia Institute of Technology, the following relation is suggested for a rectangular sharp-crested weir

$$C_d = 0.602 + 0.075\frac{H}{P} \quad [\text{dimensionless}] \tag{10.5}$$

where P is the weir height. The limitation of the approach is that $H/P < 2$.

When the face of a sharp-crested weir is inclined, the coefficient of discharge is increased by the factor in Table 10.3. When a C_d value of 0.62 is used, this results in the following Francis formula in FPS units:

$$Q = 3.33LH^{3/2} \quad [\text{L}^3\text{T}^{-1}] \quad (\text{English units}) \tag{10.6}$$

Figure 10.4 Flow over free-discharging sharp-crested weir.

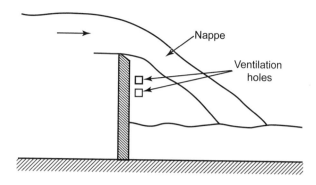

Figure 10.5 Thin plate weir: (a) free-discharging profile, (b) weir section.

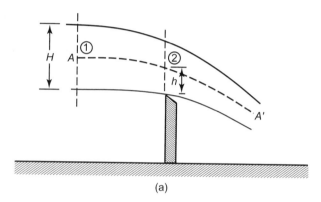

(a)

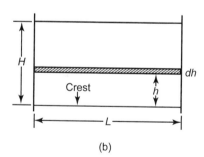

(b)

10.4.4 Rectangular Sharp-Crested Weir with End Contractions

When the weir length is less than the width of the channel, it is known as a weir with end contractions. The effective length of the weir in this case is less than the actual weir length, due to contraction of the flow jet caused by the sidewalls. The formula is given by

$$Q = \frac{2}{3} C_d \sqrt{2g}\left(L - 0.1nH\right)H^{3/2} \quad [\mathrm{L^3 T^{-1}}] \tag{10.7}$$

where n is the number of end contractions. If both ends are contracted, $n = 2$.

Table 10.3 Slope Factor for Inclined Sharp-Crested Weir

$\dfrac{H}{P}$	Slope (horizontal: vertical)		
	1:3	2:3	3:3
0	1.04	1.08	1.12
0.5	1.03	1.07	1.10
1.0	1.03	1.06	1.09
1.5	1.02	1.06	1.08
2.0	1.03	1.06	1.07
3.0	1.05	1.09	1.12

When a weir is divided into sections by N piers, each section will have an end contraction effect. For the abutments flush with the bank, $n = 2N$. If the two end abutments also are contracted, then the total number of end contractions, $n = 2(N + 1)$. Weir length, L, is the clear span that subtracts for the width of N piers from the total weir length.

10.4.5 Rectangular Sharp-Crested Weir with Velocity of Approach

The effective head responsible for the discharge is the sum of water head and the head due to velocity of approach. If the latter is not very small, the limits of integration in eq. (d) should be modified between $v_0^2/2g$ and $\left(H + v_0^2/2g\right)$. The resulting equation is

$$Q = \frac{2}{3} C_d \sqrt{2g}\,(L - 0.1nH)\left[\left(H + \frac{v_0^2}{2g}\right)^{3/2} - \left(\frac{v_0^2}{2g}\right)^{3/2}\right] \quad [\text{L}^3\text{T}^{-1}] \qquad (10.8)$$

In eq. (10.8), v_0 cannot be found unless Q is known. Thus the equation is first solved by neglecting v_0. The approximate discharge thus determined is used to find v_0. A revised value of Q is determined using this v_0. The process is repeated until the final discharge is within 1% of the preceding discharge.

EXAMPLE 10.4

An end-contracted weir of total length 286 ft and crest height 5 ft is used to discharge water without exceeding a head of 2.5 ft from a tank 300 ft wide. The weir carries piers that are 10 ft clear distance apart and 2 ft wide, to support a footway. Determine the discharge.

SOLUTION

1. From eq. (10.5), $C_d = 0.602 + 0.075\dfrac{(2.5)}{5} = 0.64$

2. Let N represent the number of piers.

3. Number of sections $= N + 1$

4. Total length, $286 = 10(N + 1) + 2N$

5. Thus, $N = 23$

6. Number of end contractions $n = 2(N + 1) = 2(24) = 48$

7. Clear weir length $= [286 - 23(2)] = 240$ ft

8. From eq. (10.7), assuming that $v_0 = 0$,

$$Q = \frac{2}{3}(0.64)\sqrt{2(32.2)}\left[240 - 0.1(48)(2.5)\right][2.5]^{3/2}$$

$$= 3085 \text{ cfs}$$

9. $v_0 = \dfrac{Q}{A} = \dfrac{3085}{(7.5)(300)} = 1.37$ ft/sec., since approach depth $= 5 + 2.5 = 7.5$ ft

10. $\dfrac{v_0^2}{2g} = \dfrac{(1.37)^2}{2(32.2)} = 0.029$ ft

11. Revised $Q = \dfrac{2}{3}(0.64)\sqrt{2(32.2)}\left[240 - 0.1(48)(2.5)\right]\left(2.529^{3/2} - 0.029^{3/2}\right)$

$$= 3135 \text{ cfs}$$

10.4.6 Triangular (V-notch) Weir

These are suitable for low discharges because the head increases more rapidly on a triangular section. In Figure 10.6, the area of elemental strip is given by

$$dA = b\, dh$$

or

$$dA = 2h \tan\frac{\theta}{2}dh \tag{a}$$

From the preceding section, at height h the velocity

$$v = \sqrt{2g(H-h)} \tag{b}$$

Discharge through the elemental area

$$dQ = C_d\left(2h \tan\frac{\theta}{2}dh\right)\sqrt{2g(H-h)} \tag{c}$$

Total discharge

$$Q = 2C_d\sqrt{2g}\,\tan\frac{\theta}{2}\int_0^H h\sqrt{H-h}\;dh \tag{d}$$

or

$$Q = \frac{8}{15}C_d\sqrt{2g}\,\tan\frac{\theta}{2}H^{5/2}\quad [L^3T^{-1}] \tag{10.9}$$

Ordinarily, V-notch weirs are not appreciably affected by the velocity of approach (U.S. Bureau of Reclamation, 1984). Typical values for the coefficient of discharge of a triangular weir are shown in Figure 10.7.

Figure 10.6 V-notch section.

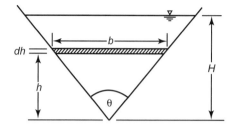

Figure 10.7 The coefficient of discharge for a triangular sharp-crested weir.

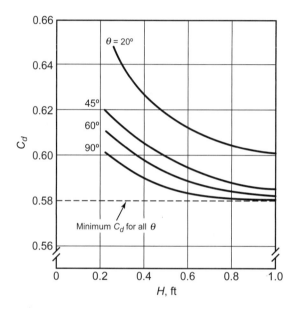

10.4.7 Trapezoidal Weir

The discharge is the sum of discharges over the rectangular section with end contractions and over the triangular section. A trapezoidal weir with a side slope of 1 horizontal to 4 vertical is known as a *Cipolletti weir*. The discharge through a Cipolletti weir is given by the Francis formula for a suppressed rectangular weir [eq. (10.6)], in which the coefficient is increased by about 1%. In FPS units:

$$Q = 3.367LH^{3/2} \quad [\text{L}^3\text{T}^{-1}] \quad (\text{English units}) \tag{10.10}$$

EXAMPLE 10.5

A trapezoidal weir has a side slope of 1 horizontal to 4 vertical. The head over the weir is 2.5 ft when the discharge is 30 cfs. What is the weir length? Compare the result with the Cipolletti formula. $C_d = 0.62$.

SOLUTION

1. Flow over the rectangular section, from eq. (10.7),

$$Q_1 = \frac{2}{3}(0.62)\sqrt{2(32.2)}\left[L-(0.1)(2)(2.5)\right]2.5^{3/2}$$
$$= 13.11(L-0.5)$$

2. Flow over the triangular section, $\tan\frac{\theta}{2} = 0.25$, from eq. (10.9),

$$Q_2 = \frac{8}{15}(0.62)\sqrt{2(32.2)}(0.25)(2.5)^{5/2}$$
$$= 6.56$$

3. Total discharge, $Q = Q_1 + Q_2$

$$30 = 13.11(L-0.5) + 6.56$$
$$\text{or } L = 2.29 \text{ ft}$$

4. By the Cipolletti eq. (10.10), $\quad 30 = 3.367\, L\,(2.5)^{3/2}$

$$\text{or } L = 2.25 \text{ ft}$$

10.4.8 Flow over Sharp-Crested, Submerged Weirs

When the downstream water level exceeds the crest height, it influences the discharge over the weir. The submergence reduces the discharge through the weir. Herschel, Villemonte, and Marvis have suggested relations between free (unsubmerged) and submerged discharge over a weir as a function of upstream and downstream heads. The Villemonte relation for various types of weirs is represented by

$$\frac{Q_s}{Q} = \left[1 - \left(\frac{H_2}{H_1}\right)^n\right]^{0.385} \qquad \text{[dimensionless]} \qquad (10.11)$$

where

$\quad Q = $ free (unsubmerged) weir discharge

$\quad Q_s = $ submerged discharge

$\quad H_1 = $ upstream head

$\quad H_2 = $ downstream head

$\quad\quad n = $ coefficient:

$\quad\quad n = 1.44$ for contracted rectangular weir,

$\quad\quad n = 1.50$ for suppressed rectangular weir, and

$\quad\quad n = 2.50$ for 90° (notch weir)

EXAMPLE 10.6

A stream has a width of 30 m, depth of 3 m, and a mean velocity of 1.25 m/s. Find the height of a weir to be built on the stream floor to raise the water level by 1 m.

SOLUTION

1. $Q = 30(3)(1.25) = 112.5$ m³/s

2. Raised water level $= 3 + 1 = 4$ m

3. Velocity of approach near weir, $v_0 = \dfrac{112.5}{30(4)} = 0.94$ m/s

4. Velocity head $= \dfrac{(0.94)^2}{2(9.81)} = 0.045$ m

5. It is not known whether the weir has a free or a submerged discharge. First assuming a free discharge and $C_d = 0.6$

$$Q = \frac{2}{3} C_d \sqrt{2g}\, L\left[\left(H + \frac{v_0^2}{2g}\right)^{3/2} - \left(\frac{v_0^2}{2g}\right)^{3/2}\right]$$

$$112.5 = \frac{2}{3}(0.6)\sqrt{2(9.81)}\,(30)\left[(H + 0.045)^{3/2} - (0.045)^{3/2}\right]$$

$$H = 1.61 \text{ m}$$

Thus the head required over the crest is 1.61 m. Since the total water depth is 4 m, this would give a crest height of 2.39 m, which is less than 3 m. It is a submerged weir.

6. Approx. $\dfrac{H}{P} = \dfrac{1.61}{2.39} = 0.67$

From eq. (10.5), $C_d = 0.602 + 0.075\,(0.67) = 0.65$

7. For a submerged weir, let the height of crest be X meters. Hence $H_1 = 4 - X$, $H_2 = 3 - X$, $Q_s = 112.5$ m³/s, disregarding approach velocity.

From eq. (10.4),

$$Q = \frac{2}{3}(0.65)\sqrt{2(9.81)}\,(30)(4 - X)^{3/2}$$

$$= 57.58(4 - X)^{3/2} \text{ m}^3/\text{s}$$

8. From eq. (10.11),

$$\frac{112.5}{57.58(4 - X)^{1.5}} = \left[1 - \left(\frac{3 - X}{4 - X}\right)^{1.5}\right]^{0.385}$$

or

$$\frac{5.70}{(4-X)^{3.9}} = 1 - \left(\frac{3-X}{4-X}\right)^{1.5}$$

By trial and error, $X = 2.3$ m

10.5 FLOW OVER BROAD-CRESTED WEIRS

When the crest of a weir has sufficient thickness, the flow becomes parallel to the crest. It is classified as a broad-crested weir. As the stream of water flows over the broad crest, the head drops from H to h_c, due to the acceleration of water as a result of a sudden reduction of sectional area, as illustrated in Figure 10.8. The acceleration raises the discharge, which attains a maximum value at $h_c = \frac{2}{3}H$ (precisely two-thirds of the energy head) when the flow is critical.

Applying the energy equation at points 1 and 2, we obtain

$$H = h_c + \frac{v^2}{2g} \tag{a}$$

or

$$v = \sqrt{2g(H - h_c)} \tag{b}$$

and

$$Q = C_d L h_c \sqrt{2g(H - h_c)} \tag{c}$$

For $H = \frac{2}{3}h_c$,

$$Q = C_d \sqrt{2g} L \left(\frac{2}{3}H\right) \sqrt{H - \frac{2}{3}H} \tag{d}$$

or

$$Q = 0.385 C_d \sqrt{2g} L H^{3/2} \quad [\mathrm{L^3 T^{-1}}] \tag{10.12}$$

H is the total head at the approach channel, including the velocity head. It is impractical to measure the energy head H directly in field measurements. A common practice is to relate discharge to upstream water head h as discussed subsequently.

For measuring flows, broad-crested weirs offer no advantages over sharp-crested weirs and thus the U.S. Bureau of Reclamation seldom uses them for measuring purposes (U.S. Bureau of Reclamation, 1984). The weir crest should be calibrated either by field tests on the actual structure or by model studies of it.

If the head of water is too high the nappe tends to spring clear of the crest and the weir does not perform as a broad-crested weir. Then it should be treated as a sharp-crested weir. The head to crest thickness ratios (h/b) that separate the broad-crested and sharp-crested weirs are shown in Figure 10.9.

Figure 10.8 Flow over broad-crested section.

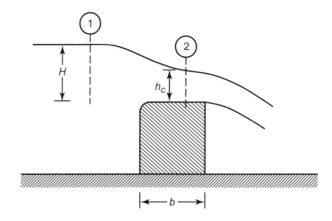

Figure 10.9 Distinction between sharp-crested and broad-crested weir.

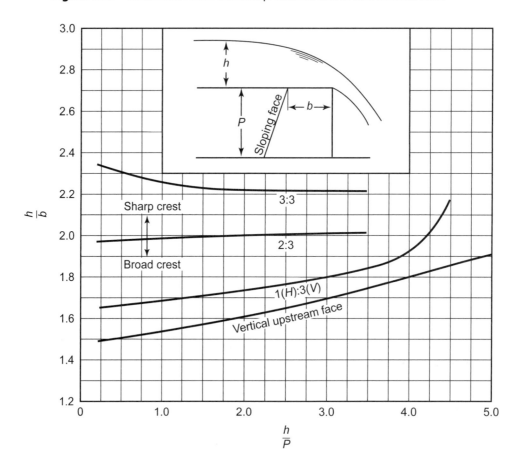

10.5.1 Coefficient of Discharge of Broad-Crested Weirs

The coefficient of discharge of a broad-crested weir is a function of (1) the ratio of head over the crest to thickness of the weir, h/b, and (2) the ratio of head to depth of water, $h/(h + P)$.

British researchers found that for the low values of these ratios as defined by the following limits, $0.08 < h/b < 0.33$ and $0.18 < h/(h + P) < 0.36$, the coefficient of discharge has a constant value of 0.85. However, when either or both of the above ratios exceed the indicated limits, C_d is higher but the position is not clear. Bos (1985) plotted 1395 data points from laboratory data of 105 broad-crested weirs from 29 different research papers and proposed a relation of C_d as a function of h/b only. Hager and Schwalt (1994) presented a different type of relation, also as a function of h/b only. The British Standards Institution (1969) adopted the coefficient as a function of h/b and $h/(h + P)$ as shown in Figure 10.10, which is widely used.

Discharge is commonly related to upstream water head h instead of the total energy head because it is impractical to measure the energy head H directly in field measurements. To correct for neglecting the velocity head, $v^2/2g$, an approach velocity coefficient, C_{av}, is introduced into eq. (10.12). Thus,

$$Q = 0.385\, C_{av}\, C_d\, \sqrt{2g}\ L\, h^{3/2} \quad [\mathrm{L^3 T^{-1}}] \tag{10.13}$$

The C_{av} values as a function of the area ratio are plotted in Figure 10.11.

A^* is the imaginary area of control if the water depth over the weir section is equal to h and A_1 is the approach area. Thus,

$$A^* = L(h)$$

and

$$A_1 = B(h + P)$$

where B is the width of the river channel. For a small approach velocity, H and h are almost equal and the value of C_{av} is one.

Many times the upstream face as well as the downstream face of a broad-crested weir are sloped. The slopes affect the coefficient of discharge. The multiplication factors that should be applied to the coefficient of discharge for sloping faces are given in Table 10.4.

The discharge coefficient is a function of the width of weir, the length of weir, the boundary layer displacement thickness over the crest, the roughness of the weir surface, and the Reynolds number. Auckers et al. (1978) provided a detailed treatment of the coefficient of discharge. Similar to the rectangular weir, the round-nosed weir with a rounded upstream corner is used for flow measurements. The rounded nose makes a weir robust and insensitive to damage and deposition upstream, in addition to having a higher coefficient of discharge. Compared to a rectangular weir, the round-nose coefficient is 5 to 10% higher for low to high value of the h/b ratio.

EXAMPLE 10.7

Determine the discharge over a broad-crested weir that is 6.5 ft wide and 100 ft long, the measured upstream level over the crest being 2.25 ft. The width of the approach channel is 120 ft and its depth below the crest of weir is 2 ft.

Figure 10.10 Coefficient of discharge for broad-crested weirs (British Standards Institution, 1969) [Scale adjustment by the author].

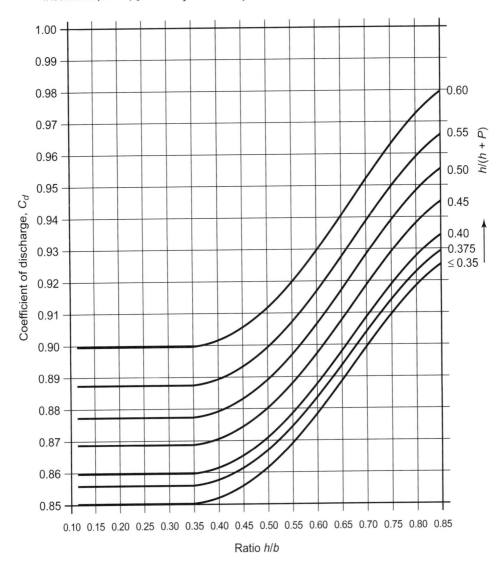

SOLUTION

1. $\dfrac{h}{b} = \dfrac{2.25}{6.5} = 0.35$, $\dfrac{h}{h+P} = \dfrac{2.25}{2.25+2} = 0.53$

From Figure 10.10, $C_d = 0.885$

Figure 10.11 C_{av} values for broad-crested weirs (with unity energy coefficient).

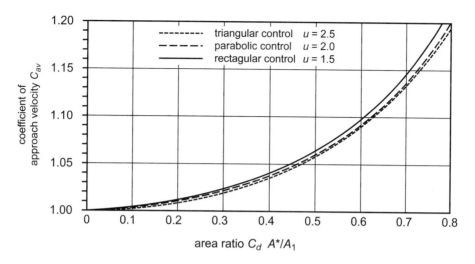

Table 10.4   Slope Factors for Broad-Crested Weirs

	Upstream slope (H:V)			Downstream slope (H:V)			
h/b	0.5:1	1:1	2:1	2:1	3:1	4:1	5:1
0.1	1.05	1.09	1.10	1.00	1.00	1.00	1.00
0.2	1.05	1.09	1.11	1.00	1.00	1.00	1.00
0.4	1.05	1.09	1.12	1.00	1.00	1.00	1.00
0.6	1.05	1.10	1.13	0.99	0.99	0.99	0.98
0.8	1.06	1.11	1.12	0.99	0.98	0.97	0.96
1.0	1.06	1.10	1.12	0.98	0.96	0.95	0.94
1.5	1.06	1.08	1.09	0.98	0.95	0.93	0.92
2.0	1.04	1.05	1.06	0.98	0.94	0.91	0.90

2. $A^* = 100 \times 2.25 = 225 \text{ ft}^2$

$A_1 = 120(2 + 2.25) = 510 \text{ ft}^2$

$$\frac{C_d A^*}{A_1} = \frac{(0.885)(225)}{510} = 0.39$$

From Figure 10.11, $C_{av} = 1.04$

3. From eq. (10.13), $Q = 0.385(0.885)(1.04)\sqrt{2(32.2)}\,100(2.25)^{3/2}$

$$= 960 \text{ ft}^3/\text{s}$$

Section 10.5   Flow over Broad-Crested Weirs

10.5.2 Broad-Crested, Submerged Weirs

The submerged broad-crested weirs have a downstream static head above the crest level. The degree of submergence is the ratio of downstream static head to upstream static head. Up to an 85% submergence ratio, there is no appreciable effect on the discharge of a broad-crested weir.

10.6 FLUMES

Flumes are devices in which the flow is locally accelerated by means of (1) a lateral contraction in the channel sides, or (2) combining the lateral contraction with a hump in the channel bed. The first type is known as a *venturi flume* or *long-throated flume*. The equation for discharge through a flume is based on the energy principle (Bernoulli's theorem). Usually, flumes are designed to achieve the critical flow in the contracted (throat) section. Flumes have four advantages: (1) they can operate with small head loss, (2) they are insensitive to the velocity of approach, (3) they make good measurements without submergence as well as under submerged conditions, and (4) there is no related sediment deposition problem.

A venturi flume section is shown in Figure 10.12. Energy at the throat is represented by

$$H = h_2 + \frac{v_2^2}{2g} \tag{a}$$

or

$$v_2 = \sqrt{2g(H - h_2)} \tag{b}$$

For a rectangular throat section,

$$Q = (Wh_2)\sqrt{2g(H - h_2)} \tag{c}$$

The flow is critical at the throat, so $h_2 = \frac{2}{3}H$. Hence:

$$Q = 0.385 C_d \sqrt{2g} \, W \, H^{3/2} \quad [\text{L}^3\text{T}^{-1}] \tag{d}$$

Using the upstream head and including the velocity of approach and coefficient of discharge:

$$Q = 0.385 \, C_{av} \, C_d \sqrt{2g} \, W \, h^{3/2} \quad [\text{L}^3\text{T}^{-1}] \tag{10.14}$$

This formula is similar to the broad-crested weir equation (10.13). The coefficient of velocity of approach is given in Figure 10.11. The coefficient of discharge for a long-throated flume is close to 1.0 for flat-bottom, smooth rounded flumes. A special group of *short flumes* has a more compact throat section than the long-throated flumes. Short flumes with standard designs, including standard converging, diverging, and throat sections, have been developed and are known as the *Parshall flumes*. The water surface profile in such flumes varies rapidly and theoretical analysis is not possible. The relations for these have been developed by the U.S. Bureau of Reclamation through extensive calibration experiments. Equations for different flume sizes are summarized in Table 10.5.

In the case of submergence, a flow-rate correction is obtained from the graphs (U.S. Bureau of Reclamation, 1984). This correction is subtracted from the value computed by the preceding equations.

Figure 10.12 Sketch of a venturi flume section.

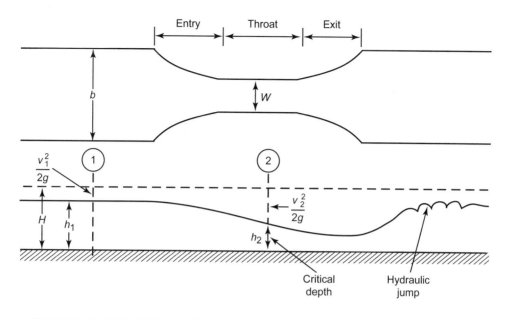

Table 10.5 Parshall Flume Discharge Relations

Throat width	Flow capacity (cfs)	Equation[a]	
3 in.	0.03–1.9	$Q = 0.992 H_a^{1.547}$	(10.15a)
6 in	0.05–3.9	$Q = 2.06 H_a^{1.58}$	(10.15b)
9 in.	0.09–8.9	$Q = 3.07 H_a^{1.53}$	(10.15c)
1–8 ft	Up to 140	$Q = 4W\, H_a^{1.522 W^{0.026}}$	(10.15d)
10–50 ft	Up to 2000	$Q = (3.687 W + 2.5) H_a^{1.6}$	(10.15e)

[a] H_a = water level in a well in the converging (approach) section (ft); W = throat width (ft); Q = discharge (cfs).

EXAMPLE 10.8

In a 6-ft Parshall flume, the gage reading in the approach well is 2 ft. The submergence is 90%, for which the correction is 14 cfs. Determine the discharge.

SOLUTION From eq. (10.15d; see Table 10.5),

$$Q_0 = 4(6)[2^{(1.522)(6)^{0.026}}] = 72.48 \text{ cfs}$$

The correction of flow rate, $Q_c = 14$ cfs, and the corrected flow, $Q = 72.48 - 14 = 58.48$ cfs.

10.7 PIPE FLOW MEASURING DEVICES

Three common devices used to measure the flow rate through pipes are

- Orifice meter
- Nozzle meter
- Venturi meter

These operate on the principle of the venturi flume in that a reduction in the flow area results in an increase of velocity in accordance with the continuity equation, and that the velocity increase causes a reduction of pressure according to the energy equation.

A typical orifice meter is constructed by inserting a plate with an orifice between two flanged pipe sections as shown in Figure 10.13.

Applying the energy principle between sections 1 and 2 with the centerline as the datum

$$\frac{p_1}{\gamma} + \frac{v_1^2}{2g} = \frac{p_2}{\gamma} + \frac{v_2^2}{2g} \tag{a}$$

Since $Q = A_1 v_1 = A_2 v_2$, the above relation results in

$$Q = A_2 \sqrt{\frac{2(p_1 - p_2)}{\rho\left[1 - \left(\dfrac{D_2}{D_1}\right)^4\right]}} \quad [\text{L}^3\text{T}^{-1}] \tag{b}$$

By including an orifice discharge coefficient C_d to account for all complexities of real-world applications,

$$Q = C_d A_o \sqrt{\frac{2\Delta p}{\rho\left[1 - \left(\dfrac{d}{D}\right)^4\right]}} \quad [\text{L}^3\text{T}^{-1}] \tag{10.16}$$

where

A_o = area of the orifice plate

Δp = pressure difference across the orifice

ρ = density of fluid through the pipe

d = diameter of the orifice

D = diameter of the pipe

From eq. (10.16), flow through the pipe can be determined by measuring the pressure difference across the orifice meter.

The coefficient of discharge is a function of the pipe restriction, d/D, and the pipe flow Reynolds number Re = $\rho v D/\mu$. Typical values are given in Figure 10.14.

The nozzle meter, shown in Figure 10.15, is similar to the orifice meter. The same eq. (10.16) is applicable. The coefficient of discharge through the nozzle meter is high due to a more streamlined flow pattern. Typical values are given in Figure 10.16.

The venturi meter, like the venturi flume, has a converging section, a throat, and a diverging section as shown in Figure 10.17. The same equation (10.16) of orifice meter is used for the venturi meter. However, the losses are minimal due to friction along the wall only. The coefficient of discharge is between 0.97 and 0.99.

Figure 10.13   Typical orifice meter.

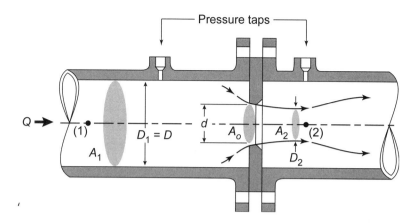

Figure 10.14   Orifice discharge coefficient.

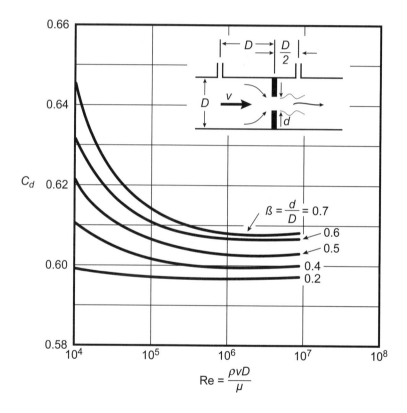

EXAMPLE 10.9

Methane gas flows through a pipe of 50 mm diameter in a refinery. The pressure drop across a 20 mm diameter nozzle is 5 kPa. What is the flow rate of methane through the pipe?

Figure 10.15 Typical nozzle meter.

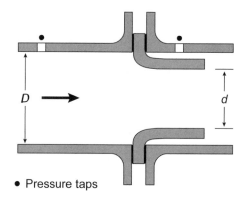

• Pressure taps

Figure 10.16 Nozzle discharge coefficient.

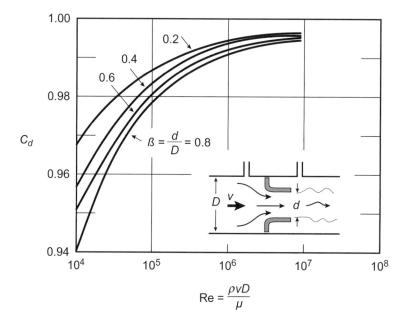

$$Re = \frac{\rho v D}{\mu}$$

Figure 10.17 Typical venturi meter.

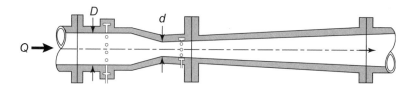

SOLUTION

1. For methane $\rho = 0.667$ kg/m^3, $\mu = 1.1 \times 10^{-5}$ N-s/m^2

$$\frac{d}{D} = \frac{20}{50} = 0.4$$

$$A_0 = \frac{\pi}{4}(0.02)^2 = 3.14 \times 10^{-4} \text{ m}^2$$

$$A = \frac{\pi}{4}(0.05)^2 = 1.96 \times 10^{-3} \text{ m}^2$$

2. From eq. (10.16), assume $C_d = 0.98$

$$Q = (0.98)\left(3.14 \times 10^{-4} \sqrt{\frac{2(5 \times 10^3)}{0.667(1 - 0.4^4)}}\right)$$

$$= 0.0382 \text{ m}^3/\text{s}$$

3. The revised C_d

$$v = \frac{Q}{A} = \frac{0.0382}{1.96 \times 10^{-3}} = 19.49 \text{ m/s}$$

$$\text{Re} = \frac{\rho v D}{\mu}$$

$$= \frac{(0.667)(19.49)(0.05)}{1.1 \times 10^{-5}}$$

$$= 59,000$$

From Figure 10.16, $C_d = 0.975$

4. Revised $Q = (0.975)(3.14 \times 10^{-4})\sqrt{\frac{2(5 \times 10^3)}{0.667(1 - 0.4^4)}}$

$$= 0.038 \text{ m}^3/\text{s}$$

10.8 PEAK-FLOW MEASURING STRUCTURES

Peak-flow measurements are commonly performed at the site of the following structures:

- A reach of stream channel (slope-area method)
- At width contractions of a bridge section
- At dams
- At culverts

A specific location at a particular structure is selected for peak-flow measurements. The hydraulic equations, derived from the energy and continuity principles, are applied in terms of the water depth or its variation. The high-water marks left by floods provide this information.

10.8.1 Slope-Area Method for a Stream Channel

A suitable reach of a stream channel is selected primarily on the basis of good high-water marks. Manning's equation is applied, in which the slope (energy gradient) term is modified for the nonuniform condition. The following set of formulas relevant to channel flow (Chapter 11) are applicable:

$$K = \frac{1.486}{n} AR^{2/3} \quad \text{(English units)} \quad [L^3 T^{-1}] \tag{10.17a}$$

$$K = \frac{1}{n} AR^{2/3} \quad \text{(metric units)} \quad [L^3 T^{-1}] \tag{10.17b}$$

$$\alpha = \frac{\sum Ki^3 / Ai^2}{K_T^3 / A^2} \quad \text{[dimensionless]} \tag{10.18}$$

$$h_v = \alpha V^2 / 2g \quad [L] \tag{10.19}$$

$$\Delta h_v = h_v(\text{upstream}) - h_v(\text{downstream}) \quad [L] \tag{10.20}$$

When Δh_v is positive (i.e., expanding reach),

$$S = \frac{\Delta h + \Delta h_v / 2}{L} \quad \text{[dimensionless]} \tag{10.21a}$$

When Δh_v is negative (i.e., contracting reach),

$$S = \frac{\Delta h + \Delta h_v}{L} \quad \text{[dimensionless]} \tag{10.21b}$$

$$Q = \sqrt{K_1 K_2 S} \quad [L^3 T^{-1}] \tag{10.22}$$

where

Q = discharge between the reach

K = conveyance of the channel; K_1 at section 1 and K_2 at section 2

A = area of cross section at a selected section

R = hydraulic radius = area/wetted perimeter

n = Manning's roughness coefficient (Table 11.4)

α = velocity head coefficient; assumed to be 1.0 if the section is not subdivided; eq. (10.18) used for subdivided section

h_v = velocity head at a section

Δh_v = difference in velocity head at two sections

Δh = difference in water surface elevations (high-water marks)

L = length of the reach

S = friction slope

The following trial-and-error procedure is used since eq. (10.22) cannot be solved directly.

1. Determine K for upstream and downstream sections by eq. (10.17).
2. Determine α for subdivided sections by eq. (10.18); otherwise, $\alpha = 1$.
3. Determine S from eq. (10.21a) or eq. (10.21b), assuming that $\Delta h_v = 0$.
4. Compute Q from eq. (10.22). Consider it as an "assumed" value.
5. For assumed Q (and $v = Q/A$), determine h_v from eq. (10.19).
6. Determine revised S and Q from eqs. (10.21) and (10.22), respectively. Repeat steps 5 and 6 until Q stabilizes.

When more than two cross sections are selected in a reach, compute the discharge between each of them.

EXAMPLE 10.10

From the following data obtained from field measurements at two sections 129.5 ft apart on Snake Creek near Connell, Washington, determine the peak discharge ($n = 0.045$).

	Section 1	Section 2
Area (ft^2)	208	209
Hydraulic radius (ft)	3.09	4.10
High-water mark (ft)	16.30	15.23

SOLUTION

(1)	(2)	(3)	(4)	(5)	(6)	(7)	(8)	(9)	(10)
							S^b	Q	
Trial	Reach	$K = \dfrac{1.486}{n} AR^{2/3}$	h	$h_v = \alpha \dfrac{v^2}{2g}$	Δh	$\Delta h_v{}^a$	[eq. (10.21)]	[eq. (10.22)]	$v = \dfrac{Q}{A}$
1	Section 1	14,577	16.3	0					7.01
					1.07	0	0.00826	1459	
	Section 2	17,688	15.23	0					6.98
2	Section 1	14,577	16.3	0.763					
					1.07	0.007	0.00829	1462	
	Section 2	17,688	15.23	0.756					

^{a} $h_{v1} - h_{v2}$
^{b} If Δh_v is positive, use (10.21a); when negative, use (10.21b).

10.8.2 Measurement at Width Contractions of a Bridge

The contraction of a stream channel at a roadway or railway crossing causes a sudden drop in the water level between an approach section and the contracted section under the bridge. From hydraulic considerations, the discharge is proportional to the square root of the head and the cross-sectional area of the contraction. The head is equal to the difference in water levels between the approach and contracted sections, to which the approach velocity head is added and the friction head loss is subtracted. Expressing the velocity head and head loss in terms of discharge and conveyance, the U.S. Geological Survey (Matthai, 1984) indicated discharge as

$$Q = CA_3 \sqrt{2g} \left[\frac{\Delta h}{1 - \alpha_1 C^2 \left(A_3 / A_1\right)^2 + 2gC^2 \left(A_3 / K_3\right)^2 \left(L + L_w K_3 / K_1\right)} \right]^{1/2} \quad [\text{L}^3\text{T}^{-1}] \quad (10.23)$$

where

Δh = difference in water levels between 1 and 3 [subscript 1 refers to the approach section, one b-width upstream (to the bridge opening of width b). Subscript 3 refers to the downstream side of the contraction]

C = coefficient of discharge

A = area of cross section, A_1 at section 1 and A_3 at section 3

α = velocity head coefficient, given by eq. (10.18)

K = conveyance, given by eq. (10.17a) or (10.17b)

L = abutment length, Figure 10.18

L_w = length of the approach reach from section 1 to water contact point on upstream of structure, Figure 10.18

The discharge coefficient, C, in eq. (10.23) is a function of many parameters related to the geometry of the bridge opening and the flow pattern. The following are the common parameters that affect the coefficient:

1. Froude number at section 3: If it exceeds 0.8, the computed discharge may not be reliable.

2. Eccentricity of opening: If the opening is very eccentric (to one side) such that $K_a / K_b \leq$ 0.12, where K_a and K_b are channel section conveyances to either side of the opening, the coefficient is reduced.

3. Piers and piles: The obstruction by piers and piles lowers the coefficient.

4. Submergence: High floods may cause the lower parts of a bridge to submerge. The additional wetted perimeter could lower the coefficient by a factor as large as 0.6.

5. Angularity: The angle between the axis of the opening and the line perpendicular to the direction of flow, known as angularity, affects the flow. A stream flowing normal to the opening has zero angularity. A 45° angularity might lower the coefficient by a factor of 0.7.

6. Rounding and sloping of embankment: Streamlining of flow pattern enhances the coefficient by a factor up to 1.2.

7. Contraction ratio: The channel contraction is defined by $m = (1 - K_3/K_1)$, where K_3 and K_1 are conveyances at section 3 (opening) and at section 1. $m = 1$ for no contraction and $m = 0$ for a fully blocked section.

8. Ratio of abutment length to width of opening (L/b).

9. Types of embankments and abutments: The USGS has divided bridge openings into four types as shown in Figure 10.18.

Figure 10.18 Classification of width contractions (Matthai, 1984).

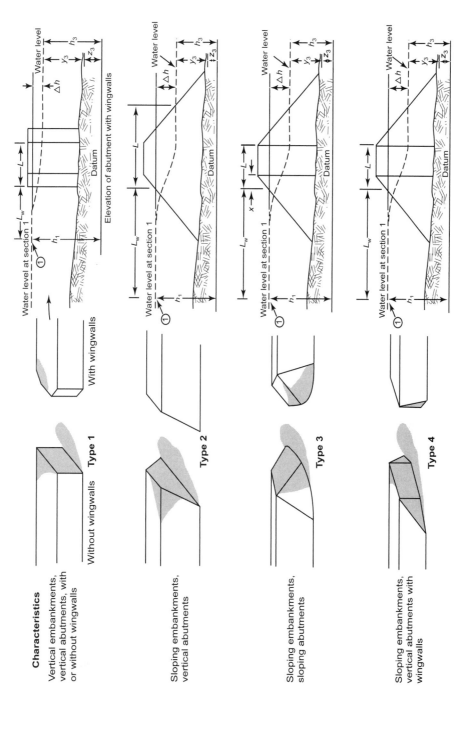

The last three items are most significant. The basic discharge coefficients under each type have been indicated by the USGS as a function of the ratio of abutment length to width of bridge opening and the channel-contraction ratio by a set of curves.

For a contraction ratio of 0.5, the basic coefficient is over 0.7 for $L/b = 0$ and around 0.9 for $L/b = 1$ or more, for all types of embankments and abutments. The basic coefficient decreases with increase in the contraction ratio and increases with the reduction of the ratio. The basic coefficients are further modified for the effect of the other factors listed above. For a detailed treatment see Matthai (1984).

EXAMPLE 10.11

The following information is obtained from a field investigation of the Rearing River in Colorado after a flood. Determine the peak discharge.

1. Water elevation at approach section = 3.00 m

2. Water elevation at contracted section = 2.74 m

3. Approach section 1 as shown in Figure 10.19

Subsection	n	Area m^2	Wetted perimeter m
1	0.050	5.18	10.67
2	0.035	10.65	11.55
3	0.048	1.54	10.97

4. Contracted section 3: $n = 0.032$. $A = 7.64$ m^2 $P_w = 11.06$ m

5. Length of approach reach, $L_w = 12.2$ m

6. Length of abutment, $L = 4.6$ m

7. Width of opening, $b = 6.4$ m

8. Sloping embankment, vertical abutment with wingwall

SOLUTION

1. $\Delta h = 3.00 - 2.74 = 0.26$ m

2.

Section	n	A	P_w	R	$R^{2/3}$	$K = \dfrac{1}{n}AR^{2/3}$	$\dfrac{K^3}{A^2}$
Approach channel							
1	0.05	5.18	10.67	0.485	0.618	64.02	9780
2	0.035	10.65	11.55	0.92	0.947	288.16	210,958
3	0.048	1.54	10.97	0.14	0.27	8.66	274
Total		17.37				360.84	221,012
Contracted section	0.032	7.64	11.06	0.69	0.78	186.22	

3. $\alpha = \dfrac{\Sigma K^3/A^2}{K_T^3/A_T^2} = \dfrac{221,012}{(360.84)^3/(17.37)^2} = 1.42$

Figure 10.19 Measurements at bridge contraction.

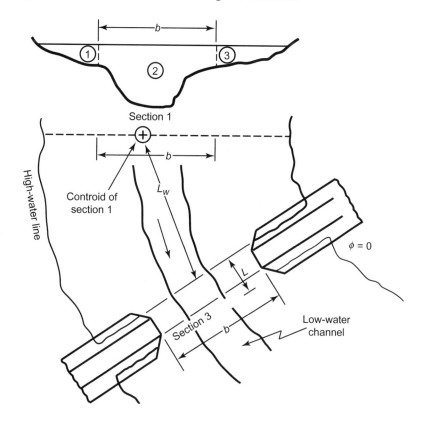

4. Contraction coefficient $m = 1 - \dfrac{K_3}{K_1} = 1 - \dfrac{186.22}{360.84} = 0.48$

$$\frac{L}{b} = \frac{4.6}{6.4} = 0.72$$

Classification of opening = type 4

$$C = 0.9 \text{ (from figure in Matthai, 1984)}$$

5. $\left(\dfrac{A_3}{A_1}\right)^2 = \left(\dfrac{7.64}{17.37}\right)^2 = 0.193$

$\left(\dfrac{A_3}{K_3}\right)^2 = \left(\dfrac{7.64}{186.22}\right)^2 = 0.00168$

$L + \dfrac{L_w K_3}{K_1} = 4.6 + \dfrac{12.2(186.22)}{360.84} = 10.90$

6. From eq. (10.23),

$$Q = (0.9)(7.64)\sqrt{(2)(9.81)}\left[\frac{0.26}{1-(1.42)(0.9)^2(0.193)+2(9.81)(0.9)^2(0.00168)(10.9)}\right]^{1/2}$$

$$= 14.7 \text{ m}^3/\text{s}$$

10.8.3 Measurement at Dams

The relations of flow over weirs are applicable to dams and embankments. The peak discharge can be ascertained for the head determined on the basis of a field survey of high water marks.

The dam sections in the field rarely have a sharp-crested profile. But many types of dams act like sharp-crested weirs. Flashboards mounted over dams may be treated as sharp-crested weirs. A dam with sharp upstream corners and level crest may act like a sharp-crested weir at a high head when the lower nappe springs clear of the downstream edge.

When the crest of a dam is designed to fit the lower nappe profile of a sharp-crested weir, the analysis is based on the spillway relations of Section 10.15. However, the majority of dams and embankments are broad-crested weirs with upstream and downstream slopes that are analyzed according to Section 10.5.

EXAMPLE 10.12

During a flood, the observations made over an embankment are shown in Figure 10.20. The embankment section is 200 m long. Determine the peak discharge.

SOLUTION

1. $\dfrac{h}{b} = \dfrac{4}{10} = 0.4$

$\dfrac{h}{h+P} = \dfrac{4}{4+15} = 0.21$

From Figure 10.10, $C_d = 0.852$

From Table 10.4, $KE_1 = 1.12$, $KE_2 = 1.00$

Revised $C_d = (0.852)(1.12)(1.00) = 0.95$

2. From eq. (10.13) with $C_{av} = 1$

$$Q = 0.385 \, C_d \, \sqrt{2g} \, L \, h^{3/2}$$

$$= 0.385(0.95)\sqrt{2(9.81)}(200)(4)^{3/2}$$

$$= 2592 \text{ m}^3/\text{s}$$

10.8.4 Measurement at Culverts

Culverts are usually used to measure flood discharges from small drainage areas. The placement of a roadway embankment and culvert inside a stream channel causes a change in the flow pattern resulting from acceleration of flow due to contraction of the cross-sectional area. Flow can be determined by measuring high-water marks that define the headwater and

Figure 10.20 Embankment section for peak-flow measurements.

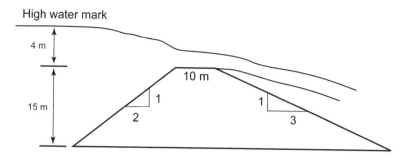

tailwater elevations and from the flow conditions inside the culvert, which can be tranquil, critical, or rapid. The continuity equation and the energy equation between the approach section and a section at the culvert are used, as described in detail in Section 14.7.

10.9 STORAGE STRUCTURES

Having estimated the water requirements for an intended project (Chapter 1) and having assessed the available water resources at a prospective site (Chapters 2 through 7), a planning engineer is faced with one of three situations:

1. The rate at which the water resources are available is always in excess of the requirements.
2. The total quantity of available resources over a period of time is equal to or in excess of the overall requirements, but the rate of requirements at times exceeds the available rate of resource supply.
3. The total available resources are less than the overall requirements.

In the first case, a *run-of-river* project can be formulated in which water can be used directly from the stream as need arises. A run-of-river project primarily incorporates a conveyance system to transport water to the location of its use. A storage reservoir is the solution to the second case. A reservoir project includes a storage structure, a control structure, and a conveyance system. Under the third condition, a supplemental source or an alternative site has to be explored. The conveyance structures are covered separately in Chapters 11 and 12.

Storage is of two types. When the demand for water can be satisfied by holding some of the high flow each year for release during a later period of low flow, it is called seasonal or *within-year* storage. However, if there is not enough high flow every year to raise the flow to a desired level, extra water must be stored during wetter years to release during dry years. This is termed *over-year* or *carryover* storage.

There are two approaches for determining the size of reservoir storage required. Simplified methods, which are commonly used in planning-stage studies, comprise mass curve analyses. Detailed methods, commonly used when developing reservoir operating plans, perform a sequential reservoir routing of the historical flow record. The theory of reservoir routing is discussed in Section 9.12. With the availability of computer program packages, the sequential routing study is also being used increasingly in the planning stage. The simplified approach is still very valuable for planning a single project when demand for water is relatively simple.

10.10 RESERVOIR STORAGE CAPACITY

There are two simplified methods of analysis of reservoir storage capacity: (1) the sequential mass-curve method, and (2) the nonsequential mass-curve method. A sequential mass-curve method known as the Rippl method considers the most critical period of recorded flow, which might be a severe drought period. The cumulative differences between inflow to the reservoir and outflow (draft) during successive periods are evaluated, the maximum value of which is the required storage:

$$S = \text{maximum } \Sigma(I_t - O_t) \quad [\text{L}^3] \tag{10.24}$$

where

$\quad S = $ required storage capacity

$\quad O_t = $ reservoir output or draft (yield) during period Δt

$\quad I_t = $ inflow during period Δt

Equation (10.24) can be solved either graphically or analytically. There are two graphic procedures. In the first method, I_t is accumulated separately as a mass inflow curve (curve a, Figure 10.21) and O_t as a mass yield curve (curve b, Figure 10.21). Yield represents the total demand for water and evaporation. For a constant draft rate, the yield curve is a straight line having a slope equal to the draft rate. At each high point on the mass inflow curve (curve a), a line is drawn parallel to the yield curve (curve b) and extended until it meets the inflow curve. The maximum vertical distance between the parallel yield line and the mass inflow curve (i.e., *FD*) represents the required storage. Assuming that the reservoir is full at *A* in Figure 10.21, going from *A* to *E* along the mass curve (*ABCDE*) represents the same volume of water as going from *A* to *E* along the straight line (*AFE*). From *A* to *B*, the draft is more than the inflow, resulting in a lowering of the reservoir; from *B* to *C*, the inflow is higher than the draft, but not enough to refill the reservoir; from *C* to *D*, the draft is more, once again causing a further drop in the level; from *D* to *E*, however, the inflow is very high, thus filling the reservoir at *E*.

The second graphical procedure plots the difference of successive accumulated values of the inflow and yield $\Sigma(I_t - O_t)$ against time as shown in Figure 10.22. The maximum vertical difference between the highest and subsequent lowest value is the storage. In the analytical procedure, the maximum accumulated difference is determined arithmetically to calculate the reservoir size.

The storage computed from the sequential analysis of historic flow data does not indicate the frequency (return period) associated with the selected size though it identifies within-year or over-year storage. Low flows of different durations can be analyzed nonsequentially to prepare flow-duration-frequency curves. From these curves, a mass inflow curve for any selected frequency can be prepared and used for determining the storage corresponding to that frequency. However, it does not make a distinction between within-year or over-year storage. These methods have been described by the HEC (1967) and Riggs and Hardison of the U.S. Geological Survey (1983).

EXAMPLE 10.13

During a critical flow period, the monthly inflows at the site of a proposed dam are given in column 2 of Table 10.6. The water supply requirements to be met from the storage are shown in column 3 of the table. A uniform release of 8 cfs has to be maintained to meet local requirements. Average monthly evaporation rate is 12 cfs. Determine the required capacity of the reservoir.

Figure 10.21 Rippl diagram for storage analysis.

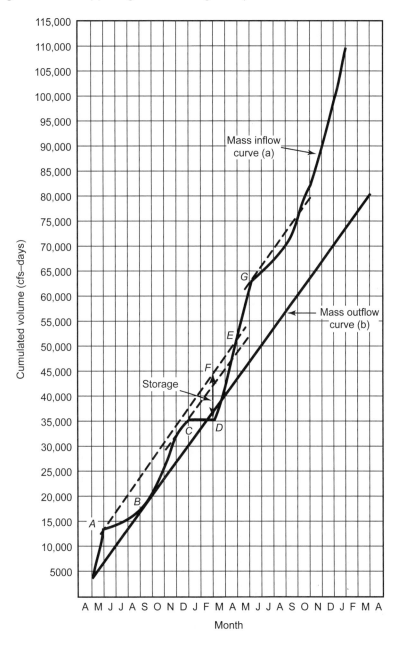

SOLUTION

1. *Rippl or mass diagram.* Inflows are cumulated in column 7 of Table 10.6 and plotted (curve a) in Figure 10.21. Total outflows are determined by adding evaporation and releases (and seepage, if any). Outflows are added in column 8 and plotted (curve b) in Figure 10.21. Lines parallel to outflow are drawn at high points.

$$\text{Storage } (FD) = 8400 \text{ cfs-days or } 16,630 \text{ acre-ft}$$

Figure 10.22 Alternative mass diagram for storage analysis.

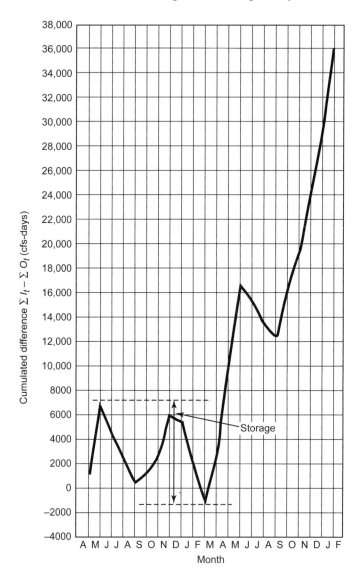

2. *Modified mass diagram.* The difference of accumulated inflows (column 7) and outflows (column 8) are computed in column 9 and plotted in Figure 10.22.

$$\text{Storage} = 8350 \text{ cfs-days or } 16{,}530 \text{ acre-ft}$$

3. *Analytical method.* The maximum difference between the highest value and the subsequent lowest value of the accumulated difference (col. 9) is the required storage.

$$\text{Storage} = 7068 - (-1262) = 8330 \text{ cfs-days or } 16{,}490 \text{ acre-ft}$$

Table 10.6   Data and Computations for the Reservoir Capacity

(1)	(2)	(3)	(4)	(5)	(6)	(7)	(8)	(9)
Month	Inflow (cfs)	Outflow (cfs)	Total Outflow[a] (cfs)	Inflow[b] Volume, I_t (cfs-day)	Outflow[c] Volume, O_t (cfs-day)	Cumulative Inflow ΣI_t (cfs-day)	Cumulative Outflow ΣO_t (cfs-day)	Difference $\Sigma I_t - \Sigma O_t$ (cfs-day)
Apr.	141	90	110	4,230	3,300	4,230	3,300	930
May	310	92	112	9,610	3,472	13,840	6,772	7,068
June	18	92	112	540	3,360	14,380	10,132	4,248
July	56	93	113	1,736	3,505	16,116	13,637	2,479
Aug.	40	90	110	1,240	3,410	17,356	17,047	309
Sept.	135	90	110	4,050	3,300	21,406	20,347	1,059
Oct.	160	90	110	4,960	3,410	26,366	23,757	2,609
Nov.	221	89	109	6,630	3,270	32,996	27,027	5,969
Dec.	85	89	109	2,635	3,379	35,631	30,406	5,225
Jan.	0	89	109	0	3,379	35,631	33,785	1,846
Feb.	0	91	111	0	3,108	35,631	36,893	−1,262
Mar.	241	90	110	7,471	3,410	43,102	40,303	2,799
Apr.	359	90	110	10,770	3,300	53,872	43,603	10,269
May	312	92	112	9,672	3,472	63,544	47,075	16,469
June	75	92	112	2,250	3,360	65,794	50,435	15,359
July	50	93	113	1,550	3,505	67,344	53,940	13,404
Aug.	82	93	113	2,542	3,505	69,886	57,445	12,441
Sept.	247	90	110	7,410	3,300	77,296	60,745	16,551
Oct.	198	90	110	6,138	3,410	83,434	64,155	19,279
Nov.	268	90	110	8,040	3,300	91,474	67,455	24,019
Dec.	266	89	109	8,246	3,379	99,720	70,834	28,886
Jan.	305	89	109	9,455	3,379	109,175	74,213	34,962

[a] Col. 3 + uniform release + evaporation.
[b] Col. 2 × number of days in respective months.
[c] Col. 4 × number of days in respective months.

10.11   STORAGE CAPACITY OF WATER SUPPLY TANKS

When widely fluctuating demands are imposed on a water supply distribution system, a distribution reservoir or a service tank is provided to accommodate an operating storage equivalent to a 24-hour demand on a maximum day. The requirements for firefighting purposes—sufficient to provide flow for 10 to 12 hours in large communities and for at least 2 hours in small communities—are added to this value. Emergency requirements for 2 or 3 days are also included. This is discussed in Section 1.8.

The operating storage is filled by a uniform 24-hour pumping at a rate of average hourly demand or at double this rate in 12 hours. In this case, the cumulated inflow is represented by a straight line corresponding to the uniform pumping into the tank and the cumulated outflow or demand is an undulating curve. To determine the storage, the construction of parallel lines on the mass curve is reversed [i.e., the lines are drawn at the lowest and highest points of the demand curve parallel to the cumulated inflow (pumping) plot]. The vertical distance between these lines is the storage. For 12-hour pumping, a parallel line on the demand curve is drawn at a point corresponding to the time at start of pumping and continued until pumping ceases. At the end of pumping time, the vertical distance between the demand curve and the parallel line is the required storage. For variable demands it is convenient to compute storage by plotting the difference of successive cumulated inflows and outflows.

EXAMPLE 10.14

The maximum daily requirement of a town is 6.18 MGD. The distribution of this requirement in a day similar to Figure 1.5 is given in column 2 of Table 10.7. Determine the required storage for **(a)** a uniform 24-hour pumping, and **(b)** a pumping period of 6 A.M. to 6 P.M.

SOLUTION

1. Referring to Table 10.7, column 3 accumulates the demands of column 2.

2. The 24-hour pumping rate $= \dfrac{\text{daily demand}}{\text{pumping period}} = \dfrac{6182.5(1000)}{24} = 257.6 \times 10^3$ gal/hr

3. The uniform pumping rate is shown in column 4. Column 5 accumulates the pumping rate of column 4.

4. The difference of the cumulated pumping and demands is given in column 6 and plotted in Figure 10.23(a).

5. Storage = maximum difference = 1319×10^3 gal or 1.32 MG.

6. The 12-hour pumping rate $= \dfrac{6182.5(1000)}{12} = 515.2 \times 10^3$ gal/hr

7. The pumping rate is shown in column 7 and the accumulated values of pumping are given in column 8.

8. The accumulated difference of pumping and demand ascertained in column 9 are plotted in Figure 10.23(b).

9. Storage = 2.57 MG.

10.12 RESERVOIR FEATURES

The storage capacity determined in Section 10.10 refers to the active storage that is required to meet the water demand for intended uses. An additional storage capacity known as the *dead storage* is provided to collect sediment and to maintain a minimum pool level. Also, if flood control is one of the purposes of the reservoir, planners provide extra space above the active storage to accommodate flood flows.

A reservoir section is shown in Figure 10.24. Many considerations are involved in the selection of a reservoir site. The suitability of the dam site and the adequacy of storage capacity are primary factors.

Table 10.7 Data and Computations for Tank Capacity [Gallons × 1000]

(1)	(2)	(3)	(4)	(5)	(6)	(7)	(8)	(9)
				24-hr Pumping			12-hr Pumping	
Time	Hourly Demand gal/hr ×1000	Cumulative Hourly Demand	Pumping Rate gal/hr ×1000	Cumulative 24-hr Pumping	Cumulative Difference (col.5 – col.3)	Pumping Rate gal/hr ×1000	Cumulative 12-hr Pumping	Cumulative Difference (col.8 – col.3)
12 A.M.	123.7	123.7	257.6	257.6	133.9	0	0	−123.7
1 A.M.	117.2	240.9	257.6	515.2	274.3	0	0	−240.9
2	113.4	354.3	257.6	772.8	418.5	0	0	−354.7
3	109.1	463.4	257.6	1030.4	567.0	0	0	−463.4
4	106.4	569.8	257.6	1288.0	718.2	0	0	−569.8
5	106.9	676.7	257.6	1545.6	868.9	0	0	−676.7
6	112.3	789.0	257.6	1803.2	1014.2	515.2	515.2	−273.8
7	196.0	985.0	257.6	2060.8	1075.8	515.2	1030.4	45.4
8	280.3	1265.3	257.6	2318.4	1053.1	515.2	1545.6	280.3
9	303.5	1568.8	257.6	2576.0	1007.2	515.2	2060.8	492.0
10	318.6	1887.4	257.6	2833.6	946.2	515.2	2576.0	688.6
11	326.2	2213.6	257.6	3091.2	877.6	515.2	3091.2	877.6
12	341.3	2554.9	257.6	3348.8	793.9	515.2	3606.4	1051.5
1 P.M.	347.8	2902.7	257.6	3606.4	703.7	515.2	4121.7	1219.0
2	344.0	3246.7	257.6	3864.0	617.3	515.2	4636.9	1390.2
3	341.3	3588.0	257.6	4121.6	533.6	515.2	5152.1	1564.1
4	342.4	3930.4	257.6	4379.2	448.8	515.2	5667.3	1736.9
5	358.6	4289.0	257.6	4636.8	347.8	515.2	6182.5	1893.5
6	395.3	4684.3	257.6	4894.4	210.1	0	6182.5	1498.2
7	504.0	5188.3	257.6	5152.0	−36.3	0	6182.5	994.2
8	449.3	5637.6	257.6	5409.6	−228.0	0	6182.5	544.9
9	272.7	5910.3	257.6	5667.2	−243.1	0	6182.5	272.2
10	138.8	6049.1	257.6	5924.8	−124.3	0	6182.5	133.4
11	133.4	6182.5	257.6	6182.5	0	0	6182.5	0
Total	6182.5		6182.5			6182.5		

10.13 DAMS

There are three common classification schemes for dams. According to the function performed, dams are classified into (1) storage dams for impounding water for developmental uses, (2) diversion dams for diverting streamflow into canals or other conveyance systems, and (3) detention dams to hold water temporarily to retard flood flows. From hydraulic design considerations, dams are classified as (1) overflow dams to carry discharge over their crests, and (2) nonoverflow dams, which are not designed to be overtopped. However, the most common classification is based on the materials of which dams are made. This

Figure 10.23 (a) Storage for 24-hour pumping; (b) storage for 12-hour pumping for Example 10.14.

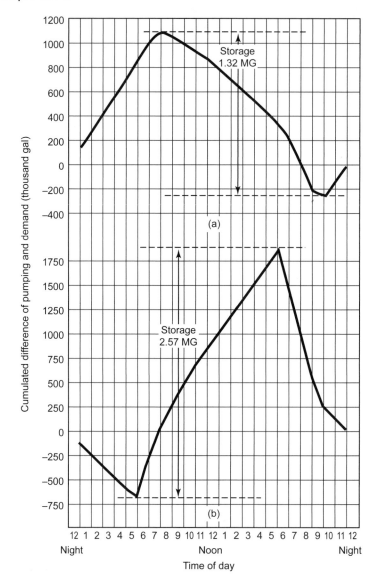

classification is further subclassified by recognizing the basic type of design, such as concrete gravity or concrete arch dams. Types of dams include:

1. Earthfill dams

2. Rockfill dams

3. Concrete dams

   a. Concrete gravity dams

   b. Concrete arch dams

   c. Concrete buttress dams

Figure 10.24 Reservoir features.

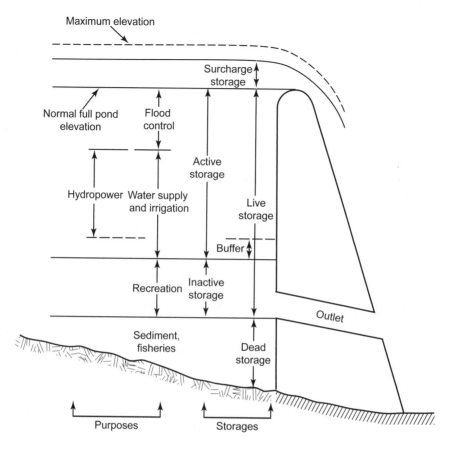

4. Stone masonry

   a. Stone-masonry gravity dams

   b. Stone-masonry arch dams

5. Timber dams

6. Steel coffer dams

10.13.1 Selection of Dam Type

Physical factors important in the choice of dam to be constructed are discussed briefly below. Topographically, a narrow stream section with high rocky walls suggests a suitable site for a concrete dam. Where the walls are strong enough to resist arch thrust, a concrete arch dam is adaptable. Low, rolling plains suggest an earthfill or rockfill dam.

When the geologic characteristics of the foundation are comprised of solid rock, any type of dam can be constructed, although concrete gravity and arch dams are favorable. Gravel foundations are suitable for earthfill, rockfill, and low concrete gravity dams. Silt and fine sand foundations are used to support earthfill and low concrete gravity dams but are not suitable for rockfill dams. Clay foundations in general are not suitable for the construction of dams. However, earthfill dams can be constructed with special treatment.

Availability of certain materials close to the dam site will reduce costs considerably if the type of dam selected utilizes these materials in sufficient quantity. Size, type, and natural restrictions in location of a spillway also influence the choice of dam. A large spillway requirement indicates the selection of a concrete gravity dam. A small spillway requirement favors the selection of an earthfill or rockfill dam. When the excavated material from a side channel spillway can be used in the dam embankment, an earthfill dam is advantageous. Side channel spillways can be used with any type of dam.

The factors listed above and others, such as the cost of diverting the stream, availability of labor, and traffic requirements on top of the dam, will usually favor one type over the others, but there is no unique choice for a given dam site. Several types have to be considered with their preliminary designs and estimates before making a final choice, mostly from cost considerations.

The structural stability of the embankment and the foundation are important considerations in dam design. Since all earth materials are pervious to some extent, the problem of water moving through the body and under the embankment is intimately associated with the design of earth dams, in addition to structural safety. The seepage flow through the embankment should be controlled so the amount lost does not interfere with the objective of the dam. There should be no erosion or sloughing of soil and the uplift pressure due to seepage underneath should not be large enough to cause piping.

10.14 SEEPAGE THROUGH AN EARTHFILL DAM

Consider a dam of homogeneous material situated on an impervious foundation. The seepage pattern through the dam is shown in Figure 10.25. This pattern is the same irrespective of the material (sand, clay, loam) of the dam. Of course, the rate of seepage depends on the type of soil. The emergence of seepage lines on the downstream slope tends to make the downstream face unstable. Either the downstream slope has to be made very flat, or the seepage must be diverted away from the downstream slope. The second alternative, being more economical, is favored and widely applied in dam design. The following two methods are used to divert the seepage line away from the downstream face:

1. A gravel filter (blanket) is provided on the base of the dam at the downstream side, to which the flow lines enter vertically. The seepage net for this case is shown in Figure 10.26. Cassagrande (1937) has shown that the phreatic line, which is the topmost seepage line, quite closely approximates a parabola.

2. An impervious core is provided. Since the outer shell material is several hundred times more pervious than the core, the shell has very little effect on the seepage lines. For seepage analysis only the central core section is considered, through which seepage lines may be drawn. The phreatic line for a central portion has been shown in Figure 10.27. The seepage line is a parabola in this method as well. However, it does not emerge at the lower end vertically as shown by the dotted line in Figure 10.27, but it tangentially meets the downstream face with an angle α. From the point D_0 on the parabola, the seepage line diverges to point D_1 by a distance Δa. As the angle α increases, the divergence from the true parabola becomes smaller. For $\alpha = 180°$, Δa becomes zero, and the two methods become the same.

The parabola intersects the water surface at A such that $AB = 0.3CB$ as shown in Figure 10.26 or Figure 10.27. Near the upstream face the phreatic line diverges from the parabola to join B perpendicularly.

Figure 10.25 Seepage pattern through a homogeneous section.

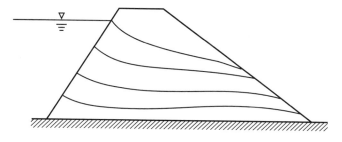

Figure 10.26 Seepage pattern through a homogeneous section with a filter blanket.

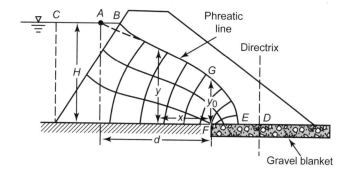

Figure 10.27 Central core portion of a composite section.

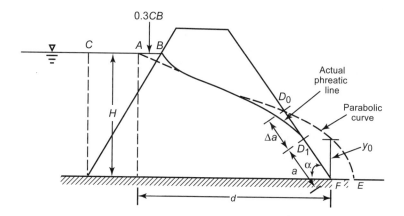

The focus of the parabola is F (Figure 10.26 or 10.27). The equation of the parabola with origin at focus F can be given by

$$x = \frac{y^2 - y_0^2}{2y_0} \quad [\text{L}] \tag{10.25}$$

where y_0 is shown in the figure.

Section 10.14 Seepage through an Earthfill Dam

At point A, $x = d$, and $y = H$. Substituting in eq. (10.25) yields

$$y_0 = \sqrt{H^2 + d^2} - d \quad [\text{L}] \tag{10.26}$$

For the location of the breakout point D_1, from F, an empirical formula is

$$a = \frac{y_0}{1 - \cos \alpha}\left(0.5 + \frac{\alpha}{360°}\right) \quad [\text{L}] \tag{10.27}$$

Another alternative approximate formula for a is

$$a = \sqrt{H^2 + d^2} - \sqrt{d^2 - H^2 \cot^2 \alpha} \quad [\text{L}] \tag{10.28}$$

The value of y_0 is determined from eq. (10.26) and $FE = 1/2\,y_0$. The parabolic curve is drawn by selecting distances x from focus F and determining y from eq. (10.25). The breakout point, D_1, is determined by eq. (10.27) or eq. (10.28) and a short transit curve from D_1 to the parabolic curve is drawn freehand. This is the phreatic line, which is the upper boundary of the flow net. The flow net is drawn according to the procedure of Section 4.1. The quantity of seepage is computed by eq. (4.1) or eq. (4.4).

For an approximate solution without reference to the flow net, apply Darcy's law at a section through D_1. The depth of the section is $a \sin \alpha$. The slope of the phreatic line is slightly flatter than the slope of the downstream face or approximately $\sin \alpha$ for α of less than 60°. Thus, for $\alpha < 60°$, by Darcy's law,

$$Q = KaL \sin^2 \alpha \quad [\text{L}^3\text{T}^{-1}] \tag{10.29}$$

where

a = breakout point from eq. (10.27) or (10.28)

L = length of dam

α = angle of downstream face

Where vertical and horizontal permeabilities differ, a transformed section of the embankment is used. All horizontal distances for flow net are distorted by a factor of $\sqrt{K_V/K_H}$, and a mean permeability value of $\sqrt{K_V K_H}$ is used in eq. (10.29).

For the horizontal drainage blanket (filter) in Figure 10.26, $\alpha = 180°$. For $\alpha > 60°$, including the horizontal blanket, the approximate seepage is given by

$$Q = Ky_0 L \quad [\text{L}^3\text{T}^{-1}] \tag{10.30}$$

where y_0 is given by eq. (10.26).

EXAMPLE 10.15

The section of a dam is shown in Figure 10.28(a). The center core has a permeability of 3×10^{-6} ft/sec in the horizontal direction and 1×10^{-6} ft/sec in the vertical direction. The core is symmetrical about the center of the dam. It has a top width of 30 ft and the two faces are at an angle of 40° from horizontal. Determine the seepage through the dam.

SOLUTION

1. The horizontal distances for the core are modified by a factor

$$\sqrt{K_V/K_H} = \sqrt{1 \times 10^{-6} / 3 \times 10^{-6}} = 0.58$$

The dimensions of the distorted core section are shown in Figure 10.28(b).

2. Revised slope angle $\tan \alpha = \dfrac{100}{69.1} = 1.45$ or $\alpha = 55.35°$

3. $CB = \dfrac{90}{\tan 55.35} = 62.2$ ft

   $AB = 0.3\, CB = 18.7$ ft and $CA = 0.7 CB = 43.5$ ft

   $d = (69.1 + 17.4 + 69.1) - 43.5 = 112.1$ ft

4. From eq. (10.26),

$$y_0 = \sqrt{(90)^2 + (112.1)^2} - 112.1 = 31.66 \text{ ft}$$

5. From eq. (10.27),

$$a = \frac{31.66}{1 - \cos 55.35}\left(0.5 + \frac{55.35}{360}\right) = 48 \text{ ft}$$

$$K = \sqrt{K_V K_H} = \sqrt{(3 \times 10^{-6})(1 \times 10^{-6})} = 1.73 \times 10^{-6} \text{ ft/sec}$$

6. From eq. (10.29) for $L = 1$ ft,

$$q = (1.73 \times 10^{-6})(48)\sin^2 55.35$$
$$= 5.6 \times 10^{-5} \text{ cfs or } 4.86 \text{ ft}^3/\text{day per foot}$$

10.14.1 Seepage under the Dam

Instead of the impervious foundation assumed in the preceding section, when the dam is located on a relatively pervious material such as alluvial sand or gravel, seepage occurs underneath the dam. If the upward seepage pressure of water near the toe is greater than the effective weight of the soil, the surface of the soil will rise at a point of least resistance, and water and soil will start flowing away from the dam. This is known as *piping* and can result in the sliding of the toe or the settling of the whole dam.

The submerged unit weight of soil of specific gravity G_s and void ratio e is given by

$$\gamma_{sub} = \frac{G_s - 1}{1 + e}\gamma_w$$

For a seepage line at a hydraulic gradient, i, the upward seepage force per unit volume is $i\gamma_w$. When the two forces are in balance,

$$i = \frac{G_s - 1}{i + e} \qquad \text{[dimensionless]} \qquad (10.31)$$

This is known as the *critical gradient*. Typically, the right side of eq. (10.31) has a value of unity. A gradient of slightly higher than one will cause piping or sand particles to be in an unstable condition known as *quicksand*. The actual gradient at the downstream end of the dam is evaluated from the flow net by dividing the head difference between the last two potential lines by the distance between these potential lines. This should be less than unity.

Figure 10.28 Composite anisotropic dam section.

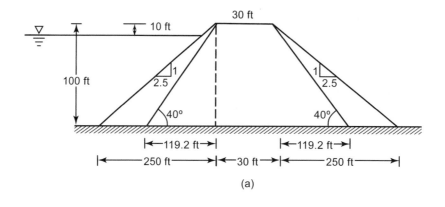

(a)

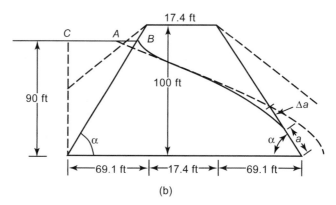

(b)

In an empirical approach, the creep ratio, L/H, is computed; here L is the length along the surface of contact between the soil and the base of the structure. This ratio is kept sufficiently large (4 for gravel and 18 for fine sand and silt) to prevent piping (Terzaghi and Peck, 1967).

When the pervious layer in the foundation is thin, one way to prevent the danger of piping is to excavate a trench to the depth of the impervious layer and to extend the center core of the dam into this trench. Alternatively, a row of sheet piles can be driven under the dam; or a combination of both can be employed.

When the pervious layer under the dam is too thick, so that a trench or pile cannot reach the impervious stratum, the upward pressure or exit gradient of the underneath seepage can be reduced by lengthening the seepage path. For this purpose, an impervious blanket is provided in front of the embankment. If the seepage pressure is still high despite the blanket, relief wells are also provided beyond the downstream toe of the dam. The approximate quantity of seepage under the dam (through the foundation) can be obtained by the application of Darcy's law.

EXAMPLE 10.16

The dam in Example 10.15 is located on a pervious stratum of 40 ft depth, having a permeability of 1.0×10^{-5} ft/sec. Determine the seepage under the dam. Assume that the average length of the seepage path is equal to the base of the dam.

SOLUTION

$$L = 530 \text{ ft}$$

$$i = \frac{H}{L} = \frac{90 \text{ ft}}{530 \text{ ft}} = 0.17$$

$$q = (1.0 \times 10^{-5} \text{ ft/s})(40 \text{ ft})(0.17)$$

$$= 6.8 \times 10^{-5} \text{ cfs/ft or } 5.88 \text{ ft}^3/\text{day per foot}$$

10.15 FLOW CONTROL STRUCTURES: SPILLWAYS

A spillway is a passageway to convey past the dam flood flows that cannot be contained in the allotted storage space or which are in excess of those released into the diversion system. Ordinarily, a reservoir is operated to release the required quantity for usage through headworks or through outlets in the dam. Thus, a spillway functions only infrequently, at times of floods or sustained high runoffs, when other facilities are inadequate. But its ample capacity is of prime importance for the safety of dams and other hydraulic structures. Determining the design flood for the spillway capacity, which comprises the peak, volume, and variation or a hydrograph of flow, is an important aspect of hydrology for which a reference is made to the study of flood flows (Chapter 8). After the hydrograph of the spillway's design flood is established, flood routing (Section 9.12) through a selected size and type of spillway establishes the maximum reservoir water level. Estimates of various combinations of spillway discharge capacity and reservoir height for alternative sizes and types of spillways provide a basis for selecting the most economical spillway. Since innumerable spillway arrangements could be considered, a judgment on the part of the designer is required to select only those alternatives that have adaptability to the site and show definite advantages.

The hydraulic aspects of spillway design extend beyond determining the inflow design flood and flood routing. These other considerations relate to design of the three spillway components: control structure, discharge channel, and terminal structure. The control structure regulates outflows from the reservoir and may consist of a sill, weir section, orifice, tube, or pipe. Design problems relate to determining the shape of the section and computing discharge through the section. The flow released through the control structure is conveyed to the streambed below the dam in a discharge channel. This can be the downstream face of the overflow section, a tunnel excavated through an abutment, or an open channel along the ground surface. The channel dimensions are fixed by the hydraulics of channel flow. An estimate of the loss of energy through the channel section is important in designing the terminal structure. Terminal structures are energy-dissipating devices that are provided to return the flow to the river without serious scour or erosion at the toe of the dam. These comprise a hydraulic jump basin, a roller bucket, a sill block apron, or a basin with impact baffles and walls. The hydraulic aspects relating only to control structures are discussed in this chapter.

10.15.1 Types of Spillways

Spillways are classified according to their most prominent feature as it pertains to the control structure or to the discharge channel. The two most common types of spillways are the concrete overflow spillway associated with gravity dams and the chute spillway often used with earthfill dams. For dams in narrow canyons, the spillway inlets are placed upstream of the dam in the form of either a side channel or a shaft (morning glory) spillway. Spillways are usually referred to as controlled or uncontrolled, depending on whether they are equipped with gates.

10.16 Overflow Spillways

This is a special form of weir whose shape is made to conform closely to the profile of the lower nappe of a ventilated sheet of water falling from a sharp-crested weir. In high-overflow spillways, the velocity of approach is negligible,* whereas low-overflow spillways have appreciable velocity of approach, which affects both the shape of the crest and the discharge coefficients. In low spillways the spillway crest curve is continuous with the toe curve, forming an S shape or ogee profile. However, the name *ogee spillway* is also applied to high spillways that have a straight tangent section between the crest curve and the toe curve.

10.16.1 Crest Shape of Overflow Spillways

The lower surface of a nappe from a sharp-crested weir is a function of (1) the head on the weir, (2) the slope or inclination of the weir face, and (3) the height of the crest, which influences the velocity of approach.

On the crest shape based on a design head H_d, when the actual head is less than H_d, the trajectory of the nappe falls below the crest profile, creating positive pressures on the crest, thereby reducing the discharge (coefficient). On the other hand, with a higher-than-design head, the nappe trajectory is higher than the crest, which creates negative pressure pockets and results in increased discharge. Accordingly, it is considered desirable to underdesign the crest shape of a high-overflow spillway for a design head, H_d[†], less than the head on the crest corresponding to the maximum reservoir level, H_e. However, with too much negative pressure, cavitation may occur. The U.S. Bureau of Reclamation (1987) recommendation has been that H_e/H_d should not exceed 1.33. Vacuum tank observations have indicated that cavitation on the crest would be incipient at an average pressure of about -25 ft of water. The Corps of Engineers has accordingly recommended that a spillway crest be designed so that the maximum expected head will result in an average pressure on the crest no lower than -15 ft of water (U.S. Army Corps of Engineers, 1986). Based on model studies by Murphy in 1970 and Maynord in 1985, design curves have been prepared that show a relationship between H_e and H_d for a pressure of -15 ft and -20 ft of water on the crest. The curves, corresponding to -15 ft pressure, can be approximated by the following equations (Reese and Maynord, 1987):

For $H_e \geq 30$ ft (9 m)

$$H_d = 0.33 H_e^{1.22} \quad \text{[without piers]} \tag{10.32a}$$

$$H_d = 0.30 H_e^{1.26} \quad \text{[with piers]} \tag{10.32b}$$

---

* The effect of the velocity of approach is negligible for a ratio of crest height to head on a weir greater than 1.33.
† Design head H_d is used to design the spillway profile. Discharge is computed with head H or H_e (without the velocity of approach head).

For $H_e < 30$ ft (9 m)

$$H_d = 0.7H_e \quad \text{[without piers]} \tag{10.32c}$$

$$H_d = 0.74H_e \quad \text{[with piers]} \tag{10.32d}$$

where

H_e = maximum total head on the crest

H_d = design head to be adopted

Numerous crest profiles have been proposed by various investigators. The U.S. Bureau of Reclamation played a leading role in investigations of the shape of the nappe. The Bureau described the complete shape of the lower nappe by separating it into two quadrants, one upstream and one downstream from the crest axis (apex), as shown in Figure 10.29. The equation for the downstream quadrant is expressed as

$$\frac{y}{H_d} = \frac{1}{K}\left(\frac{x}{H_d}\right)^n \quad \text{[dimensionless]} \tag{10.33}$$

where

H_d = design head excluding the velocity of approach head

x, y = coordinates of the crest profile, with the origin at the

highest point (O), as shown in Figure 10.29

K, n = constants that depend on upstream inclination and velocity of approach

Murphy (1973) suggested that n can be set equal to 1.85 for all cases and K can be varied from 2.0 for a deep approach to 2.2 for a very shallow approach, as shown in Figure 10.30(a).

In a high-overflow section, the crest profile merges with the straight downstream section of slope α, as shown in Figure 10.29 (i.e., $dy/dx = \alpha$). Differentiating eq. (10.33) and expressing that in terms of x yield the distance to the position of downstream tangent as follows:

$$\frac{X_{DT}}{H_d} = 0.485(K\alpha)^{1.176} \quad \text{[dimensionless]} \tag{10.34}$$

where

X_{DT} = horizontal distance from the apex to the downstream tangent point

α = slope of the downstream face

The discharge efficiency of a spillway is highly dependent on the curvature of the crest immediately upstream of the apex. To fit a single equation to the upstream quadrant had proven more difficult. Many compound curves have been proposed, including the tricompound circle for which the Bureau of Reclamation provided a tabular solution (1987). Investigations by Murphy (1973), as confirmed on model studies by Maynord (Reese and Maynord, 1987), suggested that an ellipse, of which both the major and minor axes vary systematically with the depth of approach, can closely approximate the lower nappe surfaces generated by the Bureau of Reclamation method. Furthermore, any sloping upstream

Figure 10.29 Definition sketch of overflow spillway section.

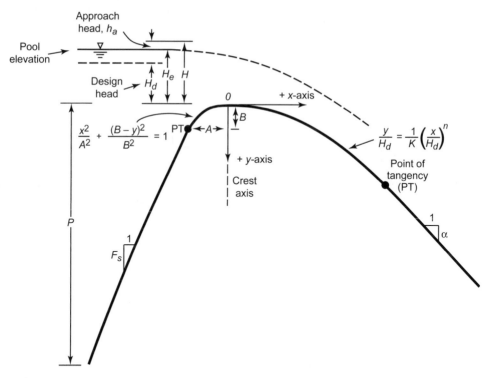

face could be used with little loss of accuracy if the slope became tangent to the ellipse calculated for a vertical upstream face.

With respect to origin at the apex, the equation of the elliptical shape for the upstream quadrant is expressed as

$$\frac{x^2}{A^2}+\frac{(B-y)^2}{B^2}=1 \quad [\text{dimensionless}] \tag{10.35}$$

where

$x =$ horizontal coordinate, positive to the right

$y =$ vertical coordinate, positive downward

$A, B =$ one-half of the ellipse axes, as given in Figure 10.30(b) and

(c) for various values of crest height and design head

For an inclined upstream face of slope F_s, the point of tangency with the elliptical shape can be determined by the following equation, obtained by differentiation of eq. (10.35):

$$X_{UT}=\frac{A^2 F_s}{\left(A^2 F_s^2 + B^2\right)^{1/2}} \quad [\text{L}] \tag{10.36}$$

where

$X_{UT} =$ horizontal distance of upstream tangent point

$F_s =$ slope of upstream face

Figure 10.30 Coordinate coefficients for spillway crest (from U.S. Army Corps of Engineers, 1986).

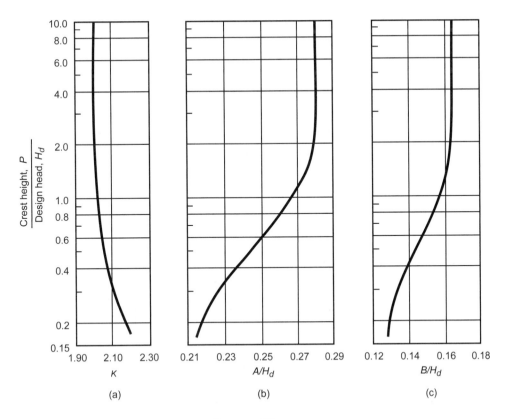

(a) (b) (c)

10.16.2 Discharge for Overflow Spillways

The following equation of flow through a weir with a consolidated coefficient, as derived in Section 10.4.2, applies to an overflow spillway.

$$Q = CL_e H^{3/2} \quad [\text{L}^3\text{T}^{-1}] \tag{10.37}$$

where

Q = discharge $\;[\text{L}^3\text{T}^{-1}]$

C = variable coefficient of discharge $\;[\text{L}^{1/2}\text{T}^{-1}]$

L_e = effective length of crest $\;[\text{L}]$

H = total head, including velocity of approach, h_a (Figure 10.29) $\;[\text{L}]$

Where crest piers and abutments cause side contractions of the overflow, the effective length is less than the crest length, as follows:

$$L_e = L - wN - 2(NK_p + K_a)H \quad [\text{L}] \tag{10.38}$$

where

L_e = effective length

L = length of crest

N = number of piers

w = width of each pier

K_p = pier contraction coefficient; K_p = 0.02 for square-nose piers, 0.01 for round-nosed piers, and 0 for pointed-nose piers (for details, see U.S. Army Corps of Engineers, 1986)

K_a = abutment contraction coefficient; K_a = 0.2 for square abutments, 0.1 for rounded abutments of radius larger than 0.15 times but less than one-half the total head, and 0 for rounded abutments of radius greater than one-half of the total head (for details, see U.S. Army Corps of Engineers, 1986)

The term $2(NK_p + K_a)$ is equivalent to $0.1n$ in eq. (10.7). The effect of the velocity of approach is negligible when the height of the crest is greater than 1.33 times the design head, H_d.

The discharge coefficient of overflow spillways is influenced by a number of factors. These include (1) crest height-to-head ratio or the velocity of approach, (2) the actual head being different from the design head, (3) the upstream face slope, and (4) the downstream submergence. When a spillway is being underdesigned, three graphs (Figure 10.31 through Figure 10.33) developed by the Bureau of Reclamation (1987) may be used to assess the coefficient of discharge, as follows:

1. Figure 10.31 provides the basic coefficient for the case when crest pressures are essentially atmospheric in a vertical-faced spillway. For the known ratio of the crest height to the design head, P/H_d, the basic coefficient, C_0, is determined from this figure. Use the scale on the right for metric units.

2. From Figure 10.32, the correction factor is determined for spillways having a sloping upstream face.

3. Figure 10.33 indicates the correction factor for the actual head being different than the design head. For the ratio of total head to design head (H/H_d), a factor is determined from this figure.

The basic coefficient multiplied by the two factors provides the corrected value of the coefficient.

EXAMPLE 10.17

Design an overflow spillway section for a design discharge of 50,000 cfs. The upstream water surface is at el. 800 and the channel floor is 680. The spillway, having a vertical face, is 180 ft long.

SOLUTION

1. Assume a high overflow spillway section ($C = 3.95$, from Figure 10.31 for $P/H_d \geq 3$).

2. From the discharge equation,

$$H^{3/2} = \frac{Q}{CL} = \frac{50,000}{3.95(180)} = 70.72 \quad \text{or} \quad H = 17.1 \text{ ft}$$

Figure 10.31 Basic discharge coefficient for vertical-faced section with the atmospheric pressure on the crest (U.S. Bureau of Reclamation, 1987).

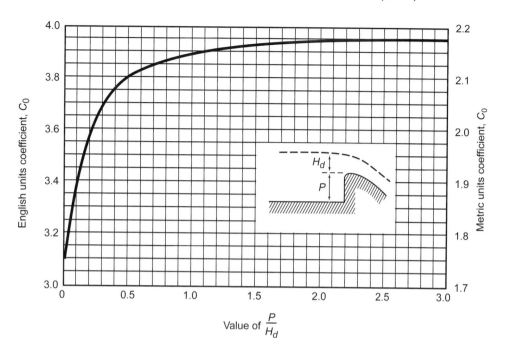

Figure 10.32 Correction factor for sloping upstream face (from U.S. Bureau of Reclamation, 1987).

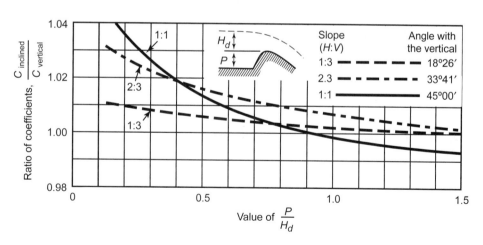

Figure 10.33 Correction factor for other than the design head (from U.S. Bureau of Reclamation, 1987).

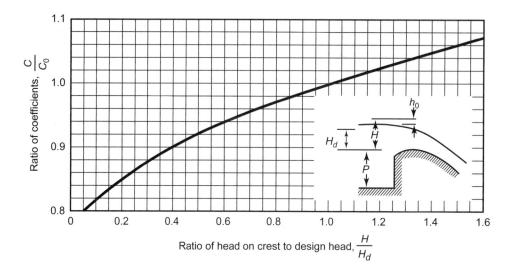

3. Depth of water upstream = 800 − 680 = 120 ft

$$\text{Velocity of approach, } v_0 = \frac{50{,}000}{120(180)} = 2.31 \text{ ft/sec}$$

$$\text{Velocity head } = \frac{(2.31)^2}{2(32.2)} = 0.08 \text{ ft}$$

4. Maximum water head, $H_e = 17.1 - 0.08 \approx 17.0$ ft
5. Height of crest, $P = 120 - 17.0 = 103$ ft
6. Since $H_e < 30$ ft, design head, $H_d = 0.7(H_e) = 0.7(17.1) = 12.0$ ft

7. $\dfrac{P}{H_d} = \dfrac{103}{12} = 8.58 > 1.33$, high overflow section.

8. *Downstream quadrant:* From eq. (10.33) of the crest shape,

$$\frac{y}{12} = \frac{1}{2}\left(\frac{x}{12}\right)^{1.85}$$

or

$$y = 0.06\,x^{1.85}$$

Coordinates of the shape computed by this equation are as follows:

x (select)(ft)	y (computed)(ft)
5	1.18
10	4.25
15	9.00
20	15.30
30	32.40

9. *Point of tangency:* Assume a downstream slope of 2:1. From eq. (10.34),

$$X_{DT} = 0.485[2(2)]^{1.176}(12) = 30 \text{ ft}$$

10. *Upstream quadrant:* From Figure 10.30(b) and (c), $A/H_d = 0.28$, $B/H_d = 0.165$.

$$A = 0.28(12) = 3.36 \text{ ft} \quad B = 0.165(12) = 2.00 \text{ ft}$$

From eq. (10.35),

$$\frac{x^2}{(3.36)^2} + \frac{(2.0 - y)^2}{(2.0)^2} = 1$$

The coordinates are computed as follows:

x (select)(ft)	y (computed)(ft)
1.0	0.09
2.0	0.39
3.0	1.10
3.36	2.00

11. The crest shape has been plotted in Figure 10.34.

10.16.3 Discharge on Submerged Overflow Spillways

The coefficient of discharge decreases under the condition of submergence. Submergence can result from either excessive tailwater depth or changed crest profile. The effect of tailwater submergence on the coefficient of discharge depends upon the degree of submergence defined by h_d/H_e and the downstream apron position, $(h_d + d)/H_e$, shown in Figure 10.35 (inset). For a value of $(h_d + d)/H_e$ up to 1.7 (approximately 2), the reduction in the coefficient depends on the factor $(h_d + d)/H_e$ and is independent of h_d/H_e as shown in Figure 10.35(a), i.e., it is subject to apron effects only. When $(h_d + d)/H_e$ is above 5, the reduction depends only on h_d/H_e, and is independent of $(h_d + d)/H_e$ as shown in Figure 10.35(b), i.e., tailwater effects control. For $(h_d + d)/H_e$ between 2 and 5, the reduction of the coefficient depends on both factors, given in Figure 10.35(c). The percent decreases in the coefficient are with respect to the unsubmerged (free discharge) condition. The effect on the discharge due to crest geometry is not well defined. A reduction of 2 to 8% has been noticed when the chute tangent intersects too close to the crest. Model studies are the best way to determine the coefficient.

Figure 10.34 Design of a spillway section (Example 10.17).

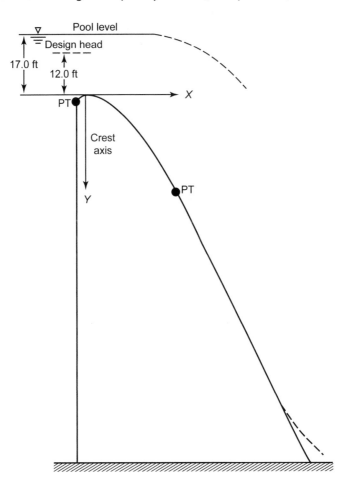

Pool level

Design head

17.0 ft

12.0 ft

PT

Crest
axis

Y

PT

X

EXAMPLE 10.18

Determine the length of an overflow spillway to pass 2000 ft³/s with a depth of flow upstream not to exceed 5 ft above the crest. The spillway is 8 ft high. The upstream face is sloped 1:1. For 2000 ft³/s, the tailwater rises 3.5 ft above the crest. The spillway is designed for the maximum head. The abutment edge adjacent to the crest is rounded to a radius of 2 ft.

SOLUTION

1. Spillway maximum head, $H_e = 5$ ft (without the approach velocity head).

2. See Figure 10.36. From the figure, $h_d = 5 - 3.5 = 1.5$ and $h_d + d = 13$. Thus,

$$\frac{(h_d + d)}{H_e} = \frac{13}{5} = 2.6, \quad \frac{h_d}{H_e} = \frac{1.5}{5} = 0.3$$

From Figure 10.35(c), % reduction = 4%

Figure 10.35 Reduction of discharge coefficient for submerged spillway: (a) apron effects, when $(h_d + d)/H_e < 2$, (b) tailwater effects, when $(h_d + d)/H_e > 5$ (modified from ASCE, 1995).

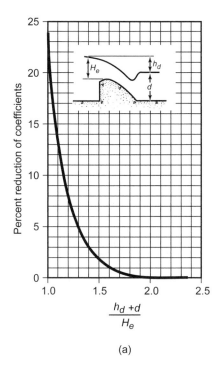

(a)

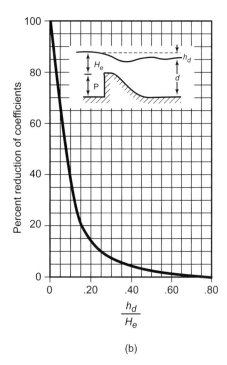

(b)

3. Since the spillway is designed for maximum head, $H_d = H_e = 5$ ft

   $P/H_d = 8/5 = 1.6$

   From Figure 10.31, $C_0 = 3.93$

   From Figure 10.32, $\dfrac{C_{\text{inclined}}}{C_0} = 0.993$

4. Hence $C = (3.93)(0.993)(96/100) = 3.75$

5. Approximate discharge/ft, $q = C\,H_e^{3/2} = (3.75)(5)^{3/2} = 41.93$ cfs/ft

   Velocity of approach, $v_0 = \dfrac{41.93}{13} = 3.22$ ft/s

   Approach velocity head $= \dfrac{v_0^2}{2g} = \dfrac{(3.22)^2}{2(32.2)} = 0.16$ ft

6. Revised computations with velocity of approach:

   $H_e = 5 + 0.16 = 5.16$ ft [this is total head, H, for eqs. (10.37) and (10.38)]

Figure 10.35 (Continued) (c) Reduction of discharge coefficient for submerged spillway when $(h_d + d)/H_e$ is between 2 and 5. Figure 10.35 (a), (b), (c) used with permission of ASCE and U.S. Army Engineers Waterways Experiment Station.

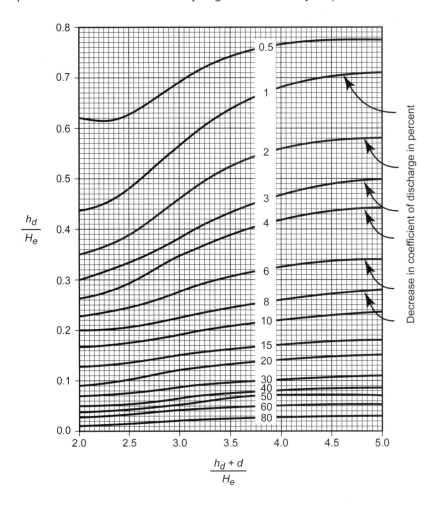

Figure 10.36 Submerged crest spillway of Example 10.18.

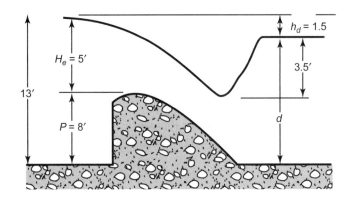

7. The revised value of $\dfrac{P}{H_e}$ will not alter the coefficients from Figures (10.31) and (10.32).

8. $\dfrac{H_d + d}{H_e} = \dfrac{13.16}{5.16} = 2.55$

$\dfrac{h_d}{H_e} = \dfrac{1.66}{5.16} = 0.32$

From Figure 10.35(c), % reduction = 3.5%

9. Revised $C = (3.93)(0.993)(96.5/100) = 3.77$

10. From eq. (10.37), $2000 = (3.77)L_e(5.16)^{3/2}$

$$L_e = 45.3 \text{ ft}$$

11. Since the abutment walls rounded to a radius larger than 0.15(5.16) or 0.77 ft, $K_a = 0.1$

From eq. (10.38),

$$L = L_e + 2Ka\, H_e$$
$$= 45.3 + 2(0.1)(5.16) = 46.3 \text{ ft}$$

10.17 CHUTE OR TROUGH SPILLWAYS

The chute spillway derives its name from the shape of the discharge channel component of the spillway. In this type of spillway, the discharge is conveyed from a reservoir to the downstream river level through a steep open channel placed either along the dam abutment or through a saddle. The designation of the channel as a chute implies that the velocity of flow is greater than critical. This name applies regardless of the control device used at the head of the chute, which can be an overflow crest, a gated orifice (acting as a large orifice), or a side-channel crest. This type of structure consists of four parts as shown in Figure 10.37: an entrance channel, a control structure or crest, the sloping chute, and a terminal structure. The entrance channel at *A* is a relatively wide channel of subcritical flow. More often, the axis of the entrance channel is curved to fit the alignment to the topography, because of the low-energy losses in the approach channel. The control section at *B* is placed in line with or upstream from the centerline of the dam. The critical velocity occurs when the water passes over the control.

Flows in the chute are ordinarily maintained at supercritical stage until the terminal structure *DE* is reached. Economy of excavation generally makes it desirable that from *B* to *C*, where a heavy cut is involved, the chute may be placed on a light slope. From *C* to *D* it follows the steep slope on the side of the river valley. An energy-dissipating device is placed at the bottom of the valley at *D*. The axis of the chute is kept straight as far as practicable; otherwise, the floor has to be superelevated to avoid piling up of the high-velocity flow around the curvature. The velocity of flow increases rapidly in the chute with drop in elevation. The curvature should be confined to the upper reaches, where the velocity is comparatively low. In the lower reaches, the alignment can be curved only if the chute floor and walls are shaped adequately to force water into a turn without overtopping the walls.

It is preferable that the width of the control section, the chute, and the stilling basins are the same. Quite often, these widths are not the same, because of the design requirements of the spillway and stilling basin. Extreme care must be taken that the transitions take place very gradually, or undesirable standing waves may develop.

Figure 10.37 Chute spillway section.

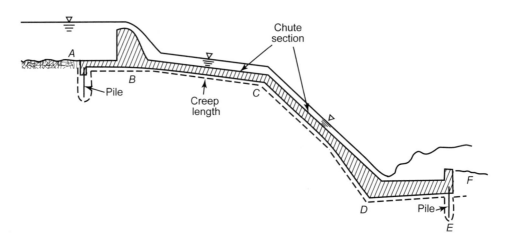

To prevent hydrostatic uplift under the chute, a cutoff wall is provided under the control structure and a drainage system of filters and pipes is provided, as shown in Figure 10.37. When the stilling basin is operating, there is a substantial uplift under the lower part of the chute and upstream part of the stilling basin floor. The floor must be made sufficiently heavy or be anchored to the foundation.

10.17.1 Slope of Chute Channel

It is important that the slope of the chute in the upstream section BC should be sufficiently steep to maintain a supercritical flow to avoid formation of a hydraulic jump in the chute. Flow through a channel is given by Manning's formula, derived in Section 11.6.1.

For a rectangular channel of depth y and width b, $A = by$ and $R \approx y$ for a wide section. Under the condition of critical flow, $y_c = (q^2/g)^{1/3}$, where $q = Q/b$ (i.e., discharge per foot width). Manning's equation in terms of slope reduces to

$$S_c = \frac{21.3n^2}{q_c^{0.222}} \quad \text{(English units)} \quad \text{[dimensionless]} \tag{10.39a}$$

$$S_c = \frac{12.6n^2}{q_c^{0.222}} \quad \text{(metric units)} \quad \text{[dimensionless]} \tag{10.39b}$$

where

q_c = critical disharge per unit width

S_c = critical slope of channel

n = roughness coefficient given in Table 11.4

Since the reliable information on the value of n is difficult, a conservative approach is indicated in the selection of n. The slope of the chute should be more than S_c for a supercritical flow. As the spillway must function properly from small to very large discharges, the critical slope has to be investigated for the entire range of flow. The required S_c from eq. (10.39) is normally a fraction of 1%.

EXAMPLE 10.19

Determine the minimum slope in the upper reach of a chute section of 100 ft width. The range of discharge is 5000 to 70,000 cfs. $n = 0.015$.

SOLUTION

1. Under minimum flow conditions, $q = \dfrac{5000}{100} = 50$ ft^3/s/ft.

2. From the slope equation (10.39a),

$$S_c = \frac{21.3(0.015)^2}{(50)^{0.222}} = 0.002$$

10.17.2 Chute Sidewalls

Except for converging and diverging sections, chute channels are designed with parallel vertical sidewalls, commonly of reinforced concrete 12 to 18 in. thick. The sidewalls are designed as the retaining walls. The height of the walls is designed to contain the depth of flow for the spillway design flood. The water surface profile from the control section downward is determined for this purpose. An allowance is made for pier end waves, roll waves, and air entrainment.

In view of uncertainties involved in the evaluation of surface roughness, pier end waves, roll waves, and air entrainment buckling, a freeboard given by the following empirical equation is added to the computed depth of the water surface profile.

$$\text{Freeboard (ft)} = 2.0 + 0.002\bar{V}d^{1/3} \quad \text{[unbalanced]} \quad \text{(English units)} \qquad (10.40)$$

where

$\bar{V}$ = mean velocity in chute section under consideration, ft/sec

d = mean depth, ft

The converging and diverging transitions in the sidewalls must be gradual. Bureau of Reclamation experiments have shown that an angular variation (flare angle) not exceeding the following value will provide an acceptable transition for either a contracting or an expanding channel.

$$\tan \alpha = \frac{1}{3\text{Fr}} \quad \text{[dimensionless]} \qquad (10.41)$$

where

α = angular variation of the sidewall with respect to the channel centerline

Fr = Froude number = $V/\sqrt{gy}$

V, y = averages of the velocities and depths at the beginning and end of the transition

10.18 SIDE-CHANNEL SPILLWAYS

In side-channel spillways, the overflow weir is placed along the side of the discharging channel, so that the flow over the crest falls into a narrow channel section (trough) opposite the weir, turns a right angle, and then continues in the direction approximately parallel to the weir crest. A plan and a cross section are shown in Figure 10.38.

A modification to the conventional side channel includes addition of a short crest length perpendicular to the channel at the upstream end resulting in an L-shaped crest. The angularity of flow at the junction of two crest sections causes the loss in effective crest length. To correct the effective length of the L-shaped crest, a graph is used which is a function of the head to crest-height ratio and head to design-head ratio (ASCE, 1995).

This type of spillway is adaptable to certain special conditions, such as when a long overflow crest is desired but the valley is narrow, or where the overflows are most economically passed through a deep narrow channel or a tunnel. The crest of the spillway is similar to an overflow or other ordinary weir section. Downstream from the side channel trough, a control section is achieved by constricting the channel sides or elevating the channel bottom to induce the critical flow. Flows upstream from this section are at subcritical stage. Downstream from this section functions similar to a chute-type spillway. Thus the side-channel design is concerned only with the hydraulic action in the trough upstream of the control section, where spatially varied flow takes place. Flow in the trough should be at a sufficient depth to carry away the accumulated flow and not to submerge the flow over the crest. The hydraulic aspect is concerned with the water surface profile in the trough, which is determined from the momentum principle (not by the energy principle as applied in the gradual flow in Section 11.12 because of excessive energy losses due to high turbulence). A trapezoidal section is a common section for the side-channel trough. The bottom width is kept to a minimum. The trough is placed on the rock foundation and a concrete lining is provided. In Figure 10.39, consider a short reach of distance Δx. The rate of change of momentum in the reach is equal to the external force acting in the reach. The momentum principle is discussed in Section 9.4.

$$\text{Rate of momentum at upper section } = \frac{\gamma}{g} Q_1 v_1 \qquad \text{(a)}$$

$$\text{Rate of momentum at lower section } = \frac{\gamma}{g} Q_2 v_2 \qquad \text{(b)}$$

$$\text{Change of rate of momentum } = \frac{\gamma}{g}\left(Q_2 v_2 - Q_1 v_1\right) \qquad \text{(c)}$$

External forces include hydrostatic pressures, weight component in the direction of flow, and the friction in the channel section. Disregarding the weight component and the channel friction that tend to compensate each other, the resultant hydrostatic force

$$P_2 - P_1 = \gamma \bullet A \bullet dy$$

Representing the average area of cross section $A = (Q_1 + Q_2)/(v_1 + v_2)$

$$\text{Hydrostatic force} = \gamma \left(\frac{Q_1 + Q_2}{v_1 + v_2}\right)\Delta y \qquad \text{(d)}$$

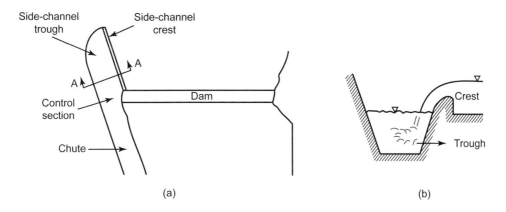

(a)   (b)

Figure 10.39   Analysis of side-channel flow.

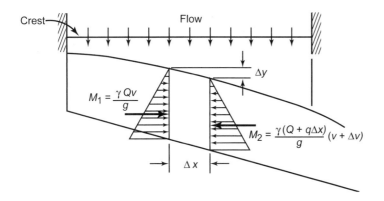

Equating the two forces (c) and (d) and rearranging, the change in water elevation can be expressed:

$$\Delta y = \frac{-Q_2}{g} \frac{v_1+v_2}{Q_1+Q_2}\left[\left(v_2-v_1\right)+\frac{v_1\left(Q_2-Q_1\right)}{Q_2}\right] \quad [L] \tag{10.42}$$

Equation (10.42) is solved by a trial-and-error procedure. For a reach of length Δx, Q_1 and Q_2 are known. Starting from the control point where the critical depth exists, a trial depth at the other end is found which will satisfy the equation.

EXAMPLE 10.20

Design a side-channel trough for a spillway of 100 ft length for a maximum discharge of 2500 cfs. The side-channel trough has a length of 100 ft and a bottom slope of 1 ft in 100 ft. A control section of 10 ft width is placed downstream from the trough with the bottom of the control at the same elevation as the bottom of the trough floor at the downstream end.

SOLUTION

1. Critical depth at the control section, $y_c = \left(\dfrac{q_c^2}{g}\right)^{1/3}$

2. $q_c = \dfrac{2500}{10} = 250$ cfs/ft

3. $y_c = \left(\dfrac{250^2}{32.2}\right)^{1/3} = 12.44$ ft

4. $v_c = \dfrac{q_c}{y_c} = \dfrac{250}{12.44} = 20.1$ ft/sec

5. Velocity head, $h_c = \dfrac{v_c^2}{2g} = \dfrac{(20.1)^2}{2(32.2)} = 6.27$ ft

6. For a side-channel trough, assume a trapezoidal section with 2 (vertical): 1 (horizontal) slope and a 10-ft bottom width. Also, assume that the transition loss from the end of the side-channel trough to the control section is equal to 0.2 of the difference in velocity heads between the ends of the transition.

7. The following energy equation may be written between the trough end and the control section, where the subscript c refers to "critical" and "100" refers to the distance of the trough from the upstream end of the spillway (i.e., at the end of the trough):

$$y_{100} + h_{100} = y_c + h_c + 0.2(h_c - h_{100})$$

or

$$y_{100} + 1.2h_{100} = 12.44 + 1.2(6.27) = 19.96 \text{ ft} \qquad (10.43)$$

Equation (10.43) is solved by trial and error. Assume that $y_{100} = 19.2$ ft; then $A = 376.3$ ft^2, $v = Q/A = 2500/376.3 = 6.64$ ft/sec, and $h_{100} = (6.64)^2/(2)(32.2) = 0.68$ ft and $1.2h_{100} = 0.8$ ft. Thus eq. (10.43) is satisfied.

8. The known values above relate to section 1 at the downstream end of the trough. Section 2 is taken at the upper end of a selected increment Δx, 25 ft in this case. A value of the change in water level, Δy, is assumed for the reach, all terms of eq. (10.42) are evaluated, and Δy is computed, as illustrated in Table 10.8. The assumed and computed values of Δy should match; otherwise, a new value is assumed for Δy.

9. The process is repeated until the upstream end of the channel is reached.

10.19 MORNING GLORY OR SHAFT SPILLWAYS

This type of spillway consists of four parts: (1) a circular weir at the entry, (2) a flared transition conforming to the shape of the lower nappe of a sharp-crested weir, (3) a vertical drop shaft, and (4) a horizontal or near-horizontal outlet conduit or tunnel. This spillway is used at dam sites in narrow canyons or where a diversion conduit/tunnel of the dam is

Table 10.8   Side-Channel Spillway Computations[a]

(1)	(2)	(3)	(4)	(5)	(6)	(7)	(8)	(9)	(10)	(11)	(12)	(13)	(14)
Δx (select)	Bottom Level[b]	Δy (Assume)	Water Level[c]	y (col. 4 – col. 2)	A[d]	$Q^e =$ $q(L - \Sigma\Delta x)$	$v = Q/A$	$Q_1 + Q_2$[f]	$Q_2 - Q_1$[g]	$v_1 + v_2$[h]	$v_2 - v_1$[i]	Δy Computed eq. (10.42)	Remarks on Assumed Δy
D/S end													
	100.0		119.2	19.20	376.3	2500	6.64						
25	100.25	1.0	120.2	19.95	398.5	1875	4.71	4375	– 625	11.35	–1.93	0.63	High
		0.62	119.82	19.57	387.2		4.84	4375	– 625	11.48	–1.80	0.61	ok
25	100.50	0.50	120.32	19.82	394.6	1250	3.17	3125	– 625	8.01	–1.67	0.41	High
		0.40	120.22	19.72	391.6		3.19	3125	– 625	8.03	–1.65	0.41	ok
25	100.75	0.25	120.47	19.72	391.6	625	1.60	1875	– 625	4.79	–1.59	0.24	ok
15	100.90	0.10	120.57	19.67	390.2	250	0.64	875	– 375	2.24	– 0.96	0.07	High
		0.07	120.54	19.64	389.3		0.64	875	– 375	2.24	– 0.96	0.07	ok

[a] $q = Q/L = 2500/100 = 25$ cfs/ft, bottom slope = 1 in 100 ft given.
[b] (Slope × channel length) + datum.
[c] Final water level at the preceding station (section) + assumed Δy. At start it is datum + y_{100}.
[d] Area of cross section of trough computed for depth y in column 5. In this case $A = (10 + 0.5y)y$.
[e] $q(L - \Sigma\Delta x)$, L = crest length, $\Sigma\Delta x$ = total of Δx up to the current step.
[f] Col. 7 + value in col. 7 at the preceding station (step).
[g] Col. 7 – value in col. 7 at preceding step.
[h] Col. 8 + col. 8 at preceding step.
[i] Col. 8 – col. 8 at preceding step.

available to be utilized as an outlet conduit. As the head increases, the control shifts from weir crest, to drop shaft, to outlet conduit. Three control conditions are indicated in Figure 10.40 and listed in Table 10.9.

Table 10.9 Discharge Characteristics of Shaft Spillway

Control Point	Condition	Characteristics	Relation
Crest	Unsubmerged flow	Weir flow	$Q = CLH^{3/2}$ (10.44)
Throat of drop shaft	Partially submerged	Orifice flow	$Q = C_d A_1 \sqrt{2gH_a}$ (10.45)
			$C_d = 0.95$
Downstream of outlet conduit	Submerged flow	Pipe flow	$Q = A_2 \sqrt{\dfrac{2gH_T}{\Sigma K}}$ (10.46)
			ΣK = loss coefficients through pipe flow

Condition 1 of a free-discharging weir prevails as long as the nappe converges into the shape of a solid jet. Under condition 2, the weir crest is drowned out. The Bureau of Reclamation (1987) indicated that this condition is approached when $H_d/R_s > 1$, where H_d is the design head and R_s is the radius of the crest. Further increase in head leads to condition 3 when the spillway is flooded out, showing only a slight depression and eddy at the surface. The relevant discharge relations under the three conditions are given by equations (10.44), (10.45), and (10.46), shown in Table 10.9. Under condition 3, the head rises rapidly for a small increase in discharge. Thus the design is not recommended under this condition (i.e., under the design head the outlet conduit should not flow more than 75% full). Thus the discharge through a shaft spillway is limited. The bureau suggested that the following weir formula may be used for the flow through the shaft spillway entrance regardless of the submergence, by adjusting the coefficient to reflect the flow conditions (U.S. Bureau of Reclamation, 1987, p. 407).

$$Q = C(2\pi R_s)H^{3/2} \quad [\text{L}^3\text{T}^{-1}] \tag{10.47}$$

Figure 10.40 Flow condition of a shift spillway: (a) condition 1: crest control; (b) condition 2: tube control; (c) condition 3: pipe control.

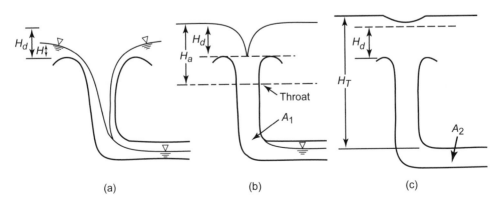

(a)   (b)   (c)

Figure 10.41 Relation for circular crest coefficient (from U.S. Bureau of Reclamation, 1987).

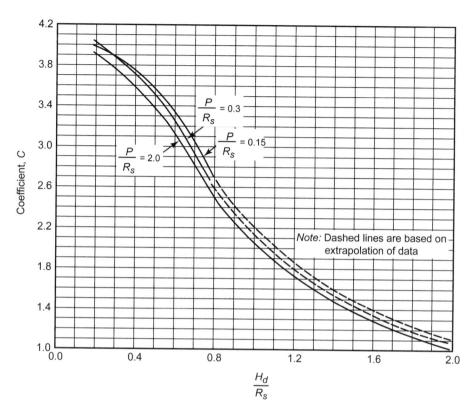

where

C = discharge coefficient related to H_d/R_s and P/R_s from model
tests, where H_d = design head and P = crest height from the
outlet pipe; values of C are given in Figure 10.41

R_s = radius of the circular crest

H = head over the weir

Alternatively, eq. (10.47) can be used to determine the crest size (radius), R_s, for a given design discharge under the maximum head. This is a minimum radius required when small subatmospheric pressures along the overflow crest can be tolerated. The Bureau of Reclamation (1987) has provided the tables to compute the crest profile. Similarly, eq. (10.45) can be used to determine the shape of the transition (drop shaft) that is required to pass the design discharge with the maximum head over the crest.

EXAMPLE 10.21

A shaft spillway is to discharge 2000 cfs under a design head of 10 ft. Determine the minimum size of the overflow crest. Also determine the shape of the transition if the control section is 4 ft below the crest level.

Section 10.19   Morning Glory or Shaft Spillways   599

1. Since the coefficient C is related to P and R_s, assume that $P/R_s > 2$ and determine R_s by trial and error.

2. Try $R_s = 7$ ft. Thus,

$$\frac{H_d}{R_s} = \frac{10}{7} = 1.43$$

From Figure 10.41, $C = 1.44$.

3. From eq. (10.47),

$$Q = C\left(2\pi R_s\right)H_d^{3/2}$$
$$= 1.44(2\pi)(7)(10)^{3/2} = 2002 \text{ cfs}$$

This is practically the same as the required discharge. Hence the crest radius = 7 ft

4. Depth at the beginning of the control section from the water surface:
$$H_a = 10 + 4 = 14 \text{ ft}$$

5. From eq. (10.45),

$$Q = 0.95\left(\pi R^2\right)\sqrt{2gH_a}$$

$$R^2 = \frac{2000}{0.95\pi\sqrt{2gH_a}}$$

or

$$\text{Drop shaft radius } R = \frac{9.14}{H_a^{1/4}}$$

H_a (select)(ft)	R (ft)
14	4.73
16	4.57
18	4.44
20	4.32

PROBLEMS

10.1 A tank has a rectangular orifice of 3 ft length and 2 ft height. The top edge of the orifice is 8 ft below the water surface. Determine the discharge through the orifice, treating it as (a) a small orifice, and (b) a large orifice. $C_d = 0.62$.

10.2 Determine the diameter of the orifice located at the bottom of a tank if the water level in the tank is kept at 15 ft above the orifice and the discharge is 20 cfs. $C_d = 0.6$.

10.3 Discharge from a sharp-edged orifice of 50 mm diameter is 9 liters/second. (a) What is the height of water over the orifice. (b) If the orifice is replaced by a free jet Borda type of mouthpiece, what is the discharge?

10.4 Under a constant head of 5 meters, the discharge from a 100 mm orifice is 0.05 m³/s. When an external mouthpiece is inserted, the discharge increases to 0.07 m³/s. The only coefficient known is that $C_c = 1$ for the mouthpiece. The coefficient of velocity remains unchanged. Determine the coefficients of contraction, velocity, and discharge for the orifice.

10.5 An orifice located at the bottom of a tank has a diameter of 14 in. The water level in the tank is 15 ft above the orifice. How much time will it take to empty the tank if the area of cross section of the tank is 10 ft × 10 ft? $C_d = 0.6$.

10.6 A portion of a swimming pool has one side vertical and the other tapered as shown in Figure P10.6. It has a length (perpendicular to the paper) of 6 m. An orifice of 150 mm diameter at the bottom of the vertical section is opened. How long will it take to empty the pool section shown in the figure? $C_d = 0.65$.

Figure P 10.6   Pool section

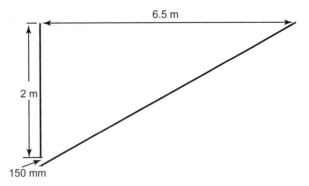

10.7 A 2.5 ft high bucket has a diameter of 2 ft at the top and 1 ft at the bottom. The bottom has a hole of 2 in. diameter. How long will it take to empty a full bucket? $C_d = 0.61$.

10.8 A cylindrical container of 2.2 m diameter and 4 m height has an inlet at the top and a 400 mm diameter outlet at the bottom. Discharge at a constant rate of 0.2 m³/s enters through the inlet. If the bottom orifice is opened when the container is full, how long will it take for the container to empty halfway? $C_d = 0.6$.

10.9 A suppressed sharp-crested weir 15 ft long has a crest of 3 ft height. Determine the discharge for a head over the crest of 25 in. Account for the velocity of approach.

10.10 Determine the length of a sharp-crested weir required to discharge 4000 cfs at a head of 2.25 ft. The weir is divided into sections by vertical posts of 2 ft width and 10 ft clear distance. The approach channel has a width of 600 ft and a depth of 3.5 ft. $C_d = 0.61$.

10.11 Determine the discharge through the stepped notch shown in Figure P10.11. $C_d = 0.6$. [*Hint:* Compute the discharge for each section (step) separately and add. For example, the head variation, and hence the range of integration on the middle section, is from 0.75 ft to 1.75 ft.]

Figure P10.11 Stepped notch

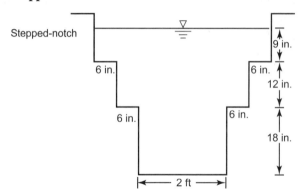

Stepped-notch

6 in. 6 in. 9 in. 12 in. 6 in. 6 in. 18 in.

|← 2 ft →|

10.12 Water passes over a rectangular weir of 10 ft width at a depth of 1 ft. If the weir is replaced by an 80° V-notch, determine the depth of water over the notch. Disregard the end contractions. C_d for notch = 0.59, C_d for rectangular weir = 0.63.

10.13 Determine the side slopes of a trapezoidal weir of weir crest length L and head H so that the discharge through it (taking into consideration the end-contraction effect on the rectangular section) is the same as that through a suppressed rectangular weir of length L and head H.

10.14 A trapezoidal weir has a length of 19 ft with side slopes of 1 horizontal to 2 vertical. What is the head over the weir for a flow of 100 cfs? $C_d = 0.62$.

10.15 A channel of 10 ft width and 3 ft depth is installed with a sharp-crested weir of crest height 1.8 ft and length 8.5 ft. The flow depth over the weir is 1.2 ft. The rectangular weir is to be replaced by a Cipolletti weir. Determine the crest length for the new weir if other conditions remain unchanged. Correct for the velocity of approach in rectangular and Cipolletti weirs. $C_d = 0.6$ for a rectangular weir.

10.16 A submerged weir in a pond is 10 ft long. The crest of the weir is 9 in. below the upstream level and 6 in. below the tailwater level. The crest height is 1 ft. Determine the discharge. Determine C_d from eq. (10.5).

10.17 A stream is 200 ft wide and 10 ft deep. It has a mean velocity of flow of 4 ft/sec. If a submerged weir of 8 ft height is installed, how much will the water upstream rise? Determine C_d from eq. (10.5). Disregard the velocity of approach. [*Hint:* Assume C_d, determine H_1 for the submerged case, calculate revised C_d from the formula, and find revised H_1.]

10.18 Determine the discharge over a broad-crested weir 6 ft wide and 100 ft long. The upstream water level over the crest is 2 ft and the crest has a height of 2.25 ft. The width of the approach channel is 150 ft.

10.19 Determine the discharge over a broad-crested weir 2 m wide and 35 m long with a rounded entrance. The measured upstream level over the crest is 0.7 m. The width of the channel is 50 m and the channel bottom is 0.6 m below the crest. Increase the discharge coefficient by 5% for the rounded entrance.

10.20 A rectangular channel 14 m wide has a uniform depth of 2 m. If the channel discharge is 10 m³/s, determine the height of a 2-m-wide broad-crested weir to be built across the channel at the end for free discharge. [*Hint:* First assume a C_d to find h, then determine the revised C_d and h.]

10.21 Calculate the discharge through a long-throated flume built in a rectangular 1-m-wide channel having a throat width of 0.5 m. The upstream head is 0.6 m.

10.22 Determine the discharge through a 15-ft Parshall flume under a head of 3 ft discharging freely.

10.23 Determine the discharge through a 4-ft Parshall flume if the approach head is 4 ft and the submergence is 80% with a flow-rate correction of 5.9 cfs.

10.24 Water flows through an orifice meter in a 50 mm pipe at a rate of 3.0×10^{-3} m^3/s. The pressure difference is 1 m head of water. Determine the diameter of the orifice.

10.25 Ethyl alcohol ($\rho = 789$ kg/m^3 and $\mu = 1.2 \times 10^{-3}$ N-s/m^2) flows through a pipe of 60 mm diameter fitted with a nozzle meter. The pressure drop is 4 kPa when the flow is 3 liters/s. Determine the diameter of the nozzle.

10.26 Water flows through a venturi meter having a pipe diameter of 6 in. and with a throat diameter of 3 in. If the flow rate is 0.25 cfs, what is the pressure difference between the pipe and the throat?

10.27 Determine the peak discharge by the slope-area method. The following field measurements were made at two sites that are 115 ft apart.

	Site 1	Site 2
Area (ft^2)	225	209
Hydraulic radius (ft)	5.20	4.05
High-water marks (ft)	15.51	15.00
Roughness coefficient	0.045	0.030

10.28 Field investigations have been made of two adjacent sites of a stream. The fall of the water surface was 0.25 m in a reach of 35 m. The cross sections at both sides were subdivided. The area and hydraulic radius of subsections are indicated below. Determine the peak flow by the slope-area method.

Site	n	A (m^2)	R (m)
1	0.075	0.60	0.25
	0.040	15.5	1.01
2	0.075	0.85	0.20
	0.045	20.5	1.20
	0.045	0.50	0.22

10.29 The following field observations were made just after a flood at the bridge opening of a stream shown in Figure 10.19 in the chapter. Determine the peak discharge. Assume that $C = 0.91$.

1. Water elevation at approach section = 9.885 ft.
2. Water elevation at contracted section = 8.995 ft.

3. Approach section 1:

Subsection	n	Area (ft^2)	Wetted Perimeter (ft)
1	0.05	15.9	30
2	0.035	116.0	38
3	0.045	12.5	25.2

4. Contracted section 3:

   $n = 0.035$ $A = 82$ ft^2 wetted perimeter $= 33.3$ ft

5. Characteristics of constriction:

   Sloping embankment

   Vertical abutment

   Length of the opening, $L = 18.5$ ft

   Length of the approach reach, $L_w = 35$ ft

10.30 The following field data were obtained at the bridge opening of a river during a flood, as shown in Figure P10.30. Determine the peak discharge. C has been found to be 0.88.

1. Water elevation at approach channel = 3.5 m

2. Water elevation at contracted section = 3.25 m

3. Vertical abutment and vertical embankment

Figure P 10.30   (a) Approach channel, (b) bridge opening

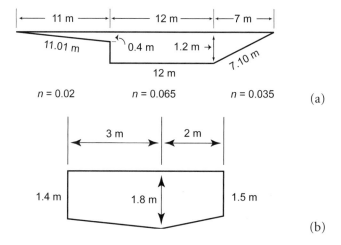

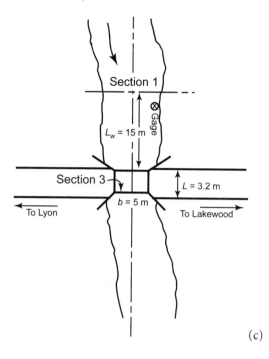

Section 1

$L_w = 15$ m

Gage

Section 3

$L = 3.2$ m

$b = 5$ m

To Lyon

To Lakewood

(c)

10.31 A 6 ft high embankment has a crest thickness of 7 ft. Both upstream and downstream faces have a slope of 2:1 (H:V). During floods, water overtops the embankment to a depth of 4 ft and flows along the embankment length of 30 ft. Determine the peak discharge. Disregard the approach velocity factor.

10.32 In Problem 10.31 the width of the approach channel is also 30 ft. Determine the peak discharge accounting for approach velocity.

10.33 At a proposed dam site for a water supply project on the Battenkill River near Greenwich, New York, the mean monthly flows are given below. If the demand for water is at a uniform rate of 60 cfs, determine the size of storage required. An allowance of 15 cfs is made to account for the seepage, evaporation losses, and downstream releases. Solve by (a) the graphic method, and (b) the analytic method.

Month	Discharge (cfs)	Month	Discharge (cfs)
Oct.	130	May	195
Nov.	70	June	100
Dec.	50	July	65
Jan.	41	Aug.	85
Feb.	30	Sept.	157
Mar.	95	Oct.	300
Apr.	230	Nov.	346

10.34 Planners propose locating an impounding reservoir on a streambed of a 150-mi^2 drainage basin. Monthly runoffs in inches for a critical year are given in the following table. The estimated demands in mgd for each month are also given. Monthly evaporation losses from the reservoir pool and the net precipitation on the pool are also indicated. If the reservoir occupies 5% of the drainage area, estimate the within-year storage requirement for the project.

Month	Runoff (in.)	Demand (mgd)	Evaporation (in.)	Net Precipitation (in.)
Oct.	1.54	125	2.25	4.10
Nov.	1.62	127	1.50	3.75
Dec.	1.99	143	1.0	4.20
Jan.	2.99	137	1.05	4.0
Feb.	2.05	150	1.75	4.6
Mar.	2.10	145	3.0	3.5
Apr.	3.20	153	4.3	3.2
May	1.50	155	5.5	3.0
June	0.35	161	5.95	4.0
July	0.20	158	5.0	4.5
Aug.	1.15	152	4.10	3.2
Sept.	2.85	153	3.20	3.9

[*Hint:* Convert all data to millions of gallons. Inflow is the sum of runoff plus net precipitation, and outflow is the sum of demand and evaporation. The runoff occurs over 95% of the drainage area excluding the reservoir pool.]

10.35 For the following hourly demand rates, determine the operating storage capacity for **(a)** uniform 24-hour pumping, and **(b)** pumping from 6 A.M. to 6 P.M. only.

Time	Demand Rate (gpm)	Time	Demand Rate (gpm)
1 A.M.	1710	1 P.M.	5850
2 A.M.	1620	2 P.M.	5814
3 A.M.	1616	3 P.M.	5787
4 A.M.	1530	4 P.M.	5850
5 A.M.	1620	5 P.M.	6030
6 A.M.	1719	6 P.M.	6407
7 A.M.	2880	7 P.M.	5000
8 A.M.	4500	8 P.M.	4836
9 A.M.	5085	9 P.M.	4092
10 A.M.	5400	10 P.M.	4803
11 A.M.	5589	11 P.M.	2500
12 noon	5670	12 midnight	3187

10.36 An earth dam has a crown (top) width of 20 ft. The height is 70 ft with a freeboard of 10 ft. The slopes are 1 (vertical):3 (horizontal). It rests on an impervious foundation having a drainage blanket extending back 50 ft from the toe. The permeability of the embankment material is 3.2×10^{-5} ft/sec. Determine the seepage per foot length of the dam.

10.37 An earth dam on an impervious foundation has a crown width of 25 ft, a total height of 80 ft, and a freeboard of 10 ft. It has an upstream slope of 20° and a downstream slope of 30°. A horizontal blanket extends back 80 ft from the back toe. The dam has a permeability of 4×10^{-4} cm/s. Determine the seepage through the dam.

10.38 In the dam of Problem 10.37, if the material has a permeability of 4×10^{-4} cm/s horizontally and 1×10^{-4} cm/s vertically, determine the seepage through the dam.

10.39 An earth dam of crown width 30 ft and height 85 ft rests on an impervious foundation. It has side slopes of 1 (vertical):4 (horizontal) and a permeability of 2×10^{-4} cm/s. A center core is provided symmetrically about the vertical axis of the dam. The core has a top width of 20 ft, side slopes of 1(vertical) to 1(horizontal), and a permeability of 2×10^{-5} cm/s. The water depth in the reservoir is 75 ft. Determine the quantity of seepage through the dam.

10.40 If the center core in Problem 10.39 has a permeability of 2×10^{-5} cm/s in the horizontal direction and 1×10^{-5} cm/s in the vertical direction, determine the seepage through the dam.

10.41 If the center core in Problem 10.40 has vertical sides (rectangular shape), determine the seepage.

10.42 An earth dam has a crown width of 25 ft and a height of 80 ft with a freeboard of 8 ft. The side slopes are at an angle of 20° with the horizontal. It has a permeability of 3×10^{-5} ft/sec horizontally and 1×10^{-5} ft/sec vertically. It rests on an impervious foundation and has no drainage blanket (the seepage line emerges on the downstream slope). Determine the seepage per foot of length of the dam.

10.43 If the dam of Problem 10.39 is located on a pervious foundation of 30 ft depth with a permeability of 1×10^{-4} cm/s, determine the quantity of seepage through the foundation assuming that the average length of seepage is equal to the base length.

10.44 For the dam of Problem 10.42, determine the seepage through the foundation if it is a pervious stratum of 40 ft depth and permeability 1.5×10^{-5} ft/sec. Assume that the seepage length is equal to the base length.

10.45 A high-overflow spillway has a maximum head of 12.0 ft. Determine the profile of the spillway crest having a vertical upstream face and a 1 (horizontal):2 (vertical) downstream slope. Take the design head to be 0.75 times the maximum head.

10.46 An overflow spillway has a design head of 2.5 m. Determine the crest profile for a spillway of 2 m height having both upstream and downstream slopes of 1:1.

10.47 A vertical-faced overflow spillway of 10 ft height is designed to discharge 8000 cfs at the pool level of 7 ft above the crest. Determine the length of the spillway. Underdesign the spillway.

10.48 An underdesigned overflow spillway of 1 m height and 40 m width has an upstream slope of 2 (horizontal):3 (vertical). Determine the discharge for a design head of 1.8 m.

10.49 A high 100-ft-wide overflow section of 1 (horizontal):3 (vertical) upstream slope is designed for a head of 10 ft. Determine the discharge when the actual total head is 12 ft.

10.50 For Problem 10.49, determine the discharge if the crest supports four square-nosed, 1-ft-wide piers and has square abutments.

10.51 An overflow is to be designed to carry a peak flow of 70,000 cfs. The upstream reservoir level is at el. 1000 ft and the average channel floor is at el. 850. The design head over the spillway is 20 ft. Determine the length of the spillway and define the crest profile for a vertical upstream section and 1 (horizontal):2 (vertical) downstream section.

10.52 An overflow spillway is to be designed having an upstream slope of 1 (horizontal):1.5 (vertical) and a downstream slope of 1 (horizontal):2 (vertical). It has to carry a peak flow of 2000 m³/s. The depth of the reservoir upstream is 40 m. The crest length is 75 m. Determine the crest height and the shape of the overflow section.

10.53 Determine the discharge on an overflow spillway designed for a maximum head of 8 ft above the crest. The overflow section is 12 ft high and 100 ft long. The tailwater can rise 6.5 ft above the crest level. The upstream face of the spillway has a slope of 1 (horizontal):3 (vertical). There are no piers and the abutment walls have square section.

10.54 Determine the crest length of an overflow spillway that will discharge 2000 ft³/s at a 5-foot maximum head. The crest height is 2 ft and the tailwater can rise 2 ft above the crest level. The upstream face of the crest is sloped 1 (horizontal): 1 (vertical). The entrance channel is 100 ft long. To support a bridge, 18-in.-wide piers with rounded nose are to be provided at a span not to exceed 20 ft. The abutment walls are rounded to a one-foot radius.

10.55 Determine the minimum slope in the upper part of a chute section of 150 ft width. The discharge range is 1000 to 100,000 cfs. $n = 0.015$.

10.56 For a 50-m-wide chute, having a discharge range of 30 to 500 m³/s, determine the minimum slope of the chute section. $n = 0.016$.

10.57 The control section of a chute spillway of design capacity 50,000 cfs is 150 ft wide, whereas the rectangular chute has a width of 100 ft. The depths of flow at the control section and at the beginning of the 100-wide section are 17.5 ft and 12.0 ft, respectively. Design the converging section.

10.58 At the bottom of the valley, where the depth is 10 ft, the chute section is to be expanded again to a width of 200 ft to join the stilling basin. Design the diverging section of Problem 10.57.

10.59 An overflow section of 120 ft length discharges 4000 cfs into a side channel. Determine the water surface profile in a side-channel trough having a bottom slope of 0.004. The channel trough and chute have a uniform rectangular section of 20 ft width. The control section is formed opposite and perpendicular to the lower end of the crest. Ignore transition losses between trough and chute.

10.60 In Problem 10.59, if the trough has a trapezoidal shape of 20 ft width and a side slope of 1 (vertical):2 (horizontal), compute the water surface profile. Assume the transition losses between the trough and the chute section at the end of the crest to be 0.2 of the difference in velocity heads in the two sections.

10.61 A 40-m-long spillway of design discharge 40 m³/s spills into a side channel of 1% bottom slope. The channel trough has a bottom width of 5 m and side slopes of 3 (vertical):2 (horizontal) and joins a chute section of 4 m width at the control. Compute the water surface profile in the channel trough. Assume the transition losses between the trough and the chute to be 0.2 of the difference in the velocity heads at two ends of the transition.

10.62 A morning glory spillway is to discharge 3500 cfs under a head of 14 ft. Determine the minimum size of the overflow crest. For a control section in the drop shaft at a depth of 20 ft below the water surface, draw the shaft profile.

10.63 A morning glory spillway has a crest diameter of 10.5 m. The design discharge is 775 m^3/s. Determine the design head over the spillway. If the control section is 2 m below the crest, determine the size of the drop shaft at the control section.

10.64 A morning glory spillway is designed such that the crest is just drowned under the maximum head of 11 ft. Compute the crest size (radius at the crest) and the shaft profile for a design discharge of 200 cfs. [*Hint:* Since the orifice control condition prevails, apply eq. (10.45) for the head versus radius computations starting at the crest level.]

11

Conveyance Systems: Open Channel Flow

🌢🌢🌢

11.1 INTRODUCTION

The two modes of transporting flowing water from one point to another are pipes and open channels. In pipes, the flow of water is under pressure, whereas open channel flow has a free water surface. The basic theory is the same for both kinds of flow, but there is an important difference in the boundary condition. Open channel flow is more difficult to deal with due to the presence of the free water surface, various possible configurations of the channel section, and the changing position of the water surface with respect to time and space. Thus, the treatment of open channel flow is somewhat empirical in nature.

An open channel can be a natural stream or a river. It also can be an artificial channel in the form of a canal, flume, chute, culvert, tunnel, ditch, partly filled pipe (conduit), or aqueduct of any shape. An artificial channel is commonly used to convey water from its source of supply to a distribution point. Further distribution beyond this point is made through a network of pipes. An artificial channel in the form of a canal is excavated either in a firm foundation such as a rock bed or in erodible materials. Irrigation canals, formed in alluvial and other granular material, are erodible unless lined with nonerodible materials. In designing erodible channels, consideration has to be given to the stability of the channel geometry such that substantial scouring does not take place. This criterion is not applicable to nonerodible channels. The two kinds of channels are described separately.

11.2 ELEMENTS OF THE CHANNEL SECTION

The flow in an open channel is due to gravitational force; hence the channel bottom should have a slope in the direction of flow. A channel having unvarying cross section and constant bed slope throughout its length is known as a *prismatic channel*. The cross section of a channel taken normal to the direction of flow at any point is referred to as the channel section. A section and a longitudinal profile of a channel are shown in Figure 11.1.

The definitions of various geometric elements are given below:

1. *Depth of flow, y*: vertical distance from the channel bottom to the free surface
2. *Depth of flow section, d*: depth of flow normal to the direction of flow; $d = y \cos \theta$, but the terms d and y are used interchangeably
3. *Top width, T*: width at the free surface

Figure 11.1 Channel section and longitudinal profile.

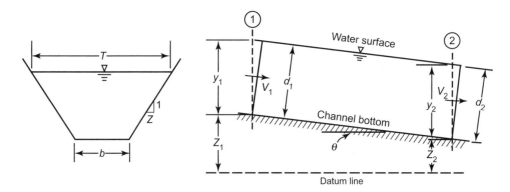

4. *Flow area, A*: cross-sectional area of the flow normal to the direction of flow; $A = bd$ for a rectangle

5. *Wetted perimeter, P_w*: across a channel section, the length of the channel surface in contact with water; $P = b + 2d$ for a rectangle

6. *Hydraulic radius*, R = ratio of the flow area to the wetted perimeter; $R = A/P_w$

7. *Hydraulic depth, D* = ratio of the flow area to the top width; $D = A/T$

8. *Section factor for critical flow,* $Z_c = A\sqrt{D}$

9. *Section factor for uniform flow,* $Z_n = AR^{2/3}$

For a circular section, the geometric elements in the dimensionless form, as a ratio with an appropriate power of the diameter, d_0, of the section, are given in Table 11.1, which provides a convenient means of determining the geometric elements for various depths of flow. For a depth of flow of $0.94d_0$, the section factor $AR^{2/3}$ has a maximum value of $0.335d_0^{8/3}$ in a circular section (i.e., discharge is maximum at this depth). Between the range of the water depth of $0.82d_0$ and d_0, there are two depths corresponding to the same level of discharge— one above $0.94d_0$ and one below it.

EXAMPLE 11.1

A circular channel section of 5 ft diameter has a water depth of 3 ft. Determine the geometric elements.

SOLUTION

1. From the given information,

$$y = 36 \text{ in. or } 3 \text{ ft} \quad \text{and} \quad d_0 = 60 \text{ in. or } 5 \text{ ft}$$

$$\frac{y}{d_0} = \frac{3}{5} = 0.6$$

Table 11.1 Geometric Elements for a Circular Section

$\dfrac{y}{d_0}$	$\dfrac{A}{d_0^2}$	$\dfrac{P_w}{d_0}$	$\dfrac{R}{d_0}$	$\dfrac{T}{d_0}$	$\dfrac{D}{d_0}$	$\dfrac{A\sqrt{D}}{d_0^{5/2}}$	$\dfrac{AR^{2/3}}{d_0^{8/3}}$
0.01	0.0013	0.2003	0.0066	0.1990	0.0066	0.0001	0.0000
0.05	0.0147	0.4510	0.0326	0.4359	0.0336	0.0027	0.0015
0.10	0.0409	0.6435	0.0635	0.6000	0.0682	0.0107	0.0065
0.15	0.0739	0.7954	0.0929	0.7141	0.1034	0.0238	0.0152
0.20	0.1118	0.9273	0.1206	0.8000	0.1398	0.0418	0.0273
0.25	0.1535	1.0472	0.1466	0.8660	0.1774	0.0646	0.0427
0.30	0.1982	1.1593	0.1709	0.9165	0.2162	0.0921	0.0610
0.35	0.2450	1.2661	0.1935	0.9539	0.2568	0.1241	0.0820
0.40	0.2934	1.3694	0.2142	0.9798	0.2994	0.1603	0.1050
0.45	0.3428	1.4706	0.2331	0.9950	0.3446	0.2011	0.1298
0.50	0.3927	1.5708	0.2500	1.0000	0.3928	0.2459	0.1558
0.55	0.4426	1.6710	0.2649	0.9950	0.4448	0.2949	0.1825
0.60	0.4920	1.7722	0.2776	0.9798	0.5022	0.3438	0.2092
0.65	0.5404	1.8755	0.2881	0.9539	0.5666	0.4066	0.2358
0.67	0.5594	1.9177	0.2917	0.9404	0.5948	0.4309	0.2460
0.70	0.5872	1.9823	0.2962	0.9165	0.6408	0.4694	0.2608
0.75	0.6318	2.0944	0.3017	0.8660	0.7296	0.5392	0.2840
0.80	0.6736	2.2143	0.3042	0.8000	0.8420	0.6177	0.3045
0.85	0.7115	2.3462	0.3033	0.7141	0.9964	0.7098	0.3212
0.90	0.7445	2.4981	0.2980	0.6000	1.2408	0.8285	0.3324
0.94[a]	0.7662	2.6467	0.2896	0.4750	1.6130	0.9725	0.3353
0.95	0.7707	2.6906	0.2864	0.4359	1.7682	1.0242	0.3349
1.00	0.7854	3.1416	0.2500	0.0000	∞	∞	0.3117

[a] Maximum flow occurs at 0.94 full depth.

2. From Table 11.1, for y/d_0 of 0.6:

$$\frac{A}{d_0^2} = 0.492, \quad A = 0.492(5)^2 = 12.3 \text{ ft}^2$$

$$\frac{P_w}{d_0} = 1.7722, \quad P_w = 1.7722(5) = 8.861 \text{ ft}$$

$$\frac{R}{d_0} = 0.2776, \quad R = 0.2776(5) = 1.39 \text{ ft}$$

$$\frac{Z_c}{d_0^{5/2}} = \frac{A\sqrt{D}}{d_0^{5/2}} = 0.3438, \quad A\sqrt{D} = 0.3438(5)^{5/2} = 1.92 \text{ ft}^{5/2}$$

$$\frac{Z_n}{d_0^{8/3}} = \frac{AR^{2/3}}{d_0^{8/3}} = 0.2092, \quad AR^{2/3} = 0.2092(5)^{8/3} = 15.29 \text{ ft}^{8/3}$$

11.3 TYPES OF FLOW

The flow in an open channel is classified according to the change in the depth of flow with respect to space and time. If the depth of flow remains the same at every section of the channel, the flow is known as *uniform* or *normal flow*. In varied or nonuniform flow, the depth changes along the length of the channel. When the change in depth occurs abruptly over a short distance, it is a rapidly varied flow; otherwise, it is a gradually varied flow.

If the depth of flow does not change during the time interval under consideration, it is referred to as *steady flow*. It is unsteady if the depth changes with time. Combining the space and time criteria, the flow in an open channel can be classified as in Table 11.2. For an unsteady uniform flow, the depth should vary from time to time while always remaining parallel to the channel bottom. This is not a practically feasible condition.

Even the condition of steady uniform flow is difficult to obtain in natural channels due to irregular section, and in artificial channels because of the existence of controls. However, steady uniform flow is a fundamental type of flow that is considered in all channel design problems. The effect of varied flow is superimposed over the uniform flow condition to determine the channel section requirements. In the computation of flow in natural streams, the steady flow condition is assumed during the time interval under consideration. Unsteady flow relates to the propagation of a wave in the channel.

Table 11.2 Types of Open Channel Flow

Type of Flow	Example
Steady uniform flow	Laboratory channel
Steady gradually varied flow	Irrigation, navigation channel
Steady rapidly varied flow	Flow over a weir, hydraulic jump
Unsteady gradually varied flow	Streamflow, flood wave
Unsteady rapidly varied flow	Surges, pulsating flow

11.4 STATE OF FLOW

Viscosity and gravity affect the state of flow in an open channel. The Reynolds number and the Froude number are both relevant in the channel flow. The Reynolds number, a ratio of the inertia force to the viscous force, is expressed as follows for an open channel:

$$\text{Re} = \frac{VR}{v} \quad \text{[dimensionless]} \tag{11.1}$$

where

$\quad$ Re = Reynolds number

$\quad$ V = mean velocity of flow, ft/sec or m/s

$\quad$ R = hydraulic radius A/P_w, ft or m

$\quad$ v = kinematic viscosity of water, ft^2/sec or m^2/s

$\quad$ $= \dfrac{\text{dynamic viscosity}}{\text{mass density}}$

The flow is *laminar* when the viscous forces are dominating, resulting in a Reynolds number of less than 500. It is *turbulent* if the viscous forces are weak and the Reynolds number is higher than 2000. The transitional range of Re is between 500 to 2000. Experiments on smooth channels and rough channels indicate the following characteristics (Chow, 1959, pp. 9–12):

1. The Darcy-Weisbach formula of flow in pipes (eq. 12.2) is applicable to uniform flow in open channels.

2. In the laminar region, the friction factor relation, $f = K/\text{Re}$ of pipe flow, is applicable to both smooth and rough channels. The value of K varies with channel shape and is higher for rough channels.

3. In the turbulent region, the friction factor relation of smooth pipes (Blasius and Prandtl-von Kármán) is approximately representative of smooth channels.

4. In the turbulent region of rough channels, the channel shape, roughness, and the Reynolds number have a pronounced effect on the friction factor. The friction factor relation deviates from the pipe flow relation.

The common type of flow pertains to item 4, thus necessitating a separate relation for the channel flow.

The gravity effect is incorporated in the Froude number, which is represented by a ratio of inertia force to gravity force, as follows:

$$\text{Fr} = \frac{V}{\sqrt{gD}} \quad \text{[dimensionless]} \tag{11.2}$$

where

$\quad$ Fr = Froude number

$\quad$ V = mean velocity of flow, ft/sec or m/s

$\quad$ D = hydraulic depth A/T, ft or m

When Fr = 1, the flow is in a *critical* state; when Fr < 1, the flow is *subcritical* or tranquil having a low velocity; and when Fr > 1, the flow is *supercritical* or shooting, having a high velocity.

The friction factor characteristics of laminar and turbulent flow as discussed above relate to the subcritical flow. In the supercritical turbulent regime of flow, the friction factor becomes larger with an increase in the Froude number. Up to a value of 3, the Froude number has a negligible effect on the friction factor.

Section 11.4   State of Flow

11.5 Critical Flow Condition

11.5.1 Concept of Specific Energy

The energy in a channel section measured with respect to the channel bottom as the datum is known as the *specific energy*. In eq. (9.6), with $Z = 0$, the specific energy is given by

$$E = y + \frac{V^2}{2g} \quad [\text{L}] \tag{11.3}$$

Since $V = Q/A$, the equation may be written as

$$E = y + \frac{Q^2}{2gA^2} \quad [\text{L}] \tag{11.4}$$

The first term on the right side relates to the static energy and the second to the kinetic energy. These have been plotted separately and then combined in Figure 11.2 for a graph of the depth against the specific energy for a constant discharge.

The combined curve (3) indicates that at point O, the specific energy is at its minimum. It will be demonstrated that this corresponds to the critical state of flow. The flow below this point is supercritical (low depth, high velocity). For a given specific energy, there are two alternate depths, one in the supercritical range and one in the subcritical range. At the critical state they merge into one depth, y_c.

For the condition of a minimum specific energy, $dE/dy = 0$. Differentiating eq. (11.4), for a constant Q,

$$\frac{dE}{dy} = 1 - \frac{2Q^2}{2gA^3} \frac{dA}{dy} = 0 \tag{a}$$

Since $dA/dy = T$ and $A/T = D$, substituting in (a) gives us

$$\frac{Q^2}{gA^2} \frac{1}{D} = 1 \quad [\text{dimensionless}] \tag{11.5}$$

Also, since $Q/A = V$,

$$\frac{V^2}{gD} = 1$$

or

$$\frac{V}{\sqrt{gD}} = 1 \quad [\text{dimensionless}] \tag{11.6}$$

The term on the left side of eq. (11.6) is the Froude number, Fr. As stated in section 11.4, Fr = 1 is the condition of critical flow. Hence the specific energy is a minimum at the critical flow.

11.5.2 Computation of Critical Flow

From eq. (11.5),

$$A\sqrt{D} = \frac{Q}{\sqrt{g}}$$

Figure 11.2 Specific energy plot.

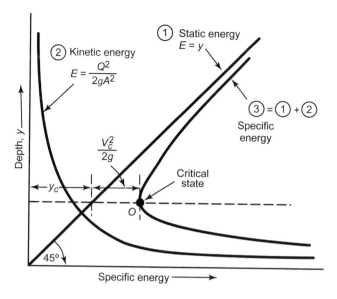

Since $Z_c = A\sqrt{D}$,

$$Z_c = \frac{Q}{\sqrt{g}} \quad [\text{L}^{5/2}] \tag{11.7}$$

Equation (11.7) is used in the following two ways:

1. *Given critical depth, y_c, to compute Q.* Compute the section factor, Z_c, for known y_c. Determine Q using eq. (11.7).

2. *To compute the critical depth for a given Q.* Calculate $Q/\sqrt{g}$, which is equal to Z_c. Express Z_c in terms of y_c in the form of an algebraic equation. For simple geometric sections, the value of y_c is solved from the equation. For a complicated section, the equation in terms of y_c can be solved by a powerful hand-held calculator, but the graphic procedure is very convenient, in which a curve of depth (y) versus $Z (= A\sqrt{D})$ is constructed. Corresponding to the Z_c value equal to $Q/\sqrt{g}$, the critical depth is obtained directly from the curve. Dimensionless curves and tables are available in handbooks (e.g., Chow, 1959, pp. 65, 625–627).

In Figure 11.2, curve (3) is almost vertical at the critical depth. A slight change in energy can cause substantial variation in depth. The flow at the critical state is thus unstable, and the water surface appears wavy. In terms of channel design, if the depth is determined to be at or near the critical state for a great length of the channel, the shape or slope of the channel should be altered (Chow, 1959). Critical flow, however, serves the purpose of a control section. It is useful in defining flow conditions and developing surface water profiles, as explained subsequently.

The slope of a channel at which the computed uniform or normal depth of flow is equal to the critical depth is known as the *critical slope*, S_c. A slope less than the critical slope is the mild or subcritical slope, and a slope greater than critical is the steep or supercritical slope. These are important in flow profiles, as discussed subsequently.

Section 11.5 Critical Flow Condition

EXAMPLE 11.2

A rectangular channel is 25 ft wide and has a flow of 500 cfs at a velocity of 5 ft/sec.
(a) Determine the specific energy of water in the channel. **(b)** What is the critical depth of water in the channel? **(c)** What is the critical velocity?

SOLUTION

(a) 1. From the continuity equation,

$$A = \frac{Q}{V} = \frac{500}{5} = 100 \text{ ft}^2$$

$$by = 100 \quad \text{or} \quad y = \frac{100}{25} = 4 \text{ ft}$$

2. Specific energy,

$$E = y + \frac{V^2}{2g} = 4 + \frac{(5)^2}{2(32.2)} = 4.39 \text{ ft}$$

(b) Section factor for critical flow,

$$Z_c = \frac{Q}{\sqrt{g}} = \frac{500}{\sqrt{32.2}} = 88.11$$

If y_c is the critical depth, then for a rectangular channel,

$$A = by_c = 25y_c \text{ ft}^2$$

$$D = \frac{A}{T} = \frac{25y_c}{25} = y_c \text{ ft}$$

$$Z_c = A\sqrt{D} = 25y_c\sqrt{y_c} = 25y_c^{3/2}$$

or

$$25y_c^{3/2} = 88.11, \quad y_c = 2.32 \text{ ft}$$

(c) At the critical flow,

$$V_c = \sqrt{gD} = \sqrt{gy_c} = \sqrt{32.2(2.32)} = 8.64 \text{ ft/sec}$$

EXAMPLE 11.3

A trapezoidal channel with a bottom width of 4 m and side slopes of 1:4 carries a discharge of 30 m³/s (Figure 11.3). Determine the **(a)** critical depth, **(b)** critical velocity, and **(c)** minimum specific energy.

Figure 11.3 Channel section for Example 11.3.

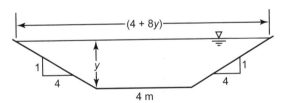

SOLUTION

(a) 1. From eq. 11.7, $Z_c = Q / \sqrt{g}$

2. $\dfrac{Q}{\sqrt{g}} = \dfrac{30}{\sqrt{9.81}} = 9.58$

3. For any depth y,

$$A = \frac{1}{2}\left[4 + (4 + 8y)\right]y = (4 + 4y)y$$

$$T = 4 + 8y$$

$$Z_c = A\sqrt{D} = A\left(\frac{A}{T}\right)^{1/2} = \frac{A^{3/2}}{T^{1/2}} = \frac{\left[(4+4y)y\right]^{3/2}}{(4+8y)^{1/2}}$$

4. Using this formula, Z_c is computed for various selected values of y in Table 11.3.

5. y versus Z_c has been plotted in Figure 11.4.

6. For Z_c corresponding to $Q / \sqrt{g}$ of 9.58, the value of y_c has been read from Figure 11.4 as $y_c = 1.22$ m

Table 11.3   Section Factor for a Trapezoidal Channel

y (m) Select:	$A = (4 + 4y)y$	$A^{3/2}$	$T = (4 + 8y)$	$T^{1/2}$	$Z_c = \dfrac{A^{3/2}}{T^{1/2}}$
1.0	8	22.63	12	3.46	6.54
1.5	15	58.09	16	4.0	14.52
2.0	24	117.58	20	4.47	26.30
2.5	35	207.06	24	4.90	42.26

(b) At critical depth, $V_c = \sqrt{gD}$:

$$D = \frac{A}{T} = \frac{(4 + 4y_c)y_c}{4 + 8y_c} = \frac{\left[4 + 4(1.22)\right]1.22}{4 + 8(1.22)} = 0.787 \text{ m}$$

$$V_c = \sqrt{9.81(0.787)} = 2.78 \text{ m/s}$$

(c) Specific energy at critical state:

$$E = y_c + \frac{V_c^2}{2g} = 1.22 + \frac{(2.78)^2}{2(9.81)} = 1.61 \text{ m}$$

Section 11.5   Critical Flow Condition

Figure 11.4 Depth versus section factor curve for critical depth for Example 11.3.

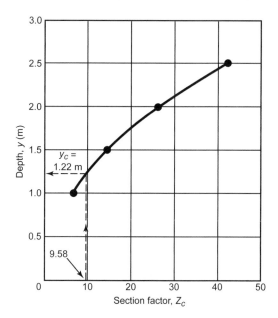

11.6 UNIFORM CHANNEL FLOW

Uniform flow occurs in a steady state only. In uniform flow, the depth, area of cross section, velocity of flow, and discharge are the same at every section of a channel. Such a condition develops when the force in the direction of the flow is fully balanced by the resistance encountered by the water as it moves downstream. This does not occur frequently. Long channels may have only small reaches of uniform flow. However, uniform flow is a basic flow in channel hydraulics. Channel designs are based on consideration of the uniform flow.

11.6.1 Hydraulics of Uniform Flow

As stated in Section 11.4, the Darcy-Weisbach (1845)[*] equation of pipe flow is applicable to uniform channel flow. The commonly encountered flow in channels is rough turbulent flow, for which the friction factor relations of pipe, by Blasius (1913)[*] and Prandtl and von Kármán (1935)[*], are not directly applicable. In 1939, Colebrook and White suggested an equation (Hydraulic Research Station, 1983) for open channel flow which is, however, not directly solvable because of its implicit form, in which the term of channel slope appears on both sides of the equation. The Hydraulic Research Station in the United Kingdom has prepared charts for the application of this equation.

The common approach in the United States is to use the Chezy or Manning formula developed in 1769 and 1889, respectively. There is conformity in the concept of Chezy and Darcy-Weisbach that the head loss varies as the square of the velocity. Chezy's formula can be derived by equating the propulsive force due to the weight of water in the direction of flow with the retarding shear force at the channel boundary. From Figure 11.5:

$$\text{propulsive force in the direction of flow,} \quad F = W \ \sin \ \theta = \gamma AL \ \sin \ \theta \tag{a}$$

$$\text{resisting force due to shear stress } \tau, \quad R = \tau PL \tag{b}$$

[*] Referenced in Chapter 12.

Figure 11.5 Derivation of Chezy's formula.

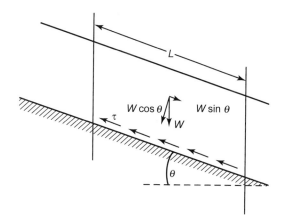

Since $F = R$,

$$\gamma AL \sin \theta = \tau PL \qquad \text{(c)}$$

For turbulent flow,

$$\tau \propto V^2 \quad \text{or} \quad \tau = KV^2 \qquad \text{(d)}$$

and for small slope,

$$\sin \theta = \tan \theta = S \qquad \text{(e)}$$

Substituting (d) and (e) in (c)

$$V = \left(\frac{\gamma}{K} \frac{A}{P_w} S \right)^{1/2} \qquad \text{(f)}$$

or

$$V = C\sqrt{RS} \quad [\text{LT}^{-1}] \qquad (11.8)$$

where

V = mean velocity, ft/sec or m/s

R = hydraulic radius, ft or m

S = slope of energy line, which is equal to channel bottom for uniform flow

C = Chezy's constant

Three formulas by Ganguillet and Kutter, Bazin, and Powell are commonly used to determine the Chezy constant, C. Of these, the first formula is most satisfactory. It uses the roughness coefficient, known as Kutter's n, which is almost equal to Manning's coefficient n.

If $C = R^{1/6}/n$ is substituted in Chezy's formula, Manning's formula results. Manning's formula has proven most reliable in practice. This empirical formula, suitable for a fully rough turbulent flow, is given by

$$V = \frac{1.486^*}{n} R^{2/3} S^{1/2} \qquad \text{(English units)} \quad [\text{LT}^{-1}] \qquad (11.9a)$$

$$V = \frac{1}{n} R^{2/3} S^{1/2} \qquad \text{(metric units)} \quad [\text{LT}^{-1}] \qquad (11.9b)$$

* This number is also commonly used as 1.49.

Section 11.6   Uniform Channel Flow

where n is Manning's roughness coefficient. It depends on channel material, surface irregularities, variation in shape and size of the cross section, vegetation and flow conditions, channel obstruction, and degree of meandering. Chow (1959) has provided a detailed table and photographs of channels for values of n in different conditions. Typical values are summarized in Table 11.4.

11.6.2 Computation of Uniform Flow

Combining the continuity equation $Q = AV$ with eq. (11.9), the Manning formula is obtained in terms of discharge as follows:

$$Q = \frac{1.486}{n} AR^{2/3}S^{1/2} \qquad \text{(English units)} \quad [L^3 T^{-1}] \qquad (11.10a)$$

$$Q = \frac{1}{n} AR^{2/3}S^{1/2} \qquad \text{(metric units)} \quad [L^3 T^{-1}] \qquad (11.10b)$$

The terms on the right side, excluding the slope, are grouped into a single term known as the conveyance, K. Thus the formula is also stated as $Q = K\sqrt{S}$. As defined earlier, $AR^{2/3}$ is called the section factor, Z_n, for uniform or normal flow, which is a function of depth for a given channel section. The depth of flow corresponding to uniform channel flow is known as the normal depth, y_n.

Three variables are involved in eq. (11.10): (1) discharge or velocity, (2) slope, and (3) section factor (a function of depth). When any two of these are known, the third one can be computed. Three cases are described below:

1. *Normal depth and slope are known; compute the discharge.* Determine the section factor, $AR^{2/3}$, for a given normal depth. Q can be computed from eq. (11.10).

2. *Discharge and normal depth are known; compute the slope.* Again the application is direct in eq. (11.10).

3. *Discharge and slope are known; compute the normal depth.* Equation (11.10) is rearranged as

$$AR^{2/3} = \frac{Qn}{1.486S^{1/2}} \qquad \text{(English units)} \quad [L^{8/3}] \qquad (11.11a)$$

$$AR^{2/3} = \frac{Qn}{S^{1/2}} \qquad \text{(metric units)} \quad [L^{8/3}] \qquad (11.11b)$$

The right side of eq. (11.11) is evaluated from the known variables. For simple geometric sections, $AR^{2/3}$ is expressed in terms of y_n and solved for directly. For other geometric shapes, $AR^{2/3}$ becomes a complicated function in terms of y_n. Today's powerful hand-held calculators are able to solve such a function to get a value of y_n. However, it is convenient to prepare a plot of y versus $AR^{2/3}$. From this the value of y_n is obtained for the computed value of $AR^{2/3}$ equal to $Qn/1.486S^{1/2}$.

For a circular channel section running full, the application of the Manning formula is direct in all three cases above. In a circular section that is flowing only partially full, the geometric relations in terms of the diameter given in Table 11.1 are very useful. For example, to determine the normal depth of flow, the section factor $AR^{2/3}$ is computed from the discharge by eq. (11.11). Looking at the last column of Table 11.1, $AR^{2/3}/d_0^{8/3}$, the value of y/d_0, hence y, is obtained. Figure 11.6 can also be used for quick computation. To use this figure it is necessary to first find the values when the section is flowing full. As stated earlier, the maximum flow in a circular section occurs at a depth of 0.94 of the diameter.

Table 11.4   Values of Manning's Roughness Coefficient[a]

Material	Manning n
1. Closed conduit or built-up channel	
1.1 Metal	
Brass	0.01
Copper	0.011
Steel—welded	0.012
Steel—riveted	0.016
Cast iron—coated	0.013
Wrought iron—galvanized	0.016
Corrugated metal (storm drain)	0.024
1.2 Nonmetal	
Glass	0.01
Cement	0.011
Cement mortar	0.013
Concrete culvert	0.013
Concrete lined channel/pipe	0.015
Wood	0.012
Clay	0.013
Brickwork	0.013
Brickwork with cement mortar	0.015
Masonry/ rubble masonry	0.025
Sanitary sewer coated with slime	0.013
Asphalt	0.013
Plastic	0.013
PVC	0.009–0.011
Polyethylene	0.009–0.015
2. Excavated or Dredged Channel	
Straight and clean	0.022
Winding and sluggish	0.025
Dredged	0.028
Rock cut/stony	0.035
Earth bottom, rubble sides	0.03
Unmaintained/uncut brush	0.08
3. Natural streams	
On plain, clean, straight, no pools	0.03
On plain, clean, winding, some pools	0.04
On plain, sluggish, weedy, deep pools	0.07
On mountain, few boulders	0.04
On mountain, large boulders	0.05

[a] For overland flow roughness coefficient see Table 13.7.

Note: Judgment must be used to determine n for channel characteristics that fall in between these categories. See Chow (1959) for a detailed reference.

Section 11.6   Uniform Channel Flow

Figure 11.6 Hydraulic elements of a circular section (ASCE, 1982).

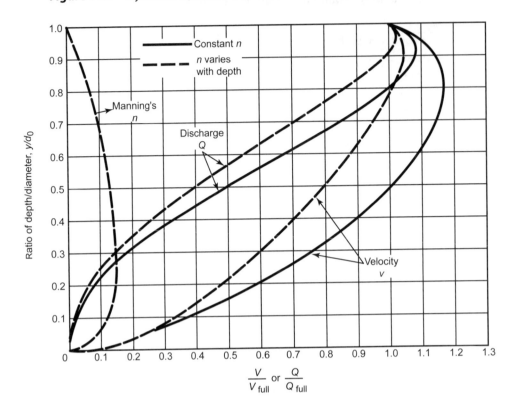

EXAMPLE 11.4

Calculate the discharge through a 3-ft-diameter circular, clean earth channel running half full. The bed slope is 1 in 4500. Manning's $n = 0.018$.

SOLUTION

1. $S = \dfrac{1}{4500} = 2.22 \times 10^{-4}$

2. $\dfrac{y_n}{d_0} = \dfrac{1.5}{3} = 0.5$

3. From Table 11.1 for $\dfrac{y}{d_0}$ of 0.5, $\dfrac{AR^{2/3}}{d_0^{8/3}} = 0.1558$

4. $AR^{2/3} = 0.1558(3)^{8/3} = 2.92$

5. $Q = \dfrac{1.49}{0.018}(2.92)\left(2.22 \times 10^{-4}\right)^{1/2} = 3.60$ cfs

ALTERNATIVE SOLUTION   From Figure 11.6 for constant n, for y/d_0 of 0.5, $Q/Q_{full} = 0.5$,

1. $A = \dfrac{\pi}{4}(3)^2 = 7.065 \text{ ft}^2$

2. $R = \dfrac{d_0}{4} = \dfrac{3}{4} = 0.75 \text{ ft}$

3. $Q_{full} = \dfrac{1.49}{0.018}(7.065)(.75)^{2/3}\left(2.22\times10^{-4}\right)^{1/2} = 7.19 \text{ cfs}$

4. $Q = 0.5 Q_{full} = 0.5(7.19) = 3.60 \text{ cfs}$

EXAMPLE 11.5

A trapezoidal channel of bottom width 25 ft and side slope 1:2.5 carries a discharge of 450 cfs with a normal depth of 3.5 ft. The elevations at the beginning and end of the channel are 685 and 650 ft, respectively. Determine the length of the channel if $n = 0.02$.

SOLUTION   Refer to Figure 11.7.

1. This is a problem of the determination of S.

2. $A = \dfrac{1}{2}(25 + 42.5)3.5 = 118.13 \text{ ft}^2$

   $P_w = 25 + 9.42 + 9.42 = 43.84 \text{ ft}$

   $R = \dfrac{A}{P_w} = \dfrac{118.13}{43.84} = 2.69 \text{ ft}$

3. $S = \left[\dfrac{Q}{(1.49/n)AR^{2/3}}\right]^2 = \left[\dfrac{450}{(1.49/0.02)(118.13)(2.69)^{2/3}}\right]^2 = 0.0007$

4. $\dfrac{H_1 - H_2}{L} = S$

   or

   $L = \dfrac{H_1 - H_2}{S} = \dfrac{685 - 650}{0.0007} = 50,000 \text{ ft}$

Figure 11.7   Channel section of Example 11.5.

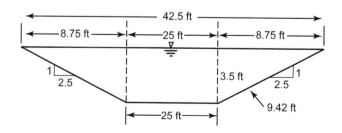

EXAMPLE 11.6

The channel of Example 11.3 has a bottom slope of 0.1% and $n = 0.025$. Determine the (a) normal depth, (b) critical slope, and (c) state of flow in the channel.

SOLUTION

(a) 1. $S = \dfrac{0.1}{100} = 0.001$

2. $\dfrac{Qn}{S^{1/2}} = \dfrac{30(0.025)}{(0.001)^{1/2}} = 23.72$

3. $A = (4 + 4y)y$

$P_w = 4 + 8.24y$

$R = \dfrac{(4 + 4y)y}{4 + 8.24y}$

4. For selected values of y, Z_n is computed below.

Selected y	$A = (4 + 4y)y$	$P_w = 4 + 8.24y$	$R = A/P_w$	$Z_n = AR^{2/3}$
1.0	8	12.24	0.65	6.00
1.5	15	16.36	0.92	14.19
2.0	24	20.48	1.17	26.65
2.5	35	24.60	1.42	44.22
3.0	48	28.72	1.67	67.57

5. y versus Z_n has been plotted in Figure 11.8. From this graph, for Z_n of 23.72 of step 2, $y_n = 1.9$ m

(b) 6. Critical depth, $y_c = 1.22$ m (from Example 11.3). For critical depth,

$$A = 10.83 \text{ m}^2$$

$$P_w = 14.05 \text{ m}$$

$$R = \frac{A}{P_w} = \frac{10.83}{14.05} = 0.77 \text{ m}$$

7. From eq. (11.10b) rearranged:

$$S_c = \left[\frac{30}{(1/0.025)(10.83)(0.77)^{2/3}} \right]^2 = 0.0068$$

(c) Since the channel bottom slope < the critical slope, it is a mild slope and subcritical flow.

Figure 11.8   Depth versus section factor for normal flow for Example 11.6.

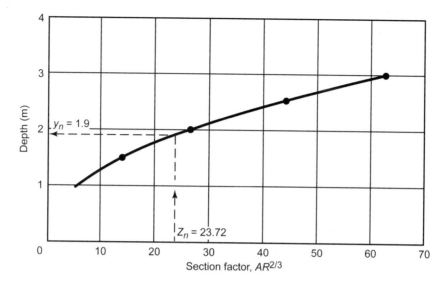

11.7 CHANNEL DESIGN

Channel design comprises determining the channel depth and other dimensions along with the bed slope. Four types of problems relate to channel design:

1. Rigid boundary channels carrying sediment-free water*
2. Rigid boundary channels carrying sediment-laden water
3. Loose boundary channels carrying sediment-free water
4. Loose boundary channels carrying sediment-laden water

11.8 RIGID CHANNELS CARRYING SEDIMENT-FREE WATER

The design of a channel that is cut in rock or constructed with a lining is based on the uniform flow condition described in the previous section. However, in order to design all channel dimensions, additional relations and criteria are needed besides Manning's equation. These additional elements are based on the following considerations.

11.8.1 Bottom Longitudinal Slope

This is governed by the topography and head requirements. When the two ends of a channel are fixed and the channel has to be laid on a predetermined alignment, the slope gets fixed accordingly. The conveyance channels for water supply, irrigation, and hydropower require a higher level at the point of delivery and therefore have a relatively small slope. In circular pipes, the slopes required for pipes flowing full at a minimum velocity of 2 ft/sec (0.6 m/s) and a value of n of about 0.015† are indicated in Table 11.5 for various flows according to a study by Pomeroy (1967).

---

* Includes colloidal material.

† For clay, concrete, cast iron, and plastic pipes, n is considered to be 0.015. For smooth joints and good construction, it is 0.013.

Table 11.5 Slopes Required for Various Flows for Circular Conduit

Flow (cfs)	Flow (m³/s)	Slope (ft/1000 ft)
0.1	2.8×10^{-3}	9.2
0.2	5.7×10^{-3}	6.1
0.3	8.5×10^{-3}	4.8
0.4	1.1×10^{-2}	4.1
0.6	1.7×10^{-2}	3.22
0.8	2.3×10^{-2}	2.73
1.0	2.8×10^{-2}	2.39
1.5	4.2×10^{-2}	1.89
2.0	5.7×10^{-2}	1.59
3.0	8.5×10^{-2}	1.26
4.0	1.1×10^{-1}	1.06

Source: Pomeroy, 1967.

11.8.2 Channel Side Slopes

The channel side slopes depend on the type of material of the channel. A nearly vertical slope is recommended for a channel comprised of rocks, while a slope of 1 (vertical): 3 (horizontal) is recommended for sandy soil. The Bureau of Reclamation prefers a 1:1.5 slope for usual sizes of lined canals.

11.8.3 Freeboard

This is the vertical distance from the water surface to the top of the channel. It should be sufficient to prevent the overtopping of the channel by waves or a fluctuating water surface. The Bureau of Reclamation has recommended the following formula:

$$u = \sqrt{cy} \text{ (English units) [unbalanced]} \tag{11.12}$$

where

u = freeboard, ft

c = coefficient varying from 1.5 for a capacity of
   20 cfs to 2.5 for a capacity of 3000 cfs or more

y = water depth in the canal, ft

11.8.4 Hydraulic Efficient Sections

For a given slope and roughness coefficient, the discharge increases with an increase in the section factor. For a given area, the section factor is highest for the least wetted perimeter. The expressions for the wetted perimeter can be written in terms of the depth for various channel shapes. Its minimization by differentiating and equating to zero provides the depth relation of the best hydraulic section. From hydraulic efficiency considerations, a semicircle is the best section for an open channel. The best closed section flowing full is a circle. Any reasonably shaped open section is more efficient than a closed (conduit) section flowing full. The dimensions of a channel are not governed entirely by the hydraulic efficiency but by practical and cost considerations as well. Trapezoidal sections are very common. The following properties are related to the best hydraulic section except where these have to be changed from practical considerations.

1. *Trapezoidal section.* The best hydraulic section never has a base width larger than the depth of water.
2. *Rectangular section.* The width is twice the depth in the best hydraulic section.
3. *Triangular section.* Side slopes are selected by practical consideration.
4. *Circular section.* A semicircle is the best section for channels open at top, and a circle is best as a closed section.

11.8.5 Design Procedure

1. Select S and estimate n from available data.
2. Substituting in the right side of eq. (11.11), determine the section factor $AR^{2/3}$.
3. Select the side slope z and assume b/y as necessary. Express $AR^{2/3}$ in terms of the depth. Solve for the depth as in the preceding section.
4. Assuming several values of the unknowns, a number of section dimensions can be obtained to make a cost comparison.
5. Add a freeboard to the water depth for an open section.

Storm sewers and wastewater sewers are designed by the procedure above, except for the computation of the quantity of flow. The storm discharge is computed based on the drainage area and wastewater flow from the quantity of water supply, as discussed in Chapter 13.

EXAMPLE 11.7

A district has a drainage area of 2500 acres with a population of 20 persons per acre. The daily water supply to the district is 40 gallons per person. It has been observed that 10% of this flow passes along the sewer between the hours of 7 and 8 A.M. If the sewer consists of vitrified clay laid to 0.1% grade, design the sewer.

SOLUTION

1. $n = 0.013$ for vitrified clay, $S = 0.1/100 = 0.001$
2. Total water supply = (acres) (person/acre) (supply/person)
$$= 2500\,(20)\,(40) = 2 \times 10^6 \text{ gpd}$$
3. Flow passing to the sewer in one hr $= 0.1(2 \times 10^6) = 0.2 \times 10^6$ gph or 7.4 cfs
4. $AR^{2/3} = \dfrac{Qn}{1.49S^{1/2}} = \dfrac{(7.4)(0.013)}{(1.49)(0.001)^{1/2}} = 2.04$
5. From Table 11.1, for maximum flow at $0.94d_0$, $AR^{2/3} = 0.3353d_0^{8/3}$
6. Hence $0.3353d_0^{8/3} = 2.04$ or $d_0 = 197$ ft.

The minimum velocity should be checked for some minimum rate of flow which is not specified. Small sewers are usually designed to flow partially full, as discussed in Section 13.6.

11.9 RIGID CHANNELS CARRYING SEDIMENT-LADEN WATER

There is an upper and a lower limit for the channel velocity. A minimum limit is necessary in order to prevent sediment deposit, aquatic growth, or sulfide formation in the case of sanitary sewers. The velocity required to transport material in sewers is only slightly dependent on conduit shape and depth of flow, and primarily dependent on the particle size and specific weight. Generally, a minimum velocity of 2 to 3 ft/sec (0.7 to 1 m/s) is used for open channels and sanitary and storm sewers. A velocity of 2 ft/sec (0.7 m/s) will be sufficient to move a 15.0-mm-diameter organic or 2.0-mm sand particle (ASCE and Water Pollution Control Federation, 1982).

An upper limit of 6 ft/s (2 m/s) has been found acceptable and prevents appreciable erosion in grass-lined channels.

To design a rigid channel with sediment load, the procedure of Section 11.8.5 is followed. Then the selection is checked for velocity limits. If the velocity is less than 2 ft/s or 0.7 m/s, the slope is increased. The slope is reduced if the velocity exceeds 6 ft/s or 2 m/s (or concrete lining is used without changing the slope).

When the sediment concentration is large, the section is further checked for the sediment carrying capacity using mostly an empirical approach. If the designed channel is not able to carry the specified sediment load, the channel slope is increased.

EXAMPLE 11.8

Design a rigid-boundary earth channel to carry a discharge of 1.08 m³ per second.

SOLUTION

1. Based on the topography and channel alignment, $S = 0.001$.

2. For a clean earth channel, $n = 0.018$.

3. Designing a trapezoidal channel: Based on the material, the side slope $z = 1$.

4. From eq. (11.11b),

$$AR^{2/3} = \frac{Qn}{S^{1/2}} = \frac{1.08(0.018)}{(0.001)^{1/2}} = 0.615$$

5. Assume that $b/y = 1$

$$A = 2y^2 \text{ m}^2$$
$$P_w = 3.83y \text{ m}$$
$$R = \frac{2y^2}{3.83y} = 0.522y \text{ m}$$

6. $AR^{2/3} = (2y^2)(0.522y)^{2/3} = 1.30y^{8/3}$

   or $1.30y^{8/3} = 0.615, \quad y = 0.75 \text{ m}$

7. Check for velocity:

$$V = \frac{Q}{A} = \frac{1.08}{2(0.75)^2} = 0.96 \text{ m/s}$$

Since V of $0.96 > V_{min}$ of 0.7, it is OK.

8. For freeboard, $y = 0.75$ m or 2.46 ft

$$u = \sqrt{1.5(2.46)} = 1.92 \text{ ft} \quad \text{or} \quad 0.59 \text{ m}$$

9. Total channel depth $= 0.75 + 0.59 = 1.34$ m

11.10 LOOSE-BOUNDARY CHANNELS CARRYING SEDIMENT-FREE WATER

The design of such channels is based on the tractive force theory developed by Lane (1955) with regard to the concept of stream power. In a channel, a force is exerted by the flowing water, which pulls on the wetted surface. This is known as *tractive force, shear force* or *drag force*. In an alluvial channel, a limiting value is a force that is just sufficient to initiate the movement of particles which would otherwise remain on the bed and banks. This limiting force is known as the *critical force* or the *permissible tractive force*. The limiting force is evaluated in different ways. By equating the exerted force to the limiting force, one can derive a basic relation to determine the channel geometry and gradient.

When the alluvial material on the bed and banks of a stream channel is in a condition just sufficient to initiate movement of this material, the channel is called a *threshold channel*. These threshold conditions are produced over a long period of time and by a large number of precipitation events. The bankfull discharge can be considered a good measure of the threshold condition, below which the particles will not move appreciably. The tractive force theory can be applied to determine the geometric relations for a threshold channel.

11.10.1 Unit Tractive Force on Channel Boundary

As noted earlier, the force exerted by moving water on a wetted surface is called tractive or drag force, which is equal to the component of the weight of water in the direction of flow. From Figure 11.9,

$$\text{tractive force, } F = W\sin\theta = \gamma ALS$$

$$\text{unit tractive force} = \frac{\text{tractive force}}{\text{contact area}}$$

$$= \frac{\gamma ALS}{P_w L}$$

$$= \gamma RS \quad [\text{FL}^{-2}] \tag{11.13}$$

The unit tractive force or stress, however, is not uniformly distributed on the bed and the banks. The U. S. Bureau of Reclamation (1952) presented the curves for the distribution on bed and banks. For common trapezoidal sections, the maximum stress, referred to as the *theoretical unit tractive force*, may be taken as

$$\text{For bed: } \tau_0 = \gamma y S \quad [\text{FL}^{-2}] \tag{11.14}$$

$$\text{For sides: } \tau_s = 0.76\gamma y S \quad [\text{FL}^{-2}] \tag{11.15}$$

Figure 11.9   Theory of tractive force.

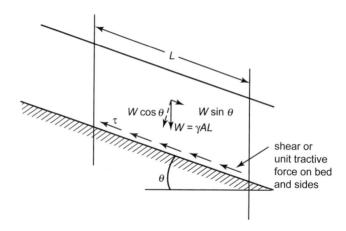

11.10.2 Critical Tractive Force

If the theoretical unit force is less than the limiting or critical tractive force, the channel will be stable. The critical tractive force is a function of the size of channel material, the sediment content of the water, and the shape of the channel. There are two distinct approaches for determining the critical tractive force: (1) the USBR method, and (2) the Stability Parameter method.

11.10.3 The USBR Method

The U.S. Bureau of Reclamation made the following recommendations for the critical tractive force for bed material.

1. For coarse noncohesive material, $\tau_{cr} = 0.4 \, d_{75}$ where τ_{cr} is in lbs/ft² and d_{75} is the diameter in inches of the particle of which 75% of bed material is finer by weight as determined by sieve analysis.

2. For fine noncohesive material, use the design curves.

3. For cohesive material, use the design curves based on void ratio and type of soil.

These approximate values have been compiled in Table 11.6.

On the sides of a channel, the particles on the slope are subject to downward gravity force in addition to the tractive force of the flowing water. This reduces the critical force by the following factor

$$K = \sqrt{1 - \frac{\sin^2 \alpha}{\sin^2 \phi}} \quad [\text{dimensionless}] \tag{11.16}$$

where

$\quad K$ = factor for permisssible stress on sides

$\quad \alpha$ = angle of channel side slope to the horizontal

$\quad \phi$ = angle of repose of material

The USBR concluded that for cohesive material, the force of cohesion is very large and the factor K can be disregarded. But for noncohesive material on the slopes, the critical

Table 11.6   Permissible or Critical Unit Tractive Force

Material	Size (mm)	Average permissible unit tractive force (lb/ft^2)
Small/medium boulders, cobbles, and shingles	64–256	1.0–2.0
Coarse gravel	8–64	0.15–1.0
Fine gravel	4–8	0.10–0.15
Coarse sand	0.5–2	0.05–0.08
Medium sand	0.25–0.5	0.05
Fine sand	0.06–0.25	0.05
Silt		0.2–0.3
Very compact clay		0.4–0.8
Compact clay		0.25–0.5
Fairly compact clay		0.15–0.25
Loose clay		0.05–0.09

tractive force (from Table 11.6) should be multiplied by factor K based upon the angle of repose of the soil.

The design procedure involves proportioning the section based on the maximum stress on the sides and checking for the maximum stress on the bottom. The steps are as follows:

1. Select a value of the critical tractive force for the channel material.
2. For noncohesive material, determine K for a predecided side slope.
3. Multiply the value of step 1 by K to obtain the permissible stress for sides of the channel.
4. Equate the permissible stress on sides to the theoretical value (i.e., $0.76\gamma yS$) and solve for y.
5. Substitute in Manning's equation to obtain the width, b, or if the width is given to determine Q.
6. Determine the theoretical shear stress on the bed = γyS. This should be less than the critical value of step 1.

EXAMPLE 11.9

Design a trapezoidal channel to carry a discharge of 1000 cfs. The channel laid on a slope of 0.0015 is excavated in coarse gravel with a 75% fine diameter of 30 mm (1.2 in.). Consider a side slope of 1:2, $n = 0.030$, and the angle of repose = $40°$.

SOLUTION

1. For coarse noncohesive material of 1.2 in., the critical stress = $0.4(1.2) = 0.48$ lb/ft^2
2. For a 1:2 slope, $\alpha = 26.57°$

3. $K = \sqrt{1 - \dfrac{(\sin 26.57)^2}{(\sin 40)^2}} = 0.72$

4. Permissible stress on sides = $0.72(0.48) = 0.346$

Section 11.10   Loose-Boundary Channels Carrying Sediment-Free Water   633

5. $0.76\gamma yS = 0.346$

or $\quad y = \dfrac{0.346}{0.76(62.4)(0.0015)} = 4.86$ ft

6. $\quad A = (b+9.72)4.86$

$P_w = (b+21.73)$

7. From Manning's equation,

$$1000 = \frac{1.49}{0.93}[b+9.72](4.86)\left[\frac{(b+9.72)4.86}{b+21.73}\right]^{2/3}(0.0015)^{1/2}$$

or

$$37.28 = \frac{(b+9.72)^{5/3}}{(b+21.73)^{2/3}}$$

By trial and error, $b = 34$ ft

8. Theoretical shear stress on the bed:

$$\tau_0 = \gamma yS = 62.4(4.86)(0.0015) = 0.45 < 0.48 \text{ OK}$$

9. Add a freeboard of 2.5 ft

11.10.4 The Stability Parameter Method

The net weight acting on a particle of diameter d_s under water is its submerged weight given by $(\gamma_s - \gamma)d_s A$. The ratio of tractive force $\tau_0 A$ to net weight is known as the *stability parameter*, commonly called the *Shields parameter* after A. Shields who introduced it first in 1936. At the inception of bed motion, the stability parameter has a critical value θ_{cr}. Thus, at critical stage

$$\theta_{cr} = \frac{\tau_0}{(\gamma_s - \gamma)d_s} \quad \text{[dimensionless]} \tag{11.17a}$$

or

$$\theta_{cr} = \frac{\gamma yS}{(\gamma_s - \gamma)d_s} \quad \text{[dimensionless]} \tag{11.17b}$$

where

θ_{cr} = critical Shields parameter

τ_0 = unit tractive force = γyS*

γ_s = specific weight of bed material

γ = specific weight of water

d_s = representative size of bed material

y = water depth

S = channel slope

* Realistically, $\tau_0 = \gamma RS$. However, the simplified use of $\tau_0 = \gamma yS$ yields the conservative design, since the computed exerted tractive force is on the higher side.

Conveyance Systems: Open Channel Flow   Chapter 11

Shields showed that the critical Shields parameter θ_{cr} is primarily a function of shear Reyolds number given by

$$Re^* = \frac{\rho v_* d_s}{\mu} \quad \text{[dimensionless]} \quad (11.18)$$

where

Re* = shear Reynolds number

v_* = shear velocity

μ = viscosity of water

The Shields diagram is presented in Figure 11.10. A few characteristics of the critical Shields parameter are as follows:

1. It never drops below 0.03.
2. For common sediments, it is never less than ¼ of the particle diameter in inches.
3. It achieves a fixed value of about 0.055 for a shear Reynolds number of over 450.
4. Likewise, when the shear Reynolds number is less than 2, it is a linear function approximately given by

$$\theta_{cr} = \frac{0.12}{Re^*}$$

Although the Shields curve is meant to represent the initiation of motion, more recent research indicates that this curve represents the permanent grain movement at all locations within the bed, and the critical values found with the Shields curve can be as much as twice the value required to cause occasional particle movement at some locations. However, the Shields parameter is widely used to determine the initiation of motion, but any other parameter based on the research can be used.

The design procedure incorporating the stability parameter is as follows:

1. Assume a critical Shields parameter, θ_{cr}.
2. From eq. (11.17b), determine depth, y.
3. Compute the shear Reynolds number from eq. (11.18).
4. Find the revised value of the Shields parameter from Fig. 11.10.
5. Repeat steps 2 through 4 until θ_{cr} stabilizes.
6. From Manning's eq. (11.10), compute b for given Q or compute Q for known b.

EXAMPLE 11.10

Solve Example 11.9. The specific gravity of sediment = 2.55. The representative grain size is 1 in.

Figure 11.10 Shields diagram.

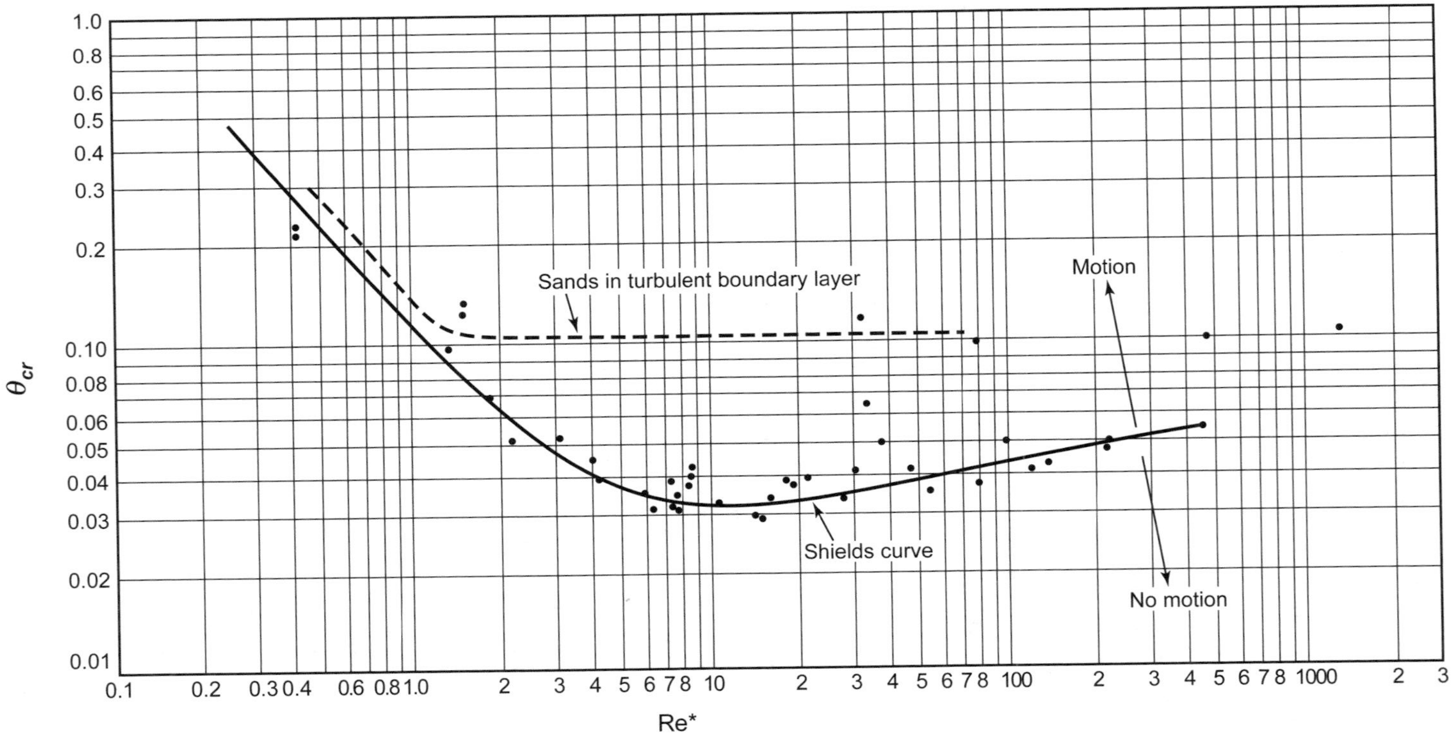

SOLUTION

1. Assume $\theta_{cr} = 0.055$

2. $\gamma_s = 2.55(62.4) = 159.12$ lbs/ft^3

$d_s = 1$ in. or 0.083 ft

$\tau_0 = \theta_{cr}(\gamma_s - \gamma)d_s$

$\quad = (0.055)(159.12 - 62.4)(0.083)$

$\quad = 0.44$ lb/ft^2

3. $\gamma y S = 0.44$

$y = \dfrac{0.44}{(62.4)(0.0015)} = 4.70$ ft

4. $v_* = \sqrt{gyS} = \sqrt{(32.2)(4.70)(0.0015)} = 0.476$

5. At 60°F, $\rho = 1.94$ slugs/ft^3, $\mu = 2.359 \times 10^{-5}$ lbs/ft^2

$$\text{Re}^* = \frac{(1.94)(0.476)(0.083)}{\left(2.359 \times 10^{-5}\right)} = 3249$$

6. From Fig. 11.10, $\theta_{cr} = 0.055$ stabilizes

7. $A = (b + 9.4)4.7$

$P_w = (b + 21.02)$

8. From Manning's equation

$$1000 = \frac{1.49}{0.03} \frac{\left[(b+9.4)(4.7)\right]^{5/3}}{(b+21.02)^{2/3}}(0.0015)^{1/2}$$

$$39.4 = \frac{(b+9.4)^{5/3}}{(b+21.02)^{2/3}}$$

$$b = 36 \text{ ft}$$

11.11 LOOSE-BOUNDARY CHANNELS CARRYING SEDIMENT-LADEN WATER

The theory of channels in erodible or alluvial material developed from the experience of irrigation engineers in the Indo-Gangetic plain in India. These channels with silty loam crust were designed based on rigid boundary hydraulics. The water was obtained from control structures on the river that carried up to 0.75% by weight of suspended load. In a few years the channels adjusted to the permanent or equilibrium or *in-regime* conditions. Engineers first tried to force the channels to run within the original designs. When these attempts repeatedly failed, they accepted the proposition that a single equation of rigid boundary was not sufficient; it had to be supplemented by a second equation representing the law of transportation of bed material, as well as a third equation representing the joint laws of the erosion of sides and bottom and the deposibility of suspended materials.

The channels had three degrees of freedom of adjustments of width, depth, and slope. Hence three equations were sufficient for the equilibrium condition. From a hydraulic point of view these three equations can be expressed by (1) Newton's second law of motion, commonly recognized as the momentum equation. The resistance equation in the form of Manning's formula or Darcy's equation is a version of this law. (2) The sediment transport equation. There is no universal formulation of the sediment transport law. Theories cover only a narrow section of the broad spectrum of conditions found in nature. (3) A morphological relation for channel characteristics, such as the ratio of width to depth.

11.11.1 Hypotheses of Stable Channel Design

Many hypotheses have been used as far as the morphological relation is concerned. For examples: (1) the channel adjusts its slope and geometry to maximize its transport capacity, (2) the shape is such that it maximizes the boundary friction factor, (3) the shape adjusts to attain the maximum flow efficiency, (4) the channel attains a form so that its mobility is minimal, (5) the channel is stable when the Froude number is minimum, (6) the channel bed adjusts for the maximum energy yield, which means the potential energy spent on excavating the bed form is minimum and the kinetic energy recovered is maximum, (7) the channel is in equilibrium when its rate of energy dissipation is the minimum needed to transport water and sediment, (8) the channel establishes a dimension so that the stream power is minimum, which means the channel has a minimal slope, (9) the channel follows the principle of least work according to which the rate of energy degradation (entropy) is minimum as permitted by the boundary conditions, (10) the channel follows the second law of thermodynamics of increase in entropy expressed by the minimum energy-characteristic equation, (11) the channel adjusts such that the sum of variances of velocity, depth, width, and slope is a minimum value.

11.11.2 The Regime Theory

A regime channel is a nonsilting, nonscouring equilibrium channel that carries its normal suspended load at a given discharge. Regime represents a long-term condition rather than an instantaneous state. Regime channels do not change over a period of several years.

The regime theory originates from the pioneering work of the irrigation engineers mentioned in Section 11.11. Lindley's regime concept (1919) constitutes the foundation, according to which the dimensions of a channel— width, depth and gradient— to carry a given supply of water (discharge) loaded with a given silt (sediment) are all fixed by nature.

Originally, the theory was developed for (1) fine-to medium-grained silts and fine sands, (2) small slopes, (3) low sediment loads, and (4) situations where the transported material and the material that formed the channel were the same or had similar physical characteristics. These conditions, however, have been relaxed, particularly in light of the extension of the regime concept into the power law theory. The regime theory presented the empirical relations for channel geometry and gradient based on field data from stable channels. Kennedy (1895), Lindley (1919), and Lacey (1930) developed the regime equations. Refinements of Lacey's equations were done by Inglis during 1941– 1947 and Blench during 1955– 1966. The regime concept was extended into the hydraulic geometry relationships for a channel in the form of power functions. Following the work of Leopold and

Maddock (1953) and Wolman (1955), voluminous studies have been done on channel hydraulic geometry. Some recent ones include Allen, Arnold and Byars (1994); Cheema, Marino, and deVries (1997); Deng and Zhang (1994); Huang and Nason (2000); Jia (1990); Rhoads (1991); and Singh, Deng and Yang (2003).

11.11.3 Lacey's Original Regime Theory

In 1930, Gerald Lacey was given the assignment of discovering equations for the design of alluvial channels. Lacey came up with an ingenious set of equations for the practical design of equilibrium or in-regime channels without explicitly specifying the relations for sediment transportation, erosion, and build-up of channels, which were not known then and even today defy exact analytical presentation.

Although many regime equations have been proposed, Lacey's theory is still the most popular. Lacey's three basic relations are:

Velocity-depth relation

$$V = 1.17 (f R)^{1/2} \quad \text{[unbalanced]} \tag{11.19}$$

Width-discharge relation

$$P_w = 2.67 Q^{1/2} \quad \text{[unbalanced]} \tag{11.20}$$

Velocity-slope relation

$$V = \frac{1.346}{n_a} R^{3/4} S^{1/2} \quad \text{[unbalanced]} \tag{11.21}$$

where

$\quad V$ = mean velocity, ft/s

$\quad R$ = hydraulic radius, ft

$\quad P_w$ = wetted perimeter, ft

$\quad S$ = longitudinal slope

$\quad n_a$ = absolute rugosity, given as follows

$$n_a = 0.0225\, f^{1/4} \quad \text{[unbalanced]} \tag{11.22}$$

$\quad f$ = Lacey's silt factor, given as follows

$$f = 1.59 D_{50}^{1/2} \quad \text{[unbalanced]} \tag{11.23}$$

$\quad D_{50}$ = mean grain diameter, mm

The values of f for various materials are indicated in Table 11.7.

11.11.4 Hydraulic Basis of the Regime Theory

From a hydraulic point of view, the momentum, sediment transportation, and channel morphology are three relations that should be satisfied. These are discussed below:

1. Equation (11.21), which is a form of Manning's equation, is known as the momentum equation. When it is compared to Manning's equation (11.9), the Lacey's rugosity can be expressed as

$$n_a = 0.9 n R^{1/12} \quad \text{[unbalanced]} \tag{11.24}$$

whereas the combination of (11.22) and (11.23) provides

$$n_a = 0.0252 D_{50}^{1/8} \quad \text{[unbalanced]} \tag{11.25}$$

In regime relations, the rugosity, n_a, being related to bed size only as given by eq. (11.25), is too simplistic a relation. According to eq. (11.24), n_a should be a function of the bed roughness and the channel section.

When n_a is substituted from eq. (11.22) into eq. (11.21) and then f is eliminated using eq. (11.19), the result is

$$V = 16 R^{2/3} S^{1/3} \quad \text{[unbalanced]} \tag{11.26}$$

This is another momentum equation, according to which the velocity is independent of roughness. This means that for any given cross section and slope, the velocity is the same for a channel of any material whether it has boulder, gravel, or sand on the bed. This is not a true condition.

2. Lacey did not use any explicit relation for sediment transport. However, if the rate of bed load is given by

$$Q_s = K P_w V^3 \quad \text{[unbalanced]} \tag{11.27}$$

where K is a constant, then the concentration of sediment in the channel can be given by

$$C = \frac{Q_s}{Q} = \frac{K P_w V^3}{AV} = \frac{K V^2}{R}$$

or

$$V = \left(\frac{C}{K} \right)^{1/2} R^{1/2} \quad \text{[unbalanced]} \tag{11.28}$$

Equation (11.28) is similar to eq. (11.19), if $f = 0.73 C/K$. Thus, eq. (11.19) can be viewed as a sediment transport equation. But then Lacey's silt factor is not only a function of sediment size, it also should relate to the concentration of the bed material.

3. Equation (11.20) is a form of the morphological relation to predict the width of an alluvial channel.

Thus, Lacey's empirical relations have theoretical backing. However, the presence of two momentum equations (11.21) and (11.26) creates a redundancy. The values of slope calculated from these two equations are not the same. Similarly, the silt factor f computed from (11.19) is not the same as computed from eq. (11.21) after substituting n_a in terms of f from eq. (11.22).

The main problem arises due to the restricted definition of the silt factor by Lacey. As stated in items (1) and (2) above, a more definitive relation is required between the silt factor and the concentration of the sediment and a better relation between rugosity, n_a and the size and bed configuration of the channel.

Table 11.7 Silt Factors and Regime Parameters

Material	Size, mm	Silt Factor, f		c		k		s	
		min	max	for f min	for f max	for f min	for f max	for f min	for f max
Massive boulders	600	39.60	39.60	0.1379	0.1379	2.7290	2.7290	0.2563	0.2563
Large stones		38.60	38.60	0.1391	0.1391	2.7058	2.7058	0.2456	0.2456
Large boulders, shingle, heavy sand		20.90	20.90	0.1706	0.1706	2.2054	2.2054	0.0884	0.0884
Small medium boulders	64–256	6.12	9.75	0.2569	0.2200	1.4645	1.7104	0.0114	0.0248
Coarse gravel	8–64	4.68	4.68	0.2810	0.2810	1.3392	1.3392	0.0073	0.0073
Fine gravel	4–8	2.00	2.00	0.3730	0.3730	1.0087	1.0087	0.0018	0.0018
Coarse sand	.5–2	1.44	1.56	0.4162	0.4052	0.9041	0.9286	0.0010	0.0012
Medium sand	.25– 5	1.31	1.31	0.4295	0.4295	0.8760	0.8760	0.0009	0.0009
Fine sand	.06–.25	1.10	1.30	0.4553	0.4306	0.8265	0.8738	0.0007	0.0009
Silt (colloidal)		1.00	1.00	0.4700	0.4700	0.8006	0.8006	0.0006	0.0006
Fine silt (colloidal)		0.40	0.90	0.6379	0.4868	0.5899	0.7730	0.0001	0.0005

11.11.5 Combining the Regime Theory with the Power Function Theory

The regime theory is a concept. The regression analysis fitting technique can be applied to this concept to obtain the power function relations. The direct reduction of Lacey's equations to power function relations can be obtained as follows:

The basic continuity equation is

$$Q = P_w R V \tag{a}$$

P_w and R in eq. (a) are substituted in terms of Q and V respectively from Lacey's equations (11.20) and (11.19). The following relation results:

$$V = \frac{f^{1/3}}{1.249} Q^{1/6} \quad \text{[unbalanced]} \tag{11.29}$$

Substituting V from eq. (11.29) into eq. (11.19)

$$R = \frac{0.47}{f^{1/3}} Q^{1/3} \quad \text{[unbalanced]} \tag{11.30}$$

and substituting V and R from eq. (11.29) and (11.30) into eq. (11.26)

$$S = \frac{f^{5/3}}{1800} Q^{-1/6} \quad \text{[unbalanced]} \tag{11.31}$$

Thus, the following power function relations of the Leopold and Maddock (1953) type emerge:

$$P_w = a\, Q^{1/2} \quad \text{[unbalanced]} \tag{11.32}$$

$$R = c\, Q^{1/3} \quad \text{[unbalanced]} \tag{11.33}$$

$$V = k\, Q^{1/6} \quad \text{[unbalanced]} \tag{11.34}$$

$$S = s\, Q^{-1/6} \quad \text{[unbalanced]} \tag{11.35}$$

where

$$a = 2.67$$

$$c = \frac{0.47}{f^{1/3}}$$

$$k = \frac{f^{1/3}}{1.249}$$

$$s = \frac{f^{5/3}}{1800}$$

The parameters a, c, k, and s are all related to the silt factor f. The values of these are computed and listed in Table 11.7 for different types of soils.

The slope computed by eq. (11.35) is a minimum slope. If the available slope is less than the computed slope, the channel should be widened or the excess head should be absorbed in a drop structure.

EXAMPLE 11.11

Design a channel so that its maximum discharge is 2000 cfs. The alluvial material is fine sand. The available slope is 1 in 4000.

SOLUTION

1. $P_w = 2.67\,(2000)^{1/2} = 119$ ft

2. From Table 11.7, $c = 0.431$
   $R = cQ^{1/3} = 0.431(2000)^{1/3} = 5.42$ ft

3. $A = P_w R = 119(5.42) = 645$ ft^2

4. Assuming side slopes of 1:1

$$A = (b + y)y = 645 \tag{a}$$

and

$$P_w = (b + 2.83y) = 119 \tag{b}$$

Solving (a) and (b)

$$b = 102 \text{ ft}, \; y = 6 \text{ ft}$$

5. From Table 11.7, $s = 0.0009$

$S = sQ^{-1/6} = 0.0009\,(2000)^{-1/6} = 0.00025$

Available S is 1 in 4000 or 0.00025

Computed slope is OK

6. Freeboard: From eq. (11.12), $u = \sqrt{(2.5)(6)} = 3.9$ ft

7. Total depth $= 6 + 3.9 = 9.9$ or 10 ft

11.12 GRADUALLY VARIED FLOW

11.12.1 Dynamic Equation of Gradually Varied Flow

When the gravity force causing the flow is not balanced with the resisting drag force, the depth varies gradually along the length of the channel. The dynamic equation of gradually varied flow is derived from the energy principle and indicates the slope of the water surface in the channel.

In Figure 11.11, the total energy at point 1, including the energy coefficient,* is

$$H = Z_1 + y + \alpha \frac{V^2}{2g} \tag{a}$$

Differentiating with respect to the channel bottom as the x-axis:

$$\frac{dH}{dx} = \frac{dZ_1}{dx} + \frac{dy}{dx} + \alpha \frac{d}{dx}\left(\frac{V^2}{2g}\right) \tag{b}$$

---

* The mean velocity is used in the relation. To account for the nonuniform distribution of velocity across a channel

section, an energy coefficient α is included, expressed as $\alpha = \dfrac{\Sigma v^3 \Delta A}{V^3 A}$

Figure 11.11 Gradually varied flow.

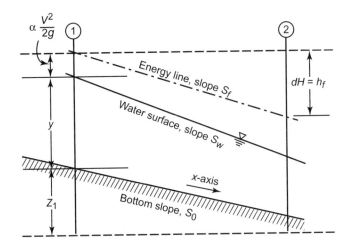

If the level increasing in the direction of flow is assumed to be positive, then $dH/dx = -S_f$, $dZ_1/dx = -S_0$, and

$$\frac{d}{dx}\left(\frac{V^2}{2g}\right) = \frac{d}{dy}\left(\frac{V^2}{2g}\right)\frac{dy}{dx}$$

Equation (b) becomes

$$\frac{dy}{dx} = \frac{S_0 - S_f}{1 + \alpha\left[d\left(V^2/2g\right)/dy\right]} \quad \text{[dimensionless]} \tag{11.36}$$

This is the equation of gradually varied flow. To reduce the equation further, it is considered that with the energy grade, S_f, used for the slope term in Manning's equation, that formula can be used for the gradually varied flow through a section, that is,

$$Q = \frac{1.49}{n}AR^{2/3}S_f^{1/2} \quad [L^3 T^{-1}] \quad \text{(English units)} \tag{11.37a}$$

$$Q = \frac{1}{n}AR^{2/3}S_f^{1/2} \quad [L^3 T^{-1}] \quad \text{(metric units)} \tag{11.37b}$$

or

$$Q = K\sqrt{S_f} \tag{c}$$

where K is the general expression for the conveyance. Also, in the case of uniform flow,

$$Q = K_n\sqrt{S_0} \tag{d}$$

where K_n is the normal flow conveyance. From eqs. (c) and (d),

$$\frac{S_f}{S_0} = \frac{K_n^2}{K^2} \tag{e}$$

The denominator term of eq. (11.36) may be developed as follows:

$$\alpha \frac{d}{dy}\left(V^2/2g\right) = \alpha \frac{d}{dy}\left(\frac{Q^2}{2gA^2}\right) = -\alpha \frac{Q^2}{g}\left(\frac{1}{A^3}\right)\frac{dA}{dy} \tag{f}$$

Since $dA/dy = T$ and in general terms, $Z = \sqrt{A^3/T}$,

$$\alpha \frac{d}{dy}\left(\frac{V^2}{2g}\right) = -\alpha \frac{Q^2}{gZ^2} \tag{g}$$

For the critical flow, $Z_c = Q/\sqrt{g/\alpha}$; hence

$$\alpha \frac{d}{dy}\left(\frac{V^2}{2g}\right) = -Z_c^2/Z^2 \tag{h}$$

Substituting eqs. (e) and (h) in eq. (11.36), we have

$$\frac{dy}{dx} = S_0 \frac{1-\left(K_n/K\right)^2}{1-\left(Z_c/Z\right)^2} \quad \text{[dimensionless]} \tag{11.38}$$

Equation (11.38) is another form of the gradually varied flow equation, which is convenient for the evaluation.

11.12.2 Types of Flow Profile Curves

The integration of eq. (11.38) will represent the surface curve of the flow. The shape or profile of the surface curve depends on (1) the slope of the channel, and (2) the depth of flow compared to the critical and normal depths. The classification is as follows:

Sign convention

1. If the water surface is rising in the direction of flow, the curve, known as the *backwater curve*, is positive.
2. If the water surface is dropping, the curve, known as the *drawdown curve*, is negative.

Channel slopes

1. When $y_n > y_c$, the slope is mild.
2. When $y_n < y_c$, the slope is steep.
3. When $y_n = y_c$, the slope is critical.
4. When $S_0 = 0$, the slope is horizontal. For horizontal slope, $y_n = \infty$.
5. When $S_0 < 0$, the slope is adverse. For adverse slope, y_n is negative or nonexistent.

Flow profiles

If the lines are drawn at the critical depth and the normal depth parallel to the channel bottom, three zones are formed. Zone 1 is the space above the upper line, zone 2 is the space between the two lines, and zone 3 is the space between the lower line and the channel bottom. The zone in which the water surface lies determines the flow profile, as shown in Figure 11.12.

Figure 11.12   Types of flow profiles.

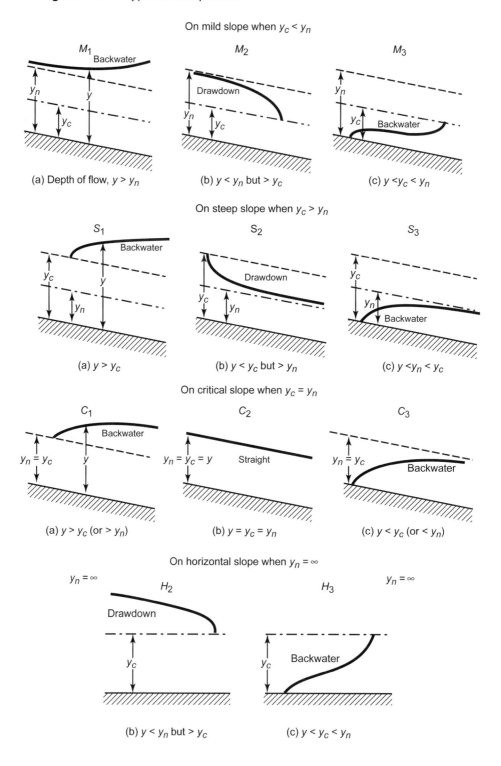

On mild slope when $y_c < y_n$

M_1

Backwater

y_n y

y_c

(a) Depth of flow, $y > y_n$

M_2

Drawdown

y_n y_c

(b) $y < y_n$ but $> y_c$

M_3

y_n

y_c Backwater

(c) $y < y_c < y_n$

On steep slope when $y_c > y_n$

S_1

Backwater

y_c y

y_n

(a) $y > y_c$

S_2

Drawdown

y_c y_n

(b) $y < y_c$ but $> y_n$

S_3

y_c

y_n

Backwater

(c) $y < y_n < y_c$

On critical slope when $y_c = y_n$

C_1

Backwater

$y_n = y_c$ y

(a) $y > y_c$ (or $> y_n$)

C_2

$y_n = y_c = y$ Straight

(b) $y = y_c = y_n$

C_3

$y_n = y_c$

Backwater

(c) $y < y_c$ (or $< y_n$)

On horizontal slope when $y_n = \infty$

$y_n = \infty$

H_2

Drawdown

y_c

(b) $y < y_n$ but $> y_c$

H_3

$y_n = \infty$

y_c Backwater

(c) $y < y_c < y_n$

Figure 11.12 (Continued)   Types of flow profiles.

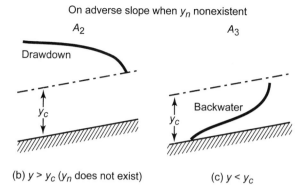

On adverse slope when y_n nonexistent

A_2 A_3

Drawdown

y_c y_c Backwater

(b) $y > y_c$ (y_n does not exist)   (c) $y < y_c$

11.12.3 Flow Profile Analysis

The analysis predicts the general shape of the flow profile in a longitudinal section of a channel without performing the quantitative analysis. For a channel of constant slope, the conditions described in the preceding section determine the type of flow profile. A break in the slope of a channel results in a change in the flow condition as well. A surface curve is often formed to negotiate the change of pattern of flow. Chow (1959, p. 232) has indicated 20 typical flow profiles for a combination of two different slopes by break in the channel slope. Certain points in the channel reach serve as a control section where the depth of flow is fixed (i.e., either it is y_c or y_n or has some other known value).

11.13 COMPUTATION OF FLOW PROFILE

The analytical determination of the shape of flow profile essentially is a solution of eq. (11.38). Since the variables on the right side of the equation cannot be expressed explicitly in terms of y, the exact integration of the equation is not practically possible. There are four approaches to computing the surface profile:

1. Graphical or numerical integration method
2. Analytical or direct integration method
3. Direct step method
4. Standard step method

   Two of these are described in detail. The water surface elevation at the start of the curve from which the computation starts is a control section. The computation should proceed upstream from the control section in subcritical flow and in the downstream direction for supercritical flow.

11.13.1 Numerical Integration Method

Consider a channel section having a depth y_1 at x_1 and y_2 at x_2 as shown in Figure 11.13(a).

$$x_2 - x_1 = \int_{x_1}^{x_2} dx$$

or

$$x_2 - x_1 = \int_{y_1}^{y_2} \frac{dx}{dy} dy$$

The right side indicates the area under the dx/dy versus y curve, as shown in Figure 11.13(b). If a to b is considered a straight line for a small difference in y_1 and y_2, then

$$x_2 - x_1 = \frac{\left[(dx/dy)_1 + (dx/dy)_2\right]}{2}(y_2 - y_1) \quad [\text{L}] \tag{11.39}$$

The procedure comprises solving eq. (11.39) by the following steps:

1. Select several values of y starting from the control point.

2. For each value of y, calculate dx/dy by the inverse of eq. (11.38).

3. Using eq. (11.39), calculate x for two successive values of y.

For a backwater curve, the y values should be selected at close intervals near the tail part of the curve.

Figure 11.13   Derivation of numerical integration method.

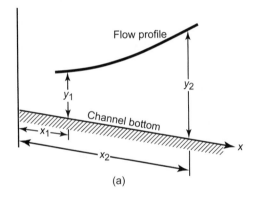

(a)

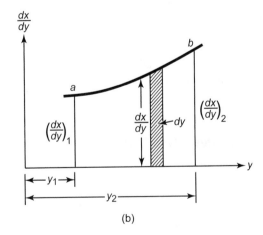

(b)

EXAMPLE 11.12

A trapezoidal channel with a bottom width of 4 m and side slopes of 1:4 carries a discharge of 30 m³/s. The channel has a constant bed slope of 0.001. A dam backs up the water to a depth of 3.0 m just behind the dam. Compute the backwater profile to a depth 5% greater than the normal channel depth. $n = 0.025$, $\alpha = 1.0$.

SOLUTION

1. The channel is the same as given in Examples 11.3 and 11.6.

2. From Example 11.3, critical depth $y_c = 1.22$ m

3. From Example 11.6, normal depth $y_n = 1.90$ m

4. Since $y_n > y_c$, the channel has a mild slope.

5. At the control point, y of 3 m is greater than y_n, thus the profile is of the M_1 type curve.

6. Section factor for critical flow,

$$Z_c = \frac{Q}{\sqrt{g/\alpha}} = \frac{30}{\sqrt{9.81}} = 9.58$$

7. Conveyance for uniform flow,

$$K_n = \frac{Q}{\sqrt{S_0}} = \frac{30}{\sqrt{0.001}} = 948.68$$

8. At the starting point of the curve, the control section depth = 3 m. The last computed point which is 5% greater than $1.90 = 1.05(1.9) = 2.0$ ft.

9. Since the flow is subcritical, computation proceeds upstream from the dam as the origin. The computations are arranged in Table 11.8.

10. The profile has been shown in Figure 11.14 by a plot between y (column 1) and x (column 10).

Figure 11.14   Backwater profile for Example 11.12.

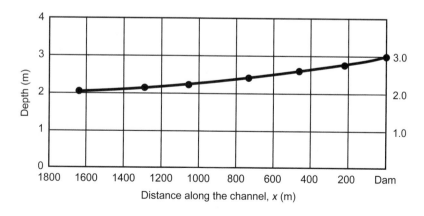

11.13.2 Direct Step Method

This method directly uses the energy principle from section to section in the entire reach of the channel. It is applicable to prismatic channels. Applying the energy principle at points 1 and 2 of Figure 11.11 gives

$$Z_1 + y_1 + \alpha_1 \frac{V_1^2}{2g} = Z_2 + y_2 + \alpha_2 \frac{V_2^2}{2g} + h_f \tag{a}$$

or

$$(Z_1 - Z_2) - h_f = \left(y_2 + \alpha_2 \frac{V_2^2}{2g} \right) - \left(y_1 + \alpha_1 \frac{V_1^2}{2g} \right) \tag{b}$$

or

$$S_0 \Delta x - S_f \Delta x = E_2 - E_1 \tag{c}$$

where Δx is the distance between the two sections, E_1 and E_2 are specific energies, and $h_f = S_f \Delta x$, or

$$\Delta x = \frac{E_1 - E_2}{S_f - S_0} \tag{d}$$

If the energy grade between the two sections is considered to be the average of the grade at sections 1 and 2, then

$$\Delta x = \frac{E_1 - E_2}{\overline{S}_f - S_0} \quad [\text{L}] \tag{11.40}$$

with

$$\overline{S}_f = \frac{S_{f1} + S_{f2}}{2} \quad [\text{dimensionless}] \tag{11.41}$$

and

$$S_f = \frac{V^2 n^2}{2.22 R^{4/3}} \quad (\text{English units}) \quad [\text{dimensionless}] \tag{11.42a}$$

$$S_f = \frac{V^2 n^2}{R^{4/3}} \quad (\text{metric units}) \quad [\text{dimensionless}] \tag{11.42b}$$

The steps of the procedure are as follows:

1. Select several values of y starting from the control point.
2. For a selected y, calculate A, R, $R^{4/3}$, and $V (= Q/A)$.
3. For a selected y, also calculate the velocity head $[\alpha(V^2/2g)]$, the specific energy, $E (= y + \alpha V^2/2g)$, and the energy slope, eq. (11.42).
4. For two successive values of y, determine the difference between the specific energy, ΔE, and the average of the energy slope, $\overline{S}_f$.
5. Compute Δx from eq. (11.40).

Table 11.8 Computation of the Flow Profile by the Numerical Integration Method

(1)	(2)	(3)	(4)	(5)	(6)	(7)	(8)	(9)	(10)
y Select:	T^a	A^b	R^c	$R^{2/3}$	$K = \frac{1}{n}AR^{2/3\,d}$	$Z = \sqrt{\dfrac{A^3}{T}}^{\,d}$	$\dfrac{dx}{dy}$ [Inverse of eq. (11.38)]^e	Δx [eq. (11.39)] (m)	x Cumulated (m)
3.0	28.0	48.0	1.67	1.41	2707.2	62.85	1113.50	228	228
2.8	26.4	42.56	1.57	1.35	2298.2	54.04	1167.50	243	471
2.6	24.8	37.44	1.47	1.29	1931.9	46.00	1260.60	270	741
2.4	23.2	32.64	1.37	1.23	1605.89	38.72	1442.03	333	1074
2.2	21.6	28.16	1.27	1.17	1317.89	32.15	1891.18	218	1292
2.1	20.8	26.04	1.22	1.14	1187.42	29.14	2466.0	332	1624
2.0	20.0	24.0	1.17	1.11	1065.60	26.29	4181.25		

^{a} $T = 4 + 8y$

^{b} $A = (4 + 4y)y$

^{c} $R = \dfrac{A}{P_w} = \dfrac{(4+4y)y}{4 + 8.24y}$

^{d} K and Z calculated for each y selected.

^{e} $\dfrac{dx}{dy} = \dfrac{1 - \left(\dfrac{Z_c}{Z}\right)^2}{S_0\left[1 - \left(\dfrac{K_n}{K}\right)^2\right]}$

651

The analytical integrating method requires use of the varied flow function tables (see Appendix D-2 of Chow, 1959). Many alternative computation procedures have been proposed under this method. The standard step method, also based on the energy principle, is a trial-and-error procedure wherein the depth of flow is determined for a given channel distance and not the inverse as in the other methods.

EXAMPLE 11.13

Determine the flow profile using the data of Example 11.12 by the direct step method.

SOLUTION   The computations are arranged in Table 11.9.

11.14 RAPIDLY VARIED FLOW

This involves a sharp change in the curvature of the water surface, sometimes producing discontinuity in the flow profile. The streamlines are so disturbed that the pressure distribution is not hydrostatic. The rapid variation in flow conditions occurs within a short reach. As a result, the energy loss due to boundary friction is negligible with rapidly varied flow, while it is dominant in gradually varied flow conditions. The overall energy losses are substantial due to turbulent conditions. The problems related to rapidly varied flow are usually studied on an individual basis, with each phenomenon given a specific treatment. Flow over spillways and hydraulic jumps are two common cases of rapidly varied flow. The former is described in Chapter 9 and the latter is described below briefly.

11.14.1 Hydraulic Jump

When a shallow stream of high velocity impinges on water of sufficient depth, the result is usually an abrupt rise in the surface in the region of impact. This phenomenon is known as *hydraulic jump*. A similar phenomenon takes place when the flow passes from a steep slope to a mild slope or when an obstruction is met in the passage of a supercritical flow. For formation of a jump, the flow should be supercritical, which converts into subcritical flow after the jump.

In the process, a substantial loss of energy takes place. Since unknown energy losses are involved in the jump, the use of the energy principle is not practical. The principle of momentum is used instead, as described in Section 9.4. By this principle, the change in the forces between two sections is equated with the change in the rate of momentum, which is equal to the mass of water multiplied by the change in velocity. Commonly, the relation is developed for horizontal or slightly inclined channels in which the weight of water between the sections and the boundary friction are disregarded. It is applicable to most field channels.

The following formulas are derived for rectangular channels from the momentum principle on the basis described above:

$$\frac{D_2}{D_1} = \frac{1}{2}\left(\sqrt{1+8\text{Fr}_1{}^2}-1\right) \quad \text{[dimensionless]} \tag{11.43}$$

where

D_1 and D_2 = depth before and after the jump

Fr_1 = Froude number before the jump = $V_1/\sqrt{gD_1}$

Table 11.9 Computation of the Flow Profile by the Direct Step Method

(1)	(2)	(3)	(4)	(5)	(6)	(7)	(8)	(9)	(10)	(11)	(12)	(13)
y Select:	A	$R=\dfrac{A}{P_w}$	$R^{4/3}$	$V=\dfrac{Q}{A}$	$\alpha\dfrac{V^2}{2g}$	$E=y+\dfrac{\alpha V^2}{2g}$	$\Delta E_s = E_1-E_2$	$S_f=\dfrac{n^2V^2}{R^{4/3}}$ $(\times 10^3)$	$\bar S_f{}^a$ $(\times 10^3)$	$\bar S_f-S_0$ $(\times 10^3)$	Δx [eq (11.40)]	x Cumulated
3.0	48.0	1.67	2.00	0.625	0.020	3.020		0.122				228
							0.195		0.147	−0.853	228	
2.8	42.56	1.57	1.82	0.705	0.025	2.825		0.171				470
							0.192		0.207	−0.793	242	
2.6	37.44	1.47	1.66	0.801	0.033	2.633		0.242				740
							0.190		0.296	−0.704	270	
2.4	32.64	1.37	1.51	0.919	0.043	2.443		0.350				1066
							0.185		0.433	−0.567	326	
2.2	28.16	1.27	1.37	1.065	0.058	2.258		0.517				1279
							0.090		0.578	−0.422	213	
2.1	26.04	1.22	1.30	1.152	0.068	2.168		0.638				1589
							0.088		0.716	−0.284	310	
2.0	24.0	1.17	1.23	1.250	0.080	2.080		0.794				

[a] Average of successive values of col. 9.

653

The depths D_1 and D_2 are referred to as conjugate depths. Equation (11.43) can be used to ascertain D_2 when D_1 is known. In the equation, subscripts 1 and 2 can be replaced by each other. Thus the equation can also be used to determine the prejump depth, D_1, for known post-jump depth, D_2 and Fr_2.

The discharge through the jump where b is the width of a rectangular channel can be given by

$$Q = b\left[(gD_1D_2)\frac{D_1+D_2}{2}\right]^{1/2} \quad [L^3T^{-1}] \tag{11.44}$$

The energy dissipated in a jump is computed from

$$E_{loss} = \frac{(D_2-D_1)^3}{4D_1D_2} \quad [L] \tag{11.45}$$

There are many applications of hydraulic jump. The main use is to dissipate energy in water flowing over spillways or weirs to prevent scouring downstream of the structure.

EXAMPLE 11.14

Water flows at a rate of 360 cfs in a rectangular channel of 18 ft width with a depth of 1 ft. (a) Is a hydraulic jump possible in the channel? (b) If so, what is the depth of flow after the jump? (c) How much energy is dissipated through the jump?

SOLUTION

(a) To determine the potential for hydraulic jump,

$$A = 18 \times 1 = 18 \text{ ft}^2$$

$$V = \frac{Q}{A} = \frac{360}{18} = 20 \text{ ft/s}$$

$$Fr_1 = \frac{V_1}{\sqrt{gD_1}} = \frac{20}{\sqrt{32.2(1)}} = 3.52$$

Since $Fr_1 > 1$, supercritical flow, jump can form.

(b) Depth of flow post-jump,

$$\frac{D_2}{D_1} = \frac{1}{2}\left(\sqrt{1+8(3.52)^2} - 1\right) = \frac{9.0}{2}$$

$$D_2 = \frac{1}{2}(9.0)(1) = 4.50 \text{ ft}$$

(c) Loss of energy,

$$E_{loss} = \frac{(D_2-D_1)^3}{4D_1D_2}$$

$$= \frac{(4.5-1.0)^3}{4(1.0)(4.5)}$$

$$= 2.38 \text{ ft-lb/lb}$$

PROBLEMS

11.1 Compute the hydraulic radius, hydraulic depth, and section factors Z_c and Z_n for the trapezoidal channel section shown in Fig. P11.1.

Figure P11.1

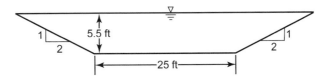

11.2 In a conduit with a diameter of 4.5 ft, the depth of flow is 4.0 ft. **(a)** Determine the hydraulic radius, hydraulic depth, and section factors for critical and normal flows. **(b)** Determine the alternate depth of flow that will carry the same discharge.

11.3 A trapezoidal channel with a side slope of 1(vertical):2(horizontal) and a bottom width of 10 ft carries a discharge of 300 cfs. **(a)** Plot the specific energy curve for the channel. **(b)** At what depth will the critical flow occur? **(c)** Determine the alternate depth to a 4.0-ft depth of flow. **(d)** What is the state of flow at the alternate depth?

11.4 Prove that for a rectangular channel at the critical state of flow, **(a)** the depth of flow is equal to two-thirds of the minimum specific energy, and **(b)** the velocity head is equal to one-third of the minimum specific energy.

11.5 A right-angled triangular channel carries a flow of 20 m^3/s. Determine the critical depth and the critical velocity of flow.

11.6 A 40-in. conduit carries a discharge of 25 cfs. Determine the critical depth using the geometric elements in Table 11.1.

11.7 A trapezoidal channel has a bed width of 3.5 m and side slope of 30° from the horizontal. Determine the critical depth and the critical velocity for a flow of 22 m^3/s.

11.8 Determine the discharge through the following sections for a normal depth of 5 ft; $n = 0.013$, and $S = 0.2\%$.

a. A rectangular section 20 ft wide.

b. A circular section 20 ft in diameter.

c. A right-angled triangular section.

d. A trapezoidal section with a bottom width of 20 ft and side slope of 1(vertical): 2 (horizontal).

e. A parabolic section having a top width of 20 ft for a 5-ft depth.

[*Hint:* $A = \frac{2}{3}Ty$, $P_w = T + \frac{8}{3}(y^2/T)$, where T is the top width and y is the depth.]

11.9 In a 3.0-m-wide rectangular channel of bed slope 0.0015, a discharge of 4 m^3/s is observed at a depth of 0.8 m. Estimate the discharge when the depth is doubled.

11.10 Determine the conveyance of the channel in Problem 11.9.

11.11 A long, rectangular channel of 15 ft width, lined with concrete, is supplied by a reservoir as shown in Fig. P11.11. Disregarding the entry losses into the channel, determine the normal depth of flow and discharge through the channel. $n = 0.015$.

Figure P11.11

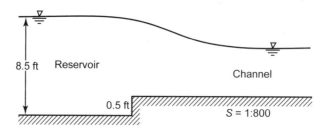

11.12 A discharge of 5.2 m³/s occurs in a rectangular channel of 2 m width having a bed slope of 1:625. Determine the **(a)** normal depth of flow, **(b)** critical depth of flow, and **(c)** state of flow. $n = 0.013$.

11.13 The channel in Problem 11.7 is excavated in smooth rock to a bed slope of 0.2%. Determine the **(a)** normal depth of flow, **(b)** critical slope, and **(c)** state of flow in the channel.

11.14 What diameter of a circular conduit flowing full would be required to carry the same quantity of flow as in a concrete trapezoidal channel of 20 ft width and 45° side slopes, running at a depth of 3.0 ft?

11.15 A concrete-lined trapezoidal channel has to be constructed to carry a discharge of 500 cfs. Design the channel. Assume that the following values were fixed based on the site conditions:

1. Bed slope = 0.002
2. $n = 0.015$
3. Side slope, $z = 1$ (vertical):1.5 (horizontal)
4. b/y ratio = 2.0

11.16 Design the channel of Problem 11.15 as a most-hydraulic-efficient channel (of semicircular section).

11.17 Design the channel of Problem 11.15 as a best-hydraulic rectangular section.

11.18 Design a storm sewer laid on a minimum grade to carry a peak flow of 10 cfs. Consider that the depth at the peak flow is 0.7 times the sewer diameter. The minimum flow to be maintained is 3.0 cfs. $n = 0.013$.

11.19 A circular sewer laid on a 1% grade is to carry 3.0 cfs when full. **(a)** Design the sewer. **(b)** At a dry weather flow of 0.6 cfs, what is the depth and velocity of flow? **(c)** What is the depth of flow at minimum velocity of 2 ft/s? ($n = 0.015$.)

11.20 A water supply conduit of 36 in. diameter was laid 20 years ago on a slope of 0.1% with $n = 0.015$. What was the discharge capacity and velocity at full flow at the time of installation? At the present time the conduit carries a flow of 15 cfs when full. Determine the present velocity and present value of n.

11.21 A sewer of vitrified clay has been laid on a gradient of 1:100. It receives the flow from 300 houses. The per capita daily water supply is 200 liters/day. The population density is 3.5 persons per house. Design the sewer. Assume that (1) the sewage quantity is equal to the water consumption; (2) the sewer is to be designed for the maximum hourly rate of flow, which is 400% of the average daily rate of flow; and (3) the minimum hourly flow rate is

50% of the average flow rate. For a minimum velocity of 0.61 m/s, change the bottom slope, if necessary.

11.22 A trapezoidal irrigation canal is excavated in silt to convey a discharge of 10 m³/s on a bed slope of 1:4000. The side slope is 1 vertical: 2 horizontal. Design the channel by the tractive force USBR method. Angle of friction = 28° and $n = 0.03$. For the permissible tractive force use the smallest value from Table 11.6.

11.23 A 35-ft-wide rectangular irrigation channel is excavated in fine gravel. Determine the total depth and discharge by the tractive force USBR method. Use the tractive force at bottom only. For the critical tractive force, use the average value from Table 11.6. The slope of the channel is 0.0008. $n = 0.03$.

11.24 Solve Problem 11.23 by the tractive force theory using the Shields parameter. The specific gravity of sediment is 2.65. The representative grain size is 8 mm. Assume water temperature of 60°F.

11.25 A trapezoidal canal has a slope of 2.5 ft per mile. The bankfull cross section has a bottom width of 50 ft, surface width of 170 ft, and depth of 12 ft. The canal material is largely sand with a representative size of 0.5 mm. Is this canal stable or erodible according to the Shields approach? Specific gravity of sand is 2.65 and the water temperature is 60°F.

11.26 Design the channel of Problem 11.22 by the regime/power function theory. The sediment load is 3000 ppm, for which the geometric parameters are increased by 10%.

11.27 For the 35-ft-wide rectangular channel in fine gravel of Problem 11.23, determine the total depth of flow, discharge, and slope by the regime/power function theory. The power function parameters c and s are 0.373 and 0.0018, respectively.

11.28 An irrigation canal is created in coarse gravel with a discharge of 1500 cfs. The side slope is 1:1. Design the canal by the regime/power function theory.

11.29 A trapezoidal concrete channel has a constant bed slope of 0.0015, a bed width of 3.0 m, and side slopes of 1:1. It carries a discharge of 20 m³/s. The channel is a tributary to a river in which the existing flood level is 3.5 m above the channel bottom. Compute the water surface profile by the numerical integration method to a depth 5% greater than the uniform flow depth. $\alpha = 1.1$ and $n = 0.025$.

11.30 Water flows under a gate opening (sluice) into a trapezoidal channel having a bed slope of 0.35%, a width of 20 ft, and side slopes of 1(vertical):2(horizontal). The sluice gate is regulated for a discharge of 400 cfs with a depth of opening of 0.8 ft. Compute the flow profile by the numerical integration method. Take $n = 0.025$ and $\alpha = 1.10$. Consider the control point at the vena contracta. (The distance from the gate opening to the vena contracta is approximately equal to the height of the opening of 0.8 ft, and the depth at vena contracta may be taken as 0.6 ft. The computations may be performed in the downstream direction from the gate.)

11.31 Determine the flow profile using the data of Problem 11.29 by the direct step method.

11.32 Determine the flow profile using the data of Problem 11.30 by the direct step method.

11.33 A rectangular channel of 20 ft width has a depth of 4 ft and velocity of 60 ft/sec. Determine **(a)** whether a jump can form in the channel, **(b)** the downstream depth needed to form the jump, and **(c)** the loss of energy through the jump.

11.34 In a rectangular channel of 12 m width, water flows at a rate of 150 m³/s. At the end of the channel there is a horizontal concrete apron of 12 m width, on which the water depth is 3 m.

Will a hydraulic jump be formed in the channel? What is the pre-jump depth? What is the loss of energy through the jump?

11.35 A rectangular section of a stream has a width of 50 ft and a depth of 5 ft. It has a slope of 1:1000 and $n = 0.015$. The flow into the stream merges from a steep channel through a sluice of 1.50 ft depth. Determine whether a hydraulic jump is going to be formed. If so, what is the pre-jump depth?

Pressure Flow Systems: Pipes and Pumps

◆◆◆

The principal components of a water system that carries water under pressure are the transmission and feeder mains from the treatment plant to the distribution system, and the distribution mains consisting of an interconnecting pipe network up to the source point. The sewage force mains that receive discharge from a pumping station also carry flows under pressure. There is a range of minimum to maximum pressure in which these components operate. The pressure is reduced as water flows through the system due to frictional resistance by the pipe walls and fittings. This is measured in terms of the energy loss. The energy equation is thus appropriate in all pipe flow problems.

12.1 ENERGY EQUATION OF PIPE FLOW

Figure 12.1 shows a pipeline segment. The total energy at any point consists of potential or elevation head, pressure head, and velocity head. The hydraulic grade line shows the elevation of pressure head along the pipe (i.e., it is a line connecting the points to which the water will rise in piezometric tubes inserted at different sections of a pipeline). This concept is similar to the water surface in open channel flow. The energy grade line represents the total head at different points of a pipe section. In a uniform pipe, the velocity head is constant. Thus the energy grade line is parallel to the hydraulic grade line.

Applying the energy equation between points 1 and 2 gives us

$$Z_1 + \frac{p_1}{\gamma} + \frac{V_1^2}{2g} = Z_2 + \frac{p_2}{\gamma} + \frac{V_2^2}{2g} + h_f \quad [\mathrm{L}] \qquad (12.1)$$

In eq. (12.1), h_f is the loss of head along the pipeline due to friction. The energy gradient $S_f = h_f/L$. Additional losses resulting from valves, fittings, bends, and so on, are known as the minor losses, h_m, and have to be included when present. Then, in eq. (12.1) the term h_f will be replaced by the total head loss, h_{loss}. Since the minor losses are localized, the energy grade line, represented by h_f/L, will have breaks wherever the minor losses occur. If mechanical energy is added to the water by a pump or removed by a turbine between the two points of interest, it should be added to or subtracted from the left side of eq. (12.1). In a uniform pipe, $V_1 = V_2$, and elevations Z_1 and Z_2 are generally known. Accordingly, to ascertain the pressure reduction, it is necessary to evaluate the head loss (and minor losses if present).

Figure 12.1 Hydraulic grade line and energy grade line in a pipe flow.

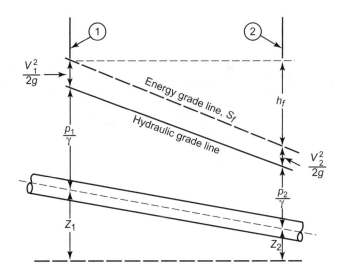

EXAMPLE 12.1

From a reservoir, water flows at a rate of 10 cfs through a pipe of 12 in. diameter, as shown in Figure 12.2. Determine the loss of head in the system.

SOLUTION

1. The area of cross section of the pipe,

$$A = \frac{\pi}{4}\left(\frac{12}{12}\right)^2 = 0.785 \text{ ft}^2$$

2. From the continuity equation, $Q = AV$ or

$$V = \frac{Q}{A} = \frac{10}{0.785} = 12.74 \text{ ft/sec}$$

3. Consider that the datum passes through point 2. Applying the energy equation between points 1 and 2 gives us

$$Z_1 + \frac{p_1}{\gamma} + \frac{V_1^2}{2g} = Z_2 + \frac{p_2}{\gamma} + \frac{V_2^2}{2g} + h_f$$

4. Since the pressure is atmospheric at points 1 and 2 and water is practically stationary (i.e., $V = 0$) at point 1,

$$40 + 0 + 0 = 0 + 0 + \frac{(12.74)^2}{2(32.2)} + h_f$$

and

$$h_f = 37.50 \text{ ft}$$

Figure 12.2 Head loss through a pipeline in Example 12.1.

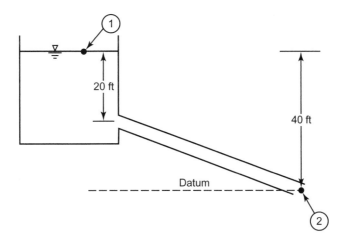

12.2 PIPE FRICTION LOSSES: DARCY-WEISBACH EQUATION

The Darcy-Weisbach equation (1845) is the most general formula for the pipe flow application. It was obtained experimentally. However, Chezy's equation (1769), derived in Section 11.6.1 from balancing the motivating and drag forces on the moving water, can be reduced to the Darcy-Weisbach equation.

According to Chezy's formula,

$$V = C\sqrt{RS} \quad [\text{LT}^{-1}] \tag{11.8}$$

Since $S = h_f/L$, $R = d/4$ for pipe, and treating $C = \sqrt{8g/f}$, eq. (11.8) reduces to

$$h_f = \frac{fL}{d}\frac{V^2}{2g} \quad [\text{L}] \tag{12.2}$$

where

h_f = loss of head due to friction in pipe, ft or m

f = friction factor, dimensionless

L = length of pipe, ft or m

d = internal diameter of pipe, ft or m

V = mean velocity of flow in pipe, ft/sec or m/s

The solution of eq. (12.2) requires the interim step of ascertaining an appropriate value of the friction factor, f, to be used in the equation.

12.2.1 Friction Factor for Darcy-Weisbach Equation

The friction factor relation depends on the state of flow, which is classified according to the Reynolds number. For pipes, the diameter is used as a characteristic dimension and the Reynolds number is given by:

$$\text{Re} = \frac{Vd}{v} \quad [\text{dimensionless}] \tag{12.3}$$

where

V = average velocity of flow, ft/sec or m/s

d = internal diameter of pipe, ft or m

v = kinematic viscosity of fluid, ft^2/s or m^2/s

The flow is classified as follows:

Type of Flow	Value of Re
Laminar	<2000
Transition to turbulent (critical region)	2000–4000
Turbulent	>4000

For laminar flow, the friction factor is a function of the Reynolds number only. It is given by the following relation:

$$\text{(For laminar flow) } f = \frac{64}{\text{Re}} \quad \text{[dimensionless]} \qquad (12.4)$$

In the critical region of Re between 2000 and 4000, the flow alternates between the laminar and turbulent regimes. Any friction factor relation cannot be applied with certainty in this region.

In turbulent regime, the friction factor is a function of the Reynolds number as well as the relative roughness of the pipe surface. In 1932 and 1933, Nikuradse published the results of now-famous experiments on smooth (uncoated) and rough pipes coated with sand grains of uniform size. The experiments' results, plotted as the friction factor versus the Reynolds number, are shown in Figure 12.3. The roughness is characterized by a parameter, k/d, where k is the average diameter of the sand grains and d is the internal diameter of the pipe. In contrast to the Nikuradse sand roughness, the roughness of commercial pipe is not uniform. As a means of differentiating, the nonuniform roughness of commercial pipe is designated ε and is given in equivalent sand roughness. Table 12.1 indicates equivalent roughness for pipe of different material. Based on the results of these experiments, the turbulent flow is further classified in three zones as shown in Figure 12.3:

1. Flow in smooth pipe, where the relative roughness ε/d is very small and plays an insignificant role in determining the friction factor.

2. Flow in fully rough pipe, where viscosity's effect is insignificant on the friction factor.

3. Flow in partially rough pipe, where both the relative roughness and viscosity are significant.

Nikuradse's experiments permitted Prandtl and von Kármán to establish the following formulas for smooth and fully rough pipes of categories 1 and 2 above. For flow in smooth pipe in a turbulent regime:

$$\frac{1}{\sqrt{f}} = -2 \log\left(\frac{2.51}{\text{Re}\sqrt{f}}\right) \quad \text{[dimensionless]} \qquad (12.5)$$

For flow in fully rough pipe in a turbulent regime:

$$\frac{1}{\sqrt{f}} = -2 \log\left(\frac{\varepsilon}{3.7d}\right) \quad \text{[dimensionless]} \qquad (12.6)$$

Figure 12.3 Nikuradse's experiment on smooth and sand-coated pipes.

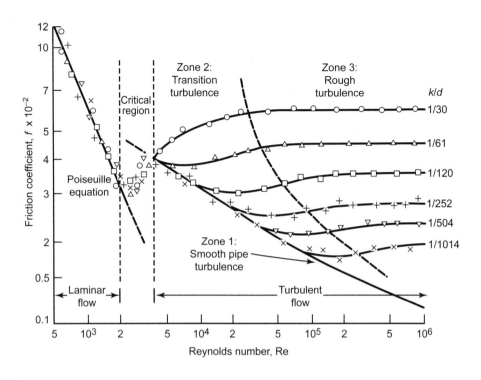

Table 12.1   Roughness Values for Pipes

Pipe Material	Equivalent Roughness, ε (ft)	Hazen-Williams Coefficient, C
PVC, plastic	Smooth	140
Brass, copper, aluminum, glass	Smooth	140
Dawn tubing	5×10^{-6}	—
Cast iron		
new	8.0×10^{-4}	130
old		100
Galvanized iron/ductile pipe	5.0×10^{-4}	120
Asphalted iron	4.0×10^{-4}	—
Wrought iron	1.5×10^{-4}	—
Commercial and		
welded steel	1.5×10^{-4}	120
Wood stave	20.0×10^{-4}	120
Concrete	40.0×10^{-4}	130
Riveted steel	60.0×10^{-4}	110
Brick sewer	—	100

Section 12.2   Pipe Friction Losses: Darcy-Weisbach Equation   663

However, most hydraulic problems and flow in commercial pipes relate to the flow of the third category of partially rough pipe. With specific reference to this transition zone between smooth and rough pipes, Colebrook, in collaboration with White in 1939, combined eqs. (12.5) and (12.6). The formula thus developed covers the entire turbulent regime.

For all types of flow in a turbulent regime,

$$\frac{1}{\sqrt{f}} = -2\log\left(\frac{\varepsilon}{3.7d} + \frac{2.51}{\mathrm{Re}\sqrt{f}}\right) \quad \text{[dimensionless]} \tag{12.7}$$

For smooth pipes, when ε/d is very small, eq. (12.7) reduces to eq. (12.5). For rough pipes at a very high Reynolds number, it takes the form of eq. (12.6). However, eq. (12.7) is implicit since the friction factor appears on both sides of the equation. As such, it involves trial-and-error solution.

By fitting the curve to the Colebrook relation of smooth pipe and combining the equation of rough pipe, Jain (1976) has suggested the following explicit equation for the entire turbulent regime, which gives results within 1% of the Colebrook equation:

$$\frac{1}{\sqrt{f}} = -2\log\left(\frac{\varepsilon}{3.7d} + \frac{5.72}{\mathrm{Re}^{0.9}}\right) \quad \text{[dimensionless]} \tag{12.8}$$

From the implicit relations of Prandtl, von Kármán, and Colebrook-White, Moody (1944) prepared a diagram between the friction factor versus the Reynolds number and the relative roughness as shown in Figure 12.4. The diagram can conveniently be used instead of eq. (12.7) to determine the friction factor. The equivalent roughness is obtained from Table 12.1.

For the application of eq. (12.7), or the Moody diagram, the velocity of flow and the diameter of the pipe should be known so that the Reynolds number can be determined.

EXAMPLE 12.2

Determine the friction factor for water flowing at a rate of 1 cfs in a cast iron pipe 2 in. in diameter at 80°F.

SOLUTION

1. Area of cross section of pipe, $\dfrac{\pi}{4} = \left(\dfrac{2}{12}\right)^2 = 0.022 \text{ ft}^2$

2. Velocity of flow, $V = \dfrac{Q}{A} = \dfrac{1}{0.022} = 45.45 \text{ ft/sec}$

3. At 80°F, kinematic viscosity, $v = 0.93 \times 10^{-5} \text{ ft}^2/\text{sec}$ (Appendix C)

4. The Reynolds number,

$$\mathrm{Re} = \frac{Vd}{v} = \frac{45.45(2/12)}{0.93 \times 10^{-5}} = 8.1 \times 10^5$$

5. Since Re > 4000, it is turbulent flow.

6. Equivalent roughness, $\varepsilon = 8.0 \times 10^{-4}$ ft (Table 12.1).

7. Relative roughness, $\varepsilon/d = \dfrac{8\times10^{-4}}{2/12} = 0.005$

8. From eq. (12.8),

$$\frac{1}{\sqrt{f}} = -2 \, \log\left[\frac{0.005}{3.7} + \frac{5.72}{\left(8.1\times10^{5}\right)^{0.9}}\right] = 5.721$$

$$f = 0.031$$

ALTERNATIVE SOLUTION

On the Moody diagram, the point of intersection of Re = 8.1×10^{5} and $\varepsilon/d = 0.005$ is projected horizontally to the left to read $f = 0.031$.

12.2.2 Extension of the Darcy-Weisbach Equation to Laminar Flow: Hagen-Poiseuille Equation

For laminar flow when the Reynolds number is less than 2000, the friction factor from eq. (12.4) is 64/Re or $64\mu/\rho Vd$. Substituting this in the Darcy-Weisbach eq. (12.2) results in the following Hagen-Poiseuille equation applicable to laminar flow

$$h_f = \frac{32\mu LV}{\gamma d^2} \quad [\text{L}] \tag{12.9}$$

where

$\qquad h_f$ = friction head

$\qquad \mu$ = dynamic viscosity of fluid

$\qquad L$ = length of pipe

$\qquad d$ = diameter of a pipe

$\qquad \gamma$ = unit (specific) weight of fluid

$\qquad V$ = velocity of flow

12.3 APPLICATION OF THE DARCY-WEISBACH EQUATION

The Darcy-Weisbach equation (12.2) is applied to solve the following three types of problems.

12.3.1 Type I: To Determine Head Loss

To compute the head loss, h_f, in a given size pipe, d, that carries a known flow, V or Q, the application of Darcy-Weisbach is direct.

1. From known d and V, or if Q is given, then $V = 4Q/\pi d^2$; determine Re from eq. (12.3).
2. For laminar flow, compute h_f by eq. (12.9).
3. For turbulent flow, find relative roughness ε/d.
4. Determine f from the Moody diagram, Figure 12.4.
5. Solve eq. (12.2) for h_f.

Figure 12.4 Moody diagram for friction factor for pipes.

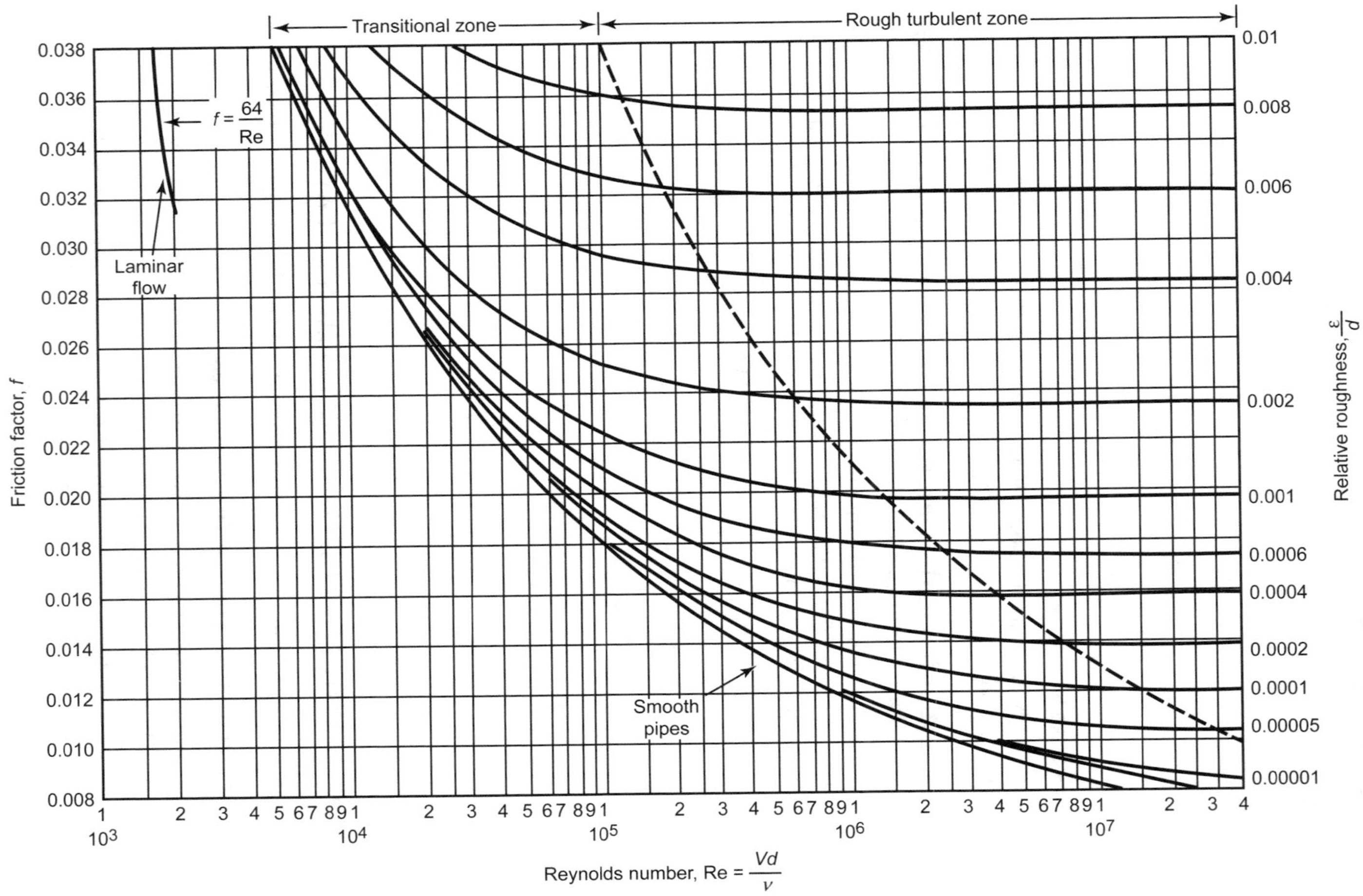

12.3.2 Type II: To Determine Velocity or Flow Rate

Equation (12.2) is used to ascertain the flow velocity, V, or flow rate, Q, through a given size pipe, d, in which the head loss, h_f, is known.

Since V (or Q) is not known, Re and hence f cannot be determined. The procedure apparently is iterative. However, the following substitutions make this a direct procedure.

Rewriting eq. (12.2),

$$f = \frac{2g\,h_f\,d}{V^2 L} \tag{a}$$

or

$$f = \frac{K^2}{V^2} \tag{b}$$

where factor

$$K = \sqrt{\frac{2g\,h_f\,d}{L}} \quad [LT^{-1}] \tag{12.10}$$

Combining eqs. (12.3) and (b), since $\nu = \mu/\rho$

$$\mathrm{Re}\sqrt{f} = \frac{K\rho d}{\mu} \tag{c}$$

The substitution of (b) and (c) in eq. (12.7) results in the following relation for turbulent flow:

$$V = -2K \log\left(\frac{\varepsilon}{3.7d} + \frac{2.51\mu}{\rho K d}\right) \quad [LT^{-1}] \tag{12.11}$$

The procedure is as follows.

1. From known h_f and d values, find factor K from eq. (12.10).
2. Compute V from eq. (12.11).
3. Determine Re from eq. (12.3) to confirm that the flow is not laminar, i.e. Re is not less than 2000.
4. If it happens to be laminar, apply Hagen-Poiseuille eq. (12.9) to compute V.

12.3.3 Type III: To Determine Diameter

Equation (12.2) is used to determine the pipe size, d, to pass a given flow velocity, V, or a rate of flow, Q, within a known limit of head loss, h_f.

Since d is unknown, Re and f cannot be ascertained. This is an iterative procedure.

1. Assume f, it is just an estimate since ε/d is not known.
2. Solve for d by eq. (12.2); if Q is given then $V = 4Q/\pi\,d^2$ in the equation.
3. Solve for Re by eq. (12.3).
4. Determine relative roughness, ε/d.
5. Read the value of f from the Moody diagram, Fig. 12.4.
6. If this value is not close to the assumed value of step 1, repeat steps (2) through (5) until the two successive values of f are about the same.

EXAMPLE 12.3

Water is delivered at a rate of 0.80 cfs by a 6-in. cast iron pipe at 80°F between two points A and B that are 1000 ft apart. If point A is 100 ft higher than point B, what is the pressure difference between the two points?

SOLUTION

1. Computing the head loss in the pipe, at 80°F, $v = 0.93 \times 10^{-5}$ ft²/sec

$$\text{Velocity of flow} = \frac{Q}{A} = \frac{0.80}{(\pi/4)(0.5)^2} = 4.08 \text{ ft/sec}$$

$$\text{Re} = \frac{Vd}{v} = \frac{4.08(0.5)}{0.93 \times 10^{-5}} = 2.2 \times 10^5$$

$$\text{Relative roughness,} \frac{\varepsilon}{d} = \frac{0.0008}{0.5} = 0.0016$$

From the Moody diagram (Figure 12.4), $f = 0.023$. Hence

$$h_f = \frac{fL}{d} \frac{V^2}{2g} = (0.023)\left(\frac{1000}{0.5}\right)\frac{(4.08)^2}{2(32.2)} = 11.9 \text{ ft}$$

2. Applying the energy equation between points A and B (velocity head is equal at both points)

$$Z_1 + \frac{p_1}{\gamma} + \frac{V_1^2}{2g} = Z_2 + \frac{p_2}{\gamma} + \frac{V_2^2}{2g} + h_f$$

$$100 + \frac{p_1}{\gamma} = 0 + \frac{p_2}{\gamma} + 11.9$$

$$\frac{p_2 - p_1}{\gamma} = 100 - 11.9 = 88.1 \text{ ft}$$

$$p_2 - p_1 = 88.1(62.4) = 5500 \text{ psf or } 38.2 \text{ psi}$$

EXAMPLE 12.4

A 2000-m-long commercial steel pipeline of 200 mm diameter conveys water at 20°C between two reservoirs, as shown in Figure 12.5. The difference in water level between the reservoirs is maintained at 50 m. Determine the discharge through the pipeline. Disregard the minor losses.

SOLUTION

1. Consider the datum at the water level of the second reservoir. Apply the energy equation at points 1 and 2:

$$Z_1 + \frac{p_1}{\gamma} + \frac{V_1^2}{2g} = Z_2 + \frac{p_2}{\gamma} + \frac{V_2^2}{2g} + h_f$$

$$50 + 0 + 0 = 0 + 0 + 0 + h_f$$

$h_f = 50$ ft, the difference in water level

Figure 12.5 Pipe connecting two reservoirs in Example 12.4.

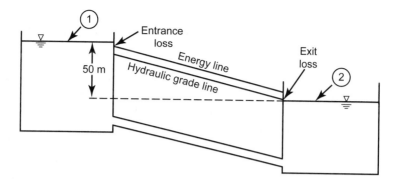

2. For commercial steel, $\varepsilon = 4.6 \times 10^{-5}$ m

   At $20°$ C, $\mu = 1 \times 10^{-3}$ N-s/m^2, $\rho = 998$ kg/m^3

3. $K = \left(\dfrac{2g\, h_f\, d}{L} \right)^{1/2}$

   $= \left[\dfrac{2(9.81)(50)(0.2)}{2000} \right]^{1/2} = 0.313$

4. $V = -2(0.313)\log\left[\dfrac{4.6 \times 10^{-5}}{3.7(0.2)} + \dfrac{2.51(1 \times 10^{-3})}{(998)(0.313)(0.2)} \right]$

   $= 2.5$ m/s

5. $\mathrm{Re} = \dfrac{\rho V d}{\mu} = \dfrac{(998)(2.5)(0.2)}{1 \times 10^{-3}} = 5 \times 10^5 > 4000$, turbulent flow, **OK**

6. $Q = AV = \dfrac{\pi}{4}(0.2)^2 (2.5) = 0.0785$ m^3/s

12.4 PIPE FRICTION LOSSES: HAZEN-WILLIAMS EQUATION

Another common formula for head loss in pipes that has found almost exclusive usage in water supply engineering is the Hazen-Williams equation:

$$V = 1.318C\, R^{0.63} S^{0.54} \quad \text{(English units)} \quad \text{[unbalanced]} \tag{12.12a}$$

$$V = 0.849C\, R^{0.63} S^{0.54} \quad \text{(metric units)} \quad \text{[unbalanced]} \tag{12.12b}$$

where

   V = mean velocity of flow, ft/sec

   C = Hazen-Williams coefficient of roughness given in Table 12.1

   R = hydraulic radius, ft

   S = slope of energy gradient = h_f / L

At $R = 1$ ft and $S = 1/1000$, the Hazen-Williams coefficient C is the same as Chezy's coefficient, C. Thus the Hazen-Williams formula is accurate within a certain range of diameters and friction slopes, although it is used indiscriminately in pipe designs. Jain et al. (1978) indicated that an error of up to 39% can be involved in the evaluation of the velocity by the Hazen-Williams formula over a wide range of diameters and slopes. Two sources of error in the Hazen-Williams formula are: (1) the multiplying factor 1.318 should change for different values of R and S to be comparable with Chezy's formula for the fixed value of C above, and (2) the Hazen-Williams coefficient C is considered to be related to the pipe material only as shown in Table 12.1, whereas it must also depend on pipe diameter, velocity, and viscosity, similar to the friction factor of Darcy-Weisbach. Jain and colleagues suggested a modified formula that incorporates the kinematic viscosity and contains a coefficient that varies with material, pipe diameter, and velocity of flow. The Hazen-Williams formula, however, has a wide application because of its simplicity.

Equation (12.12) can be written as follows in terms of discharge for a circular pipe by substituting $V = Q/A$, $A = (\pi/4)d^2$, and $R = d/4$:

$$Q = 0.432 C d^{2.63} S^{0.54} \qquad \text{(English units)} \qquad \text{[unbalanced]} \qquad \text{(12.13a)}$$

$$Q = 0.278 C d^{2.63} S^{0.54} \qquad \text{(metric units)} \qquad \text{[unbalanced]} \qquad \text{(12.13b)}$$

A nomogram based on eq. (12.13) is given in Figure 12.6 to facilitate the solution. Equation (12.13) or the nomogram provides a direct solution to all three cases of pipe problems mentioned in Section 12.3: computation of head loss, assessment of flow, and determination of pipe size.

The nomogram in Figure 12.6 is based on the coefficient $C = 100$. For pipes of a different coefficient, adjustments are made as follows:

$$\text{To adjust discharge: } Q = Q_{100}\left(\frac{C}{100}\right) \qquad [L^3 T^{-1}] \qquad \text{(12.14a)}$$

$$\text{To adjust diameter: } d = d_{100}\left(\frac{100}{C}\right)^{0.38} \qquad [L] \qquad \text{(12.14b)}$$

$$\text{To adjust friction slope: } S = S_{100}\left(\frac{100}{C}\right)^{1.85} \qquad \text{[dimensionless]} \qquad \text{(12.14c)}$$

where the subscript 100 refers to the value obtained from the nomogram.

EXAMPLE 12.5

Compute the head loss in Example 12.3 by the Hazen-Williams formula.

SOLUTION

1. For a new cast iron pipe, $C = 130$.

2. From eq. (12.13a), $0.80 = 0.432(130)(0.5)^{2.63} S^{0.54}$ or $S = 0.011$.
   Hence $h_f = SL = 0.011(1000) = 11.0$ ft.

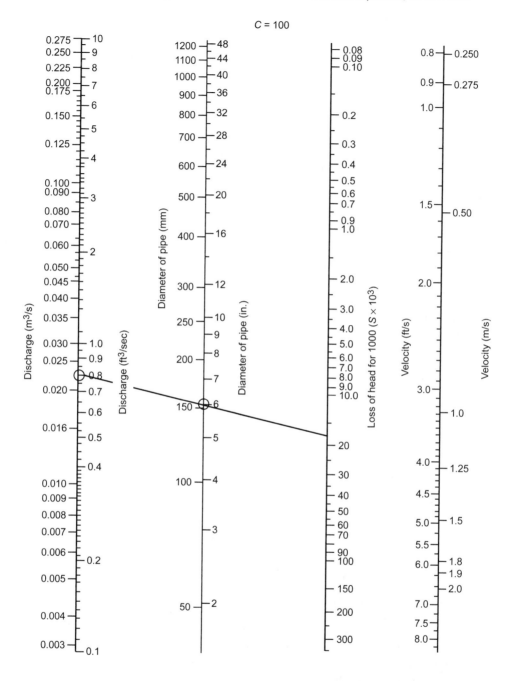

1. A point is marked at 0.80 cfs on the discharge scale.

2. Another point is marked at 6 in. on the diameter scale.

3. The straight line joining these points meets the head loss for 1000 scale at 18. Thus $S_{100} \times 10^3 = 18$ or $S_{100} = 0.018$.

4. Adjustment of S for $C = 130$. From eq. (12.14c),

$$S = S_{100}\left[\frac{100}{C}\right]^{1.85} = (0.018)\left[\frac{100}{130}\right]^{1.85} = 0.011$$

$$h_f = SL = 0.011(1000) = 11.0 \text{ ft}$$

12.5 MINOR HEAD LOSSES

In addition to the continuous head loss along the pipe length due to friction, local head losses occur at changes in pipe section, bends, valves, and fittings. These losses may be disregarded for long pipes but are significant for pipes that are less than 100 feet long. Since pipe lengths in water supply and wastewater plants are generally short, minor losses are important. There are two ways to compute these losses. In the equivalent-length technique, a fictitious length of pipe is estimated that will cause the same pressure drop as any fitting or change in a pipe cross section. This length is added to the actual pipe length. In the second method, the loss is considered proportional to the kinetic energy head given by the following formula:

$$h_m = \Sigma K \frac{V^2}{2g} \quad [\text{L}] \tag{12.15}$$

where

h_m = minor loss of head, ft or m

ΣK = summation of all loss coefficients

V = mean velocity of flow, ft/sec or m/s

Some typical values of the loss coefficient are given in Table 12.2.

12.6 PIPELINE ANALYSIS AND DESIGN

The analysis involves determining the head loss or rate of flow through a pipeline of a given size. The design decision involves selecting a pipe size that will carry a design discharge between two points with specified reservoir elevations or a known pressure difference. The problems can be solved by the Darcy-Weisbach equation as explained in Section 12.2. If minor losses are disregarded, the Hazen-Williams equation (Section 12.4) leads to the direct solution of both the analysis and design problems. The problems presented by various pipeline systems are discussed subsequently.

Table 12.2   Minor Head Loss Coefficients

Item	Loss Coefficient, K
Entrance from tank to pipe	
Flush connection	0.5
Projecting connection	0.8
Exit from pipe to tank	1.0
Sudden contraction $\left(h_m = KV_2^2/2g \right)$	
$d_2/d_1 = 0.2$	0.48
$d_2/d_1 = 0.4$	0.42
$d_2/d_1 = 0.6$	0.32
$d_2/d_1 = 0.8$	0.20
$d_2/d_1 = 0.9$	0.05
Sudden enlargement $\left(h_m = KV_1^2/2g \right)$	
$d_1/d_2 = 0.9$	0.04
$d_1/d_2 = 0.8$	0.13
$d_1/d_2 = 0.6$	0.41
$d_1/d_2 = 0.4$	0.71
$d_1/d_2 = 0.2$	0.92
90° bend and 180° return—threaded	1.5
45° bend—threaded	0.4
90° bend and 180° return—flanged	0.3
45° bend—flanged	0.3
Tee—threaded	
through flow	0.9
branched flow	2.0
Tee—flanged	
through flow	0.2
branched flow	1.0
Gate valve (open)	0.19
Check valve (open)	2.0
Globe valve (open)	10.0
Angle valve (open)	2.0
Butterfly valve (open)	0.3

Note: Subscript 1 refers to upstream and 2 refers to downstream section.

12.7 SINGLE PIPELINES

Application is made of the (1) energy equation, (2) Darcy-Weisbach or Hazen-Williams equation, and (3) minor losses relation, as demonstrated in Example 12.6.

EXAMPLE 12.6

Two reservoirs are connected by a 200-ft-long cast iron pipeline, as shown in Figure 12.7. If the pipeline is to convey a discharge of 2 cfs at 60°F, what diameter of pipeline is required?

SOLUTION

1. Apply the energy equation between points 1 and 2 with respect to point 2 as the datum:

$$Z_1 + \frac{p_1}{\gamma} + \frac{V_1^2}{2g} = Z_2 + \frac{p_2}{\gamma} + \frac{V_2^2}{2g} + h_{loss}$$

$$20 + 0 + 0 = 0 + 0 + 0 + h_{loss}$$

$$h_{loss} = 20 \text{ ft} \tag{a}$$

2. Friction loss,

$$h_f = \frac{f}{d} \frac{L}{2g} \frac{V^2}{2g} = \frac{f}{d} \frac{L}{\left[(\pi/4)d^2\right]^2} \frac{Q^2}{2g}$$

$$= \frac{f L Q^2}{39.68 d^5} \text{ (in FPS units)} \tag{b}$$

3. Minor losses,

$$h_m = \sum \frac{K V^2}{2g} = \sum K \frac{Q^2}{\left[(\pi/4)d^2\right]^2 2g} = \frac{\sum K Q^2}{39.68 d^4} \text{ (in FPS units)} \tag{c}$$

Item	K
Entrance loss	0.5
Exit loss	1.0
Two 90° bends at 1.5	3.0
Globe valve	10.0
Total	14.5

4. $h_{loss} = h_f + h_m$ (d)

5. Substituting eqs. (a), (b) and (c) in (d) above:

$$\frac{f L Q^2}{39.68 d^5} + \frac{\sum K Q^2}{39.68 d^4} = 20$$

$$\frac{f(200)(2)^2}{(39.68)d^5} + \frac{14.5(2)^2}{(39.68)d^4} = 20$$

$$20 d^5 - 1.46 d - 20.16 f = 0 \tag{e}$$

Figure 12.7 Pipe system connecting two reservoirs in Example 12.6.

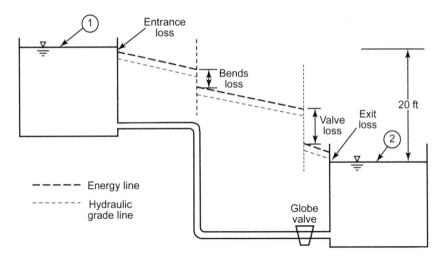

6. In the first trial, assume that $f = 0.03$. Substitute f in eq. (e): $20d^5 - 1.46d - 0.605 = 0$.
Solve by trial and error: $d = 0.59$ ft. Thus

$$A = \frac{\pi}{4}(.59)^2 = 0.273 \text{ ft}^2$$

$$V = \frac{Q}{A} = \frac{2}{0.273} = 7.33 \text{ ft/s}$$

$$Re = \frac{Vd}{\nu} = \frac{7.33(0.59)}{1.217 \times 10^{-5}} = 3.5 \times 10^5$$

$$\frac{\varepsilon}{d} = \frac{8 \times 10^{-4}}{0.59} = 0.0014$$

$f = 0.0215$ from the Moody diagram

7. First revision: Substitute f in eq. (e): $20d^5 - 1.46d - 0.433 = 0$. Solve by trial and
error: $d = 0.57$ ft. Thus

$$A = \frac{\pi}{4}(0.57)^2 = 0.255 \text{ ft}^2$$

$$V = \frac{2}{0.255} = 7.84 \text{ ft/s}$$

$$Re = \frac{7.84(0.57)}{1.217 \times 10^{-5}} = 3.7 \times 10^5$$

$$\frac{\varepsilon}{d} = \frac{8 \times 10^{-4}}{0.57} = 0.0014$$

$f = 0.0215$ (from the Moody diagram)

Since f stabilizes, $d = 0.57$ ft or 6.8 in.

Section 12.7   Single Pipelines

675

12.8 SINGLE PIPELINES WITH PUMPS

In waterworks and wastewater systems, pumps are common at the source to lift the water level and at intermediate points to boost the pressure. Figure 12.8 illustrates a situation where water is supplied from a lower reservoir to an upper-level reservoir.

To analyze the system, the energy equation is applied between downstream and upstream ends of the pipe:

$$Z_1 + \frac{p_1}{\gamma} + \frac{V_1^2}{2g} + H_p = Z_2 + \frac{p_2}{\gamma} + \frac{V_2^2}{2g} + h_f + h_m \qquad \text{(a)}$$

Treating $V_1 = V_2$ yields

$$H_p = -\left(Z_1 + \frac{p_1}{\gamma}\right) + \left(Z_2 + \frac{p_2}{\gamma}\right) + h_f + h_m \qquad \text{(b)}$$

$$H_p = \Delta Z + h_{loss} \quad [\text{L}] \qquad (12.16)$$

where

H_p = energy added by the pump, ft or m

ΔZ = difference between piezometric heads or water levels after pumping and before pumping, or total static head, ft or m

h_f = friction head loss = $(f\,L/d)(V^2/2g)$, ft or m

h_m = minor head losses = $\Sigma K V^2/2g$, ft or m

h_{loss} = total of friction and minor head losses

The energy head, H_p, and the brake horsepower of the pump are related as

$$\text{BHP} = \frac{\gamma Q H_p}{550\eta} \quad [\text{FLT}^{-1}] \quad \text{(English units)} \qquad (12.17a)$$

$$\text{BHP} = \frac{\gamma Q H_p}{\eta} \quad [\text{FLT}^{-1}] \quad \text{(metric units)} \qquad (12.17b)$$

where

BHP = pump brake power, horsepower or Watt

Q = discharge through pipe

H_p = pump head

η = overall pump efficiency

EXAMPLE 12.7

Water has to be transported at a rate of 1 cfs from a reservoir of water elevation 1000 ft to a reservoir at water elevation of 1100 ft through a 4000-ft-long, 6-in.-diameter steel pipeline at 50°F. Determine the horsepower of the pump required having an efficiency of 70%. Disregard the minor losses.

Figure 12.8 Pumped pipeline system.

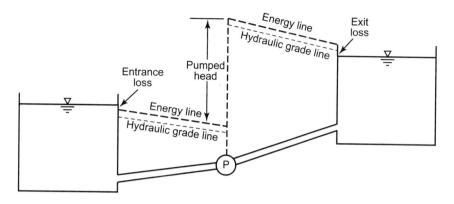

SOLUTION

1. $\Delta Z = 1100 - 1000 = 100$ ft

2. $h_m = 0$

3. $V = \dfrac{Q}{A} = \dfrac{1}{(\pi/4)(0.5)^2} = 5.1$ ft/sec

4. $\text{Re} = \dfrac{Vd}{v} = \dfrac{5.1(0.5)}{1.41\times10^{-5}} = 1.81\times10^5$

5. $\dfrac{\varepsilon}{d} = \dfrac{0.00015}{0.5} = 0.0003$

6. $f = 0.0265$ (from the Moody diagram, Fig. 12.4)

7. $h_f = (0.0265)\left(\dfrac{4000}{0.5}\right)\dfrac{(5.1)^2}{2(32.2)} = 85.6$ ft

8. From eq. (12.16), $H_p = \Delta Z + h_{\text{loss}}$

$$H_p = 100 + 85.6 + 0 = 185.6 \text{ ft}$$

9. From eq. (12.17a),

$$\text{BHP} = \dfrac{\gamma\, Q\, H_p}{550\eta} = \dfrac{62.4(1)(185.6)}{550(0.70)} = 30 \text{ hp}$$

12.9 PIPES IN SERIES

Pipes in series or a compound pipeline consists of several pipes of different sizes connected together as shown in Figure 12.9. According to the continuity and the energy equations, the following relations apply to the pipes in series:

$$Q = Q_1 = Q_2 = Q_3 = \cdots \quad [\text{L}^3\text{T}^{-1}] \tag{12.18a}$$

$$h_f = h_{f1} + h_{f2} + h_{f3} + \cdots \quad [\text{L}] \tag{12.18b}$$

Figure 12.9 Compound pipeline.

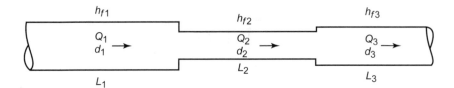

For analysis purpose, the different-sized pipes are replaced by a pipe of a uniform diameter of a length that will pass a discharge, Q, with the total head loss, h_f, given by eq. (12.18). This is known as the equivalent pipe. The procedure will be illustrated by an example.

EXAMPLE 12.8

In Figure 12.10, cast iron pipes 1, 2, and 3 are 1000 ft of 6-in. diameter, 500 ft of 3-in. diameter, and 1800 ft of 4-in. diameter, respectively. If the difference in head is 50 ft, determine the discharge at 50°F.

SOLUTION   The nomogram of Figure 12.6 for $C = 100$ can be used for converting a series of pipes of any material (any value of C) into an equivalent length. The steps are as follows:

1. Assume a discharge through the series of pipes: say, $Q = 0.5$ cfs.
2. For each pipe, for the assumed discharge and known diameter, compute the friction slope by the Hazen-Williams equation (12.13a)* or the nomogram, as shown in column 4 of Table 12.3.
3. Multiply the friction slope (column 4) by the pipe length (column 5) to obtain the head loss (column 6).

Figure 12.10 Compound pipe system connecting two reservoirs.

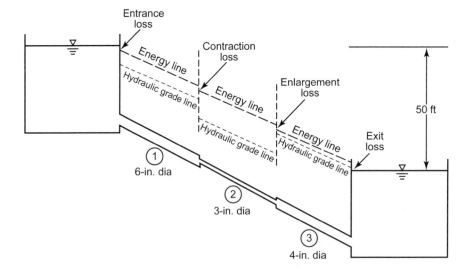

---

* Alternatively, the Darcy-Weisbach formula can be used to determine h_f for steps 2 and 3 above.

Table 12.3 Computation for Pipes in Series

(1) Pipe	(2) Pipe Size (ft)	(3) Discharge, Q Assumed	(4) Friction Slope, S Computed	(5) Pipe Length, L (ft)	(6) Head Loss, h_f, (ft) (SL)
1	0.5	0.5	0.008	1000	8.00
2	0.25	0.5	0.222	500	111.00
3	0.333	0.5	0.055	1800	99.00
EQV	0.333 (select)	0.5 (assumed)	0.055	?	218

4. The addition of col. 6 provides the total head loss per eq. (12.18b). In this case, $h_f = 218$ ft.

5. Select any desired size of uniform pipe. Selected $d = 4$ in.

6. For the selected diameter and the assumed Q of step 1, compute S, again by eq. (12.13a) or the nomogram. In this case $S = 0.055$.

7. The required length of uniform pipe, $L =$ step 4/step 6:

$$L = \frac{h_f}{S} = \frac{218}{0.055} = 3964 \text{ ft}$$

Thus a uniform pipe of 4 in. diameter and 3964 ft length is equivalent to the three pipes in series. The discharge can be determined by the method for a single pipe using the Darcy-Weisbach or Hazen-Williams equation.

8. Given $h_f = 50$ ft, thus $S = \dfrac{50}{3964} = 0.0126$

For $d = 0.333$, $C = 100$ for cast iron, and $S = 0.0126$; by the Hazen-Williams equation (12.13a), $Q = 0.23$ cfs.

12.10 PIPES IN PARALLEL

For the parallel or looping pipes of Figure 12.11, the continuity and energy equations provide the following relations:

$$Q = Q_1 + Q_2 + Q_3 + \cdots \quad [\text{L}^3\text{T}^{-1}] \tag{12.19a}$$

$$h_f = h_{f1} = h_{f2} = h_{f3} = \cdots \quad [\text{L}] \tag{12.19b}$$

A procedure similar to that used for pipes in series is also used in this case, as illustrated in the following example.

EXAMPLE 12.9

A welded steel pipeline of 2 ft diameter is 1 mile long. To augment the supply, a pipe of the same diameter is attached in parallel to the first in the middle half of the length as shown in Figure 12.12. The head above the outlet is 100 ft. Find the discharge through the pipe. Disregard the minor losses.

Figure 12.11   Parallel pipe system.

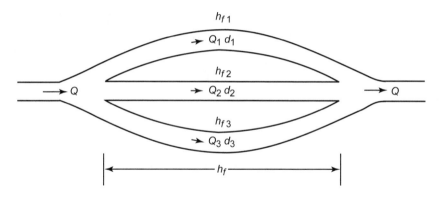

Figure 12.12   Pipes in parallel for Example 12.9.

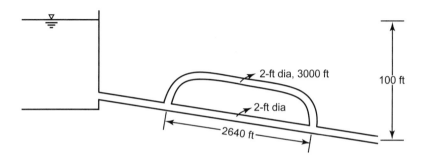

SOLUTION   The following steps are followed to convert the parallel pipes into a single pipe of a uniform size.

1. Assume a head loss through the system: say, $h_f = 100$ ft.

2. For each pipe, compute $S = h_f/L$ (column 5, Table 12.4).

3. From the known values of diameter (column 2) and computed slope (column 5) for each pipe, compute the discharge (column 6) by the Hazen-Williams equation (12.13a) or the nomogram. $C = 100$ can be used.

4. The addition of column 6 is the total discharge, Q, per eq. (12.19a). In this case, Q = 88.26 cfs.

5. Select any desired size of a uniform pipe. The selected $d = 2$ ft.

6. For the selected diameter and total Q of step 4, compute the friction slope, S, again by eq. (12.13a) or the nomogram, which is 0.128.

7. The required length of the uniform pipe

$$L = \frac{\text{step 1}}{\text{step 6}} = \frac{h_f}{S} = \frac{100}{0.128} = 781.3 \text{ ft}$$

Table 12.4 Computation of Parallel Pipe System

(1)	(2)	(3)	(4)	(5)	(6)
Pipe	Pipe Diameter (ft)	h_f (ft), Assumed	Length (ft)	$S = h_f/L$ (col. 3/col.4)	Discharge Q (cfs) Computed
1	2	100	2640	0.0379	45.67
2	2	100	3000	0.0333	42.59
EQV	2	100	?	0.128	88.26
	(selected)	(assumed)		(computed)	

A single pipe of 2 ft diameter and 781.3 ft length is equal to the parallel portion of the pipe.

8. Total length of uniform pipe of 2 ft diameter = 2640 + 781.3 = 3421.3 ft. Given h_f = 100 ft,

$$S = \frac{100}{3421.3} = 0.029$$

9. For steel, C = 120. From eq. (12.13a),

$$Q = 0.432(120)(2)^{2.63}(0.029)^{0.54} = 47.43 \text{ cfs}$$

12.11 PIPES NETWORK

An extension of pipes in parallel is a system in which the pipes are interconnected to form a complex loop configuration. The flow to an outlet comes from several paths. The analytical solution of such systems, referred to as pipes networks, is quite complicated. Three simple methods are the Hardy Cross method, the linear theory method, and the Newton-Raphson method. Of these, the Hardy Cross method, which involves a series of successive approximations and corrections to flows in individual pipes, is a popular procedure of analysis.

According to the Darcy-Weisbach equation,

$$h_f = \frac{fL}{d}\frac{V^2}{2g} = \frac{16}{\pi^2}\frac{fL}{d^5}\frac{Q^2}{2g} \tag{a}$$

As per the Hazen-Williams equation,

$$Q = 0.432Cd^{2.63}\left(\frac{h_f}{L}\right)^{0.54} \tag{b}$$

or

$$h_f = \frac{4.727L}{C^{1.85}d^{4.87}}Q^{1.85} \tag{c}$$

Both (a) and (c) can be expressed in the general form

$$h_f = KQ^n \quad [\text{L}] \tag{12.20}$$

where

K = equivalent resistance as given in Table 12.5

$n = 2.0$ for the Darcy-Weisbach equation and 1.85 for the Hazen-Williams equation

The sum of head losses around any closed loop is zero: that is,

$$\Sigma\, h_f = 0 \tag{d}$$

Consider that Q_a is an assumed pipe discharge that varies from pipe to pipe of a loop to satisfy the continuity of flow. If δ is the correction made in the assumed flow of all pipes of a loop to satisfy eq. (d), then by substituting eq. (12.20) in eq. (d),

$$\Sigma\, K(Q_a + \delta)^n = 0 \tag{e}$$

Expanding eq. (e) by the binomial theorem and retaining only the first two terms yields

$$\delta = -\frac{\Sigma\, K\, Q_a^n}{n\, \Sigma\, K\, Q_a^{n-1}} \tag{f}$$

or

$$\delta = -\frac{\Sigma\, h_f}{n\, \Sigma\, \left| h_f\, / Q_a \right|} \quad [\mathrm{L^3 T^{-1}}] \tag{12.21}$$

Equations (12.20) and (12.21) are used in the Hardy Cross procedure. The values of n and K are obtained based on the Darcy-Weisbach or Hazen-Williams equations from Table 12.5. The procedure is summarized as follows:

1. Divide the network into a number of closed loops. The computations are made for one loop at a time.

2. Compute K for each pipe using the appropriate expression from Table 12.5 (column 3 of Table 12.6).

3. Assume a discharge, Q_a and its direction in each pipe of the loop (column 4). At each joint (node), the total flow *in* should equal the flow *out*. Consider the clockwise flow to be positive and the counterclockwise flow to be negative.

4. Compute h_f in column 5 for each pipe by eq. (12.20), retaining the sign of column 4. The algebraic sum of column 5 is $\Sigma\, h_f$.

5. Compute h_f / Q_a (column 5/column 6) for each pipe without regard to the sign. The sum of column 6 is $\Sigma\, |h_f / Q_a|$.

6. Determine the correction, δ, by eq. (12.21). Apply the correction algebraically to the discharge of each member of the loop.

7. For common members among two loops, both δ corrections should be made, one for each loop.

8. For the adjusted Q, steps 4 through 7 are repeated until δ becomes very small for all loops.

Lyle and Weinberg (1957) and Watters (1984) have created computer programs for the Hardy Cross analysis.

Table 12.5   Equivalent Resistance, K, for Pipe

Formula	Units of Measurement	K
Hazen-Williams	Q, cfs; L, ft; d, ft; h_f, ft	$\dfrac{4.73L}{C^{1.85}d^{4.87}}$
	Q, gpm; L, ft; d, in.; h_f, ft	$\dfrac{10.44L}{C^{1.85}d^{4.87}}$
	Q, m³/s; L, m; d, m; h_f, m	$\dfrac{10.70L}{C^{1.85}d^{4.87}}$
Darcy-Weisbach	Q, cfs; L, ft; d, ft; h_f, ft	$\dfrac{fL}{39.70d^{5}}$
	Q, gpm; L, ft; d, in.; h_f, ft	$\dfrac{fL}{33.15d^{5}}$
	Q, m³/s; L, m; d, m; h_f, m	$\dfrac{fL}{12.10d^{5}}$

EXAMPLE 12.10

Find the discharge in each pipe of the welded steel network shown in Figure 12.13. The pressure head at A is 100 ft.

SOLUTION

1. From Table 12.5,

$$K_{AB} = \frac{4.73(1000)}{(120)^{1.85}(1)^{4.87}} = 0.67$$

$$K_{AC} = \frac{4.73(1500)}{(120)^{1.85}(1)^{4.87}} = 1.01$$

$$K_{BC} = \frac{4.73(1700)}{(120)^{1.85}(0.667)^{4.87}} = 8.23$$

$$K_{BD} = \frac{4.73(2300)}{(120)^{1.85}(1)^{4.87}} = 1.55$$

$$K_{DE} = \frac{4.73(1500)}{(120)^{1.85}(0.667)^{4.87}} = 7.26$$

$$K_{CE} = \frac{4.73(1100)}{(120)^{1.85}(1)^{4.87}} = 0.74$$

2. In the first trial, the flow in each pipe is assumed as indicated in Figure 12.14(a). The corrections δ are computed in Table 12.6. After applying these corrections, the discharges after first iteration are indicated in Figure 12.14(b), which are also the assumed discharges for second iteration.

Section 12.11   Pipes Network

683

Figure 12.13 Pipes network of Examples 12.10 and 12.11.

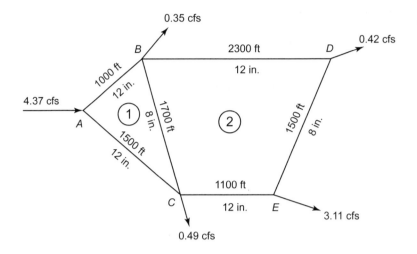

Figure 12.14 (a) Assumed discharges; (b) discharges after first iteration.

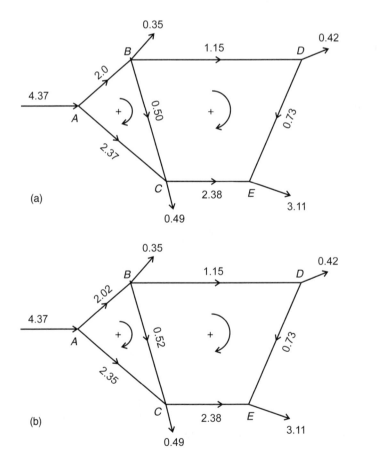

Table 12.6 Iteration 1 of Hardy Cross Procedure

(1)	(2)	(3)	(4)	(5)	(6)	(7)
Loop	Pipeline	K	Q_a (cfs)	$h_f = KQ_a^{1.85}$ (ft)	$\left\lvert\dfrac{h_f}{Q_a}\right\rvert$	Q corrected = $Q_a + \delta$ (cfs)
1	AB	0.67	+2.0	+2.42	1.21	+2.02
	BC	8.23	+0.50	+2.28	4.56	+0.52
	CA	1.01	− 2.37	− 4.98	2.10	− 2.35
				$\overline{-0.28}$	$\overline{7.87}$	
2	BD	1.55	+1.15	+2.01	1.75	+1.15
	DE	7.26	+0.73	+4.06	5.56	+0.73
	EC	0.74	− 2.38	− 3.68	1.55	− 2.38
	CB	8.23	− 0.50	− 2.28	4.56	− 0.52
				$\overline{+0.11}$	$\overline{13.42}$	

For loop 1

$$\delta_1 = -\frac{-0.28}{1.85(7.87)} = +0.02$$

$$\text{adjusted } Q_{AB} = 2.0 + (+0.02) = 2.02$$

For loop 2

$$\delta_2 = -\frac{0.11}{1.85(13.42)} = -0.00$$

$$\text{adjusted } Q_{BC} = +0.5 + (+0.02) - (-0.00) = 0.52$$

3. The corrections δ for the second iteration are computed in Table 12.7. The corrections are negligible. The final discharges are indicated in col. 2 of Table 12.8.

4. The pressure heads are computed in Table 12.8.

12.12 DISTRIBUTION SYSTEM DESIGN

12.12.1 System Configuration

Pipes, valves, and hydrants are the basic elements of a distribution network. The pipe system consists of the following:

1. Primary or arterial lines, which are the main lines that carry water from pumping stations or storage tanks through the distribution district. These lines are laid in interlocking loops with the mains not more than 3000 ft or 1 km apart.

2. Secondary lines that run between primary mains spaced two to four blocks apart.

3. Small distribution mains that form a grid over the entire service area. Fire hydrant connections and service connections to individual homes are made from small distribution mains. They have a maximum spacing of 600 ft.

Table 12.7 Iteration 2 of Hardy Cross Procedure

(1)	(2)	(3)	(4)	(5)	(6)	(7)
				$h_f = KQ_a^{1.85}$	$\left\|\dfrac{h_f}{Q_a}\right\|$	Q corrected =
Loop	Pipeline	K	Q_a (cfs)	(ft)		$Q_a + \delta$ (cfs)
1	AB	0.67	+2.02	+2.46	1.22	+2.02
	BC	8.23	+0.52	+2.45	4.71	+0.52
	CA	1.01	− 2.35	− 4.91	2.09	− 2.35
				0	8.02	
2	BD	1.55	+1.15	+2.01	1.75	+1.15
	DE	7.26	+0.73	+4.06	5.56	+0.73
	EC	0.74	− 2.38	− 3.68	1.55	− 2.38
	CB	8.23	− 0.50	− 2.45	4.71	− 0.52
				−0.06	13.57	

For loop 1

$$\delta_1 = -\frac{0}{1.85(8.02)} = 0.00$$

$$\text{adjusted } Q_{AB} = 2.02 + (0.00) = 2.02$$

For loop 2

$$\delta_2 = -\frac{-0.06}{1.85(13.57)} = 0.00$$

$$\text{adjusted } Q_{BC} = 0.52 + (+0.00) - (-0.00) = 0.52$$

Table 12.8 Final Flows and Pressure Heads

(1)	(2)	(3)	(4)	(5)
Link	Flow (cfs)	Head loss $h_f = KQ^{1.85}$ (ft)	Node	Pressure head (ft)
AB	2.02	2.46	A	100 ft (given)
BC	0.52	2.45	B	$100 - h_{AB} = 97.54$
AC	2.35	4.91	C	$100 - h_{AC} = 95.09$
BD	1.15	2.01	D	$h_B - h_{BD} = 97.54 - 2.01$
DE	0.73	4.06		$= 95.53$
CE	2.38	3.68	E	$h_C - h_{CE} = 95.09 - 3.68$
				$= 91.41$

Note: A slight discrepancy may exist in balancing of the heads for all nodes.

Pumping with on-line storage is a common method of distribution in which the water is pumped at a fairly uniform rate to a storage tank from which variable demands are met as discussed in Section 10.11. The other methods of distribution are gravity distribution and pumping without storage. Parts of communities at different elevations are supplied through separate, though interconnected, distribution systems, each with its own service storage.

Hydrants can be spaced 300 ft apart when the fire flow is less than 1000 gpm but should be only 200 ft apart for flow exceeding 5000 gpm. The valves are located not more than 800 ft apart in high value districts to allow shutdown of a section for service or repairs.

12.12.2 Design Flow Estimation

Table 1.1 indicated that a distribution grid should be designed for a capacity that is the greater of (1) maximum daily demand plus fire demand or (2) maximum hourly demand. The demands are based on per person use and the projected population at the end of the design period which, for the distribution system, is the community's full development since the life of the pipe system may exceed 100 years.

In the distribution system, small flows from individual connections are withdrawn from a large number of points. For network analysis, the withdrawals in a segment of the distribution line, usually between the intersections of streets, are combined. The distribution system consists of nodes, corresponding to pipe intersections, and links representing pipe segments. The segment-wise flows are estimated based on the industrial, commercial, and residential composition of the area representing the segment and the population density within the segment. The combined flow pertaining to a segment is applied at the end node of that segment. To optimize design flow, it is important to know the demand patterns of different usages because the peak demands of those uses may not coincide in time.

Fire flow is assessed as in Section 1.9.3 for an entire service district. The takeoff point could be any node within the network of the district. The fire demand could be placed at the farthest node for design and then checked by applying the fire demand at other nodes. To determine the design basis, the overall demand for the network in terms of the maximum daily plus fire flow and the maximum hourly flow should be determined. When peak hourly flow controls, both design flow bases should be considered since there still may be some segments within the network for which the maximum daily plus fire flow could be critical because of the high concentration of fire demands. The network, with withdrawals at nodes and the total demand as the input, is then analyzed by the Hardy Cross method of Section 12.11 to determine the distributed flows in the pipe segments and the pressure heads at the nodes.

12.12.3 Velocity and Pipe Sizes

Once discharges in various pipe sections are determined, their sizes are selected by the continuity equation ($Q = AV$) by assuming a velocity of 2 to 5 ft/sec. The minimum recommended sizes are: (1) primary/major streets, 12-in., (2) secondary/branching pipes, 8-in., and (3) small distribution, 6-in. In smaller communities, the lines that provide only domestic supplies may be of 4-in. diameter.

12.12.4 Pressure Requirements

The adequacy of a distribution system is judged from the pressure maintained in the pipelines. The recommended range for residential areas is from 40 to 50 psi, and it is 60 to 75 psi for commercial areas. A minimum desirable pressure in any area is considered to be 30 psi, which can supply buildings of up to four stories. During a serious fire, the pressure in a vicinity may be allowed to drop to about 20 psi.

For firefighting purposes, a pressure of 75 psi is desired at the hydrant and 100 psi within the system. The normal pressures are, however, increased at such times by turning on special high-pressure fire pumps at the pumping station or by motor pumper trucks. To account for the losses within the system, main lines should be designed for pressures between 40 and 75 psi. A distribution system should be designed to maintain a reasonably uniform pressure throughout the system. High head losses indicate a deficient system and warrant replacement of existing pipes with larger sized pipes where excessive losses take place.

12.12.5 A Distribution System Project

The first step in distribution design is to sketch a development plan; that is, an arrangement of the pipelines that will be needed to serve the area. The extent of detail depends upon the level of analysis and the size of the network. Nonessential lines are always eliminated and the less important ones are combined using the equivalent pipe method. Minor losses are usually disregarded; they can be included via equivalent pipe lengths whenever necessary. In a large system, it may be worthwhile to consider the network comprising main feeders only. It also is expeditious in analysis of a large system to balance portions successively instead of analyzing the entire system at once.

The centerpiece of the design is a pipe network analysis similar to that of Section 12.11, for which several commercial computerized programs are available. The designs are invariably performed on computer. The steps of the procedure are explained below with a simple example.

EXAMPLE 12.11

Figure 12.13 represents a simplified network component of a water district. Design the network.

SOLUTION

1. Skeletonize the distribution network of the existing or proposed system. In this example the small distribution lines lying within Figure 12.13 have not been shown.

2. Project water-use patterns for domestic, commercial, and industrial purposes. Estimate the maximum daily and maximum hourly flows to the end of the design period. For hourly flows, use the specific hour in which the summation of domestic, commercial, and industrial demand is greatest. Separately, estimate the fire flows for the area covered by the network. Disaggregate flows to the various links or segments of the network as indicated in Table 12.9.

3. Concentrate the disaggregated flows to the node at the end of the concerned link. These are computed below and shown in Figure 12.13. The fire flow has been applied on the far node E.

   Node B: Flow of Link $AB = 0.35$ cfs

   Node C: Flow of Link AC and Link $BC = 0.21 + 0.28 = 0.49$ cfs

   Node D: Flow of Link $BD = 0.42$ cfs

   Node E: Flow of Link $CE +$ Link $DE +$ Fire $= 0.46 + 0.42 + 2.23 = 3.11$ cfs

   Input at node A = Total flow = 4.37 cfs

4. Assume a discharge and its direction in each pipe such that at each node, the total flow entering equals the flow coming out. Select the initial pipe sizes using a flow velocity of 5 ft/s subject to the minimum size limits of Section 12.12.3. These are shown in Figure 12.13.

Table 12.9 Estimates of Design Flows for Example 12.11

(1)	(2)	(3)	(4)	(6)
Link	Water use	Area or population served	Maximum daily unit consumption	Flow cfs
AB	commercial	5 acres	4,500 gal/acre/day	0.35
AC	industrial	6.6 acres	21,000 gal/acre/day	0.21
BC	commercial	4 acres	45,000 gal/acre/day	0.28
BD	residential—single family	1000 persons	270 gal/person/day	0.42
DE	residential—single family	1000 persons	270 gal/person/day	0.42
CE	residential—multi-family	1100 persons	270 gal/person/day	0.46
Total				2.14
Fire Demand			1000 gal/min	2.23

Maximum daily + fire demand = 2.14 + 2.23 = 4.37 cfs ← controls

Maximum hourly = 150% of maximum daily = 1.5 (2.14) = 3.21 cfs

5. Compute factor K for each pipe. Solve for flows in pipes through the iterative procedure of Section 12.11. This has been done in Example 12.10. Confirm that each pipe carries the quantities required for that segment. Recheck the velocities in each pipe.

6. Calculate pressures at the nodes, as shown in Table 12.8. Compare the calculated pressures to the desired standard pressures.

7. Where velocities are more than 5 ft/s in step 5 or pressures are low in step 6, increase the pipe sizes and repeat the procedures of steps 5 and 6 until a satisfactory solution is obtained. In this example adjustment of the pipe size is not required.

8. Perform the above analysis with fire flows assigned to other nodes to select the final design.

12.13 HYDRAULIC TRANSIENTS IN PIPES

The terms *transient flow* and *water hammer* are used synonymously to describe unsteady flow in pipes. This unsteady phenomenon associated with sudden increase or decrease in flow is accompanied by pressure fluctuations. If the pressure induced exceeds the rating of a pipe, a rupture might occur unless a protection device is installed. Hydroelectric plants are particularly vulnerable to water hammer. The protection devices available include standpipes, surge tanks, flywheels, pressure relief valves, control valves not to be closed too rapidly, and air chambers that use gas to cushion water hammer blow. Prior to selecting a protection device, an adequate analysis of water hammer has to be made. This has been presented in Section 9.14.

12.14 PUMPS

In the design of a pumping station, a water resources engineer is concerned with the selection of a pump based on its performance information. Therefore, the components of a pump and its design details have not been considered. Karassik et al. (1986) is recommended reading. In water and wastewater works, centrifugal pumps are most common in application. In the following sections the characteristics of this type of pump are specifically described.

12.15 PUMP CLASSIFICATION: SPECIFIC SPEED

The pump performance parameters comprise (1) rotational speed, (2) discharge capacity, (3) pumping head, (4) power applied, and (5) efficiency. For each pump, these parameters have certain relationships to each other that vary from pump to pump. By combining the three main parameters of speed, discharge, and head, a single term known as the *specific speed* has been created that is fixed for all pumps operating under dynamic conditions that are geometrically similar (homologous) to one another. The term is, thus, suitable to group the pumps with respect to the similarity of their design and to compare the performance of pumps of different designs.

The specific speed, expressed as follows, is measured in inconsistent but standard units in the United States and thus is used as an index.

$$N_S = \frac{N\sqrt{Q}}{H^{3/4}} \quad [\ ]^* \tag{12.22}$$

where

N_S = specific speed

N = rotational speed, rpm

Q = discharge capacity, gpm

H = total head, ft

Under similar operating conditions of head and capacity (similar conditions established by the laws of similarity explained in the next section), the specific speed is the same for all pumps of geometrically similar designs. For any pump, however, the value of the specific speed changes under different operating conditions. In the classification of a pump, the specific speed corresponding to the operating condition at the maximum efficiency, called the *type specific speed*, is used.

Certain features of the specific speed are as follows:

1. The efficiency starts dropping drastically when lowering the specific speed below 1000.

2. High specific speeds (above 5000) also have a lower efficiency than the medium specific speed range.

3. At all specific speeds, smaller pump capacities have lower efficiencies than pumps with higher capacities.

Radial-flow centrifugal pumps have a specific speed between 500 and 3500 and are suitable for low discharge under relatively high pressure. Mixed-flow pumps with a specific speed in the range of 3500 to 7500 are used for flows of more than 1000 gpm. Axial-flow pumps with a specific speed between 7500 and 15,000 deliver a high discharge of over 5000 gpm.

---

* In principle, the term is dimensionless in the form $NQ^{1/2}/(gH)^{3/4}$.

12.16 RELATIONS FOR GEOMETRICALLY SIMILAR PUMPS

The relations of parameters of geometrically similar pumps are known as the *affinity laws*. These are useful in predicting the performance of a pump from pumping tests on a model pump or homologous* pump.

Since $Q = AV$, where the area is proportional to the square of the impeller diameter, D^2, and the velocity is proportional to the impeller diameter, D, and angular speed, N, $Q \propto ND^3$. Since $V = \sqrt{2gH}$, $H \propto V^2$ or $H \propto D^2 N^2$. Also, the power is the multiplication of Q and H and hence $P \propto N^3 D^5$.

Thus, from dimensional analysis considerations, the affinity laws are

$$\frac{Q_2}{Q_1} = \frac{N_2}{N_1} \left(\frac{D_2}{D_1}\right)^3 \quad \text{[dimensionless]} \tag{12.23}$$

$$\frac{H_2}{H_1} = \left(\frac{N_2}{N_1}\right)^2 \left(\frac{D_2}{D_1}\right)^2 \quad \text{[dimensionless]} \tag{12.24}$$

$$\frac{P_2}{P_1} = \left(\frac{N_2}{N_1}\right)^3 \left(\frac{D_2}{D_1}\right)^5 \quad \text{[dimensionless]} \tag{12.25}$$

where

$Q =$ capacity

$H =$ head

$P =$ power

and where the subscript 1 refers to the parameters at which characteristics are known and the subscript 2 refers to the unit for which values are to be predicted.

EXAMPLE 12.12

A model pump of 5 in. diameter develops 0.25 hp at a speed of 800 rpm under a head of 2.5 ft. A geometrically similar pump 15 in. in diameter is to operate at the same efficiency at a head of 49.0 ft. What speed and power should be expected?

SOLUTION   From eq. (12.24),

$$\frac{H_2}{H_1} = \left(\frac{N_2}{N_1}\right)^2 \left(\frac{D_2}{D_1}\right)^2$$

$$\frac{49}{2.5} = \left(\frac{N_2}{800}\right)^2 \left(\frac{15}{5}\right)^2$$

$$N_2 = 1181 \text{ rpm}$$

---

* Two units that are geometrically similar and have similar vector diagrams are said to be homologous.

From eq. (12.25),

$$\frac{P_2}{P_1} = \left(\frac{N_2}{N_1}\right)^3 \left(\frac{D_2}{D_1}\right)^5$$

$$P_2 = \left(\frac{1181}{800}\right)^3 \left(\frac{15}{5}\right)^5 (0.25) = 195 \text{ hp}$$

12.17 RELATIONS FOR ALTERATIONS IN THE SAME PUMP

For a given pump operating at a given speed, there are definite relationships among parameters, known as the *performance characteristics*. If the pump size is altered or the speed is changed, the same relations do not hold. However, in the same pump there is a geometric similarity of velocity which means that at any speed or diameter, the velocities are similarly directed and ratios of all velocities are the same. The velocity triangle due to peripheral and flow velocities at the exit from the impeller is similar before and after the alteration in the pump diameter. For this condition, the following relations apply:

$$\frac{Q_2}{Q_1} = \frac{N_2}{N_1}\frac{D_2}{D_1} \qquad \text{[dimensionless]} \qquad (12.26)$$

$$\frac{H_2}{H_1} = \left(\frac{N_2}{N_1}\right)^2 \left(\frac{D_2}{D_1}\right)^2 \qquad \text{[dimensionless]} \qquad (12.27)$$

$$\frac{P_2}{P_1} = \left(\frac{N_2}{N_1}\right)^3 \left(\frac{D_2}{D_1}\right)^3 \qquad \text{[dimensionless]} \qquad (12.28)$$

The efficiency is considered constant, with change in speed and diameter in the relations above.

Equations (12.26), (12.27), and (12.28) are used to determine the revised characteristics of a pump for a desired change in speed, diameter, or both. Alternatively, the equations are used to determine the speed or diameter to produce a desired change in the discharge capacity or the head of the pump without changing the efficiency.

EXAMPLE 12.13

A pump tested at 1800 rpm gives the following results: capacity = 4000 gpm, head = 157 ft, power = 190 hp. **(a)** Obtain the performance of this pump at 1600 rpm. **(b)** If along with the speed, the diameter of the impeller is reduced from 15 in. to 14 in., obtain the revised pump characteristics.

SOLUTION

(a) $N_2/N_1 = 1600/1800 = 0.89$ and $D_2/D_1 = 1$

From eq. (12.26),

$$\frac{Q_2}{Q_1} = 0.89 \text{ or } Q_2 = 0.89(4000) = 3560 \text{ gpm}$$

From eq. (12.27),

$$\frac{H_2}{H_1} = (0.89)^2 \text{ or } H_2 = (0.89)^2(157) = 124 \text{ ft}$$

From eq. (12.28),

$$\frac{P_2}{P_1} = (0.89)^3 \text{ or } P_2 = (0.89)^3(190) = 134 \text{ hp}$$

(b) $D_2/D_1 = 14/15 = 0.933$

From eq. (12.26),

$$\frac{Q_2}{Q_1} = (0.933)(0.89) = 0.83 \text{ or } Q_2 = 0.83(4000) = 3320 \text{ gpm}$$

From eq. (12.27),

$$\frac{H_2}{H_1} = (0.933)^2(0.89)^2 = 0.69 \text{ or } H_2 = 0.69(157) = 108 \text{ ft}$$

From eq. (12.28),

$$\frac{P_2}{P_1} = (0.933)^3(0.89)^3 = 0.57 \text{ or } P_2 = 0.57(190) = 108 \text{ hp}$$

12.18 HEAD TERMS IN PUMPING

For a diagram that encompasses the following definitions, refer to Figure 12.15.

Static suction lift. The vertical distance from the water level in the source tank to the centerline of the pump. If the pump is located at a lower level than the source tank [Figure 12.15(c)], the static suction lift is negative.

Static discharge head. The vertical distance from the centerline of the pump to the water level in the discharge tank or to the exit end of the pipe, whichever is higher.

Total static head. The sum of the static suction lift and the static discharge head, which is equal to the difference between the water levels of discharge end and source tank.

Total dynamic head (TDH). The sum of the total static head and the friction and minor losses. This term is known as the total head. The relation for the total head was developed in Section 12.8, given by eq. (12.16):

$$H_p = DZ + h_{\text{loss}} \quad [\text{L}] \tag{12.16}$$

The total dynamic head, H_p, is used to calculate the horsepower requirement for the pump as given by eq. (12.17).

Figure 12.15 Head terms in pumping.

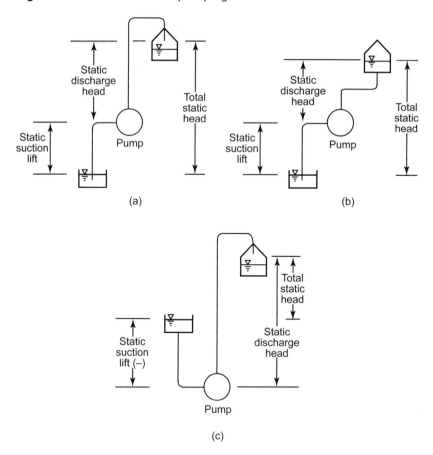

(a)

(b)

(c)

12.19 SYSTEM HEAD CURVE

For any piping system, the friction loss,

$$h_f \left(= \frac{f}{d} \frac{L}{2g} V^2 \right)$$

and the minor loss, $h_m \left(= \Sigma KV^2/2g \right)$, can be expressed in terms of the flow through the system. Thus eq. (12.16) can be expressed as

$$H_p = \Delta Z + \frac{0.81}{g} \left(\frac{f\, LQ^2}{d^5} + \frac{\Sigma KQ^2}{d^4} \right) \quad [\text{L}] \tag{12.29}$$

The plot of eq. (12.29) between H_p versus Q, as shown in Figure 12.16, is known as the *system head curve.* This curve, representing the behavior of the piping system, is important in the selection of a pump.

Figure 12.16   Typical system head curve.

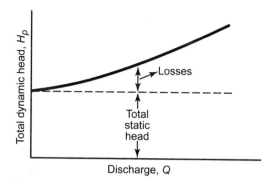

12.20 PUMP CHARACTERISTIC CURVES

As stated earlier, for a given pump at a given speed, there are definite relationships among the pump discharge capacity, head, power, and efficiency. These relations are derived from actual tests on a given pump or a similar unit and are usually depicted graphically by the *pump characteristics* or *performance curves*, comprising the following:

- Pumping head versus discharge
- Brake horsepower versus discharge
- Efficiency versus discharge

Figure 12.17 illustrates typical characteristic curves. The general shape of the curves varies with the size, speed, and design of a particular pump. The important feature of the curves is that an increase in the head reduces the capacity. These curves are supplied by the manufacturer of the pump. In fact, since a pump casing can accommodate impellers of several sizes, the manufacturer supplies a series of sets of curves drawn on the same graph, corresponding to various sizes of the impellers, which can be derived by use of the laws explained in Section 12.17. A set of characteristic curves represents the behavior of a given-size pump operating at a given speed, in the same manner as a system head curve represents the behavior of a piping system. At a given speed, a pump is rated at the head and discharge, which gives the maximum efficiency, referred to as the *best efficiency point*, shown by point A in Figure 12.17. The characteristic curves, particularly the head-discharge curve, are important in pump selection.

EXAMPLE 12.14

The characteristic data as supplied by the manufacturer for an 8-in. pump rotating at 1750 rpm are given in Table 12.10. **(a)** Plot the pump characteristic curves. **(b)** Determine the type specific speed. **(c)** If the pump speed is reduced to 1450 rpm, determine the pump characteristics. **(d)** If the diameter of the pump is reduced to 6 in. but the speed remains 1750 rpm, determine the pump characteristics. **(e)** If a similar unit of 6 in. diameter is used at 1750 rpm, determine its characteristics.

Figure 12.17 Pump characteristic curves.

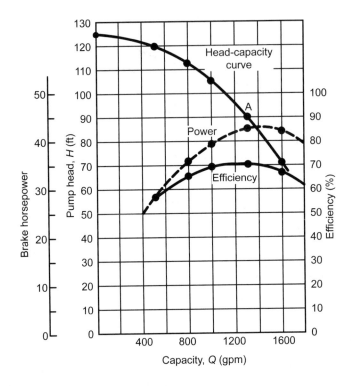

Table 12.10   Data for 8-in. Pump at 1750 rpm

Q (gpm)	H (ft)	P (hp)	Efficiency (%)
0	124		0
500	119	27.75	54
800	112	35.34	64
1000	104	38.69	68
1300	90	42.30	70
1600	70	42.20	67

SOLUTION

(a) The characteristic curves are plotted in Figure 12.17.

(b) At best (maximum) efficiency point, $Q = 1300$ gpm, $H = 90$ ft.

$$N_S = \frac{N\sqrt{Q}}{H^{3/4}} = \frac{1750\sqrt{1300}}{90^{3/4}} = 2159$$

(c) For a change in the pump speed without diameter change, from eqs. (12.26) to (12.28),

$$Q_2 = \frac{N_2}{N_1} Q_1 = \frac{1450}{1750} Q_1 = 0.83 Q_1$$

$$H_2 = \left(\frac{N_2}{N_1}\right)^2 H_1 = (0.83)^2 H_1 = 0.69 H_1$$

$$P_2 = \left(\frac{N_2}{N_1}\right)^3 P_1 = (0.83)^3 P_1 = 0.57 P_1$$

The values of Table 12.10 are adjusted by these factors in columns 1, 2, and 3 of Table 12.11.

(d) For the reduced diameter without speed change of the same pump, from eqs. (12.26) to (12.28),

$$Q_2 = \frac{D_2}{D_1} Q_1 = \frac{6}{8} Q_1 = 0.75 Q_1$$

$$H_2 = \left(\frac{D_2}{D_1}\right)^2 H_1 = (0.75)^2 H_1 = 0.56 H_1$$

$$P_2 = \left(\frac{D_2}{D_1}\right)^3 P_1 = (0.75)^3 P_1 = 0.42 P_1$$

The values are given in columns 4, 5, and 6 of Table 12.11.

(e) For a homologous unit, from eqs. (12.23) to (12.25),

$$Q_2 = \left(\frac{D_2}{D_1}\right)^3 Q_1 = (0.75)^3 Q_1 = 0.42 Q_1$$

$$H_2 = \left(\frac{D_2}{D_1}\right)^2 H_1 = (0.75)^2 H_1 = 0.56 H_1$$

$$P_2 = \left(\frac{D_2}{D_1}\right)^5 P_1 = (0.75)^5 P_1 = 0.237 P_1$$

The final values are computed in columns 7, 8, and 9 in Table 12.11 by the factors above.

12.21 SINGLE PUMP AND PIPELINE SYSTEM

The suitability of a given pump for a certain known piping system is determined by super-imposing the system head curve of the piping system on the characteristic curve of the pump. The intersection point of the two curves indicates the operating point (i.e., the head and discharge of the given pump). If the efficiency of the pump is too low at this point, another pump should be considered. This is illustrated in Example 12.15.

Table 12.11 Adjusted Pump Characteristics

8-in. at 1450 rpm Same Pump			6-in. at 1750 rpm Same Pump			6-in. at 1750 rpm Similar Unit		
(1)	(2)	(3)	(4)	(5)	(6)	(7)	(8)	(9)
Q	H	P	Q	H	P	Q	H	P
$0.83Q_1$	$0.69H_1$	$0.57P_1$	$0.75Q_1$	$0.56H_1$	$0.42P_1$	$0.42Q_1$	$0.56H_1$	$0.237P_1$
(cfs)	(ft)	(hp)	(cfs)	(ft)	(hp)	(cfs)	(ft)	(hp)
0	85.6		0	69.4		0	69.4	
415	82.1	15.82	375	66.6	11.66	210	66.6	6.58
664	77.3	20.14	600	62.7	14.84	336	62.7	8.38
830	71.8	22.05	750	58.2	16.25	420	58.2	9.17
1079	62.1	24.11	975	50.4	17.77	546	50.4	10.03
1328	48.3	24.05	1200	39.2	17.72	672	39.2	10.00

EXAMPLE 12.15

A pump having the characteristic curves given in Table 12.10 is to be used in the pipeline system shown in Figure 12.18. Determine the **(a)** operating head and discharge, **(b)** efficiency of the pump and hence its suitability, and **(c)** input power (brake horsepower) for the pump. $f = 0.02$.

SOLUTION

1. For the system head curve, from eq. (12.29),

$$H_p = \Delta Z + \frac{0.81}{g}\left(\frac{f\,LO^2}{d^5} + \frac{\Sigma KO^2}{d^4}\right)$$

	K
One entrance	0.5
One exit	1.0
Four 90° bends at 1.5	6.0
One globe valve	10.0
	17.5

$$H_p = (100-60) + \frac{0.81}{32.2}\left[\frac{0.02(1000)Q^2}{(8/12)^5} + \frac{(17.5)Q^2}{(8/12)^4}\right]$$

or

$$H_p = 40 + 6.05Q^2 \quad [\text{L}] \tag{12.30}$$

2. From eq. (12.30), H_p for various assumed Q values are computed in Table 12.12 and plotted in Figure 12.19.

3. On the same graph, the head-capacity curve of the pump characteristics has been plotted from the data in Table 12.10.

Figure 12.18 Piping system for Example 12.15.

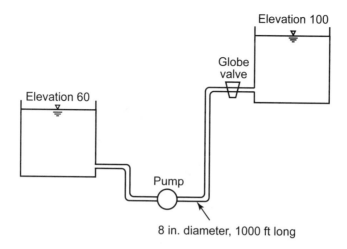

Table 12.12   Computation for System Head Curve

Capacity (gpm) (Select)	$Q = \dfrac{\text{Capacity}}{449}$ (cfs)	H_p [by eq. (12.30)] (ft)
0	0	40
500	1.11	47.45
800	1.78	59.17
1000	2.23	70.08
1500	3.34	107.49

(a) **4.** The operating head and capacity at the intersection point: $Q = 1300$ gpm, $H = 90$ ft.

(b) **5.** At this point, the efficiency = 70%, which is the maximum efficiency; hence the pump is OK.

(c) **6.** Input power $= \dfrac{\gamma QH}{550\eta} = \dfrac{(62.4)(1300/449)(90)}{550(0.70)} = 42.2$ hp

12.22 MULTIPLE PUMP SYSTEM

A single pump is suitable within a narrow range of head and discharge in proximity of the optimum pump efficiency. However, in a piping system the discharge and head requirements may vary considerably at different times. Within a certain range, these fluctuations in head and discharge can be accommodated by adopting variable-speed motors. Pump characteristic curves can be altered by suitable adjustment of the speed, as discussed in Section 12.17. When the fluctuations are considerable, or either the head or capacity requirement is too high for a single pump, two or more pumps are used in series or in parallel. It is advantageous both from hydraulic and economic considerations to use pumps of identical size to match their performance characteristics. The pumps are used in series in a system

Figure 12.19 Determination of pump operating condition.

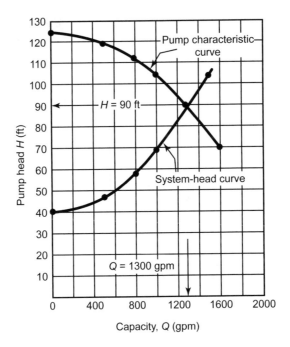

where substantial head changes take place without appreciable difference in the discharge (i.e., the system head curve is steep). In series, each pump has the same discharge. The parallel pumps are useful for systems with considerable discharge variations with no appreciable head change. In parallel, each pump has the same head.

12.23 PUMPS IN SERIES

The following relations apply:

$$H = H_A + H_B + \cdots \qquad [L] \qquad (12.31)$$

$$Q = Q_A = Q_B = \cdots \qquad [L^3 T^{-1}] \qquad (12.32)$$

$$\eta = \frac{H_A + H_B + \cdots}{H_A / \eta_A + H_B / \eta_B + \cdots} \qquad [\text{dimensionless}] \qquad (12.33)$$

$$P = \frac{\gamma Q (H_A + H_B + \cdots)}{550\eta} \qquad [FLT^{-1}] \quad (\text{English units}) \qquad (12.34a)$$

$$P = \frac{\gamma Q (H_A + H_B + \cdots)}{\eta} \qquad [FLT^{-1}] \quad (\text{metric units}) \qquad (12.34b)$$

where $A, B, \ldots$ refer to different pumps.

The composite characteristic curves of pumps in series can be prepared by the equations above. The composite head characteristic curve is prepared by adding the ordinates (heads) of all the pumps for the same values of discharge [eq. (12.31)], as shown in Figure 12.20. The intersection point of the composite head characteristic curve and the system head curve provides the operating condition.

EXAMPLE 12.16

Pumps A and B have the following characteristics:

Pump A: 8-in., 1450 rpm			Pump B: 10-in., 1750 rpm		
Q (gpm)	H (ft)	Efficiency	Q (gpm)	H (ft)	Efficiency
0	186	0	0	172	0
500	179	54	400	166	59
1000	158	70	800	140	77
1500	112	67	1200	90	74

These pumps are arranged in series in a system having a static lift of 80 ft. The pipeline comprises 6-in-diameter pipe of 1200 ft length, with minor losses 20 times the velocity head. Determine the operating condition and the power input. $f = 0.022$.

SOLUTION

1. The characteristic curves for pumps A and B are plotted in Figure 12.20. The composite head characteristics are computed below.

Capacity (gpm)	Head for Pump A (ft)	Head for Pump B (ft)	Total head $H_A + H_B$ (ft)
0	186	172	358
400	182	166	348
800	168	140	308
1200	144	90	234

The composite pump characteristics curve is illustrated in Figure 12.20.

2. The system head curve is computed below:

$$H_p = \Delta Z + \frac{0.81}{32.2}\left[\frac{0.022(1200)Q^2}{(0.5)^5} + \frac{20Q^2}{(0.5)^4}\right]$$

$$= 80 + 29.3Q^2$$

Capacity (gpm)	Q = Capacity/449 (cfs)	$H_p = 80 + 29.3Q^2$ (ft)
0	0	80.0
500	1.11	116.1
1000	2.23	225.7
1200	2.67	288.9

The system head curve has been plotted in Figure 12.20.

Figure 12.20 Head characteristics and operating condition for pumps in series.

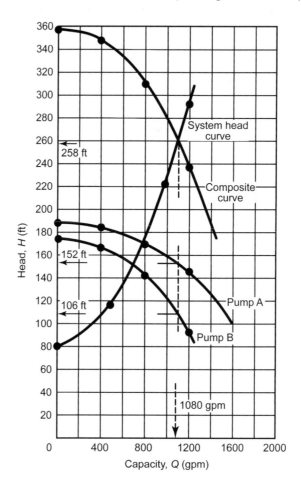

3. The discharge and head at the operating condition from the intersection of the characteristic and the system curves are $Q = 1080$ gpm, $H = 258$ ft.

4. Corresponding to $Q = 1080$ gpm:

For Pump A: $H_A = 152$ ft, $\eta_A = 69\%$

For Pump B: $H_B = 106$ ft, $\eta_B = 75\%$

5. $\eta = \dfrac{H_A + H_B}{H_A/\eta_A + H_B/\eta_B} = \dfrac{260}{152/69 + 106/75} = 71.9\%$

6. Input power $= \dfrac{\gamma Q(H_A + H_B)}{550\eta} = \dfrac{62.4(1080/449)(258)}{550(0.719)} = 97.9$ hp

12.24 PUMPS IN PARALLEL

For parallel pumps, the relations are as follows:

$$H = H_A = H_B = \cdots \qquad [\text{L}] \qquad (12.35)$$

$$Q = Q_A + Q_B + \cdots \qquad [\text{L}^3\text{T}^{-1}] \qquad (12.36)$$

$$\eta = \frac{Q_A + Q_B + \cdots}{Q_A / \eta_A + Q_B / \eta_B + \cdots} \qquad [\text{dimensionless}] \qquad (12.37)$$

$$P = \frac{\gamma H (Q_A + Q_B + \cdots)}{550\eta} \qquad [\text{FLT}^{-1}] \quad (\text{English units}) \qquad (12.38a)$$

$$P = \frac{\gamma H (Q_A + Q_B + \cdots)}{\eta} \qquad [\text{FLT}^{-1}] \quad (\text{metric units}) \qquad (12.38b)$$

The composite head characteristic curve is obtained by summing up the abscissas (discharges) of all the pumps for the same values of head [eq. (12.36)], as shown in Figure 12.21.

EXAMPLE 12.17

The two pumps of Example 12.16 are arranged in parallel. The static head is 40 ft. The pipeline system consists of 9 in.-diameter pipe of 1200 ft length with minor losses 20 times the velocity head. Determine the operating condition and the power input. $f = 0.022$.

SOLUTION

1. The characteristic curves for pumps A and B are plotted in Figure 12.21. The composite characteristics are computed below.

Head (ft)	Capacity of Pump A (gpm)	Capacity of Pump B (gpm)	Total Capacity $Q_A + Q_B$ (gpm)
172	720	0	720
160	960	500	1460
140	1240	800	2040
120	1440	980	2420

The composite curve is plotted in Figure 12.21.

2. The system head curve is computed as follows:

$$H_p = 40 + \frac{0.81}{32.2} \left[\frac{0.022(1200)Q^2}{(0.75)^5} + \frac{20Q^2}{(0.75)^4} \right]$$

$$= 40 + 4.39Q^2$$

The system curve is plotted in Figure 12.21.

Figure 12.21 Head characteristics and operating condition for pumps in parallel.

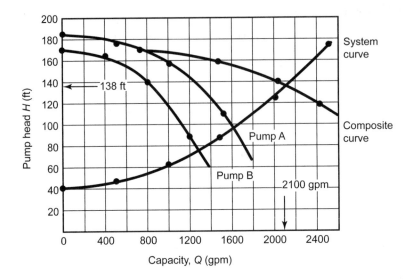

Capacity, Q (gpm)	Q = capacity/449 (cfs)	$H_p = 40 + 4.39Q^2$ (ft)
0	0	40
500	1.11	45.4
1000	2.23	61.8
1500	3.34	89.0
2000	4.45	126.9
2500	5.57	176.2

3. The operating condition at the intersection of the characteristic and system curves is Q = 2100 gpm, H = 138 ft.

4. Corresponding to the head of 138 ft:

For Pump A: Q_A = 1280 gpm, η_A = 68%
For Pump B: Q_B = 840 gpm, η_B = 77%

5. Overall $\eta = \dfrac{Q_A + Q_B}{Q_A/\eta_A + Q_B/\eta_B} = \dfrac{2100}{1280/68 + 840/77} = 70.6\%$

6. Input power $= \dfrac{(62.4)(2100/449)(138)}{550(0.706)} = 104$ hp

12.25 LIMIT ON PUMP LOCATION

The pressure on the suction side of a pump which is located above the supply tank is below the atmospheric (vacuum) pressure. If the absolute pressure at the suction inlet (point S in Figure 12.22) falls below the vapor pressure of the water at the operating temperature, vapor pockets are formed which can damage the pump. This phenomenon is known as *cavitation*.

The absolute pressure at the suction intake of a pump is referred to as the net positive suction head (NPSH). It should not fall below a certain minimum value that is influenced by the further reduction in pressure within the impeller. For each pump, the manufacturer indicates the *required* NPSH based on the pump performance test. The *available* NPSH for a given system should be more than the required NPSH. This places a limitation on the elevation for the location of a pump.

In terms of the pressure head at the suction inlet, the available NPSH is

$$\text{NPSH} = \frac{p_s}{\gamma} + \frac{V_s^2}{2g} - \frac{e_w}{\gamma} \quad \text{[L]} \tag{12.39}$$

where the subscript refers to values at suction inlet and e is the vapor pressure, given in Appendix C.

It is not convenient in many instances to measure p_s and V_s at the suction inlet. In such cases it is preferable to express NPSH in terms of pressure on the reservoir surface.

$$\text{NPSH} = \frac{p_0}{\gamma} - Z - h_L - \frac{e_w}{\gamma} \quad \text{[L]} \tag{12.40}$$

where

p_0 = absolute pressure at the reservoir surface; atmospheric for open reservoir

Z = elevation of suction intake from the reservoir surface

h_L = friction and local head losses up to the suction inlet

The cavitation parameter, σ, is given by

$$\sigma = \frac{\text{NPSH}}{\text{total pump head}} \quad \text{[dimensionless]} \tag{12.41}$$

The computed σ should be higher than the critical σ furnished by the manufacturer. The cavitation parameters range from 0.05 for a specific speed of 1000 to 1.0 for a specific speed of 8000.

Figure 12.22 Pressure head at suction inlet.

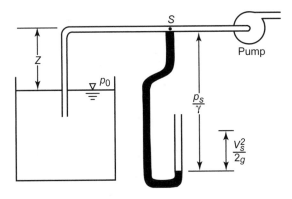

EXAMPLE 12.18

The pumping system shown in Figure 12.22 is to deliver 1 cfs of water at 60°F. The suction line is 6 in. in diameter in a 300-ft-long cast-iron pipe. The suction inlet is 18 ft above the reservoir level. The atmospheric pressure of 14.7 psi absolute exists over the reservoir. The required NPSH of the pump is 7. Determine whether the system will have a cavitation problem.

SOLUTION

$$V = \frac{Q}{A} = \frac{1}{(\pi/4)(0.5)^2} = 5.1 \text{ ft/sec}$$

$$Re = \frac{Vd}{v} = \frac{5.1(0.5)}{1.217 \times 10^{-5}} = 2.1 \times 10^5$$

$$\frac{\varepsilon}{d} = \frac{0.0008}{0.5} = 0.0016$$

From the Moody diagram, Fig. 12.4, $f = 0.023$:

$$h_f = \frac{f L V^2}{d \; 2g} = (0.023)\left(\frac{300}{0.5}\right)\frac{(5.1)^2}{2(32.2)} = 5.57 \text{ ft}$$

Entrance loss $K = 0.8$

Bend $K = 1.5$

$\overline{\quad \Sigma K = 2.3 \quad}$

$$h_m = \frac{2.3(5.1)^2}{2(32.2)} = 0.93 \text{ ft}$$

At 60°F, e_w (vapor pressure) = 0.26 psi (Appendix C), p_0 = 14.7 psi (atmospheric). From eq. (12.40),

$$\text{NPSH} = \frac{p_0}{\gamma} - Z - h_L - \frac{e_w}{\gamma}$$

$$= \frac{14.7(144)}{62.4} - 18 - (5.57 + 0.93) - \frac{0.26(144)}{62.4}$$

$$= 8.82 \text{ ft}$$

Since the available NPSH of 8.82 is greater than the required NPSH of 7.0, there is no cavitation problem.

PROBLEMS

12.1 Water flows from a reservoir at a rate of 2.5 cfs through a system of pipes as shown in Fig. P12.1. Determine the total energy loss in the system.

Figure P 12.1

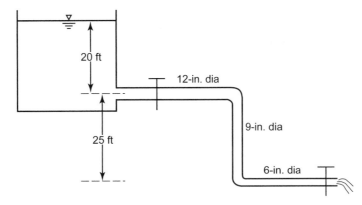

12.2 Crude oil of density 925 kg/m^3 flows from a closed tank that has a pressure 70 kPa above the atmosphere to an open tank. If the oil level in the open tank is 2 m higher than the closed tank, determine the total loss of energy between the two tanks.

12.3 At the inlet point of a 1000-ft-long pipeline 6 in. in diameter, the energy head is 80 ft from a reference datum. The pipe carries a flow of 0.6 cfs. The pressure at the outlet is 18 psi and the elevation of the pipe at the end is 12 ft from the datum. Determine the loss of energy through the pipe.

12.4 Determine the slope of the energy grade line for Problem 12.3. Determine the piezometric height (height of the hydraulic grade line) above the reference datum at a point 300 ft from the inlet.

12.5 Water flows at a velocity of 0.3 ft/sec in a cast-iron pipe of 1 in. inside diameter at 40°F. Determine the friction factor.

12.6 Determine the friction factor for a 100-mm-inside-diameter commercial steel pipe in which water flows at 60°C at a rate of 0.1 m^3/s.

12.7 The discharge through a concrete pipe of 1 ft diameter and 3000 ft length is 1.0 cfs. What is the energy loss if the water temperature is 60°F? Apply the Darcy-Weisbach equation.

12.8 Water at 80°F flows from a storage tank through an 80-ft-long galvanized iron pipe of 4 in. diameter laid horizontally. Calculate the depth of water required in the tank above the centerline of the pipe to carry a discharge of 2 cfs. Disregard the minor losses. Apply the Darcy-Weisbach equation.

12.9 A fluid of density 900 kg/m^3 and viscosity $\mu = 0.28$ N-s/m^2 flows steadily down a vertical 0.1-m-diameter pipe and exits at the lower end. Determine the maximum pressure in kPa at 8 m above the pipe exit if the flow is to be laminar.

12.10 A 25-m-long, 10-mm diameter hose of roughness $\varepsilon = 0.02$ mm is fastened to a water faucet where the pressure is 6 kPa. Determine the rate of flow at 20°C.

12.11 A fire hydrant is supplied through a 6-in. cast-iron pipe of 16,000 ft length. The total drop in pressure is limited to 35 psi. What is the discharge through the pipe for a water temperature of 80°F? Apply the Darcy-Weisbach equation.

12.12 Water is transferred at a rate of 2.1 cfs from an upper reservoir to a lower reservoir by a concrete pipe 3000 ft long. The difference in water level is 60 ft. Determine the size of pipe for a water temperature of 60°F. Apply the Darcy-Weisbach equation.

12.13 Air at standard temperature (15°C) and pressure (101 kPa) flows at a rate of 0.05 m³/s through a galvanized iron pipe that slopes down 2%. Determine the pipe diameter if the pressure drop is not to exceed 1 kPa per 100 m of pipe.

12.14 An upper reservoir A is connected to a lower reservoir B by a 10,000-ft-long, 2-ft diameter cast-iron pipe. The difference in water elevations of the two reservoirs is 100 ft. There is a hill between the two reservoirs whose summit is 15 ft above the upper reservoir. The pipeline has to cross the hill when its length is 3000 ft. Determine the **(a)** discharge of water through the pipeline, and **(b)** minimum depth below the summit to which the pipeline should be laid if the pressure in the pipe is not to fall below atmospheric pressure. Assume a water temperature of 50°F. Disregard the minor losses. Apply the Darcy-Weisbach equation.

12.15 For Problem 12.7, determine the loss of energy by the Hazen-Williams formula. $C = 120$.

12.16 Rework Problem 12.8 by the Hazen-Williams formula. $C = 120$.

12.17 For Problem 12.11, determine the discharge by the Hazen-Williams formula.

12.18 For Problem 12.12, determine the pipe size by the Hazen-Williams formula.

12.19 Air flows through a schedule 80 steel pipe (internal diameter 0.546 in.) from a closed storage tank to a second closed storage tank 100 ft above the first storage tank. The threaded piping system contains ten 90° elbows, two 180° bends, three tees with line flow and two tees with branch flow, one globe valve and two gate valves in a 200-ft-long pipe. The pressure in the first tank is 50 psi. If the air flow is 5 ft³/min and the air has a specific weight of 0.1 lb/ft³ and dynamic viscosity of 3×10^{-7} lb-s/ft², what is the pressure in the second tank? If the pressure variation in two tanks is more than 10%, then the Darcy-Weisbach equation of incompressible flow cannot be applied. Is the equation valid in this case?

12.20 A pipeline 200 m long delivers water from an impounding reservoir to a service reservoir with the difference in water levels of 20 m. The pipeline of commercial steel is 400 mm in diameter. It has two 90° flanged elbows, a check valve, and an orifice ($K = 2.5$). Determine the flow through the pipe. Use a water temperature of 20°C.

12.21 In a cast-iron piping system of 450 ft length shown in Fig. P12.21, the rate of flow is 1 cfs at 70°F. Determine the diameter of the pipe.

Figure P 12.21

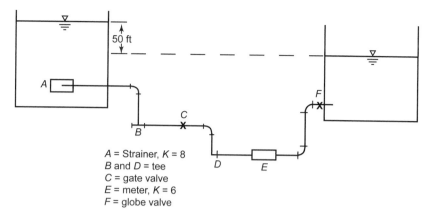

A = Strainer, K = 8
B and D = tee
C = gate valve
E = meter, K = 6
F = globe valve

12.22 A 300-mm-diameter cast-iron pipeline 1000 m long delivers water from an upper reservoir to a lower reservoir. The difference in water levels of the two reservoirs is 30 m. **(a)** Determine the discharge through the pipe. **(b)** If the discharge has to be maintained at 0.3 m³/s without a change in the pipe, determine the horsepower of the pump to be installed at an efficiency of 70%. Consider the entrance and exit losses and the water temperature of 10°C.

12.23 In the supply system shown in Fig. P12.23, the pressure required at the delivery end is 60 psi. Determine **(a)** the pumping head, and **(b)** the power delivered by the pump. Consider the entrance and exit losses and a water temperature of 70°F.

Figure P 12.23

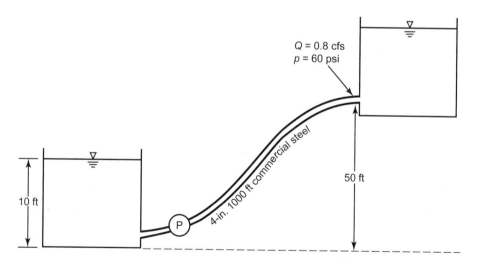

12.24 A fire hydrant is supplied through three welded steel pipelines arranged in series. The total drop in pressure due to friction in the pipeline is limited to 50 psi. What is the discharge through the hydrant? Disregard the minor losses.

Pipe	Diameter (in.)	Length (ft)
1	6	500
2	8	2,000
3	12	16,000

12.25 Find the elevation of the downstream reservoir for a flow of 5 cfs in the system shown in Fig. P12.25. Consider the minor losses. The water temperature = 50°F.

Figure P12.25

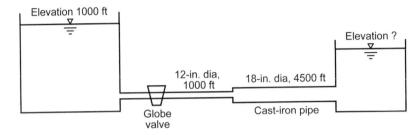

12.26 A flow of 1 m³/s is divided into three parallel cast-iron pipes of diameters 500, 250, and 400 mm and lengths 400 m, 100 m, and 250 m, respectively. Determine the head loss and flow through each pipe.

12.27 Two reservoirs having a difference of 40 ft in water elevations are connected by a 1.5-mile-long cast-iron pipe of 1 ft diameter. A second pipe 1.5 ft in diameter and one-half mile in length is laid alongside the first one for the last half-mile. How is the discharge affected?

12.28 For the pipe system shown in Fig. P12.28, determine the rate of flow. Disregard the minor losses.

Figure P12.28

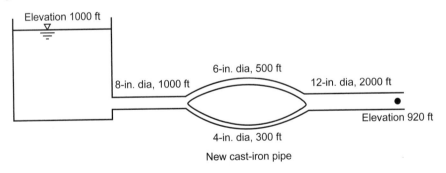

12.29 In the network shown in Fig. P12.29, determine the **(a)** flows in the pipe, and **(b)** pressure heads at the nodes. C = 130.

Figure P 12.29

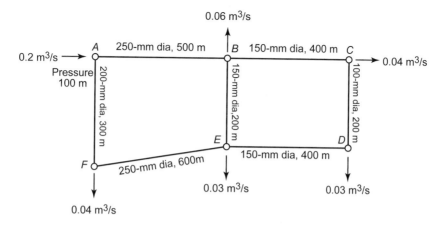

12.30 The length of each pipe in the network shown in Fig. P12.30 is 1000 ft. Determine the (a) flows in the pipe, and (b) pressure heads at the nodes. The pressure at point C is 40 ft. The elevations of different points are given below. $C = 100$.

Figure P 12.30

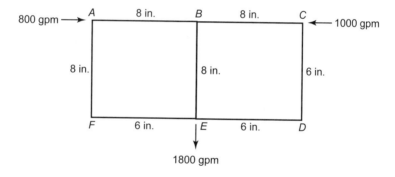

Node	A	B	C	D	E	F
Elevation (ft)	200	150	300	150	200	150

12.31 In the pipe network shown in Fig. P12.31, determine the discharge in each pipe. Assume that $f = 0.015$.

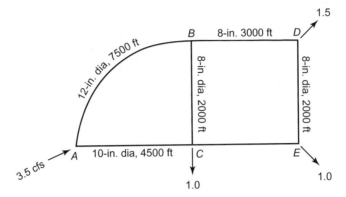

12.32 A water district network is shown in Figure P12.32. The types of water needs, the tribu-tary area or the population served, and the maximum daily unit consumption assigned to each pipe segment are given in the accompanying table. The initial sizes of PVC pipes are shown in the figure. Water enters the district at node A. The fire demand is initially assigned at node D. Assess the adequacy of the network by analyzing for discharge in each pipe and pressure head at each node. (Assume the pressure head at A to be 175 ft).

Figure P 12.32

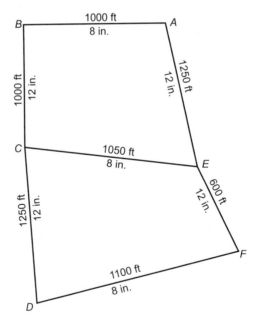

Segment	Water use	Area/population served	Maximum daily unit consumption
AB	industrial	6.77 acres	21,000 gal/day/acre
BC	commercial	2.90 acres	45,000 gal/day/acre
AF	commercial	2.90 acres	45,000 gal/day/acre
CF	residential	120 persons	270 gal/day/person
CD	residential	502 persons	270 gal/day/person
ED	residential	598 persons	270 gal/day/person
EF	commercial	3.30 acres	45,000 gal/day/acre
Fire demand			0.94 cfs

12.33 For the water district network shown in Figure P12.33, the tributary area or the population served, and the maximum daily unit consumption assigned to each pipe segment are given in the accompanying table. The initial sizes of PVC pipes are shown in the figure. Water enters the district at node *A*. The fire demand is assigned at node *D*. Determine the pressure at node *D* if the pressure at node A is 75 psi.

Figure P 12.33

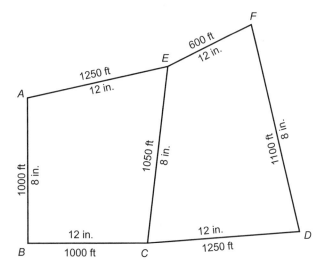

Segment	Water use	Area/population served	Maximum daily unit consumption
AB	industrial	16.06 acres	21,000 gal/day/acre
BC	commercial	6.73 acres	46,000 gal/day/acre
AF	commercial	6.88 acres	45,000 gal/day/acre
CF	residential	308 persons	250 gal/day/person
CD	residential	1191 persons	270 gal/day/person
ED	residential	1419 persons	270 gal/day/person
EF	commercial	7.83 acres	45,000 gal/day/acre
Fire demand			2.23 cfs

12.34 A 6-in. pump operating at 1760 rpm delivers 1480 gpm for a head of 132 ft at its maximum efficiency. Determine the specific speed.

12.35 A pump operating at 25 hertz delivers 60 liters of water at 30 m head at maximum efficiency. Determine the type specific speed of the pump. Is it a low or high discharge pump?

12.36 A centrifugal pump discharges 300 gpm against a head of 55 ft when the speed is 1500 rpm. The diameter of the impeller is 12.5 in. and the brake horsepower is 6.0. **(a)** Determine the efficiency of the pump. **(b)** What should be the speed of a geometrically similar pump of 15 in. diameter running at a capacity of 400 gpm?

12.37 If the rotational speed of a pump motor is reduced by 35%, what is the effect on the pump performance in terms of capacity, head, and power requirements?

12.38 The following performance characteristics are obtained for a pump tested at 1800 rpm. Determine the performance data of this pump operating at 1400 rpm.

Capacity (gpm)	Head (ft)	Power (hp)	Efficiency (%)
3000	200	175	87
2000	221	143	78.5
1000	229	107	54

12.39 The speed of the pump in Problem 12.38 is not changed, but the impeller diameter is reduced from 14.75 in. to 14.0 in. Determine the performance data of the pump.

12.40 If both the speed and the impeller diameter are changed for the pump as stated in Problems 12.38 and 12.39, determine the pump characteristic data.

12.41 From the manufacturer's data, a pump of 10 in. impeller diameter has a capacity of 1200 gpm at a head of 61 ft when operating at a speed of 900 rpm. It is desired that the capacity be about 1500 gpm at the same efficiency. Determine the adjusted speed of the pump and the corresponding head.

12.42 The characteristic data for a 14.75-in.-diameter pump rotating at 1800 rpm are given below. **(a)** Plot the pump characteristic curves. **(b)** Determine the type specific speed. **(c)** If the pump speed is changed to 1600 rpm, determine the performance characteristics of the pump. **(d)** If the diameter of the pump is reduced to 14 in. (speed retained at 1800 rpm), determine the performance characteristics. **(e)** If another similar pump of 14 in. diameter is used, determine the pump characteristics.

Capacity (gpm)	Head (ft)	Power (hp)	Efficiency (%)
0	230	76.5	0
1000	228.5	107	54
2000	221.0	142.3	78.4
3000	200.5	174.5	87.0
3500	183.5	185.0	87.6
4000	157.0	189.5	83.7

12.43 The pump with the characteristics given in Problem 12.42 is to be used for the piping system shown in Fig. P12.43. Determine the **(a)** pump operation point in terms of head and discharge, **(b)** operation efficiency, **(c)** suitability of the pump, and **(d)** input horsepower to the pump. $f = 0.022$.

Figure P 12.43

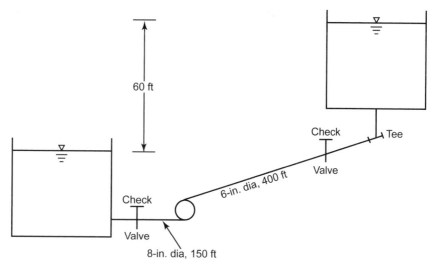

12.44 Water is pumped from a lower tank to a higher tank having a difference in elevation of 10 m. The piping system comprises a 200-mm-diameter pipe of 2000 m length with minor losses = 6.2 times the velocity head. Determine the rate of flow. $f = 0.02$.

Pump characteristics:

Discharge (liters/s)	0	10	20	30	40	50
Total head (m)	25	23.2	20.8	17.0	12.4	7.3
Efficiency (%)	—	45	65	71	65	45

12.45 Each of two identical pumps has the following characteristics:

Q (gpm)	H (ft)	Efficiency (%)
0	124	0
500	119	54
800	112	64
1000	104	68
1300	90	70
1600	70	67

The pumps are used in series to supply water between two tanks with a total static head of 40 ft. The pipeline is 6 in. in diameter and 1200 ft in length, with minor losses 20 times the velocity head. Determine the **(a)** operating condition of head and discharge, **(b)** head developed by each pump, and **(c)** input power. $f = 0.022$.

12.46 The two pumps of Problem 12.45 are arranged in parallel to supply water for a total static head of 40 ft. The pipeline is 10 in. in diameter and 1200 ft long with minor losses 20 times the velocity head. Determine the operating condition of head and discharge and the input power. $f = 0.022$.

12.47 The suction side of a pipe system is as shown in Fig. P12.47. The pump discharges 495 gpm of water under a total head of 100 ft at 40°F. Atmospheric pressure = 32.7 ft of water. Determine the **(a)** available NPSH, and **(b)** cavitation parameter. If the required NPSH is 9, will there be a cavitation problem?

Figure P 12.47

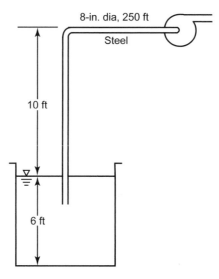

8-in. dia, 250 ft

Steel

10 ft

6 ft

12.48 A pump delivers water under a head of 130 ft at 100°F. The barometric (atmospheric) pressure is 14.3 psi absolute. At the suction intake the pressure is a vacuum (below atmosphere) of 17 in. Hg and the velocity is 12 ft/sec. Determine the NPSH and the cavitation parameter. [*Hint:* Head of water = head of Hg × sp. gr. of Hg.]

13

Urban Drainage Systems

◆◆

13.1 TYPES OF DRAINAGE SYSTEMS

The term *drainage* applies to the process of removing excess water to prevent public inconvenience and to provide protection against loss of property and life. In an undeveloped area, drainage occurs naturally as a part of the hydrologic cycle. This natural drainage system is not static but is constantly changing with environmental and physical conditions.

Development of an area interferes with nature's ability to accommodate severe storms without significant damage, and an artificial drainage system becomes necessary. A drainage system can be classified according to the following categories:

1. urban drainage system
2. agriculture land drainage system
3. roadway drainage system
4. airport drainage system

Urban drainage systems are the topic of this chapter; the other three drainage systems are covered in the following chapter.

In an urbanized area, runoff is contributed by (1) excess surface water after a rainfall, from roofs, yards, streets, and so forth and (2) wastewater* from households, commercial establishments, and industries. Past practice was to convey the entire runoff through a single system known as the *combined sewer system*. Combined sewers are no longer built; the present practice is to construct a system to discharge rainfall excess only, and a separate system to transport wastewater, also referred to as the *dry-weather flow*. The former is known as the *stormwater sewer system* and the latter as the *sanitary sewer system*.

In a combined sewer system the ratio of maximum flow rate (stormwater) to minimum flow rate (dry-weather flow) is over 20, even exceeding 100 during severe storms. The flow through the sewer system is carried to treatment facilities. During severe storms, however, it is not practical to carry the entire quantity to the treatment works. Excess diluted flow is passed into a stream at the nearest discharge point through *storm sewer overflows*. The combined overflows introduce large quantities of polluting materials into receiving waters.

---

* *Wastewater* is the spent or used water of a community, comprising water-carried wastes from residences, institutions, commercial buildings, and industries. *Sewage* is the liquid waste of a community conveyed by a sewer. Thus both terms have the same meaning. In recent usage the word "wastewater" has taken precedence. *Sewerage* implies collection, treatment, and disposal of sewage.

These should be kept to a minimum and the "first flushes," which are more polluted, should be passed entirely to the treatment facilities. Combined sewers tend to get silted up during dry-weather flow because the combined flow capacity is large and the dry-weather flow is comparatively small. These are flushed at the time of a storm.

With separate stormwater and sanitary systems, sewers are designed to maintain a self-cleansing velocity at higher discharges. For storm sewers, comparatively shorter lengths are needed because the stormwater can be discharged directly to the nearest point in a water course. There is no need for overflow structures. Combined sewers are less costly to construct but have additional operational cost for treatment of larger quantities. Generally, separate systems are more favorable.

13.2 Layout of an Urban Drainage System

There are perpendicular, zonal, fan, and ring patterns of sewer systems, as shown in Figure 13.1. Among the sewer components, a *lateral* sewer is the unit into which no other common sewer discharges. A *submain* receives the discharge of a number of laterals. A *main* or *trunk* sewer receives the discharge from one or more submains. In a storm system, main sewers outfall to receiving waters. In sanitary or combined sewer systems, an intercepting sewer receives flow from a number of mains and conducts it to a point of treatment. Excess water is allowed to overflow into the water course through outfalls.

Figure 13.1   Layout of sanitary sewers: (a) perpendicular pattern; (b) fan pattern; (c) zone pattern. *p*, pumping station; TP, treatment plant.

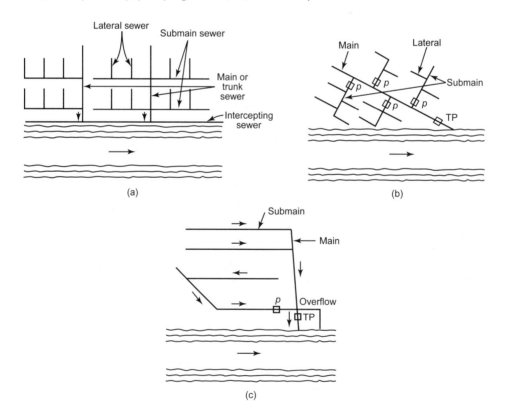

For the layout in a horizontal plane, the sewer lines are laid alongside the streets or utility easement. A layout is produced of the main sewer leaving the area at its lowest point and submains and laterals radiating to the outlying areas. The tributary area to each sewer is drawn from the ground contour map.

The vertical layout is influenced by the ground surface profile. A cover of about 10 ft is provided on sewer pipes in northern states to prevent freezing. A minimum cover of 3 ft is adopted in the South from imposed load considerations. For a minimum excavation, a sewer line is laid parallel to the ground surface. However, this is not practical in very steep or in flat terrain. For sanitary sewers, a slope has to be provided to maintain a self-cleansing velocity of a minimum of 2 ft/sec (0.6 m/s). Table 11.5 indicates the required slopes for various flows. In flat areas all sewer lines drain to a collection point for pumping to a gravity main. A separate sewer system is constructed for a higher-elevation area.

Manholes, for service and maintenance of sewers, are located at (1) all sewer intersections, (2) major changes in slope of sewers, (3) changes in size of sewers, (4) changes in horizontal layout of sewers, (5) drop in vertical layout of sewers, and (6) along straight sewer runs at a spacing of 300 to 500 ft (90 to 150 m); this is 500 to 1000 ft (150 to 300 m) for large-diameter sewers.

A section of sewer invert entering a manhole and another section leaving the manhole are not always at the same elevation. A drop as shown at manhole 102 in Figure 13.2 is sometimes needed to adjust the slope of a sewer section without resorting to a steep gradient or deep cut. In a sewer on a short radius curve or where there is a bend or change in the sewer direction, a noticeable loss of head (energy) is involved. For the usual velocities, this justifies a drop of about 0.1 ft (30 mm) in the invert levels of sewers entering and leaving the manhole. When the sewer increases in size to overcome the head loss, a drop in the invert elevation is made so that either the tops, or crowns, of the two sewer sections or the 0.8-depth points of the two sewers remain at the same elevation. The invert elevation of a subsequent sewer segment should never be higher than the previous section.

For each sewer line, a profile is drawn that shows the (1) ground level, (2) location of borings, (3) rock levels, (4) underground structures, (5) elevations of foundations and cellars, (6) cross streets, (7) location and number of manholes, (8) the elevations of the sewer invert entering and leaving the manhole at each manhole, and (9) slopes and sizes of sewer lines. A profile is illustrated in Figure 13.2. The profile assists in the design, cost estimation, and construction of a sewer.

Figure 13.2 Profile of a sewer section (modified from McGhee, 1991).

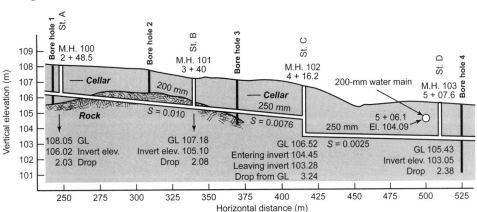

13.3 DESIGN OF A SANITARY SEWER SYSTEM

A design period throughout which a sewer capacity will be adequate is established prior to the design of a sewer system. This is typically between 25 and 50 years. A sanitary sewer performs two main functions: carrying the peak sanitary discharge, and transporting suspended solids without deposition. It is designed according to the following considerations:

1. Determine the hydraulic capacity of the sewer for the estimated peak flow at the end of the design period, known as the *design discharge.*
2. Check for a minimum cleansing velocity of 2 ft/sec (0.6 m/s) for the peak rate of flow at the beginning of the design period (at the present time).

Sanitary sewers are designed with one-half to full depth of flow at design discharge. Usually, pipes up to 16 in. (400 mm) in diameter are designed to flow half full, pipes between 16 and 35 in. (900 mm) to flow two-thirds full, and pipes over 35 in. to flow at three-fourths to full depth of flow. An upper limit on the velocity is often taken to be about 10 ft/sec (3 m/s) in sanitary sewers.

The design process involves determining the appropriate diameter and slope of the sewer sections. Since the trench excavation is a major cost component, the aim is to lay the sewers to a minimum grade while still meeting the criteria above.

Two steps in the design are (1) determining the peak rate of sewage flow, and (2) selecting the sewer for the estimated flow. The second step is based on Manning's equation and has been described in Chapter 11. Peak-flow computations are discussed in the following section.

13.4 QUANTITY OF WASTEWATER

For purposes of sewer design, an estimate must be made of present and future quantities of domestic, commercial, institutional, and industrial wastewaters, groundwater infiltration, and any stormwater entering the system. The assessment of wastewater quantity is made from the quantity of water consumed in a city. It is frequently assumed that the average rate of wastewater flow, including a moderate amount of infiltration, is equal to the average rate of water consumption in the city. The domestic component of water consumption is estimated by multiplying the population by the per capita rate of consumption. To this are added the contributions from commercial buildings, institutions, industries, and extraneous sources. The accuracy of population numbers at present and projected at the end of the design period is important. Population forecasting techniques are discussed in Chapter 1. Since the system is designed in sections to serve segments of area, population density is used to determine the quantity of flow as follows:

$$Q = pDA \quad [\text{L}^3\text{T}^{-1}] \tag{13.1}$$

where

$\quad\quad Q$ = domestic average daily flow

$\quad\quad p$ = water usage per person per day

$\quad\quad D$ = population density, persons per unit area

$\quad\quad A$ = tributary area

The density is estimated by dividing the total population by the total area of the community with proper deductions for parks, playgrounds, swamps, lakes, ponds, and rivers. Main and intercepting sewers are designed for future density at the end of the design period. Laterals and submains are designed on the basis of saturation density for the area.

Domestic water usage per capita depends on the living conditions of consumers. On an average it is 101 gallons (384 liters) per person per day. The average projected domestic consumption rate in the future is not expected to rise.

The quantity of commercial usage in smaller communities is averaged to 30 gpd per person. In large cities, an allowance from 4500 to 160,000 gpd per acre of floor area per building served is included.

Industrial wastewater quantities vary over a wide range depending on the type and size of industry. Data on wastes from various industries are available in certain guides. The average contribution from industry may vary from 8 to 25 gpd per person per shift.

Institutions such as hospitals, jails, and schools, and public services such as fire protection agencies, have a fixed pattern of demand. Their requirements amount to 13 to 26 gpd per person.

For infiltration, a moderate allowance has been included by assuming a rate of sewage equal to the rate of water consumption. When a separate estimate has to be made, an allowance of 30,000 gpd per mile (71,000 liters/day per kilometer) is made for infiltration in small to medium-sized sewers up to 24 inches.

In the average city, the total water consumption is expected to be about 200% of the domestic consumption [i.e., 180 gpd per person (700 liters/day per person)].

The overall capacity is designed for the peak wastewater flow, which varies from about 150% of the average daily flow for high-population areas to about 450% for low-population areas, as shown in Figure 13.3. Thus, for average conditions, the peak flow is estimated at 270 to 800 gpd (1000 to 3000 liters/day) per capita. Many state regulatory agencies have set a lowest design rate of 400 gpd per capita (1500 liters/day per capita) for laterals and 250 gpd per capita (950 liters/day per capita) for mains where no actual measurements and other pertinent data are available (American Society of Civil Engineers and Water Pollution Control Federation, 1982).

EXAMPLE 13.1

Determine the maximum hourly (peak) rate of dry-weather flow from an area of 25 km^2 if the present population density is 8000 persons per square kilometer. Assume a domestic water consumption of 300 liters per day per capita and total consumption, including infiltration, to be 200% of domestic use.

SOLUTION

1. Population = density × area = 8000(25) = 200×10^3 persons
2. Domestic rate of flow = $300(200 \times 10^3) = 60 \times 10^6$ liters/day or 0.69 m^3/s
3. Total average daily flow = $2(60 \times 10^6) = 120 \times 10^6$ liters/day or 1.39 m^3/s
4. From Figure 13.3, ratio of peak to average flow = 2.0
5. Peak wastewater flow = 2(1.39) = 2.78 m^3/s

Figure 13.3 Ratio of peak flow to average daily flow (modified from American Society of Civil Engineers and Water Pollution Control Federation, 1982).

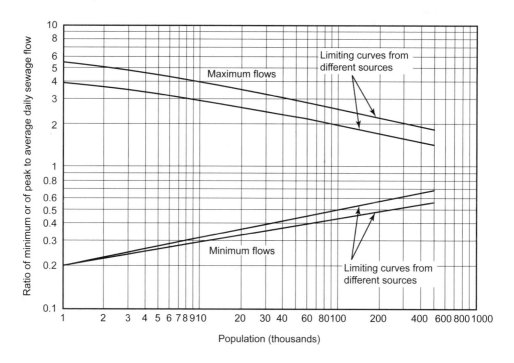

EXAMPLE 13.2

The future population density at the end of 15 years for Example 13.1 is projected to be 11,000 persons per square kilometer. Calculate the design flow (peak discharge at the end of the design period).

SOLUTION

1. Future population = $11,000(25) = 275 \times 10^3$ persons

2. Domestic flow = $300(275 \times 10^3) = 82.5 \times 10^6$ liters/day or 0.95 m³/s

3. Total average daily flow = $2(0.95) = 1.9$ m³/s

4. Ratio of maximum to average, from Figure 13.3 = 1.9

5. Peak flow in future = $1.9(1.9) = 3.61$ m³/s

13.5 FRICTION COEFFICIENT FOR SANITARY SEWERS

Experiments on small-diameter pipes of different materials indicate that after a short period of use, the capacity of a sewer depends on the characteristics of the slime grown on the pipe wall. Thus, to a large extent, roughness values become independent of the nature of the pipe material. The friction coefficient, n, of 0.012 in Manning's formula is considered satisfactory for pipes up to 35 in. (900 mm) in diameter. For larger pipes, the values of n are based on the type of material of the pipe, as given in Table 11.4.

13.6 DESIGN PROCEDURE FOR SANITARY SEWERS

Based on the criteria stated in Section 13.3, the design procedure is as follows:

1. Compute the present peak sewage flow, Q_{peak}.
2. Compute the ultimate (at the end of the design period) peak sewage flow, Q_{design}.
3. Determine the minimum slope from Table 11.5 for Q_{design}; ascertain the ground-level gradient. The critical (steeper) of the two slopes is taken as the sewer slope.
4. For Q_{design}, calculate the pipe diameter by Manning's formula for the slope of step 3 and the partial flow condition at 1/2 full for a lateral, 2/3 full for a submain, and 3/4 full for a main or an interceptor.
5. From Figure 11.6 on hydraulic elements, determine the full discharge capacity for the pipe size of step 4; from Table 11.1, obtain the depth of flow corresponding to the flow ratio Q_{peak}/Q_{full}.
6. For this depth of flow and a velocity of 2 ft/sec, recompute the required slope from Manning's formula. Select the steepest of the slopes from steps 3 and 6 as the design slope.
7. If the slope of step 3 is in considerable excess (over 10 times higher) than that of step 6,[*] the velocity of flow for design discharge should be checked to confirm that it is less than the scour velocity of 10 ft/sec. If necessary, a larger diameter should be selected.
8. Fix the invert levels accordingly.

EXAMPLE 13.3

The ground-surface elevations at the location of two manholes 300 ft apart are 101.2 ft and 100 ft, respectively. The present rate of peak sewage flow is 1 mgd, which is expected to increase to 1.5 mgd at the end of 25 years. Design the sewer.

SOLUTION

1. $Q_{design} = (1.5 \text{ mgd})\left(1.547\dfrac{\text{cfs}}{\text{mgd}}\right) = 2.32 \text{ cfs}$

2. $Q_{peak} = 1 \times 1.547 = 1.55 \text{ cfs}$

3. Surface grade $= \dfrac{101.2 - 100.0}{300} = 0.004$

4. For 2.32 cfs, minimum slope (from Table 11.5) = 0.0016

5. Critical slope = 0.004

6. Designing for 1/2-full condition. From Table 11.1, for $y/d_0 = 0.50$, $AR^{2/3} = 0.156 d_0^{8/3}$

---

* When the slope from step 6 is steeper than that from step 3 and is selected as the design slope, the velocity of flow at Q_{design} will exceed 2 ft/sec, but it is unlikely to exceed 10 ft/sec, since the velocity variation beyond 0.5 depth is not substantial.

7. Using Manning's equation (11.10),

$$2.32 = \left(\frac{1.49}{0.012}\right)\left(0.156 d_0^{8/3}\right)\left(0.004\right)^{1/2}$$

$$d_0 = 1.27 \text{ ft or } 15.25 \text{ in.}$$

Use d_0 = 15 in. or 1.25 ft

8. From Figure 11.6, y/d_0 = 0.5 (half-full condition), Q/Q_{full} = 0.5

$$Q_{full} = \frac{Q_{design}}{0.5} = \frac{2.32}{0.5} = 4.64 \text{ cfs}$$

9. Q_{peak}/Q_{full} = 1.55/4.64 = 0.33. From Figure 11.16, for Q/Q_{full} of 0.33, y/d_0 = 0.4. From Table 11.1, for y/d_0 of 0.4, R/d_0 = 0.214 or $R = 0.214 d_0$

10. From Manning's equation (11.9),

$$V = \frac{1.49}{n} R^{2/3} S^{1.2}$$

or

$$2 = \frac{1.49}{0.012}(0.214 \times 1.25)^{2/3} S^{1/2}$$

$$S = 0.0015 < 0.004, \text{ adopt } 0.004$$

11. Invert elevations

$$\text{Upstream end} = \text{ground level} - \text{cover} - \text{diameter}$$

$$= 101.2 - 10 - 1.25 = 89.95 \text{ ft}$$

$$\text{Downstream end} = 100 - 10 - 1.25 = 88.75 \text{ ft}$$

13.7 A SANITARY SEWER PROJECT

An example illustrates the procedure. This may be varied to suit the requirements of local and state regulatory agencies.

EXAMPLE 13.4

Design a sanitary sewer system for a part of the city shown in Figure 13.4. The present density of population is 8000 persons per square kilometer. It is estimated that the maximum density will be 12,000 persons per square kilometer. Assume that the maximum rate of sewage flow is 1500 liters/day per person for all the sewers. The ground-level gradient for various sewer segments is given in column 15 of Table 13.1.

SOLUTION   Guided by contours, the directions of flow are indicated with arrows on the map. The manholes, provided at intersections, at changes in direction, and at intervals not exceeding 150 m, have been numbered from the upper end of the tributary lateral. The area tributary to each line segment from manhole to manhole is sketched by the dashed lines on the map and indicated in column 6 of Table 13.1. Design proceeds from the uppermost point of the system downward through the laterals. Where a branch joins a line already

Figure 13.4 Sewer system for a portion of a city.

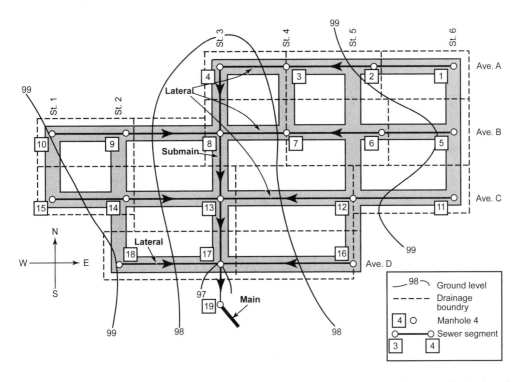

designed, the computation is made for the branch at the intersection (i.e., manhole 4) and restarted from the uppermost lateral meeting the other end of the branch (i.e., from manhole 5).

The procedure has been explained in Tables 13.1 and 13.2. The first table computes the sewage quantity and the second completes the hydraulic design. Computation is started from manhole 1 to 2 in Table 13.1. The tributary area of 0.007 km^2 is multiplied by the present and maximum density to obtain the incremental population. The total population is obtained by adding the incremental population from each manhole, until the submain 4– 8 is met. The total population at the submain at manhole 4 equals the total of the laterals from manholes 1 through 4. After this, the laterals from manholes 5 to 8 and 10 to 8 are investigated as shown in lines 5 through 9 of Table 13.1. Their total is added to that of line 4 of the table to obtain the total population contributing to the submain at manhole 8 (line 10). This procedure is repeated until the point of connection to the main is reached. The quantity of wastewater flow is obtained by multiplying the total population by per capita consumption. Surface elevation data are known from a land survey. The street slope is the difference in elevation at the tops of the upper and lower manholes divided by the length of the line. Table 13.1 (column 15) indicates the slopes.

Once the design flow (column 12 of Table 13.1) and present peak flow (column 11) have been determined, the hydraulic design is accomplished by the method of Example 13.3. This is arranged in Table 13.2. For the stipulated depth of flow of column 6, the value of parameter $AR^{2/3}/d_0^{8/3}$ is read from Table 11.1 and substituted in Manning's equation. In the equation, using Q_{design} (column 12 of Table 13.1) for the discharge, and the larger of the values in columns 4 and 5 for the slope, and $n = 0.012$, the diameter, d_0, is

Table 13.1 Computation of Sewage Quantity

(1)	(2)	(3)	(4)	(5)	(6)	(7)	(8)	(9)	(10)	(11)	(12)	(13)	(14)	(15)
		Manhole		Length of Line (m)	Area Drained (km²)	Increment of Population		Total Population		Wastewater Flow (m³/s)		Surface Elevation (m)		Street Slope
No.	Location	From	To			Present[a]	Ultimate[b]	Present	Ultimate	Present[c]	Ultimate	Upper Manhole	Lower Manhole	
1	Avenue A	1	2	100	0.007	56	84	56	84	.0010	.0015	99.96	99.26	0.007
2	Avenue A	2	3	100	0.007	56	84	112	168	.0019	.0029	99.26	98.46	0.008
3	Avenue A	3	4	90	0.005	40	60	152	228	.0026	.0040	98.46	97.65	0.009
4	Street 3	4	8	90	—	—	—	152	228	.0026	.0040	97.65	97.47	0.002
5	Avenue B	5	6	100	0.014	112	168	112	168	.0019	.0029	99.11	98.61	0.005
6	Avenue B	6	7	100	0.014	112	168	224	336	.0039	.0058	98.61	98.01	0.006
7	Avenue B	7	8	90	0.010	80	120	304	456	.0053	.0079	98.01	97.47	0.006
8	Avenue B	10	9	100	0.006	48	72	48	72	.0008	.0013	98.99	98.19	0.008
9	Avenue B	9	8	120	0.008	64	96	112	168	.0019	.0029	98.19	97.47	0.006
10	Street 3	8	13	97	—	—	—	568	852	.0099	.0148	97.47	97.28	0.002
11	Avenue C	11	12	140	0.008	64	96	64	96	.0011	.0017	100.04	98.78	0.009
12	Avenue C	12	13	150	0.016	128	192	192	288	.0033	.0050	98.78	97.28	0.010
13	Avenue C	15	14	100	0.006	48	72	48	72	.0008	.0013	99.56	98.36	0.012
14	Avenue C	14	13	120	0.014	112	168	160	240	.0028	.0042	98.36	97.28	0.009
15	Street 3	13	17	93	—	—	—	920	1,380	.0160	.0240	97.28	97.0	0.003
16	Avenue D	16	17	150	0.008	64	96	64	96	.0011	.0017	98.35	97.0	0.009
17	Avenue D	18	17	130	0.007	56	84	56	84	.0010	.0015	98.95	97.0	0.015
18	Street 3	17	19	95	—	—	—	1,040	1,560	.0181	.0272	97.0	96.62	0.004

[a] [Area (km²) (col. 6) × 8000 persons/km²]

[b] [Area (km²) (col. 6) × 12000 persons/km²]

[c] Population (col. 9) × rate $\left(1500\dfrac{\text{Liters}}{\text{person} \times \text{day}}\right)\left[\dfrac{1\,\text{m}^3}{1000\,\text{L}}\right]\left[\dfrac{1\,\text{day}}{24 \times 60 \times 60\,\text{s}}\right]$

Table 13.2 Hydraulic Design of Sanitary Sewers

(1)	(2)	(3)	(4)	(5)	(6)	(7)	(8)	(9)	(10)	(11)	(12)	(13)	(14)	(15)
	Manhole		Slope		Design for Q_{design}					Check for $V = 0.6$ m/s (2 ft/sec) at Q_{peak}				
					Flow									Adopted
No.	From	To	Street	Minimum (Table 11.5)	Depth, $\frac{y}{d_0}$	$\frac{AR^{2/3}}{d_0^{8/3}}$ (Table 11.1)	Diameter d_0 (mm)[a]	$\frac{Q_{design}}{Q_{full}}$ (Fig. 11.6)	Q_{full} m^3/s	$\frac{Q_{peak}}{Q_{full}}$	$\frac{y}{d_0}$ (Fig. 11.6)	$\frac{R}{d_0}$ (Table 11.1)	Slope, S Required[b]	Sewer Slope
1	1	2	0.007	0.009	0.5	0.156	85[c]	0.5	.003	0.33	0.4	0.214	0.011	0.011
2	2	3	0.008	0.009	0.5	0.156	105	0.5	.0058	0.33	0.4	0.214	0.008	0.009
3	3	4	0.009	0.007	0.5	0.156	120	0.5	.0080	0.33	0.4	0.214	0.007	0.009
4	4	8	0.002	0.007	0.67	0.24	105	0.8	.005	0.52	0.51	0.253	0.007	0.007
5	5	6	0.005	0.009	0.5	0.156	105	0.5	.0058	0.33	0.4	0.214	0.008	0.009
6	6	7	0.006	0.006	0.5	0.156	145	0.5	.0116	0.34	0.41	0.218	0.005	0.006
7	7	8	0.006	0.005	0.5	0.156	160	0.5	.0158	0.34	0.41	0.218	0.005	0.006
8	10	9	0.008	0.009	0.5	0.156	80	0.5	.0026	0.31	0.4	0.214	0.012	0.012
9	9	8	0.006	0.009	0.5	0.156	105	0.5	.0058	0.33	0.4	0.214	0.008	0.009
10	8	13	0.002	0.004	0.67	0.24	190	0.8	.0185	0.54	0.53	0.259	0.003	0.004
11	11	12	0.009	0.009	0.5	0.156	85	0.5	.0034	0.32	0.39	0.210	0.011	0.011
12	12	13	0.010	0.007	0.5	0.156	125	0.5	.0100	0.33	0.4	0.214	0.006	0.010
13	15	14	0.012	0.009	0.5	0.156	75	0.5	.0026	0.31	0.38	0.206	0.013	0.013
14	14	13	0.009	0.007	0.5	0.156	120	0.5	.0082	0.34	0.41	0.218	0.007	0.009
15	13	17	0.003	0.003	0.67	0.24	240	0.8	.0300	0.53	0.52	0.256	0.002	0.003
16	16	17	0.009	0.009	0.5	0.156	85	0.5	.0034	0.32	0.39	0.210	0.011	0.011
17	18	17	0.015	0.009	0.5	0.156	75	0.5	.003	0.33	0.4	0.214	0.013	0.015
18	17	19	0.004	0.002	0.67	0.24	240	0.8	.034	0.53	0.52	0.256	0.002	0.004

[a] $d_0 = \left[\dfrac{n \cdot (\text{col. 12 of Table 13.1})}{(\text{col. 7})(\text{col. 4 or 5})^{1/2}} \right]^{3/8}$, $n = .012$

[b] $S = 0.36 n^2 / [(\text{col. 13})(\text{col. 8 in m})]^{4/3}$
For English units, 1.49 appears in denominator within parenthesis in both equations for d_0 and S above.

[c] There may be a minimum limit on the size of sewer, say 200 mm. In the worst case it will slightly reduce the velocity at Q_{peak} which can be offset by increasing the sewer slope 33% as compared to col. 14.

727

computed as shown in column 8. For the depth ratio of flow in column 6, the discharge ratio of column 9 is obtained from Figure 11.6. Full-pipe flow (column 10) is thereby obtained by dividing the design flow (column 12 of Table 13.1) by the ratio of column 9. The ratio in column 11 is determined by dividing peak flow (column 11 of Table 13.1) by column 10. For the discharge ratio in column 11, the depth ratio is ascertained in column 12 from Figure 11.6. Table 11.1 provides the hydraulic radius parameter of column 13 corresponding to the depth ratio of column 12. Using Manning's equation in terms of the velocity [eq. (11.9)], the required slope for a velocity of 0.6 m/s (2 ft/sec) is computed. The steeper of the slopes in columns 4, 5, and 14 is adopted to lay the sewer line. This is a general procedure, in which variations are made to suit the local conditions and preferences of a designer.

Another tabulation, as shown in Table 13.3, is added to determine the elevations of the sewers. Columns 1 through 8 of the table are completed from information in Tables 13.1 and 13.2. Column 9 is equal to column 5 times column 8. Invert elevations are computed from surface elevation, fall of sewer, cover, and sewer size. Recommendations regarding matching of the crown levels and the drops to accommodate losses, as discussed earlier, are kept in mind.

Table 13.3 Sewer Elevation Analysis

(1)	(2)	(3)	(4)	(5)	(6)	(7)	(8)	(9)	(10)	(11)
	Manhole			Length of Sewer (m)	Surface Elevation (m)		Grade of Sewer	Fall of [a] Sewer	Invert Elevation (m)	
			Diameter of Sewer (mm)							
No.	From	To			Upper Manhole	Lower Manhole			Upper Manhole	Lower Manhole
1	1	2	85	100	99.96	99.26	0.011	1.10	97.88	96.78
2	2	3	105	100	99.26	98.46	0.009	0.9	96.76	95.86
3	3	4	120	90	98.46	97.65	0.009	0.81	95.85	95.04
4	4	8	105	90	97.65	97.47	0.007	0.63	95.01	

[a] Col. 9 = col. 8 × col. 5.

EXAMPLE 13.5

In the layout of Figure 13.4, determine the sewer arrangement for manholes 1 through 4. Provide a minimum cover of 2 m.

SOLUTION   Invert elevations:

1. Manhole 1:

   Surface elevation − cover diameter = 99.96−2−0.085 = 97.88 m

2. Manhole 2:

   Lower end of sewer 1−2 = upper end − sewer fall

   $$= 97.88 - 1.10 = 96.78 \text{ m}$$

   Upper end of sewer 2−3: The crown levels of sewers at the manhole should match. The crown level of sewer 1−2 = 96.78 + 0.085 = 96.86 m. The invert of sewer 2−3 = 96.86 − 0.105 = 96.76 m. Hence there is a drop of 0.02 m in manhole 2.

3. Manhole 3:

Lower end of sewer 2–3 = 96.76−0.9 = 95.86 m

Upper end of sewer 3–4: The crown level of sewer 2–3 = 95.86 + 0.105 = 95.97 m

The invert of sewer 3–4 = 95.97 −0.12 = 95.85 m

4. Manhole 4:

Lower end of sewer 3–4 = 95.85 − 0.81 = 95.04 m

Upper end of sewer 4–8: Since the size of sewer 4–8 is smaller than the size of sewer 3–4, no drop on this count is required. However, since a change in direction is involved, a minimum drop of 30 mm should be provided. To provide a drop of 30 mm at manhole 4, the invert at the upper end of sewer 4–8 = 95.04−0.03 = 95.01 meters.

These invert elevations are shown in columns 10 and 11 of Table 13.3.

13.8 DESIGN OF A STORM SEWER SYSTEM

Similar to the sanitary sewer system, the design process involves two steps: (1) determining the quantity of stormwater, and (2) establishing a sewer capacity to pass this quantity. The quantity pertains to the peak rate of runoff within the urban or residential drainage area produced by a precipitation storm of a certain specified return period referred to as the design frequency. For this design rate of flow, the storm sewers are designed for just flowing full at the grade of ground surface slope by applying Manning's equation. A minimum velocity of 3 ft/sec (0.9 m/s) is maintained when flowing full in order to produce a minimum nondepositing slope for silt and grit particles that are heavier than the sewage solids. For this purpose, the slope at times is increased in excess of the surface grade. The upper limit on velocity is about 15 ft/sec (5 m/s) from scour considerations.

13.9 QUANTITY OF STORMWATER

There are two common methods of computing peak stormwater flows from small urban watersheds (usually less than 20 mi^2). The rational method, used for the first time in the United States by Kuichling in 1889, is still very popular for estimating stormwater quantity. The Natural Resources Conservation Service has developed the other procedure to determine the peak discharge by using the soil cover complex curves. With respect to the rational method, McPherson (1969) demonstrated that extensive variability of results is inherent in applying this method due to considerable variations in the interpretation of the variables and methodology of use. A wide latitude of subjective judgment is involved in the method. Improved methods are evolving very slowly because of a dearth of the rainfall-runoff field measurements that have national transferability of results. McPherson pointed out that until improved methods are developed, the rational method is as satisfactory as any other oversimplified empirical approach.

13.10 RATIONAL METHOD

The basic equation in the rational method has the form

$$Q = C_f CiA \quad [\text{L}^3\text{T}^{-1}] \tag{13.2}$$

where

Q = peak rate of flow

C_f = frequency factor

C = runoff (rational) coefficient

i = intensity of precipitation for a duration equal to time of concentration, t_c, and a return period, T

A = drainage area

Equation (13.2) is dimensionally homogeneous (i.e., if A is in ft^2, i is in ft/sec, and Q is in cfs). However, this equation also yields correct values for Q in cfs, i in in./hr, and A in acres. C and C_f are dimensionless coefficients. Their product should not exceed 1.

13.10.1 Frequency Correction factor, C_f

In a common form of eq. (13.2), C_f is taken as unity, which applies to design storms of a 2- to 10-year recurrence interval— a representative frequency for residential sewers. For storms of higher return periods, the coefficients are higher because of smaller infiltration and other losses, as shown in Table 13.4.

The U.S. Department of Transportation (1979) has proposed the curves of Figure 13.5 to determine the frequency correction factor. Application of the curves is explained in the following example.

Table 13.4 Frequency Factor

Return Period (years)	C_f
2– 10	1.0
25	1.1
50	1.2
100	1.25

EXAMPLE 13.6

For a drainage area comprising a paved surface ($C = 0.9$), determine the runoff coefficient corresponding to a 50-year design frequency.

SOLUTION

1. From Table 13.4, $C_f = 1.2$ and $CC_f = 0.9(1.2) = 1.08$. Since CC_f cannot be greater than 1.0, $CC_f = 1.00$

2. Alternatively, from Figure 13.5 at imperviousness (C value) of 90% and for a 10-year frequency,

$$\frac{C}{C_{max}} = 0.83$$

Figure 13.5 Correction for design storm frequency (from U.S. Department of Transportation, 1979).

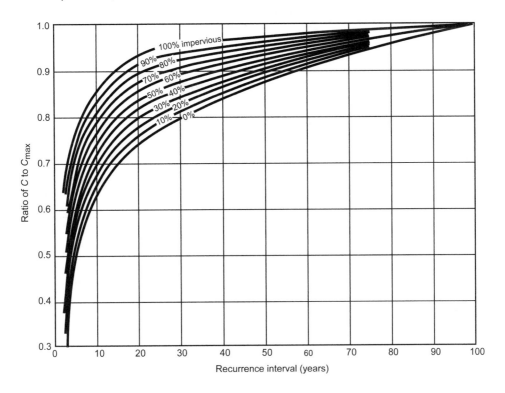

For a 50-year frequency,

$$\frac{C}{C_{max}} = 0.96$$

$$C_f = \frac{\text{ratio for 50 yr}}{\text{ratio for 10 yr}} = \frac{0.96}{0.83} = 1.16$$

$$CC_f = 1.16(0.9) = 1.04 > 1.0$$

Hence $CC_f = 1.00$.

13.10.2 Runoff Coefficient, C

This is a highly critical element that serves the function of converting the average rainfall rate of a particular recurrence interval to the peak runoff intensity of the same frequency. Therefore, it accounts for many complex phenomena of the runoff process. Its magnitude will be affected by antecedent moisture condition, ground slope, ground cover, depression storage, soil moisture, shape of drainage area, overland flow velocity, intensity of rain, and so on. Yet its value is generally considered fixed for any drainage area, depending only on the surface type. This simplistic approach is a major cause of criticism for the rational method. The values of the coefficient are given in Table 13.5.

Table 13.5　Rational Runoff Coefficient

Urban Catchments

General Description	C	Surface	C
City	0.7–0.9	Asphalt paving	0.7–0.9
Suburban business	0.5–0.7	Roofs	0.7–0.9
Industrial	0.5–0.9	Lawn heavy soil	
		>7° slope	0.25–0.35
Residential multiunits	0.6–0.7	2–7°	0.18–0.22
Housing estates	0.4–0.6	<2°	0.13–0.17
Bungalows	0.3–0.5	Lawn sandy soil	
		>7°	0.15–0.2
Parks, cemeteries	0.1–0.3	2–7°	0.10–0.15
		<2°	0.05–0.10

Rural Catchments (less than 10 km²)

Ground Cover	Basic Factor	Corrections: Add or Subtract
Bare surface	0.40	Slope < 5%: –0.05
Grassland	0.35	Slope > 10%: +0.05
Cultivated land	0.30	Recurrence interval < 20 yr: –0.05
Timber	0.18	Recurrence interval > 50 yr: +0.05
		Mean annual precipitation < 600 mm: –0.03
		Mean annual precipitation > 900 mm: +0.03

Source: Stephenson (1981).

For an area having different types of surfaces, a composite coefficient is determined by estimating the fraction of each type of surface within the total area, multiplying each fraction by the appropriate coefficient for that type of surface, and then summing the products for all types of surfaces. The coefficients are selected so as to reflect the conditions that are expected at the end of the design period.

The U.S. Department of Transportation (1979) has included a set of curves based on the following formula of Mitci, which relates the runoff coefficient to the degree of imperviousness (type of surface) and the antecedent rainfall (time from beginning of a rainfall to the occurrence of the design intensity rain of the duration of the time of concentration within the overall rainfall period).

$$C = \frac{0.98t}{4.54+t}P + \frac{0.78t}{31.17+t}(1-P) \quad \text{[unbalanced]} \tag{13.3}$$

where

C = runoff coefficient corrected for antecedent rain condition

t = time, in minutes, from beginning of rainfall to the occurrence of the design intensity rain of short duration

P = percent of impervious surface (coefficient from Table 13.5)

Within the long-period rainfall, since the time sequence of the design rainfall intensity used in the rational method is not fixed, eq. (13.3) or the curves of the Department of Transportation cannot be used directly. Usually, the short-duration intense design storm (conforming to the time of concentration at the point under consideration) is placed at the midpoint of the longer-duration storm. This gives the time position to determine C for known imperviousness (type of surface). However, a common practice is to use the runoff coefficient from Table 13.5 and assume it does not vary through the duration of the storm.

<hr>

EXAMPLE 13.7

A drainage area consists of 42% turf ($C = 0.3$) and 58% paved surface ($C = 0.9$). The point under design has a time of concentration of 20 minutes. The total duration of the rainstorm is 3 hours. Determine the value of C corrected for antecedent rainfall.

SOLUTION

1. Composite $C = \dfrac{0.42(0.3) + 0.58(0.9)}{0.42 + 0.58} = 0.65$

2. Midpoint of longer-duration storm = 1.5 hr or 90 min.

3. Time to start of 20-minute design rainfall from beginning of 3-hour storm,

$$t = 90 - \frac{1}{2}(20) = 80 \text{ min.}$$

4. Time to end of 20-minute design rainfall from beginning of 3-hour storm,

$$t = 90 + \frac{1}{2}(20) = 100 \text{ min.}$$

5. From eq. (13.3),

$$C = \frac{0.98(80)}{4.54 + 80}(0.65) + \frac{0.78(80)}{31.17 + 80}(1 - 0.65) = 0.80$$

$$C = \frac{0.98(100)}{4.54 + 100}(0.65) + \frac{0.78(100)}{31.17 + 100}(1 - 0.65) = 0.82$$

6. Average $C = 0.81$

<hr>

13.10.3 Drainage Area, A

Area represents the drainage area for a site under consideration. For a natural system it represents the watershed. For a sewer system it is the area tributary to a point of inlet. If a system consists of a number of sewers, the complete area is subdivided into component parts, separating a tributary area to each inlet point of every sewer segment. Many arrangements of sewer layout and inlet locations are tried before adopting a final one.

13.10.4 Rainfall Intensity, i

Intensity of rainfall is dependent on the duration of rainfall (short-duration storms are more intense) and the storm frequency or recurrence interval (less frequent storms are more intense). The rainfall intensity-duration-frequency (IDF) relation for a gaging site is developed from the data of a recording rain gage. The procedure has been explained in Section 2.8. Since the point rainfalls or observations at a gaging site are considered representative of a

10-mi^2 drainage area, IDF analysis for a station or combined for two stations is adequate for application in the small-area urban drainage design. Typical IDF curves are shown in Figure 13.6.

The collection and analysis of rainfall data and preparation of IDF curves for the site condition are done only for extensive projects. Generally, the rainfall data and the maps prepared by the National Weather Service and other government agencies are used in place of local statistical analysis to prepare the IDF curves for the locality selected.

In the absence of data and maps to prepare the curves, empirical relations are used for duration of less than 2 hours. For any given frequency, the intensity is related to the duration by eq. (2.15) (reproduced below):

$$i = \frac{A}{t+B} \quad \text{[unbalanced]} \tag{2.15}$$

where

i = intensity, in./hr

t = duration, min

A, B = constants that depend on the frequency and climatic conditions

The values for constants A and B are obtained using observed rainfall or the National Weather Service data for the locality selected. General values of the constants for the different parts of the country (Figure 2.7) are given in Table 2.8 for frequency levels of 2, 5, 10, 25, 50, and 100 years.

For application in the rational formula, the extreme (probable maximum, etc.) value of the rainfall intensity is not used because nearly complete protection of the area is not justified. The following range of design frequency is commonly used:

1. 2 to 15 years for sewers in residential areas, most commonly 10 years

2. 10 to 50 years for sewers in commercial and high-value areas

3. 50 years or more for flood-protection works

Figure 13.6   Intensity-duration-frequency (IDF) curves for Bridgewater, CT.

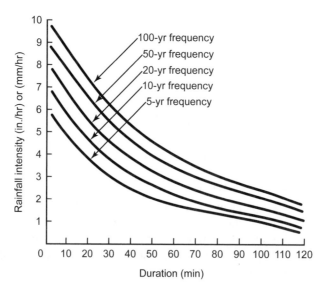

Urban Drainage Systems   Chapter 13

13.10.5 Time of Concentration, t_c

With regard to storm duration to be considered for runoff assessment, *time of concentration*, t_c, is relevant. It is defined as the time required for runoff from the hydraulically most remote part of the drainage area to reach the point of reference. There is another definition of this term as well, as stated in Section 7.8.1. For various routes of flow, t_c is taken as the longest time of travel to the point of reference. There are many ways to estimate t_c. Some of these methods are designed primarily for overland flow, some primarily for channel flow, and a few for both overland and channel flows. Many formulas are summarized in Table 13.6, which can be used when overland flow conditions dominate. The specific condition for which a formula applies is indicated in the table. To apply the Izzard formula, rainfall intensity must be known. A suitable procedure is to assume a time of concentration, determine the intensity from eq. (2.15), and calculate the time of concentration from the Izzard formula. If the initially assumed value was inconsistent, the process above is repeated.

Table 13.6 Empirical Relations for Time of Overland Flow, t_i

Name	Formula for t_i	Remarks	Eq. Number
1. Kirpich	$0.0078\dfrac{L^{0.77}}{S^{0.385}}$		(13.4)
2. Kerby	$0.828\left(\dfrac{rL}{S^{0.5}}\right)^{0.467}$	Applicable to $L < 1300$ ft $r = 0.02$ smooth pavement 0.1 bare packed soil 0.3 rough bare or poor grass 0.4 average grass 0.8 dense grass, timber	(13.5)
3. Izzard	$\dfrac{41.025\left(0.007i + K\right)L^{0.33}}{S^{0.333}i^{0.667}}$	Applicable to $iL < 500$ $K = 0.007$ smooth asphalt 0.012 concrete pavement 0.017 tar and gravel pavement 0.046 closely clipped sod 0.060 dense bluegrass turf	(13.6)
4. Bransby-Williams	$\dfrac{0.00765L}{S^{0.2}A^{0.1}}$		
5. Federal Aviation Agency	$\dfrac{0.388\left(1.1 - C\right)L^{0.5}}{S^{0.333}}$	$C = $ Rational coefficient	(13.7)
6. Kinematic Wave	$\dfrac{0.94L^{0.6}n^{0.6}}{i^{0.4}S^{0.3}}$	$n = $ Manning's coefficient for overland flow	(13.8)
7. NRCS (SCS)	see eqs. (13.9) and (13.10)		

where: $i = $ rainfall intensity, in./hr; $L = $ Length of flow path, ft; $S = $ slope of flow path, ft/ft; $A = $ drainage area, acres; and $t_i = $ overland flow time, min.

Since rainfall intensity reduces with increase in storm duration, the duration should be as short as possible. However, if the rainfall duration is less than t_c, then only a part of the drainage area will be contributing to the runoff. For an entire area to contribute, the shortest storm duration should equal t_c. Thus the time of concentration is used as a unit duration for which the rainfall intensity is determined.

In storm sewer design, in addition to the time required for the rain falling on the most remote point of the tributary area to flow across the ground surface, along streets and gutters, to the point of entry to a sewer, the time of flow through the sewer line is also important. Either the surface and sewer flow times are added together (rational method) or they are considered separately [NRCS (SCS) TR-55 method].

According to the Natural Resources Conservation Service (1986), water moves through a watershed as sheet flow, shallow concentrated flow, open channel flow, or some combination of these before it enters the sewer line. The types that occur depend on the drainage area and can best be determined by field inspection. Time of concentration is the sum of sheet flow, shallow concentrated flow, and channel flow, whichever occur.

Sheet flow in the form of a thin layer can occur for a maximum length of 300 ft. The travel time is given by Manning's kinematic solution (Overton and Meadows, 1976) as follows:

$$t_{t1} = \frac{0.42(nL)^{0.8}}{(P_2)^{0.5} S^{0.4}} \quad \text{[unbalanced]} \tag{13.9}$$

where

t_{t1} = sheet flow travel time, min

n = Manning's roughness coefficient (Table 13.7)

L = Flow length, ft

P_2 = 2-yr 24-hr rainfall, in.

S = land slope

The Manning roughness coefficient n, as presented in Table 13.7, is taken to be a constant value for a particular surface. This holds for a large Reynolds number and a fully developed turbulent condition that exists in an open channel flow. Comparison of the Darcy-Weisbach equation and the Manning equation has revealed that the value of n actually increases for low Reynolds numbers. Engman (1986) and others have assessed the values of n utilizing data from controlled experiments and observations on small experimental watersheds. The value tends to be higher for overland flow than for channel flow for rough surfaces.

Table 13.7   Overland Flow Roughness Coefficient

Surface	Manning's n
Concrete, asphalt, bare soil	0.01–0.016
Gravel, clay-loam eroded	0.012–0.03
Sparse vegetation, cultivated soil	0.053–0.13
Short grass	0.1–0.2
Dense grass, bluegrass, Bermuda grass	0.17–0.48
Woods	0.4 –0.8

After a maximum of 300 ft, sheet flow usually becomes shallow concentrated flow. The average velocity for this flow can be determined from Figure 13.7 using the land slope and the type of soil cover. The travel time for shallow concentrated flow is the length divided by the average velocity.

$$t_{t2} = \frac{L}{V} \times \frac{1}{60} \quad [\text{T}^{-1}] \tag{13.10}$$

where

t_{t2} = shallow concentrated flow travel time, min

L = concentrated flow length, ft

V = flow velocity, ft/s (Figure 13.7)

Open channel is assumed to begin where a channel form is visible from field investigations or on aerial photographs. Manning's equation of open channel flow is used to determine the average velocity and the travel time therefrom. The coefficient n for channel flow is obtained from Table 11.4.

Whenever a drainage area consists of several types of surfaces, the time of concentration is determined by adding the times for different surfaces.

EXAMPLE 13.8

An urbanized watershed in Providence, Rhode Island, is shown in Figure 13.8. Determine the time of concentration to point C by the various methods. The average velocity of flow in the storm drain = 3 ft/s.

SOLUTION

(a) Time of overland flow:

1. Kirpich method

$$t_i = \frac{0.0078(1000)^{0.77}}{(0.02)^{0.385}} = 7.18 \text{ min}$$

2. Kerby method

$$r = 0.02$$

$$t_i = 0.828 \left[\frac{(0.02)(1000)}{(0.02)^{0.5}} \right]^{0.467} = 8.36 \text{ min}$$

3. Izzard method. Assume that the time of concentration = 10 min. For Providence, RI (area 3 in Fig. 2.7) and 5-year frequency

$$i = \frac{131.1}{t+19} = \frac{131.1}{10+19} = 4.52 \text{ in./hr}$$

$$iL = (4.52)(1000) = 4520 > 500; \text{ thus the formula is not applicable}$$

4. Bransby-Williams method

$$t_i = \frac{0.00765(1000)}{(0.02)^{0.2}(375)^{0.1}} = 9.25 \text{ min}$$

Figure 13.7 Average velocity of overland flow (from U.S. SCS 1975b).

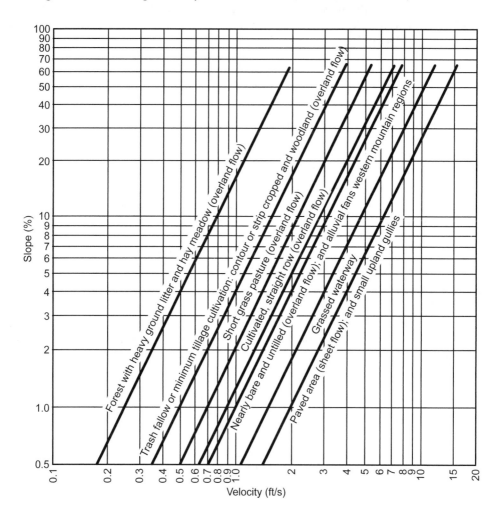

Figure 13.8 Urbanized watershed for Example 13.8.

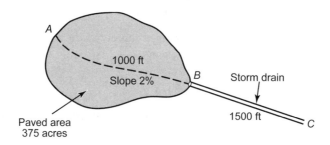

Urban Drainage Systems   Chapter 13

5. Federal Aviation Agency method

$$C = 0.9 \text{ for asphalt paving}$$

$$t_i = \frac{0.388(1.1 - 0.9)(1000)^{0.5}}{(0.02)^{0.333}} = 9.03 \text{ min}$$

6. Kinematic wave method

$$n = 0.011 \text{ (Table 13.7)}$$
$$i = 4.52 \text{ in./hr (method 3 above)}$$

$$t_i = \frac{0.94(1000)^{0.6}(0.011)^{0.6}}{(4.52)^{0.4}(0.02)^{0.3}} = 7.00 \text{ min}$$

7. NRCS (SCS) method. Sheet flow for first 300 ft

$$P_2 = 3.5 \text{ in./hr (from Fig. B-3 in the TR-55, NRCS, 2006)}$$

$$t_{t1} = \frac{0.42\left[(0.011)(300)\right]^{0.8}}{(3.5)^{0.5}(0.02)^{0.4}} = 2.79 \text{ min}$$

Shallow concentrated flow for remaining length of 700 ft

$$V = 2.8 \text{ ft/s from Fig. 13.7}$$

$$t_{t2} = \frac{700}{2.8\,(60)} = 4.17 \text{ min}$$

$$t_i = 2.79 + 4.17 = 6.96 \text{ min}$$

(b) Sewer flow time:

$$t_f = \frac{\text{sewer length}}{\text{velocity}} = \frac{1500}{3} = 500 \text{ sec or } 8.3 \text{ min}$$

Adding time to inlet and sewer flow time, t_c varies from 15.26 min to 17.55 min, depending on the method of computation.

13.11 APPLICATION OF THE RATIONAL METHOD

The drainage area usually consists of more than one type of surface. Equation (13.2) is then applied in the following form:

$$Q = iC_f \sum_{j=1}^{n} C_j a_j \quad [\text{L}^3\text{T}^{-1}] \tag{13.11}$$

where

Q = peak discharge

C_f = frequency factor

C_j = runoff coefficient of subdrainage area a_j

$$A = \sum_{j=1}^{n} a_j$$

i = rainfall intensity for the time of concentration,
which is equal to the longest total time to the
point where the value of Q is desired

Equation 13.11 provides the design flow at the mouth of the composite drainage area, as shown in Example 13.9. However, a storm drainage system consists of many segments of sewer drains. The tributary area and the amount of flow entering each drain is different. To determine the peak discharge, not only at the outlet but at interim points of entry to each drain, a step-by-step application of eq. (13.11), known as the Lloyd-Davies method, is made as illustrated in the following example. C_f is usually taken as 1.0 in urban storm studies.

EXAMPLE 13.9

An urban watershed is shown in Figure 13.9 along with the travel paths from the most remote points in each subarea. The details of the subareas are given in Table 13.8. Assuming that Figure 13.6 reflects the intensity-duration-frequency in in./hr for the site, determine the 20-year peak flow at the drainage outlet G.

Figure 13.9 Watershed for Example 13.9.

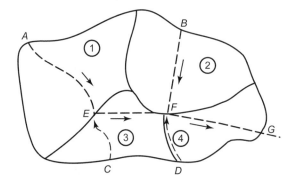

Table 13.8 Details of Subareas of Watershed in Figure 13.9

No.	Area Drained (acres)	Type of Surface	Path	Length (ft)	Slope (%)
1	14	Lawn	AE	1600	4.0
2	12.5	Bare surface	BF	1490	3.0
3	11.1	Asphalt paved	CE	1280	2.0
			EF	1300	1.5
4	8.5	Concrete paved	DF	1250	2.0
			FG	1510	1.5

SOLUTION

1. Travel time by paths:

Path	Length (ft)	Slope (%)	Average Velocity (ft/sec) (Fig. 13.7)	$\text{Time} = \dfrac{\text{Length}}{\text{Velocity}} \times \dfrac{1}{60}$ (min)	Time (hr)
AE	1600	4	3.0	8.9	0.15
BF	1490	3	1.8	13.8	0.23
CE	1280	2	2.9	7.4	0.12
EF	1300	1.5	2.5	8.7	0.14
DF	1250	2	2.9	7.2	0.12
FG	1510	1.5	2.5	10.1	0.17

2. Possible routes:

    a. $AE + EF + FG = 8.9 + 8.7 + 10.1 = 27.7 \text{ min} \leftarrow t_c$

    b. $CE + EF + FG = 7.4 + 8.7 + 10.1 = 26.2 \text{ min}$

    c. $BF + FG = 13.8 + 10.1 = 23.9 \text{ min}$

    d. $DF + FG = 7.2 + 10.1 = 17.3 \text{ min}$

3. From Figure 13.6, for 20-year frequency and a t_c of 27.7 min, $i = 4.7$ in./hr

4.

Area	Area Drained (acres)	C	aC (acres)	$\Sigma \, aC$ (acres)
1	14	0.2	2.8	2.8
2	12.5	0.35	4.38	7.18
3	11.1	0.8	8.88	16.06
4	8.5	0.9	7.65	23.71

5. $Q = i\Sigma \, aC = 4.7(23.71) = 111.44 \text{ cfs}$

EXAMPLE 13.10

A storm drainage system comprises four areas (Figure 13.10) with the data given in Table 13.9. Determine the 5-year design flow for each section of the sewer line. Average velocity in sewer = 4.5 ft/sec. Five-year rainfall intensity, $i = 105/(t_c + 15)$.

SOLUTION

1. Flow time in sewers:

$$\text{Sewer } 1-2 = \frac{300}{4.5} = 66.6 \text{ sec or } 1.1 \text{ min}$$

$$\text{Sewer } 2-3 = \frac{400}{4.5} = 88.89 \text{ sec or } 1.5 \text{ min}$$

$$\text{Sewer } 3-4 = \frac{400}{4.5} = 88.89 \text{ sec or } 1.5 \text{ min}$$

Section 13.11   **Application of the Rational Method**   **741**

Figure 13.10 Storm drainage system.

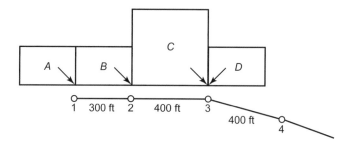

2. Computations are arranged in Table 13.10.

Consider manhole 3: Areas C and D are contributing at the manhole as listed in column 2 of the table. The coefficients for these areas and the product aC are given in the next two columns. Column 5 accumulates the product; it sums up the value of column 4 to the value in the previous line of column 5 if the flow of the previous area passes through the manhole in question. That is, all areas whose flow passes through the manhole in consideration are added together. In column 6 all possible routes for the flow to reach the manhole in question (i.e., manhole 3) are listed along with their overland flow and sewer flow time in columns 7 and 8, respectively. For the route of flow from the previous manhole (i.e., B– 2– 3), the highest value of the total time (column 9) of the preceding manhole computation is taken as the overland flow time. The flow time in a sewer section is determined from the average velocity. In a design problem, this is established from the design of the pipe size of the preceding section. The maximum of the total time in column 9 from all routes is underlined and denotes the time of concentration, t_c. Corresponding to this t_c, intensity is determined either from the intensity-duration-frequency curve or the empirical relation. Peak discharge is the multiplication of the last value in column 5 with column 10 and C_f if it is greater than 1.0.

13.12 THE NRCS (SCS) TR-55 METHOD

In 1964 the NRCS (then SCS) developed a computerized watershed model known as the TR-20 which has since been updated. It is a very versatile model that has the capability of solving many hydrologic problems involving the formulation of runoff hydrographs; routing hydrographs through channels and reservoirs, thus providing discharges at selected locations; combining or separating hydrographs at confluences; and determining peak discharges and their time of occurrences for individual storm events. The model is widely used in small watershed projects and floodplain studies. When the sole purpose is to assess the peak discharge or peak flow hydrograph for drainage design, a method simplified from TR-20 is used. This is referred to as TR-55 (Technical Release 55). TR-55, *Urban Hydrology for Small Watersheds*, was first released in 1975. Major revisions were made in 1986. The new WIN TR-55 is a completely rewritten Windows-based model (2002). The technical framework remains the same as in the 1986 documentation. There are two approaches in the TR-55 method, known as the graphical method and the tabular method.

The graphical method uses the following equation to determine the peak flow:

$$q_p = q_u A_m Q F_p \quad [\text{L}^3\text{T}^{-1}] \tag{13.12}$$

Table 13.9 Details of Drainage Area

Unit	Area (acres)	Runoff Coefficient	Overland flow time (min)
A	12	0.8	10
B	12	0.8	10
C	34	0.6	28
D	10	0.9	8

Table 13.10 Step Method of Computation

(1)	(2)	(3)	(4)	(5)	(6)	(7)	(8)	(9)	(10)	(11)
						Travel Time (min)			Intensity (in./hr)	
Manhole	Area, a (acres)	Coefficient, C	aC (acres)	ΣaC (acres)	Routes	Overland	In Sewer	Total Time	from formula	Q_{peak}[a] (cfs)
1	12	0.8	9.6	9.6	A–1	10	—	10	4.20	40.32
2	12	0.8	9.6	19.2	B–2	10	—	10		
					A–1–2	10	1.1	11.1	4.02	77.18
3	34	0.6	20.4	39.6						
	10	0.9	9.0	48.6	C–3	28	—	28	2.44	118.58
					D–3	8	—	8		
					B–2–3	11.1	1.5	12.6		

[a] Col. 5 × col. 10.

where

q_p = peak discharge, cfs

q_u = unit peak discharge, cfs/mi^2/in.

A_m = drainage area, mi^2

Q = runoff corresponding to 24-hr rainfall, in., of a desired design frequency

F_p = pond or swamp adjustment factor

Runoff (Q) is determined by the procedure of Section 2.15 from the data of 24-hour rainfall of the desired design frequency. Generalized 24-hour rainfall data for various frequencies are available in TR-55 (NRCS, 1986). Adjustment factor (F_p) is obtained from Table 13.11. Unit peak discharge (q_u) is determined from the graphs contained in TR-55. A sample graph is shown in Figure 13.11. The information required to use the graph to ascertain q_u is as follows:

1. Time of concentration as calculated from Section 13.10.5, which is the sum of overland flow and channel flow, if any, within the area denoting the highest value from all possible routes.
2. The ratio I_a/P, I_a being obtained from Table 13.12 and P is the 24-hr rainfall
3. Rainfall distribution type

The NRCS has developed four synthetic 24-hour rainfall distributions from the National Weather Service duration-frequency data to represent various regions of the United States:

Type I: Hawaii, Alaska
Type IA: coastal side of Sierra Nevada and Cascade Mountains in southern and northern California, Oregon, and Washington
Type III: Gulf of Mexico and east coast area from Maryland to Maine
Type II: rest of the United States

Table 13.11 Adjustment Factor (F_p) for Pond and Swamp Areas that Occur Throughout the Watershed

Percentage of Pond and Swamp Areas	F_p
0	1.00
0.2	0.97
1.0	0.87
3.0	0.75
5.0	0.72

Source: NRCS (1986).

Figure 13.11 Unit peak discharge (q_u) for NRCS (SCS) type I rainfall distribution (from NRCS, 1986).

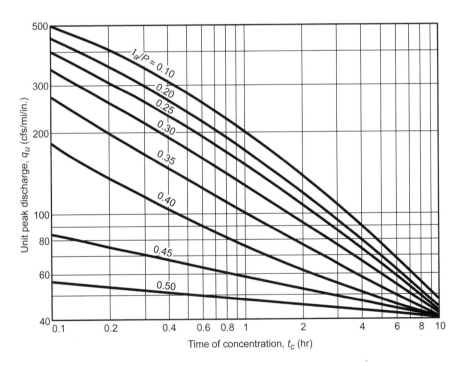

Table 13.12 I_a Values for Runoff Curve Numbers

Curve Number	I_a (in.)	Curve Number	I_a (in.)	Curve Number	I_a (in.)	Curve Number	I_a (in.)
40	3.000	55	1.636	70	0.857	85	0.353
41	2.878	56	1.571	71	0.817	86	0.326
42	2.762	57	1.509	72	0.778	87	0.299
43	2.651	58	1.448	73	0.740	88	0.273
44	2.545	59	1.390	74	0.703	89	0.247
45	2.444	60	1.333	75	0.667	90	0.222
46	2.348	61	1.279	76	0.632	91	0.198
47	2.255	62	1.226	77	0.597	92	0.174
48	2.167	63	1.175	78	0.564	93	0.151
49	2.082	64	1.125	79	0.532	94	0.128
50	2.000	65	1.077	80	0.500	95	0.105
51	1.922	66	1.030	81	0.469	96	0.083
52	1.846	67	0.985	82	0.439	97	0.062
53	1.774	68	0.941	83	0.410	98	0.041
54	1.704	69	0.899	84	0.381		

Source: NRCS (1986).

The tabular method is suitable when a complete hydrograph is desired instead of peak flow only or when a watershed is subdivided into subareas. For each subarea, the following information is ascertained: (1) weighted curve number; (2) runoff, Q; (3) ratio I_a/P; (4) time of concentration within the subarea, t_c; and (5) travel time downstream of the subarea to the outlet, T_t.* For the selected rainfall distribution type and known I_a/P, one of the tables in TR-55 provides the hydrograph ordinates for the subarea that correspond to the time of concentration t_c and the travel time T_t. Once the ordinates for all subareas at different times are tabulated, their summation at each time yields the composite hydrograph.

EXAMPLE 13.11

Solve Example 13.9 by the NRCS (SCS) TR-55 method. The rainfall distribution is type II and the 20-year 24-hour rainfall is 4 in.

SOLUTION   Computations are given in Tables 13.13 and 13.14.

Table 13.13   Computation of Runoff and Initial Abstraction

Area	Drainage Area, A_m (mi²)	24-Hr Rainfall (in.)	Curve Number, CN (Table 2.22)	Runoff, Q (in.) (Table 2.25)	Area × Runoff, A_mQ (mi²·in.)	I_a (in.) (Table 13.12)	I_a/P
1	0.0219	4	68	1.20	0.026	0.94	0.24
2	0.0195	4	75	1.67	0.033	0.67	0.17
3	0.0173	4	98	3.77	0.065	0.04	0.01
4	0.0133	4	98	3.77	0.050	0.04	0.01

* In the graphic procedure, the time of concentration, t_c, and travel time T_t are added for the entire area.

Table 13.14 Hydrograph Computation

Area	Time of Conc., t_c (hr) (Example 13.9)	Downstream Travel Route	D/S Travel Time, $\Sigma T_t{}^a$ (hr) (Example 13.9)	I_a/P (rounded)	$A_m Q$ (mi² in.)	Hydrograph Times (hr)										
						11.9	12.0	12.1	12.2	12.3	12.4	12.5	12.6	12.7	12.8	13.0
						Hydrograph Ordinates in cfs = (Value from TR-55^b) × ($A_m Q$)										
1	AE = 0.15	EF + FG	0.31	0.2	0.026	0.6^c	0.8	1.6	3.6	8.0	12.7	13.9	12.0	9.2	6.9	3.9
2	BF = 0.23	FG	0.17	0.2	0.033	0.8^c	1.5	3.8	9.3	16.3	19.4	16.6	12.2	8.7	6.3	3.9
3	CE = 0.12	EF + FG	0.31	—	0.065	3.6^d	6.0	11.3	21.9	37.8	43.0	35.4	23.3	17.5	12.4	7.1
4	DF = 0.12	FG	0.17	—	0.050	3.1^d	5.5	10.8	20.9	35.2	35.1	24.3	15.6	10.5	7.6	4.7
						8.1	13.8	27.5	55.7	97.3	110.2	90.2	63.1	45.9	33.2	19.6

^{a} Add travel time for the route indicated in previous column.
^{b} From Exhibit 5-II (NRCS, 1986, pp. 5–29 and 5–30). See table below.
^{c} Table values at I_a/P of 0.1 and 0.3 are averaged.
^{d} Table values at I_a/P of 0.1 are used.
Rounding of t_c and T_t

Area	t_c	T_t	Sum
1	0.2	0.3	0.5
2	0.2	0.2	0.4
3	0.1	0.3	0.4
4	0.1	0.2	0.3

13.13 A STORM SEWER DESIGN PROJECT

Design flows Q_{design} for various sewer sections are estimated by the rational method or the NRCS (SCS) method. The sewer design sequence is as follows:

1. For Q_{design}, and a velocity of 3 ft/sec (0.9 m/s), determine the diameter of the sewer by the continuity equation, $Q = VA$. This is the maximum size.

2. For Q_{design} and the surface grade, determine the diameter by Manning's equation. For a roughness coefficient, use Table 11.4.

3. If the diameter in step 2 is smaller than step 1, select it (rounded to a next-higher standard size) as the sewer size and the surface grade as the sewer slope.

4. If the diameter of step 2 is bigger than step 1, recompute the slope by Manning's formula, adopting the sewer diameter of step 1. In this case a steeper slope than street grade is required. In certain circumstances it is necessary to use a slope lower than this. A provision should then be made for flushing of the pipe.

An example illustrates the procedure.

EXAMPLE 13.12

Design a storm drainage system for the section of a city shown in Figure 13.12 (the city for which a sanitary sewer system is designed in Example 13.4). Design for the following conditions.

1. The coefficients of runoff, C, at the time of maximum development are given in Figure 13.12.

2. The overland flow time, which can be calculated by the method of Section 13.10.5, is assumed to be 15 minutes for each inlet.

3. The system is to be designed for 5-year peak flows. The rainfall intensity in mm/hr is given by $i = 3330/(t + 19)$, where t is in minutes.

4. Manning's n is 0.013.

SOLUTION   Unlike sanitary sewers, the storm sewer line need not run through individual lots because the connections from housing units are not required. Thus the drains can be laid by the shortest route. However, the arrangement will be governed by the topography (contour pattern) of the drainage area that dictates the direction of runoff and hence the positioning of the inlets and laying of sewer lines. Also, relatively larger areas can be covered by each section of the storm drain. For the section of the city in Figure 13.12, the arrow indicates the lowest point in each block.

The general direction of flow of the block will be toward the arrow. A layout of drains has been arranged keeping this in mind. A minimum number of manholes are included. The manhole number is shown within a ❑. The drainage area tributary to each manhole intercept point, determined on the basis of the contours, is indicated in the figure along with the runoff coefficient, which has been taken as 0.6 for the commercial district and 0.4 for the residential area.

The design flows at each intercept point are computed in Table 13.15 by the step method of Section 13.11. To calculate the flow at manhole 3, the time of concentration needs to be determined, which involves computation of the time of travel in sewer 1– 3 and sewer 2– 3 (column 9). This requires a determination of the size and the flow velocity

Figure 13.12 Storm drains layout for a section of a city.

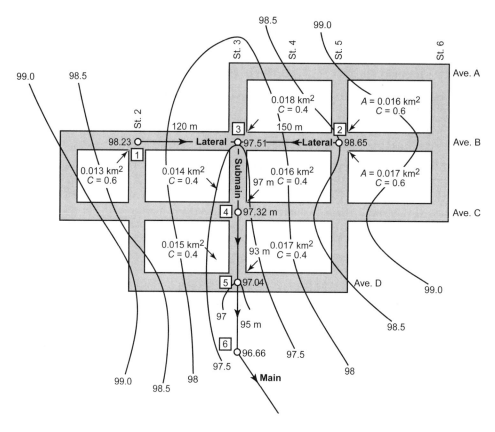

(design) of the two sewer sections. Thus the design of each section proceeds simultaneously with estimation of design flow. The design flows at the head of section 1– 3 (manhole 1) and section 2– 3 (manhole 2) are computed in the first two lines of column 12 in Table 13.15. For these flows, pipe sections 1– 3 and 2– 3 are designed in lines 1 and 2 in Table 13.16. This provides the average velocity of flow (column 13) and the travel time through the sewer (column 14) in Table 13.16. This time of flow is included in column 9 of Table 13.15 for manhole 3 to determine the time of concentration. The design flow is then determined for the next pipe to be designed. Thus the computation alternates in Tables 13.15 and 13.16.

The design procedure of Table 13.16 is as follows. The value in column 4 is taken from column 12 of Table 13.15. Columns 5, 6, and 7 are based on the layout plan. Column 8 is the difference between columns 6 and 7, divided by column 5. In column 9, the maximum sewer size for a minimum velocity of 0.9 m/s is determined using the continuity equation, $Q = AV$. In column 10, the diameter corresponding to the street slope of column 8 is computed from Manning's equation. The design diameter in column 11 is the minimum of columns 9 and 10 (rounded to a standard size). If this pertains to column 10, the sewer grade in column 12 is equivalent to the street grade. If the design diameter is based on column 9, the sewer grade is computed from Manning's equation. The velocity of flow in column 13 is determined by the continuity equation for known flow (column 4) and

Table 13.15 Computation of Peak Discharge

(1) Manhole	(2) Location	(3) Tributary Area, a (km²)	(4) Coefficient, C	(5) aC (m²)	(6) ΣaC (m²)	(7) Route	Travel Time (min)			(11) Intensity (mm/hr)	(12) Q (m³/s)
							(8) Overland	(9) In Sewer	(10) Total		
1	Avenue B	0.013	0.6	7,800	7,800	TA–1ᵃ	15	0	15	97.9	0.212
2	Avenue B	0.016	0.6	9,600	9,600						
		0.017	0.6	10,200	19,800	TA–2	15	0	15	97.9	0.538
3	Street 3	0.018	0.4	7,200	34,800ᵇ	TA–3	15	0	15		
						1–3	15	1.47	<u>16.47</u>	93.9	0.908
						2–3	15	1.32	16.32		
4	Street 3	0.014	0.4	5,600	40,400	TA–4	15	0	15		
		0.016	0.4	6,400	46,800	3–4	16.47	1.23	<u>17.70</u>	90.7	1.180
5	Street 3	0.015	0.4	6,000	52,800	TA–5	15	0	15		
		0.017	0.4	6,800	59,600	4–5	17.70	0.95	<u>18.65</u>	88.4	1.464

ᵃ TA–1 = Tributary area to manhole 1.
ᵇ Col. 5 for TA + col. 6 for manhole 1 via route 1–3 + col. 6 for manhole 2 via route 2–3 = 7200 + 7800 + 19,800 = 34,800.

Table 13.16 Storm Sewer Design Computations

(1)	(2)	(3)	(4)	(5)	(6)	(7)	(8)	(9)	(10)	(11)	(12)	(13)	(14)
Sewer Line					Surface Elevation (ft)					Design Parameters			
Line	From	To	Design Flow, Q_{design} (m³/s)	Length of Sewer (m)	Upstream	Downstream	Street Slope	Maximum Diameter for Velocity of 0.9 m/s[a] (mm)	Diameter for Street Grade[b] (mm)	Diameter[c] (mm)	Sewer Grade	Velocity at Full[d] (m/s)	Travel Time (min) $\left(\dfrac{\text{col. 5}}{\text{col. 13}} \times \dfrac{1}{60}\right)$
1	1	3	0.212	120	98.23	97.51	0.006	550	445	445	0.006	1.36	1.47
2	2	3	0.538	150	98.65	97.51	0.0076	875	600	600	0.0076	1.90	1.32
3	3	4	0.908	97	97.51	97.32	0.002	1135	940	940	0.002	1.31	1.23
4	4	5	1.180	93	97.32	97.04	0.003	1290	960	960	0.003	1.63	0.95
5	5	6	1.464	95	97.04	96.66	0.004	1440	990	990	0.004	1.90	0.83

[a] $D = (1.274 Q/v)^{1/2} \times 1000$ (continuity equation), Q is Q_{design}

[b] $D = \left[\dfrac{(3.211)nQ}{s^{1/2}} \right]^{0.375} \times 1000$ (Manning's equation).

[c] Smaller of col. 9 or col. 10. If col. 9 is smaller, recompute the slope for this diameter by the Manning eq.

[d] $v = 1.274 \left(\dfrac{Q}{D^2} \right)$; D in m (continuity equation).

diameter (column 11). When the velocity is excessive, it is reduced to a limiting value of 5.0 m/s and for the known design flow, the diameter is recomputed by the continuity equation and the slope from Manning's equation.

13.14 DETENTION BASIN STORAGE CAPACITY

Urbanization of rural areas increases peak discharges that adversely affect downstream floodplains. Many local governments are adopting ordinances that require that the postdevelopment discharge not exceed the predevelopment discharge, i.e., *zero excess runoff*, for a defined storm frequency at a development area. A detention basin is the most widely used measure to control the peak discharge. When a detention basin is installed, the reservoir routing procedure can be used to estimate the effect on hydrographs. The size of the detention basin can be adjusted to maintain a required level of outflow discharge. A quick method of estimation has been included in TR-55 that relates the ratio of peak outflow to peak inflow discharge (q_o/q_i) with the ratio of detention storage volume to runoff volume (V_s/V_r), as illustrated in Figure 13.13. This figure is used to estimate the detention storage volume (V_s) from the known information of runoff volume (V_r), peak outflow discharge (q_o), and peak inflow discharge (q_i) or to estimate q_o from the known values of V_r, V_s, and q_i. The value q_o is the predevelopment level of peak flow or a desired level of discharge from the drainage area. The value q_i is the peak discharge from the developed area computed by the TR-55 method of Section 13.12. While using the tabular method to estimate q_i for a subarea, the peak discharge associated with travel time $T_t = 0$ is used. V_r is the drainage area times the runoff Q, which is determined when computing q_i. The computed V_s is adequate for preliminary designs.

13.14.1 Rational-Method-Based Procedure

The detention storage is the maximum difference between the cumulated inflow volume and the cumulated outflow volume, $V = \max \{ V_{in} - V_{out} \}$ where V_{in} is the peak rate of flow from the watershed times the storm duration and V_{out} is the maximum flow from basin outlet times the storm duration, thus:

$$V_{in} = i \sum aCT \quad [\mathrm{L}^3] \tag{13.13}$$

and

$$V_{out} = Q_o T \quad [\mathrm{L}^3] \tag{13.14}$$

where

       i = rainfall intensity from the IDF curve

       T = storm duration

      Q_o = maximum outflow rate

Various storm durations are assumed. For each assumed value, i is obtained from the IDF relation, V_{in} and V_{out} are computed from the above equations. The maximum difference is the required storage volume.

Figure 13.13 Detention basin storage volume (from NRCS, 1986).

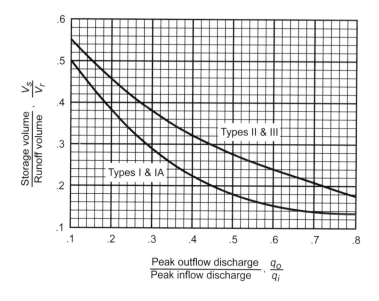

Peak outflow discharge / Peak inflow discharge, $\dfrac{q_o}{q_i}$

EXAMPLE 13.13

For the watershed in Figure 13.9, if the outflow of a 5-year frequency should not exceed 50 cfs, determine the size of the detention basin required. The intensity-duration-frequency relation is given by $i = 315/(t_c + 25)$, t_c is in minutes.

SOLUTION

1. From Example 13.9, $\sum aC = 23.71$ acres
2. Refer to Table 13.17
3. Detention basin capacity (col. 6 of Table 13.17) = 156.4×10^3 ft³

Table 13.17 Computation of Detention Storage

(1) Rainfall Duration min	(2) Intensity $i = 315/(t_c + 25)$ in. per hr	(3) Peak Inflow $Q_i = i\sum aC$ cfs	(4) Peak Outflow Q cfs	(5) Rate of Flow Detained[a] cfs	(6) Detention Capacity[b] 1000 ft³
10	9.0	213.39	50	163.39	98.0
20	7.0	165.97	50	115.97	139.2
30	5.73	135.86	50	85.86	154.5
35	5.25	124.48	50	74.48	156.4
40	4.85	114.99	50	64.99	156.0
45	4.50	106.70	50	56.70	153.1

[a] Col. 5 = col. 3 − col. 4

[b] Col. 6 = col. 5 × col. 1 × $\left[\dfrac{60\ \text{sec}}{1\ \text{min}}\right]$

PROBLEMS

13.1 From an area of 10 km² having a present population density of 8000 persons/km², determine the peak dry-weather flow if the domestic water consumption is 670 liters/person per day. The domestic consumption is expected to be 50% of the average total consumption. Determine the design flow if the density increases to 13,000 persons/km² at the end of the design period.

13.2 Wastewater to a sewer is contributed by two 500-acre areas. One area is sparsely populated, with 15 persons/acre, and the other is an apartment district with a heavy population of 150 persons/acre. For the total consumption rate of 160 gal/person per day in both areas, calculate the peak rate of sewage flow.

13.3 The composition of a district of 30 km² size is 60% residential area, 25% commercial zone, and 15% industrial zone. Twenty percent of the commercial zone is estimated to be covered by buildings. The residential section has a density of 10,000 persons/km² and the industrial zone of 50,000 persons/km². Determine the peak rate of flow. Assume the following parameters:

1. Domestic consumption = 300 liters/person per day
2. Commercial consumption = 12.2 liters/day per sq. meter of building area
3. Industrial consumption = 60 liters/person per shift
4. Number of shifts = 2
5. Length of sewer line = 10.2 km

13.4 A sewer has to be laid in a place where the ground has a slope of 1.75 in 1000 m. If the present and ultimate peak sewage discharge rates are 40 and 165 liters/sec, respectively, design the sewer section. Design for 2/3 full condition.

13.5 Between two manholes 500 ft apart, the ground elevations are 100 ft and 99.25 ft, respectively. The present peak rate of sewage flow is 2.75 cfs, which is estimated to go up to 10.0 cfs at the end of 25 years. Design the sewer section.

13.6 The layout of a sanitary sewer system is shown in Fig. P13.6. Data on area, length, and elevations are given below. The present population density, 40 persons/acre, is expected to rise to 100 per acre by conversion of the dwellings to apartments. The peak rate of sewage flow is 400 gpd per person. Design the sewer system.

Block	Area (acres)	Length (ft)	Elevation (ft) Upstream	Downstream
A	2.0	390	101.50	97.17
B	2.5	350	100.67	97.17
C	1.5	330	97.17	93.29
D	1.3	230	98.69	97.54
E	1.2	295	100.5	97.54
F	5.7	650	97.54	94.29
G	2.1	300	94.29	93.29
H	3.5	550	93.29	86.42

Problems   **753**

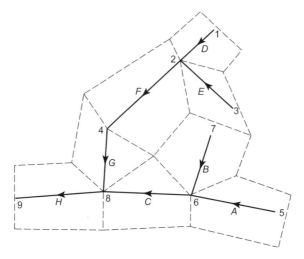

13.7 Design the sanitary sewer system for the city apartment district shown in Fig. P13.7. The length of the sewer segments, the tributary area of each, and the ground-level elevations are shown in the figure. The present density of population is 100,000 persons/km², which is expected to rise to 150,000/km² by the end of the design period. The maximum rate of sewage flow is 1500 liters/person per day.

Figure P 13.7

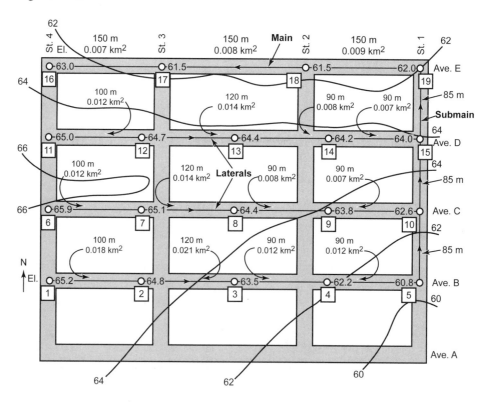

13.8 Determine the sewer arrangement of manholes 5 through 8 of Problem 13.6. Provide a cover of 10 ft.

13.9 Determine the arrangement of sewers at manholes 1 through 5 in Problem 13.7, allowing a cover of 2 m.

13.10 A drainage area consists of 30% turf ($C = 0.3$), 35% bare surface ($C = 0.4$), and 35% paved surface ($C = 0.9$). The time of concentration at the inlet point under consideration is 12 minutes. The total duration of the rainstorm is 3 hours. Determine the value of the runoff coefficient corrected for antecedent rainfall condition.

13.11 A storm drain system is shown in Fig. P13.11. For the flow conditions indicated, determine the time of concentration by the different methods. Assume that $A = 5230$ and $B = 30$ for the intensity relation of eq. (2.15).

Figure P 13.11

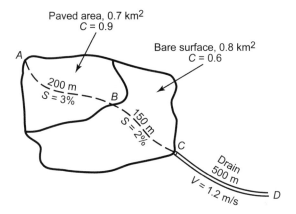

13.12 For the storm drain system shown in Fig. P13.12, determine the time of concentration at point C using Figure 13.7.

Figure P 13.12

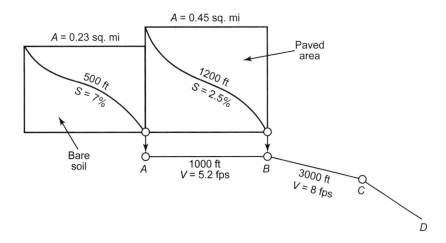

13.13 An urban watershed has a main ditch as shown in Fig. P13.13. Paths from remote points to the main ditch are also indicated. The details of each subarea are listed below. Determine the peak flow at the outlet ditch by the rational method, assuming that the 5-year rainfall intensity is given by $170/(t_c + 23.0)$, where t_c is in minutes and i in in./hr. Assume shallow concentrated flow through each subarea.

Area	Drainage Area (acres)	Type of Surface	Path	Length (ft)	Slope (%)
1	12.0	Bare surface, $C = 0.4$	AB	1300	6.0
2	13.5	Asphalt paved, $C = 0.8$	BD	1250	1.5
			CD	1420	2.0
3	11.8	Lawn, $C = 0.3$	ED	1800	1.0
4	14.1	Concrete paved, $C = 0.9$	DG	1510	1.5
			FG	1660	2.5

Figure P 13.13

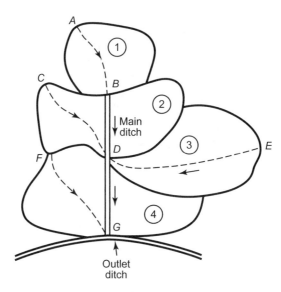

13.14 Assuming that Figure 13.6 reflects the intensity-duration-frequency relation for the watershed of Problem 13.13, determine the peak rate of runoff for 5-, 10-, and 20-year frequency.

13.15 Determine by the rational method the peak flow at the outfall of the watershed shown in Fig. P13.15. The 5-year intensity relation is $190/(t_c + 25.0)$, t_c in minutes, i in in./hr. Given:

1. Area, $A_1 = 13.0$ acres, $C_1 = 0.6$.
    Area, $A_2 = 20.0$ acres, $C_2 = 0.4$.
    Area, $A_3 = 18.5$ acres, $C_3 = 0.5$.

2. Time of travel to inlet points I_1, I_2, I_3 within each area = 10 min.
3. Inlet to manhole times (min): $I_1 M_1 = 6.5$, $I_2 M_2 = 12.1$, $I_3 M_4 = 13.5$ min.
4. Average velocity of flow between manholes = 4.1 ft/sec.

Figure P 13.15

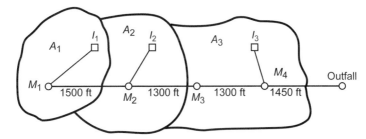

13.16 Solve Problem 13.13 by the NRCS (SCS) TR-55 tabular hydrograph method to determine the peak flow hydrograph. Assume type II rainfall distribution and a 5-year 24-hour rainfall of 3.5 in. Hydrologic soil group is A. Obtain tables of hydrograph unit discharges from the NRCS publication *Urban Hydrology for Small Watersheds*, Technical Release 55, on the World Wide Web at http://www.wcc.nrcs.usda.gov/hydro/hydro-tools-models-tr55.html.

13.17 Solve Problem 13.15 by the NRCS (SCS) TR-55 tabular hydrograph method to determine the peak flow hydrograph. The entire area comprises the urban business district of hydrologic group A. Assume type III rainfall distribution and a 5-year 24-hour rainfall of 4 in. Obtain the tables from the source indicated in Problem 13.16.

13.18 A storm system consists of four areas with details as shown in Fig. P13.18. The direction of flow from each area and between the manholes is given by an arrow. Determine the 10-year peak rate of flow for each sewer section by the rational method, assuming that Figure 13.6 reflects the intensity-duration-frequency relation for the area. The travel time between each manhole = 5 min.

Figure P 13.18

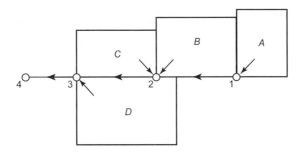

| | | | Overland flow time |
Drainage Unit	Area (acres)	Coefficient, C	(min)
A	5	0.6	8.0
B	5.4	0.7	9.2
C	2.5	0.7	7.0
D	6.5	0.5	14.5

13.19 For the drainage system shown in Fig. P13.19, determine the design flow for each sewer section by the rational method. The rainfall intensity (in./hr) is represented by $i = 100/(t_c + 15)$; t_c is in minutes. Flow from each area is shown by an arrow. Average velocity through the pipes = 4 ft/sec.

Unit	Area (acres)	C	Overland flow time (min)
A	0.4	0.7	10
B	0.5	0.7	10
C	0.3	0.80	5
D	0.3	0.80	5
E	0.4	0.65	10
F	0.5	0.60	10
G	0.6	0.60	10

Pipe Lengths	Feet	Slope
1– 3	500	0.008
2– 3	1500	0.01
3– 5	600	0.0075
4– 5	750	0.01
5– 6	600	0.0075

Figure P 13.19

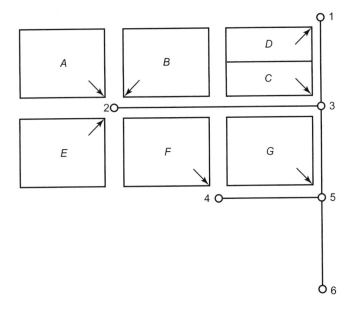

13.20 Determine the design discharge for the storm sewers between the manholes as shown in Fig. P13.20. The design rainfall intensity in in./hr is given by $i = 96/(t_c + 16)$ (t_c in min) and the average velocity of flow is 3.5 ft/sec.

Figure P 13.20

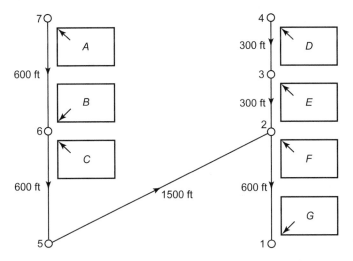

Block:	A	B	C	D	E	F	G
Area (acres)	10	9	8	10	9	8	12
C	0.65	0.7	0.55	0.6	0.5	0.45	0.4
Overland flow time (min)	10	9	7	11	10	9	10

13.21 Design the storm sewer system of Problem 13.19. Assume that $n = 0.013$.

13.22 For the part of a city in Problem 13.7, design the storm sewer system. The layout is shown in Fig. P13.22. Assume the following conditions.

1. The tributary areas and ground surface elevations are given in the figure.

2. The runoff coefficients are indicated in the figure.

3. The overland flow time from each area is 20 minutes.

4. The design frequency is 2 years, for which the rainfall intensity (mm/hr) is given by $i = 2590/(t + 17)$, where t is in minutes.

5. Manning's $n = 0.013$.

Figure P 13.22

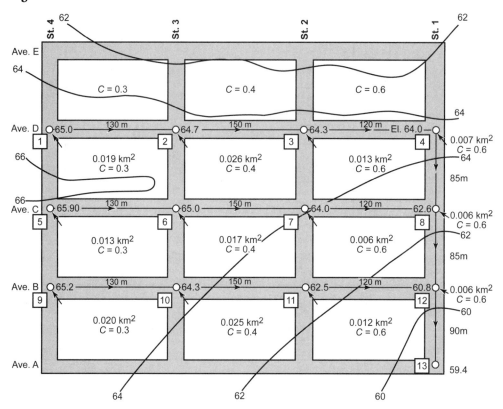

13.23 The developed-condition peak discharge for the 25-year frequency storm from a 75-acre area is 400 cfs and the corresponding runoff (Q) is 4.2 in. If the maximum discharge from the area should be retained at the predevelopment level of 176 cfs, how much storage will be required by the NRCS (SCS) method? The area is in the type II distribution region.

13.24 A detention basin is planned as part of a development for a 15-acre watershed. The peak discharge from the developed area for a 2-year frequency is 93.5 cfs and the runoff is 1.6 in. The outflow structure consists of a 2.5-foot long rectangular weir. The maximum stage (head) over the weir should not exceed 3.5 ft. The watershed is in type I distribution. Determine the detention basin storage size by the NRCS (SCS) method. Assume the flow over the weir, $q_o = 3.2LH^{3/2}$.

13.25 For a 100-acre watershed, determine the size of a detention basin if the maximum discharge from the watershed (outflow) is not to exceed 36 cfs. The composite runoff coefficient is 0.4. The IDF relation is given by $i = 103/(13.5 + t)$, where i is in in./hr and t is storm duration (concentration time) in min.

13.26 Determine the capacity of a detention basin if the peak rate of outflow for the watershed in Problem 13.13 is limited to 50 cfs.

13.27 Determine the capacity of a detention basin in Problem 13.14 if the outflow from the basin is not to exceed 50 cfs at all levels of frequency.

13.28 Determine the size of a detention basin for the watershed of Problem 13.15 for restricting the outflow to 40 cfs.

14

Other Drainage Systems

♦♦

14.1 AGRICULTURAL DRAINAGE SYSTEMS

The removal of water from the surface of land is *surface drainage*. Similarly, the removal or control of water beneath the land surface is termed *subsurface drainage*. Urban storm drainage is concerned with surface drainage. Agricultural drainage, on the other hand, deals both with the removal of excess precipitation and irrigation surface waste as a part of surface drainage and with removal and control of groundwater percolated from precipitation or irrigation or leaked from canals through subsurface drainage. The latter component, termed *land drainage*, is achieved by the flow of water through the porous soil medium by gravity to the natural outlet. If water is added through irrigation or heavy precipitation at a faster rate than it can travel to the outlets, the water table rises and can approach the surface to waterlog the land. In such cases, additional constructed outlets are provided in the form of drains. The installation of drains or constructed subdrainage systems has been found to be essential for agricultural land because of the rapid buildup of the water table. For highway and airport pavement structures, some provision of subsurface drains also is required, in addition to major surface drainage facilities.

14.2 SURFACE DRAINAGE FOR AGRICULTURAL LAND

Surface flow that should be carried away from agricultural lands includes precipitation excess and farm irrigation surface waste (excess). In a humid region, the former constitutes almost the entire surface flow. In arid regions, irrigation waste is the major constituent. Surface runoff from agricultural land is much less than urban runoff because of the perviousness of the land surface. The procedures to determine surface flows due to precipitation and irrigation are described below. Once the quantity of runoff (from storm and irrigation) at various points of interception is known, the surface drainage system is designed (1) as a separate system for a large land area along the line of an urban storm drain system, or (2) as a system of open drains comprising laterals or field drains, submains, and mains. The submains and mains include contributions from the subsurface drains. The drains are sized for the combined surface and subsurface quantity of flow by Manning's equation.

The peak rate of surface runoff due to a rainstorm of specified frequency can be determined by the rational method using an appropriate value of the runoff coefficient for the agricultural area. Surface drains are designed to handle flows from 5- to 15-year storm frequencies. Where damages can be expensive, a more conservative design frequency of 25 years should be used. The NRCS (SCS) procedure of soil and cover conditions, as described in Section 13.12, is more appropriate for agricultural land.

Surface runoff produced by waste from irrigation varies with many factors, including soil texture, land slope, length of irrigation run, and irrigation efficiency. This may amount to as much as 50% of the water applied to any farm unit. The total amount of farm (irrigation) waste at any point depends on the amount that is wasted from a unit area times the total irrigable area up to that point. From the data of an irrigation canal in a particular location, a canal capacity curve can be prepared as shown in Figure 14.1 that indicates the required capacity of the canal to irrigate various sizes of areas. Unless a better estimate of farm waste is available, a standard factor is applied to the canal capacity to determine the farm waste from the irrigated area. The drain is located on a topographic map. For any point on the drain, the total irrigated acreage is determined. The canal capacity for that acreage is read from Figure 14.1. By applying a factor between 15 and 25%, the irrigation waste for drainage design is computed. For example, assume that a topographic map shows an irrigable area of 500 acres at a certain point on the drain in question. From Figure 14.1, the canal capacity is 14 cfs for 500 acres. The irrigation waste will be 15% of this, or 2.1 cfs.

Figure 14.1 Typical canal capacity curve (from U.S. Bureau of Reclamation, 1984).

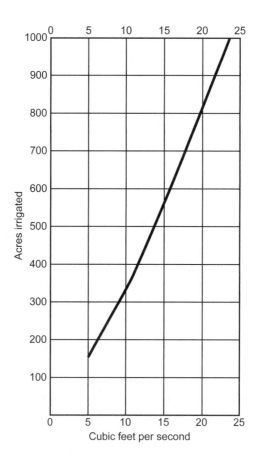

Other Drainage Systems   Chapter 14

14.3 Subsurface Drainage for Agricultural Land

In terms of agricultural requirements, the major objectives of subsurface drainage are (1) to maintain the water table below the plant root zone, which will otherwise rise close to the land surface due to excess infiltrated precipitation in humid regions or application of irrigation water in arid areas, and (2) to leach an adequate quantity of water through the root zone of the plants to keep the salinity from exceeding a specified limit, since irrigation water contains salt that gets deposited in the root zone.

The drainage system consists of either open drains or pipe drains or their combination. Open drains or ditches, with an exposed water surface, are used both for surface drainage and subsurface drainage. They are used as field drains, branches, mains, and intercepting drains. Their main advantages are the ability to carry a large quantity of water and low initial cost, which is, however, partly offset by a high maintenance cost. Their principal disadvantages arise from the loss of the land they occupy, which could otherwise be cultivated, and difficulty in farming operations. The size is determined from the theory of open channel flow by using Manning's formula to carry the subsurface drained flow alone or along with the surface flow, depending on the intended use. The shape, depth of flow, and grade of ditch enter in the design. A semicircle is an efficient section for which the channel properties listed in Table 11.1 up to one-half full depth can be used. A trapezoidal section is also commonly used. The best form for a trapezoidal section is when the top width is equal to twice the length of the sloping sides. Depth of ditches is usually 6 to 12 ft. Ditches should be deep enough to receive the discharge from the drains emptying into them. The slope is determined by the topography of the land. It is very small. A minimum slope should be 0.005%. The maximum grade should not induce velocity more than scour velocity, which ranges from 2 to 6 ft/sec, depending on the soil. Lateral ditches are rarely placed closer than 0.3 mile (0.5 km) apart. A spacing of 0.6 mile (1 km) is satisfactory for the laterals for favorable slopes.

Pipe drains or tube drains are buried beneath the surface of the soil. The modern tendency is in favor of pipe drains. Ditches are used for main and intercepting drains into which the pipe drains empty. Pipe drains are also designed by Manning's formula (open channel flow) for carrying the design flow with just flowing full condition. The smallest size in general use in the United States is 4 in. (100 mm) [plastic tubing of 3 in. (80 mm) has been installed in some places]. The grade for the pipe is set to maintain a velocity of 1 to 1.5 ft/sec when running full to carry the small sediment that enters into the pipe. A minimum grade of 0.15% is recommended for 4-in. pipe and 0.005% for 12-in. or larger pipe. Spacing of 50 to 150 ft is usually adopted. The spacing requirement for a parallel drain system is discussed subsequently. Where very close spacing is required, mole drains are provided, which are the round channels formed by pulling a steel lug (cutting edge) through the subsoil at shallow depth.

Pipe drains can be made of unglazed clay tile, concrete pipe, or corrugated plastic pipe. The pipes are placed with the ends of pipes butted together. Water enters through the space between abutted sections. If the space between pipe sections is 1/8 in. or larger, the joints are covered with a filter material or special joints are used.

Based on the function performed, there are five types of drains: relief, interceptor, collector, suboutlet, and outlet drains. Relief drains are used to lower groundwater over relatively large flat areas where the gradient of both the water table and subsurface strata do not permit sufficient lateral movement of the groundwater (U.S. Bureau of Reclamation, 1984). Interceptor drains cut off or intercept groundwater moving at a steeper slope down the hill. Collector drains receive water from subsurface relief and interceptor drains and from surface drains carrying irrigation surface waste and precipitation surface runoff. These can be open or pipe drains. Suboutlets receive inflow from a number of collector drains and convey it to outlets. They are located in topographic lows such as draws and creeks. Outlet drains take away water from the drainage area. They are usually natural water courses but can be constructed structures.

EXAMPLE 14.1

The subsurface flow into a plastic pipe drain is 0.5 cfs (computation of discharge into subsurface drains is discussed in Section 14.4.2). The drain empties into a semicircular open ditch. In addition to the subsurface flow, the ditch captures the surface flow from an irrigable area of 500 acres. The irrigable waste constitutes 15% of the water applied. The farmland has a slope of 0.05%. The canal capacity curve of Figure 14.1 is applicable. Design the **(a)** pipe drain and **(b)** the intercepting ditch. $n = 0.011$ for pipe and 0.025 for open ditch.

SOLUTION

(a) Pipe Drain

1. Size to maintain a velocity of 1.5 ft/s; $A = Q/v$

$$\text{or } \frac{\pi}{4}d^2 = \frac{0.5 \text{ cfs}}{1.5 \text{ ft/s}} = 0.33 \text{ ft}^2$$

$$d = 0.65 \text{ ft or } 7.8 \text{ in. (use 8 in.)}$$

2. From Manning's eq. (11.10) rearranged,

$$S = \left[\frac{Q \cdot n}{1.486 AR^{2/3}}\right]^2$$

$$A = \frac{\pi}{4}\left(\frac{8}{12}\right)^2 = 0.349 \text{ ft}^2 \text{ and } R = \frac{d}{4} = 0.167 \text{ ft}$$

$$S = \left[\frac{(0.5)(0.011)}{(1.486)(0.349)(0.167)^{2/3}}\right]^2$$

$$= 0.0012 \text{ or } 0.12\% \text{ ok (minimum between .005\% and 0.15\%}$$

(b) Open ditch surface grade $= 0.0005$

3. From Figure 14.1, for 500 acres, capacity $= 14$ cfs
4. Irrigation waste flow $= 0.15(14) = 2.1$ cfs
5. Including pipe drain flow, total flow $= 2.1 + 0.5 = 2.6$ cfs
6. From Manning's eq. (11.10)

$$2.6 = \frac{1.486}{0.025} AR^{2/3} (0.0005)^{1/2}$$

$$AR^{2/3} = 1.956 \text{ ft}^{8/3}$$

7. From Table 11.1, for y/d_0 of 0.5 (half-full), $AR^{2/3}/d_0^{8/3} = 0.1558$, and $A/d_0^2 = 0.3927$

$$d_0^{8/3} = \frac{AR^{2/3}}{0.1558} = \frac{1.956}{0.1558} = 12.55 \text{ ft}^{8/3}$$

$$d_0 = 2.58 \text{ ft (use 2.6 ft)}$$

8. Check for maximum velocity not to exceed 6 ft/s

$$A = 0.3927(2.6)^2 = 2.65 \text{ ft}^2, \text{ since } A/d_0^2 = 0.3927$$

$$v = Q/A = \frac{2.6}{2.65} = 1 \text{ ft/s} \quad \text{OK}$$

14.3.1 Layout of Pipe (Tube) Drainage System

The arrangement of drains is mostly determined by topography. The common types of layout are shown in Figure 14.2 and described briefly below.

1. *Random system.* This system is used where the topography is undulating and drainage is required in isolated areas or in small swales and valleys.

2. *Gridiron system.* The parallel laterals enter the submain from one side. It is suitable for flat land or where the land slopes away on one side.

3. *Herringbone system.* The parallel laterals enter the submain at an angle, usually from both sides. It is suitable for a valley-shaped land where the submain is placed in the depression and better grades for laterals are obtained by angling them upslope.

4. *Interception drains.* Seepage moving down a slope (hill) is collected by drains placed along the toe of the slope.

Figure 14.2 Arrangements of tile (subsurface) drains: (a) random; (b) gridiron; (c) herringbone; (d) intercepting (from Linsley and Franzini, 1979).

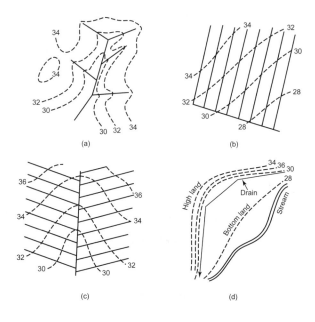

The design of a subsurface drainage system involves the layout of drains and the determination of depth, spacing, and size of drains, together with outlet and appurtenant works. Several layouts and tentative designs are worked out before adopting a final arrangement that is most suitable technically and economically.

14.4 DEPTH AND SPACING OF DRAINS

Methods for estimating the depth and spacing of drains have been developed based on drainage theory, which is essentially the theory of movement of groundwater through a porous medium. As discussed in Chapter 3, the governing partial differential equation of groundwater flow is too complex to be solved for real field conditions. In drainage theory, an idealized soil-water system is considered and a practical judgment is used in the application. The depth and spacing requirements as determined from mathematical analyses are verified from operating systems in similar conditions, if possible. Where wide variations exist between field observations and mathematical solutions, field data should be checked to justify adoption of field-observed values. The simplest drainage theory, developed by the Dutch engineer Hooghoudt in 1940 and still very popular, considers a steady-state condition of a stabilized water table. It applies Darcy's law to derive an expression for the spacing of drains. Kirkham, in 1958, derived the spacing equation based on the steady-state groundwater equation. The U.S. Bureau of Reclamation extended the theory to unsteady-state conditions of falling and rising water table by applying the linearized differential equation of groundwater flow. Some other theories considered the nonlinear form of the groundwater equation. Unsaturated flow theories have also been developed. The extensive research in the theory of drainage is evidenced by the fact that over 60 drain-spacing formulas have been reported in the category of steady- and unsteady-state flows.

The Bureau of Reclamation is a leader in the field of irrigation drainage. The validity of the Bureau's method has been demonstrated from field tests. According to the Bureau, the height of the water table at the midpoint between the drains is given by

$$\frac{y}{y_0} = \frac{192}{\pi^3} \sum_{n=1,3,5,\ldots}^{\infty} (-1)^{(n-1)/2} \frac{n^2 - 8/\pi^2}{n^5} e^{-\pi^2 n^2 \alpha t / L^2} \quad [L] \tag{14.1}$$

where

$\alpha = KD/S$

K = hydraulic conductivity

D = average depth of flow region = $d + y_0/2$

S = specific yield (% by volume) from Figure 14.3

L = drain spacing

y = water-table height above drain at midpoint
 at the end of drain period, t (Figure 14.4)

y_0 = water-table height above drain at midpoint
 at the beginning of drainout

d = depth from drain to barrier (impermeable surface)

The Bureau of Reclamation presented the solution in the form of the curves given in Figure 14.4 in terms of dimensionless parameters, y/y_0 versus KDt/SL^2 for the case where drains are located above a barrier and in terms of y/y_0 versus Ky_0t/SL^2 for drains located on the barrier.

Figure 14.3 General relationship between specific yield and hydraulic conductivity (from U.S. Bureau of Reclamation, 1984).

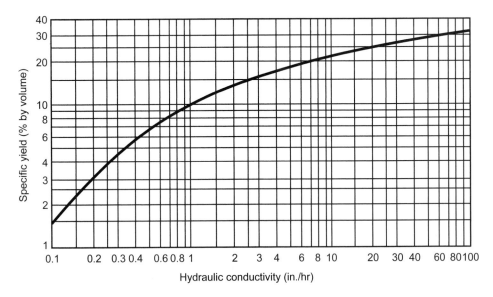

The water level reaches its highest position after the last irrigation (at the end of the peak period of irrigation) or after recharge. The water table recedes during a nonirrigation or slack period. It starts rising again with the beginning of irrigation or recharge. If annual discharge (drainage) from an area is less than annual recharge (from precipitation and irrigation), the water table will progressively rise upward from year to year. When the annual discharge and recharge are about equal, the range of the cyclic annual water table fluctuation becomes reasonably constant. This condition is referred to as *dynamic equilibrium*. The Bureau's method determines the drain spacing that will produce a dynamic equilibrium for a specified water-table depth.

For application of the method, it is necessary that the initial water-table condition be known. When the drains are being planned on an operating project, field measurements will provide information on buildup in the water table due to irrigation/precipitation. For a new project, the amount of deep percolation reaching the drain is determined as a percentage of the net irrigation input of water into the soil. These percentages are given in Table 14.1. The buildup in the water table is computed by dividing the amount of deep percolation by the specific yield of the soil within the zone of fluctuation of the water table, as given in Figure 14.3. In humid and semihumid areas, the infiltration due to rainfall should also be considered. Due to rainfall, the fraction of infiltration going into deep percolation and contributing to the water-table buildup is computed by the same procedure.

14.4.1 Application of Bureau of Reclamation Method

The flow converges toward the drain, resulting in the loss of head. To account for this convergency, the depth from drain to barrier, d, is converted to an equivalent depth by applying the Hooghoudt correction, as given in Figure 14.5 This equivalent depth is used to determine D. This correction is not required when the drains are located on the barrier.

Figure 14.4 Calculation of drain spacing by the transient-flow theory (from U.S. Bureau of Reclamation, 1984).

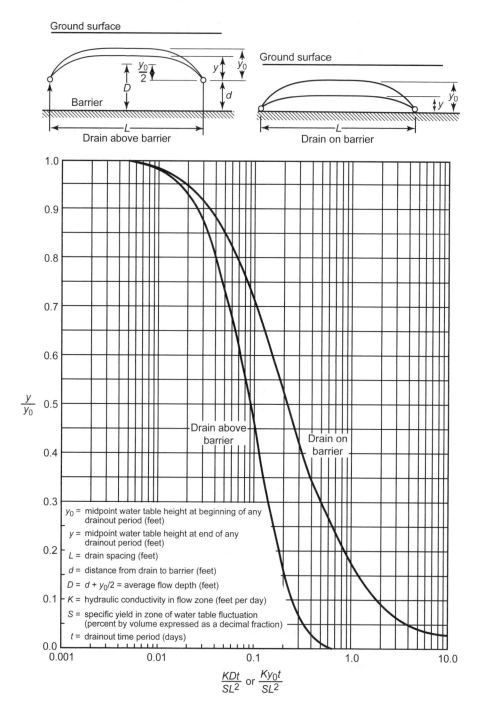

Table 14.1   Deep Percolation of Irrigation Input of Water

By Texture

Texture	Percent	Texture	Percent
LS	30	CL	10
SL	26	SiCL	6
L	22	SC	6
SiL	18	C	6
SCL	14		

By Infiltration Rate

Infiltration Rate (in./hr)	Deep Percolation (%)	Infiltration Rate (in./hr)	Deep Percolation (%)
0.05	3	1.00	20
0.10	5	1.25	22
0.20	8	1.50	24
0.30	10	2.00	28
0.40	12	2.50	31
0.50	14	3.00	33
0.60	16	4.00	37
0.80	18		

Source: U.S. Bureau of Reclamation (1984).

Figure 14.5   Hooghoudt's correction for convergency for drains of radius 0.6 ft (from U.S. Bureau of Reclamation, 1984).

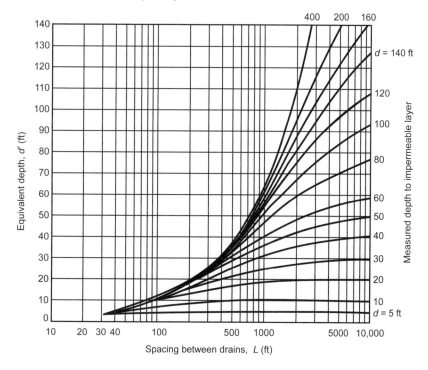

The method involves a trial-and-error procedure as follows:

1. Estimate the initial (maximum) water-table height, y_0, after the last irrigation of the season or recharge.
2. Assume a drain spacing L.
3. Calculate the successive positions of the water table during the nonirrigation (drainout) period.
4. Calculate the buildup and drainout of the water table from each irrigation for the next season.
5. If the water-table height at the end of the series of calculations is not the same as the initial height, y_0, repeat the procedure with a different L until a dynamic equilibrium is achieved.

EXAMPLE 14.2

Determine the drain spacing for the following conditions:

1. Depth from the surface to impervious layer = 30 ft.
2. Depth of the drain from the ground surface = 10 ft.
3. Root zone (water table below surface) requirement = 5 ft.
4. Average hydraulic conductivity = 3.9 in./hr or 7.8 ft/day.
5. The spring snowmelt and irrigation schedule is as follows. The runoff is about 20% and the infiltration rate in the root zone is 1.25 in./hr.

Event	Water Application (in.)	Date	Time between Events (days)
Snowmelt	6	April 15	
			47
First irrigation	5	June 1	
			30
Second irrigation	7	July 1	
			29
Third irrigation	7.1	July 30	
			29
Fourth irrigation	7	August 28	
		Total	135

SOLUTION

1. Maximum allowable water-table height above the drain, $y_0 = 10 - 5 = 5$ ft.
2. For hydraulic conductivity of 3.9 in./hr, from Figure 14.3, $S = 17\%$.
3. Assume that the drainage spacing, $L = 1600$ ft.
4. Drain to barrier depth = 20 ft. The corrected depth d from Figure 14.5 = 18.5 ft.
5. For an infiltration rate of 1.25 in./hr, the deep percolation from Table 14.1 is 22%.
6. Water-table buildup from the snowmelt and each irrigation is computed in Table 14.2.

Table 14.2 Water-Table Buildup

(1)	(2)	(3)	(4)	(5)	(6)	(7)
				Deep	Water Table Increment	
Event	Water Applied, in.	Runoff, in. (0.2 × col.2)	Net Input, in. (col.2 – col. 3)	Percolation, in. (0.22 × col. 4)	(in.) (col. 5/S)	(ft)
Snowmelt	6	1.20	4.80	1.06	6.24	0.52
First irrigation	5	1.00	4.00	0.88	5.18	0.43
Second irrigation	7	1.40	5.60	1.23	7.24	0.60
Third irrigation	7.1	1.42	5.68	1.25	7.35	0.61
Fourth irrigation	7	1.40	5.60	1.23	7.24	0.60

7. The starting point of the computation is the end of the last irrigation season when the water table is at its maximum allowable height of 5 ft above the drain. Then during the non-irrigation period of 230 days (365 – 135), the water table recedes. It builds up again with recharge/irrigation as per the schedule indicated. If the assumed spacing is correct, the water table should again rise to 5 ft after application of the last irrigation.

8. The nonirrigation period is divided into two periods of 115 days each.

9. The computations are arranged in Table 14.3.

14.4.2 Design Discharge for Determining Subsurface Drain Pipe Size

Discharge into drains takes place due to (1) deep percolation of water from the surface, and (2) underground flow from upslope irrigated areas, and from canals, streams, and other water bodies. Thus

$$q = q_p + q_u \quad [\text{L}^2\text{T}^{-1}] \tag{14.2}$$

where

q = total flow per unit length of drain

q_p = flow due to deep percolation

q_u = flow from upslope sources

Flow of parallel spaced drains can be computed using the following formulas:

$$q_p = \frac{2\pi K y_0 D}{L} \quad \text{(for drains above a barrier)} \quad [\text{L}^2\text{T}^{-1}] \tag{14.3}$$

$$q_p = \frac{4 K y_0^2}{L} \quad \text{(for drains on a barrier)} \quad [\text{L}^2\text{T}^{-1}] \tag{14.4}$$

All terms are as defined previously.

Subsurface flow from upslope is given by Darcy's law for the saturated portion above the drain. Hence

$$q_u = K i A \frac{y_0}{y_0 + d} \quad [\text{L}^2\text{T}^{-1}] \tag{14.5}$$

Table 14.3 Bureau of Reclamation Method

(1)	(2)	(3)	(4)[a]	(5)	(6)[b]	(7)[c]	(8)[d]
Event	Period, t days	Water-Table Buildup (ft) (Table 14.2)	Initial Height, y_0 (ft)	$D = d' + \dfrac{y_0}{2}$ (ft)	$\dfrac{kDt}{SL^2}$	$\dfrac{y}{y_0}$ (Fig. 14.4)	Height after the Period, y (ft)
Last season							5.00
Nonirrigation 1	115		5.00	21.00	0.043	0.74	3.70
Nonirrigation 2	115		3.70	20.35	0.042	0.75	2.78
Snowmelt		0.52					
	47		3.30	20.15	0.017	0.93	3.07
First irrigation		0.43					
	30		3.50	20.25	0.011	0.97	3.40
Second irrigation		0.60					
	29		4.00	20.50	0.011	0.97	3.88
Third irrigation		0.61					
	29		4.49	20.75	0.011	0.97	4.36
Fourth irrigation		0.60					
			4.96				

[a] Col. 4 = col. 8 of preceding period plus col. 3 during the period.
[b] Col. 6 = $K/SL^2 \times$ col. 2 $\times$ col. 5.
[c] Col. 7 = From Figure 14.4, corresponding to the value in col. 6.
[d] Col. 8 = col. 7 $\times$ col. 4.

where

i = slope of water table obtained from a water-table contour map along a line normal to the contours

A = saturated area along the plane parallel to the contours or normal to the direction of flow, for a unit length of drain

The total design discharge is obtained from $Q = qX$, where a pipe is X units long. The formula is applied for a length X which serves an area that can be irrigated probably within about 2 days (U.S. Bureau of Reclamation, 1984). Once the discharge is determined, the size of parallel subsurface drains is determined by the Manning equation according to the procedure of Example 14.1.

Flow q in the preceding equations is the maximum rate of discharge. For a collector drain receiving water from a number of drains, each branch will not deliver at the maximum rate at the same time. The Bureau of Reclamation has suggested that the following equations will provide a reasonable design capacity for most collector drains:

$$Q = C\frac{2\pi K y_0 D}{L}\left(\frac{A}{L}\right) \quad \text{(drains above barrier)} \quad [\mathrm{L}^3\mathrm{T}^{-1}] \tag{14.6}$$

$$Q = C\frac{4K y_0^2}{L}\left(\frac{A}{L}\right) \quad \text{(drains on barrier)} \quad\quad [\mathrm{L}^3\mathrm{T}^{-1}] \tag{14.7}$$

where

A = area drained

C = area discharge factor given in Table 14.4

Table 14.4 Area Discharge Factors

Area Drained (acres)	Factor C
0–40	1.0
40–160	1.0–0.82
160–320	0.82–0.72
320–640	0.72–0.60
640–960	0.60–0.54
960–1280	0.54–0.50
1280–5000	0.50

Source: U.S. Bureau of Reclamation (1984).

EXAMPLE 14.3

Determine the peak discharge for parallel drains of Example 14.2 for the maximum water table. The length of pipe is 6075 ft. The average ground surface gradient is 1% and the water table is generally parallel to the ground surface. Also design the drain. $n = 0.012$.

SOLUTION

1. From Example 14.2: L = 1600 ft., max $y_0 = 5$ ft, $d' = 18.5$ ft, $K = 7.8$ ft/day

$$D = d' + y_0 = 18.5 + \frac{5}{2} = 21.0$$

2. From eq. (14.3),

$$q_p = \frac{2\pi(7.8)(5)(21)}{1600}$$

$$= 3.21 \text{ ft}^2/\text{day or } 3.72 \times 10^{-5} \text{ cfs/ft}$$

3. $A = (d + y_0)1 = (20 + 5)1 = 25 \text{ ft}^2/\text{ft}$

$i = 1\%$ or 0.01 (given)

From eq. (14.5),

$$q_u = (7.8)(0.01)(25)\left(\frac{5}{5+20}\right)$$

$$= 0.39 \text{ ft}^2/\text{day or } 0.45 \times 10^{-5} \text{ cfs/ft}$$

4. $q = 3.72 \times 10^{-5} + 0.45 \times 10^{-5} = 4.17 \times 10^{-5} \text{ cfs/ft}$

$Q = (4.17 \times 10^{-5})6075 = 0.25 \text{ cfs}$

5. Size for a velocity of 1.5 ft/s:

$$\frac{\pi}{4}d^2 = Q/v = \frac{0.25}{1.5} = 0.17 \text{ ft}^2$$

or

$$d = 0.456 \text{ ft or } 5.58 \text{ in. (use 6 in.)}$$

6. $A = \dfrac{\pi}{4}(0.5)^2 = 0.196 \text{ ft}^2$, $R = \dfrac{d}{4} = \dfrac{0.5}{4} = 0.125 \text{ ft}$

From the Manning equation,

$$S = \left[\frac{(0.25)(0.012)}{(1.486)(0.196)(0.125)^{2/3}} \right]^2$$

$$= 0.0017 \text{ or } 0.17\% \text{ ok}$$

14.5 ROADWAY DRAINAGE SYSTEMS

Roadways occupy a narrow strip of land but stretch lengthwise through many watersheds of different characteristics. Two different types of drainage problems are associated with roadways:

1. It is necessary to take away precipitation falling on the road surface and to divert storm-water approaching the road. The facility alongside the road or the longitudinal system takes care of this.

2. A roadway crosses many natural water courses and channels in valley areas. Water carried by these channels has to be conveyed across the road smoothly. Cross-drainage works comprising culverts and bridges are provided for this purpose.

14.6 LONGITUDINAL DRAINAGE SYSTEMS

Roads are designed with a crown in the center and cross slopes in both directions away from the centerline. On rural roads, water falling on roads flows laterally off the road surface into the countryside or into shoulder drains. On city streets and urban highways, water falling on or near pavements and sidewalks is directed by the cross slopes to the gutters formed between the edge of the road surface and the vertical curb. It flows along gutters to curb or gutter inlets and from them into underground storm drains. Thus the longitudinal system consists of (1) collector structures such as gutters, gutter inlets, curb opening inlets, grate for inlets, and so on, and (2) underground drains (conduits) that conduct water to the out-fall. A detailed treatment of the design of collector structures is presented by the U.S. Department of Transportation (1979). The design of drains is discussed here.

The drainage of city streets is mostly a part of the city storm sewer system, which is based on a 5- or 10-year frequency. Highways use high drainage standards since they need urgent water removal from high-traffic pavements; the main storm drains of freeways are based on 50- to 100-year storms. For highway sections that traverse through city areas, a coordination of local drainage authority and highway authority is required since they utilize the local drainage facilities, especially the outfall facilities. For a highway section at grade, the primary need is to provide an adequate number of suitably located inlets for rapid removal of water into the existing urban drainage system facilities. When the section is in a cut, there will be sumps or low points at which excess runoff will collect and pumping will be needed. On an elevated highway section, connection to the existing facility is easier.

Highways traversing outside city areas need separate drainage systems. The provision of detention storage to handle runoff from highways is considered part of a stormwater management plan. The acquisition of suitable sites and the cooperation of other agencies are required for such storage facilities.

14.6.1 Design Flows for Longitudinal Drainage

Drains are engineered to carry the design flows by the procedure of storm drains of Section 13.13. The design flows for longitudinal drainage components—gutters, inlets, spillways, underground drains—are determined by the rational method (Section 13.10). For a direct application to roadways, the rational formula is expressed as follows:

$$q = C_f CiL \quad [L^2 T^{-1}] \tag{14.8}$$

where

q = peak flow per unit length of pavement

C_f = frequency correction factor (Section 13.10.1)

C = runoff coefficient (Section 13.10.2)

i = rainfall intensity of the design frequency for the time of transverse flow across the pavement

L = length of overland flow normal to contours

Like the original rational equation, eq. (14.8) is dimensionally homogeneous (i.e., if i is in ft/sec and L is in ft, then q is in cfs/ft). Usually, the unit for i of in./hr is used; then eq. (14.8) is divided by 43,200 to get q in cfs/ft.

The length of overland flow, L, is approximated by the formula

$$L = W\sqrt{r^2 + 1}/r \quad [L] \tag{14.9}$$

where

W = roadway width from the center (one-half of total width)

r = ratio of cross slope to longitudinal slope

The time of overland flow for ascertaining the rainfall intensity is determined by the methods of Section 13.10.5, commonly by the Izzard method, eq. (13.6).

For short distances and steep slopes, the total discharge at an inlet is taken to be q multiplied by the length between inlets. For long distances and flat slopes, the discharge at an inlet is determined either by the routing procedure to provide allowance for the channel storage, or the original rational eq. (13.2) is used, incorporating the entire area tributary up to the point of the inlet. The original rational equation is also used when many interconnecting drains are involved and the travel time through them has to be considered. The U.S. Department of Transportation (DOT, 1979) has designed underground drains to run under pressure (applying the pipe flow formula). However, the drains are commonly designed as a nonpressure system by the method of Section 13.13.

DOT further recommends that any pipe wholly or partly under a roadbed should have a minimum diameter of 18 in. Elsewhere, it should be a minimum of 15 in. in diameter. Erosion control of the slopes of highway embankments requires serious consideration. An easy grade for the slope, sod and grass covers, and intercepting dikes or ditches are some erosion-control measures discussed in Section 14.8 in the context of airport drainage.

A subsurface (flow) drainage system to remove the infiltrated water and to lower the high water table from all important highway pavement structures is an essential part of a highway design. The requirements and guidelines for the design of subsurface drainage for highways are given by DOT (1973).

EXAMPLE 14.4

A 30-ft-wide road section has a longitudinal slope of 0.013. It has a cross slope of 1/4 in./ft. Determine the peak flow at the gutter inlet if the spacing of inlets is 150 ft. The 10-year rainfall intensity in in./hr is given by $i = 170/(t + 23)$, when t is in minutes.

SOLUTION

1. Using eq. (14.9),

$$\text{cross slope} = \frac{1/4}{12} = 0.0208$$

$$r = \frac{0.0208}{0.013} = 1.60$$

$$L = \frac{15\left[(1.6)^2 + 1\right]^{1/2}}{1.6} = 17.7 \text{ ft}$$

2. To determine the time of overflow by the Izzard method:

a. Assume that $t_c = 5$ min.

b. $i = \dfrac{170}{5 + 23} = 6.07$ in./hr

c. Using eq. (13.6),

$$L = 17.7 \text{ ft}$$

$$S = \frac{0.25}{12} = 0.0208$$

$$iL = (6.07)(17.7) = 107.4 < 500 \text{ ok}$$

$$t_i = \frac{41.025\left[(.0007)(6.07) + 0.017\right](17.7)^{0.33}}{(0.0208)^{0.333}(6.07)^{0.667}}$$

$$= 2.45 \text{ min}$$

d. Repeat with $t = 2.4$ min

$$i = \frac{170}{2.4 + 23} = 6.69 \text{ in./hr}$$

$$t_i = \frac{41.025\left[(.0007)(6.69) + 0.017\right](17.7)^{0.33}}{(0.0208)^{0.333}(6.69)^{0.667}}$$

$$= 2.35 \approx 2.4 \text{ ok}$$

3. Using eq. (14.8),

$$q = \frac{(1)(0.8)(6.69)}{43,200}(17.7) = 0.0022 \text{ cfs/ft}$$

4. $Q = 0.0022(150) = 0.33$ cfs.

14.7 Cross-Drainage Systems: Culverts

Bridges and culverts are two cross-drainage works that pass stream channels under roadways. The hydraulics of bridge openings have been discussed in Section 10.8.2. The distinction between a bridge and a culvert on the basis of size is arbitrary, with a structure whose span is in excess of 20 ft being classed as a bridge. However, a distinctive feature is that culverts can be designed to flow with a submerged inlet. A culvert acts as a control structure. In the hydraulic sense, a device is said to control flow if it limits the flow of water which would otherwise be exceeded under existing upstream and downstream conditions. In a control device, the head adjustment across the control section takes place until a balance is achieved between the inflow and the discharge through the section. In the case of culverts, a difficulty arises because the control section can be at the inlet or at the outlet, depending on the type of flow. In supercritical flow the flow velocity is faster than the velocity of a wave, so that the water waves cannot travel upstream, and hence control cannot be exercised from downstream (i.e., there is inlet control). In the subcritical flow condition, control from downstream will back up water until an equilibrium profile is achieved upstream of the control (i.e., outlet control exists).

Inlet control means that conditions at the entrance—depth of headwater and entrance geometry—control the capacity of the culvert. An orifice type of flow takes place at the entrance. A culvert runs part full (atmospheric pressure). Thus the barrel size beyond the inlet can be reduced without affecting the discharge, or the capacity can be increased by improving the inlet conditions. The detailed design of improved inlets has been discussed by DOT (1972). These improvements comprise provision of wingwalls; beveling or rounding of culvert edges; tapering the sides of the inlet, including slope tapering; and providing a drop inlet. The geometry of the top and sides of the inlet is important, but not as important as that of the culvert floor. The inlet geometry and channel contraction affect the coefficient of discharge as discussed by Bodhaine (1982).

In outlet control, the culvert can flow full or part full, depending on headwater and tailwater levels. The friction head in the barrel of a culvert affects the headwater or the total energy to pass the discharge through the culvert.

Discharge through culverts depends not only on the type of control but on different types of flow under each control. A general classification of flow through culverts is shown in Table 14.5, separated into two groups: unsubmerged and submerged flow. For submerged flow, the headwater-to-barrel diameter ratio should exceed approximately 1.2. The features of each type of flow with respect to culvert slope, flow depth, and control section are indicated in the table. The first three types relate to unsubmerged flow, with the first one relating to inlet control conditions. The other three types in the table relate to submerged culvert flow. The discharge equations given in the table for each type disregard entrance losses.

Table 14.5 Classification of Culvert Flow

(1) Category	(2) Type	(3) Culvert Slope	(4) Flow	(5) Control Section	(6) Discharge, Q	(7) Eq. Number	(8) Illustration
Unsubmerged, $H/D \leq 1.2$	1	Steep	Part full	Inlet	$C_d A_c \sqrt{2g\left(H + V_1^2/2g - dc - h_{1,2}\right)}$	(14.10)	
	2	Mild	Part full	Outlet	$C_d A_c \sqrt{2g\left(H + z + V_1^2/2g - dc - h_{1,2} - h_{2,3}\right)}$	(14.11)	
	3	Mild	Part full	Outlet	$C_d A_3 \sqrt{2g\left(H + z + V_1^2/2g - h_3 - h_{1,2} - h_{2,3}\right)}$	(14.12)	

Table 14.5 Classification of Culvert Flow (Continued)

(1)	(2)	(3)	(4)	(5)	(6)	(7)	(8)
Category	Type	Culvert Slope	Flow	Control Section	Discharge, Q	Eq. Number	Illustration
Submerged, $H/D > 1.2$	4	Any	Full	Outlet	$C_d A_0 \sqrt{\dfrac{2g(H + z - h_4)}{1 + 2\alpha C_d^2 n^2 L / R_0^{4/3}}}$	(14.13)	
	5	Any	Full	Outlet	$C_d A_0 \sqrt{\dfrac{2g(H + z - D)}{1 + \alpha C_d^2 n^2 L / R_0^{4/3}}}$	(14.14)	
	6	Any	Full	Intlet	$C_d A_0 \sqrt{2gH}$	(14.15)	

$\alpha = 29$ for FPS units and 19.6 for metric units.
A_c = area of section of flow at critical depth.
A_0 = area of culvert barrel.
A_3 = area of section of flow at exit end of culvert.
R_0 = hydraulic radius of culvert barrel.
V_1 = mean velocity in the approach section.
Other variables are shown on figures.

14.7.1 Design of Culverts

Some box culverts are designed such that their top forms the base of the roadway. These are unsubmerged culverts that belong to types 1, 2, and 3. For a trial selected size, the type of flow can be determined as follows:

1. For the design flow, determine the critical depth, d_c (Section 11.5.2), and the normal depth, d_n (Section 11.6.2).

2. Compare the depths above with the tailwater, h_4. When:

$d_n < d_c$ and $h_4 < d_c$	type 1 flow	critical depth at inlet
$d_n > d_c$ and $h_4 < d_c$	type 2 flow	critical depth at outlet
$d_n > d_c$ and $h_4 > d_c$	type 3 flow	subcritical throughout

   The appropriate discharge equation of Table 14.5 is used to confirm the size and type. If not adequate, then guided by the computed size, the trial may be repeated. In eqs. (14.10) through (14.15) in Table 14.5, the friction head loss between indicated sections is determined by Manning's equation, arranged as follows:

$$h_{ab} = \frac{n^2 L V^2}{2.22 R^{4/3}} \quad [\text{L}] \quad \text{(English units)} \tag{14.16a}$$

or

$$h_{ab} = \frac{n^2 L V^2}{R^{4/3}} \quad [\text{L}] \quad \text{(metric units)} \tag{14.16b}$$

where

   h_{ab} = friction head loss between two points a and b
   V = velocity of flow
   L = length of section ab
   R = hydraulic radius, A/P

The majority of culverts are designed for submerged conditions (types 4, 5, and 6), since the entrance is submerged at least with the peak rate of flow.

When a culvert is submerged by both headwaters and tailwaters, it is a type 4 condition in which eq. (14.13) is applicable. However, the distinction between types 5 and 6 when the tailwater is low is not as obvious.

To classify type 5 or 6 flow, the curves of Figures 14.6 and 14.7, which are adapted from Bodhaine (1982), are used. Figure 14.6 is applicable to a concrete barrel box or pipe culverts of square, rounded, or beveled entrances with or without wingwalls. Figure 14.7 is for rough (corrugated) pipes of circular or arch sections mounted flush in a vertical headwall with or without wingwalls. The procedure to classify type 5 or 6 flow is as follows:

1. Compute the ratios L/D, r/D or w/D, S_0, and (for rough pipes), $\alpha n^2 H / R_0^{4/3}$ where r is the radius of rounding, w is the effective bevel, and $\alpha = 29$ for FPS units and 19.6 for metric units, shown in Figures 14.6 and 14.7.

2. For concrete pipes, select the curve of Figure 14.6 corresponding to r/D or w/D for the culvert.

Figure 14.6 Criterion for classifying type 5 and type 6 flow in box or pipe culverts with concrete barrels and square, rounded, or beveled entrances, either with or without wingwalls (from Bodhaine, 1982).

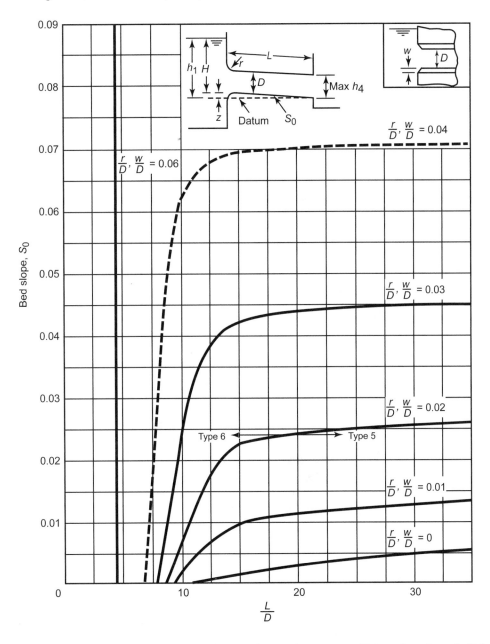

3. For rough pipes, select from Figure 14.7 the graph corresponding to the value of r/D for the culvert and then select the curve corresponding to the $\alpha n^2 H / R_0^{4/3}$ computed for the culvert.

4. Plot the point defined by the computed values of S_0 and L/D for the culvert.

Figure 14.7 Criterion for classifying type 5 and type 6 flow in pipe culverts with rough barrels (from Bodhaine, 1982).

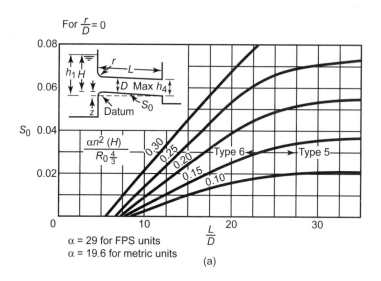

For $\frac{r}{D} = 0$

$\alpha = 29$ for FPS units
$\alpha = 19.6$ for metric units

(a)

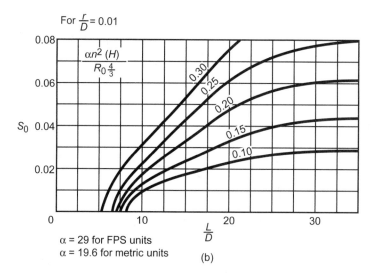

For $\frac{r}{D} = 0.01$

$\alpha = 29$ for FPS units
$\alpha = 19.6$ for metric units

(b)

5. If the point plots to the right of the curve in step 2 or 3, the flow is type 5. If it plots to the left, the flow is type 6.

As in the case of bridge openings, the coefficient of discharge, C_d, is a function of many variables relating to type of flow, degree of channel contraction, and the geometry of the culvert entrance. The coefficient varies from 0.4 to 0.98. A systematic presentation has been made by Bodhaine (1982).

Other Drainage Systems   Chapter 14

Figure 14.7 (Continued) Criterion for classifying type 5 and type 6 flow in pipe culverts with rough barrels (from Bodhaine, 1982).

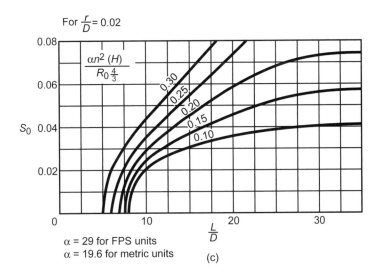

For $\frac{r}{D} = 0.02$

$$\frac{\alpha n^2 (H)}{R_0^{\frac{4}{3}}}$$

$\alpha = 29$ for FPS units
$\alpha = 19.6$ for metric units

(c)

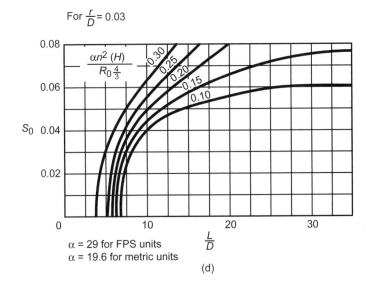

For $\frac{r}{D} = 0.03$

$$\frac{\alpha n^2 (H)}{R_0^{\frac{4}{3}}}$$

$\alpha = 29$ for FPS units
$\alpha = 19.6$ for metric units

(d)

EXAMPLE 14.5

A culvert section is shown in Figure 14.8 with upstream and downstream water levels. Design the culvert for a peak discharge of 120 cfs. For the culvert section, corrugated metal pipe ($n = 0.024$) is to be used without rounding. $C_d = 0.5$.

SOLUTION

1. Consider a pipe section 4 ft in diameter.
2. $H/D = 6/4 = 1.5 > 1.2$; submerged flow

Figure 14.8 Submerged culvert section of Example 14.5.

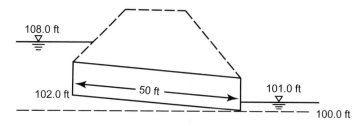

3. Since not submerged by tailwaters, it is either type 5 or 6.

4. Thus,

$$r/D = 0, \quad S_0 = \left(\frac{102.0 - 100.0}{50} \right) = 0.04, \quad L/D = 50/4 = 12.5$$

$$R_0 = D/4 = \frac{4}{4} = 1, \quad \frac{29n^2 H}{R_0^{4/3}} = \frac{29(0.024)^2 (6)}{(1)^{4/3}} = 0.10$$

From Figure 14.7(a), flow is type 6, as the point plots to the left of curve of 0.10.

5. Apply eq. (14.15):

$$A_0 = \frac{\pi}{4}(4)^2 = 12.56 \text{ ft}^2$$

$$H = 108 - 102 = 6 \text{ ft}$$

$$Q = 0.5(12.56)\sqrt{2(32.2)(6)}$$

$$= 123 \text{ cfs} \approx 120 \text{ cfs ok}$$

EXAMPLE 14.6

Design a box culvert of concrete section ($n = 0.015$) to carry a discharge of 520 cfs for the condition shown in Figure 14.9. It has a square-edged entrance. The approach stream has a rectangular section of width 40 ft. $C_d = 0.93$.

Figure 14.9 Unsubmerged box culvert section.

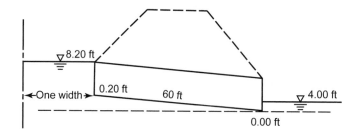

SOLUTION

1. Assume an 8-ft square section. The approach section will be one width (8 ft) upstream.

2. $H/D = 8/8 = 1.0 < 1.2$; unsubmerged case.

3. Determine the critical depth. From eq. (11.7),

$$Z_c = \frac{Q}{\sqrt{g}} = 520/\sqrt{32.3} = 91.64 \text{ ft}$$

Since

$$A_c = 8d_c$$

$$Z_c = A_c^{3/2}/T^{1/2} = \left(8d_c\right)^{3/2}/(8)^{1/2} = 8.0d_c^{3/2}$$

$$8.0d_c^{3/2} = 91.64 \text{ or } d_c = 5.09 \text{ ft}$$

4. Determine the normal depth. From eq. (11.10a),

$$AR^{2/3} = \frac{Qn}{1.49S^{1/2}} = \frac{520(0.015)}{1.49(0.0033)^{1/2}} = 91.13$$

But

$$AR^{2/3} = (8d_n)^{5/3}/(8+2d_n)^{2/3}$$

Hence

$$\frac{\left(8d_n\right)^{5/3}}{\left(8+2d_n\right)^{2/3}} = 91.13$$

Solving either by trial and error or by plotting d versus $AR^{2/3}$, $d_n = 6.3$ ft.

5. Since $d_n > d_c$ but $h_4 < d_c$ it is type 2 flow.

6. For velocity of approach,

$$V_1 = \frac{Q}{A} = \frac{520}{8.2(40)} = 1.59 \text{ ft/sec}$$

$$\frac{V_1^2}{2g} = \frac{(1.59)^2}{2(32.2)} = 0.04 \text{ ft}$$

7. For head losses between 1 and 2,

$$h_{1,2} = \frac{n^2 L V^2}{2.22R_1^{4/3}}$$

$$R_1 = \frac{40(8.2)}{40+2(8.2)} = 5.82 \text{ ft}$$

$$h_{1,2} = \frac{(0.013)^2 (8)(1.59)^2}{2.22(5.82)^{4/3}} = \text{ negligible}$$

8. For head losses between 2 and 3,

$$A_2 = 8(6.3) = 50.4 \text{ ft}^2$$

$$R_2 = \frac{50.4}{8 + 2(6.3)} = 2.45 \text{ ft}$$

$$V_2 = \frac{Q}{A_2} = \frac{520}{50.4} = 10.32 \text{ ft/sec}$$

$$h_{2-3} = \frac{(0.015)^2 (60)(10.32)^2}{2.22(2.45)^{4/3}} = 0.20 \text{ ft}$$

9. From eq. (14.11), $A_c = 8(5.09) = 40.72 \text{ ft}^2$

$$Q = C_d A_c \sqrt{2g\left(H + z + \frac{V_1^2}{2g} - d_c - h_{1-2} - h_{2-3}\right)}$$

$$= 0.93(40.72)\sqrt{2(32.2)(8 + 0.2 + 0.04 - 5.09 - 0 - 0.20)}$$

$$= 522 \approx 520 \text{ cfs ok}$$

14.8 Airport Drainage Systems

The objectives of airport drainage systems are (1) to collect and drain surface water runoff, (2) to remove excess groundwater and lower the water table where it is too high, and (3) to protect all slopes from erosion.

The first objective is met by (1) properly grading the airport area so that all shoulders and slopes drain away from runways, taxiways, and all paved areas; (2) providing a field storm drainage system serving all the depressed areas; and (3) constructing peripheral and other ditches to convey the outfall from the drainage system, to collect surface flows from the airport and adjoining sites, and to intercept groundwater flow from higher adjacent areas. Proper coordination of grading and draining is most desirable since a drainage system cannot function effectively unless the area is graded correctly to divert the surface flow into the drainage system. Similarly, ditches form an integral part of the drainage system.

Subsurface drainage is provided to take care of the second objective of diverting subterranean flows, lowering the water table, and controlling the moisture in the base and subbase of the pavements. Intercepting ditches or intercepting drainlines are provided to collect flows through the porous water-bearing stratum. For draining off the moisture pocketed in pervious soils over an impervious stratum or in the low-lying areas of an undulating impervious stratum, the subsurface drains are placed within wet masses of soil. It is desirable to place the best drainable soils adjacent to and beneath the paved areas to provide drainage away from the pavement. Less-drainable soils are placed in nontraffic areas. The draining of large areas through subsurface drainage systems is usually not required on airports since it can be done more efficiently by grading properly and installing surface drainage (Federal Aviation Agency, 1965).

Cut-and-fill slopes address the problem of erosion. As a first step of protection, these slopes are made as flat as possible. Deep-cut slopes of over 10 ft, with higher ground above them, are provided with a cutoff ditch running back to the top-of-cut line and set back a few feet from the top of the bank to intercept the water flowing down from the higher ground. A ditch is constructed at the base of the bank to collect runoff. The cut slopes are protected by riprap, sod, grass, or vegetation. The fill slopes above 5 ft high are protected by constructing beams and gutters along the top of the slope to prevent water from running down the slope.

Only the surface storm drainage is discussed here. The design starts with a comprehensive study of the topography of the site and surrounding areas to identify surface and subsurface direction of flow, natural water courses, and outfalls. The topography affects the layout of the runways, taxiways, aprons, and buildings. The outline of the boundary of the airport is superimposed on the map. A plan is prepared from the topographic map, showing the contours of the finished grade and the location of such features as runways, taxiways, aprons, buildings, and roads. This is known as the *drainage working drawing*.

On the plan, the entire surface drainage system is sketched, showing all laterals, submains, and main storm sewers; direction of flow; gradients; and identifying each subarea, catch basin, inlet, gutter, shallow channel, manhole, and peripheral and outfall ditches.

The layout should cover all depressed areas in which overland flow will accumulate. Inlet structures are located at the lowest points within each field area. Each inlet is connected to the drainage line. The pipelines lead to the major outfalls.

Once a layout of inlets, manholes, and storm pipes has been made, determination of the area contributing to each inlet, tabulation of data, and computations of peak flows and drain capacities proceed in exactly the manner described in Section 13.13 for urban storm sewer design. This is illustrated in Example 14.7. Several different drainage layouts are necessary to select the most economic and effective system.

The rate of outflow from a drainage area is controlled by the capacity of the drainpipe. Whenever the rate of runoff to an inlet exceeds the drain capacity, ponding or temporary storage occurs. Where considerable low-lying flat field areas exist away from the pavements, the desirability of using a ponding facility should be considered. This will reduce the size and/or number of drains. Also, this will act as a safety factor in the case of heavier-than-design storms. The volume of storage in ponding is determined by the method of Section 13.14.

EXAMPLE 14.7

A surface (storm) drainage system for a part of an airport is shown in Figure 14.10. The finished contours, drainage layout, and length and slope of drains are marked on the figure. The computed tributary area, the composite runoff coefficient, and the time of overland flow to each intercept are given in Table 14.6. The 5-year rainfall intensity in in./hr is given by $190/(t + 25)$, where t is in minutes. Manning's coefficient $n = 0.015$. Design the drainage system.

SOLUTION

1. The design is performed in exactly the same manner as for the storm design in Section 13.13.

2. Computations for peak flows by the rational method and size of drains are arranged in Tables 14.7 and 14.8, respectively.

Figure 14.10 Section of an airport.

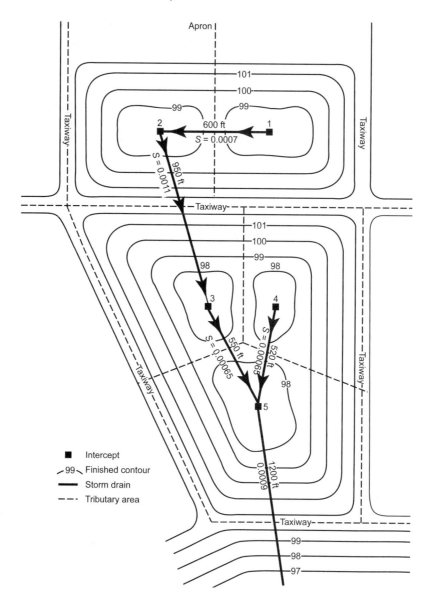

Table 14.6   Drainage Data for Example 14.7

Intercept	Tributary Area (acres)	Weighted Runoff Coefficient	Time of Overland Flow (min)
1	14.2	0.65	25.0
2	16.3	0.65	28.0
3	20.7	0.35	35.0
4	13.5	0.35	35.0
5	25.0	0.35	40.0

Table 14.7 Computation of Peak Discharge

(1)	(2)	(3)	(4)	(5)	(6)	(7)	(8)	(9)	(10)	(11)	(12)
							Travel Time (min)				
Intercept	Location	Tributary Area, a (acres)	Coefficient, C	aC (acres)	ΣaC (acres)	Route	Overland Flow	In Sewer[a]	Total	Intensity (in./hr)	Q (cfs)
1		14.2	0.65	9.23	9.23	TA-1	25	0	25	3.8	35.1
2		16.3	0.65	10.60	19.83[b]	TA-2	28	0	28		
						1–2	25	3.18	<u>28.18</u>	3.57	70.8
3		20.7	0.35	7.25	27.08	TA-3	35	0	<u>35</u>	3.17	85.8
						2–3	28.18	4.14	32.32		
4		13.5	0.35	4.73	4.73	TA-4	35	0	35	3.17	15.0
5		25.0	0.35	8.75	40.56[c]	TA-5	40	0	<u>40</u>	2.92	118.4
						4–5	35	2.82	37.82		
						3–5	35	2.85	37.85		

[a] From column 14 of Table 14.8.

[b] Col. 5 for intercept 2 + col. 6 for intercept 1 via route 1–2.

[c] Col. 5 for intercept 5 (TA–5) + (col. 6 for intercept 3 via route 3–5) + (col. 6 for intercept 4 via route 4–5) = 8.75 + 27.08 + 4.73 = 40.56.

Table 14.8 Storm Sewer Design Computations

(1)	(2)	(3)	(4)	(5)	(6)	(7)	(8)	(9)	(10)	(11)	(12)	(13)	(14)
	Drain Line		Design Flow (cfs)	Length of Sewer (ft)	Surface Elevation		Street Slope	Maximum Diameter for Velocity of 3 ft/s[a] (in.)	Diameter for Street Grade[b] (in.)	Diameter (in.)	Design Parameters		
Line	From	To			Upstream	Downstream					Sewer[c] Grade	Velocity at Full[d] (ft/sec)	Travel Time (min)[e]
1	1	2	35.1	600			0.0007	46	49	45	0.0011	3.14	3.18
2	2	3	70.8	950			0.0011	66	58	60	0.0011	3.82	4.14
3	3	5	85.8	550			0.00065	72	69	69	0.00065	3.22	2.85
4	4	5	15.0	520			0.00065	30	36	30	0.0018	3.07	2.82
5	5	outlet	118.4	1200			0.0009	85	74	75	0.0008	3.78	5.29

[a] $D=(1.274Q/v)^{1/2}$ ft$\times 12\frac{in.}{ft}$ (continuity eq.)

[b] $D=\left(\frac{2.155nQ}{S^{1/2}}\right)^{0.375}$ ft$\times 12\frac{in.}{ft}$ (Manning's eq.)

[c] Recompute S from Manning's eq. $S=\left[\frac{2.155nQ}{D^{8/3}}\right]^2$, D in ft

[d] $v=\frac{0.59}{n}D^{2/3}S^{1/2}$ (Manning's eq.), D in ft

[e] Col. 5/col. 13 $\left(\frac{1}{60}\right)$

14.9 COMPUTER APPLICATIONS FOR DRAINAGE

The Hydrology Web of the Pacific Northwest National Laboratory of the U. S. Department of Energy hosts a comprehensive list of links to hydrology and related hydrology resources. Under the computer applications category it maintains a detailed listing of hydrological software, including the programs for stormwater management. The site also provides the supporting resources for software downloading and documentation.

The Surface-Water Quality and Flow Modeling Interest Group (SMIG) of the U.S. Geological Survey maintains an archives of commercial water resources and storm water models. The site provides links to modeling software.

Dodson and Associates have compiled hydrology and hydraulic software in HydroCD, which is a source of 60 stormwater management programs.

The Natural Resources Conservation Service has formulated software based on its TR-55 document. The software has DOS as well as Windows versions. It has a menu-driven interface to calculate the peak flow by the graphical method or to generate a peak inflow hydrograph by the tabular method, as discussed in section 13.12, for small watersheds from 1 to 200 acres.

Haestad Methods offers three salient programs—SewerCAD, StormCAD, and Water-CAD—which they have now combined into a single unified platform. The program provides analysis of peak flows and allows users to size pipes, offset inverts, drop manholes, and design catch basins in addition to analysis and design of water distribution systems and sanitary sewer systems.

The Texas Department of Transportation has created a program, THYSYS, that can analyze storm drain layouts, compute discharges to inlets using the rational method, design inlets, and compute hydraulic grade lines for storm drain lines in a network with up to 100 junctions.

The University of Central Florida has developed the Stormwater Management and Design Aid (SMADA) that allows users to create runoff hydrographs by multiple methods and to perform routing of these hydrographs through ponds, canals, and pipes. The other routines within the program perform optimal sewer design and retention system design.

The EPA's Storm Water Management Model (SWMM) has a diverse array of routines. It can develop runoff hydrographs from rain or snowfall. These hydrographs are for analysis and design of a storm sewer network. SWMM can also estimate the rate of sewage flow from land-use and population statistics and analyze a sanitary sewer network. It can model runoff pollutants and the treatment of sewage through the system. SWMM5 is a completely revised release (November 2004) of SWMM with GIS and CAD interfacing, while SWMM 4.4H has been revised most recently in 2005.

PROBLEMS

14.1 The irrigable area covered up to a point by an open drain is 450 acres. The irrigation waste constitutes 25% of the water applied. For canal capacity, the curve of Figure 14.1 applies. The land topography is flat, having a slope of 0.008%. Determine the drain of a semicircular section. $n = 0.03$.

14.2 A subsurface flow of 0.6 cfs drains through a plastic tube drain ($n = 0.011$) into the open drain of Problem 14.1. Design the tube drain and redesign the open drain for a best trapezoidal section when the top width is twice that of the sloping sides. The side slope is 1:1.

14.3 In an agricultural area, 6 in. of water are applied per irrigation application, 20% of which runs off. The infiltration rate of the soil in the upper root zone is 1.25 in./hr. If the hydraulic conductivity of the soil is 0.8 in./hr, determine the rise in the water table after the irrigation.

14.4 The irrigation application in an area is 5.0 in. The soil in the root zone has a sandy loam texture. About 30% of the irrigation water runs off. For a hydraulic conductivity of 0.4 in./hr, determine the water-table buildup.

14.5 For an agricultural drainage system, design the drains (determine the drain spacing) for the following conditions:

1. Depth from surface to impervious layer = 30 ft

2. Depth of drains from surface = 8 ft

3. Root zone or water-table requirement = 4 ft below the surface

4. Uniform hydraulic conductivity = 10 ft/day

5. Water application is as follows:

Event	Date	Time between Events (days)
Snowmelt	Apr. 22	
		45
First irrigation	June 6	
		25
Second irrigation	July 1	
		20
Third irrigation	July 21	
		14
Fourth irrigation	Aug. 4	
		14
Fifth irrigation	Aug. 18	
		14
Sixth irrigation	Sept. 1	

6. Each irrigation application, as well as spring snowmelt, contributed to a deep percolation of 1 in. (Adapted from the Bureau of Reclamation, 1984.)

14.6 In Problem 14.5, if the depth from the surface to the impervious layer is only 8 ft (i.e., the drains are located on the barrier), determine the drain spacing.

14.7 The following conditions were observed for an agricultural drainage system.

1. Vertical distance from impervious layer to drains = 30 ft

2. Maximum water table above drains = 6 ft

3. Average hydraulic conductivity = 7.5 in./hr

4. Infiltration rate = 3 in./hr

5. Runoff = 40%

6. Irrigation schedule as per the following table

Event	Amount, in.	Time between events, days
First irrigation	5	
		35
Second irrigation	4.5	
		30
Third irrigation	6	
		25
Fourth irrigation	6.5	
		36
Fifth irrigation	3.75	

Determine the drain spacing so that the water table does not build up more than 1 foot above the drain.

14.8 If the drains in Problem 14.7 are located on the impervious surface, determine the spacing of the drains so that the water table does not rise more than 2 ft above the drains.

14.9 For the parallel pipe drains of Problem 14.5, the length of drain is 7230 ft. The surface gradient is 0.5%. The water table is generally parallel to the ground surface. Determine the peak discharge due to deep percolation and upstream slope. Also design the drain. $n = 0.011$.

14.10 For the pipe drains of Problem 14.8 on the impervious surface, the length of drain is 8300 ft. The average water-table gradient is 0.008. Determine the peak discharge due to deep percolation and upland slope and the size of the drain for $n = 0.011$.

14.11 For a collector drain receiving water from the parallel drains of Problem 14.9, determine the discharge from a drained area of 100 acres. Use the median value for the C factor from Table 14.4.

14.12 If the drained area is 50 acres, determine the discharge for a collector drain receiving water in Problem 14.10. $C = 0.91$.

14.13 A 60-ft-wide asphalt-paved highway section has a longitudinal slope of 1% and a cross slope of 1/4 in. to 1 ft. If the gutter inlet spacing is 180 ft, determine the peak flow at the inlet. The rainfall intensity (in./hr) is given by $i = 180/(t + 25)$, where t is in minutes. $C = 0.8$.

14.14 A highway section of 100 ft width traverses a suburb, where the intensity-duration frequency curves of Figure 13.6 apply. The maximum 10-year intensity is 7 in./hr. A grated inlet located at station 285 + 95 has an elevation of 551.45 ft. Another inlet is located at station 284 + 05, with an elevation of 549.60 ft. The cross slope is 0.0208. Determine the inlet peak flow of 10-year frequency. $C = 0.7$.

14.15 A highway section is as shown in Figure P14.15. The runoff is caught by the grated inlets 1, 2, and 7 at station 204 + 00 and at the gutter sump 8 at station 205 + 78.1. In addition, the runoff from the drainage area to the south of the highway is collected in two inlets, 3 and 5. Inlet 3 connects to inlet 2 and then to manhole 4 at 205 + 95 and inlet 5 directly connects to manhole 4. The runoff from the south side of the highway is then conveyed under the highway to manhole 6, where the north side inlets are also picked up. The accumulated runoff is discharged from manholes 6 to 9 and into a natural water course.

The tributary areas and their breakdown between pervious and impervious portions, length of drains, and surface slopes are indicated on the figure. The overland flow time to the intercepts has been considered to be 5 min. The 10-year rainfall intensity (in./hr) is given by $i = 149/(t + 15.7)$, where t is in minutes. Design the longitudinal drainage system. $C = 0.3$ for pervious areas and 0.95 for impervious areas. Use concrete pipe ($n = 0.013$).

Figure P14.15

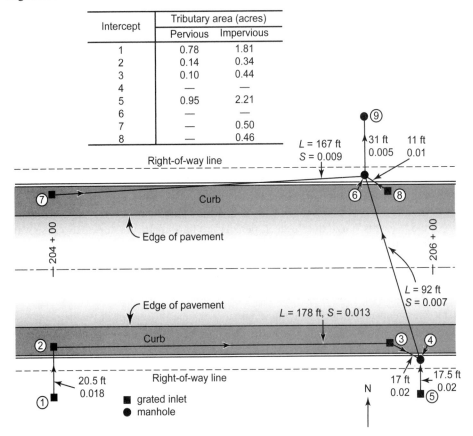

Intercept	Tributary area (acres)	
	Pervious	Impervious
1	0.78	1.81
2	0.14	0.34
3	0.10	0.44
4	—	—
5	0.95	2.21
6	—	—
7	—	0.50
8	—	0.46

14.16 A culvert section is as shown in Fig. P14.16. Design the culvert of a beveled concrete pipe section ($n = 0.012$) to carry a peak discharge of 210 cfs. $C_d = 0.96$, $w = 0.15$ ft.

Figure P14.16

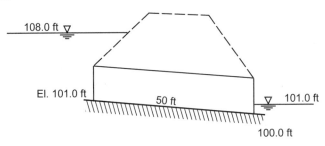

14.17 A culvert section is as shown in Fig. P14.17. Design a culvert of a rounded corrugated metal pipe ($n = 0.024$) to carry a peak flow of 125 cfs. $C_d = 0.5$, $r = 0.05$ ft.

Figure P 14.17

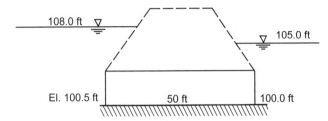

14.18 Design a circular corrugated-metal pipe culvert ($n = 0.024$) to carry a discharge of 250 cfs for the condition shown in Fig. P14.18. The approach stream has a width of 40 ft. $C_d = 0.90$.

Figure P 14.18

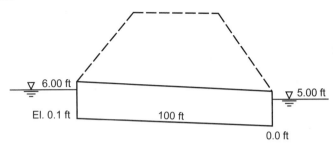

14.19 Design a corrugated-metal pipe culvert ($n = 0.024$) set in a vertical concrete headwall, to carry a flow of 725 cfs for the condition shown in Figure P14.19. The approach stream has a width of 80 ft and $n = 0.013$. $C_d = 0.9$.

Figure P 14.19

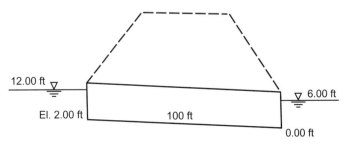

14.20 Determine the flow through an 8-ft-square concrete box culvert ($n = 0.015$) of square-edged entrance for the conditions given in Figure P14.20. The approach channel is 50 ft wide and $n = 0.013$. $C_d = 0.95$. [*Hint:* Assume a discharge, determine the type of flow, use the appropriate discharge equation to verify the assumed flow. Repeat with changed discharge until it verifies. As a first guess, $Q = C_d A_3 \sqrt{2g\Delta H}$, ΔH is the difference of head and tailwaters and A_3 is the area at the culvert exit.]

Figure P14.20

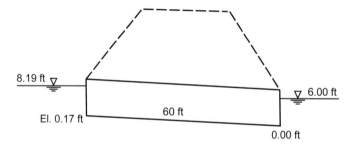

8.19 ft

6.00 ft

El. 0.17 ft

60 ft

0.00 ft

14.21 Design a surface drainage system for the part of the airport shown in Fig. P14.21. The finished contours, drains layout, and length and slope of the drains are as shown. The tributary area, weighted runoff coefficient, and time of overland flow to each intercept point are listed below. The 5-year rainfall intensity (in./hr) is given by $i = 96/(t + 16)$, where t is in minutes. Use $n = 0.015$.

Intercept	Tributary Area (acres)	Weighted Coefficient, C	Time to Overland Flow (min)
1	15.0	0.40	25.0
2	16.5	0.40	26.0
3	25.0	0.40	30.0
4	12.0	0.40	20.0
5	30.0	0.35	40.0

Figure P 14.21 Portion of an airport showing final contouring and drainage layout.

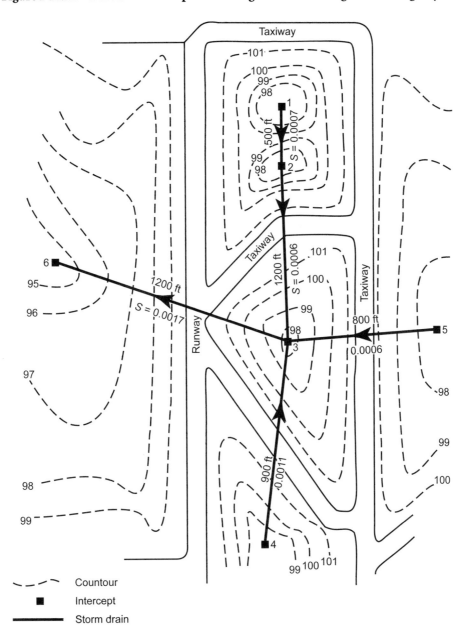

Appendix

◆◆

Table A.1 Length Equivalents

Unit	mm	m	in.	ft	yd	mi
			Equivalent			
Millimeter	1	10^{-3}	0.0394	0.00328	0.00109	6.214×10^{-7}
Meter	10^3	1	39.37	3.281	1.0936	6.214×10^{-4}
Inch	25.4	0.0254	1	0.0833	0.02778	1.578×10^{-5}
Foot	304.8	0.3048	12	1	0.333	1.894×10^{-4}
Yard	914.4	0.9144	36	3	1	5.682×10^{-4}
Mile	1.609×10^6	1.609×10^3	6.336×10^4	5280	1760	1

Table A.2 Area Equivalents

Unit	$in.^2$	ft^2	m^2	acre	mi^2
			Equivalent		
Square inch	1	6.944×10^{-3}	6.452×10^{-4}	1.59×10^{-7}	2.491×10^{-10}
Square foot	144	1	0.0929	2.30×10^{-5}	3.587×10^{-8}
Square meter	1550	10.764	1	2.50×10^{-4}	3.861×10^{-7}
Acre	6.270×10^6	43,560	4047	1	1.56×10^{-3}
Square mile	4.014×10^9	2.788×10^7	2.59×10^6	640	1

Table A.3 Volume Equivalents

Unit	$in.^3$	gal	ft^3	m^3	acre-ft	cfs-day
			Equivalent			
Cubic inch	1	0.00433	5.79×10^{-4}	1.64×10^{-5}	1.33×10^{-8}	6.70×10^{-9}
Gallon	231	1	0.134	0.00379	3.07×10^{-6}	1.55×10^{-6}
Cubic foot	1728	7.48	1	0.0283	2.30×10^{-5}	1.16×10^{-5}
Cubic meter	61,000	264	35.3	1	8.11×10^{-4}	4.09×10^{-4}
Acre-foot	7.53×10^7	3.26×10^5	43,560	1233	1	0.504
Cubic foot per second-day	1.49×10^8	6.46×10^5	86,400	2447	1.98	1

Table A.4   Velocity Equivalents

Unit	ft/sec	mi/hr	m/sec	km/hr	kn
			Equivalent		
Feet per second	1	0.6818	0.3048	1.097	0.5925
Miles per hour	1.467	1	0.4470	1.609	0.8690
Meters per second	3.281	2.237	1	3.600	1.944
Kilometers per hour	0.9113	0.6214	0.2778	1	0.5400
Knots	1.688	1.151	0.5144	1.852	1

Table A.5   Discharge Equivalents

Unit	gal/day	ft^3/day	gal/min	acre-ft/day	cfs	m^3/s
				Equivalent		
U.S. gallons per day	1	0.134	6.94×10^{-4}	3.07×10^{-6}	1.55×10^{-6}	4.38×10^{-8}
Cubic feet per day	7.48	1	5.19×10^{-3}	2.30×10^{-5}	1.16×10^{-5}	3.28×10^{-7}
U.S. gallons per minute	1440	193	1	4.42×10^{-3}	2.23×10^{-3}	6.31×10^{-5}
Acre-feet per day	3.26×10^5	43,560	226	1	0.504	0.0143
Cubic feet per second	6.46×10^5	86,400	449	1.98	1	0.0283
Cubic meters per second	2.28×10^7	3.05×10^6	15,800	70.0	35.3	1

Table A.6   Pressure Equivalents

Unit	ft H$_2$O	in. Hg	mm Hg	mbar	kPa	psi	kg/m^2
				Equivalent			
Foot of water (32°F)	1	0.883	22.42	29.89	2.989	0.4335	304.8
Inch of mercury (32°F)	1.133	1	25.40	33.86	3.386	0.4912	345.3
Millimeter of mercury (0°C)	0.0446	0.03937	1	1.333	0.1333	0.01934	13.60
Millibar	0.0335	0.02953	0.7501	1	0.1000	0.01450	10.20
Kilopascal (N/m$^2 \times 10^3$)	0.335	0.2953	7.501	10.00	1	0.1450	102.0
Pounds per square inch	2.307	2.036	51.71	68.95	6.895	1	703.1
Kilograms per square meter	.000328	0.002896	0.07356	0.09807	0.009807	0.001422	1

Table A.7 Energy Equivalents

Unit	Btu	cal	J	kW-hr	ft-lb	hp-hr
			Equivalent			
British thermal unit (60°F)	1	252.0	1055	0.0002930	777.9	0.0003929
Calorie (15°C)	0.003969	1	4.186	1.163×10^{-6}	3.087	1.559×10^{-6}
Joule	0.0009482	0.2389	1	2.778×10^{-7}	0.7376	3.725×10^{-7}
Kilowatt-hour	3413	860,100	3.600×10^{6}	1	2.655×10^{6}	1.341
Foot-pound	0.001286	0.3239	1.356	3.766×10^{-7}	1	5.051×10^{-7}
Horsepower-hour	2545	641,300	2.685×10^{6}	0.7457	1.980×10^{6}	1

Table A.8 Power Equivalents

Unit	W or J/sec	kW	ft-lb/sec	hp	Btu/hr
			Equivalent		
Watts (or Joules per second)	1	0.001	0.737	0.00134	3.412
Kilowatts	1,000	1	737.6	1.314	3,412
Foot-pounds per second	1.356	0.001356	1	0.001818	4.63
Horsepower	745.5	0.7455	550	1	2,545
British thermal units per hour	0.293	2.93×10^{-4}	0.216	3.93×10^{-4}	1

Table A.9 Dynamic Viscosity Equivalents

Unit	N-sec/m^2	g/cm-sec (poise)	lb-sec/ft^2	kg/m-hr
			Equivalent	
Newtons-seconds per square meter	1	10.0	0.0209	3600
Grams per centimeter-second	0.1	1	2.089×10^{-3}	360
Pounds-seconds per square foot	47.88	478.80	1	1.724×10^{5}
Kilograms per meter-hour	2.778×10^{-4}	2.778×10^{-3}	5.80×10^{-6}	1

1 poise = 100 centipoise (cp)

B

Appendix

◆◆

Table B.1   Some Other Useful Conversion Factors

Multiply:	By:	To obtain:
Pound (m)	453.6	Grams (g) mass
Mass (kg)	9.81	Weight in Newtons
Pound (f)	4.448	Newton (N)
	0.4536	Kilogram
Liter	1000	Cubic centimeter
Pounds per ft^2	47.88	N/m^2 *or* pascal
Horsepower	745.7	Watt
	550	Foot-lb/sec
Standard atmosphere	101.325	Kilopascal (kPa)
U.S. or short ton	2000	Pound
Metric ton or tonne	1000	Kilogram
Short ton	0.907	Metric ton
	0.892	Long ton
Nautical mile	1852	Meter
U.S. mile	1609	Meter
Square mile	2.59	Square kilometer
Square kilometer	100	Hectare (ha)
°F	$5/9(°F - 32)$	°C
Log to base e (i.e., $\log_e$, where $e = 2.718$)	0.434	Log to base 10 (i.e., $\log_{10}$)

Appendix

Table C.1 Physical Properties of Water in English Units

Temp. (°F)	Specific gravity	Specific weight (lb/ft³)	Surface Tension (lb/ft)	Heat of vaporization (Btu/lb)	Viscosity Dynamic (lb-sec/ft²)	Viscosity Kinematic (ft²/sec)	Bulk modulus of elasticity (psi)	Vapor pressure in. Hg	Vapor pressure Millibar	Vapor pressure lb/in.²
32	0.99986	62.418	0.518×10^{-2}	1075.5	3.746×10^{-5}	1.931×10^{-5}	293×10^{3}	0.180	6.11	0.089
40	0.99998	62.426[a]	0.514	1071.0	3.229	1.664	294	0.248	8.39	0.122
50	0.99971	62.409	0.509	1065.3	2.735	1.410	305	0.362	12.27	0.178
60	0.99902	62.366	0.504	1059.7	2.359	1.217	311	0.522	17.66	0.256
70	0.99798	62.301	0.500	1054.0	2.050	1.058	320	0.739	25.03	0.363
80	0.99662	62.216	0.492	1048.4	1.799	0.930	322	1.032	34.96	0.507
90	0.99497	62.113	0.486	1042.7	1.595	0.826	323	1.422	18.15	0.698
100	0.99306	61.994	0.480	1037.1	1.424	0.739	327	1.933	65.47	0.950
120	0.98856	61.713	0.473	1025.6	1.168	0.609	333	3.448	116.75	1.693
140	0.98321	61.379	0.454	1014.0	0.981	0.514	330	5.884	119.26	2.890
160	0.97714	61.000	0.441	1002.2	0.838	0.442	326	9.656	326.98	4.742
180	0.97041	60.580	0.426	990.2	0.726	0.386	318	15.295	517.95	7.512
200	0.96306	60.121	0.412	977.9	0.637	0.341	308	23.468	794.72	11.526
212	0.95837	59.828	0.404	970.3	0.593	0.319	300	29.921	1013.25	14.696

[a] Maximum specific weight is 62.427 lb/ft³ at 39.2°F.

Table C.2   Physical Properties of Water in Metric Units[a]

Temp. (°C)	Specific gravity	Density (g/cm³)	Surface Tension (N/m)	Heat of vaporization (cal/g)	Viscosity		Bulk modulus of elasticity (N/m²)	Vapor pressure		
					Dynamic (poise)[b]	Kinematic (stokes)[c]		mm Hg	kPa	g/cm²
0	0.99987	0.99984	75.6×10^{-3}	597.3	1.79×10^{-2}	1.79×10^{-2}	2.02×10^{9}	4.58	0.611	6.23
5	0.99999	0.99996	74.9	594.5	1.52	1.52	2.06	6.54	0.873	8.89
10	0.99973	0.99970	74.2	591.7	1.31	1.31	2.10	9.20	1.228	12.51
15	0.99913	0.99910	73.5	588.9	1.14	1.14	2.14	12.78	1.706	17.38
20	0.99824	0.99821	72.8	586.0	1.00	1.00	2.18	17.53	2.337	23.83
25	0.99708	0.99705	72.0	583.2	0.890	0.893	2.22	23.76	3.169	32.30
30	0.99568	0.99565	71.2	580.4	0.798	0.801	2.25	31.83	4.243	43.27
35	0.99407	0.99404	70.4	577.6	0.719	0.723	2.27	42.18	5.625	57.34
40	0.99225	0.99222	69.6	574.7	0.653	0.658	2.28	55.34	7.378	75.23
50	0.98807	0.98804	67.9	569.0	0.547	0.554	2.29	92.56	12.340	125.83
60	0.98323	0.98320	66.2	563.2	0.466	0.474	2.28	149.46	19.926	203.19
70	0.97780	0.97777	64.4	557.4	0.404	0.413	2.25	233.79	31.169	317.84
80	0.97182	0.97179	62.6	551.4	0.355	0.365	2.20	355.28	47.367	483.01
90	0.96534	0.96531	60.8	545.3	0.315	0.326	2.14	525.89	70.113	714.95
100	0.95839	0.95836	58.9	539.1	0.282	0.294	2.07	760.00	101.325	1033.23

[a] SI units:

Density: $kg/m^3 = g/cm^3 \times 10^3$

Specific weight: N/m^3 = density in $kg/m^3 \times 9.81$

Dynamic viscosity: $N \cdot s/m^2 = poise \times 10^{-1}$

Kinematic viscosity: $m^2/s = stokes \times 10^{-4}$

Vapor pressure: $N/m^2 = millibar \times 10^2$ or $g/cm^2 \times 98.1$

[b] poise = (g/cm-s)

[c] stokes = (cm^2/s)

D

Appendix

🔹🔹

Table D.1 Physical Properties of Air at Standard Atmospheric Pressure in English Units

Temperature (°F)	Density, ρ (slugs/ft^3)	Specific Weight[b], γ (lb/ft^3)	Dynamic Viscosity, μ (lb-s/ft^2)	Kinematic Viscosity, ν (ft^2/s)	Specific Heat Ratio, k (—)	Speed of Sound, c (ft/s)
−40	2.939×10^{-3}	9.456×10^{-2}	3.29×10^{-7}	1.12×10^{-4}	1.401	1004
−20	2.805	9.026	3.34	1.19	1.401	1028
0	2.683	8.633	3.38	1.26	1.401	1051
10	2.626	8.449	3.44	1.31	1.401	1062
20	2.571	8.273	3.50	1.36	1.401	1074
30	2.519	8.104	3.58	1.42	1.401	1085
40	2.469	7.942	3.60	1.46	1.401	1096
50	2.420	7.786	3.68	1.52	1.401	1106
60	2.373	7.636	3.75	1.58	1.401	1117
70	2.329	7.492	3.82	1.64	1.401	1128
80	2.286	7.353	3.86	1.69	1.400	1138
90	2.244	7.219	3.90	1.74	1.400	1149
100	2.204	7.090	3.94	1.79	1.400	1159
120	2.128	6.846	4.02	1.89	1.400	1180
140	2.057	6.617	4.13	2.01	1.399	1200
160	1.990	6.404	4.22	2.12	1.399	1220
180	1.928	6.204	4.34	2.25	1.399	1239
200	1.870	6.016	4.49	2.40	1.398	1258
300	1.624	5.224	4.97	3.06	1.394	1348
400	1.435	4.616	5.24	3.65	1.389	1431
500	1.285	4.135	5.80	4.51	1.383	1509
750	1.020	3.280	6.81	6.68	1.367	1685
1000	0.845	2.717	7.85	9.30	1.351	1839
1500	0.629	2.024	9.50	15.1	1.329	2114

Table D.2   Physical Properties of Air at Standard Atmospheric Pressure in Metric Units

Temperature (°C)	Density, ρ (kg/m^3)	Specific Weight[b], γ (N/m^3)	Dynamic Viscosity, μ (N-s/m^2)	Kinematic Viscosity, ν (m^2/s)	Specific Heat Ratio, k (—)	Speed of Sound, c (m/s)
−40	1.514	14.85	1.57×10^{-5}	1.04×10^{-5}	1.401	306.2
−20	1.395	13.68	1.63	1.17	1.401	319.1
0	1.292	12.67	1.71	1.32	1.401	331.4
5	1.269	12.45	1.73	1.36	1.401	334.4
10	1.247	12.23	1.76	1.41	1.401	337.4
15	1.225	12.01	1.80	1.47	1.401	340.4
20	1.204	11.81	1.82	1.51	1.401	343.3
25	1.184	11.61	1.85	1.56	1.401	346.3
30	1.165	11.43	1.86	1.60	1.400	349.1
40	1.127	11.05	1.87	1.66	1.400	354.7
50	1.109	10.88	1.95	1.76	1.400	360.3
60	1.060	10.40	1.97	1.86	1.399	365.7
70	1.029	10.09	2.03	1.97	1.399	371.2
80	0.9996	9.803	2.07	2.07	1.399	376.6
90	0.9721	9.533	2.14	2.20	1.398	381.7
100	0.9461	9.278	2.17	2.29	1.397	386.9
200	0.7461	7.317	2.53	3.39	1.390	434.5
300	0.6159	6.040	2.98	4.84	1.379	476.3
400	0.5243	5.142	3.32	6.34	1.368	514.1
500	0.4565	4.477	3.64	7.97	1.357	548.8
1000	0.2772	2.719	5.04	18.2	1.321	694.8

E

Appendix

◆◆

Table E.1   Values of the Error Function

z	0.0	.01	.02	.03	.04	.05	.06	.07	.08	.09
0.0		.01128	.02256	.03384	.04511	.05637	.06762	.07886	.09008	.10128
.1	.11246	.12362	.13476	.14587	.15695	.16800	.17901	.18999	.20094	.21184
.2	.22270	.23352	.24430	.25502	.26570	.27633	.28690	.29742	.30788	.31828
.3	.32863	.33891	.34913	.35928	.36936	.37938	.38933	.39921	.40901	.41874
.4	.42839	.43797	.44747	.45689	.46623	.47548	.48466	.49375	.50275	.51167
.5	.52050	.52924	.53790	.54646	.55494	.56332	.57162	.57982	.58792	.59594
.6	.60386	.61168	.61941	.62705	.63459	.64203	.64938	.65663	.66378	.67084
.7	.67780	.68467	.69143	.69810	.70468	.71116	.71754	.72382	.73001	.73610
.8	.74210	.74800	.75381	.75952	.76514	.77067	.77610	.78144	.78669	.79184
.9	.79691	.80188	.80677	.81156	.81627	.82089	.82542	.82987	.83423	.83851
1.0	.84270	.84681	.85084	.85478	.85865	.86244	.86614	.86977	.87333	.87680
1.1	.88021	.88353	.88679	.88997	.89308	.89612	.89910	.90200	.90484	.90761
1.2	.91031	.91296	.91553	.91805	.92051	.92290	.92524	.92751	.92973	.93190
1.3	.93401	.93606	.93807	.94002	.94191	.94376	.94556	.94731	.94902	.95067
1.4	.95229	.95385	.95538	.95686	.95830	.95970	.96105	.96237	.96365	.96490
1.5	.96611	.96728	.96841	.96952	.97059	.97162	.97263	.97360	.97455	.97546
1.6	.97635	.97721	.97804	.97884	.97962	.98038	.98110	.98181	.98249	.98315
1.7	.98379	.98441	.98500	.98558	.98613	.98667	.98719	.98769	.98817	.98864
1.8	.98909	.98952	.98994	.99035	.99074	.99111	.99147	.99182	.99216	.99248
1.9	.99279	.99309	.99338	.99366	.99392	.99418	.99443	.99466	.99489	.99511
2.0	.99532	.99552	.99572	.99591	.99609	.99626	.99642	.99658	.99673	.99688
2.1	.99702	.99715	.99728	.99741	.99753	.99764	.99775	.99785	.99795	.99805
2.2	.99814	.99822	.99831	.99839	.99846	.99854	.99861	.99867	.99874	.99880
2.3	.99886	.99891	.99897	.99902	.99906	.99911	.99915	.99920	.99924	.99928
2.4	.99931	.99935	.99938	.99941	.99944	.99947	.99950	.99952	.99955	.99957
2.5	.99959	.99961	.99963	.99965	.99967	.99969	.99971	.99972	.99974	.99975
2.6	.99976	.99978	.99979	.99980	.99981	.99982	.99983	.99984	.99985	.99986
2.7	.99987	.99987	.99988	.99989	.99989	.99990	.99991	.99991	.99992	.99992
2.8	.99992	.99993	.99993	.99994	.99994	.99994	.99995	.99995	.99995	.99996
2.9	.99996	.99996	.99996	.99997	.99997	.99997	.99997	.99997	.99997	.99998
3.0	.99998	.99999	.99999	1.0000						

F

Appendix

◆◆◆

Table F.1   Coefficients a_{n-i+1} for the Shapiro-Wilk Test of Normality

i/n	2	3	4	5	6	7	8	9	10
1	0.7071	0.7071	0.6872	0.6646	0.6431	0.6233	0.6052	0.5888	0.5739
2	—	.0000	.1677	.2413	.2806	.3031	.3164	.3244	.3291
3	—	—	—	.0000	.0875	.1401	.1743	.1976	.2141
4	—	—	—	—	—	.0000	.0561	.0947	.1224
5	—	—	—	—	—	—	—	.0000	.0399

i/n	11	12	13	14	15	16	17	18	19	20
1	0.5601	0.5475	0.5359	0.5251	0.5150	0.5056	0.4968	0.4886	0.4808	0.4734
2	.3315	.3325	.3325	.3318	.3306	.3290	.3273	.3253	.3232	.3211
3	.2260	.2347	.2412	.2460	.2495	.2521	.2540	.2553	.2561	.2565
4	.1429	.1586	.1707	.1802	.1878	.1939	.1988	.2027	.2059	.2085
5	.0695	.0922	.1099	.1240	.1353	.1447	.1524	.1587	.1641	.1686
6	0.0000	0.0303	0.0539	0.0727	0.0880	0.1005	0.1109	0.1197	0.1271	0.1334
7	—	—	.0000	.0240	.0433	.0593	.0725	.0837	.0932	.1013
8	—	—	—	—	.0000	.0196	.0359	.0496	.0612	.0711
9	—	—	—	—	—	—	.0000	.0163	.0303	.0422
10	—	—	—	—	—	—	—	—	.0000	.0140

i/n	21	22	23	24	25	26	27	28	29	30
1	0.4643	0.4590	0.4542	0.4493	0.4450	0.4407	0.4366	0.4328	0.4291	0.4254
2	.3185	.3156	.3126	.3098	.3069	.3043	.3018	.2992	.2968	.2944
3	.2578	.2571	.2563	.2554	.2543	.2533	.2522	.2510	.2499	.2487
4	.2119	.2131	.2139	.2145	.2148	.2151	.2152	.2151	.2150	.2148
5	.1736	.1764	.1787	.1807	.1822	.1836	.1848	.1857	.1864	.1870
6	0.1399	0.1443	0.1480	0.1512	0.1539	0.1563	0.1584	0.1601	0.1616	0.1630
7	.1092	.1150	.1201	.1245	.1283	.1316	.1346	.1372	.1395	.1415
8	.0804	.0878	.0941	.0997	.1046	.1089	.1128	.1162	.1192	.1219
9	.0530	.0618	.0696	.0764	.0823	.0876	.0923	.0965	.1002	.1036
10	.0263	.0368	.0459	.0539	.0610	.0672	.0728	.0778	.0822	.0862

Table F.1 Coefficients a_{n-i+1} for the Shapiro-Wilk Test of Normality (Continued)

i/n	21	22	23	24	25	26	27	28	29	30
11	0.0000	0.0122	0.0228	0.0321	0.0403	0.0476	0.0540	0.0598	0.0650	0.0697
12	—	—	.0000	.0107	.0200	.0284	.0358	.0424	.0483	.0537
13	—	—	—	—	.0000	.0094	.0178	.0253	.0320	.0381
14	—	—	—	—	—	—	.0000	.0084	.0159	.0227
15	—	—	—	—	—	—	—	—	.0000	.0076

i/n	31	32	33	34	35	36	37	38	39	40
1	0.4220	0.4188	0.4156	0.4127	0.4096	0.4068	0.4040	0.4015	0.3989	0.3964
2	.2921	.2898	.2876	.2854	.2834	.2813	.2794	.2774	.2755	.2737
3	.2475	.2463	.2451	.2439	.2427	.2415	.2403	.2391	.2380	.2368
4	.2145	.2141	.2137	.2132	.2127	.2121	.2116	.2110	.2104	.2098
5	.1874	.1878	.1880	.1882	.1883	.1883	.1883	.1881	.1880	.1878
6	0.1641	0.1651	0.1660	0.1667	0.1673	0.1678	0.1683	0.1686	0.1689	0.1691
7	.1433	.1449	.1463	.1475	.1487	.1496	.1503	.1513	.1520	.1526
8	.1243	.1265	.1284	.1301	.1317	.1331	.1344	.1356	.1366	.1376
9	.1066	.1093	.1118	.1140	.1160	.1179	.1196	.1211	.1225	.1237
10	.0899	.0931	.0961	.0988	.1013	.1036	.1056	.1075	.1092	.1108
11	0.0739	0.0777	0.0812	0.0844	0.0873	0.0900	0.0924	0.0947	0.0967	0.0986
12	.0585	.0629	.0669	.0706	.0739	.0770	.0798	.0824	.0848	.0870
13	.0435	.0485	.0530	.0572	.0610	.0645	.0677	.0706	.0733	.0759
14	.0289	.0344	.0395	.0441	.0484	.0523	.0559	.0592	.0622	.0651
15	.0144	.0206	.0262	.0314	.0361	.0404	.0444	.0481	.0515	.0546
16	0.0000	0.0068	0.0131	0.0187	0.0239	0.0287	0.0331	0.0372	0.0409	0.0444
17	—	—	.0000	.0062	.0119	.0172	.0220	.0264	.0305	.0343
18	—	—	—	—	.0000	.0057	.0110	.0158	.0203	.0244
19	—	—	—	—	—	—	.0000	.0053	.0101	.0146
20	—	—	—	—	—	—	—	—	.0000	.0049

i/n	41	42	43	44	45	46	47	48	49	50
1	0.3940	0.3917	0.3894	0.3872	0.3850	0.3830	0.3808	0.3789	0.3770	0.3751
2	.2719	.2701	.2684	.2667	.2651	.2635	.2620	.2604	.2589	.2574
3	.2357	.2345	.2334	.2323	.2313	.2302	.2291	.2281	.2271	.2260
4	.2091	.2085	.2078	.2072	.2065	.2058	.2052	.2045	.2038	.2032
5	.1876	.1874	.1871	.1868	.1865	.1862	.1859	.1855	.1851	.1847

(continued)

i/n	41	42	43	44	45	46	47	48	49	50
6	0.1693	0.1694	0.1695	0.1695	0.1695	0.1695	0.1695	0.1693	0.1692	0.1691
7	.1531	.1535	.1539	.1542	.1545	.1548	.1550	.1551	.1553	.1554
8	.1384	.1392	.1398	.1405	.1410	.1415	.1420	.1423	.1427	.1430
9	.1249	.1259	.1269	.1278	.1286	.1293	.1300	.1306	.1312	.1317
10	.1123	.1136	.1149	.1160	.1170	.1180	.1189	.1197	.1205	.1212
11	0.1004	0.1020	0.1035	0.1049	0.1062	0.1073	0.1085	0.1095	0.1105	0.1113
12	.0891	.0909	.0927	.0943	.0959	.0972	.0986	.0998	.1010	.1020
13	.0782	.0804	.0824	.0842	.0860	.0876	.0892	.0906	.0919	.0932
14	.0677	.0701	.0724	.0745	.0775	.0785	.0801	.0817	.0832	.0846
15	.0575	.0602	.0628	.0651	.0673	.0694	.0713	.0731	.0748	.0764
16	0.0476	0.0506	0.0534	0.0560	0.0584	0.0607	0.0628	0.0648	0.0667	0.0685
17	.0379	.0411	.0442	.0471	.0497	.0522	.0546	.0568	.0588	.0608
18	.0283	.0318	.0352	.0383	.0412	.0439	.0465	.0489	.0511	.0532
19	.0188	.0227	.0263	.0296	.0328	.0357	.0385	.0411	.0436	.0459
20	.0094	.0136	.0175	.0211	.0245	.0277	.0307	.0335	.0361	.0386
21	0.0000	0.0045	0.0087	0.0126	0.0163	0.0197	0.0229	0.0259	0.0288	0.0314
22	—	—	.0000	.0042	.0081	.0118	.0153	.0185	.0215	.0244
23	—	—	—	—	.0000	.0039	.0076	.0111	.0143	.0174
24	—	—	—	—	—	—	.0000	.0037	.0071	.0104
25	—	—	—	—	—	—	—	—	.0000	.0035

G

Appendix

◆◆◆

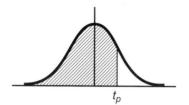

t_p

Table G.1   Cumulative Student t Distribution

v	$t_{.995}$	$t_{.975}$	$t_{.95}$	$t_{.90}$	$t_{.75}$	$t_{.70}$	$t_{.60}$	$t_{.55}$
1	63.66	12.71	6.31	3.08	1.000	0.727	0.325	0.158
3	5.84	3.18	2.35	1.64	0.765	0.584	0.277	0.137
5	4.03	2.57	2.02	1.48	0.727	0.559	0.267	0.132
8	3.36	2.31	1.86	1.40	0.706	0.546	0.262	0.130
10	3.17	2.23	1.81	1.37	.700	.542	.260	.129
11	3.11	2.20	1.80	1.36	.697	.540	.260	.129
12	3.06	2.18	1.78	1.36	.695	.539	.259	.128
13	3.01	2.16	1.77	1.35	.694	.538	.259	.128
14	2.98	2.14	1.76	1.34	.692	.537	.258	.128
15	2.95	2.13	1.75	1.34	.691	.536	.258	.128
16	2.92	2.12	1.75	1.34	.690	.535	.258	.128
17	2.90	2.11	1.74	1.33	.689	.534	.257	.128
18	2.88	2.10	1.73	1.33	.688	.534	.257	.127
19	2.86	2.09	1.73	1.33	.688	.533	.257	.127
20	2.84	2.09	1.72	1.32	.687	.533	.257	.127
21	2.83	2.08	1.72	1.32	.686	.532	.257	.127
22	2.82	2.07	1.72	1.32	.686	.532	.256	.127
23	2.81	2.07	1.71	1.32	.685	.532	.256	.127
24	2.80	2.06	1.71	1.32	.685	.531	.256	.127
25	2.79	2.06	1.71	1.32	.684	.531	.256	.127
26	2.78	2.06	1.71	1.32	.684	.531	.256	.127
27	2.77	2.05	1.70	1.31	.684	.531	.256	.127
28	2.76	2.05	1.70	1.31	.683	.530	.256	.127
29	2.76	2.04	1.70	1.31	.683	.530	.256	.127
30	2.75	2.04	1.70	1.31	.683	.530	.256	.127
40	2.70	2.02	1.68	1.30	.681	.529	.255	.126
60	2.66	2.00	1.67	1.30	.679	.527	.254	.126
120	2.62	1.98	1.66	1.29	.677	.526	.254	.126
∞	2.58	1.96	1.645	1.28	.674	.524	.253	.126

Appendix

◆◆◆

Table H.1   Cumulative *F* Distribution (*m* Numerator and *n* Denominator Degrees of Freedom)

a	n	m →10	12	15	20	30	60	120	∞
.90		2.32	2.28	2.24	2.20	2.15	2.11	2.08	2.06
.95		2.98	2.91	2.84	2.77	2.70	2.62	2.58	2.54
.975	10	3.72	3.62	3.52	3.42	3.31	3.20	3.14	3.08
.99		4.85	4.71	4.56	4.41	4.25	4.08	4.00	3.91
.995		5.85	5.66	5.47	5.27	5.07	4.86	4.75	4.64
.90		2.19	2.15	2.10	2.06	2.01	1.96	1.93	1.90
.95		2.75	2.69	2.62	2.54	2.47	2.38	2.34	2.30
.975	12	3.37	3.28	3.18	3.07	2.96	2.85	2.79	2.72
.99		4.30	4.16	4.01	3.86	3.70	3.54	3.45	3.36
.995		5.09	4.91	4.72	4.53	4.33	4.12	4.01	3.90
.90		2.06	2.02	1.97	1.92	1.87	1.82	1.79	1.76
.95		2.54	2.48	2.40	2.33	2.25	2.16	2.11	2.07
.975	15	3.06	2.96	2.86	2.76	2.64	2.52	2.46	2.40
.99		3.80	3.67	3.52	3.37	3.21	3.05	2.96	2.87
.995		4.42	4.25	4.07	3.88	3.69	3.48	3.37	3.26
.90		1.94	1.89	1.84	1.79	1.74	1.68	1.64	1.61
.95		2.35	2.28	2.20	2.12	2.04	1.95	1.90	1.84
.975	20	2.77	2.68	2.57	2.46	2.35	2.22	2.16	2.09
.99		3.37	3.23	3.09	2.94	2.78	2.61	2.52	2.42
.995		3.85	3.68	3.50	3.32	3.12	2.92	2.81	2.69

Table H.1   Cumulative *F* Distribution (*m* Numerator and *n* Denominator Degrees of Freedom) (Continued)

a	n	m ⟶ 10	12	15	20	30	60	120	∞
.90		1.82	1.77	1.72	1.67	1.61	1.54	1.50	1.46
.95		2.16	2.09	2.01	1.93	1.84	1.74	1.68	1.62
.975	30	2.51	2.41	2.31	2.20	2.07	1.94	1.87	1.79
.99		2.98	2.84	2.70	2.55	2.39	2.21	2.11	2.01
.995		3.34	3.18	3.01	2.82	2.63	2.42	2.30	2.18
.90		1.71	1.66	1.60	1.54	1.48	1.40	1.35	1.29
.95		1.99	1.92	1.84	1.75	1.65	1.53	1.47	1.39
.975	60	2.27	2.17	2.06	1.94	1.82	1.67	1.58	1.48
.99		2.63	2.50	2.35	2.20	2.03	1.84	1.73	1.60
.995		2.90	2.74	2.57	2.39	2.19	1.96	1.83	1.69
.90		1.65	1.60	1.54	1.48	1.41	1.32	1.26	1.19
.95		1.91	1.83	1.75	1.66	1.55	1.43	1.35	1.25
.975	120	2.16	2.05	1.94	1.82	1.69	1.53	1.43	1.31
.99		2.47	2.34	2.19	2.03	1.86	1.66	1.53	1.38
.995		2.71	2.54	2.37	2.19	1.98	1.75	1.61	1.43
.90		1.60	1.55	1.49	1.42	1.34	1.24	1.17	1.00
.95		1.83	1.75	1.67	1.57	1.46	1.32	1.22	1.00
.975	∞	2.05	1.94	1.83	1.71	1.57	1.39	1.27	1.00
.99		2.32	2.18	2.04	1.88	1.70	1.47	1.32	1.00
.995		2.52	2.36	2.19	2.00	1.79	1.53	1.36	1.00

Appendix

◆◆◆

Table I.1   Values of $H_\alpha = H_{0.05}$ for Computing a One-Sided Lower 95% Confidence Limit on a Lognormal Mean

s_y	3	5	7	10	12	15	21	31	51	101
0.10	−2.130	−1.806	−1.731	−1.690	−1.677	−1.666	−1.655	−1.648	−1.644	−1.642
0.20	−1.949	−1.729	−1.678	−1.653	−1.646	−1.640	−1.636	−1.636	−1.637	−1.641
0.30	−1.816	−1.669	−1.639	−1.627	−1.625	−1.625	−1.627	−1.632	−1.638	−1.648
0.40	−1.717	−1.625	−1.611	−1.611	−1.613	−1.617	−1.625	−1.635	−1.647	−1.662
0.50	−1.644	−1.594	−1.594	−1.603	−1.609	−1.618	−1.631	−1.646	−1.663	−1.683
0.60	−1.589	−1.573	−1.584	−1.602	−1.612	−1.625	−1.643	−1.662	−1.685	−1.711
0.70	−1.549	−1.560	−1.582	−1.608	−1.622	−1.638	−1.661	−1.686	−1.713	−1.744
0.80	−1.521	−1.555	−1.586	−1.620	−1.636	−1.656	−1.685	−1.714	−1.747	−1.783
0.90	−1.502	−1.556	−1.595	−1.637	−1.656	−1.680	−1.713	−1.747	−1.785	−1.826
1.00	−1.490	−1.562	−1.610	−1.658	−1.681	−1.707	−1.745	−1.784	−1.827	−1.874
1.25	−1.486	−1.596	−1.662	−1.727	−1.758	−1.793	−1.842	−1.893	−1.949	−2.012
1.50	−1.508	−1.650	−1.733	−1.814	−1.853	−1.896	−1.958	−2.020	−2.091	−2.169
1.75	−1.547	−1.719	−1.819	−1.916	−1.962	−2.015	−2.088	−2.164	−2.247	−2.341
2.00	−1.598	−1.799	−1.917	−2.029	−2.083	−2.144	−2.230	−2.318	−2.416	−2.526
2.50	−1.727	−1.986	−2.138	−2.283	−2.351	−2.430	−2.540	− 2.654	−2.780	−2.921
3.00	−1.880	−2.199	−2.384	−2.560	−2.644	−2.740	−2.874	−.014	−3.169	−3.342
3.50	−2.051	−2.429	−2.647	−2.855	−2.953	−3.067	−3.226	−3.391	−3.574	−3.780
4.00	−2.237	−2.672	−2.922	−3.161	−3.275	−3.406	−3.589	−3.779	−3.990	−4.228
4.50	−2.434	−2.924	−3.206	−3.476	−3.605	−3.753	−3.960	− 4.176	−4.416	−4.685
5.00	−2.638	−3.183	−3.497	−3.798	−3.941	−4.107	−4.338	−4.579	−4.847	−5.148
6.00	−3.062	−3.715	−4.092	−4.455	−4.627	−4.827	−5.106	−5.397	−5.721	−6.086
7.00	−3.499	−4.260	−4.699	−5.123	−5.325	−5.559	−5.886	−6.227	−6.608	−7.036
8.00	−3.945	−4.812	−5.315	−5.800	−6.031	−6.300	−6.674	−7.066	−7.502	−7.992
9.00	−4.397	−5.371	−5.936	−6.482	−6.742	−7.045	−7.468	−7.909	−8.401	−8.953
10.00	−4.852	−5.933	−6.560	−7.168	−7.458	−7.794	−8.264	−8.755	−9.302	−9.918

Source: After Land, 1975

Appendix

Table J.1   Coefficients for Calculating Normal Distribution One-Sided 100(1 – α)% Tolerance Intervals and Confidence Intervals for Percentiles

	Percentile or coverage $p = 0.95$									
	$(1 - \alpha)$									
n	0.010	0.025	0.050	0.100	0.200	0.800	0.900	0.950	0.975	0.990
2	0.000	0.273	0.475	0.717	1.077	6.464	13.090	26.260	52.559	131.426
3	0.295	0.478	0.639	0.840	1.126	3.604	5.311	7.656	10.927	17.370
4	0.443	0.601	0.743	0.922	1.172	2.968	3.957	5.144	6.602	9.083
5	0.543	0.687	0.818	0.982	1.209	2.683	3.400	4.203	5.124	6.578
6	0.618	0.752	0.875	1.028	1.238	2.517	3.092	3.708	4.385	5.406
7	0.678	0.804	0.920	1.065	1.261	2.407	2.984	3.399	3.940	4.728
8	0.727	0.847	0.958	1.096	1.281	2.328	2.754	3.187	3.640	4.285
9	0.768	0.884	0.990	1.122	1.298	2.268	2.650	3.031	3.424	3.972
10	0.804	0.915	1.017	1.144	1.313	2.220	2.568	2.911	3.259	3.738
11	0.835	0.943	1.041	1.163	1.325	2.182	2.503	2.815	3.129	3.556
12	0.862	0.967	1.062	1.180	1.337	2.149	2.448	2.736	3.023	3.410
13	0.887	0.989	1.081	1.196	1.347	2.122	2.402	2.671	2.936	3.290
14	0.909	1.008	1.098	1.210	1.356	2.098	2.363	2.614	2.861	3.189
15	0.929	1.026	1.114	1.222	1.364	2.078	2.329	2.566	2.797	3.102
16	0.948	1.042	1.128	1.234	1.372	2.059	2.299	2.524	2.742	3.028
17	0.965	1.057	1.141	1.244	1.379	2.043	2.272	2.486	2.693	2.963
18	0.980	1.071	1.153	1.254	1.385	2.029	2.249	2.453	2.650	2.905
19	0.995	1.084	1.164	1.263	1.391	2.016	2.227	2.423	2.611	2.854
20	1.008	1.095	1.175	1.271	1.397	2.004	2.208	2.396	2.576	2.808
21	1.021	1.107	1.184	1.279	1.402	1.993	2.190	2.371	2.544	2.766

	Percentile or coverage $p = 0.95$									
	$(1 - \alpha)$									
n	0.010	0.025	0.050	0.100	0.200	0.800	0.900	0.950	0.975	0.990
22	1.033	1.117	1.193	1.286	1.407	1.983	2.174	2.349	2.515	2.729
23	1.044	1.127	1.202	1.293	1.412	1.973	2.159	2.328	2.489	2.694
24	1.054	1.136	1.210	1.300	1.416	1.965	2.145	2.309	2.465	2.662
25	1.064	1.145	1.217	1.306	1.420	1.957	2.132	2.292	2.442	2.633
26	1.074	1.153	1.225	1.311	1.424	1.949	2.120	2.275	2.421	2.606
27	1.083	1.161	1.231	1.317	1.427	1.943	2.109	2.260	2.402	2.581
28	1.091	1.168	1.238	1.322	1.431	1.936	2.099	2.246	2.384	2.558
29	1.099	1.175	1.244	1.327	1.434	1.930	2.089	2.232	2.367	2.536
30	1.107	1.182	1.250	1.332	1.437	1.924	2.080	2.220	2.351	2.515
35	1.141	1.212	1.276	1.352	1.451	1.900	2.041	2.167	2.284	2.430
40	1.169	1.236	1.297	1.369	1.462	1.880	2.010	2.125	2.232	2.364
50	1.212	1.274	1.329	1.396	1.480	1.852	1.965	2.065	2.156	2.269
60	1.245	1.303	1.354	1.415	1.493	1.832	1.933	2.022	2.103	2.202
120	1.352	1.395	1.433	1.478	1.533	1.772	1.841	1.899	1.952	2.015
240	1.431	1.463	1.492	1.525	1.565	1.773	1.780	1.819	1.854	1.896
480	1.491	1.514	1.535	1.558	1.588	1.706	1.738	1.766	1.790	1.818
∞	1.645	1.645	1.645	1.645	1.645	1.645	1.645	1.645	1.645	1.645

References

◆◆

Chapter 1

American Society of Civil Engineers, *Consumptive Use of Water and Irrigation Water Requirements*, Technical Committee of the Irrigation and Drainage Division, ASCE, New York, 1973.

American Society of Civil Engineers, "Describing Irrigation Efficiency and Uniformity," On-Farm Irrigation Committee of the Irrigation and Drainage Division, *ASCE J. Irrig. Drain. Div.*, v. 104, n. IR 1, pp. 35–42, 1978.

American Society of Civil Engineers, *Principles of Project Formulation for Irrigation and Drainage Projects*, Technical Committee of the Irrigation and Drainage Division, ASCE, New York, 1982.

American Water Works Association, *Energy and Water Use Forecasting*, AWWA, Denver, CO, 1980.

Blaney, H. F., "Monthly Consumptive Use Requirements for Irrigated Crops," *ASCE J. Irrig. Drain. Div.*, v. 85, n. IR1, pp. 1–12, 1959.

Blaney, H. F., and Criddle, W. D., *Determining Water Requirements in Irrigated Areas from Climatological Data*, U.S. Dept. of Agriculture, Soil Conservation Service (now the NRCS), Washington, DC, 1945.

Blaney, H. F., and Criddle, W. D., *Determining Consumptive Use and Irrigation Water Requirements*, Tech. Bull. 1275, U.S. Dept. of Agriculture, Washington, DC, 1962.

Carl, K. J., Young, R. A., and Anderson, G. C., "Guidelines for Determining Fire-Flow Requirements," *J. Am. Water Works Assoc.*, v. 65, May 1973.

Criddle, W. D., "Methods of Computing Consumptive Use of Water," *ASCE J. Irrig. Drain. Div.*, v. 84, n. IR1, pp. 1–27, 1958.

Davis, C. V., and Sorenson, K. E. (eds.), Handbook of Applied Hydraulics, 3rd ed., Sections 30, 31, 33, and 36, McGraw-Hill, New York, 1969.

Davis, D. W., *Technical Factors in Small Hydropower Planning*, Tech. Paper 61, Hydrologic Engineering Center, U.S. Army Corps of Engineers, Davis, CA, 1979.

Davis, D. W., and Smith, B. W., *Feasibility Analysis in Small Hydropower Planning*, Technical Committee of the Irrigation and Drainage Division, ASCE, New York, 1982.

Fair, G. M., Geyer, J. C., and Okun, D. A., *Water and Wastewater Engineering*, vol. 1, *Water Supply and Wastewater Removal*, John Wiley & Sons, New York, 1966.

FAO/UNESCO, *Irrigation, Drainage, and Salinity: An International Source Book*, Hutchinson Publishing, London, 1973.

Fredrich, A. J., *Hydroelectric Power Analysis in Reservoir Systems*, Tech. Paper 24, Hydrologic Engineering Center, U.S. Army Corps of Engineers, Davis, CA, 1970.

Fritz, J. J., *Small and Mini Hydropower Systems*, McGraw-Hill, New York, 1984.

Goodman, A. S., *Principles of Water Resources Planning*, Prentice-Hall, Englewood Cliffs, NJ, 1984.

Hansen, V. E., Israelson, O. W., and Stringham, G. E., *Irrigation Principles and Practices*, 4th ed., John Wiley & Sons, New York, 1980.

Hargreaves, G. H., "Water Requirements and Man-Induced Climate Change," *ASCE J. Irrig. Drain. Div.* v. 107, n. IR3, pp. 247–255, 1981.

Hart, F. C., "Experience with Seepage Control in the Pacific Northwest," *Proc. of the Seepage Symposium*, ARS 41–90, pp. 93–97, Phoenix, AZ, 1963.

Helweg, O. J., and Alvarez, D., "Estimating Irrigation Water Quantity and Quality," *ASCE J. Irrig. Drain. Div.*, v. 106, n. IR3, pp. 175–188, 1980.

Holtz, D., and Sebastian, S., *Municipal Water Systems*, Indiana Univ. Press, Bloomington, IN, 1978.

Kollar, K. L., and Macauley, P., "Water Requirements for Industrial Development," *J. Amer. Water Works Assoc.*, v. 72, n. 1, 1980.

Kuiper, E., *Water Resources Development: Planning, Engineering and Economics*, Butterworth & Company Ltd., London, 1965.

Linsley, R. K., Franzini, J. B. Freyberg, D. L., and Tchobanoglous, G., *Water Resources Engineering*, 4th ed., McGraw-Hill, New York, 1991.

McClean, J. E., "More Accurate Population Estimates by Means of Logistic Curves," *Civ. Eng.*, pp. 35–37, Feb. 1952.

McGhee, T. J., *Water Supply and Sewerage*, McGraw-Hill, New York, 1991.

McJunkin, F. E., "Population Forecasting by Sanitary Engineers," *ASCE J. Sanit. Eng. Div.*, v. 90, n. SA4, pp. 31–55, 1964.

Munson, W. C., "Method for Estimating Consumptive Use of Water for Agriculture," *ASCE J. Irrig. Drain. Div.*, v. 88, n. IR4, pp. 45–57, 1960.

National Research Council, *Estimating Water Use in the U. S.—A New Paradigm for the National Water Use*, National Academy Press, Washington, DC, 2002.

United Nations, *The Demand for Water: Procedures and Methodologies for Projecting Water Demands in the Context of Regional and National Planning*, Water Resources Series No. 3, U.N., New York, 1976.

U.S. Geological Survey, *Guidelines for Preparation of State Water Use Estimates for 2005*, Tech. and Methods Bk. 4, Ch. E1, Reston, VA, 2007.

U.S. National Water Commission, *Forecasting Water Demands*, compiled by R. G. Thompson and others, Report PB 206 491, U.S. NWC, Springfield, VA, 1971.

U.S. Soil Conservation Service (now the NRCS), *Irrigation Water Requirements*, Tech. Release 21(revised), U.S. Dept. of Agriculture, Washington, DC, 1970.

U.S. Soil Conservation Service (now the NRCS), *Crop Consumptive Irrigation Requirements and Irrigation Efficiency Coefficients for the U.S.*, U.S. Dept. of Agriculture, Washington, DC, 1976.

U.S. Water Resources Council, *The Nation's Water Resources 1975–2000*, Second National Water Assessment, U.S. GPO, Washington, DC, 1978.

Viessman, W., and Hammer, M. J., *Water Supply and Pollution Control*, Prentice-Hall, Englewood Cliffs, NJ, 2004.

Warnick, C. C., *Hydropower Engineering*, Prentice-Hall, Englewood Cliffs, NJ, 1984.

Worstell, R. V., "Estimating Seepage Losses from Canal Systems," *ASCE J. Irrig. Drain. Div.*, v. 102, n. IR1, pp. 137–147, 1975.

Chapter 2

Allen, R. G., Jensen, M. E., Wright, J. L., and Burman, R. D., "Operational Estimates of Evapotranspiration," *Agron. Jour.*, v. 81, pp. 650–662, 1989.

Allen, R. G., et al. (eds.), *Lysimeters for Evapotranspiration and Environmental Measurements*, ASCE, New York, 1991.

Alley, W. M., "Treatment of Evapotranspiration, Soil Moisture Accounting and Aquifer Recharge in Monthly Water Balance Models," *Water Resour. Res.*, v. 20, n. 8, pp. 1137–1149, Aug. 1984.

American Society of Agriculture Engineers, *Advances in Evapotranspiration*, Publication #74–85, ASAE, St. Joseph, MI, 1985.

American Society of Civil Engineers, *Hydrology Handbook*, 2nd ed., Manuals and Reports on Engineering Practice n. 28, ASCE, New York, 1996.

Bell, F. C., "Generalized Rainfall-Duration-Frequency Relationships," *J. Hydraul. Div.*, v. 95, n. HY1, pp. 311–326, 1969.

Chow, V. T. (ed.), *Handbook of Applied Hydrology*, Secs. 9–12, 14, McGraw-Hill, New York, 1964.

Chow, V. T., Maidment, D. R., and Mays, L. W., *Applied Hydrology*, McGraw-Hill, New York, 1988.

Chu, S. T., "Infiltration During Unsteady Rain," *Water Resour. Res.*, v. 14, n. 3, pp. 461–466, 1978.

Cook, H. L., "The Infiltration Approach to the Calculation of Surface Runoff," *Trans. Am. Geophys. Union*, v. 27, n. V, pp. 726–743, Oct. 1946.

Criddle, W. D., "Methods of Computing Consumptive Use of Water," *ASCE J. Irrig. Drain. Div.*, v. 84, n. IR1, 1958.

Doorenbos, J., and Pruitt, W. O., *Crop Water Requirements*, Irrigation and Drainage Paper n. 24, U.N. Food and Agriculture Organization, Rome, Italy, 1977.

Gray, D. M. (ed.), *Handbook on the Principles of Hydrology*, National Research Council of Canada, Water Information Center, Inc., Port Washington, NY, 1973.

Green, W. H., and Ampt, G., "Studies of Soil Physics, Part I: The Flow of Air and Water through Soils," *J. Agric. Sci.*, v. 4, n. 1, pp. 1–24, 1911.

Haan, C. T., "A Water Yield Model for Small Watersheds," *Water Resour. Bull.*, v. 8, n. 1, pp. 58–69, 1972.

Hanks, R. J., "Model for Predicting Plant Yield as Influenced by Water Use," *Agron. J.*, v. 66, pp. 600–665, 1974.

Harbeck, G. E., *A Practical Field Technique for Measuring Reservoir Operation Utilizing Mass Transfer Theory*, Professional Paper 272-E, U.S. Geological Survey, Washington, DC, 1962.

Harbeck, G. E., and Cruse, R. R., *Evaporation Control Research*, 1955–58, Water Supply Paper 1480, U.S. Geological Survey, Washington, DC, 1960.

Hargreaves, G. H., and Samani, Z. A., "Estimating Potential Evapotranspiration," *ASCE J. Irrig. Drain., Div.*, v. 108, n. 3., pp. 225–230, 1982.

Hjelmfelt, A. T., and Cassidy, J. J., *Hydrology for Engineers and Planners*, Iowa State Univ. Press, Ames, IA, 1975.

Holton, H. N., et al., *USDAHL-74 Revised Model of Watershed Hydrology*, Tech. Bull. 1518, Agriculture Research Service, U.S. Dept. of Agriculture, Washington, DC, 1975.

Horton, R. E., "Analyses of Runoff-Plat Experiments with Varying Infiltration Capacity," *Trans. Am. Geophys. Union*, pt. 4, pp. 693–711, 1939.

Horton, R. E., "A Simplified Method of Determining the Constants in the Infiltration-Capacity Equation," *Trans. Am. Geophys. Union*, pp. 575–577, 1942.

Huggins, L. F., and Monke, E. J., *The Mathematical Simulation of the Hydrology of Small Watersheds*, Tech. Rep. 1, Water Resources Research Center, Purdue Univ., West Lafayette, IN, 1966.

Hydrologic Engineering Center (HEC), *Hydrograph Analysis*, Hydrologic Engineering Methods for Water Resources Development, Vol. 4, U.S. Army Corps of Engineers, Davis, CA, 1973.

Jensen, M. E. (ed.), *Consumptive Use of Water and Irrigation Water Requirements*, Report Tech. Comm. on Irrig. Water Requirements, ASCE, New York, 1974.

Jensen, M. E., Burman, R. D., and Allen, R. G., *Evapotranspiration and Irrigation Water Requirements*, ASCE Manuals and Report on Eng. Practice n. 70, New York, 1990.

Jensen, M. E., and Haise, H. R., "Estimating Evapotranspiration from Solar Radiation," *ASCE J. Irrig. Drain. Div.*, v. 89, n. IR4, pp. 15–41, 1963.

Jones, F. E., *Evaporation of Water with Emphasis on Applications and Measurements*, Lewis Publishers, Chelsea, MI, 1992.

References

Kohler, M. A., and Parmele, L. H., "Generalized Estimates of Free Water Evaporation," *Water Resour. Res.*, v. 3. n. 4, pp. 997–1005, 1967.

Kohler, M. A., Nordenson, T. J., and Fox, W. E., *Evaporation From Pans and Lakes*, Research Paper 38, U.S. Weather Bureau, Washington, DC, 1955.

Larson, L.W., and Peck, E. L., "Accuracy of Precipitation Measurements for Hydrologic Modeling," *Water Resour. Res.*, v. 10, n. 4, Aug. 1974.

Linsley, R. K., "A Simple Procedure for the Day-to-day Forecasting of Runoff from Snow Melt," *Trans. Am. Geophys. Union*, pp. 62–66, 1943.

Linsley, R. K., Kohler, M. A., and Paulhus, J. L. H., *Hydrology for Engineers*, 3rd ed., McGraw-Hill, New York, 1982.

List, R. J., *Smithsonian Meteorological Tables*, 6th revised ed., Smithsonian Institute, Washington, DC, 1984.

Maidment, D. R., *Handbook of Hydrology*, McGraw-Hill, New York, 1993.

Martin, D. L., et al., "SCS Methods for Determining Irrigation Water Requirements," *Proc. ASCE National Conference on Irrig. and Drain. Eng.*, pp. 1013–1038, 1993.

Mather, J. R., "Using Computed Streamflow in Watershed Analysis," *Water Resour. Bull.*, v. 17, n. 3, pp. 474–482, 1981.

McGhee, T. J., *Water Supply and Sewerage*, 5th ed., McGraw-Hill, New York, 1991.

Mein, R. G., and Larson, C. L., *Modeling Infiltration Component of the Rainfall-Runoff Process*, Bull. 43, Water Resources Research Center, Univ. of Minnesota, Minneapolis, MN, 1971.

Mein, R. G., and Larson, C. L., "Modeling Infiltration during a Steady Rain," *Water Resour. Res.*, v. 9, n. 2, pp. 384–394, 1973.

Mishra, S. K., and Singh, V. P., "Another Look at SCS-CN Method," *ASCE J. of Hydrol. Eng.*, v. 4, n. 3, July 1999.

Monteith, J. L., "Evaporation and the Environment," in *The State and Movement of Water in Living Organisms*, XIXth Symposium, Soc. for Exp. Biol., Swansea, Cambridge Univ. Press, London, 1965.

Monteith, J. L., and Unsworth, M. H., *Principles of Environmental Physics*, 2nd ed., Edward Arnold, UK, 1990.

Morel-Seytoux, H. J., "Two Phase Flows in Porous Media," *Adv. Hydrol.*, v. 9, 1973.

Morel-Seytoux, H. J., and Khanji, J., "Derivation of an Equation of Infiltration," *Water Resour. Res.*, v. 10, n. 4, pp. 795–800, 1974.

Mustonen, S. E., and McGuinness, J. L., "Lysimeter and Evapotranspiration," *Water Resour. Res.*, v. 3, n. 4, pp. 989–996, Feb. 1967.

Osborn, H. B., "Estimating Precipitation in Mountainous Regions," *ASCE J. Hydraul. Div.*, v. 110, n. 12, pp. 1859–1863, 1984.

Palmer, W. C., *Meteorological Drought*, Research Paper 45, U.S. Weather Bureau, Washington, DC, 1965.

Penman, H. L., "Natural Evaporation from Open Water, Bare Soil and Grass," *Proc. Royal Society*, v. A193, pp. 120–145, London, 1948.

Philip, J. R., "Theory of Infiltration," *Adv. Hydrosci.*, v. 5, 1969.

Rawls, W. J., Brakensiek, D. L., and Miller, N., "Green-Ampt Infiltration Parameters from Soil Data," *ASCE J. Hydraul. Div.*, v. 109, n. 1, pp. 62–70, 1983.

Reinhart, K. G., and Taylor, R. E., "Infiltration and Available Water Storage Capacity in the Soil," *Trans. Am. Geophys. Union*, v. 35. n. 5, pp. 791–794, 1954.

Ripple, C. D., Rubin, J., and Van Hylckama, T. E. A., *Estimating Steady-State Evaporation Rates from Bare Soils under Conditions of High Water Table*, Water Supply Paper 2019-A, U.S. Geological Survey, Washington, DC, 1972.

Robinson, T. W., and Johnson, A. I., *Selected Bibliography on Evaporation and Transpiration*, Water Supply Paper 1539-R, U.S. Geological Survey, Washington, DC, 1961.

Searcy, J. K., and Hardison, C. H., *Double Mass Curves*, Water Supply Paper 1541-B, U.S. Geological Survey, Washington, DC, 1960.

Sherman, L. K., "Derivation of Infiltration Capacity from Average Loss Rates," *Trans. Am. Geophys. Union*, v. 21. pp. 541–550, 1940.

Sherman, L. K., and Musgrave, G.W., "Infiltration" in *Physics of the Earth, IX: Hydrology*, ed. O. E. Meinzer, McGraw-Hill, New York, 1942.

Skaggs, R. W., and Khaleel, R., "Infiltration," in *Hydrologic Modeling of Small Watersheds*, ed. C. T Haan, H. P. Johnson, and D. L. Brakensiek, American Society of Agricultural Engineers, St. Joseph, MI, 1982.

Smith, R. E., and Parlange, J. Y., "A Parameter- Efficient Hydrologic Infiltration Model," *Water Resour. Res.*, v. 14, n. 3, pp. 533–538, 1978.

Snyder, R. L., et al., *Using Reference Evapotranspiration and Crop Coefficients to Estimate Crop Evapotranspiration*, Leaflets n. 21427 and n. 21428, Cooperative Extension, Univ. of California, Berkeley, 1989.

Sokolov, A. A., and Chapman, T.G., *Methods for Water Balance Computations*, Studies and Reports in Hydrology, UNESCO, Paris, 1974.

Thomas, H. A., *Improved Methods for National Water Assessment*, Report, Harvard Water Resources Group, Harvard Univ., Cambridge, MA, 1981.

Thornthwaite, C.W., and Mather, J. R., *The Water Balance*, publication of the Climatology Lab., Climatol. Drexel Inst. Technol., v. 8, n. 1, 1955.

U.S. Army Corps of Engineers, *Runoff from Snowmelt*, Engineering Manual 1110-2-1406, Washington, DC, 1960.

U.S. Army Corps of Engineers, *Snow Hydrology*, North Pacific Div., Portland, OR, 1965.

U.S. Bureau of Reclamation and U.S. Forest Service, *Factors Affecting Snowmelt and Streamflow*, U.S. Dept. of Interior, Washington, DC, 1958.

U.S. Geological Survey, *National Handbook of Recommended Methods for Water Data Acquisition*, Chap. 8, "Evaporation and Transpiration," 1982; prepared by agencies of the U.S. government under the sponsorship of the Water Data Coordinator, U.S. Geological Survey, 1977.

U.S. Nat. Resources Conservation Service, *National Engineering Handbook*, 210-630 (NEH Part 630), Chapters from 1971–2004, U. S. Dept. of Agriculture, Washington, DC, 2004.

U.S. Soil Conservation Service, *Engineering Field Manual for Soil Conservation Service Practice*, U.S. Dept. of Agriculture, Washington, DC, 1975.

U.S. Soil Conservation Service, *Urban Hydrology for Small Watersheds*, Tech. Release 55, U.S. Dept. of Agriculture, Washington, DC, 1986.

U.S. Weather Bureau, *Rainfall Frequency Atlas of the United States*, Tech. Paper 40, Washington, DC, 1961.

Webb, E. K., "A Pan-Lake Evaporation Relationship," *J. Hydrol.*, v. 4, pp. 1–11, 1966.

Weiss, L. L., and Wilson, W. T., "Evaluation of Significance of Slope Changes in Double-Mass Curves," *Trans. Am. Geophys. Union*, v. 34, n. 6, pp. 893–896, Dec. 1953.

Winter, T. C., "Uncertainties in Estimating the Water Balance of Lakes," *Water Resour. Bull.*, v. 17, n. 1, pp. 82–115, Feb. 1981.

World Meteorological Organization, *Manual for Depth-Area-Duration Analysis of Storm Precipitation*, WMO No. 237, Tech. Paper 129, Geneva, 1969.

World Meteorological Organization, *Preparation of Maps of Precipitation and Evaporation with Special Regard to Water Balance*, WMO, Geneva, 1970.

Wright, J. L., "New Evapotranspiration Crop Coefficients," *ASCE J. Irrig. and Drain. Eng.*, v. 108, n. IR2, pp. 57–74, 1982.

References

Chapter 3

Bennett, G. D., *Introduction to Groundwater Hydraulics*, Techniques of Water Resources Investigations, Chap. B2, Book 3, U.S. Geological Survey, Washington, DC, 1976.

Bowen, R., *Groundwater*, Elsevier Applied Science, London, 1986.

Cooper, H. H., "The Equation of Groundwater Flow in Fixed and Deforming Coordinates," *J. Geophys. Res.*, v. 71, n. 20, pp. 4785–4790, 1966.

Davis, S. N., and DeWiest, R. J. M., *Hydrogeology*, John Wiley & Sons, New York, 1966.

Delleur, J. W. (ed.), *The Handbook of Groundwater Engineering*, 2nd ed., CRC Press, Boca Raton, FL, 2006.

DeWiest, R. J. M., *Geohydrology*, John Wiley & Sons, New York, 1965.

DeWiest, R. J. M., "On the Storage Coefficient and Equation of Groundwater Flow," *J. Geophys. Res.*, v. 71, n. 4, pp. 1117–1122, 1966.

DeWiest, R. J. M. (ed.), *Flow through Porous Media*, Academic Press, New York, 1969.

Fetter, C. W., *Applied Hydrogeology*, 4th ed., Prentice-Hall, Englewood Cliffs, NJ, 2000.

Freeze, A. R., and Cherry, J. A., *Groundwater*, Prentice-Hall, Englewood Cliffs, NJ, 1979.

Gupta, R. S., *Groundwater Reservoir Operation for Drought Management*, Ph.D. dissertation, Polytechnic Univ. of New York, 1983.

Hantush, M. S., "Depletion of Storage, Leakage, and River Flow in Gravity Wells in Sloping Sand," *J. of Geophys. Res.*, v. 69, n. 12, p. 2551–2560, 1964.

Heath, R. C., *Basic Ground-Water Hydrology*, Water Supply Paper 2220, U.S. Geological Survey, Washington, DC, 1983.

Hermance, J. F., *A Mathematical Primer on Groundwater Flow*, Prentice-Hall, Upper Saddle River, NJ, 1999.

Jacob, C. E., "Flow of Groundwater" in *Engineering Hydraulics*, ed. H. Rouse, John Wiley & Sons, New York, 1950.

Jorgensen, D. G., Relationship between Basic Soil Engineering Equations and Basic Ground Water Flow Equations, Water Supply Paper 2064, U.S. Geological Survey, Washington, DC, 1980.

Linsley, R. K., Kohler, M. A., and Paulhus, J. L. H., *Hydrology for Engineers*, 3rd ed., McGraw-Hill, New York, 1982.

Lohman, S. W., *Ground-Water Hydraulics*, Professional Paper 708, U.S. Geological Survey, Washington, DC, 1972.

Marino, M.A., and Luthin, J.N., *Seepage and Groundwater*, Developments in Water Science Series No. 13, Elsevier Science, New York, 1982.

McWhorter, D. B., and Sunada, D. K., *Ground-Water Hydrology and Hydraulics*, Water Resources Publications, Fort Collins, CO, 1981.

Meinzer, O. E., *Outline of Groundwater Hydrology*, Water Supply Paper 494, U.S. Geological Survey, Washington, DC, 1923 (reprint 1960).

Pinneker, E. V. (ed.), *General Hydrogeology*, Cambridge Univ. Press, Cambridge, 1983.

Todd, D. K., and Mays, L. W., *Groundwater Hydrology*, 3rd ed., John Wiley & Sons, New York, 2004.

UNESCO/WMO, *International Glossary of Hydrology*, World Meteorological Organization, 1974.

U.S Department of Interior, Water and Water Resources Service, *Groundwater Manual*, U.S. GPO, Washington, DC, 1977 (reprint 1981).

Verruijit, A., *Theory of Groundwater Flow*, 2nd ed., Palgrave Macmillan, London, 1982.

Viessman, W., and Lewis, G. L., *Introduction to Hydrology*, 5th ed., HarperCollins, New York, 2002.

Walton, W. C., *Groundwater Resources Evaluation*, McGraw-Hill, New York, 1970.

Chapter 4

Batu, V., *Aquifer Hydraulics: A Comprehensive Guide to Hydrogeologic Data Analysis*, John Wiley & Sons, New York, 1998.

Bennett, G. D., *Introduction to Ground-Water Hydraulics*, Techniques of Water Resources Investigations, Chap. B2. Book 3, U.S. Geological Survey, Washington, DC, 1976.

Bentall, Ray (compiler), *Methods of Determining Permeability, Transmissibility and Drawdown*, Water-Supply Paper 1536–I, U.S. Geological Survey, Washington, DC, 1963.

Bierschenk, W. H., "Determining Well Efficiency by Multiple Step-Drawdown Tests," *International Assoc. of Scientific Hydrology*, n. 64, 1964.

Birsoy, Y. K., and Summers, W. K., "Determination of Aquifer Parameters from Step Tests and Intermittent Pumping Data," *Ground Water*, v. 18, n. 2, 1980.

Boulton, N. S., "Analysis of Data from Non-equilibrium Pumping Tests Allowing for Delayed Yield from Storage," *Proc. Inst. of Civil Engineers*, London, v. 26, pp. 269–282, 1963.

Boulton, N. S., "The Influence of the Delayed Drainage on Data from Pumping Tests in Unconfined Aquifers," *J. Hydrol.*, v. 19, pp. 157–169, 1973.

Brown, R. H., et al. (eds.), *Ground-Water Studies: An International Guide for Research and Practice*, UNESCO, Paris, 1977.

Bruggeman, G. A., *Analytical Solution of Geohydrological Problems*, Elsevier Science, New York, 1999.

Chow, V. T., "On the Determination of Transmissibility and Storage Coefficients from Pumping Test Data," *Trans. Am. Geophys. Union*, v. 33, pp. 397–404, 1952.

Davis, S. N., and DeWiest, R. J. M., *Hydrogeology*, John Wiley & Sons, New York, 1966.

De Vries, J. J., "Some Calculation Methods for Determination of the Travel Time of Groundwater," *Aqua-Vu*, ser. A, no. 5, Comm. of the Institute of Earth Sciences, Free Reformed Univ., Amsterdam, 1972.

De Vries, J. J., "Groundwater Hydraulics," *Aqua-Vu*, ser. A, no. 6, Comm. of the Institute of Earth Sciences, Free Reformed Univ., Amsterdam, 1975.

DeWiest, R. J. M., *Geohydrology*, John Wiley & Sons, New York, 1965.

DeWiest, R. J. M. (ed.), *Flow through Porous Media*, Academic Press, New York, 1969.

Driscoll, F. G., *Groundwater and Wells*, 2nd ed., Johnson Division, St. Paul, MN, 1986.

Fetter, C. W., *Applied Hydrogeology*, 4th ed., Macmillan, New York, 2000.

Freeze, A. R., and Cherry, J. A., *Groundwater*, Prentice-Hall, Englewood Cliffs, NJ, 1979.

Glover, R. E., *Transient Groundwater Hydraulics*, Water Resources Publications, Littleton, CO, 1977.

Gupta, R. S., *Groundwater Reservoir Operation for Drought Management*, Ph.D. dissertation, Polytechnic Univ. of New York, 1983.

Hantush, M. S., "Analysis of Data from Pumping Tests in Leaky Aquifers," *Trans. Amer. Geophys. Union*, v. 37, n. 6, pp. 702–714, 1956.

Hantush, M. S., "Modification of the Theory of Leaky Aquifers," *J. Geophys. Res.*, v. 65, pp. 3713–3725, 1960.

Hantush, M. S., "Hydraulics of Wells," in *Advances in Hydroscience*, vol. I, ed. V. T. Chow, Academic Press, New York, 1964a.

Hantush, M. S., "Depletion of Storage, Leakage and River Flow by Gravity Wells in Sloping Sand," *J. Geophys. Res.*, v. 69, n. 12, pp. 2551–2560, 1964b.

Hantush, M. S., and Jacob, C. E., "Nonsteady Radial Flow in an Infinite Leaky Aquifer," *Trans. Amer. Geophys. Union*, v. 36, n. 1, pp. 95–100, 1955.

Harr, M. E., *Groundwater and Seepage*, McGraw-Hill, New York, 1962.

Heath, R. C., *Basic Ground-Water Hydrology*, Water Supply Paper 2220, U.S. Geological Survey, Washington, DC, 1983.

Jacob, B., *Hydraulics of Groundwater*, Dover Publications, New York, 2007.

Jacob, C. E., "Radial Flow in a Leaky Artesian Well," *Trans. Amer. Geophys. Union*, v. 27, n. 2, 1946.

Jacob, C. E., "Drawdown Test to Determine Effective Radius of Artesian Well," *Trans. Amer. Soc. of Civil Engineers*, v. 112, 1947.

Jacob, C. E., "Flow of Ground Water," in *Engineering Hydraulics*, ed. H. Rouse, John Wiley & Sons, New York, 1950.

Jacob, C. E., *Determining the Permeability of Water Table Aquifers*, USGS Water Supply Paper 1536–1, 1963.

Jenkins, C. T., *Computation of Rate and Volume of Stream Depletion by Wells*, Techniques of Water Resources Investigations, Chap. D1, Book 4, U.S. Geological Survey, Washington, DC, 1968.

Kasenow, M., *Introduction to Aquatic Analysis*, 4th ed., Water Resources Publications, Highland Ranch, CO, 1997.

Kasenow, M., *Applied Ground-Water Hydrology and Well Hydraulics*, 2nd ed., Water Resources Publications, Highland Ranch, CO, 2001.

Kasenow, M., and Pare, P., *Leaky Confined Aquifers and the Hantush Inflection Point*, Water Resources Publications, Highland Ranch, CO, 1994.

Kawecki, M. W., "Meaningful Interpretation of Step Drawdown Tests," *Ground Water*, v. 33, n. 1, 1995.

Lang, S. M., *Methods for Determining the Proper Spacing of Wells in Artesian Aquifers*, Water Supply Paper 1545-B, U.S. Geological Survey, Washington, DC, 1961.

Li, W. H., *Differential Equations of Hydraulic Transients, Dispersion and Groundwater Flow*, Prentice-Hall, Englewood Cliffs, NJ, 1972.

Linsley, R. K., Kohler, M. A., and Paulhus, J. L. H., *Hydrology for Engineers*, 3rd ed., McGraw-Hill, New York, 1982.

Lohman, S. W., *Ground-Water Hydraulics*, Professional Paper 708, U.S. Geological Survey, Washington, DC, 1972.

Marino, M. A., and Luthin, J.N., *Seepage and Groundwater*, Developments in Water Science Series No. 13, Elsevier Science, New York, 1982.

McWhorter, D. B., and Sunada, D. K., *Ground-Water Hydrology and Hydraulics*, Water Resources Publications, Fort Collins, CO, 1977.

Neuman, S. P., "Supplementary Comments on Theory of Flow in Unconfined Aquifers Considering Delayed Response of the Water Table," *Water Resour. Res.*, v. 8, pp. 1031–1045, 1972.

Neuman, S. P., "Supplementary Comments on Theory of Flow in Unconfined Aquifers Considering Delayed Response of the Water Table," *Water Resour. Res.*, v. 9, pp. 1102–1103, 1973.

Neuman, S. P., "Analysis of Pumping Test Data from Anisotropic Unconfined Aquifers Considering Delayed Gravity Response," *Water Resour. Res.*, v. 11, pp. 329–342, 1975.

Neuman, S. P., and Witherspoon, P. A., "Applicability of Current Theories of Flow in Leaky Aquifers," *Water Resour. Res.*, v. 5, pp. 817–829, 1969.

Polubarinova-Kochina, P. Ya., *Theory of Groundwater Movement* (translated from Russian by R. J. M. DeWiest), Princeton Univ. Press, Princeton, NJ, 1962.

Prickett, T. A., and Lonnguist, C. G., *Selected Digital Computer Techniques for Groundwater Resource Evaluation*, Bulletin 55, Illinois State Water Survey, Dept. of Registration and Education, Urbana, 1971.

Ramsahoye, L. E., and Lang, S. M., *A Simple Method for Determining Specific Yield from Pumping Tests*, Water-Supply Paper 1536-C, U.S. Geological Survey, Washington, DC, 1961.

Reed, J. E., *Type Curves for Selected Problems of Flow to Wells in Confined Aquifers*, Techniques of Water Resources Investigations, Chap. B3, Book 3, U.S. Geological Survey, Washington, DC, 1980.

References

Roscoe Moss Company, *Handbook of Groundwater Development*, John Wiley & Sons, New York, 2001.

Sen, Z., *Applied Hydrogeology for Scientists and Engineers*, Lewis Publishers, Boca Raton, FL, 1995.

Shehan, N. T., "A Nongraphical Method of Determining U and $W(u)$," *Ground Water*, v. 5, n. 2, 1976.

Stallman, R. W., *Aquifer-Test Design Observation and Data Analysis*, Techniques of Water Resources Investigations, Chap. B1, Book 3, U.S. Geological Survey, Washington, DC, 1971.

Todd, D. K., "Groundwater," in *Handbook of Applied Hydrology*, ed. V. T. Chow, McGraw-Hill, New York, 1964.

Todd, D. K., *Groundwater Hydrology*, 3rd ed., John Wiley & Sons, New York, 2004.

Trescott, P. C., *Documentation of Finite-Difference Model for Simulation of Three-Dimensional Groundwater Flow*, Open File Report 75–438, U.S. Geological Survey, Washington, DC, 1975 (reprint July 1976).

Trescott, P. C., Pinder, G. F., and Larson, S. P., *Finite-Difference Model for Aquifer Simulation in Two Dimensions with Results of Experiments*, Techniques of Water Resources Investigations, Chap. C1, Book 7, U.S. Geological Survey, Washington, DC, 1976.

U.S. Bureau of Reclamation, *Groundwater Movement*, Engineering Monograph # 31, U.S. Dept. of Interior, Denver, CO, 1966.

U.S. Department of Interior, Water and Power Resources Service, *Ground Water Manual*, U.S. GPO, Washington, DC, 1977 (reprint 1981).

Verruijit, A., *Theory of Groundwater Flow*, 2nd ed., Palgrave Macmillan, London, 1982.

Viessman, W., and Lewis, G. L., *Introduction to Hydrology*, 5th ed., Prentice-Hall, Englewood Cliffs, NJ, 2002.

Walton, W. C., *Selected Analytical Methods for Well and Aquifer Evaluation*, Illinois State Water Survey Bull. 49, 1962.

Walton, W. C., *Groundwater Resources Evaluation*, McGraw-Hill, New York, 1970.

Walton, W. C., *Groundwater Pumping Tests*, Lewis Publishers, Chelsea, MI, 1988.

Wang, H. F., and Anderson, M. P., *Introduction to Groundwater Modeling: Finite Difference and Finite Element Methods*, Academic Press, NY, 1995.

Williams, D. E., "Modern Techniques in Well Design," *J. Amer. Water Works Assoc.*, v. 77, n. 9, 1985.

Wright, C. E. (ed.), *Surface Water and Ground-Water Interaction*. UNESCO, Paris, 1980.

Chapter 5

Adriano, D. C., et al. (ed.), *Contamination in Groundwater*, Science Reviewers, Northwood, UK, 1994.

Anderson, M. P. *Studies in Geophysics, Groundwater Contamination*, National Academy Press, Washington, DC, 1984.

Arial, M. M., and Liao, B., "Analytical Solution for Two dimensional Transport Equation with Time-Dependent Dispersion Coefficients," *ASCE J. Hydro. Div.*, v. 1, n. 1, pp. 20–32, Jan., 1996.

Barry, D. A., and Sposito, G., "Analytical Solution of a Convection-Dispersion Model with Time-Dependent Transport Coefficients," *Water Resour. Res.*, v. 25, n. 12, pp. 2407–2416, 1989.

Basha, H. A., and El-Habel, F. S., "Analytical Solution of the One-Dimensional Time Dependent Transport Equation," *Water Resour. Res.*, v. 29, n. 9, pp. 3209–3214, 1993.

Bear, J., *Dynamics of Fluids in Porous Media*, Dover Publications, New York, 1988.

Bear, J., *Hydraulics of Groundwater*, Dover Publications, New York, 2007.

Corey, A. T., *Mechanics of Immiscible Fluids in Porous Media*, Water Resources Publications, Highlands Ranch, CO, 1994.

Crank, J., *The Mathematics of Diffusion*, 2nd ed., Oxford Press, New York, 1980.

Crow, E. L., and Shimizu, K. (eds.), *Lognormal Distribution: Theory and Applications*, Marcel Dekker, New York, 1988.

Cussler, E. L., *Diffusion: Mass Transfer in Fluid Systems*, 2nd ed., Cambridge Univ. Press, Cambridge, UK, 1997.

Dagan, G., *Flow and Transport in Porous Formations*, Springer-Verlag, Berlin, 1989.

Demond, A. H., and Roberts, P. V., "An Examination of Relative Permeability Relations for Two-Phase Flow in Porous Media," *Water Resour. Bull.*, v. 23, n. 4, pp. 617–628, 1987.

Dixon, W. J., and Massey, F. J., Jr., *Introduction to Statistical Analysis*, 4th ed., McGraw-Hill, New York, 1983.

Domenico, P. A., and Robbins, G. A., "A New Method of Contaminant Plume Analysis," *Groundwater*, v. 23, n. 4, pp. 476–485, 1985.

Domenico, P. A., and Schwartz, F.W., *Physical and Chemical Hydrogeology*, 2nd ed., John Wiley & Sons, New York, 2001.

Dougherty, E. R., *Probability and Statistics for the Engineering, Computing, and Physical Sciences*, Prentice-Hall, Englewood Cliffs, NJ, 1990.

Douglas, J. and Hornung, U. (ed.), *Flow in Porous Media*, Birkhauser Verlag, Basel, Switzerland, 1993.

Farr, A. M., Houghtalen, R. J., and McWhorter, D. B., "Volume Estimation of Light Nonaqueous Phase Liquids in Porous Media," *Groundwater*, v. 28, n. 1, pp. 48–56, 1990.

Fetter, C. W., *Contaminant Hydrogeology*, 2nd ed., Prentice-Hall, Englewood Cliffs, NJ, 1998.

Freund, J. E., and Smith, R. M., *Statistics: A First Course*, 4th ed., Prentice-Hall, Englewood Cliffs, NJ, 1986.

Geankoplis, C. J., *Transport Processes: Momentum, Heat and Mass*, Allyn and Bacon, Boston, MA, 1983.

Gehhar, L. W., Gutjahr, A. L., and Naff, R. L. "Stochastic Analysis of Macrodispersion in a Stratified Aquifer," *Water Resour. Res.*, v. 15, n. 6, pp. 1387–1397, 1979.

Gibbons, R. D., *Statistical Methods for Groundwater Monitoring*, John Wiley & Sons, New York, 1994.

Gilbert, R.O., *Statistical Methods for Environmental Pollution Monitoring*, Van Nostrand Reinhold, New York, 1987.

Gupta, R. S., *An Introduction to Environmental Engineering and Science*, ABS Consulting, Rockville, MD, 2004.

Haan, C. T., *Statistical Methods in Hydrology*, 2nd ed., Iowa State Univ. Press, Ames, IA, 2002.

Hahn, G. J., and Meeker, W. Q., *Statistical Intervals: A Guide for Practitioners*, John Wiley & Sons, New York, 2001.

Hamond, H. F., and Fechner, E. J., *Chemical Fate and Transport in the Environment*, Academic Press, San Diego, CA, 1994.

Hewitt, G. F., Shires, G. L., and Polezhaev, Y. V. (eds.), *International Encyclopedia of Heat and Mass Transfer*, CRC Press, 1997.

Holley, E. R., "Unified View of Diffusion and Dispersion," *J. Hydraul. Div.*, v. 95. n. HY2, pp. 621–632, March, 1969.

Hunt, J. C. R., (ed.), *Turbulence and Diffusion in Stable Environment*, Clarendon Press, Oxford, UK, 1985.

Incropera, F. P., and Dewitt, D. P., *Fundamentals of Heat and Mass Transfer*, 6th ed., John Wiley & Sons, New York, 2006.

Jost, W., *Diffusion in Solids, Liquids, Gases*, Academic Press, New York, 1960.

Keely, J. F., Piwoni, M. D., and Wilson, J. T., "Evolving Concepts of Subsurface Contaminant Transport," *J. Water Poll. Control Fed.*, v. 58, pp. 349–357, 1986.

Kent, D. C., Pettyjohn, W. A., and Prickett, T. A., "Analytical Method for the Prediction of Leachate Plume Migration," *Groundwater Monitoring Review*, v. 5, n. 2, pp. 46–59, 1985.

Kobus, H., and Kinzelbach, W. K. H. (ed.), *Contaminant Transport in Ground Water*, Balkerna, Rotterdam, 1989.

Land, C. E., "Confidence Intervals for Linear Functions of the Normal Mean and Variance," *The Annals of Mathematical Statistics*, v. 42, n. 4, pp. 1187–1205, 1971.

Land, C. E., "An Evaluation of Approximate Confidence Interval Estimation Methods for Lognormal Means," *Technometrics*, v. 14, n. 1, pp. 145–158, 1972.

Land, C. E., "Standard Confidence Limits for Linear Functions of the Normal Mean and Variance," *Journal of the American Statistical Association*, v. 68, n. 344, pp. 960–963, 1973.

Land, C. E., "Tables of Confidence Limits for Linear Functions of the Normal Mean and Variance," in *Selected Tables in Mathematical Statistics*, v. III, pp. 385–419, American Mathematical Society, Providence, RI, 1975.

Lindstorm, F. T., and Boersma, L., "Analytical Solution for Convective-Dispersive Transport in Confined Aquifers with Different Initial and Boundary Conditions," *Water Resour. Res.*, v. 25, n. 2, pp. 241–256, 1989.

Lothar, S., *Applied Statistics: A Handbook of Techniques*, 2nd ed., Springer-Verlag, New York, 1984.

McWhorter, D. B., and Sunada, D. K., "Exact Integral Solutions for Two-Phase Flow," *Water Resour. Res.*, v. 26, n. 3, pp. 399–413, 1990.

Ogata, A., and Banks, R. B., *A Solution of the Differential Equation of Longitudinal Dispersion in Porous Media*, Professional Paper 411-A, USGS, 1961.

Palmer, C. M., *Principles of Contaminant Hydrogeology*, 2nd ed., Lewis Publishers, Boca Raton, FL, 1996.

Perkins, T. K., and Johnston, O. C., "A Review of Diffusion and Dispersion in Porous Media," *J. of the Society of Petroleum Engineers*, v. 17, pp. 70–83, 1963.

Schnoor, J. L., *Environmental Modeling—Fate and Transport of Pollutants in Water, Air and Soil*, John Wiley & Sons, New York, 1996.

Schwartz, F.W., "Macroscopic Dispersion in Porous Media: The Controlling Factors," *Water Resour. Res.*, v. 13, n. 4, pp. 743–752, 1977.

Serrano, S. E., "Hydrologic Theory of Dispersion Heterogeneous Aquifers," *ASCE J. Hydrol. Eng.*, v. 1, n. 4, pp. 144–151, Oct., 1996.

Stephan, K., and Park, N. J., *Heat and Mass Transfer*, trans. H. D. Baehr, Springer Verlag, Berlin, 1998.

Stroock, D. W., *Lectures on Stochastic Analysis: Diffusion Theory*, Cambridge Univ. Press, Cambridge, UK, 1987.

Taigbenu, A., and Liggett, J. A., "An Integral Solution for the Diffusion Advection Equation," *Water Resour. Res.*, v. 22, n. 8, pp. 1237–1246, 1986.

U.S. Environmental Protection Agency, *Handbook of Groundwater*, EPA/625/6–87/016, Cincinnati, OH, 1987.

U.S. Environmental Protection Agency, *Transport and Fate of Contaminants in the Subsurface*," EPA /625/4–89/019, Cincinnati, OH, 1989.

U.S. Environmental Protection Agency, *Statistical Analysis of Groundwater Monitoring Data at RCRA Facilities*, Interim Final Guidance, Washington, DC, 1989.

U.S. Environmental Protection Agency, *Statistical Training Course for Groundwater Monitoring Data Analysis*, Office of Solid Waste, EPA530-R-93–003, Washington, DC, 1992.

U.S. Environmental Protection Agency, *RCRA Groundwater Monitoring: Draft Technical Guidelines*, Government Institutes, Rockville, MD, 1994.

Van Genuchten, M. T., and Alves, W. J., "Analytical Solution of the One-Dimensional Convective-Dispersive Solute Transport Equation," *Tech. Bull.* 1661, U.S. Dept. of Agriculture, 1982.

References

Wadsworth, H. M., *Handbook of Statistical Methods for Engineers and Scientists*, 2nd ed., McGraw-Hill, New York, 1997.

Weber, W. J., and Digiano, F. A., *Process Dynamics in Environmental Systems*, John Wiley & Sons, New York, 1996.

Yates, S. R., "An Analytical Solution for One-Dimensional Transport in Porous Media with an Experimental Dispersion Function," *Water Resour. Res.*, v. 28, n. 8, pp. 2149–2158, 1992.

Chapter 6

Boyer, M. C., "Streamflow Measurement," in *Handbook of Applied Hydrology*, ed. V. T. Chow, McGraw-Hill, New York, 1964.

Buchanan, T. J., and Somers, W. P., *Discharge Measurements at Gaging Stations*, Techniques of Water Resources Investigations, Chap. A8, Book 3, U.S. Geological Survey, Washington, DC, 1969.

Carter, R.W., and Davidian, J., *General Procedure for Gaging Streams*, Techniques of Water Resources Investigations, Chap. A6, Book 3, U.S. Geological Survey, Washington, DC, 1968.

Chow, V. T., *Open-Channel Hydraulics*, McGraw-Hill, New York, 1959.

Corbett, D.M., et al., *Stream-Gaging Procedure*, Water Supply Paper 888, U.S. Geological Survey, Washington, DC, 1943.

Dickinson, W. T., *Accuracy of Discharge Determinations*, Hydrology Paper 20, Colorado State Univ., Fort Collins, CO, 1967.

Fitzgerald, M. G., and Karlinger, M. R., *Daily Water and Sediment Discharges from Selected Rivers of the Eastern United States: A Time-Series Modeling Approach.* Water Supply Paper 2216, U.S. Geological Survey, Washington, DC, 1983.

Gils, H., "Discharge Measurement in Open Water by Means of Magnetic Induction," *Hydrometry*, vol. I, Proceedings of the Koblenz Symposium, Sept. 1970, pp. 374–381, UNESCO, Paris, 1973.

Gonzalez, D. D., Scott, C. H., and Culbertson, J. K., *Stage-Discharge Characteristics of a Weir in a Sand-Channel Stream*, Water Supply Paper 1898-A, U.S. Geological Survey, Washington, DC, 1969.

Gupta, R. S., *Hydrology of Liberia*, Ministry of Lands and Mines, Government of Liberia, Dec. 1978.

Gupta, R. S., *Hydrologic Study of the Saint John River Basin in Liberia, West Coast of Africa*, Polytechnic Univ. of New York, 1980.

Haan, C. T., *Statistical Methods in Hydrology*, 2nd ed., Iowa State Univ. Press, Ames, IA, 2002.

Herschy, R. W., *Streamflow Measurement*, Taylor Francis, Oxford, UK, 1995.

Herschy, R. W. (ed.), *New Technology in Hydrometry: Development in the Acquisition and Management of Streamflow Data*, Adam Higler, 1985.

Herschy, R. W. (ed.), *Hydrometry: Principles and Practices*, John Wiley & Sons, New York, 1999.

Hiranandani, M. G., and Chitale, S. V., *Stream Gauging: A Manual*, Government of India, Ministry of Irrigation and Power, Central Water and Power Research Station, Poona, India, 1960.

Kennedy, E. J., *Computation of Continuous Records of Streamflow*, Techniques of Water Resources Investigations, Chap. A13, Book 3, U.S. Geological Survey, Washington, DC, 1983.

Linsley, R. K., Franzini, J. B., Freyberg, D. L., and Tchobanoglous, G., *Water Resources Engineering*, 4th ed., McGraw-Hill, New York, 1991.

Linsley, R. K., Kohler, M. A., and Paulhus, J. L. H., *Hydrology for Engineers*, 3rd ed., McGraw-Hill, New York, 1982.

Mays, L. W., *Water Resources Engineering*, John Wiley & Sons, New York, 2005.

Mitchell, W. D., *Stage-Fall-Discharge Relations for Steady Flow in Prismatic Channels*, Water Supply Paper 1164, U.S. Geological Survey, Washington, DC, 1954.

Moss, M.E., et al., *Design of Surface-Water Data Networks for Regional Information*, Water Supply Paper 2178, U.S. Geological Survey, Washington, DC, 1982.

Rantz, S. E., "Characteristics of Logarithmic Rating Curves," in *Selected Techniques in Water Resources Investigations, 1966–67*, compiled by E. G. Chase and F. N. Payne, Water-Supply Paper 1982, U.S. Geological Survey, Washington, DC, 1968.

Rantz, S. E., et al., *Measurement and Computation of Streamflow*, vol. 1, *Measurement of Stage and Discharge*, Water Supply Paper 2175, U.S. Geological Survey, Washington, DC, 1982a.

Rantz, S. E., et al., *Measurement and Computation of Streamflow*, vol. 2, *Computation of Discharge*, Water Supply Paper 2175, U.S. Geological Survey, Washington, DC, 1982b.

Riggs, H. C., *Some Statistical Tools in Hydrology*, Techniques of Water Resources Investigations, Chap. Al, Book 4, U.S. Geological Survey, Washington, DC, 1968.

Riggs, H. C., *Regional Analyses of Streamflow Characteristics*, Techniques of Water Resources Investigations, Chap. B3, Book 4, U.S. Geological Survey, Washington, DC, 1973.

Savini, J., and Bodhaine, G. L., *Analysis of Current-meter Data at Columbia River Gaging Stations*, Water Supply Paper 1869-F, U.S. Geological Survey, Washington, DC, 1971.

Searcy, J. K., "Graphical Correlation of Gaging-Station Records," in *Manual of Hydrology*, Part 1, *General Surface-Water Techniques*, Water Supply Paper 1541-C, U.S. Geological Survey, Washington, DC, 1960.

Searcy, J. K., and Hardison, C. H., "Double-Mass Curves," in *Manual of Hydrology*, Part 1, *General Surface-Water Techniques*, Water Supply Paper 1541-B, U.S. Geological Survey, Washington, DC, 1960.

Smith, W., *Techniques and Equipment Required for Precise Stream Gaging in Tide-Affected Fresh-Water Reaches of the Sacramento River, California*, Water-Supply Paper 1869-G, U.S. Geological Survey, Washington, DC, 1971.

Smoot, G. F., and Novak, C. E., *Measurement of Discharge by the Moving-Boat Method*, Techniques of Water Resources Investigations, Chapter A11, Book 3, U.S. Geological Survey, Washington, DC, 1969.

Speigel, M. R., *Theory and Problems of Statistics*, Schaum's Outline Series, 3rd ed., McGraw-Hill, New York, 1998.

Thomas, D. M., and Benson, M. A., *Generalization of Streamflow Characteristics from Drainage-Basin Characteristics*, Water Supply Paper 1975, U.S. Geological Survey, Washington, DC, 1970.

UNESCO, *Hydrologic Information Systems*, ed. G. W. Whetstone and V. J. Grigoriev, Studies and Reports in Hydrology, UNESCO/WMO, Paris/Geneva, 1972.

UNESCO, *Three Centuries of Scientific Hydrology*, UNESCO/WMO/IAHS, Paris, 1974.

U.S. Bureau of Reclamation, *Water Measurement Manual*, revised reprint, U.S. Dept. of Interior, Washington, DC, 1984.

U.S. Geological Survey, *National Handbook of Recommended Methods for Water-Data Acquisition*, Chap. 1, "Surface Water," Office of the Water Data Coordinator, U.S. Geological Survey, Reston, VA, 1977.

Vanoni, V. A., "Velocity Distribution in Open Channels," *Civ. Eng.*, v. 11, n. 6, pp. 350–357, June 1941.

Vanoni, V. A., and Nomicos, G. A., "Resistance Properties of Sediment-Laden Streams," *Trans. ASCE*, Paper 3055, pp. 1140–1175, 1960.

Viessman, W., and Lewis, G. L., *Introduction to Hydrology*, 5th ed., Prentice-Hall, Englewood Cliffs, NJ, 2002.

Villemonte, J. R., "Submerged-Weir Discharge Studies," *Eng. News Rec.*, v. 130, n. 26, Dec. 25, 1947.

Winter, T. C., "Uncertainties in Estimating the Water Balance of Lakes," *Water Resour. Bull.*, v. 17. n. 1, pp. 82–115, Feb. 1981.

Yevdjevich, V. M., "Statistical and Probability Analysis of Hydrologic Data, Part II: Regression and Correlation Analysis," in *Handbook of Applied Hydrology*, ed. v. T. Chow, McGraw-Hill, New York, 1964.

Chapter 7

American Society of Civil Engineers, *Hydrology Handbook*, 2nd ed., ASCE, New York, 1996.

American Society of Civil Engineers, *Flood Runoff Analysis*, adapted from the U.S. Army Corps of Engineers, n. 19, ASCE, New York, 1997.

Anderson, M. G., and Burt, T. P. (eds.), *Hydrological Forecasting*, John Wiley & Sons, Chichester, West Sussex, England, 1985.

Barndorff-Nielson, O. E., *Stochastic Methods in Hydrology: Rain, Landforms and Floods*, v. 7, World Scientific Publishing, River Edge, NJ, 1998.

Beard, L. R., *Simulation of Daily Streamflow*, Tech. Paper 6, Hydrologic Engineering Center, U.S. Army Corps of Engineers, Davis, CA (no date).

Beard, L. R., *Statistical Methods in Hydrology*, Hydrologic Engineering Center, U.S. Army Corps of Engineers, Davis, CA, 1962.

Beard, L. R., *Use of Interrelated Records to Simulate Streamflow*, Tech. Paper 1, Hydrologic Engineering Center, U.S. Army Corps of Engineers, Davis, CA, 1965.

Beard, L. R., *Streamflow Synthesis for Ungaged Rivers*, Tech. Paper 5, Hydrologic Engineering Center, U.S. Army Corps of Engineers, Davis, CA, 1967.

Beard, L. R., et al., *Estimating Monthly Streamflows within a Region*, Tech. Paper 18, Hydrologic Engineering Center, U.S. Army Corps of Engineers, Davis, CA, 1970.

Bender, D. L., and Roberson, J. A., "The Use of Dimensionless Unit Hydrograph to Derive Unit Hydrograph for Some Pacific Northwest Basins," *J. Geophys. Res.*, v. 66, n. 2, pp. 521–527, 1961.

Bernard, M., "An Approach to Determinate Streamflow," *Trans. ASCE*, v. 100, 1935.

Beven, K. J., *Rainfall-Runoff Modeling: The Primer*, John Wiley & Sons, NY, 2004.

Box, G. E. P., and Jenkins, G. W., *Time Series Analysis, Forecasting and Control*, Holden-Day, San Francisco, 1976.

Bras, R. L., and Rodriguez-Iturbe, I., *Random Functions and Hydrology*, Dover Publications, Mineola, NY, 1993.

Brown, S. (ed.), *Hydrology, Hydraulics and Water Quality*, National Research Council, Washington, DC, 1994.

Carlson, R. E., MacCormick, A. J. A., and Watts, D. G., "Application of Linear Random Models to Four Annual Streamflow Series," *Water Resour. Res.*, v. 6, pp. 1070–1078, 1970.

Cheremisinoff, N. P., *Practical Statistics for Engineers and Scientists*, Technomic, Lancaster, PA, 1987.

Chow, V. T., "Runoff," in *Handbook of Applied Hydrology*, ed. V. T. Chow, McGraw-Hill, New York, 1964.

Chow, V. T., Maidment, D. R., and Mays, L. W., *Applied Hydrology*, McGraw-Hill, New York, 1988.

Clark, C. O., "Storage and Unit Hydrograph," *Trans. ASCE*, v. 1 10, 1943.

Commons, G., "Flood Hydrographs," *Civil Eng.*, v. 12, pp. 571–572, 1942.

Davis, M. L., and Cornwell, D. A., *Introduction to Environmental Engineering*, 3rd ed., McGraw-Hill, New York, 1998.

Dunne, T., "Models of Runoff Processes and Their Significance," in *Studies in Geophysics: Scientific Basis of Water Resources Management*, prepared by Geophysics Study Committee, Geophysics Research Board, Assembly of Math and Physics Sciences, National Research Council, National Academy Press, Washington, DC, 1982.

Dunne, T., and Black, R. D., "Partial Area Contributions to Storm Runoff in Small New England Watershed," *Water Resour. Res.*, v. 6, n. 5, pp. 1296–1311, 1970.

Eichert, B. S., et al., *Methods of Hydrological Computations for Water Projects*, International Hydrological Program, UNESCO, Paris, 1982.

Fiering, M. B., and Jackson, B. B., *Synthetic Streamflows*, Water Resources Monograph 1, American Geophysicists Union, Washington, DC, 1971.

Fitzgerald, M. G., and Karlinger, M. R., *Daily Water and Sediment Discharges from Selected Rivers of the Eastern United States: A Time-Series Modeling Approach*, Water Supply Paper 2216, U.S. Geological Survey, Washington, DC, 1983.

Fleming, G., *Computer Simulation Techniques in Hydrology*, American Elsevier Publishing, New York, 1975.

Freeze, R. A., "Role of Subsurface Flow in Generating Surface Runoff, 1: Baseflow Contribution to Channel Flow," *Water Resour. Res.*, v. 8, n. 3, pp. 609–623, 1972a.

Freeze, R. A., "Role of Subsurface Flow in Generating Surface Runoff, 2: Upstream Source Areas," *Water Resour. Res.*, v. 8, n. 3, pp. 1272–1283, 1972b.

Freeze, R. A., "Reply to Comments by Knisel," *Water Resour. Res.*, v. 9, n. 4, p. 1101, 1973.

Gray, D. M., "Synthetic Unit Hydrograph for Small Watersheds," *ASCE J. of Hydraul. Div.*, v. 87, n. HY4, pp. 33–53, 1961.

Gray, D. M. (ed.), *Handbook on the Principles of Hydrology*, Water Information Center, New York, 1973.

Haan, C. T., *Statistical Methods in Hydrology*, 2nd ed., Iowa State Univ. Press, Ames, IA, 2002.

Hedman, E. R., *Mean Annual Runoff as Related to Channel Geometry of Selected Streams in California*, Water Supply Paper 1999-E, U.S. Geological Survey, Washington, DC, 1970.

Hedman, E. R., and Osterkamp, W. R., *Streamflow Characteristics Related to Channel Geometry of Streams in Western United States*, Water Supply Paper 2193, U.S. Geological Survey, Washington, DC, 1982.

Hewlett, J. D., "Comments on Letters Relating to Role of Subsurface Flow in Generating Surface Runoff, 2: Upstream Source Areas," by A. Freeze, *Water Resour. Res.*, v. 10, n. 3, June 1974.

Hewlett, J. D., *Principles of Forest Hydrology*, Univ. of Georgia Press, Athens, GA, 1982.

Hewlett, J. D., and Hibbert, A. R., "Factors Affecting the Response of Small Watersheds to Precipitation in Humid Areas," in *International Symposium on Forest Hydrology*, ed. W. E. Sopper and H. W. Lull, Pergamon Press, Oxford, 1967.

Hickok, R. B., Keppel, R. V., and Rafferty, B. R., "Hydrograph Synthesis for Small Arid-land Watersheds," *Agr. Eng.*, v. 40, pp. 608–615, 1959.

Horton, R. E., "The Role of Infiltration in the Hydrologic Cycle," *Trans. Am. Geophys. Union*, v. 1 4, pp. 446–460, 1933.

Hromadka, T. V., Durbin, T. J., and DeVries, J. J., *Computer Methods in Water Resources*, Lighthouse Publications, Mission Viejo, CA, 1985.

Hromadka, T. V., Beech, B. L., and Clements, J. M., *Computational Hydraulics for Civil Engineers*, Lighthouse Publications, Mission Viejo, CA, 1986.

Hydrologic Engineering Center, *Hydrologic Engineering Methods for Water Resources Development: Hydrograph Analysis*, vol. 4, U.S. Contribution to I.H.D., U.S. Army Corps of Engineers, Davis, CA, 1973.

Hydrologic Engineering Center, *Hydrologic Engineering Methods for Water Resources Development: Hydrologic Frequency Analysis*, vol. 3, U.S. Contribution to I.H.D., U.S. Army Corps of Engineers, Davis, CA, 1975.

Interagency Advisory Committee on Water Data, *Guidelines for Determining Flood Flow Frequency*, Bull. 17B, USGS, Office of Water Data Coordination, Reston, VA, 1982.

International Hydrological Decade, *Design of Water Resources Projects with Inadequate Data*, vols. 1and 2, Studies and Reports in Hydrology, UNESCO, Paris, 1974.

Kirkby, M. J. (ed.), *Hillslope Hydrology*, Wiley-Interscience, New York, 1978.

Kirkby, M. J., and Chorley, R. J., "Throughflow, Overland Flow, and Erosion," *Bull of Internat. Assoc. for Scientific Hydr.*, Special Publication 3, 1967.

References

Klemes, V., "Physically Based Stochastic Hydrologic Analysis," in *Advances in Hydrosciences*, vol. 11, ed. V. T. Chow, Academic Press, New York, 1978.

Knapp, B. J., *Elements of Geographical Hydrology*, George Allen & Unwin, London, 1979.

Knisel, W. G., "Comments on Role of Subsurface Flow in Generating Surface Runoff, 2: Upstream Source Areas," by A. Freeze, *Water Resour. Res.*, v. 9, n. 4, pp. 1097–1100, Aug. 1973.

Kottegoda, N. T., *Stochastic Water Resources Technology*, Halsted Press, New York, 1980.

Lane, W. L., *Applied Stochastic Techniques, User's Manual*, U.S. Bureau of Reclamation, Denver, 1979.

Linsley, R. K., Franzini, J. B., Freyberg, D. L., and Tchobanoglous, G. *Water Resources Engineering*, 4th ed., McGraw-Hill, New York, 1991.

Linsley, R. K., Kohler, M. A., and Paulhus, J. L. H., *Hydrology for Engineers*, 3rd ed., McGraw-Hill, New York, 1982.

Maass, A., et al., *The Design of Water Resources Systems*, Harvard Univ. Press, Cambridge, MA, 1962.

Maidment, D. R. (ed.), *Handbook of Hydrology*, McGraw-Hill, New York, 1993.

Matalas, N. C., "Mathematical Assessment of Synthetic Hydrology," *Water Resour. Res.*, v. 3, n. 4, pp. 937–945, 1967.

Matalas, N. C., and Jacobs, B., *A Correlation Procedure for Augmenting Hydrological Data*, Professional Paper 434-E, U.S. Geological Survey, Washington, DC, 1964.

Matalas, N. C., and Wallis, J. R., "Generation of Synthetic Flow Sequence," in *Systems Approach to Water Management*, ed. A. K. Biswas, McGraw-Hill, New York, 1976.

Meinzer, O. E. (ed.), *Physics of the Earth: Hydrology*, McGraw-Hill, New York, 1942.

Mejia, J. M., and Rousselle, J., "Disaggregation Models in Hydrology Revisited," *Water Resour. Res.*, v. 12, n. 2, pp. 185–186, 1976.

Mitchell, W. D., *Unit Hydrographs in Illinois*, Division of Waterways, State of Illinois, Springfield, 1948.

Musgrave, G. W., and Holton, H. N., "Infiltration," in *Handbook of Applied Hydrology*, ed. V. T. Chow, McGraw-Hill, New York, 1964.

Pilgrim, D. H., Huff, D. D., and Steele, T. D., "A Field Evaluation of Subsurface and Surface Runoff, II: Runoff Process," *J. Hydrol.*, v. 38, pp. 319–341, 1978.

Riggs, H. C., *Some Statistical Tools in Hydrology*, Techniques of Water Resources Investigations, Chap. Al, Book 4, U.S. Geological Survey, Washington, DC, 1968.

Sabol, G. V., "Clark Unit Hydrograph and R-Parameter Estimation," *ASCE J. Hydraul. Eng.*, v. 114, n. 1, pp. 103–111, 1988.

Salas, J. D., Delleur, J. W., Yevjevich, V., and Lane, W. L., *Applied Modeling of Hydrologic Time Series*, Water Resources Publications, Littleton, CO, 1980.

Searcy, J. K., *Graphical Correlation of Gaging-Station*, Water Supply Paper 1541-C, U.S. Geological Survey, Washington, DC, 1960.

Searcy, J. K., and Hardison, C. H., *Double-Mass Curves*, Water Supply Paper 1541-B, U.S. Geological Survey, Washington, DC, 1960.

Sen, Z., "Wet and Dry Periods of Annual Flow Series," *ASCE, J. Hydraul. Div.*, v. 102, n. HY10, pp. 1503–1514, 1976.

Sen, Z., "Autorun Analysis of Hydrologic Times Series," *ASCE J. Hydraul.*, v. 36, pp. 75–85, 1978.

Sen, Z., "Autorun Model for Synthetic Flow Generation," *ASCE J. Hydraul.*, v. 81, pp. 157–170, 1985.

Sherman, L. K., "Streamflow from Rainfall by the Unit-graph Method," *Eng. News Record*, v. 1 08, pp. 501–502, 1932.

Sherman, L. K., "The Unit Hydrograph Method," Ch. X-IE, in *Physics of the Earth—Hydrology*, ed. O. E. Meinzer, McGraw-Hill, New York, 1942.

Singh, V. P., *Hydrologic Systems: Rainfall-Runoff Modeling*, vol. 1, Prentice-Hall, Englewood Cliffs, NJ, 1988.

Singh, V. P., "On the Theories of Hydraulic Geometry," *International Jour. of Sediment Research*, v. 18, n. 3, pp 196–218, 2003.

Snyder, F. F., "Synthetic Unit Graphs," *Trans. Amer. Geophys. Union*, v. 19, pp. 447–454, 1938.

Srikanthan, R., "Sequential Generation of Monthly Streamflows," *ASCE J. Hydraul.*, v. 38, pp. 71–80, 1978.

Stedinger, J. R., and Vogel, R. M., "Disaggregation Procedures for Generating Serially Correlated Flow Vectors," *Water Resour. Res.*, v. 20, n. 1, pp. 47–56, 1984.

Steinger, J. R., Lettenmaier, D. P., and Vogel, R. M., "Multisite ARMA (1, 1) and Disaggregation Models for Annual Streamflow Generation," *Water Resour. Res.*, v. 21, n. 4, pp. 497–509, 1985.

Svanidize, G., *Mathematical Modeling of Hydrologic Series*, Water Resources Publications, Littleton, CO, 1980.

Taylor, A. B., and Schwarz, H. E., "Unit Hydrograph Lag and Peak Flow Related to Basin Characteristics," *Trans. Amer. Geophys. Union*, v. 33, pp. 235–246, 1952.

Thomas, D. M., and Benson, M. A., *Generalization of Streamflow Characteristics from Drainage-Basin Characteristics*, Water Supply Paper 1975, U.S. Geological Survey, Washington, DC, 1970.

Thomas, H. A., *Improved Methods for National Assessment*, Water Resources Contract WR15249270, Water Resources Group, Harvard Univ., Cambridge, MA, 1981.

U.S. Nat. Resources Conserv. Service, *National Engineering Handbook*, 210-630 (NEH Part 630), Chapters from 1971–2004, U. S. Dept of Agriculture, Washington, DC, 2004.

Valencia, R. D., and Schaake, J. C., "Disaggregation Process in Stochastic Hydrology," *Water Resour. Res.*, v. 9, n. 3, pp. 580–585, 1973.

Viessman, W., and Lewis, G. L., *Introduction to Hydrology*, 5th ed., Prentice-Hall, Englewood Cliffs, NJ, 2002.

Wetherill, G. B., et al., *Regression Analysis with Applications*, Chapman & Hall, London, 1986.

Williams, H. M., "Discussion on Military Airfields: Design of Drainage Facilities," *Trans. ASCE*, v. 110, pp. 820–826, 1945.

World Meteorological Organization, *Hydrological Network Design and Information Transfer*, WMO No. 433, Report 8, Geneva, 1976.

Yevdjevich, V. M., "Regression and Correlation Analysis," in *Handbook of Applied Hydrology*, ed. V. T. Chow, McGraw-Hill, New York, 1964.

Chapter 8

Adamowski, K., "Plotting Formula for Flood Frequency," *Water Resour. Bull.*, v. 17, n. 2, pp. 197–202, 1981.

Altman, D. G., Espey, W. H., and Feldman, A. D., *Investigation of Soil Conservation Service Urban Hydrology Techniques*, Tech. Paper 77, Hydrologic Engineering Center, U.S. Army Corps of Engineers, Davis, CA, 1980.

American Society of Civil Engineers, *Hydrology Handbook*, 2nd ed. ASCE, NY, 1996.

American Society of Civil Engineers, *Flood Runoff Analysis*, adapted from U.S. Army Corps of Engineers, n. 19, ASCE, NY, 1997.

Barrows, H. K., *Floods: Their Hydrology and Control*, McGraw-Hill, New York, 1948.

Beard, L. R., *Statistical Methods in Hydrology*, Civil Works Investigation Project CW-151, U.S. Army Corps of Engineers, Sacramento, CA, 1962.

Beard, L. R., *Hypothetical Flood Computation for a Stream System*, Tech. Paper 12, Hydrologic Engineering Center, U.S. Army Corps of Engineers, Davis, CA, 1968.

Benson, M. A., "Characteristics of Frequency Curves Based on a Theoretical 1000-Year Record," in *Flood-Frequency Analysis*, by T. Dalrymple, Water Supply Paper 1543-A, U.S. Geological Survey, Washington, DC, 1960.

Brater, E. F., and King, H. W., *Handbook of Hydraulics*, 6th ed., McGraw-Hill, New York, 1976.

Brater, E. F., King, H. W., Lindell, J. F., and Wei, C. Y., *Handbook of Hydraulics*, 7th ed., McGraw-Hill, New York, 1996.

Burnham, M. W., *Adoption of Flood Flow Frequency Estimates at Ungaged Locations*. Training Doc. 11, Hydrologic Engineering Center, U.S. Army Corps of Engineers, Davis, CA, 1980.

Chow, V. T., Discussion on "Annual Floods and the Partial Duration Flood Series" by W. B. Langbein, *Trans. Amer. Geophys. Union*, v. 31, pp. 939–941, 1950.

Chow, V. T., "A General Formula for Hydrologic Frequency Analysis," *Trans. Am. Geophys. Union*, v. 32, pp. 231–237, 1951.

Chow, V. T., "Frequency Analysis," in *Handbook of Applied Hydrology*, ed. V. T. Chow, McGraw-Hill, New York, 1964.

Creager, W. P., Justin, J. D., and Hinds, J., *Engineering for Dams*, v. I, John Wiley & Sons, New York, 1945.

Crippen, J. R., "Envelope Curves for Extreme Flood Events," *ASCE J. of Hydraul. Eng.*, v. 108, n. HY 10, pp. 1208–1212, 1982.

Crippen, J. R., and Bue, C. D., *Maximum Flood Flows in the Conterminous United States*, Water Supply Paper, 1887, U.S. Geological Survey, Washington, DC, 1977.

Cudworth, A. G., *Flood Hydrology Manual*, Water Resour. Tech. Publ., U.S. Dept. of the Interior, Bureau of Reclamation, Denver, CO, 1992.

Cunnane, C., "Unbiased Plotting Positions—A Review," *J. Hydrol.*, v. 37, pp. 205–222, 1978.

Dalrymple, T., *Flood-Frequency Analysis*, Water Supply Paper 1543-A, U.S. Geological Survey, Washington, DC, 1960.

Dalrymple, T., "Flood Characteristics and Flow Determination," in *Handbook of Applied Hydrology*, ed. V. T. Chow, McGraw-Hill, New York, 1964.

Ely, P. B., and Peters, J. C., *Probable Maximum Flood Estimation—Eastern United States*, Tech. Paper 100, Hydrologic Engineering Center, U.S. Army Corps of Engineers, Davis, CA, 1984.

Feldman, A. D., *Flood Hydrograph and Peak Flow Frequency Analysis*, Tech. Paper 62, Hydrologic Engineering Center, U.S. Army Corps of Engineers, Davis, CA, 1979.

Gray, D. M., "Statistical Methods—Fitting Frequency Curves, Regression Analysis," in *Handbook on the Principles of Hydrology*, ed. D. M. Gray, Water Information Center, Port Washington, NY, 1973 (published by the National Committee for the IHD in 1970).

Haan, C. T., *Statistical Methods in Hydrology*, 2nd ed., Iowa State Univ. Press, Ames, IA, 2002.

Hershfield, D. M., "Estimating the Probable Maximum Precipitation," *ASCE J. Hydraul. Div.*. v. 87, n. HY5, pp. 99–116, Sept. 1961.

Hinds, J., Creager, W. P., and Justin, J. D., *Engineering for Dams*, vol. 2, John Wiley & Sons, New York, 1945.

Hjelmfelt, A. T., and Cassidy, J. J., *Hydrology for Engineers and Planners*, Iowa State Univ. Press, Ames, 1975.

Hoggan, D. H., *Computer Assisted Flood Plain Hydrology and Hydraulics*, 2nd ed., McGraw-Hill, New York, 1996.

Horn, D. R., "Graphic Estimation of Peak Flow Reduction in Reservoirs," *ASCE J. Hydraul. Eng*, v. 113, n. 11, 1987.

Hudson, H. E., and Hazen, R., "Droughts and Low Streamflows," in *Handbook of Applied Hydrology*, ed. V. T. Chow, McGraw-Hill, New York, 1964.

Hydrologic Engineering Center, *Hydrologic Frequency Analysis*, Hydrologic Engineering Methods for Water Resources Development, vol. 3, U.S. Army Corps of Engineers, Davis, CA, 1975.

Hydrologic Engineering Center, *Feasibility Studies for Small Scale Hydropower Additions*, vol. 3, U.S. Army Corps of Engineers, Davis, CA, 1979.

Hydrologic Engineering Center, *Comparative Analysis of Flood Routing Methods*, Research Doc. n. 24 , U.S. Army Corps of Engineers, Davis, CA, 1980.

Interagency Advisory Committee on Water Data (IACWD) (formerly U.S. Water Resources Council), "Guidelines for Determining Flood Flow Frequency," Bull. 17B, U.S. Dept. of the Interior, Office of Water Data Coordination, Reston, VA, 1982.

Langbein, W. B., "Annual Floods and the Partial-Duration Flood Series," *Trans. Amer. Geophys. Union*, v. 30, pp. 879–881, 1949.

Lawler, E. A., "Flood Routing," in *Handbook of Applied Hydrology*, ed. V. T. Chow, McGraw-Hill, NY, 1964.

Maidment, D. R., *Handbook of Hydrology*, McGraw-Hill, New York, 1993.

Mays, L. W., *Water Resources Handbook*, McGraw-Hill, New York, 1996.

Morris, E. C., *Mixed Population Frequency Analysis*, Training Document 17, Hydrologic Engineering Center, U.S. Army Corps of Engineers, Davis, CA, 1982.

Raudkivi, A. J., *Hydrology: An Advanced Introduction to Hydrological Processes and Modeling*, Pergamon Press, Oxford, 1979.

Riggs, H. C., *Frequency Curves*, Techniques of Water Resources Investigations, Book 4, Chap. A2, U.S. Geological Survey, Washington, DC, 1968.

Riggs, H. C., *Low-Flow Investigations*, Techniques of Water Resources Investigations, Book 4, Chap. B1, U.S. Geological Survey, Washington, DC, 1972.

Riggs, H. C., *Streamflow Characteristics*, Developments in Water Science Series No. 22, Elsevier Science, New York, 1985.

Sauer, V. B., et al., *Flood Characteristics of Urban Watersheds in the United States*, Water Supply Paper 2207, U.S. Geological Survey, Washington, DC, 1983.

Singh, V. P. (ed.), *Computer Models of Watershed Hydrology*, Water Resources Publications, Highlands Ranch, CO, 1995.

Singh, V. P., and Frevert, D. K., *Watershed Models*, CRC Press, Boca Raton, FL, 2005.

Singh, V. P., and Singh, K., "Parameter Estimation for Log-Pearson Type III Distribution by POME," *ASCE J. Hydraul. Eng.*, v. 114, n. 1, pp. 112–122, 1988.

Sokolov, A. A., Rantz, S. E., and Roche, M., *Floodflow Computation: Methods Compiled from World Experience*, Studies and Reports in Hydrology, No. 22, UNESCO, Paris, 1976.

Speigel, M. R., *Statistics*, Schaum's Outline Series, 3rd ed., McGraw-Hill, New York, 1998.

Srikanthan, R., and McMahon, T. A., "Recurrence Interval of Long Hydrologic Events," *ASCE J. Hydraul. Eng.*, v. 112, n. 6, 1986.

Tung, Y. K., and Mays, L. W., "Generalized Skew Coefficients for Flood Frequency Analysis," *Water Resour. Bull.*, v. 17, n. 2, pp. 262–269, 1981.

UNESCO, *Flood Studies: An International Guide for Collection and Processing of Data*, Technical Papers in Hydrology 8, UNESCO, Paris, 1971.

U.S. Army Corps of Engineers, *Flood Hydrograph Analyses and Computations*, Engineering Manual 1110-2-1405, U.S. GPO, Washington, DC, 1959.

U.S. Army Corps of Engineers, *Hydrologic Frequency Analysis*, Engineering Manual 1110-2-1415, U.S. GPO, Washington, DC, 1993.

U.S. National Weather Service, *Seasonal Variation of the Probable Maximum Possible Precipitation, East of the 105 Meridian for Areas from 10 to 1000 Square Miles and Durations of 6, 12, 24, and 48 Hours*, Hydrometeorology Report 35, Washington, DC, 1956.

References

U.S. National Weather Service, *Generalized Estimates of Maximum Possible Precipitation for the United States West of the 105 Meridian*, Tech. Paper 38, U.S. Weather Bureau, Washington, DC, 1960.

U.S. National Weather Service, *Probable Maximum Precipitation Estimates—U.S. East of the 105 Meridian*, Hydrometeorology Report 51, National Oceanic and Atmospheric Administration, U.S. Dept. of Commerce, Washington, DC, 1978 (reprinted 1986).

U.S. National Weather Service, *Seasonal Variation of 10-Sq. Mi. Probable Maximum Precipitation Estimates—U.S. East of the 105 Meridian*, Hydrometeorology Report 53, National Oceanic and Atmospheric Administration, U.S. Dept. of Commerce, Washington, DC, 1980.

U.S. National Weather Service, *Application of Probable Maximum Precipitation Estimates—U.S. East of the 105 Meridian*, Hydrometeorology Report 52, National Oceanic and Atmospheric Administration, U.S. Dept. of Commerce, Washington, DC, 1982.

U. S. National Weather Service, *Probable Maximum Precipitation, U. S. West of 105 Meridian*, Hydrometeorological Reports 49, 55A, 57, 58, 59, National Oceanic and Atmospheric Administration, U.S. Dept. of Commerce, Washington, DC, 1984–1998.

U.S. Water Resources Council, *Guidelines for Determining Flood Flow Frequency*, Bulletin 17B, Hydrology Committee of the Water Resources Council, Washington, DC, 1981.

Viessman, W., and Lewis, G. L., *Introduction to Hydrology*, 5th ed., Prentice-Hall, Englewood Cliffs, NJ, 2002.

Wandle, S. W., *Estimating Peak Discharges of Small Rural Streams in Massachusetts*, Water Supply Paper 2214, U.S. Geological Survey, Washington, DC, 1983.

Wang, B. H., and Jawed, K., "Transformation of PMP to PMF: Case Studies," *ASCE J. Hydraul. Eng.*, v. 112, n. 7, pp. 541–561, 1986.

World Meteorological Organization, *Estimation of Maximum Floods*, Technical Note 98, Report of a Working Group of the Commission for Hydrometeorology, WMO, Geneva, 1969.

Yevdjevich, V., *Probability and Statistics in Hydrology*, Water Resources Publications, Fort Collins, CO, 1972.

Chapter 9

Abbott, M. B., *An Introduction to the Methods of Characteristics*, American Elsevier, New York, 1966.

Abbott, M. B., *Computational Hydraulics: Elements of the Theory of Free Surface Flows*, Ashgate Publishing Ltd., London, 1992.

Allievi, L., *Theory of Water Hammer*, two vol., (trans. E. E. Halmos), sponsored by the ASCE and ASME, 1925.

Afouda, A. A., *A Generalized Kinematic Approach to Basin Modeling*, Publ. 129, International Assoc. of Hydrological Sciences, Wallingford, UK, 1980.

Akan, A. O., and Yen, B. C., "Mathematical Model of Shallow Water Flow over Porous Media," *J. Hydraul. Div.*, ASCE, v. 107, n. HY4, pp. 479—494, 1981.

ASCE Task Committee, "Turbulence Modeling of Surface Water Flow and Transport," Parts I–V, *J. Hydraul. Eng.*, v. 114, n. 9, pp. 970–1073, 1988.

Auckers, P., "Alluvial Channel Hydraulics," *J. Hydrol.*, v. 100, pp. 177–204, 1988.

Beven, K., "On the Generalized Kinematic Routing Method," *Water Resour. Res.*, v. 15, n. 5, pp. 1238–1242, 1979.

Brakensiek, D. L., "Kinematic Flood Routing," *Trans. ASCE*, v. 10, n. 3, 1967.

Carter, R. W., and Godfrey, R. G., *Storage and Flood Routing*, Water Supply Paper 1543-B, U.S. Geological Survey, Washington, DC, 1960.

Chadwick, A., Morfett, J. and Martin, B. , *Hydraulics in Civil and Environmental Engineering*, 4th ed., Taylor Francis, UK, 2004.

Chalfen, M., and Niemiec, A., "Analytical and Numerical Solution of Saint-Venant Equations," *J. Hydrol.*, v. 86, pp 1–13, 1986.

Chang, C. N., et al., "On the Mathematics of Storage Routing," *J. Hydrol.*, v. 61, pp. 357–370, 1983.

Chaudhry, M. H., *Applied Hydraulic Transients*, 2nd ed., Van Nostrand Reinhold, New York, 1987.

Chaudhry, M. H., *Open Channel Flow*, Prentice-Hall, Upper Saddle River, NJ, 1993.

Crowe, C. T., Roberson, J. A., and Elger, D. F., *Engineering Fluid Mechanics*, John Wiley & Sons, NY, 2004.

Cundy, T. W., and Tento, S. W., "Solution to the Kinematic Wave Approach to Overland Flow Routing with Rainfall Excess Given by Philip's Equation," *Water Resour. Res.*, v. 21, n. 8, pp. 1132–1140, 1985.

Cunge, J. A., "On the Subject of a Flood Propagation Computation Method," *J. Hydraul. Research*, v. 7, n. 2, pp. 205–230, 1969.

Cunge, J. A., "Applied Mathematical Modeling of Open Channel Flow," Chap. 14 in *Unsteady Flow in Open Channels*, ed. K. Mahmood, and V. Yevjevich, Water Resources Publications, Fort Collins, CO, 1975.

Davis, C. V., *Handbook of Applied Hydraulics*, 2nd ed., McGraw-Hill, NY, 1952.

Davis, J. L., *Wave Propagation in Solids and Fluids*, Springer-Verlag, New York, 1988.

Duchateau, P., and Zachmann, D. W., *Partial Differential Equations*, Schaum's Outline Series, McGraw-Hill, New York, 1986.

Elliot, R. C., and Chaudhry, M. H., "A Wave Propagation Model for Two-Dimensional Dam-Break Flows," *J. Hydraul. Res.*, v. 30, n. 4, pp. 467–483, 1992.

Ferrick, M. G., "Analysis of River Wave Types," *Water Resour. Res.*, v. 21, n. 2, pp. 209–20, 1985.

Field, W. G., and Williams, B. J., "A Generalized One-Dimensional Kinematic Catchment Model," *J. Hydrol.*, v. 60, pp. 25, 1983.

Fread, D. L., *Applicability Criteria for Kinematic and Diffusion Routing Models*, Laboratory of Hydrology, National Weather Service, NOAA, U.S. Department of Commerce, Silver Spring, MD, 1985.

Fox, J. A., *Hydraulic Analysis of Unsteady Flow in Pipe Networks*, Halsted Press, New York, 1977.

Giles, R. V., Liu, C., and Evett, J. B., *Fluid Mechanics and Hydraulics*, Schaum's Outline Series, McGraw-Hill, NY, 1994.

Govindaraju, R. S., Jones, S. E., and Kavvas, M. L., "On the Diffusion Wave Model for Overland Flow: 2 Steady-State Analysis," *Water Resour. Res.*, v. 24, n. 5, pp. 745–754, 1988.

Govindaraju, R. S., Kavvas, M. L., and Jones, S. E., "Approximate Analytical Solution for Overland Flows," *Water Resour. Res.*, v. 26, n. 12, pp. 2903–2912, 1990.

Gunlach, D. L., and Thomas, W. A., *Guidelines for Calculating and Routing a Dam-Break Flood*, Res. Note 5, Hydrologic Engineering Center, U.S. Army Corps of Engineers, Davis, CA, 1977.

Hager, W. H., and Hager, K., "Application Limits for the Kinematic Wave Approximation," *Nordic Hydrol.*, v. 16, pp. 203–212, 1985.

Henderson, F. M., "Flood Waves in Prismatic Channels," *ASCE J. Hydraul. Div.*, v. 89, n. HY4, pp. 39–67, 1963.

Holden, A. P., and Stephenson, D., "Finite Difference Formulations of Kinematic Equations," *J. Hydraul. Eng.*, v. 121, n. 5, pp. 423–26, 1995.

Holmes, H., Whirlow, D. K., and Wright, L. G., "The Flowmeter—A Unique Device for Open Channel Discharge Measurement," in *Hydrometry*, vol. I, *Studies and Report in Hydrology*, UNESCO, Paris, 1973.

Katopodes, N. D., "On Zero-Inertia and Kinematic Waves," *ASCE J. Hydraul. Div.*, v. 108, n. 11, pp. 1380–87, 1982.

Kreyszig, E., *Advanced Engineering Mathematics*, 9th ed., John Wiley & Sons, New York, 2006.

Lighthill, M. J., and Whitham, G. B., "On Kinematic Waves: 1, Flood Movement in Long Rivers," *Proc. Royal Society*, v. 229, n. 1178, 1955.

References

Lighthill, M. J., *Waves in Fluids*, Cambridge Univ. Press, London, 1978.

Morris, E. M., and Woolhiser, D. A., "Unsteady One-Dimensional Flow Over a Plane: Partial Equilibrium and Recession Hydrographs," *Water Resour. Res.*, v. 16, n. 2, pp. 355–366, 1980.

Munson, B. R., Young, D. F., and Okiishi T. H., *Fundamentals of Fluid Mechanics*, 5th ed., John Wiley & Sons, New York, 2006.

Nalluri, C., *Civil Engineering Hydraulics: Essential Theory with Worked Examples*, 4th ed., Blackwell Science, Oxford, 2001.

Overton, D. E., and Meadows, M. E., *Storm Water Modeling*, Academic Press, New York, 1976.

Pearson, C. P., "One-Dimensional Flow over a Plane: Criteria for Kinematic Wave Modeling," *J. Hydrol.*, v. 3, pp. 30–48, 1989.

Ponce, V. M., "The Kinematic Wave Controversy," *ASCE J. Hydraul. Eng.*, v. 117, n. 4, pp. 511–25, 1991.

Ponce, V. M., *Engineering Hydrology: Practices and Applications*, 2nd ed., Prentice-Hall, Englewood Cliffs, NJ, 1994.

Ponce, V. M., Li, R. M., and Simons, D. B., "Applicability of Kinematic and Diffusion Wave Models," *ASCE J. Hydraul. Div.*, v. 104, n. HY3, pp. 353–360, 1978.

Ponce, V. M., and Simons, D. B., "Shallow Wave Propagation in Open Channel Flow," *ASCE J. Hydraul. Div.*, v. 103, n. HY12, pp. 1461–1476, 1977.

Popescu, M., *Applied Hydraulic Transients: For Hydropower Plants and Pumping Stations*, Taylor Francis, UK, 2003.

Price, R. K., "Comparison of Four Numerical Methods for Flood Routing," *ASCE J. Hydraul. Div.*, v. 100, n. HY7, July 1974.

Rahman, M., *Applied Differential Equations for Scientists and Engineers: Partial Differential Equations*, v. 2, Computational Mechanics Publications, Southampton, UK, 1991.

Renardy, M., and Rogers, R. C., *An Introduction to Partial Differential Equations*, 2nd ed., Princeton Univ. Press, NJ, 1995.

Rich, G. R., *Hydraulic Transients*, 2nd ed., Dover Publications, New York, 1963.

Roberson, J. A., Cassidy, J. J., and Chaudhry, H. M., *Hydraulic Engineering*, John Wiley & Sons, New York, 2001.

Sharp, D. B., *Water Hammer: Practical Solutions*, John Wiley & Sons, New York, 1996.

Singh, V. P., *Kinematic Wave Modeling in Water Resources*, John Wiley & Sons, New York, 1997.

Singh, V. P., and Mahmood, K., "Kinematic Watershed Runoff," *Proc. 3rd World Congr. on Water Resour.*, Mexico City, v. 4, 1979.

Smith, A. A., "A Generalized Approach to Kinematic Flood Routing," *J. Hydrol.*, v. 45, pp. 71–89, 1980.

Stephenson, D., and Meadows, M. E., *Kinematic Hydrology and Modeling*, Development in Water Sciences n. 26, Elsevier, Amsterdam, 1986.

Strekoff, T., et al., *Comparative Analysis of Flood Routing Methods*, Res. Doc. 24, Hydrologic Engineering Center, U.S. Army Corps of Engineers, Davis, CA, 1980.

Subramanya, K., *Engineering Hydrology*, Tata McGraw-Hill, India, 1999.

U.S. Army Corps of Engineers, *Routing of Floods through River Channels*, Engineering Manual EM 1110-2-1405, U.S. GPO, Washington, DC, 1959.

Vieira, J. H. D., "Conditions Governing the Use of Approximations for the Saint Venant Equations for Shallow Surface Water Flow," *J. Hydrol.*, v. 60, pp. 43–58, 1983.

Watters, G. Z., *Analysis and Control of Unsteady Flow in Pipe Lines*, 2nd ed., Butterworth Publishers, Woburn, MA, 1984.

Weinmann, P. E., and Laurenson, E. M., "Approximate Flood Routing Methods," *ASCE J. Hydraul. Div.*, v. 105, HY12, pp. 1521–1536, 1979.

Wooding, R. A., "A Hydraulic Model for the Catchment-Stream Problem: I. Kinematic Theory," *J. Hydrol.*, v. 3, n. 3–4, pp. 254–267, 1965a.

Wooding, R. A., "A Hydraulic Model for the Catchment-Stream Problem: II. Numerical Solutions," *J. Hydrol.*, v. 3, n. 3–4, pp. 268–282, 1965b.

Woolhiser, M. H., and Liggett, J. A., "Unsteady One-Dimensional Flow over a Plane: Rising Hydrograph," *Water Resour. Res.*, v. 3, n. 3, pp. 753–771, 1967.

Wylie, E. B., and Streeter, V. L., *Fluid Transients in Systems*, Prentice-Hall, Englewood Cliffs, NJ, 1993.

Chapter 10

Bos, M. G., *Long Throated Flumes and Broad-Crested Weirs*, Martinus Nijhoff/Dr. W. Junk Publishers, Dordrecht, The Netherlands, 1985.

Bos, M., *Discharge Measurement Structures*, Water Resources Publications, Littleton, CO, 1989.

Ackers P., White W. R., and Harrison, A. J. M., *Weirs and Flumes for Measurement*, John Wiley & Sons, NY, 1978.

American Society of Civil Engineers, *Report on Reevaluating Adequacy of Existing Dams*, Committee on Hydrometeorology, ASCE, New York, 1973.

American Society of Civil Engineers, *Hydraulic Design of Spillways*, Tech. Eng. and Design Guides as adapted from the U.S. Army Corps of Engineers No. 12, ASCE, New York, 1995.

Beard, L. R., *Methods for Determination of Safe Yield and Compensation Water from Storage Reservoirs*, Technical Paper 3, Hydrologic Engineering Center, U.S. Army Corps of Engineers, Davis, CA, 1965.

Bodhaine, G. L., *Measurement of Peak Discharge at Culverts by Indirect Methods*, Techniques of Water Resources Investigations, Book 3, Chap. A3, U.S. Geological Survey, Washington, DC, 1968 (reprint 1982).

Brater, E. F., King, H. W., Lindell, J. E., and Wei C. Y., *Handbook of Hydraulics*, 7th ed., McGraw-Hill, New York, 1996.

Brune, G. M., "Trap Efficiency of Reservoirs," *Trans. Am. Geophys. Union*, v. 34, n. 3, 1953.

Brune, G. M., and Allen, R. E., "A Consideration of Factors Influencing Reservoir Sedimentation in the Ohio Valley Region," *Trans. Am. Geophys. Union*, v. 22, pp. 649–655, 1941.

Cassagrande, A., "Seepage through Dams," *J. New Engl. Water Works Assoc.*, June 1937.

Chadwick, A. J., and Morfett, J. C., *Hydraulics in Civil Engineering*, George Allen Unwin Ltd., London, 1986.

Cheremisinoff, P. N., Cheremisinoff, N. P., and Su Ling Cheng, (eds.), *Handbook of Civil Engineering Practice*, Technomic, Lancaster, PA, 1988.

Clemmens, A. J., Wahl, T. L., Bos, M. G., and Replogle, J. A., *Water Measurement with Flumes and Weirs*, International Inst. for Land Reclamation & Improvement, Wageningen, The Netherlands, 2001.

Chow, V. T., *Open-Channel Hydraulics*, McGraw-Hill, New York, 1959.

Chow, V. T. (ed.), *Handbook of Applied Hydrology*, McGraw-Hill, New York, 1964.

Creager, W. P., Justin, J. D., and Hinds, J., *Engineering for Dams*, vol. I, John Wiley & Sons, New York, 1945.

Crowe, C. T., Roberson, J. A., and Elger, D. F., *Engineering Fluid Mechanics*, John Wiley & Sons, NY, 2004.

Dalrymple, T., and Benson, M. A., *Measurement of Peak Discharge by the Slope-Area Method*, Techniques of Water Resources Investigations, Book 3, Chap. A2, U.S. Geological Survey, Washington, DC, 1967 (reprint 1984).

Davis, C. V., and Sorenson, K. E. (eds.), *Handbook of Applied Hydraulics*, 3rd ed., McGraw-Hill, New York, 1969.

Fredrich, A. J., *Techniques for Evaluating Long-Term Reservoir Yields*, Technical Paper 14, Hydrologic Engineering Center, U.S. Army Corps of Engineers, Davis, CA, 1969.

Golze, A. R. (ed.), *Handbook of Dam Engineering*, Van Nostrand Rheinhold, New York, 1977.

Grzywienski, A., "Anti-vacuum Profiles for Spillways of Large Dams," *Trans. 4th Congress on Large Dams*, vol. 2, pp. 105–124, International Commission on Large Dams of the World-Power Conference, New Delhi, India, 1951.

Hager, W. H., "Lateral Outflow over Side Weirs," *ASCE J. Hydraul. Eng.*, v. 113, n. 4, pp. 491–504, 1987a.

Hager, W. H., "Continuous Crest Profile for Standard Spillway," *ASCE J. Hydraul. Eng.*, v. 113, n. 11, pp. 1453–1456, 1987b.

Hinds, J., Creager, W. P., and Justin, J. D., *Engineering for Dams*, vol. 2, John Wiley & Sons, New York, 1945.

Hulsing, H., *Measurement of Peak Discharge at Dams by Indirect Method*, Techniques of Water Resources Investigations, Book 3, Chap. A5, U.S. Geological Survey, Washington, DC, 1967 (reprint 1984).

Hwang, N. H. D., and Houghtalen, R. J., *Fundamentals of Hydraulic Engineering Systems*, 3rd ed., Prentice-Hall, Englewood Cliffs, NJ, 1995.

Hydrologic Engineering Center, *Feasibility Studies for Small Scale Hydropower Additions*, vol. 3, U.S. Army Corps of Engineers, Davis, CA, 1979.

Justin, J. D., Hinds, J., and Creager, W. P., *Engineering for Dams*, vol. 3, John Wiley & Sons, New York, 1945.

Kilpatrick, F. A., and Schneider, V. R., *Use of Flumes in Measuring Discharge*, Techniques of Water Resources Investigations, Book 3, Chap. A14, U.S. Geological Survey, Washington, DC, 1983.

Koelzer, V. A., and Bitouri, M., "Hydrology of Spillway Design: Large Structures—Limited Data," *ASCE J. Hydraul. Div.*, v. 90, n. HY3, pp. 261–293, May 1964.

Kulin, G., and Compton, P. R., *Guide to Methods and Standards for the Measurement of Water Flow*, National Bureau of Standards, spl. publ. # 421, 1975.

Martin, R. O. R., and Hanson, R. L., *Reservoirs in the United States*, Water Supply Paper 1838, U.S. Geological Survey, Washington, DC, 1966.

Matthai, H. F., *Measurement of Peak Discharge at Width Contractions by Indirect Methods*, Techniques of Water Resources Investigations, Book 3, Chap. A4, U.S. Geological Survey, Washington, DC, 1967 (reprint 1984).

McCarthy, D. F., *Essentials of Soil Mechanics and Foundations*, 7th ed., Prentice-Hall, Englewood Cliffs, NJ, 2006.

Morris, H. M., and Wiggert, J. M., *Applied Hydraulics in Engineering*, 2nd ed., John Wiley & Sons, New York, 1972.

Munson, B. R., Young, D. F., and Okiishi T. H., *Fundamentals of Fluid Mechanics*, 5th ed., John Wiley & Sons, New York, 2006.

Murphy, T. E., *Spillway Crest Design, Misc. Paper H-73–5*, Waterways Experimentation Station, U.S. Army Corps of Engineers, Vicksburg, MS, 1973.

Nalluri, C., *Civil Engineering Hydraulics: Essential Theory with Worked Examples*, 4th ed., Blackwell Science, Oxford, 2001.

National Academy of Science, *Safety of Existing Dams: Evaluation and Improvement*, Washington, DC, 1983.

Ogrosky, H. O., "Hydrology of Spillway Design: Small Structures—Limited Data," *ASCE J. Hydraul. Div.*, v. 90, n. HY3, pp. 265–310, May 1964.

Reese, A. J., and Maynord, S. T., "Design of Spillway Crest," *ASCE J. Hydraul. Eng.*, v. 113, n. 4, pp. 476–490, 1987.

Riggs, H. C., and Hardison, C. H., *Storage Analyses for Water Supply*, Techniques of Water Resources Investigations, Book 4, Chap. B2, U.S. Geological Survey, Washington, DC, 1983.

Senturk, F., *Hydraulics of Dams and Reservoirs*, Water Resources Publications, Highlands Ranch, CO, 1996.

Simon, A. L., *Hydraulics*, 3rd ed., John Wiley & Sons, New York, 1986.

Snyder, F. F., "Hydrology of Spillway Design: Large Structures—Adequate Data," *ASCE J. Hydraul. Div.*, v. 90, n. HY3, pp. 239–259, May 1964.

Terzaghi, K., and Peck, R. B., *Soil Mechanics in Engineering Practice*, 2nd ed., John Wiley & Sons, Inc., New York, 1967.

Thomas, H. H., *The Engineering of Dams*, parts I and II, John Wiley & Sons, Chichester, West Sussex, England, 1976.

U.S. Army Corps of Engineers, *Hydraulic Design of Spillways*, Engineering Manual 1110-2-1603, U.S. GPO, Washington, DC, 1965.

U.S. Army Corps of Engineers, *Hydraulic Design of Spillways*, *Draft of Chapter 3 on Spillway Crest*, Engineering Manual 1110-2-1603, U.S. GPO, Washington, DC, 1986.

U.S. Bureau of Reclamation, *Dams and Control Works*, 3rd ed., U.S. GPO, Washington, DC, 1954.

U.S. Bureau of Reclamation, *Design of Small Canal Structures*, U.S. Dept. of Interior, Denver, CO, 1978.

U.S. Bureau of Reclamation, *Water Measurement Manual*, revised reprint, U.S. Dept. of Interior, Denver, CO, 1984.

U.S. Bureau of Reclamation, *Design of Small* Dams, 3rd ed., U.S. Dept. of Interior, Washington, DC, 1987.

Chapter 11

American Society of Civil Engineers, Task Force on Friction Factors in Open Channels, "Friction Factors in Open Channels," *ASCE J. Hydraul. Div.*, v. 89, n. HY4, March 1963.

American Society of Civil Engineers and Water Pollution Control Federation, *Design and Construction of Sanitary and Storm Sewers*, 5th printing, ASCE, New York, 1982.

Asawa, G. L., *Irrigation Engineering*, 2nd ed., New Age International (P), New Delhi, India, 1996.

Auckers, P., "Alluvial Channel Hydraulics," *J. Hydrol.*, v. 100, pp. 177–204, 1988.

Bakhmeteff, B. A., *Hydraulics of Open Channels*, McGraw-Hill, New York, 1932.

Benefield, L. D., Judkin, J. F., and Parr, A. D., *Treatment Plant Hydraulics for Environmental Engineers*, Prentice-Hall, Englewood Cliffs, NJ, 1984.

Blench, T., "Regime Theory for Self-Formed Sediment Bearing Channels," *Trans. ASCE*, v. 117, pp. 383–400, 1952.

Blench, T., "Regime Theory Design of Canals with Sand Beds," *ASCE, Jour. of Irrig. and Drainage Div.*, IR2, pp. 205–213, 1970.

Brater, E. F., King, H. W., Lindell, J. F., and Wei, C. Y., *Handbook of Hydraulics*, 7th ed., McGraw-Hill, New York, 1996.

Chang, H. H., "Stable Alluvial Canal Design," *Proc. ASCE, Hydrau. Div.*, v.106, HY5, 1980.

Chanson, H., *Hydraulics of Open Channel Flow*, Butterworth Heinemann, Oxford, UK, 2004.

Charleton, F. G., Brown, P. M., and Benson, R. W., "The Hydraulic Geometry of Some Gravel Rivers in Britain," *Hydraulic Research Station*, Report IT 180, Wallingford, UK, 1978.

Chaudhry, M. H., *Open-Channel Flow*, Prentice-Hall, Englewood Cliffs, NJ, 1993.

Cheremisinoff, N. P., *Fluid Flow—Pumps, Pipes and Channels*, Ann Arbor Science Publishers, Ann Arbor, MI, 1981.

Chow, V. T., *Open-Channel Hydraulics*, McGraw-Hill, New York, 1959.

Crowe, C. T., Elger, D. F., and Roberson, A., *Engineering Fluid Mechanics*, 8th ed., John Wiley & Sons, New York, 2007.

Davidian, J., *Computation of Water Surface Profiles in Open Channels*, Techniques of Water Resources Investigations, Chap. A15, Book 3, U.S. Geological Survey, Washington, DC, 1984.

Davis, C. V., and Sorenson, K. E. (eds.), *Handbook of Applied Hydraulics*, Secs. 2, 6, 7, McGraw-Hill, New York, 1969.

Henderson, F. M., *Open Channel Flow*, Macmillan, New York, 1966.

Hydraulic Research Station, *Charts for the Hydraulic Design of Channels and Pipes*, 5th ed., Hydraulic Research, Wallingford, England, 1983.

Inglis, Sir Claude, *Annual Report (Technical)*, Central Irrigation and Hydropower Research Station, Poona, India, 1940–41.

Kellerhals, R., "Stable Channel with Gravel Paved Beds," *J. WW, ASCE*, v. 93, pp. 63–68, 1967.

Kennedy, R. G., "The Prevention of Silting in Irrigation Canals," *Proceedings of Institution of Civil Engineers*, London, v. 119, pp. 281–290, 1895.

Koutitas, C. G., *Elements of Computational Hydraulics*, Chapman & Hall, London, 1983.

Lacey, G., "Stable Channels in Alluvium," *Proceedings of Institution of Civil Engineers*, London, v. 229, pp. 259–384, 1930.

Lacey, G., *Regime Flow in Incoherent Alluvium*, Central Board of Irrigation and Power, Publication 20, Government of India, Simla, India, 1940.

Lal, J., *Hydraulics*, Metropolitan Book Co., New Delhi, 1958.

Lane, E. W., "Stable Channels in Erodible Material," *Trans. ASCE*, v. 102, pp. 123–142, 1937.

Lane, E. W., "Design of Stable Channels," *Trans. ASCE*, v. 120, pp. 1234–1260, 1955.

Leopold, L. B., and Maddock, T., *Hydraulic Geometry of Stream Channels and Some Physiographic Implications*, USGS Prof. Paper # 252, U.S. GPO, Washington, DC, 1953.

Limerinos, J. T., *Determination of the Manning Coefficient from Measured Bed Roughness in Natural Channels*, Water Supply Paper 1898-B, U.S. Geological Survey, Washington, DC, 1970.

Lindley, E. S., "Regime Channels," *Minutes of Proceedings, Punjab Engineering Congress*, Lahore, India, v. 7, pp. 63–74, 1919.

Linsley, R. K., Franzini, J. B., Freyberg, D. L., and Tchobanoglous, G., *Water Resources Engineering*, 4th ed., McGraw-Hill, New York, 1991.

Nalluri, C., *Civil Engineering Hydraulics: Essential Theory with Worked Examples*, 4th ed., Blackwell Science, Oxford, UK, 2001.

Ouazar, D., Barthet, H., and Brebbia, C. A., *Computational Hydraulics*, Springer-Verlag, New York, 1988.

Pickard, W. F., "Solving the Equations of Uniform Flow," *Proc. ASCE*, v. 89, n. HY4, Part I, July 1963.

Pomeroy, R. D., "Flow Velocities in Small Sewers," *J. Water Poll. Control Fed.*, 39, 1525, 1967.

Raudkivi, A. J., *Loose Boundary Hydraulics*, Taylor Francis, UK, 1998.

Richards, K., *Rivers: Form and Process in Alluvial Channels*, The Blackburn Press, Caldwell, NJ, 2004.

Rouse, H. (ed.), *Engineering Hydraulics*, Chap. 9, John Wiley & Sons, New York, 1950.

Simon, A. L., *Practical Hydraulics*, John Wiley & Sons, New York, 1976.

Singh, V. P., "On the Theories of Hydraulic Geometry," *International Jour. of Sediment Research*, v. 18, n. 3, pp 196–218, 2003.

Streeter, V. L., Wylie, E. B., and Bedford, K. W., *Fluid Mechanics*, 9th ed., McGraw-Hill, New York, 1998.

Stevens, M. A., and Nordin, C. F., "Critique of the Regime Theory for Alluvial Channel," *ASCE Jour. of Hydrau. Eng.*, v.113, n.11, pp. 1359–1380, 1987.

Terrell, P. W., and Whitney, M. B., "Design of Stable Canals and Channels in Erodible Material," *Trans. ASCE*, v. 123, pp. 101–115, 1958.

U.S. Bureau of Reclamation, *Design of Small Canal Structures*, U.S. Dept. of Interior, Denver, CO, 1978 (reprint 1983).

U.S. Soil Conservation Service (now the NRCS), *Handbook of Channel Design for Soil and Water Conservation*, SCS-Technical Paper 61, U.S. Dept. of Agriculture, Washington, DC, 1954.

U.S. Soil Conservation Service (now the NRCS), *Design of Open Channels*, SCS-Technical Release # 25, U.S. Dept. of Agriculture, Washington, DC, 1977.

Whitaker, S., *Introduction to Fluid Mechanics*, Krieger, 1992.

White, W. R., Bettes, R., and Paris, E., "Analytical Approach to River Regime," *Proc. ASCE Hydrau. Div.*, v.108, HY10, 1982.

Woodward, S. M., and Posey, C., *Hydraulics of Steady Flow in Open Channels*, John Wiley & Sons, New York, 1962.

Chapter 12

Abbott, M. B., *An Introduction to the Method of Characteristics*, American Elsevier, New York, 1966.

American Society of Civil Engineers, *Pipeline Design for Water and Wastewater*, Committee on Pipeline Planning, ASCE, New York, 1975.

American Society of Civil Engineers and Water Pollution Control Board, *Design and Construction of Sanitary and Storm Sewers*, Manual on Engineering Practice 37, ASCE, New York, 1982.

Asthana, K. C., "Transformation of Moody Diagram," *ASCE J. Hydraul. Div.*, v. 100, n. HY6, pp. 797–808, June 1974.

Barr, D. I. H., "Explicit Working for Turbulent Pipe Flow Problems," *ASCE J. Hydraul. Div.*, v. 102, n. HY5, pp. 667–673, May 1976.

Bauer, W. J., Louis, D. S., and Voorduin, W. L., "Basic Hydraulics, Part 1: Close Conduit," in *Handbook of Applied Hydraulics*, ed. C. V. Davis and K. E. Sorenson, McGraw-Hill, New York, 1969.

Benefield, L. D., Judkins, J. F., and Parr, A. D., *Treatment Plant Hydraulics for Environmental Engineers*, Prentice-Hall, Englewood Cliffs, NJ, 1984.

Blasius, H., "Das Ähnlichkeitsgesetz bei Reibungsvorgängen in Flüssigkeiten" (The Law of Simulitude for Frictional Motions in Fluids), Forschungsheft Des Vereins Detscher Ingenieure, n. 131, Berlin, 1913.

Brater, E. F., King, H. W., Lindel, J. F., and Wei, C. Y., *Handbook of Hydraulics*, 7th ed., McGraw-Hill, New York, 1996.

Cheremisinoff, N. P., *Fluid Flow: Pumps, Pipes, and Channels*, Ann Arbor Science Publishers, MI, 1981.

Colebrook, C. F., and White, C. M., "Experiments with Fluid Friction in Roughened Pipes," *Royal Society Proceedings*, v. 161, London, 1937.

Colebrook, C. F., "Turbulent Flows in Pipes with Particular Reference to the Transition Region Between the Smooth and Rough Pipe Laws," *J. Inst. Civil Engrs.*, v. 11, London, 1939.

Darcy, H., "Sur des recherches expérimentales relatives au mouvement des eaux dans les tuyaux" ("Experimental Research on the Flow of Water in Pipes"), *Comptes Rendus Des Séances De L'Académie Des Sciences*, v. 38, pp. 1109–1121, 1854 (Darcy's name is associated with the equation for his research on flow in pipes although Weisbach first formulated the equation.)

Franzini, J. B., and Finnemore, E. J., *Fluid Mechanics with Engineering Applications*, 10th ed., McGraw-Hill, New York, 2001.

Halliwell, A. R., "Velocity of a Waterhammer Wave in an Elastic Pipe," *ASCE J. Hydraul. Div.*, v. 89, n. HY4, pp. 1–21, 1963.

Hwang, N. H. C., and Houghtalen, R. J., *Fundamentals of Hydraulic Engineering Systems*, 3rd ed., Prentice-Hall, Englewood Cliffs, NJ, 1995.

Hydraulic Research Station, *Charts for the Hydraulic Design of Channels and Pipes*, 5th ed., Hydraulic Research Station Ltd., London, 1983.

Jain, A. K., "Accurate Explicit Equation for Friction Factor," *ASCE J. Hydraul. Div.*, v. 102, n. HY5, pp. 674–677, May 1976.

Jain, A. K., Mohan, D. M., and Khanna, P., "Modified Hazen-Williams Formula," *ASCE J. Environ. Eng. Div.*, v. 104, n. EE1, pp. 137–146, Feb. 1978.

Karassik, I. J., and McGuire, T. J., *Centrifugal Pumps*, 2nd ed., Springer, 1997.

Karassik, I. J., Messina, J., Cooper, P., and Heald, C., *Pump Handbook*, 3rd ed., McGraw-Hill, New York, 2000.

Kerr, S. L., "New Aspects of Maximum Pressure Rise in Closed Conduits," *Trans, Am. Soc. Mech. Eng.*, v. HYD-51–3, pp. 13–30, 1929.

Lai, R. Y. S., and Lee, K. K., "Moody Diagram for Direct Pipe Diameter Calculation," *ASCE J. Hydraul. Div.*, v. 101, n. HY10, pp. 1377–1379, Oct. 1975.

Lal, J., *Hydraulics*, Metropolitan, New Delhi, 1958.

Li, W. H., "Direct Determination of Pipe Size," *ASCE Civ. Eng.*, p. 74, June 1974.

Linsley, R. K., Franzini, J. B., Freyberg, D. L., and Tchobanoglous, G., *Water Resources Engineering*, 4th ed., McGraw-Hill, New York, 1991.

Lyle, N. H., and Weinberg, G., "Pipeline Network Analysis by Electronic Digital Computer," *J. Am. Water Works Assoc.*, v. 49, pp. 517–529, 1957.

McClain, C. H., *Fluid Flow in Pipes*, 2nd ed., Industrial Press, New York, 1963.

McGhee, T. J., *Water Supply and Sewage*, 6th ed., McGraw-Hill, New York, 1991.

Moody, L. F., "Friction Factors for Pipe Flow," *Trans. ASME*, v. 66, p. 671, 1944.

Mott, R. L., *Applied Fluid Mechanics*, 6th ed., Prentice-Hall, Englewood Cliffs, NJ, 2005.

Nalluri, C., *Civil Engineering Hydraulics: Essential Theory with Worked Examples*, 4th ed., Blackwell Science, Oxford, UK, 2001.

Ouazar, D., Barthet, H., and Brebbia, C. A., *Computational Hydraulics*, Springer-Verlag, New York, 1988.

Powell, R. W., "Diagram Determines Pipe Sizes Directly," *ASCE Civ. Eng.*, pp. 45–46, Sept. 1950.

Prandtl, L., "The Mechanics of Viscous Fluids," v. III, div. G, in *Aerodynamics Theory* by Durand, W. F. (editor-in-chief), Springer-Verlag, Berlin, 1935 (Prandtl modified the expression developed by von Kármán in 1930.)

Simon, A. L., *Practical Hydraulics*, John Wiley & Sons, New York, 1976.

Smith, P. D., *Basic Hydraulics*, Butterworth & Company, London, 1982.

Streeter, V. L., "Steady Flow in Pipes and Conduits," in *Engineering Hydraulics*, ed. H. Rouse, John Wiley & Sons, New York, 1950.

Streeter, V. L., *Fluid Mechanics*, 5th ed., McGraw-Hill, New York, 1971.

Streeter, V. L., Wylie, E. B., and Bedford, K. W., *Fluid Mechanics*, 9th ed., McGraw-Hill, New York, 1998.

Swamee, P. K., and Jain, A. K., "Explicit Equations for Pipe-Flow Problems," *ASCE J. Hydraul. Div.*, v. 102, n. HY5, pp. 657–664, May 1976.

Weisbach, J., *Lehrbuch Der Ingenieur-und Maschinenmechanik (Textbook of Engineering Mechanics)*, Brunswick, Germany, 1845.

Whitaker, S., *Introduction to Fluid Mechanics*, Krieger, 1992.

Wood, D. J., "An Explicit Friction Factor Relationship," *ASCE Civ. Eng.*, pp. 60–61, Dec. 1966.

Wood, D. J., and Rayes, A. G., "Reliability of Algorithms for Pipe Network Analysis," *ASCE J. Hydraul. Div.*, v. 107, n. 10, 1981.

Chapter 13

Al-Layla, M. A., Ahmad, S., and Middlebrooks, E. J., *Handbook of Wastewater Collection and Treatment: Principles and Practice*, Garland Publishing, New York, 1980.

American Society of Civil Engineers, *Some Notes on the Rational Method of Storm Drain Design*, ASCE Urban Water Resources Research Program, TMN.6, ASCE, New York, 1969.

American Society of Civil Engineers, *Pipeline Design for Water and Wastewater*, Committee on Pipeline Planning, Pipeline Division, ASCE, New York, 1975.

American Society of Civil Engineers, *Urban Subsurface Drainage*, Manual #95, Urban Drainage Standards Committee, ASCE, Reston, VA, 1998.

American Society of Civil Engineers and Water Environment Federation, *Design and Construction of Urban Stormwater Management Systems*, ASCE, New York, 1992.

American Society of Civil Engineers and Water Pollution Control Federation, *Gravity Sanitary Sewer Design and Construction*, Manual on Engineering Practice No. 60, ASCE, New York, 1982a.

American Society of Civil Engineers and Water Pollution Control Federation, *Design and Construction of Sanitary and Storm Sewers*, Manual on Engineering Practice No. 37, ASCE, New York, 1982b.

Bartlett, R. E., *Surface Water Sewerage*, Applied Science Pub., Essex, UK, 1976.

Bell, F. C., "Generalized Rainfall-Duration-Frequency Relationships," *ASCE J. Hydraul. Div.*, v. 95, n. HY1, pp. 311–327, 1969.

Bowman, C. C., "Manning's Equation for Shallow Flow," *Proc. ARS-SCS Workshop on Hydraulics of Surface Irrig.*, ARS-41–43, pp. 21–23, U.S. Department of Agriculture, Washington, DC, 1960.

Bras, R. L., and Perkins, F. E., "Effects of Urbanization on Catchment Response," *ASCE J. Hydraul. Div.*, v. 101, n. HY3, 1975.

Butler, D., Davies, J. W., and Davis, J., *Urban Drainage*, Brunner-Routledge, London, 2000.

Engman, E. T., "Roughness Coefficients for Routing Surfaces Runoff," *ASCE J. Irrig. and Drainage*, v. 112, n. 1, pp. 39–53, 1986.

Escritt, L. B., *Sewerage and Sewage Disposal*, 4th ed., vol. II, Macdonald & Evans, Plymouth, England, 1972.

Fair, G. M., Geyer, J. C., and Okun, D. A., *Water and Wastewater Engineering*, vol. 1, John Wiley & Sons, New York, 1966.

Geyer J. C., and Lentz, J. J., "An Evaluation of the Problems of Sanitary Sewer System Design," *J. Water Poll. Control Fed.*, v. 38, pp. 1138–1147, 1966.

Ghosh, S. N., *Flood Control and Drainage Engineering*, 3rd ed., Taylor Francis, UK, 2006.

Gupta, B. R. N., *Design Aids for Public Health Engineers*, John Wiley & Sons, New York, 1986.

Hromadka, T. V., Clements, J. M., and Saluja, H., *Computer Methods in Urban Watershed Hydraulics*, Lighthouse Publications, Mission Viejo, CA, 1984.

Hromadka, T. V., Durbin, T. J., and DeVries, J. J., *Computer Methods in Water Resources*, Lighthouse Publications, Mission Viejo, CA, 1985.

International Conference on Urban Drainage, *Urban Stormwater Hydraulics and Hydrology*, Water Resources Publications, Highlands Ranch, CO, 1982.

Johnson, R. E. L., "Development of Sanitary Sewer Design Criteria," *J. Water Poll. Control Fed.*, v. 37, pp. 1597–1606, 1965.

Kinori, B. Z., *Manual of Surface Drainage Engineering*, vol. I, Elsevier Science Publishing, Amsterdam, 1970.

Linsley, R. K., Franzini, J. B., Freyberg, D. L., and Tchobanoglous, G. *Water Resources Engineering*, 4th ed., McGraw-Hill, New York, 1991.

Marsalek, J., Watt, W. E., Zeman, E., and Sieker, H. (eds.), *Advances in Stormwater and Agriculture Runoff Source Controls*, Kluwer Academic Pub., Dordrechy, The Netherlands, 2001.

McCuen, R. H., *A Guide to Hydrologic Analysis Using SCS Methods*, Prentice-Hall, Englewood Cliffs, NJ, 1982.

McCuen, R. H., Wong, S. L., and Walter, J. R., "Estimating Urban Time of Concentration," *ASCE J. Hydraul. Div.*, v. 110, n. 7, 1984.

McGhee, T. J., *Water Supply and Sewarage*, 6th ed., McGraw-Hill, New York, 1991.

McPherson, M. B., *Some Notes on the Rational Method of Storm Drain Design*, Tech Memo. No. 6, ASCE, Water Resources Research Program, Harvard Univ., Cambridge, MA, 1969.

National Association of Home Builders, *Residential Wastewater Systems*, NAHB, Washington, DC, 1980.

Okun, D. A., and Ponghis, G., *Community Wastewater Collection and Disposal*, World Health Organization, Geneva, 1975.

Overton, D. E., and Meadows, M. E., *Storm Water Modeling*, Academic Press, New York, 1976.

Sauer, V. B., et al., *Flood Characteristics of Urban Watersheds in the United States*, Water Supply Paper 2207, U.S. Geological Survey, Washington, DC, 1983.

Schaake, J. C., Geyer, J. C., and Knapp, J. W., "Experimental Examination of the Rational Method," *ASCE J. Hydraul. Div.*, v. 93, n. HY6, pp. 353–363, 1967.

Sieker, F., and Verworn, H. R., *Urban Storm Drainage*, Pergamon, New York, 1998.

Singh, V. P., *Hydrologic Systems: Rainfall-Runoff Modeling*, vol. 1, Prentice-Hall, Englewood Cliffs, NJ, 1988.

Sokolov, A. A., Rantz, S. E., and Roche, M., *Floodflow Computation: Methods Compiled from World Experience*, Studies and Report in Hydrology No. 22, UNESCO, Paris, 1976.

Stanley, W. E., and Kaufman, W. J., "Sewer Capacity Design Practice," *J. Boston Soc. Civ. Eng.*, v. 317, 1953.

Stephenson, D., *Stormwater Hydrology and Drainage*, Developments in Water Science Series No. 14, Elsevier Science Publishing, Amsterdam, 1981.

Tholin, A. L., and Keifer, C. J., "The Hydrology of Urban Runoff," *Trans. ASCE*, v. 125, p. 1308, 1960.

Urban Land Institute, National Association of Home Builders, *Residential Storm Water Management: Objective, Principles and Design Considerations*, NAHB, Washington, DC, 1975.

U.S. Bureau of Reclamation, *Design of Small Dams*, 2nd ed., U.S. Dept. of Interior, U.S. GPO, Washington, DC, 1977.

U.S. Soil Conservation Service (now the NRCS), *Engineering Field Manual for Soil Conservation Practices*, Chap. 14, U.S. Dept. of Agriculture, Washington, DC, 1975a.

U.S. Soil Conservation Service, *Urban Hydrology for Small Watersheds*, Technical Release 55, U.S. Dept. of Agriculture, Washington, DC, 1975b.

U.S. Soil Conservation Service (now the NRCS), *Urban Hydrology for Small Watersheds*, Revised Technical Release 55, Dept. of Agriculture, Washington, DC, 1986.

Walesh, S. G., *Urban Surface Water Management*, John Wiley & Sons, New York, 2001.

Wandle, S. W., Jr., *Estimating Peak Discharges of Small, Rural Streams in Massachusetts*, Water Supply Paper 2214, U.S. Geological Survey, Washington, DC, 1983.

Whipple, W., et al., *Stormwater Management in Urbanizing Areas*, Prentice-Hall, Englewood Cliffs, NJ, 1983.

White, J. B., *Wastewater Engineering*, 3rd ed., Hodder Arnold, UK, 1978.

Yen, B. C. (ed.), *Catchment Runoff and Rational Formula*, Water Resources Publications, Littleton, CO, 1991.

Chapter 14

American Association of State Highway Transportation Officials (AASHTO), *Highway Drainage Guidelines*, AASHTO, Washington, DC, 1994.

American Society of Agriculture Engineers, *Drainage and Water Table Control*, Saint Joseph, MI, 1993.

American Society of Civil Engineering, *Urban Subsurface Drainage*, Urban Drainage Standards Committee, ASCE, Reston, VA, 1998.

Bartlett, R. E., *Surface Water Sewerage*, Applied Science Publishers Ltd., Barking, Essex, England, 1976.

Beers, van W. F. J., *Some Nomograms for the Calculation of Drainage Spacings*, Bull. 8, International Inst. for Land Reclama. & Improvement, Wageningen, The Netherlands, 1965.

Bodhaine, G. L., *Measurement of Peak Discharge at Culverts by Indirect Methods*, Techniques of Water Resources Investigations, Chap. A3, Book 3, U.S. Geological Survey, Washington, DC, 1982.

Bras, R. L., and Perkins, F. E., "Effects of Urbanization on Catchment Response," *ASCE J. Hydraul. Div.*, v. 101, n. HY3, 1975.

Cedergren, H. R., *Drainage of Highway and Airfield Pavements*, Krieger Publishing, Melbourne, FL, 1987.

Chow, V. T., "Hydrologic Design of Culverts," *ASCE J. Hydraul. Div.*, v. 88, n. HY2, pp. 39–55, 1962.

Chow, V. T. (ed.), *Handbook of Applied Hydrology*, Sec. 20, McGraw-Hill, New York, 1964.

Federal Aviation Adm., *Airport Drainage*, Advisory Circular 150/5320–5A, U.S. GPO, Washington, DC, 1965.

Federal Highway Adm., *Urban Drainage Design Manual*, Univ. Press of the Pacific, 2005.

Kinori, B. Z., *Manual of Surface Drainage Engineering*, vol. I, Elsevier Science Publishing, Amsterdam, 1970.

Luthin, J. N., *Drainage Engineering*, Krieger, New York, 1978.

Marsalek, J., Watt, W. E., Zeman, E., and Sieker, H. ed., *Advances in Stormwater and Agriculture Runoff Source Controls*, Kluwer Academic Pub., Dordrechy, The Netherlands, 2001.

McCuen, R. H., *A Guide to Hydrologic Analysis Using SCS Methods*, Prentice-Hall, Englewood Cliffs, NJ, 1982.

Ritzema, H. P. (ed.), *Drainage Principles and Applications*, 2nd ed., Water Resources Publications, Highlands Ranch, CO, 1994.

Ritzema, H. P., et al., *Drainage of Irrigated Lands: A Manual*, n. 9 Food and Agriculture Organization, 1996.

Schilfgaarde, J. V. (ed.), *Drainage for Agriculture*, American Society of Agronomy, Madison, WI, 1974.

Schultz, B., *Guidelines on the Construction of Horizontal Subsurface Systems*, U.S. Committee on Irrigation and Drainage Systems, Denver, CO, 1990.

Sieker, F., and Verworn, H. R., *Urban Storm Drainage*, Pergamon, New York, 1998.

Smart, P., and Hebertson, J. G., *Drainage Design*, Van Nostrand Reinhold, 1992.

Smith, K. V. H., *Hydraulic Design in Water Resources Engineering: Land Drainage*, Springer-Verlag, NY, 1986.

U.S. Bureau of Public Roads, *Hydraulic Charts for the Selection of Highway Culverts*, Hydraulic Engineering Circular # 5, U.S. Dept of Commerce, Washington, DC, 1965.

U.S. Bureau of Reclamation, *Design of Small Canal Structures*, U.S. Dept. of the Interior, Denver, CO, 1978.

U.S. Bureau of Reclamation, *Drainage Manual*, U.S. Dept. of Interior, U.S. GPO, Denver, CO, 1984.

U.S. Department of Transportation, *Hydraulic Design of Improved Inlets for Culverts*, Hydraulic Engineering Circular 13, Federal Highway Administration, Washington, DC, 1972.

U.S. Department of Transportation, *Guidelines for the Design of Subsurface Drainage Systems for Highway Structural Sections*, Federal Highway Administration, Washington, DC, 1973.

U.S. Department of Transportation, *Design of Urban Highway Drainage, The State-of-the-Art*, Federal Highway Administration, Washington, DC, 1979.

U.S. Department of Transportation, *Urban Drainage Design Manual*, Federal Highway Administration, Hydraulic Engineering Circular 22, Washington, DC, 1996.

References

Answers to Selected Problems

◆◆◆

Chapter 1

1.2   (**a**) and (**b**) 45.8 thousand

1.4   39.8 years

1.6   112.5 years

1.7   (**a**) 284.1 million, (**b**) $P_t = \dfrac{284.1}{1+28.59}e^{-0.02t}$ (**c**) 235.5 million

1.8   (**a**) 118.69 thousand (**b**) 110 thousand

1.10 $P_{2000} = 22.3$ thousand, $P_{2020} = 24$ thousand

1.12   Flow 1.44 mgd, duration 2 hr

1.13   0.24 mgd

1.14   0.18 mgd

1.15   0.39 mgd

1.16   (**a**) 15.75 mgd (**b**) 15.75 mgd (**c**) 15.75 mgd (**d**) low 21.0 mgd, high 31.5 mgd
           (**e**) 23.63 mgd (**f**) 28.93 mgd

1.17   municipal 95.7 mgd, manufacturing 132.3 mgd, thermal 238.8 mgd,
           dilution 7445.2 mgd

1.18   29 in.

1.20   18.56 in.

1.21

Month	Jan	Feb	Mar	Apr	May	June	July	Aug	Sept	Oct	Nov	Dec	Total
Gross demand acre-ft $\times 10^3$	9.95	17.05	31.53	27.15	10.26	14.62	31.67	55.96	25.65	13.88	14.02	9.65	261.39

1.22   (**a**) 4214 kW (**b**) 25.7×10^6 kWh (**c**) 0.7

1.23   (**a**) 2595.7 kW (**b**) 8.89×10^6 kWh (**c**) 0.39

1.24   66%

Chapter 2

2.2   (−) 3 in.

2.4   138.9 m^3/s

2.6   no runoff

2.8

Month	Jan.	Feb.	Mar.	Apr.	May
Moisture storage, in.	6.96	7.82	9.13	8.98	7.86
Groundwater storage, in.	2.276	2.083	1.932	1.795	1.658
Groundwater recharge, in.	0.026	0.036	0.062	0.060	0.045
Direct runoff, cfs	0.144	0.204	0.350	0.340	0.255
Groundwater discharge, cfs	6.62	6.71	5.62	5.40	4.82

2.10   4.77 in.

2.11   Breakpoint at 2000; data from 1990 to 1999 are adjusted by a factor of 0.78.

2.12   **(b)** Breakpoint at 1987, **(c)** 367 mm, **(d)** 312 mm

2.13   **(a)** 4.29 in. **(b)** 4.41 in.

2.14   4.38 in.

2.15   **(a)** 800 mm **(b)** 881mm

2.16   880 mm

2.17   Intensity (in./hr) of duration

Period	5 min.	10 min.	15 min.	20 min.	30 min.
5-yr	4.24	3.6	2.96	2.50	1.80
3-yr	3.78	3.0	2.50	2.16	1.56

2.18

	Duration, hr				
	6	12	24	48	72
Area			Depth, in.		
10	26.0	30.3	33.5	37.5	38.0
200	18.3	21.8	25.0	29.3	30.8
1000	13.0	15.8	20.0	23.3	24.5
5000	8.3	11.3	14.0	17.5	18.5

2.19   **(a)** 10.5 mm **(b)** 10.1 mm

2.20   **(a)** 14.25 cm/mo, **(b)** 14.55 cm/mo

2.21   10.5 mm/day

2.22   14.54 cm/mo

2.23

Date	Jun 8	Jun 15	Jun 21	Jun 27	Average
K	0.69	0.70	0.75	0.75	0.72

2.24   7.63 mm/day

2.25   13.6 cm/mo

2.26   8.4 mm/day

2.27   13.93 cm/mo

2.28 6.34×10^{-3} m/day

2.29 5.98×10^{-3} m/day

2.30 $D_1 = 16\%$, $D_2 = 27\%$, $D_3 = 33\%$, $D_4 = 24\%$, $K_{C1} = .4$, $K_{C2} = 1.00$, $K_{C3} = 0.55$

2.31 $D_1 = 21\%$, $D_2 = 25\%$, $D_3 = 33\%$, $D_4 = 21\%$, $K_{C1} = .3$, $K_{C2} = 1.15$, $K_{C3} = 0.75$

2.32

Month	June	July	Aug.	Sept.	Oct.	Nov.
mm/day	1.4	4.23	5.5	3.0	1.16	0.57

2.33

Month	May	June	July	Aug.	Sept.
mm/day	1.5	3.19	7.48	6.90	5.16

2.34   **(c)** Linear decrease from 9 in./hr at $t = 0$ to 5.35 in./hr at $t = 40$ min, thereafter
$f_p = 1.2 + (5.35 - 1.2)\, e^{-0.076t'}$ where $t' = t + 40$

t, min	40	60	80	100	120
f_p, in./hr	5.35	2.11	1.40	1.24	1.21

2.35

Period, min	40–60	60–80	80–100
RO, in.	0.76	0	0.30

2.36

Period, hr	0.56–1	1–2	2–3	3–4
RO, mm	1.01	5	6.5	7.0

2.37

Period, min	0–20	20–40	40–60	60–80	80–100	100–120	120–140
RO, in.	0	0	0.75	0.78	0	0	0.16

2.38

Period, min	up to 100.5	100.5–120	120–150	150–180
RO, in.	0	0.3	1.3	0.4

2.39

Period, min	0–45	45–60	60–80	80–100	100–120	120–140
RO, in.	0	0.19	0.69	0.10	0.08	0.58

2.40

Period, min	0–100	100–120	120–150	150–180
RO, in.	0	0	1.16	0.65

2.41

Period, min	0–40	40–60	60–80	80–100	100–120	120–140
RO, in.	0	0.51	0.72	0.20	0.18	0.74

2.42

Period, min	0–90	90–120	120–150	150–180
RO, in.	0	1.96	2.46	1.49

2.44   20.8 acre-ft

2.46   (a) 1.68 in./day (b) 0.92 in./day

2.47

Snowmelt, in.	Zone 1	Zone 2	Zone 3
First day	0.46	0.31	0.15
Second day	0.46	0.29	0.12
Third day	0.73	0.56	0.38

Chapter 3

3.2

	at *A*	at *B*	at *C*
a. Pressure head, m	65	15	88
b. Hydraulic head, m	90	90	88
e. Hydraulic gradient = 0.0015			

3.4   (a) 211.2×10^3 ft^3/day (b) 3300 days

3.6 6.6×10^{-9} ft^2

3.8   0.025 cm^2/s

3.9   4.04 m/day

3.10   18.58 ft

3.12   0.65 ft^3

3.14   (a) 1.3 g/cc (b) 0.51 (c) 1.04

3.16   6.0 cm^3/cm^2 area

3.17 **(a)**

Negative head, cm	0	10	20	30	40	50	60	70	80	90	100
Vol. moist. content %	53.2	52.92	52.5	51.66	50.40	48.58	46.20	44.24	43.12	42.56	42.28

(c) 5.45 cm^3/cm^2 area

3.18 **(a)** 53.2% **(b)** 42.3% **(c)** 10.9%

3.20 25%

3.22 $67.2 \times 10^3 \, m^3$

3.24 3.56×10^{-6} in.^{2}/lb, 23%, 77%

3.26 **(a)** $\bar{v} = 3.8 \times 10^{-3} \, \bar{j} + 17.4 \times 10^{-3} \, \bar{k}$ **(c)** $v = 17.81 \times 10^{-3}$ ft/s, $\theta = 78°$ from horizontal

3.27 **(a)** $\bar{v} = 6.3 \times 10^{-3} \, \bar{j} + 2.46 \times 10^{-3} \, \bar{k}$ **(c)** $v = 6.76 \times 10^{-3}$ ft/min

Chapter 4

4.2 **(a)** 109 m^3/day/m

(b)

Dist., m	0	1	6	9.5	13.6	17.3	22	26	28.6	30
u, kN/m^2	99.1	90.2	81.2	72.3	63.4	54.4	45.5	36.6	27.7	18.7

4.3

Dist., m	0	15	30
u, kN/m^2	112	80.6	31.5

4.4 **(a)** 0.057 ft^3/min/ft

(b)

At	A	B	C
u, lbs/ft^2	749	686	437

4.6 6.75 m^3/day/m

4.7 8.6 m^3/day/m

4.8 0.051 m^3/s

4.10 0.2 m^3/s

4.12 12.15 yrs

4.14 39.8 days

4.15 13,700 ft^2/day, 19×10^{-5}

4.16 707 ft^2/day, 6.7×10^{-3}

4.17 3070 ft^2/day, 5.3×10^{-4}

4.18 $0.033 \text{ m}^3/\text{s}, 2.1 \times 10^{-4}$

4.19 $119.4 \text{ m}^2/\text{day}, 0.01$

4.20   18 ft

4.21 $13,765 \text{ ft}^2/\text{day}, 18.8 \times 10^{-5}$

4.22 $0.032 \text{ m}^3/\text{s}, 2 \times 10^{-4}$, valid after 2 min

4.23 $93,180 \text{ gpd/ft}, 5.1 \times 10^{-4}$

4.24 $67,000 \text{ gpd/ft}, 0.0032$
valid for vertical flow and delayed yield but drawdown over 25%

4.26 $0.091 \text{ m}^3/\text{s}$

4.27   2 m

4.28 $4.4 \times 10^{-3} \text{ m}^2/\text{min}, 6.4 \times 10^{-5}, 8.4 \times 10^{-5} \text{ m/min}$

4.29 $0.1 \text{ ft}^2/\text{min}, 4 \times 10^{-4}, 1.5 \times 10^{-4} \text{ ft/min}$

4.30   **(a)** 0.34 m **(b)** 6.86 m

4.31   6.13 m

4.32   8.12 m

4.34

20	16.69	19.15	19.03	
20	16.69	19.15	19.08	19.08
20	19.01	19.35	17.76	17.76
20	19.75	19.83	19.43	19.43

4.36 $B = 3.1 \text{ s/ft}^2, C = 10.59 \text{ s}^2/\text{ft}^5$

4.37   6 wells at 350 ft spacing

4.38 $25.87 \text{ ft}, 1.55 \text{ ft}^2/\text{min}$

Chapter 5

5.2 $4.71 \times 10^{-6} \text{ mg/s}$

5.4

Distance, cm	± 0	± 1	± 3	± 5
At $t = 1$ min				
C, mg/cm^3	0.236	0.153	0.005	0.00
At $t = 5$ min				
C, mg/cm^3	0.106	0.097	0.049	0.012
At $t = 10$ min				
C, mg/cm^3	0.075	0.072	0.051	0.026

5.5

Distance, cm	± 0	± 1	± 2	± 3	± 5
At $t = 1$ min					
C, mg/cm^3	*	0.22	0.102	0.02	0
At $t = 5$ min					
C, mg/cm^3	*	*	*	0.106	0.075
At $t = 10$ min					
C, mg/cm^3	*	*	*	*	*

* tracer moved out of this position due to advection

5.6 The values are twice that of Problem 5.5 in the positive direction of X only

5.7

Time, min	0	10	30	60	90	120	150
C/C_o	0	0	0	0.11	0.69	0.96	1.00

5.8 0.51 cm/min, 0.91 cm^2/min

5.9

x, m	10	20	30	40
C, mg/L	37.8	4.5	0.11	0

5.10 0.99×10^{-5} cm^2/s, diffusion controlled

5.11 (a) 85.86 cm (b) 1

5.12 $Q_{water} = 4.78$ ft^3/day, $Q_{CTC} = 2.6$ ft^3/day

5.14 5.9 cm

5.16 5.6 ft

5.17 0.89 ft^3/s

5.18 p-plot curvature indicates non-normality

5.19 $W_{comp} = 0.992$, $W_{critical} = 0.897$, lognormal data

5.20 p-plot curvature indicates non-normality

5.21 $W_{comp} = 0.993$, $W_{critical} = 0.915$, lognormal data

5.22 $W_{comp} = 0.978$, $W_{critical} = 0.955$, normal data

5.24 $X_{mean} = 13.65$, $S = 6.19$, $r_1 = 0.42$ (low), independent data

5.25 $CI_L = 2.687$ ppm, average GWPS = 1.668, evidence of contamination

5.26 Well # 1, $CI_L = 2.44$, GWPS < CI_L, evidence of contamination

Well # 6, $CI_L = 0.74$, GWPS > CI_L, no evidence of contamination

5.27 Well # 1, $CI_L = 3.28$ ppm, ln MCL = 3.91, no evidence of contamination

Well # 6, $CI_L = 1.64$ ppm, ln MCL = 3.91, no evidence of contamination

5.28 $CI_L = 5$, MCL = 5, at threshold

5.29 $TI_U = 137$ ppb, no evidence of contamination

5.30 $TI_U = 5.14$ ppm,   Well # 1 $TI_U >$ con., no evidence of contamination

Well # 2 $TI_U <$ con., evidence of contamination

5.31 $TI_U = 5.14$ ppm, all wells ln con. $< TI_U$, no evidence of contamination

5.32 $PI_U = 92$ ppb, compl. well con. $> PI_U$, evidence of contamination

5.33   Well # 1, $PI_U = 14$ ppb, well conc. = 10, no evidence of contamination

Well # 2, $PI_U = 14$ ppb, well conc. = 30, evidence of contamination

5.34 $PI_U = 72.8$ ppb, month 11 has evidence of contamination

5.35

Month	9	10	11	12
Z	0.822	2.334	3.312	1.108
S	0	1.334	3.646	3.754

no evidence of contamination

5.36

	2004				2005			
Quarter	1	2	3	4	1	2	3	4
Z	0	−1	0.83	−0.83	2	1	5	2.5
S	0	0	0	0	1	1	5	6.5

Out of control in third and fourth quarter of 2005.

Chapter 6

6.1   1.27 m/s, 2.02 Pa

6.2   5.3 Pa, 1.05 m/s, 0.015 m^3/s

6.3   0.944 m/s, 3.6×10^{-4}, 14.16 m^3/s

6.4   1.99 ft/s, 6.97 ft^3/ft

6.6   1.97 ft/s, 6.9 ft^3/ft

6.8   3.16 ft/s

6.9

Method	v, ft/s	Error, %
1. Vertical	2.02	0
2. Graphic	2.03	0.5
3. Two-point	2.03	0.5
4. Six-tenth	2.0	1.0
5. Three-point	2.01	0.5
6. 0.2-depth	2.0	1.0
7. Surface	2.0	1.0
8. Five-point	2.03	0.5
9. Six-point	2.04	1.0

6.10   (**a**) 4.27 m (**b**) 4.42 m, 6.98 m

6.12   2.95 m^3/s

6.13   110.89 cfs

6.14   3.02 m^3/s

6.15   109.02 cfs

6.16   2.95 m^3/s

6.17   110.4 cfs

6.18   12.4 m^3/s

6.19   2 m^3/s

6.20   1.017 m^3/s

6.21   8.52 m^3/s

6.22   0.318 m^3/s

6.23   (**a**) 0 and 1 (**b**) 0.6

6.24   (**a**) 24 cm (**b**) 25 cm

6.25   (**a**) 0 (**b**) 0.1

6.26   (**a**) $2.65\,(h-0.6)^{1.47}$ (**b**) $2.61(h-0.6)^{1.48}$

6.27   (**a**) $\dfrac{1}{1430}(h-25)^2$ (**b**) $\dfrac{1}{1486}(h-25)^{2.04}$

6.28 $90.9\,(h-0.1)^{2.41}$

6.30   478 m^3/s

Chapter 7

7.2

Time, hr	10	15	20	25	30	35 …	70	75	80	85
DR, m^3/s	0	3.83	10.80	14.35	15.85	13.03	1.94	1.25	0.5	0

7.46 mm

7.4

Time, days	2	3	4	5	6	7	8	9	10	11	12
Method 1, m^3/s	0	11.5	35.5	71.7	98.8	75.8	35.9	14	7	3	0
Method 2, m^3/s	0	8	37.5	74	103	75	29	0	0		
Method 3, m^3/s	0	7	28	60	83.5	56.5	21	3	0		

Depth: Method 1, 47 mm; Method 2, 43.35 mm; Method 3, 34.45 mm

7.6

Time, hr	10	20	30	40	50	60	70	80	85
Method 1, m^3/s	0	10.95	16.15	11.55	7.9	4.7	1.95	0.5	0
Method 2, m^3/s	0	11.1	16.45	11.85	8.15	4.9	2.15	0.6	0
Method 3, m^3/s	0	10.65	15.65	11.10	7.60	4.45	1.9	0.45	0

7.8

Time, hr	10	15	20	25	30	35 …	65	70	75	80
UHG, m^3/s	0	0.51	1.45	1.92	2.12	1.74	0.43	0.26	0.17	0.07

7.9

Time	midnight	6 AM	noon	6 PM	midnight …	6 PM	midnight
UHG, cfs	0	33.56	5671.4	4060.4	2986.6	1308.7	906.04

7.10

Time, hr	0	1	2	3	4	5	6	7	8	9	10
IUH, cfs	0	260	960	2369	3205	3553	3970	4273	3870	3378	2983
UHG, cfs	0	130	610	1665	2787	3379	3762	4122	4072	3634	3191

Time, hr	11	12	13	14	15	16	17	18	19
IUH, cfs	2619	2299	2019	1773	1557	1367	1200	1054	925
UHG, cfs	2801	2459	2159	1896	1665	1462	1284	1127	990

7.11

Time, hr	0	5	10	15	20	25	30	35	40
IUH, m^3/s	0	0.23	0.51	1.37	1.65	1.54	1.31	1.11	0.94
UHG, m^3/s	0	0.12	0.37	0.94	1.51	1.60	1.42	1.21	1.02

7.12

Time, hr	0	1	2	3	4	5	6	7	8	9 ...	12	13	14
UHG, cfs	0	25	50	125	125	175	150	188	150	162	25	13	0

7.14

Time, hr	0	1	2	3	4	5	6	7	8	9 ...	12	13	14
UHG, cfs	0	25	50	125	125	175	150	188	150	162	25	13	0

7.15

Time, hr	0	1	2	3	4	19	20	21	22
UHG, cfs	0	133.3	833.3	1466.7	2000	200	200	66.7	0

7.16 $t_{pR} = 9.5$ hr, $Q_{pR} = 5893$ cfs, $T = 99.5$ hr, $W_{50} = 13.56$ hr, $W_{75} = 7.75$ hr

7.18

Time, hr	0	5.35	10.7	21.4	32.1	42.8	53.5
UHG, cfs	0	2984	6350	1778	349	70	0

7.20

Time, hr	0	6	12	24	36	48	60
UHG, m^3/s	0	4.07	8.67	2.43	0.50	0.10	0

7.21

Time (time unit)	0	1	2	3	4	5	6	7
DRH (rainfall unit)	0	0.75	0.50	1.50	1.23	0.68	0.14	0

7.22

Time, hr	1200	1500	1800 ...	0900	1200	1500 ...	2100	2400	0300
Flow, m^3/s	10	59.4	197.9	232.4	232.4	162.1	41	20	13

7.23

Time, clock hr	0	1000	1500	2000	0100 ...	2100	0200	0700	1200	...
Time, hr	0	10	15	20	25	45	50	55	60	
Flow, m^3/s	1.37	1.12	1.17	1.20	8.90	41.76	42.93	33.78	33.78	
						2300	0400	0900	1400	
						95	100	105	110	
						3.16	2.26	1.26	0.56	

7.24

July	6	7	8 ...	14	15	16	18	19	20	21
Flow m^3/s	62	345	236	28	80	60	105	583	393	263

7.25 $P_e = 0.8P_0 + 0.2P_1$

7.26 $P_e = 0.7P_0 + 0.3P_1$

7.27 $Q = 0.0056P_e - 2.405$, P_e mm, Q m^3/s

7.28 $Q = 1.88P_e - 12.64$, P_e in., Q mill acre-ft

7.29   1900 cfs

7.30   (a) $r = 0.96$, excellent correlation (b) $Q_{St.John} = 0.566Q_{St.Paul} - 1036$

7.31   (a) $r = 0.95$, excellent correlation (b) $Q_{Quinebaug} = 4.9Q_{LittleRiver} + 8.15$

7.32 $\bar{X} = 4955.3$ cfs, $S = 1794$ cfs, $g = -0.21$, $r_1 = -0.195$, $r_2 = -0.02$, normally distributed

7.33 $\bar{X} = 26.52$ m^3/s, $S = 7.75$, $g = 1.08$, $r_1 = 0.368$, $r_2 = 0.239$

7.34

i	1	2	3	4 ...	7	8	9
qi, cfs	4034.8	6209.6	4078.7	4434.4	5236.5	6304.3	7633.9

7.35

i	1	2	3	4	5 ...	8	9	10
qi, m^3/s	20.71	25.58	32.21	23.19	22.03	20.04	25.60	27.12

7.36 $\phi_{1,1} = 0.80$, $\theta_{1,1} = 0.366$, $Se^2 = 0.81$

       $X_1 = 49.4$ m^3/s, $X_2 = 38.1$ m^3/s, $X_3 = 38.3$ m^3/s

7.37 $\phi_{1,1} = 0.65$, $\theta_{1,1} = 0.33$, $Se = 0.849$

i	1	2	3	4	5	6	7
X_i, m^3/s	20.78	25.90	31.71	22.57	22.49	11.13	26.60

7.38

Month	Oct.	Nov.	Dec.	Jan. ...	July	Aug.	Sept.
Q, cfs	392	408	307	258	373	240	174

7.40   126.2 cfs

7.41 $w = 13.18 Q^{0.38}$, $d = 0.1 Q^{0.39}$, $v = 0.836 Q^{0.22}$

7.42 $w = 49.43 Q^{0.286}$, $d = 0.077 Q^{0.55}$, $v = 0.26 Q^{0.20}$

7.43   85%, 254 cfs, 215 cfs

7.44 $Q_{90} = 170$ m^3/s, $Q_{mean} = 295$ m^3/s, $Q_{median} = 235$ m^3/s

Chapter 8

8.2   3.5%

8.4   87%

8.6   9950 yrs

8.7 $ 9748

8.8   45 cfs, 11%.

8.9 $F_{computed} = 1.11$, $F_{theor.} = 2.1$, data homogeneous
$t_{computed} = 0.9$, $t_{theor.} = 1.68$, data homogeneous

8.10 $F_{computed} = 1.5$, $F_{theor.} = 3.17$, data homogeneous
$t_{computed} = 0.85$, $t_{theor.} = 2.06$, data homogeneous

8.12   330 m^3/s, 700 m^3/s

8.14   **(a)** 14,000 cfs **(b)** 97%

8.15   **(a)** 16,910 cfs **(b)** 98%

8.16

$P\%$	0.5	1 ...	10	50	80	90	99
Q, m^3/s	575	518	332	187	126	102	61

8.17

$P\%$	1	2 ...	10	50	80	90	99
Q, m^3/s	1363	1126	675	321	211	173	114

8.18

$P\%$	0.5	1	2	10	50	80	90	95	99
Q, cfs	11,427	10,810	10,145	9413	5625	4157	3497	3008	2222

8.19

$P\%$	0.1	1	2	5	10	20
Q, m^3/s	717	531	475	400	342	282

8.20

$P\%$	1	5	10	50	80	90
Q, m^3/s	1155	803	661	333	213	168

8.21

$P\%$	0.1	1	5	10	50	80	90	95
Q, cfs	11,127	9802	8621	7992	5769	4309	3546	2917

8.22

P_N %	1.33	10.6	50	89.4	98.67
Q_5, m^3/s	683	401	210	117	74
Q_{95}, m^3/s	427	289	166	84	46

8.23

P_N %	1.45	10.8	50	89.2	98.55
Q_5, cfs	13,934	9878	6256	3943	2628
Q_{95}, cfs	9137	7379	5057	2945	1723

8.24

Q, cfs	26,500	25,000	20,000	17,500	15,000	12,500	10,000
$P_{combined}$, %	5	6	20.7	35.8	56	75.4	89.7

500 yrs

8.26 25,260 cfs.

8.28

Area, mi.^{2}	10	200	1000	5000	10,000	20,000
Duration 6-hr, depth, in.	24.96	17.52	12.48	7.92	6.72	4.80
Duration 12-hr, depth, in.	29.04	20.88	15.12	10.8	8.88	7.92

8.29 **(a)** PMP = 13 in., **(b)** $\bar{p}$ = 10.83 in.

8.30 Q_{peak} = 10,790 cfs, Q_{50} = 5400 cfs, Q_{75} = 8090 cfs, W_{50} = 4.5 hrs, W_{75} = 2.5 hrs

8.32 Q = 1,298,356 cfs, error 17%

8.33 3,304,700 cfs

8.34 **(a)** 90% **(b)** 3.1 yrs

8.35 126 cfs, 42%

8.36

P (X <)%	10	20	40	60	80	90	95	98	99.9
Q, m^3/s	3.87	4.16	4.65	5.23	6.38	7.56	8.91	11.01	21.83

(a) 90% **(b)** 2.9 yrs

8.37

P (X <)%	10	20	40	60	80	90	95	98	99	99.9
Q, cfs	134.9	143.8	159.2	176.8	211.5	246.6	285.7	345.8	398.9	640.8

(a) 135 cfs **(b)** 40%

Chapter 9

9.2   **(a)** 191 mm **(b)** 10 m/s, 17.14 m/s **(c)** 2.52 kPa, 105 kPa

9.3 $0.162 \text{ m}^3/\text{s}$

9.4 $0.162 \text{ m}^3/\text{s}$

9.5   2.8 ft, 13.86 ft/s

9.6   2.7 ft, 14.35 ft/s

9.7   7.53 ft, 2.97 ft, 19.9 ft/s

9.8   7.53 ft, 3 ft, 19.7 ft/s

9.9   46.22 kN, 165°

9.10   3.93 m

9.11 $x = 1 - e^{-0.15t}$, $v = e^{0.2t}$

9.12 $\ln x = 0.47t^{3/2} + 2.30$, $A = 3.35t + 500$

9.13

Time, min	y, mm	x, m
1	0.42	1.81
5	2.08	26.53

9.14

Time, min	0	1	2	3	5
q, m^3/s/m ($\times 10^{-5}$)	0	1.26	4.0	7.86	18.43

9.15 $t_c = 10$ min

	Rising					Receding	
Time, s	0	1	2	5	10	15	20
q, cfs/ft ($\times 10^{-4}$)	0	2.84	9.01	41.49	131.74	52	23

9.16 $t_c = 8.37$ min

	Rising					Receding			
Time, s	0	1	3	5	8.37	12	15	20	25
q, m^3/s/m ($\times 10^{-5}$)	0	1.01	6.32	14.82	34.95	34.95	18.6	6.6	1.3

9.17 $t_c = 11.77$ min

	Rising					Receding		
Time, s	0	2	5	8	12.3	15	20	25
q, m^3/s/m ($\times 10^{-5}$)	0	3.06	14.09	30.84	30.84	20	9.6	5.2

9.18 $KFr^2 = 133$, kinematic approximation valid

9.20

Time, hr	12	24	36	48	60	72	84	96
O, cfs	300	348	575	823	799	678	514	326

9.21

Time, hr	6 AM	Noon	6 PM	Midnight	6 AM	Midnight...	6 AM	Midnight
O, cfs	150	159	211	418	749	383	279	104

9.22

Time, hr	0	1	2	3	4	5	6	7	8	9	10
O, m^3/s	15	18.19	38.71	93.35	168.24	208.75	190.41	144.42	84.22	47.34	24.12

9.24   x = 0.1, K = 10 hrs

9.26

Time, hr	12	24	36	48	60	72	84	96
Q, cfs	300	307.5	390.9	567.1	686.7	705.3	644.8	520.8

9.27

Time	6 AM	Noon	6 PM	Midnight ... 3rd	6 PM	Midnight	6 AM	Noon
O, cfs	150	155.4	192.0	338.9	178.5	130.6	108.3	85.6

9.28

Time, hr	—	0	0.4	0.8	1.2 ...	3.2	3.6	4.0	4.4
I, cfs	0	0	600	2100	2500	80	0	0	0
O, cfs	0	0	50	415	1120	650	440	240	200
Elev., ft	60	60.3	61.6	63.1	63.8	61.65	61.25	61.0	60.8

9.29

Time, hr	—	0	0.5	1.0	1.5	2.0 ...	4.5	5.0	5.5	6.0
I, cfs	0	0	20	70	160	280	0	0	0	0
O, cfs	0	0	2	17	52	105	98	65	42	27
Elev., ft	15	15.3	16.1	16.95	18.0	18.9	17.25	16.75	16.35	16.0

9.30   16.6 m, 11.4 m

9.32   410.5 m

9.34

Time, s	0	2.5	5	7.5	10.0	12.5	15.0	17.5	20.0
Head, m	10	16.03	28.87	48.40	55.6	51.2	48.1	41.8	43.0

9.35

Time, s	0	1.9	3.8	5.7	7.6	9.5	11.4	13.3
Head, ft	50	63.62	79.48	97.45	116.66	135.81	153.29	166.70
velocity, m/s	8.15	8.04	7.71	7.11	6.22	5.04	3.57	1.86

9.36   (a) $V_2 = 5.34$ m/s, $V_w = 9.17$ m/s

(b) $y_2 = 1.266$ m, $V_2 = 2.54$ m/s

(c) $y_1 = 0.386$ m, $V_2 = -1.456$ m/s

(d) $V_1 = 0.529$ m/s, $V_2 = 2.015$ m/s

9.37   0.85 m, 4.95 m

9.38   2.3 m

y, m	1.76	2	2.5	3
x, m	7.72	8.54	10.11	11.53

9.40   6.61 m

Chapter 10

10.2   14 in.

10.4   0.71, 0.9, 0.643

10.6   721 s

10.7   43.4 s

10.8   20.58 s

10.10   442 ft

10.11   47.11 cfs

10.12   3.02 ft

10.14   1.35 ft

10.16   16.94 ft³/s

10.17   3.4 ft

10.18   778.15 ft³/s

10.20   1.38 m

10.21   0.42 m³/s

10.22   335.2 ft³/s

10.24   35 mm

10.26   24.66 psf

10.27   1450 cfs

10.28   44.3 m³/s

10.29   567 cfs

10.30 15.08 m³/s

10.31 726.7 ft³/s

10.32 752 ft³/s

10.33 (**a**) 6730 acre-ft (**b**) 66,845 acre-ft

10.34 5603 mill. gal

10.35 (**a**) 1.127 mill. gal (**b**) 2.335 mill. gal

10.36 18.6 ft³/day/ft

10.37 15.8 ft³/day/ft

10.38 14.3 ft³/day/ft

10.40 1.05 ft³/day/ft

10.42 21.6 ft³/day/ft

10.44 8 ft³/day/ft

10.45

Downstream quadrant

x, ft	5	10	15	20	22.3
y, ft	1.52	5.46	11.57	19.70	24.10

Upstream quadrant

x, ft	1	2	2.52
y, ft	0.12	0.58	1.48

10.46

Downstream quadrant

x, m	0.5	1.0	1.5	2.0
y, m	0.06	0.23	0.48	0.82

Upstream quadrant

x, m	0.2	0.4	0.56
y, m	0.02	0.08	0.19

10.47 97 ft

10.48 369.46 m³/s

10.49 16,836 ft³/s

10.50 15,031 ft³/s

10.52 34.9 m

Downstream quadrant

x, m	1	3	5	7	8	8.9
y, m	0.168	1.28	3.30	6.15	7.87	9.60

Upstream quadrant

x, m	0.3	0.6	0.9
y, m	0.027	0.115	0.322

10.53 8257 ft³/s

10.54 50.73 ft

10.56 $S = 0.0036$, Fr $= 1.32$, stable

10.57 $\alpha = 13.5°$, $L = 104$ ft

10.58 $\alpha = 11.70°$, $L = 266$ ft

10.59

x from downstream end, ft	30	60	90	110
Δy, ft	5.82	1.49	0.74	0.2

10.60

x from downstream end, ft	30	60	90	110
Δy, ft	0.27	0.19	0.11	0.03

10.61

x from downstream end, m	10	20	30	38
Δy, m	0.13	0.09	0.05	0.02

10.62 $R_s = 8.5$ ft

H_a, ft	20	22	24	26	28	30
R, ft	5.72	5.59	5.47	5.36	5.26	5.17

10.64 $R_s = 1.59$ ft

H_a, ft	11	13	15	17	20
R, ft	1.59	1.52	1.47	1.42	1.37

Chapter 11

11.2 (a) $R = 1.35$ ft, $D = 5.31$ ft, $Z_c = 34.43$, $Z_n = 18.26$ (b) 4.39 ft

11.3 (a)

y, ft	1	2	3	4	5
E, ft	10.7	3.78	3.61	4.27	5.14

(b) 2.55 ft (c) 1.7 ft (d) supercritical

11.5 2.41 m, 3.44 m/s

11.6 1.57 ft

11.7 1.27 m, 3 m/s

11.8 (a) 1143 cfs (b) 645.13 cfs (c) 187.5 cfs (d) 1786 cfs (e) 688.6 cfs

11.10 268.79 m^3/s

11.11 6.95 ft, 856.94 ft^3/s

11.12 (a) 1.25 m (b) 0.85 m (c) subcritical

11.14 9.45 ft

11.16 $y_n = 5.9$ ft, channel depth = 9.1 ft.

11.18 $d = 2.5$ ft, $S = 0.00084$

11.20 18.33 ft^3/s, 2.59 ft/s, 2.12 ft/s, 0.0183

11.21 $d_o = 130$ mm, $S = 0.0166$

11.22 $y = 5.12$ ft, $b = 26$ ft

11.23 2.5 ft, 175 ft^3/s

11.24 2.96 ft, 228 ft^3/s

11.25 $\tau_0 = 0.35$ lbs/ft^2, $\tau_{allow} = 0.007$ lbs/ft^2, erosion occurs

11.26 $y = 4.77$ ft, $b = 28.9$ ft

11.27 2.4 ft, 170 ft^3/s

11.28 $y = 3.43$ ft, $b = 93.7$ ft

11.29

Toward upstream, depth, m	3.5	3.3	3.0	2.7	2.5	2.4	2.3	2.26
distance, m		154	402	689	931	1087	1314	1448

11.30

Toward downstream, depth, ft	0.6	0.8	1.0	1.2	1.4	1.6	1.8	2.0	2.2
distance, ft		20	41	63	85	106	125	140	146

11.31

Toward upstream, depth, m	3.5	3.3	3.0	2.7	2.5	2.26
distance, m		154	401	685	923	1406

11.32

Toward downstream, depth, ft	0.6	0.8	1.0	1.2	1.4	1.6	1.8	2.0	2.2
distance, ft		18	38	59	80	100	118	132	140

11.34 $F_2 = 0.77$, $F_1 = 1.31$, $D_1 = 2.1$ m, $E = 0.03$ m

Chapter 12

12.2   5.72 m

12.4   0.026, 72.05 ft

12.6   0.017

12.8   50.6 ft

12.10 33.2×10^{-6} m^3/s

12.11   0.52 cfs

12.12   0.71 ft

12.13   91 mm

12.14   **(a)** 27.95 ft^3/s **(b)** 46.23 ft

12.16   49.2 ft

12.18   0.64 ft

12.19   45 psi, $\Delta_p = 9.9\%$, valid

12.20   0.66 m^3/s

12.21   0.415 ft

12.22   **(a)** 0.21 m^3/s **(b)** 125 kW

12.24   2.05 cfs

12.26   4.01 m, 0.48 m^3/s, 0.17 m^3/s, 0.35 m^3/s

12.28   3.19 cfs

12.29

Pipe	Q m^3/s	h_f, m	Node	Pressure head, m
AB	0.116	10.4	A	100
BC	0.037	12.1	B	89.6
CD	0.003	0.4	C	77.5
DE	0.033	9.8	D	77.9
EF	0.044	2.1	E	87.7
AF	0.084	10.2	F	89.8
BE	0.019	1.8		

12.30

Pipe	Q, gpm	h_f, ft	Node	Pressure head, ft
AB	359.7	4.5	A	132.4
BC	617.9	12.1	B	177.9
CD	382.1	20.2	C	40 (given)
DE	382.1	20.2	D	169.8
EF	440.3	26.3	E	99.6
AF	440.3	6.5	F	175.9
BE	977.6	28.3		

12.31

Pipe	AB	BC	AC	BD	DE	EC
Q, cfs	1.85	0.54	1.65	1.31	0.19	1.19

12.32

Pipe	Q, cfs	h_f, ft	Node	Pressure head, ft
AB	0.78	2.02	A	175 (assumed)
BC	0.56	0.15	B	172.98
AF	1.52	1.22	C	172.83
FC	0.52	1.0	D	172.43
FE	0.8	0.18	E	173.6
ED	0.57	1.24	F	173.78
DC	0.83	0.40	D	172.36

12.33

Pipe	Q, cfs	h_f, ft	Node	Pressure head, ft
AB	1.86	10.18	A	173.1 ft or 75 psi
BC	1.34	0.78	B	162.92
AF	3.61	6.0	C	162.14
FC	1.24	4.93	D	160.16 or 69.4 psi
FE	1.89	0.88		
ED	1.34	6.05		
DC	1.98	1.98		

12.34 1739

12.35 1480, low discharge

12.36 (a) 69% (b) 1157 rpm

12.38

Changed Q, gpm	Changed H, ft	Changed P, hp
2334	121	82
1556	133.8	67.3
778	138.6	50.4

12.40

Changed Q, gpm	Changed H, ft	Changed P, hp
2214	109.0	70.4
1476	120.4	57.5
738	124.8	43.0

12.42 **(b)** 1980

(c) Changed speed			(d) Changed diameter			(e) Similar pump		
Q, gpm	H, ft	P, hp	Q, gpm	H, ft	P, hp	Q, gpm	H, ft	P, hp
0	181.8	53.7	0	207.4	65.4	0	207.4	58.9
890	180.6	75.2	950	206.0	91.5	855	206.0	82.4
⋮								
3556	124.0	133.1	3797	141.5	162.0	3420	141.5	145.9

12.44 $Q = 27.5$ L/s, $H = 18$ m, $n = 70\%$, BHP $= 6.9$ kW

12.45 **(a)** 204 ft, 1040 gpm **(b)** 102 ft **(c)** 77.7 hp

12.46 $Q = 2130$ gpm, $H = 101$ ft, $n = 69\%$, $P = 78.8$ hp

12.48 13.97 ft, 0.106

Chapter 13

13.2 29.4 mgd

13.4 540 mm diameter

13.6

Block	A	B	C	D	E	F	G	H
Diameter, in.	4	5	5	4	4	7	8	8
Grade	0.011	0.010	0.012	0.012	0.012	0.005	0.003	0.012

13.7

Manhole	1–2	2–3	3–4	4–5	5–10	6–7	7–8	8–9	9–10	10–15
Diameter, mm	340	380	400	420	605	260	365	390	375	730
Grade	.004	.011	.014	0.016	.001	.008	.006	.007	.013	.001

Manhole	11–12	12–13	13–14	14–15	15–19	19–18	18–17	17–16
Diameter, mm	310	415	495	530	460	645	810	820
Grade	.003	.003	.002	.002	.0235	.003	.001	.001

13.8 Invert elevation, ft

MH 5 91.17 → MH 6 86.88

MH 7 90.29 → MH 6 86.79, MH 6 86.69 → MH 8 82.73

13.10 0.775

13.11

Method	Kirpich	Kerby	Izzard	Bransby-Williams	Federal Aviation	Kinematic Wave
Time, min	15.54	25.92	NA	17.83	29.16	17.61

13.12 12.7 min

13.14

Frequency, yr	5	10	20
Q, cfs	95.5	121	143

13.16

Time, hr	11.9	12.0	12.1	12.2 …	12.7	12.8	13.0	13.2
Q, cfs	28.7	54.8	88.9	77.9	36.0	27.5	18.0	13.6

13.18

At manhole #	1	2	3
Q, cfs	18	46.9	58.9

13.20

At manhole #	7	6	5	4	3	2	1
Q, cfs	24	57	52	21.4	35.5	77.3	83

13.21

Between manhole	1–3	2–3	3–5	4–5	5–6
Diameter, in.	8	12	15	8	18
Grade	.008	0.01	.0075	.01	.0075

13.22

Manhole	1–2	2–3	3–4	4–8	5–6	6–7	7–8	8–12
Q, m³/s	0.111	0.294	0.413	0.466	0.076	0.198	0.253	0.753
d, mm	400	580	670	495	295	420	420	565

Manhole	9–10	10–11	11–12	12–13
Q, m³/s	0.117	0.297	0.417	1.18
d, mm	345	445	490	705

13.24 0.32 acre-ft

13.26 94.05×10^3 ft³

13.28 83.56×10^3 ft³

Chapter 14

14.2 Tube drain d = 0.7 ft, open drain b = 1.73 ft, y = 2.11 ft.

14.4 15 in.

14.5 spacing 1400 ft

14.6 400 ft

14.7 1000 ft

14.8 300 ft

14.10 $q = 12.04 \times 10^{-6}$ cfs/ft, $d = 0.36$ ft

14.12 0.06 cfs

14.14 1.24 cfs

14.15

Intercept	1–2	2–3	3–4	5–4	4–6	7–6	8–6	6–9
Q, cfs	13.96	16.52	19.42	17.06	36.10	3.45	3.17	41.66
d, in.	18	21	21	21	30	15*	15*	36

* minimum recommended

14.16 4.5 ft diameter

14.18 10 ft diameter

14.20 525 cfs

14.21

Manhole	1–2	2–3	4–3	5–3	3–6
Q, cfs	14.05	27.63	12.80	18.0	60.20
d, in.	30	42	30	33	51

Index

E

Earthfill dams 574–576

Eddy dispersion 235

Eddy shedding current meter 333

Effective porosity 138, 143–144. *See also* Specific yield

Effective rainfall 25–26, 29

Elastic wave theory 515, 518

Electromagnetic discharge measurement 318–319

Electromagnetic velocity meter 333

Electronic data acquisition and storage 288

Embankments 562, 564–565, 574, 604

Empirical deterministic hydrology 345

Energy balance method 66–71

Energy equation 484–487

Energy, specific 616–619

Engman, E. T. 736

Erosion, and airport drainage systems 786

Error limits/confidence limits 451–454

Estimation, rational and statistical 459–460

Evaporation

    defined 61

    energy balance method 66–71

    from free-water bodies 62–64

    transpiration and 61–62

    weighted 72

Evaporation measurement

    aerodynamic estimation method 64–66

    energy balance method 66–71

    using pans 63–64

    Penman's combination method 72

Evapotranspiration

    actual, from any surface 76–78

    Blaney-Criddle method 22, 80

    computation of irrigation demands and 29

    in cropped areas 22

    defined 61–62

    from a drainage basin 72–80

    potential 73

    precipitation and 61–62

    reference crop, Penman-Monteith method 73–76

    remote-sensing techniques and 113

    soil moisture content and 44–45

Evapotranspirometers 73

F

False negative/positive errors 259, 268

Fetter, C. W. 199

Fick's First Law of Diffusion 234–235

Fick's Second Law of Diffusion 237

Field capacity/water-holding capacity 140

Field data curve 182–184

Finite-difference/finite-element method 209–210, 491

Finite-Element Surface-Water Modeling System 475

Fire demands/requirements and duration 15–18, 687

Fisher distribution 51, 268, 436–438

Fisher, R. A. 443

Float gages/sensors 285

Floats, discharge measurement by 312

Flood flow

    computation procedures 427–430

    genetic and empirical equations for computing 468–469

    hydrograph 465, 467

    100-year exceedance 431

    risk basis for design flood 431–433

    series 434–435, 440–441

    See also Design flood; Peak discharge/flow

Flood-frequency analysis

    analytical method 442–451

    combined-population (composite) 454–456

    confidence limits and probability adjustments 451–454

    damage-frequency relation or curve of damage 433

    empirical method 440–442

    of flood-frequency curve 439–451, 454–456

    of flood volume 457

    graphical method 439

    models of 474–475

Exceedance (flood) probability 427, 430–434, 447, 452

Expected probability adjustment 446

Explicit approximation method 210

Extreme flow. *See* Flood-frequency analysis

Extreme value distribution 443–445

I

Ice cover, measurement under 292
Images, theory of 204–207, 209
Immiscible liquids 250–256
Implicit approximation method 210
Impulse system 289
Index station analysis 49–52
Industrial wastewater 720–721
Industrial water requirements 18–21
Infiltration
    cumulative 91, 95–96
    distinct cases of 80–81
    Green-Ampt model 90–97
    sewer design and 721
    Richards equation of, 83
Infiltration capacity
    curve approach 83–85, 87
    definition of 341
    empirical models for 84
    rainfall intensity and 81, 84–86, 88
Infiltration-index approach 104, 107
Instantaneous unit hydrograph 354, 356, 358–359
Intensity-duration-frequency analysis 59–60
Interception, definition of 41
Interceptor drains 763–765
Interflow 339, 346–347
Interim Final Guidance to Statistical Analysis (EPA) 262
International Glossary of Hydrology (WMO/UNESCO) 121
Interpolation 58, 332, 343, 401–403, 460, 470
Interstitial water 122
Interval estimates 269, 274
Intrinsic permeability 130–131, 252
Inverse procedure 355–356
Irrigation demands
    computation of 29, 31
    factors influencing 21
    farm delivery requirements 29
    monthly, computation of 30–31
Irrigation drainage systems 766–769
Isochronal IUH computation method 358–360
Isohyetal precipitation computation method 54, 56–57
Isotropic soil
    flow net in 163–165

vector notation (Darcy's law) 149
Iterative alternate direction implicit (IADI) method 213
Izzard formula 735, 737, 776

J

Jacob, C. E. 146, 155, 189, 198, 219
Jain, S. K. 664
Jia, Y. 639
Johnston, R. A. 248

K

Karassik, I. J. 690
Kasenow, M. 219
Kennedy, A. B. W. 638
Kerby method 735, 737
Khaleel, R. 91
Khanji, J. 90
Kinematic routing 499–503
Kinematic wave theory
    empirical relations for time of overland flow 735
    hydrograph formulation in 492–497
    Manning's equation 490–491
    methods of solving equations in 491–492
    Muskingum-Cunge routing method 500–502
    streamflow routing by 499–500
    time of concentration for urban watershed 739
    validity for hydrographs 497–498
    validity of routing in 503
    wave analysis, definition of 489
King, H. W. 536
Kirkby, M. J. 341
Kirkham, D. 766
Kirpich method 735, 737
K-lag serial correlation coefficient 266
Kohler, M. A. 63
Kostiakov, A. N. 83, 84
K-T relationship/table 446–449
Kuichling, E. 729
Kutter, W. R. 621

L

Lacey's original regime theory/rugosity 639–640

Lag time 352, 365, 368

Lagging method of duration adjustment 361–363

Laminar flow 150, 218, 235, 615, 662, 665

Lane, E. W. 631

Langbein, W. B. 435, 457

Lange, R. T. 123

Laplace equation 157, 163, 168

Larson, C. L. 91, 92

Larson, L. W. 49

Latent heat of vaporization of water 66–68

Leakage, coefficient of 160

Leakance 136–137

Leaky aquifers
    steady-state flow in 200–202
    theory of 198–199
    unsteady-state flow in 202–204

Least-squares line 326

Leopold, L. B. 405, 638

Level of significance 271, 435, 445–446, 462

Liggett, J. A. 497

Light nonaqueous phase liquids (LNAPL) 250, 253–254

Lindley's regime concept 638

Linear regression 325–329

Lloyd-Davies method 740

Logarithmic rating curve 321–322

Lognormal distribution 442–443

Lognormal probability graph 439

Lognormality/normality, tests for 262–266

Log-Pearson type III (gamma-type) distribution 445

Longitudinal drainage systems 774–776

Long-term population forecasting
    component methods 11–12
    graphic comparison method 8
    mathematical logistic curve method 8–10
    ratio and correlation methods 10–11

Long-throated flume 552

Loose-boundary channels
    carrying sediment-free water 631–637
    carrying sediment-laden water 637–643
    critical tractive force 632–633
    hypotheses of stable channel design 638
    power function theory and 642
    regime theory and 638–640, 642
    stability parameter method 634–636

unit tractive force on channel boundary 631

Loss-rate function 98–100

Low flow
    computation of 470, 472
    frequency analysis 470, 472–473
    regionalization/application to ungaged sites 473

Luthin, J. N. 183

Lysimeter 73

M

Macrodispersion 248

Maddock, T. 405, 639, 642

Manning, R. 330, 489–490
    equation of open channel flow 621–623, 737, 763
    roughness coefficient 621–622

Marino, C. T. 183, 639

Markov model/process (autoregressive model) 388–396

Marvis, F. T. 545

Mass curve analysis 407, 438

Mass transport equations 236–238
    advection-diffusion equation 237–238
    advection-diffusion-dispersion equation 238
    Fick's Second Law of Diffusion 237
    mass transport with reaction 238
    solutions of 238–245

Mass-transfer coefficient 65

Mathematical hydrology, defined 345

Mather, J. R. 44

Matthai, H. F. 562

Maximum contaminant levels (MCLs) 259, 268–270, 272

Maynord, J. T. 580, 581

Mays, L. W. 447, 450

McCarthy, D. F. 505

McJunkin, F. E. 8

McPherson, M. B. 729

Meadows, M. E. 497

Mean vertical velocity 294–295

Mechanical dispersions 235–236

Mein, R. G. 91, 92

Meinzer, O. E. 122

Meinzers 131–133

rainfall transformation to. *See* Unit
hydrograph
routing by kinematic theory 499–500
runoff and 339–340
short, augmentation of 374
simple stage-discharge relation 320–322
slope-stage-discharge relation 330–331
stage-discharge curve equation 324–329
synthetic 386–387, 389, 391
unit hydrograph estimation 370–372, 374
variability of 406–411
velocity distribution 292–293
velocity measurement 295–300
zero flow determination 322–324, 326
See also Discharge computation/
measurement; Stream gaging; Water
stage measurement
Streamlines, definition of 163
Streltsova, T. D. 183
Strongly implicit procedure 213
Student distribution 435–436
Subcritical/supercritical flow 615
Submergence 587
Subsurface drainage systems 761, 763–764,
786
Subsurface runoff, definition of 339
Subsurface stormflow 339, 341–342
Subsurface water, classification of 121–123,
125
Surface drainage systems 761–762, 787–788
Surface resistance, definition of 74
Surface runoff, definition of 339
Surface tension 250
Surface water flow estimation/computation
correlation technique 374–386
hydrograph analysis 339–374
methods of measurement 283–333,
427–475. *See also* Discharge
computation/measurement; Stage
measurement; Streamflow; Stream
gaging
synthetic technique 386–400
ungaged sites 401–415
Surface-Water Quality and Flow Modeling
Interest Group (SMIG) 791
Surge
absolute/relative velocity of 520, 522–523

positive and negative 520–526
Synthetic technique 386–400
Synthetic unit hydrograph 364
estimation of streamflow from 371–372,
374
Natural Resources Conservation Service
(NRCS) Method 369, 371
Snyder's method 364–365, 368, 370–371
System head curves 694, 697–699, 701, 703
Systems hydrology, defined 345

T

Taylor dispersion 235
Technical Release (TR)-20 program 412
Telemark/telemetry systems 289, 332
Theis curve/equation/method 180–182,
187–189, 195, 198
Theis, C. V. 180
Theoretical unit tractive force 631
Theory of images 204–209
Thiessen polygon precipitation conversion
method 54–55
Thomas, H., model of water balance equation
44–45
Thornthwaite, C. W. 44, 65
Three-dimensional flow, velocity components
for 148
Threshold channel 631
Throughflow 339
Time of concentration 352, 358, 495, 733, 735,
737
Time series, hydrologic 387–388
Tippett, L. H. C. 443
Tolerance interval technique 272–274, 276
Total dynamic head (TDH) 693
Total head loss 164, 659, 678–679
Total net flux 153, 156
Total static head 693
Tractive force 631
Transients, hydraulic 511–512
in open channels 520–526
in pipes 513–519, 689
Transmissivity 131, 135, 157, 184, 188,
209–210
Transpiration 61–62
Transport function, definition of 73

Transport processes
  advective transport 233–234
  aqueous phase or soluble contaminants
    247–248
  diffusive transport 234–235
  dispersive transport 235–236
  immiscible liquids 250–256
  mass transport equations 236–238
  nonaqueous phase liquids (NAPL)
    250–256
  of sediment 639–641
  solutions of mass transport equation
    238–245
Trapezoidal channel 618–619
Trapezoidal weirs 544–545
Trescott, P. C. 213
Triangular (V-notch) weirs 544
Trough/chute spillways 591–593
T-tests 268, 269
Tung, Y. K. 447, 450
Turbulent dispersion 235
Turbulent flow 150, 615, 620–621, 662–665,
  667
Two-phase groundwater flow 157
Type curves 182–188, 195, 199, 202–203
Type I and II errors 259, 268
Type specific speed 690

U

Ultrasonic discharge measurement 317–318
Ultrasonic sensors 287
Unconfined aquifers 194–197
Ungaged sites, estimation of flow at 401–415
Uniform channel flow
  computation of 622–623, 625–627
  definition of 614, 620
  formula 490
  hydraulics of 620–622
Unit hydrograph 352–355
  changing of duration 361–364
  Clark/instantaneous method 354, 356,
    358–359, 362
  computation of 357
  definition of 352
  derivation of 355–356, 358, 360–361
  dimensionless 371
  distribution graphs 354

duration adjustment, lagging method of
  361–363
flood flow 466
instantaneous 354, 356, 358–359
inverse procedure derivation 356,
  358–359, 361
preparing from runoff data 340
probable maximum flood and 459
s-curve computation and duration
  conversion 366–367
storm hydrograph derivation 357
streamflow estimation from 370–372, 374
summation-curve method of adjustment
  364–367
synthetic. *See* Synthetic unit hydrograph
time parameters of 352
width and peak discharge relation of 467
Unsaturated groundwater flow 157, 766
Unsteady-Flow Network Model 475
Unsteady-state analysis
  of confined aquifers 189–194
  of unconfined aquifers 183–189, 194–197
Unsteady-state flow
  analytical solution of equations for
    180–183
  aquifer-test analysis and 182–183
  depth and spacing of drains, 766
  flood hydraulics models 474
  in leaky aquifers 202–204
  production well drawdown 218
Upconing 256–258
Urban drainage systems 717–752. *See also*
  Sewer systems

V

Valve closure 515–520
Vanoni, V. A. 294
Vaporization of water, latent heat of 66–68
Variable source area concept 342
Variates 427–428, 435
Vector
  head gradient 148–149, 151, 162
  unit 148, 150
  velocity 149, 151–152, 162
Velocity
  of approach 542–543
  Darcy 126–130, 149–150

distribution 292–293
flow 125, 667
mean vertical 294–295
measurement of 295–300
meters, new types of 333
of overland flow 738
pipe size and 687
potential 149–150
seepage 127–129
surge 520, 522–523
vector 149, 151–152, 162
Vena contracta 535–536
Venturi flume/meter 552–554, 556
Vertical-axis cup meter 290
Vertical-velocity curve 296, 299–300
Vieira, D. 497
Villemonte, J. R. 545
V-notch weirs 543–544
Void ratio 138–139
Volumetric moisture content 140, 142–143, 161
Von Kármán, T. 620, 662
logarithmic velocity distribution 293

W
Wading rods 301
Walton, W. C. 219
Wastewater
conveyance losses and 27, 29
dilution requirements 19–21
quantity, assessment of 720–722
sewage/sewerage/wastewater, definition of 717
sewage quantity, computing 726
Water
availability of 41–113
conveyance losses and waste 27–29
demand for 1–35
gross water requirements, computation of 29
loss. See Evaporation; Evapotranspiration; Transpiration
pressure flow systems 659–706
vapor transfer, aerodynamic resistance to 73–74
variable of available 44–45
See also Groundwater; Groundwater flow;

Groundwater monitoring
Water balance equation 42–48
definition of 42
errors of computation of 47–48
for direct runoff within a basin 44–47
hydrologic cycle and 42
for larger river basins for long duration 43
precipitation and 48
for water bodies for short duration 42–43
Water hammer 513, 689
Water requirements
assessment of 2
consumptive use of crops 22–25
effective rainfall and 25–26
farm losses and 27
fire demands 15–18
hydropower demand 32–35
industrial 18–21
irrigation 21–29
long-term population forecasting of 8–12
municipal 3–4
navigation demands 35
on-farm and off-farm efficiency 27–29
per capita usage 12–15
population forecasting and 4
resource development for 1–2
short-term estimates of 5–8
waste dilution 19–21
Water stage measurement
automatic gages 285–287
converting records into discharge 331–332
data recording mechanisms for 287–288
data storage methods for 288
definition of 284–285
remote transmission of data 289
sensor devices for 285–287
simple stage-discharge relation 320–321
station analysis of stage data 331
Water supply tanks, storage capacity of 569–572
Water Surface Profile Model (WSPRO) 474
Water usage, per capita 12–15
Water-bearing formations 123–125
Watershed hydrology models 412–414, 474–475
NRCS (SCS) TR-55 method 742–746
surface water hydrology 410–415

Water-table aquifers
    aquitards and 128–129
    specific retention of 137, 140–143
    specific storage in 137, 145–146
    specific yield of 137–138, 143–145, 767
Wave celerity 490, 499–500, 520, 523
Weibull, W. 440
Weirs 539–552
    broad-crested. *See* Broad-crested weir
    classification of 539
    sharp-crested. *See* Sharp-crested weir
    trapezoidal 544–545
    triangular (V-notch) 544
    Villemonte relation for 545
    See also Spillways
Wells
    confined flow to 172–174
    efficiency and loss relationship 220
    equality of variances among 268
    field design 222–225
    flow near boundaries 204–209
    fully penetrating 172
    function, or $W(u)$ 181–182, 184–185, 201–202
    groundwater flow unconfined to 176–178
    leaky well function $(u, r/B)$ 198–199
    multiple, drawdown-time or -distance

    analysis 191–193
    near a stream 204–206
    near impermeable boundaries 207, 209
    production, analysis of 218–222
    specific capacity analysis 221–222
    statistical procedures for groundwater monitoring 268
    stilling 285
    unsteady flow to (Theis equation) 180–182
Wetting fluid, definition of 251
Wetting front 81, 90–93
White, C. M. 620, 664
Winter, T. C. 47, 48
Witherspoon, P. A. 182, 199
Wolman, M. G. 639
Woolhiser, D. A. 497

Y
Yang, C. T. 639

Z
Zero excess runoff 751
Zero flow, determining stage of 322–324, 326
Zhang, K. 639
Zonation, principle of 121–123